MECHANICAL METALLURGY

Third Edition

George E. Dieter

University of Maryland

McGraw-Hill Book Company

New York St. Louis San Francisco Auckland Bogotá Hamburg
Johannesburg London Madrid Mexico Montreal New Delhi
Panama Paris São Paulo Singapore Sydney Tokyo Toronto

MECHANICAL METALLURGY
INTERNATIONAL EDITION

1st printing 1986.

This book was set in Times Roman by Science Typographers Inc.
The editor was Sanjeev Rao.
The production supervisor was Diane Renda.
Project supervision was done by Science Typographers Inc.

Library of Congress Cataloging-in-Publication Data

Dieter, George Ellwood.
Mechanical metallurgy.

(McGraw-Hill series in materials science and engineering)
Includes bibliographies and indexes.
1. Strength of materials. 2. Physical metallurgy.
I. Title. II. Series.
TA405.D53 1986 620.1′63 85-18229
ISBN 0-07-016893-8

When ordering this title use ISBN 0-07-100178-6

Printed in Singapore by
Singapore National Printers (Pte) Ltd.

ABOUT THE AUTHOR

George E. Dieter is currently Dean of Engineering and Professor of Mechanical Engineering at the University of Maryland. The author received his B.S. Met.E. degree from Drexel University, and his D.Sc. degree from Carnegie-Mellon University. After a career in industy with DuPont Engineering Research Laboratory, he became Head of the Metallurgical Engineering Department at Drexel University, where he later became Dean of Engineering. Professor Dieter later joined the faculty of Carnegie-Mellon University, as Professor of Engineering and Director of the Processing Research Institute. He moved to the University of Maryland four years later.

A former member of the National Materials Advisory Board, Professor Dieter is a fellow of the American Society for Metals, and a member of AAAS, AIME, ASEE, NSPE, and SME.

CONTENTS

MECHANICAL METALLURGY

McGraw-Hill Series in Materials Science and Engineering

Brick, Pense, and Gordon: *Structure and Properties of Engineering Materials*
Dieter: *Engineering Design: A Materials and Processing Approach*
Dieter: *Mechanical Metallurgy*
Drauglis, Gretz, and Jaffe: *Molecular Processes on Solid Surfaces*
Flemings: *Solidification Processing*
Fontana: *Corrosion Engineering*
Gaskell: *Introduction to Metallurgical Thermodynamics*
Guy: *Introduction to Materials Science*
Kehl: *The Principles of Metallographic Laboratory Practice*
Leslie: *The Physical Metallurgy of Steels*
Rhines: *Phase Diagrams in Metallurgy: Their Development and Application*
Rozenfeld: *Corrosion Inhibitors*
Shewmon: *Transformations in Metals*
Smith: *Principles of Materials Science and Engineering*
Smith: *Structure and Properties of Engineering Alloys*
Vander Voort: *Metallography: Principles and Practice*
Wert and Thomson: *Physics of Solids*

PREFACE TO THE THIRD EDITION

The objective of *Mechanical Metallurgy* continues to be the presentation of the entire scope of mechanical metallurgy in a single comprehensive volume. Thus, the book starts with a continuum description of stress and strain, extends this background to the defect mechanisms of flow and fracture of metals, and then considers the major mechanical property tests and the basic metalworking processes. As before, the book is intended for the senior or first-year graduate-student level. Emphasis is on basic phenomena and relationships in an engineering context. Extensive references to the literature have been included to assist students in digging deeper into most topics.

Since the second edition in 1976 extensive progress has been made in all research areas of the mechanical metallurgy spectrum. Indeed, mechanical behavior is the category of research under which the greatest number of papers are published in *Metallurgical Transactions*. Since 1976 the field of fracture mechanics has grown greatly in general acceptance. In recognition of this a separate chapter on fracture mechanics has been added to the present edition, replacing a chapter on mechanical behavior of polymeric materials. Other topics added for the first time or greatly expanded in coverage are deformation maps, finite element methods, environmentally assisted fracture, and creep-fatigue interaction.

As an aid to the student, numerous illustrative examples have been included throughout the book. Answers have been provided to selected problems for the student, and a solutions manual is available for instructors. In this third edition, major emphasis is given to the use of SI units, as is common with most engineering texts today. However, engineering units have been retained to some extent.

I would like to express my thanks for the many useful comments and suggestions provided by Ronald Scattergood, North Carolina State University, and Oleg Sherby, Stanford University, who reviewed this text during the course of its development.

Acknowledgment is given to Professor Ronald Armstrong, University of Maryland, for providing many stimulating problems, and Dr. A. Pattniak, Naval Research Laboratory, for assistance in obtaining the fractographs. Special thanks goes to Jean Beckmann for her painstaking efforts to create a perfect manuscript.

George E. Dieter

PREFACE TO THE SECOND EDITION

In the 12 years since the first edition of *Mechanical Metallurgy* at least 25 textbooks dealing with major segments of the book have appeared in print. For example, at least 10 books dealing with the mechanics of metalworking have been published during this period. However, none of these books has dealt with the entire scope of mechanical metallurgy, from an understanding of the continuum description of stress and strain through crystalline and defect mechanisms of flow and fracture, and on to a consideration of the major mechanical property tests and the basic metalworking processes.

Important advances have been made in understanding the mechanical behavior of solids in the period since the first edition. The dislocation theory of plastic deformation has become well established, with excellent experimental verification for most of the theory. These advances have led to a better understanding of the strengthening mechanisms in solids. Developments such as fracture mechanics have matured to a high level of technical sophistication and engineering usefulness. An important development during this period has been the "materials science movement" in which crystalline solids, metals, ceramics, and polymers are considered as a group whose properties are controlled by basic structural defects common to all classes of crystalline solids.

In this revision of the book the emphasis is as before. The book is intended for the senior or first-year graduate-student level. Extensive revisions have been made to up-date material, to introduce new topics which have emerged as important areas, and to clarify sections which have proven difficult for students to understand. In some sections advanced material intended primarily for graduate students has been set in smaller type. The problems have been extensively revised and expanded, and a solutions manual has been prepared. Two new chapters, one dealing with the mechanical properties of polymers and the other with the machining of metals, have been added, while the chapters of statistical methods and residual stresses have been deleted. In total, more than one-half of the book has been rewritten completely.

George E. Dieter

PREFACE TO THE FIRST EDITION

Mechanical metallurgy is the area of knowledge which deals with the behavior and response of metals to applied forces. Since it is not a precisely defined area, it will mean different things to different persons. To some it will mean mechanical properties of metals or mechanical testing, others may consider the field restricted to the plastic working and shaping of metals, while still others confine their interests to the more theoretical aspects of the field, which merge with metal physics and physical metallurgy. Still another group may consider that mechanical metallurgy is closely allied with applied mathematics and applied mechanics. In writing this book an attempt has been made to cover, in some measure, this great diversity of interests. The objective has been to include the entire scope of mechanical metallurgy in one fairly comprehensive volume.

The book has been divided into four parts. Part One, Mechanical Fundamentals, presents the mathematical framework for many of the chapters which follow. The concepts of combined stress and strain are reviewed and extended into three dimensions. Detailed consideration of the theories of yielding and an introduction to the concepts of plasticity are given. No attempt is made to carry the topics in Part One to the degree of completion required for original problem solving. Instead, the purpose is to acquaint metallurgically trained persons with the mathematical language encountered in some areas of mechanical metallurgy. Part Two, Metallurgical Fundamentals, deals with the structural aspects of plastic deformation and fracture. Emphasis is on the atomistics of flow and fracture and the way in which metallurgical structure affects these processes. The concept of the dislocation is introduced early in Part Two and is used throughout to provide qualitative explanations for such phenomena as strain hardening, the yield point, dispersed phase hardening, and fracture. A more mathematical treatment of the properties of dislocations is given in a separate chapter. The topics covered in Part Two stem from physical metallurgy. However, most topics are discussed in greater detail and with a different emphasis than when they are first covered in the usual undergraduate course in physical metallurgy. Certain topics that are more physical metallurgy than mechanical metallurgy are included to provide continuity and the necessary background for readers who have not studied modern physical metallurgy.

Part Three, Applications to Materials Testing, deals with the engineering aspects of the common testing techniques of mechanical failure of metals. Chapters are devoted to the tension, torsion, hardness, fatigue, creep, and impact tests. Others take up the important subjects of residual stresses and the statistical analysis of mechanical-property data. In Part Three emphasis is placed on the interpretation of the tests and on the effect of metallurgical variables on mechanical behavior rather than on the procedures for conducting the tests. It is assumed that the actual performance of these tests will be covered in a concurrent laboratory course or in a separate course. Part Four, Plastic Forming of Metals, deals with the common mechanical processes for producing useful metal shapes. Little emphasis is given to the descriptive aspects of this subject, since this can best be covered by plant trips and illustrated lectures. Instead, the main attention is given to the mechanical and metallurgical factors which control each process such as forging, rolling, extrusion, drawing, and sheet-metal forming.

This book is written for the senior or first-year graduate student in metallurgical or mechanical engineering, as well as for practicing engineers in industry. While most universities have instituted courses in mechanical metallurgy or mechanical properties, there is a great diversity in the material covered and in the background of the students taking these courses. Thus, for the present there can be nothing like a standardized textbook on mechanical metallurgy. It is hoped that the breadth and scope of this book will provide material for these somewhat diverse requirements. It is further hoped that the existence of a comprehensive treatment of the field of mechanical metallurgy will stimulate the development of courses which cover the total subject.

Since this book is intended for college seniors, graduate students, and practicing engineers, it is expected to become a part of their professional library. Although there has been no attempt to make this book a handbook, some thought has been given to providing abundant references to the literature on mechanical metallurgy. Therefore, more references are included than is usual in the ordinary textbook. References have been given to point out derivations or analyses beyond the scope of the book, to provide the key to further information on controversial or detailed points, and to emphasize important papers which are worthy of further study. In addition, a bibliography of general references will be found at the end of each chapter. A collection of problems is included at the end of the volume. This is primarily for the use of the reader who is engaged in industry and who desires some check on his comprehension of the material.

The task of writing this book has been mainly one of sifting and sorting facts and information from the literature and the many excellent texts on specialized aspects of this subject. To cover the breadth of material found in this book would require parts of over 15 standard texts and countless review articles and individual contributions. A conscientious effort has been made throughout to give credit to original sources. For the occasional oversights that may have developed during the "boiling-down process" the author offers his apologies. He is indebted to many authors and publishers who consented to the reproduction of illustrations. Credit is given in the captions of the illustrations.

Finally, the author wishes to acknowledge the many friends who advised him in his work. Special mention should be given to Professor A. W. Grosvenor, Drexel Institute of Technology, Dr. G. T. Horne, Carnegie Institute of Technology, Drs. T. C. Chilton, J. H. Faupel, W. L. Phillips, W. I. Pollock, and J. T. Ransom of the du Pont Company, and Dr. A. S. Nemy of the Thompson-Ramo-Wooldridge Corp.

George E. Dieter

LIST OF SYMBOLS

A area; amplitude
a linear distance; crack length
a_0 interatomic spacing
B constant; specimen thickness
b width or breadth
$\mathbf{b}$ Burgers vector of a dislocation
C generalized constant; specific heat
C_{ij} elastic coefficients
c length of Griffith crack
D diameter, grain diameter
E modulus of elasticity for axial loading (Young's modulus)
e conventional, or engineering, linear strain
exp base of natural logarithms (= 2.718)
F force per unit length on a dislocation line
G modulus of elasticity in shear (modulus of rigidity)
$\mathcal{G}$ crack-extension force
H activation energy
h distance, usually in thickness direction
(h, k, l) Miller indices of a crystallographic plane
I moment of inertia
J invariant of the stress deviator; polar moment of inertia
K strength coefficient
K_f fatigue-notch factor
K_t theoretical stress-concentration factor
K_{Ic} fracture toughness
k yield stress in pure shear
L length

l, m, n direction cosines of normal to a plane
ln natural logarithm
log logarithm to base 10
M_B bending moment
M_T torsional moment, torque
m strain-rate sensitivity
N number of cycles of stress or vibration
n strain-hardening exponent
n' generalized constant in exponential term
P load or external force
Q activation energy
p pressure
q reduction in area; plastic-constraint factor; notch sensitivity index in fatigue
R radius of curvature; stress ratio in fatigue; gas constant
r radial distance
S total stress on a plane before resolution into normal and shear components
S_{ij} elastic compliance
s engineering stress
T temperature
T_m melting point
t time; thickness
t_r time for rupture
U elastic strain energy
U_0 elastic strain energy per unit volume
u, v, w components of displacement in x, y, and z directions
$[uvw]$ Miller indices for a crystallographic direction
V volume
v velocity
W work
Z Zener-Hollomon parameter
α linear coefficient of thermal expansion; phase angle
$\alpha, \beta, \theta, \phi$ generalized angles
Γ line tension of a dislocation
γ shear strain
Δ volume strain or cubical dilatation; finite change
δ deformation or elongation; deflection; logarithmic decrement; Kronecker delta
ε general symbol for strain; natural or true strain
$\bar{\varepsilon}$ significant, or effective, true strain
$\dot{\varepsilon}$ true-strain rate

$\dot{\varepsilon}_s$ minimum creep rate
η efficiency; coefficient of viscosity
θ Dorn time-temperature parameter
κ bulk modulus or volumetric modulus of elasticity
λ Lamé's constant; interparticle spacing
μ coefficient of friction
ν Poisson's ratio
ρ density
σ normal stress; true stress
σ_0 yield stress or yield strength
σ_0' yield stress in plane strain
$\bar{\sigma}$ significant, or effective, true stress
$\sigma_1, \sigma_2, \sigma_3$ principal stresses
σ' stress deviator
σ'' hydrostatic component of stress
σ_a alternating, or variable, stress
σ_m average principal stress; mean stress
σ_r range of stress
σ_u ultimate tensile strength
σ_w working stress
τ shearing stress; relaxation time

PART ONE

MECHANICAL FUNDAMENTALS

CHAPTER
ONE
INTRODUCTION

1-1 SCOPE OF THIS BOOK

Mechanical metallurgy is the area of metallurgy which is concerned primarily with the response of metals to forces or loads. The forces may arise from the use of the metal as a member or part in a structure or machine, in which case it is necessary to know something about the limiting values which can be withstood without failure. On the other hand, the objective may be to convert a cast ingot into a more useful shape, such as a flat plate, and here it is necessary to know the conditions of temperature and rate of loading which minimize the forces that are needed to do the job.

Mechanical metallurgy is *not* a subject which can be neatly isolated and studied by itself. It is a combination of many disciplines and many approaches to the problem of understanding the response of materials to forces. On the one hand is the approach used in strength of materials and in the theories of elasticity and plasticity, where a metal is considered to be a homogeneous material whose mechanical behavior can be rather precisely described on the basis of only a very few material constants. This approach is the basis for the rational design of structural members and machine parts. The topics of strength of materials, elasticity, and plasticity are treated in Part One of this book from a more generalized point of view than is usually considered in a first course in strength of materials. The material in Chaps. 1 to 3 can be considered the mathematical framework on which much of the remainder of the book rests. For students of engineering who have had an advanced course in strength of materials or machine design, it probably will be possible to skim rapidly over these chapters. However, for most students of metallurgy and for practicing engineers in industry, it is worth spending the time to become familiar with the mathematics presented in Part One.

The theories of strength of materials, elasticity, and plasticity lose much of their power when the structure of the metal becomes an important consideration and it can no longer be considered a homogeneous medium. Examples of this are in the high-temperature behavior of metals, where the metallurgical structure may continuously change with time, or in the ductile-to-brittle transition, which occurs in carbon steel. The determination of the relationship between mechanical behavior and structure (as detected chiefly with microscopic and x-ray techniques) is the main responsibility of the mechanical metallurgist. When mechanical behavior is understood in terms of metallurgical structure, it is generally possible to improve the mechanical properties or at least to control them. Part Two of this book is concerned with the metallurgical fundamentals of the mechanical behavior of metals. Metallurgical students will find that some of the material in Part Two has been covered in a previous course in physical metallurgy, since mechanical metallurgy is part of the broader field of physical metallurgy. However, these subjects are considered in greater detail than is usually the case in a first course in physical metallurgy. In addition, certain topics which pertain more to physical metallurgy than mechanical metallurgy have been included in order to provide continuity and to assist nonmetallurgical students who may not have had a course in physical metallurgy.

The last three chapters of Part Two are concerned primarily with atomistic concepts of the flow and fracture of metals. Many of the developments in these areas have been the result of the alliance of the solid-state physicist with the metallurgist. This has been an area of great progress. The introduction of transmission electron microscopy has provided an important experimental tool for verifying theory and guiding analysis. A body of basic dislocation theory is presented which is useful for understanding the mechanical behavior of crystalline solids.

Basic data concerning the strength of metals and measurements for the routine control of mechanical properties are obtained from a relatively small number of standardized mechanical tests. Part Three, Applications to Materials Testing, considers each of the common mechanical tests, not from the usual standpoint of testing techniques, but instead from the consideration of what these tests tell about the service performance of metals and how metallurgical variables affect the results of these tests. Much of the material in Parts One and Two has been utilized in Part Three. It is assumed that the reader either has completed a conventional course in materials testing or will be concurrently taking a laboratory course in which familiarization with the testing techniques will be acquired.

Part Four considers the metallurgical and mechanical factors involved in forming metals into useful shapes. Attempts have been made to present mathematical analyses of the principal metalworking processes, although in certain cases this has not been possible, either because of the considerable detail required or because the analysis is beyond the scope of this book. No attempt has been made to include the extensive specialized technology associated with each metalworking process, such as rolling or extrusion, although some effort has been made to give a general impression of the mechanical equipment required and to familiarize the reader with the specialized vocabulary of the metalworking field. Major emphasis has been placed on presenting a fairly simplified picture of the forces involved in each process and of how geometrical and metallurgical factors affect the forming loads and the success of the metalworking process.

1-2 STRENGTH OF MATERIALS—BASIC ASSUMPTIONS

Strength of materials is the body of knowledge which deals with the relation between internal forces, deformation, and external loads. In the general method of analysis used in strength of materials the first step is to assume that the member is in equilibrium. The equations of static equilibrium are applied to the forces acting on some part of the body in order to obtain a relationship between the external forces acting on the member and the internal forces resisting the action of the external loads. Since the equations of equilibrium must be expressed in terms of forces acting external to the body, it is necessary to make the internal resisting forces into external forces. This is done by passing a plane through the body at the point of interest. The part of the body lying on one side of the cutting plane is removed and replaced by the forces it exerted on the cut section of the part of the body that remains. Since the forces acting on the "free body" hold it in equilibrium, the equations of equilibrium may be applied to the problem.

The internal resisting forces are usually expressed by the *stress*[1] acting over a certain area, so that the internal force is the integral of the stress times the differential area over which it acts. In order to evaluate this integral, it is necessary to know the distribution of the stress over the area of the cutting plane. The stress distribution is arrived at by observing and measuring the strain distribution in the member, since stress cannot be physically measured. However, since stress is proportional to strain for the small deformations involved in most work, the determination of the strain distribution provides the stress distribution. The expression for the stress is then substituted into the equations of equilibrium, and they are solved for stress in terms of the loads and dimensions of the member.

Important assumptions in strength of materials are that the body which is being analyzed is continuous, homogeneous, and isotropic. A *continuous body* is one which does not contain voids or empty spaces of any kind. A body is *homogeneous* if it has identical properties at all points. A body is considered to be *isotropic* with respect to some property when that property does not vary with direction or orientation. A property which varies with orientation with respect to some system of axes is said to be *anisotropic*.

[1] For present purposes *stress* is defined as force per unit area. The companion term *strain* is defined as the change in length per unit length. More complete definitions will be given later.

While engineering materials such as steel, cast iron, and aluminum may appear to meet these conditions when viewed on a gross scale, it is readily apparent when they are viewed through a microscope that they are anything but homogeneous and isotropic. Most engineering metals are made up of more than one phase, with different mechanical properties, such that on a micro scale they are heterogeneous. Further, even a single-phase metal will usually exhibit chemical segregation, and therefore the properties will not be identical from point to point. Metals are made up of an aggregate of crystal grains having different properties in different crystallographic directions. The reason why the equations of strength of materials describe the behavior of real metals is that, in general, the crystal grains are so small that, for a specimen of any macroscopic volume, the materials are statistically homogeneous and isotropic. However, when metals are severely deformed in a particular direction, as in rolling or forging, the mechanical properties may be anisotropic on a macro scale. Other examples of anisotropic properties are fiber-reinforced composite materials and single crystals. Lack of continuity may be present in porous castings or powder metallurgy parts and, on an atomic level, at defects such as vacancies and dislocations.

1-3 ELASTIC AND PLASTIC BEHAVIOR

Experience shows that all solid materials can be deformed when subjected to external load. It is further found that up to certain limiting loads a solid will recover its original dimensions when the load is removed. The recovery of the original dimensions of a deformed body when the load is removed is known as *elastic behavior*. The limiting load beyond which the material no longer behaves elastically is the *elastic limit*. If the elastic limit is exceeded, the body will experience a permanent set or deformation when the load is removed. A body which is permanently deformed is said to have undergone *plastic deformation*.

For most materials, as long as the load does not exceed the elastic limit, the deformation is proportional to the load. This relationship is known as Hooke's law; it is more frequently stated as *stress is proportional to strain*. Hooke's law requires that the load-deformation relationship should be linear. However, it does not necessarily follow that all materials which behave elastically will have a linear stress-strain relationship. Rubber is an example of a material with a nonlinear stress-strain relationship that still satisfies the definition of an elastic material.

Elastic deformations in metals are quite small and require very sensitive instruments for their measurement. Ultrasensitive instruments have shown that the elastic limits of metals are much lower than the values usually measured in engineering tests of materials. As the measuring devices become more sensitive, the elastic limit is decreased, so that for most metals there is only a rather narrow range of loads over which Hooke's law strictly applies. This is, however, primarily of academic importance. Hooke's law remains a quite valid relationship for engineering design.

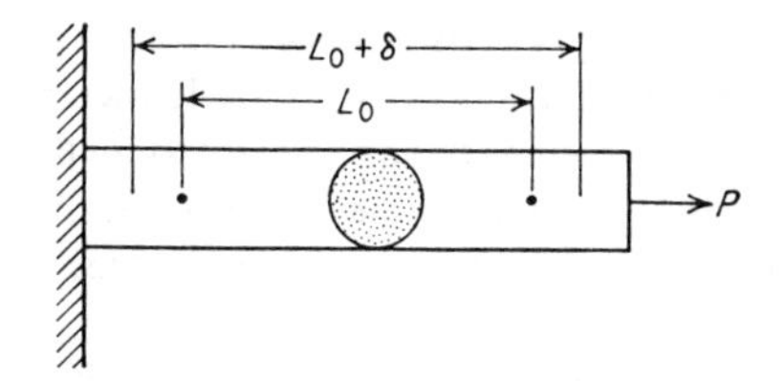

Figure 1-1 Cylindrical bar subjected to axial load.

Figure 1-2 Free-body diagram for Fig. 1-1.

1-4 AVERAGE STRESS AND STRAIN

As a starting point in the discussion of stress and strain, consider a uniform cylindrical bar which is subjected to an axial tensile load (Fig. 1-1). Assume that two gage marks are put on the surface of the bar in its unstrained state and that L_0 is the gage length between these marks. A load P is applied to one end of the bar, and the gage length undergoes a slight increase in length and decrease in diameter. The distance between the gage marks has increased by an amount δ, called the deformation. The *average linear strain* e is the ratio of the change in length to the original length.

$$e = \frac{\delta}{L_0} = \frac{\Delta L}{L_0} = \frac{L - L_0}{L_0} \tag{1-1}$$

Strain is a dimensionless quantity since both δ and L_0 are expressed in units of length.

Figure 1-2 shows the free-body diagram for the cylindrical bar shown in Fig. 1-1. The external load P is balanced by the internal resisting force $\int \sigma\, dA$, where σ is the stress normal to the cutting plane and A is the cross-sectional area of the bar. The equilibrium equation is

$$P = \int \sigma\, dA \tag{1-2}$$

If the stress is distributed uniformly over the area A, that is, if σ is constant, Eq. (1-2) becomes

$$P = \sigma \int dA = \sigma A$$

$$\sigma = \frac{P}{A} \tag{1-3}$$

In general, the stress will not be uniform over the area A, and therefore Eq. (1-3) represents an *average stress*. For the stress to be absolutely uniform, every longitudinal element in the bar would have to experience exactly the same strain, and the proportionality between stress and strain would have to be identical for each element. The inherent anisotropy between grains in a polycrystalline metal rules out the possibility of complete uniformity of stress over a body of macroscopic size. The presence of more than one phase also gives rise to nonuniformity of stress on a microscopic scale. If the bar is not straight or not centrally loaded, the strains will be different for certain longitudinal elements and the stress will not be uniform. An extreme disruption in the uniformity of the stress pattern occurs when there is an abrupt change in cross section. This results in a stress raiser or stress concentration (see Sec. 2-15).

Below the elastic limit Hooke's law can be considered valid, so that the average stress is proportional to the average strain,

$$\frac{\sigma}{e} = E = \text{constant} \tag{1-4}$$

The constant E is the *modulus of elasticity*, or *Young's modulus.*

1-5 TENSILE DEFORMATION OF DUCTILE METAL

The basic data on the mechanical properties of a ductile metal are obtained from a tension test, in which a suitably designed specimen is subjected to increasing axial load until it fractures. The load and elongation are measured at frequent intervals during the test and are expressed as average stress and strain according to the equations in the previous section. (More complete details on the tension test are given in Chap. 8.)

The data obtained from the tension test are generally plotted as a stress-strain diagram. Figure 1-3 shows a typical stress-strain curve for a metal such as aluminum or copper. The initial linear portion of the curve OA is the elastic region within which Hooke's law is obeyed. Point A is the elastic limit, defined as the greatest stress that the metal can withstand without experiencing a permanent strain when the load is removed. The determination of the elastic limit is quite tedious, not at all routine, and dependent on the sensitivity of the strain-measuring instrument. For these reasons it is often replaced by the *proportional limit*, point A'. The proportional limit is the stress at which the stress-strain curve deviates from linearity. The slope of the stress-strain curve in this region is the modulus of elasticity.

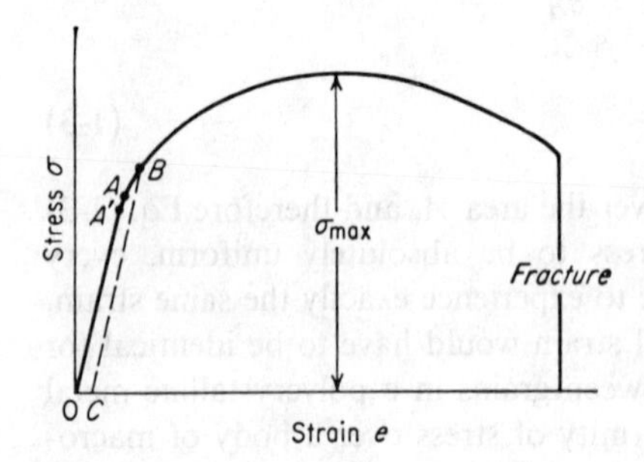

Figure 1-3 Typical tension stress-strain curve.

For engineering purposes the limit of usable elastic behavior is described by the *yield strength*, point B. The yield strength is defined as the stress which will produce a small amount of permanent deformation, generally equal to a strain of 0.002. In Fig. 1-3 this permanent strain, or offset, is OC. Plastic deformation begins when the elastic limit is exceeded. As the plastic deformation of the specimen increases, the metal becomes stronger (strain hardening) so that the load required to extend the specimen increases with further straining. Eventually the load reaches a maximum value. The maximum load divided by the original area of the specimen is the *ultimate tensile strength*. For a ductile metal the diameter of the specimen begins to decrease rapidly beyond maximum load, so that the load required to continue deformation drops off until the specimen fractures. Since the average stress is based on the original area of the specimen, it also decreases from maximum load to fracture.

1-6 DUCTILE VS. BRITTLE BEHAVIOR

The general behavior of materials under load can be classified as ductile or brittle depending upon whether or not the material exhibits the ability to undergo plastic deformation. Figure 1-3 illustrates the tension stress-strain curve of a ductile material. A completely brittle material would fracture almost at the elastic limit (Fig. 1-4*a*), while a brittle metal, such as white cast iron, shows some slight measure of plasticity before fracture (Fig. 1-4*b*). Adequate ductility is an important engineering consideration, because it allows the material to redistribute localized stresses. When localized stresses at notches and other accidental stress concentrations do not have to be considered, it is possible to design for static situations on the basis of average stresses. However, with brittle materials, localized stresses continue to build up when there is no local yielding. Finally, a crack forms at one or more points of stress concentration, and it spreads rapidly over the section. Even if no stress concentrations are present in a brittle material, fracture will still occur suddenly because the yield stress and tensile strength are practically identical.

It is important to note that brittleness is not an absolute property of a metal. A metal such as tungsten, which is brittle at room temperature, is ductile at an elevated temperature. A metal which is brittle in tension may be ductile under hydrostatic compression. Furthermore, a metal which is ductile in tension at room

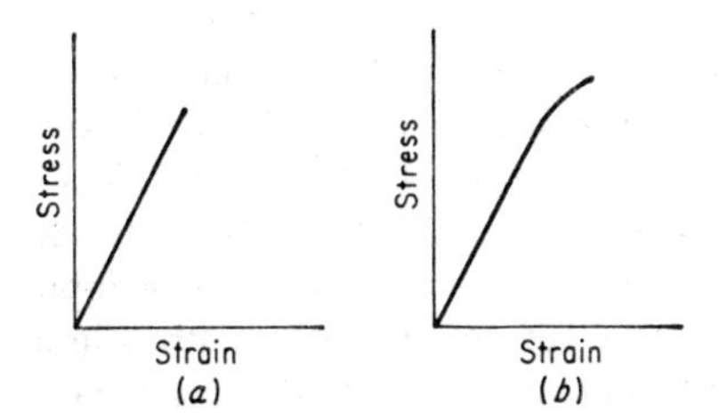

Figure 1-4 (a) Stress-strain curve for completely brittle material (ideal behavior); (b) stress-strain curve for brittle metal with slight amount of ductility.

temperature can become brittle in the presence of notches, low temperature, high rates of loading, or embrittling agents such as hydrogen.

1-7 WHAT CONSTITUTES FAILURE?

Structural members and machine elements can fail to perform their intended functions in three general ways:

1. Excessive elastic deformation
2. Yielding, or excessive plastic deformation
3. Fracture

An understanding of the common types of failure is important in good design because it is always necessary to relate the loads and dimensions of the member to some significant material parameter which limits the load-carrying capacity of the member. For different types of failure, different significant parameters will be important.

Two general types of excessive elastic deformation may occur: (1) excessive deflection under condition of stable equilibrium, such as the deflection of beam under gradually applied loads; (2) sudden deflection, or *buckling*, under conditions of unstable equilibrium.

Excessive elastic deformation of a machine part can mean failure of the machine just as much as if the part completely fractured. For example, a shaft which is too flexible can cause rapid wear of the bearing, or the excessive deflection of closely mating parts can result in interference and damage to the parts. The sudden buckling type of failure may occur in a slender column when the axial load exceeds the Euler critical load or when the external pressure acting against a thin-walled shell exceeds a critical value. Failures due to excessive elastic deformation are controlled by the modulus of elasticity, not by the strength of the material. Generally, little metallurgical control can be exercised over the elastic modulus. The most effective way to increase the stiffness of a member is usually by changing its shape and increasing the dimensions of its cross section.

Yielding, or excessive plastic deformation, occurs when the elastic limit of the metal has been exceeded. Yielding produces permanent change of shape, which may prevent the part from functioning properly any longer. In a ductile metal under conditions of static loading at room temperature yielding rarely results in fracture, because the metal strain hardens as it deforms, and an increased stress is required to produce further deformation. Failure by excessive plastic deformation is controlled by the yield strength of the metal for a uniaxial condition of loading. For more complex loading conditions the yield strength is still the significant parameter, but it must be used with a suitable failure criterion (Sec. 3-4). At temperatures significantly greater than room temperature metals no longer exhibit strain hardening. Instead, metals can continuously deform at constant stress in a time-dependent yielding known as *creep*. The failure criterion under creep conditions is complicated by the fact that stress is not proportional to strain and the further fact that the mechanical properties of the material may change appreciably during service. This complex phenomenon will be considered in greater detail in Chap. 13.

The formation of a crack which can result in complete disruption of continuity of the member constitutes fracture. A part made from a ductile metal which is loaded statically rarely fractures like a tensile specimen, because it will first fail by excessive plastic deformation. However, metals fail by fracture in three general ways: (1) sudden brittle fracture; (2) fatigue, or progressive fracture; (3) delayed fracture. In the previous section it was shown that a brittle material fractures under static loads with little outward evidence of yielding. A sudden brittle type of fracture can also occur in ordinarily ductile metals under certain conditions. Plain carbon structural steel is the most common example of a material with a ductile-to-brittle transition. A change from the ductile to the brittle type of fracture is promoted by a decrease in temperature, an increase in the rate of loading, and the presence of a complex state of stress due to a notch. This problem is considered in Chap. 14. A powerful and quite general method of analysis for brittle fracture problems is the technique called *fracture mechanics.* This is treated in detail in Chap. 11.

Most fractures in machine parts are due to *fatigue*. Fatigue failures occur in parts which are subjected to alternating, or fluctuating, stresses. A minute crack starts at a localized spot, generally at a notch or stress concentration, and gradually spreads over the cross section until the member breaks. Fatigue failure occurs without any visible sign of yielding at *nominal* or average stresses that are well below the tensile strength of the metal. Fatigue failure is caused by a critical *localized* tensile stress which is very difficult to evaluate, and therefore design for fatigue failure is based primarily on empirical relationships using nominal stresses. Fatigue of metals is discussed in greater detail in Chap. 12.

One common type of delayed fracture is *stress-rupture* failure, which occurs when a metal has been statically loaded at an elevated temperature for a long period of time. Depending upon the stress and the temperature there may be no yielding prior to fracture. A similar type of delayed fracture, in which there is no warning by yielding prior to failure, occurs at room temperature when steel is statically loaded in the presence of hydrogen.

All engineering materials show a certain variability in mechanical properties, which in turn can be influenced by changes in heat treatment or fabrication. Further, uncertainties usually exist regarding the magnitude of the applied loads, and approximations are usually necessary in calculating stresses for all but the most simple member. Allowance must be made for the possibility of accidental loads of high magnitude. Thus, in order to provide a margin of safety and to protect against failure from unpredictable causes, it is necessary that the allowable stresses be smaller than the stresses which produce failure. The value of stress for a particular material used in a particular way which is considered to be a safe stress is usually called the *working stress* σ_w. For static applications the working stress of ductile metals is usually based on the yield strength σ_0 and for brittle metals on the ultimate tensile strength σ_u. Values of working stress are established by local and federal agencies and by technical organizations such as the American Society of Mechanical Engineers (ASME). The working stress may be considered as either the yield strength or the tensile strength divided by a number called the *factor of safety*.

$$\sigma_w = \frac{\sigma_0}{N_0} \quad \text{or} \quad \sigma_w = \frac{\sigma_u}{N_u} \tag{1-5}$$

where σ_w = working stress
σ_0 = yield strength
σ_u = tensile strength
N_0 = factor of safety based on yield strength
N_u = factor of safety based on tensile strength

The value assigned to the factor of safety depends on an estimate of all the factors discussed above. In addition, careful consideration should be given to the consequences, which would result from failure. If failure would result in loss of life, the factor of safety should be increased. The type of equipment will also influence the factor of safety. In military equipment, where light weight may be a prime consideration, the factor of safety may be lower than in commercial equipment. The factor of safety will also depend on the expected type of loading. For static loading, as in a building, the factor of safety would be lower than in a machine, which is subjected to vibration and fluctuating stresses.

1-8 CONCEPT OF STRESS AND THE TYPES OF STRESSES

Stress is defined as force per unit area. In Sec. 1-4 the stress was considered to be uniformly distributed over the cross-sectional area of the member. However, this is not the general case. Figure 1-5*a* represents a body in equilibrium under the action of external forces $P_1, P_2, \ldots, P_5$. There are two kinds of external forces which may act on a body: surface forces and body forces. Forces distributed over the surface of the body, such as hydrostatic pressure or the pressure exerted by one body on another, are called *surface forces*. Forces distributed over the volume of a body, such as gravitational forces, magnetic forces, or inertia forces (for a body in motion), are called *body forces*. The two most common types of body forces encountered in engineering practice are centrifugal forces due to high-speed rotation and forces due to temperature differential over the body (thermal stress).

In general the force will not be uniformly distributed over any cross section of the body illustrated in Fig. 1-5*a*. To obtain the stress at some point O in a plane such as *mm*, part 1 of the body is removed and replaced by the system of external forces on *mm* which will retain each point in part 2 of the body in the same position as before the removal of part 1. This is the situation in Fig. 1-5*b*. We then take an area ΔA surrounding the point O and note that a force ΔP acts

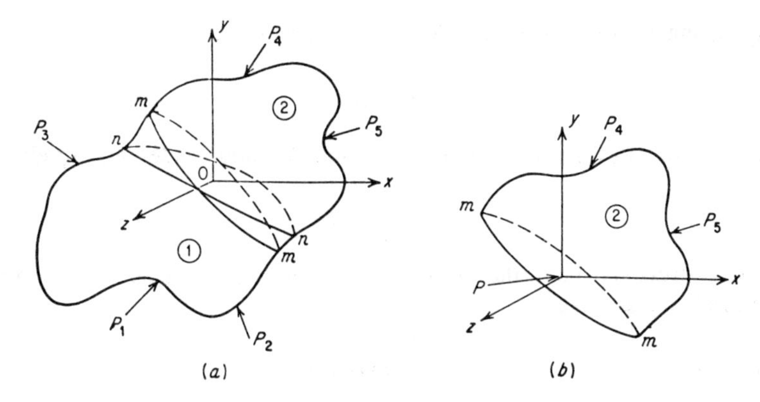

Figure 1-5 (a) Body in equilibrium under action of external forces $P_1, \ldots, P_5$; (b) forces acting on parts.

on this area. If the area ΔA is continuously reduced to zero, the limiting value of the ratio $\Delta P/\Delta A$ is the stress at the point O on plane mm of body 2.

$$\lim_{\Delta A \to 0} \frac{\Delta P}{\Delta A} = \sigma \tag{1-6}$$

The stress will be in the direction of the resultant force P and will generally be inclined at an angle to ΔA. The same stress at point O in plane mm would be obtained if the free body were constructed by removing part 2 of the solid body. However, the stress will be different on any other plane passing through point O, such as the plane nn.

It is inconvenient to use a stress which is inclined at some arbitrary angle to the area over which it acts. The total stress can be resolved into two components, a *normal stress* σ perpendicular to ΔA, and a *shearing stress* (or shear stress) τ lying in the plane mm of the area. To illustrate this point, consider Fig. 1-6. The force P makes an angle θ with the normal z to the plane of the area A. Also, the plane containing the normal and P intersects the plane A along a dashed line that

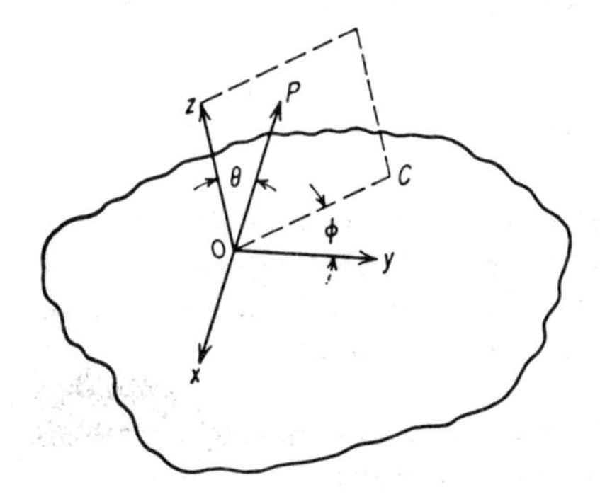

Figure 1-6 Resolution of total stress into its components.

makes an angle ϕ with the y axis. The normal stress is given by

$$\sigma = \frac{P}{A}\cos\theta \tag{1-7}$$

The shear stress in the plane acts along the line OC and has the magnitude

$$\tau = \frac{P}{A}\sin\theta \tag{1-8}$$

This shear stress may be further resolved into components parallel to the x and y directions lying the plane.

x direction
$$\tau = \frac{P}{A}\sin\theta\sin\phi \tag{1-9}$$

y direction
$$\tau = \frac{P}{A}\sin\theta\cos\phi \tag{1-10}$$

Therefore, in general a given plane may have one normal stress and two shear stresses acting on it.

1-9 CONCEPT OF STRAIN AND THE TYPES OF STRAIN

In Sec. 1-4 the average linear strain was defined as the ratio of the change in length to the original length of the same dimension.

$$e = \frac{\delta}{L_0} = \frac{\Delta L}{L_0} = \frac{L - L_0}{L_0}$$

where e = average linear strain
δ = deformation

By analogy with the definition of stress at a point, the strain at a point is the ratio of the deformation to the gage length as the gage length approaches zero.

Rather than referring the change in length to the original gage length, it often is more useful to define the strain as the change in linear dimension divided by the instantaneous value of the dimension.

$$\varepsilon = \int_{L_0}^{L_f} \frac{dL}{L} = \ln\frac{L_f}{L_0} \tag{1-11}$$

The above equation defines the *natural*, or *true*, *strain*. True strain, which is useful in dealing with problems in plasticity and metal forming, will be discussed more fully in Chap. 3. For the present it should be noted that for the very small strains for which the equations of elasticity are valid the two definitions of strain give identical values.

Not only will the elastic deformation of a body result in a change in length of a linear element in the body, but it may also result in a change in the initial angle

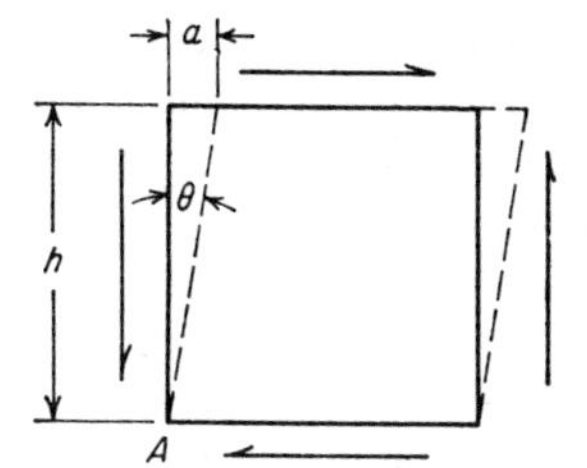

Figure 1-7 Shear strain.

between any two lines. The angular change in a right angle is known as *shear strain*. Figure 1-7 illustrates the strain produced by the pure shear of one face of a cube. The angle at A, which was originally 90°, is decreased by the application of a shear stress by a small amount θ. The shear strain γ is equal to the displacement a divided by the distance between the planes, h. The ratio a/h is also the tangent of the angle through which the element has been rotated. For the small angles usually involved, the tangent of the angle and the angle (in radians) are equal. Therefore, shear strains are often expressed as angles of rotation.

$$\gamma = \frac{a}{h} = \tan\theta = \theta \qquad (1\text{-}12)$$

1-10 UNITS OF STRESS

Stress has dimensions of force per unit area. Conventional practice among engineers in the English speaking world for many years was to express force (load) in pounds and area in square inches, so that stress had units of pounds per square inch (psi). Since it is common to deal with loads in the thousands of pounds, an obvious simplification is to work with units of 1,000 lb called *kips*. Thus, stress may be expressed in units of kips per square inch (ksi). 1 ksi = 1,000 psi.

With the adoption of the International System of Units, usually called SI (for Système International), the official unit of stress is the newton per square meter, N/m^2. SI units have been adopted worldwide, except in American engineering practice. However, SI units have been adopted as the primary units by most engineering and scientific societies in the United States. Therefore, American engineering students must be "bilingual" in this respect and in this book we shall use both sets of units.

In the SI system there are six basic units: meter (m) for length; kilogram (kg) for mass; second (s) for time; ampere (A) for electric current; degree Kelvin (K) for thermodynamic temperature; and candela (cd) for intensity of illumination. The unit of force is the newton (N). It is derived as the force required to accelerate a mass of one kilogram (kg) to an acceleration of one meter per second per second (m/s^2). $N = kg - m/s^2$. The unit of stress or pressure, N/m^2 has been designated a pascal (Pa).

A system of prefixes is used to indicate the magnitude of a unit; see Appendix A. Only the prefixes which vary by a factor of 1,000 are commonly used. Since the N/m^2 represents a very small stress ($1\ N/m^2 = 0.000145$ psi), it is more common to express stress in units of meganewtons per square meter, MN/m^2. $1\ MN/m^2 = 10^6\ N/m^2 = MPa = 145$ psi.

Certain conventions that apply to SI style and usage in written form have been adopted to keep the units as simple as possible. Prefixes are to be used in their simplest form, e.g., use GPa not kMPa. Prefixes are to be used only in the numerator, e.g., use MN/m^2 instead of N/mm^2. However, because the kilogram is a basic SI unit it should be used instead of the gram, e.g., use MJ/kg instead of kJ/g. This does not violate the rule against prefixes in the denominator because kg is a basic unit. Remember that when the unit is raised to a power the prefix is also raised to that power. thus, mm^3 is $10^{-9}\ m^3$ not $10^{-3}\ m^3$. SI units are always written in singular form, e.g., 60 newtons is 60 N not 60 Ns. A single space is left between the number and the first symbol, e.g., 100 MPa not 100MPa. There is no period after the symbol of a SI unit except when it ends a sentence.

Example The shear stress required to nucleate a grain boundary crack in high-temperature deformation has been estimated to be

$$\tau = \left(\frac{3\pi \gamma_b G}{8(1-\nu)L} \right)^{1/2}$$

where γ_b is the grain boundary surface energy, let us say $2\ J/m^2$; G is the shear modulus, 75 GPa; L is the grain boundary sliding distance, assumed equal to the grain diameter 0.01 mm, and ν is Poisson's ratio, $\nu = 0.3$. To calculate τ we need to be sure the units are consistent and that the prefixes have been properly evaluated.

To check the equation express all units in newtons and meters.

$$\tau = \left(\frac{\dfrac{N-m}{m^2} \times \dfrac{N}{m^2}}{m} \right)^{1/2} = \left(\frac{N^2}{m^4} \right)^{1/2} = \frac{N}{m^2}$$

Note that a joule (J) is a unit of energy; J = N − m (see Appendix A)

$$\tau = \left(\frac{3\pi \times 2 \times 75 \times 10^9}{8(1-0.3) \times 10^{-2} \times 10^{-3}} \right)^{1/2} = (252.4 \times 10^{14})^{1/2}$$

$$= 15.89 \times 10^7\ N/m^2 \quad \text{(this shear stress is about 23,000 psi)}$$

$$= 158.9\ MN/m^2 = 158.9\ MPa$$

BIBLIOGRAPHY

Caddell, R. M.: "Deformation and Fracture of Solids," Prentice-Hall Inc., Englewood Cliffs, N.J., 1980.

Felbeck, D. K., and A. G. Atkins: "Strength and Fracture of Engineering Solids," Prentice-Hall Inc., Englewood Cliffs, N.J., 1984.

Gordon, J. E.: "Structures—or Why Things Don't Fall Through the Floor," Penguin Books, London, 1978.

Hertzberg, R. W.: "Deformation and Fracture Mechanics of Engineering Materials," 2d ed., John Wiley & Sons, New York, 1983.

LeMay, I.: "Principles of Mechanical Metallurgy," Elsevier North-Holland Inc., New York, 1981.

Meyers, M. A., and K. K. Chawla: "Mechanical Metallurgy: Principles and Applications," Prentice-Hall Inc., Englewood Cliffs, N.J., 1984.

Polakowski, N. H., and E. J. Ripling: "Strength and Structure of Engineering Materials," Prentice-Hall Inc., Englewood Cliffs, N.J., 1964.

CHAPTER
TWO

STRESS AND STRAIN RELATIONSHIPS FOR ELASTIC BEHAVIOR

2-1 INTRODUCTION

The purpose of this chapter is to present the mathematical relationships for expressing the stress and strain at a point and the relationships between stress and strain in a solid which obeys Hooke's law. While part of the material covered in this chapter is a review of information generally covered in strength of materials, the subject is extended beyond this point to a consideration of stress and strain in three dimensions. The material included in this chapter is important for an understanding of most of the phenomenological aspects of mechanical metallurgy, and for this reason it should be given careful attention by those readers to whom it is unfamiliar. In the space available for this subject it has not been possible to carry it to the point where extensive problem solving is possible. The material covered here should, however, provide a background for intelligent reading of the more mathematical literature in mechanical metallurgy.

It should be recognized that the equations describing the state of stress or strain in a body are applicable to any solid continuum, whether it be an elastic or plastic solid or a viscous fluid. Indeed, this body of knowledge is often called *continuum mechanics*. The equations relating stress and strain are called *constitutive equations* because they depend on the material behavior. In this chapter we shall only consider the constitutive equations for an elastic solid.

2-2 DESCRIPTION OF STRESS AT A POINT

As described in Sec. 1-8, it is often convenient to resolve the stresses at a point into normal and shear components. In the general case the shear components are at arbitrary angles to the coordinate axes, so that it is convenient to resolve each

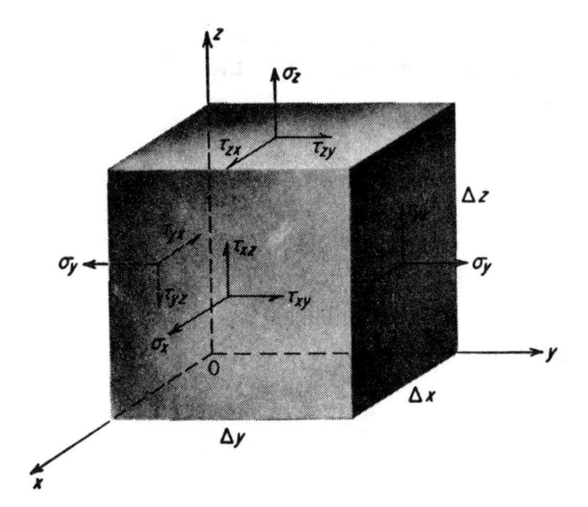

Figure 2-1 Stresses acting on an elemental cube.

shear stress further into two components. The general case is shown in Fig. 2-1. Stresses acting normal to the faces of the elemental cube are identified by the subscript which also identifies the direction in which the stress acts; that is σ_x is the normal stress acting in the x direction. Since it is a normal stress, it must act on the plane perpendicular to the x direction. By convention, values of normal stresses greater than zero denote tension; values less than zero indicate compression. All the normal stresses shown in Fig. 2-1 are tensile.

Two subscripts are needed for describing shearing stresses. The first subscript indicates the plane in which the stress acts and the second the direction in which the stress acts. Since a plane is most easily defined by its normal, the first subscript refers to this normal. For example, τ_{yz} is the shear stress on the plane perpendicular to the y axis in the direction of the z axis. τ_{yx} is the shear stress on a plane normal to the y axis in the direction of the x axis.

A shear stress is positive if it points in the positive direction on the positive face of a unit cube. (It is also positive if it points in the negative direction on the negative face of a unit cube.) All of the shear stresses in Fig. 2-2a are positive shear stresses regardless of the type of normal stresses that are present. A shear stress is negative if it points in the negative direction of a positive face of a unit cube and vice versa. The shearing stresses shown in Fig. 2-2b are all negative stresses.

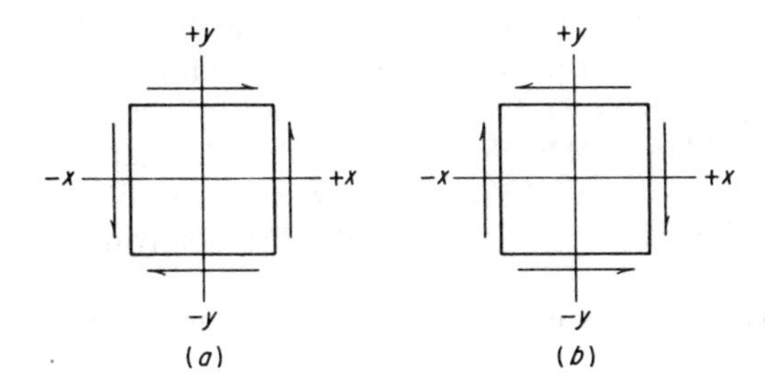

Figure 2-2 Sign convention for shear stress. (a) Positive; (b) negative.

The notation for stress given above is the one used by Timoshenko[1] and most American workers in the field of elasticity. However, many other notations have been used, some of which are given below.

$$\begin{array}{ccccc}
\sigma_x & \sigma_{11} & X_x & \widehat{xx} & p_{xx} \\
\sigma_y & \sigma_{22} & Y_y & \widehat{yy} & p_{yy} \\
\sigma_z & \sigma_{33} & Z_z & \widehat{zz} & p_{zz} \\
\tau_{xy} & \sigma_{12} & X_y & \widehat{xy} & p_{xy} \\
\tau_{yz} & \sigma_{23} & Y_z & \widehat{yz} & p_{yz} \\
\tau_{zx} & \sigma_{31} & Z_x & \widehat{zx} & p_{zx}
\end{array}$$

It can be seen from Fig. 2-1 that nine quantities must be defined in order to establish the state of stress at a point. They are σ_x, σ_y, σ_z, τ_{xy}, τ_{xz}, τ_{yx}, τ_{yz}, τ_{zx}, and τ_{zy}. However, some simplification is possible. If we assume that the areas of the faces of the unit cube are small enough so that the change in stress over the face is negligible, by taking the summation of the moments of the forces about the z axis it can be shown that $\tau_{xy} = \tau_{yx}$.

$$\begin{gathered}(\tau_{xy}\,\Delta y \Delta z)\,\Delta x = (\tau_{yx}\,\Delta x\,\Delta z)\,\Delta y \\ \therefore \quad \tau_{xy} = \tau_{yx}\end{gathered} \tag{2-1}$$

and in like manner

$$\tau_{xz} = \tau_{zx} \qquad \tau_{yz} = \tau_{zy}$$

Thus, the state of stress at a point is completely described by six components: three normal stresses and three shear stresses, σ_x, σ_y, σ_z, τ_{xy}, τ_{xz}, τ_{yz}.

2-3 STATE OF STRESS IN TWO DIMENSIONS (PLANE STRESS)

Many problems can be simplified by considering a two-dimensional state of stress. This condition is frequently approached in practice when one of the dimensions of the body is small relative to the others. For example, in a thin plate loaded in the plane of the plate there will be no stress acting perpendicular to the surface of the plate. The stress system will consist of two normal stresses σ_x and σ_y and a shear stress τ_{xy}. A stress condition in which the stresses are zero in one of the primary directions is called *plane stress*.

Figure 2-3 illustrates a thin plate with its thickness normal to the plane of the paper. In order to know the state of stress at point O in the plate, we need to be able to describe the stress components at O for any orientation of the axes through the point. To do this, consider an oblique plane normal to the plane of the paper at an angle θ between the x axis and the outward normal to the oblique plane. Let the normal to this plane be the x' direction and the direction lying in

[1] S. P. Timoshenko, and J. N. Goodier, "Theory of Elasticity," 2d ed., McGraw-Hill Book Company, New York, 1951.

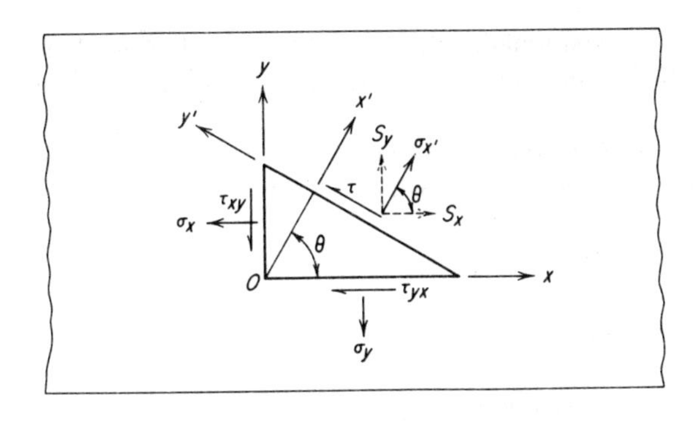

Figure 2-3 Stress on oblique plane (two dimensions).

the oblique plane the y' direction. It is assumed that the plane shown in Fig. 2-3 is an infinitesimal distance from O and that the element is so small that variations in stress over the sides of the element can be neglected. The stresses acting on the oblique plane are the normal stress σ and the shear stress τ. The direction cosines between x' and the x and y axes are l and m, respectively. From the geometry of Fig. 2-3, $l = \cos\theta$ and $m = \sin\theta$. If A is the area of the oblique plane, the areas of the sides of the element perpendicular to the x and y axes are Al and Am.

Let S_x and S_y denote the x and y components of the total stress acting on the inclined face. By taking the summation of the *forces* in the x direction and the y direction, we obtain

$$S_xA = \sigma_xAl + \tau_{xy}Am$$

$$S_yA = \sigma_yAm + \tau_{xy}Al$$

or

$$S_x = \sigma_x\cos\theta + \tau_{xy}\sin\theta$$

$$S_y = \sigma_y\sin\theta + \tau_{xy}\cos\theta$$

The components of S_x and S_y in the direction of the normal stress σ are

$$S_{xN} = S_x\cos\theta \quad \text{and} \quad S_{yN} = S_y\sin\theta$$

so that the normal stress acting on the oblique plane is given by

$$\begin{aligned}\sigma_{x'} &= S_x\cos\theta + S_y\sin\theta \\ \sigma_{x'} &= \sigma_x\cos^2\theta + \sigma_y\sin^2\theta + 2\tau_{xy}\sin\theta\cos\theta\end{aligned} \tag{2-2}$$

The shearing stress on the oblique plane is given by

$$\begin{aligned}\tau_{x'y'} &= S_y\cos\theta - S_x\sin\theta \\ \tau_{x'y'} &= \tau_{xy}(\cos^2\theta - \sin^2\theta) + (\sigma_y - \sigma_x)\sin\theta\cos\theta\end{aligned} \tag{2-3}$$

The stress $\sigma_{y'}$ may be found by substituting $\theta + \pi/2$ for θ in Eq. (2-2), since $\sigma_{y'}$ is orthogonal to $\sigma_{x'}$.

$$\sigma_{y'} = \sigma_x\cos^2(\theta + \pi/2) + \sigma_y\sin^2(\theta + \pi/2) + 2\tau_{xy}\sin(\theta + \pi/2)\cos(\theta + \pi/2)$$

and since $\sin(\theta + \pi/2) = \cos\theta$ and $\cos(\theta + \pi/2) = -\sin\theta$, we obtain

$$\sigma_{y'} = \sigma_x\sin^2\theta + \sigma_y\cos^2\theta - 2\tau_{xy}\sin\theta\cos\theta \tag{2-4}$$

Equations (2-2) to (2-4) are the transformation of stress equations which give the stresses in an $x'y'$ coordinate system if the stresses in an xy coordinate system and the angle θ are known.

To aid in computation, it is often convenient to express Eqs. (2-2) to (2-4) in terms of the double angle 2θ. This can be done with the following identities:

$$\cos^2\theta = \frac{\cos 2\theta + 1}{2}$$

$$\sin^2\theta = \frac{1 - \cos 2\theta}{2}$$

$$2\sin\theta\cos\theta = \sin 2\theta$$

$$\cos^2\theta - \sin^2\theta = \cos 2\theta$$

The transformation of stress equations now become

$$\sigma_{x'} = \frac{\sigma_x + \sigma_y}{2} + \frac{\sigma_x - \sigma_y}{2}\cos 2\theta + \tau_{xy}\sin 2\theta \tag{2-5}$$

$$\sigma_{y'} = \frac{\sigma_x + \sigma_y}{2} - \frac{\sigma_x - \sigma_y}{2}\cos 2\theta - \tau_{xy}\sin 2\theta \tag{2-6}$$

$$\tau_{x'y'} = \frac{\sigma_y - \sigma_x}{2}\sin 2\theta + \tau_{xy}\cos 2\theta \tag{2-7}$$

It is important to note that $\sigma_{x'} + \sigma_{y'} = \sigma_x + \sigma_y$. Thus the sum of the normal stresses on two perpendicular planes is an *invariant* quantity, that is, it is independent of orientation or angle θ.

Equations (2-2) and (2-3) and their equivalents, Eqs. (2-5) and (2-7), describe the normal stress and shear stress on any plane through a point in a body subjected to a plane-stress situation. Figure 2-4 shows the variation of normal stress and shear stress with θ for the biaxial-plane-stress situation given at the top of the figure. Note the following important facts about this figure:

1. The maximum and minimum values of normal stress on the oblique plane through point O occur when the shear stress is zero.
2. The maximum and minimum values of both normal stress and shear stress occur at angles which are 90° apart.
3. The maximum shear stress occurs at an angle halfway between the maximum and minimum normal stresses.
4. The variation of normal stress and shear stress occurs in the form of a sine wave, with a period of $\theta = 180°$. These relationships are valid for any state of stress.

For any state of stress it is always possible to define a new coordinate system which has axes perpendicular to the planes on which the maximum normal stresses act and on which no shearing stresses act. These planes are called the *principal planes*, and the stresses normal to these planes are the *principal stresses*. For two-dimensional plane stress there will be two principal stresses σ_1 and σ_2 which occur at angles that are 90° apart (Fig. 2-4). For the general case of stress

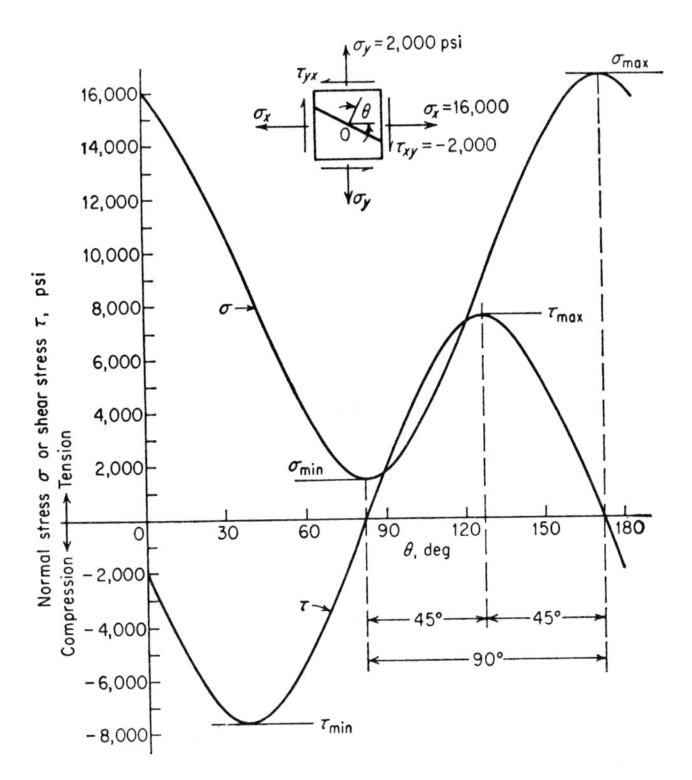

Figure 2-4 Variation of normal stress and shear stress on oblique plane with angle θ.

in three dimensions there will be three principal stresses σ_1, σ_2, and σ_3. According to convention, σ_1 is the algebraically greatest principal stress, while σ_3 is the algebraically smallest stress. The directions of the principal stresses are the *principal axes* 1, 2, and 3. Although in general the principal axes 1, 2, and 3 do not coincide with the cartesian-coordinate axes x, y, z, for many situations that are encountered in practice the two systems of axes coincide because of symmetry of loading and deformation. The specification of the principal stresses and their direction provides a convenient way of describing the state of stress at a point.

Since by definition a principal plane contains no shear stress, its angular relationship with respect to the xy coordinate axes can be determined by finding the values of θ in Eq. (2-3) for which $\tau_{x'y'} = 0$.

$$\tau_{xy}(\cos^2\theta - \sin^2\theta) + (\sigma_y - \sigma_x)\sin\theta\cos\theta = 0$$

$$\frac{\tau_{xy}}{\sigma_x - \sigma_y} = \frac{\sin\theta\cos\theta}{\cos^2\theta - \sin^2\theta} = \frac{\frac{1}{2}(\sin 2\theta)}{\cos 2\theta} = \frac{1}{2}\tan 2\theta$$

$$\tan 2\theta = \frac{2\tau_{xy}}{\sigma_x - \sigma_y} \qquad (2\text{-}8)$$

Since $\tan 2\theta = \tan(\pi + 2\theta)$, Eq. (2-8) has two roots, θ_1 and $\theta_2 = \theta_1 + n\pi/2$. These roots define two mutually perpendicular planes which are free from shear.

Equation (2-5) will give the principal stresses when values of $\cos 2\theta$ and $\sin 2\theta$ are substituted into it from Eq. (2-8). The values of $\cos 2\theta$ and $\sin 2\theta$ are found from Eq. (2-8) by means of the pythagorean relationships.

$$\sin 2\theta = \pm \frac{\tau_{xy}}{\left[(\sigma_x - \sigma_y)^2/4 + \tau_{xy}^2\right]^{1/2}}$$

$$\cos 2\theta = \pm \frac{(\sigma_x - \sigma_y)/2}{\left[(\sigma_x - \sigma_y)^2/4 + \tau_{xy}^2\right]^{1/2}}$$

Substituting these values into Eq. (2-5) results in the expression for the maximum and minimum principal stresses for a two-dimensional (biaxial) state of stress.

$$\left.\begin{aligned}\sigma_{max} &= \sigma_1 \\ \sigma_{min} &= \sigma_2\end{aligned}\right\} = \frac{\sigma_x + \sigma_y}{2} \pm \left[\left(\frac{\sigma_x - \sigma_y}{2}\right)^2 + \tau_{xy}^2\right]^{1/2} \tag{2-9}$$

The direction of the principal planes is found by solving for θ in Eq. (2-8), taking special care to establish whether 2θ is between 0 and $\pi/2$, π, and $3\pi/2$, etc. Figure 2-5 shows a simple way to establish the direction of the largest principal stress σ_1. σ_1 will lie between the algebraically largest normal stress and the shear diagonal. To see this intuitively, consider that if there were no shear stresses, then $\sigma_x = \sigma_1$. If only shear stresses act, then a normal stress (the principal stress) would exist along the shear diagonal. If both normal and shear stresses act on the element, then σ_1 lies between the influences of these two effects.

To find the maximum shear stress we return to Eq. (2-7). We differentiate the expression for $\tau_{x'y'}$ and set this equal to zero.

$$\frac{d\tau_{x'y'}}{d\theta} = (\sigma_y - \sigma_x)\cos 2\theta - 2\tau_{xy}\sin 2\theta = 0$$

$$\tan 2\theta_s = \frac{\sigma_y - \sigma_x}{2\tau_{xy}} = -\frac{\sigma_x - \sigma_y}{2\tau_{xy}} \tag{2-10}$$

Comparing this with the angle at which the principal planes occur, Eq. (2-8), $\tan 2\theta_n = 2\tau_{xy}/(\sigma_x - \sigma_y)$, we see that $\tan 2\theta_s$ is the negative reciprocal of $\tan 2\theta_n$.

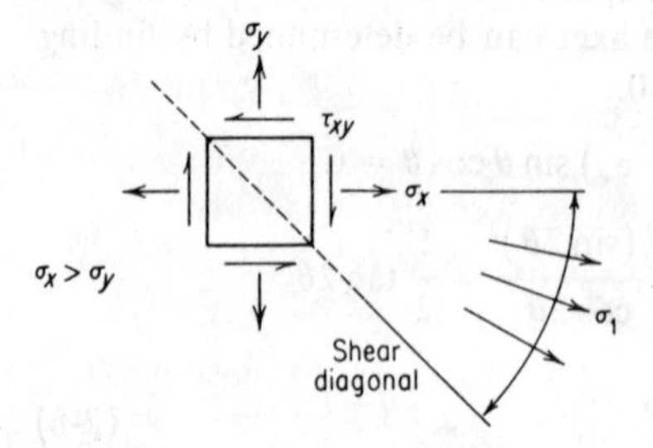

Figure 2-5 Method of establishing direction of σ_1.

This means that $2\theta_s$ and $2\theta_n$ are orthogonal, and that θ_s and θ_n are separated in space by 45°. The magnitude of the maximum shear stress is found by substituting Eq. (2-10) into Eq. (2-7).

$$\tau_{max} = \pm\left[\left(\frac{\sigma_x - \sigma_y}{2}\right)^2 + \tau_{xy}^2\right]^{1/2} \tag{2-11}$$

Example The state of stress is given by $\sigma_x = 25p$ and $\sigma_y = 5p$ plus shearing stresses τ_{xy}. On a plane at 45° counterclockwise to the plane on which σ_x acts the state of stress is 50 MPa tension and 5 MPa shear. Determine the values of $\sigma_x, \sigma_y, \tau_{xy}$.

From Eqs. (2-5) and (2-7)

$$\sigma_{x'} = \frac{\sigma_x + \sigma_y}{2} + \frac{\sigma_x - \sigma_y}{2}\cos 2\theta + \tau_{xy}\sin 2\theta \qquad \text{Eq. (2-5)}$$

$$50 \times 10^6 = \frac{25p + 5p}{2} + \frac{25p - 5p}{2}\cos 90° + \tau_{xy}\sin 90°$$

$$15p + \tau_{xy} = 50 \times 10^6\ \text{N/m}^2$$

$$\tau_{x'y'} = \frac{\sigma_y - \sigma_x}{2}\sin 2\theta + \tau_{xy}\cos 2\theta \qquad \text{Eq. (2-7)}$$

$$5 \times 10^6 = \left(\frac{5p - 25p}{2}\right)\sin 90° + \tau_{xy}\cos 90$$

$$-10p = 5 \times 10^6 \qquad p = -5 \times 10^5\ \text{N/m}^2$$

$$\therefore \quad \sigma_x = 25(-5 \times 10^5) = -12.5\ \text{MPa}$$

$$\sigma_y = 5(p) = -2.5\ \text{MPa}$$

$$\tau_{xy} = 50 \times 10^6 - 15(-5 \times 10^5)$$

$$= 50 \times 10^6 + 7.5 \times 10^6 = 57.5\ \text{MPa}$$

We also can find $\sigma_{y'}$, orthogonal to $\sigma_{x'} = 50$ MPa, since $\sigma_x + \sigma_y = \sigma_{x'} + \sigma_{y'}$

$$-12.5 - 2.5 = 50 + \sigma_{y'}$$

$$\sigma_{y'} = -65\ \text{MPa}$$

2-4 MOHR'S CIRCLE OF STRESS—TWO DIMENSIONS

A very useful graphical method for representing the state of stress at a point on an oblique plane through the point was suggested by O. Mohr. The transformation of stress equations, Eqs. (2-5) and (2-7), can be rearranged to give

$$\sigma_{x'} - \frac{\sigma_x + \sigma_y}{2} = \frac{\sigma_x - \sigma_y}{2} \cos 2\theta + \tau_{xy} \sin 2\theta$$

$$\tau_{y'x'} = \frac{\sigma_y - \sigma_x}{2} \sin 2\theta + \tau_{xy} \cos 2\theta$$

We can solve for $\sigma_{x'}$ in terms of $\tau_{x'y'}$ by squaring each of these equations and adding

$$\left(\sigma_{x'} - \frac{\sigma_x + \sigma_y}{2}\right)^2 + \tau_{x'y'}^2 = \left(\frac{\sigma_x - \sigma_y}{2}\right)^2 + \tau_{xy}^2 \tag{2-12}$$

Equation (2-12) is the equation of a circle of the form $(x - h)^2 + y^2 = r^2$. Thus, Mohr's circle is a circle in $\sigma_{x'}$, $\tau_{x'y'}$ coordinates with a radius equal to $\tau_{\max}$ and the center displaced $(\sigma_x + \sigma_y)/2$ to the right of the origin.

In working with Mohr's circle there are only a few basic rules to remember. An angle of θ on the physical element is represented by 2θ on Mohr's circle. The same sense of rotation (clockwise or counterclockwise) should be used in each case. A different convention to express shear stress is used in drawing and interpreting Mohr's circle. This convention is that a shear stress causing a clockwise rotation about any point in the physical element is plotted above the horizontal axis of the Mohr's circle. A point on Mohr's circle gives the magnitude and direction of the normal and shear stresses on any plane in the physical element.

Figure 2-6 illustrates the plotting and use of Mohr's circle for a particular stress state shown at the upper left. Normal stresses are plotted along the x axis, shear stresses along the y axis. The stresses on the planes normal to the x and y

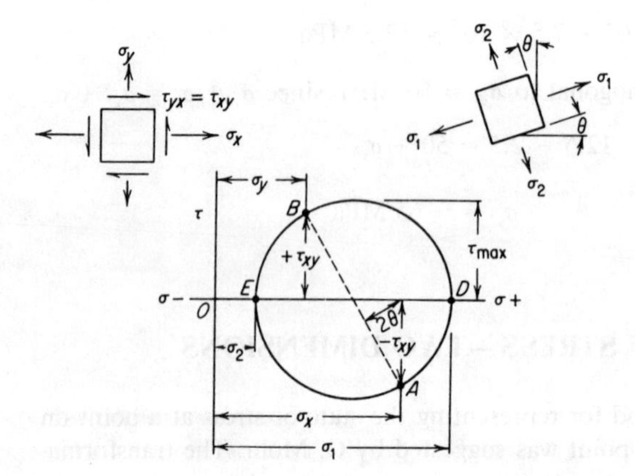

Figure 2-6 Mohr's circle for two-dimensional state of stress.

axes are plotted as points A and B. The intersection of the line AB with the σ axis determines the center of the circle. At points D and E the shear stress is zero, so these points represent the values of the principal stresses. The angle between σ_x and σ_1 on Mohr's circle is 2θ. Since this angle is measured counterclockwise on Mohr's circle on the physical element, σ_1 acts counterclockwise from the x axis at an angle θ (see sketch, upper right). The stresses on any other plane whose normal makes an angle of θ with the x axis could be found from Mohr's circle in the same way.

2-5 STATE OF STRESS IN THREE DIMENSIONS

The general three-dimensional state of stress consists of three unequal principal stresses acting at a point. This is called a *triaxial state of stress*. If two of the three principal stresses are equal, the state of stress is known as *cylindrical*, while if all three principal stresses are equal, the state of stress is said to be *hydrostatic*, or *spherical*.

The determination of the principal stresses for a three-dimensional state of stress in terms of the stresses acting on an arbitrary cartesian-coordinate system is an extension of the method described in Sec. 2-3 for the two-dimensional case. Figure 2-7 represents an elemental free body similar to that shown in Fig. 2-1 with a diagonal plane JKL of area A. The plane JKL is assumed to be a principal plane cutting through the unit cube. σ is the principal stress acting normal to the plane JKL. Let l, m, n be the direction cosines of σ, that is, the cosines of the angles between σ and the x, y, and z axes. Since the free body in Fig. 2-7, must be in equilibrium, the forces acting on each of its faces must

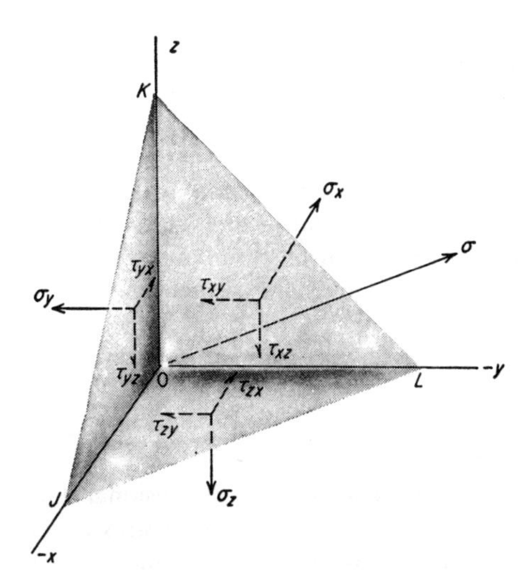

Figure 2-7 Stresses acting on elemental free body.

balance. The components of σ along each of the axes are S_x, S_y, and S_z.

$$S_x = \sigma l \qquad S_y = \sigma m \qquad S_z = \sigma n$$

$$\text{Area } KOL = Al \qquad \text{Area } JOK = Am \qquad \text{Area } JOL = An$$

Taking the summation of the forces in the x direction results in

$$\sigma Al - \sigma_x Al - \tau_{yx} Am - \tau_{zx} An = 0$$

which reduces to

$$(\sigma - \sigma_x)l - \tau_{yx}m - \tau_{zx}n = 0 \tag{2-13a}$$

Summing the forces along the other two axes results in

$$-\tau_{xy}l + (\sigma - \sigma_y)m - \tau_{zy}n = 0 \tag{2-13b}$$

$$-\tau_{xz}l - \tau_{yz}m + (\sigma - \sigma_z)n = 0 \tag{2-13c}$$

Equations (2-13) are three homogeneous linear equations in terms of l, m, and n. The only nontrivial solution can be obtained by setting the determinant of the coefficients of l, m, and n equal to zero, since l, m, and n cannot all be zero.

$$\begin{vmatrix} \sigma - \sigma_x & -\tau_{yx} & -\tau_{zx} \\ -\tau_{xy} & \sigma - \sigma_y & -\tau_{zy} \\ -\tau_{xz} & -\tau_{yz} & \sigma - \sigma_z \end{vmatrix} = 0$$

Solution of the determinant results in a cubic equation in σ.

$$\sigma^3 - (\sigma_x + \sigma_y + \sigma_z)\sigma^2 + \left(\sigma_x\sigma_y + \sigma_y\sigma_z + \sigma_x\sigma_z - \tau_{xy}^2 - \tau_{yz}^2 - \tau_{xz}^2\right)\sigma$$
$$-\left(\sigma_x\sigma_y\sigma_z + 2\tau_{xy}\tau_{yz}\tau_{xz} - \sigma_x\tau_{yz}^2 - \sigma_y\tau_{xz}^2 - \sigma_z\tau_{xy}^2\right) = 0 \tag{2-14}$$

The three roots of Eq. (2-14) are the three principal stresses σ_1, σ_2, and σ_3. To determine the direction, with respect to the original x, y, z axes, in which the principal stresses act, it is necessary to substitute, σ_1, σ_2, and σ_3 each in turn into the three equations of Eq. (2-13). The resulting equations must be solved simultaneously for l, m, and n with the help of the auxiliary relationship $l^2 + m^2 + n^2 = 1$.

Note that there are three combinations of stress components in Eq. (2-14) that make up the coefficients of the cubic equation. Since the values of these coefficients determine the principal stresses, they obviously do not vary with changes in the coordinate axes. Therefore, they are invariant coefficients.

$$\sigma_x + \sigma_y + \sigma_z = I_1$$

$$\sigma_x\sigma_y + \sigma_y\sigma_z + \sigma_x\sigma_z - \tau_{xy}^2 - \tau_{xz}^2 - \tau_{yz}^2 = I_2$$

$$\sigma_x\sigma_y\sigma_z + 2\tau_{xy}\tau_{yz}\tau_{xz} - \sigma_x\tau_{yz}^2 - \sigma_y\tau_{xz}^2 - \sigma_z\tau_{xy}^2 = I_3$$

The first invariant of stress I_1 has been seen before for the two-dimensional state of stress. It states the useful relationship that the sum of the normal stresses for any orientation in the coordinate system is equal to the sum of the normal stresses for any other orientation. For example

$$\sigma_x + \sigma_y + \sigma_z = \sigma_{x'} + \sigma_{y'} + \sigma_{z'} = \sigma_1 + \sigma_2 + \sigma_3 \tag{2-15}$$

Example Determine the principal stresses for the state of stress

$$\begin{bmatrix} 0 & -240 & 0 \\ -240 & 200 & 0 \\ 0 & 0 & -280 \end{bmatrix} \text{MPa}$$

From Eq. (2-14)

$$\sigma^3 - (200 - 280)\sigma^2 + \left[(200)(-280) - (-240)^2\right]\sigma - (-280)(-240)^2 = 0$$

$\sigma = -280$ MPa is a principal stress because $\tau_{zx} = \tau_{xz} = 0$ and $\tau_{zy} = \tau_{yz} = 0$

$$[\sigma - (-280)]\sqrt{\sigma^3 - I_1\sigma^2 + I_2\sigma - I_3} = \sigma^2 - 200\sigma - (240)^2$$

$$\sigma = \frac{200 \pm \left[(-200)^2 + 4(240)^2\right]^{1/2}}{2} = 100 \pm 260$$

$$\sigma_1 = 360 \text{ MPa}; \qquad \sigma_2 = -160 \text{ MPa}; \qquad \sigma_3 = -280 \text{ MPa}$$

In the discussion above we developed the equation for the stress on a particular oblique plane, a principal plane in which there is no shear stress. Let us now develop the equations for the normal and shear stress on *any* oblique plane whose normal has the direction cosines l, m, n with the x, y, z axes. We can use Fig. 2-7 once again if we realize that for this general situation the total stress on the plane S will not be coaxial with the normal stress, and that $S^2 = \sigma^2 + \tau^2$. Once again the total stress can be resolved into components S_x, S_y, S_z, so that

$$S^2 = S_x^2 + S_y^2 + S_z^2 \tag{2-16}$$

Taking the summation of the *forces* in the x, y, and z directions, we arrive at the expressions for the orthogonal components of the total stress:

$$S_x = \sigma_x l + \tau_{yx} m + \tau_{zx} n \tag{2-17a}$$

$$S_y = \tau_{xy} l + \sigma_y m + \tau_{zy} n \tag{2-17b}$$

$$S_z = \tau_{xz} l + \tau_{yz} m + \sigma_z n \tag{2-17c}$$

To find the normal stress σ on the oblique plane, it is necessary to determine the components of S_x, S_y, S_z in the direction of the normal to the oblique plane. Thus,

$$\sigma = S_x l + S_y m + S_z n$$

or, after substituting from Eqs. (2-17) and simplifying with $\tau_{xy} = \tau_{yx}$, etc.

$$\sigma = \sigma_x l^2 + \sigma_y m^2 + \sigma_z n^2 + 2\tau_{xy} lm + 2\tau_{yz} mn + 2\tau_{zx} nl \tag{2-18}$$

The magnitude of the shear stress on the oblique plane can be found from $\tau^2 = S^2 - \sigma^2$. To get the magnitude and direction of the two shear stress components lying in the oblique plane it is necessary to resolve the stress components S_x, S_y, S_z into the y' and z' directions lying in the oblique plane.[1] This development will not be carried out here because the pertinent equations can be derived more easily by the methods given in Sec. 2-6.

Since plastic flow involves shearing stresses, it is important to identify the planes on which the *maximum* or *principal shear stresses* occur. In our discussion of the two-dimensional state of stress we saw that τ_{max} occurred on a plane halfway between the two principal planes. Therefore it is easiest to define the principal shear planes in terms of the three principal axes 1, 2, 3. From $\tau^2 = S^2 - \sigma^2$ it can be shown that

$$\tau^2 = (\sigma_1 - \sigma_2)^2 l^2 m^2 + (\sigma_1 - \sigma_3)^2 l^2 n^2 + (\sigma_2 - \sigma_3)^2 m^2 n^2 \qquad (2\text{-}19)$$

where l, m, n are the direction cosines between the normal to the oblique plane and the principal axes.

The principal shear stresses occur for the following combinations of direction cosines that bisect the angle between two of the three principal axes:

$$\begin{array}{cccc} l & m & n & \tau \\ 0 & \pm\sqrt{\frac{1}{2}} & \pm\sqrt{\frac{1}{2}} & \tau_1 = \frac{\sigma_2 - \sigma_3}{2} \\ \pm\sqrt{\frac{1}{2}} & 0 & \pm\sqrt{\frac{1}{2}} & \tau_2 = \frac{\sigma_1 - \sigma_3}{2} \\ \pm\sqrt{\frac{1}{2}} & \pm\sqrt{\frac{1}{2}} & 0 & \tau_3 = \frac{\sigma_1 - \sigma_2}{2} \end{array} \qquad (2\text{-}20)$$

Since according to convention σ_1 is the algebraically greatest principal normal stress and σ_3 is the algebraically smallest principal stress, τ_2 has the largest value of shear stress and it is called the *maximum shear stress* τ_{max}.

$$\tau_{max} = \frac{\sigma_1 - \sigma_3}{2} \qquad (2\text{-}21)$$

The maximum shear stress is important in theories of yielding and metal-forming operations. Figure 2-8 shows the planes of the principal shear stresses for a cube whose faces are the principal planes. Note that for each pair of principal stresses there are two planes of principal shear stress, which bisect the directions of the principal stresses.

[1] P. C. Chou and N. J. Pagano, "Elasticity," p. 24, D. Van Nostrand Company, Inc., Princeton, N.J., 1967.

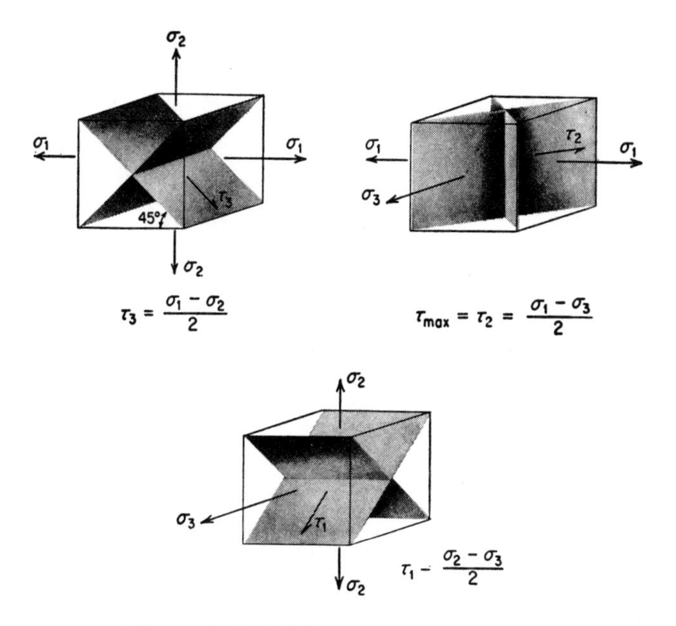

Figure 2-8 Planes of principal shear stresses.

2-6 STRESS TENSOR

Many aspects of the analysis of stress, such as the equations for the transformation of the stress components from one set of coordinate axes to another coordinate system or the existence of principal stresses, become simpler when it is realized that stress is a second-rank tensor quantity. Many of the techniques for manipulating second-rank tensors do not require a deep understanding of tensor calculus, so it is advantageous to learn something about the properties of tensors.

We shall start with the consideration of the transformation of a vector (a first-rank tensor) from one coordinate system to another. Consider the vector $\mathbf{S} = S_1 i_1 + S_2 i_2 + S_3 i_3$, when the unit vectors i_1, i_2, i_3 are in the directions x_1, x_2, x_3. (In accordance with convention and convenience in working with tensor quantities, the coordinate axes will be designated x_1, x_2, etc., where x_1 is equivalent to our previous designation x, x_2 is equivalent to the old y, etc.) S_1, S_2, S_3 are the components of $\mathbf{S}$ referred to the axes x_1, x_2, x_3. We now want to find the components of $\mathbf{S}$ referred to the x_1', x_2', x_3' axes, Fig. 2-9. S_1' is obtained by resolving S_1, S_2, S_3 along the new direction x_1'.

$$S_1' = S_1 \cos(x_1 x_1') + S_2 \cos(x_2 x_1') + S_3 \cos(x_3 x_1')$$

or

$$S_1' = a_{11} S_1 + a_{12} S_2 + a_{13} S_3 \tag{2-22a}$$

where a_{11} is the direction cosine between x_1' and x_1, a_{12} is the direction cosine

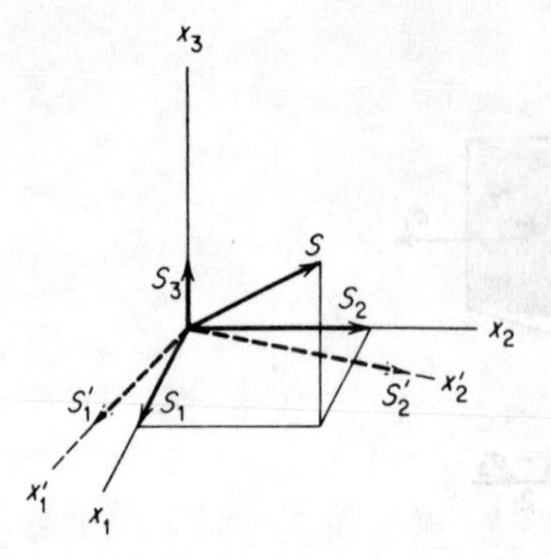

Figure 2-9 Transformation of axes for a vector.

between x_1' and x_2, etc. Similarly,

$$S_2' = a_{21}S_1 + a_{22}S_2 + a_{23}S_3 \tag{2-22b}$$

$$S_3' = a_{31}S_1 + a_{32}S_2 + a_{33}S_3 \tag{2-22c}$$

We note that the leading suffix for each direction cosine in each equation is the same, so we could write these equations as

$$S_1' = \sum_{j=1}^{3} a_{1j}S_j \qquad S_2' = \sum_{j=1}^{3} a_{2j}S_j \qquad S_3' = \sum_{j=1}^{3} a_{3j}S_j$$

These three equations could be combined by writing

$$S_i' = \sum_{j=1}^{3} a_{ij}S_j(i = 1, 2, 3) = a_{i1}S_1 + a_{i2}S_2 + a_{i3}S_3 \tag{2-23}$$

Still greater brevity is obtained by writing Eq. (2-23) in the Einstein suffix notation

$$S_i' = a_{ij}S_j \tag{2-24}$$

The suffix notation is a very useful way of compactly expressing the systems of equations usually found in continuum mechanics. In Eq. (2-24) it is understood that when a suffix occurs twice in the same term (in this case the suffix j), it indicates *summation* with respect to that suffix. Unless otherwise indicated, the summation of the other index is from 1 to 3.

In the above example, i is a *free suffix* and it is understood that in the expanded form there is one equation for each value of i. The repeated index is called a *dummy suffix*. Its only purpose is to indicate summation. Exactly the same three equations would be produced if some other letter were used for the dummy suffix, for example, $S_i' = a_{ir}S_r$ would mean the same thing as Eq. (2-24).

We saw in Sec. 2-5 that the complete determination of the state of stress at a point in a solid requires the specification of nine components of stress on the orthogonal faces of the element at the point. A vector quantity only requires the specification of three components. Obviously, stress is more complicated than a vector. Physical quantities that transform with coordinate axes in the manner of Eq. (2-18) are called *tensors* of the *second rank*. Stress, strain, and many other physical quantities are second-rank tensors. A scalar quantity, which remains unchanged with transformation of axes, requires only a single number for its specification. Scalars are tensors of zero rank. Vector quantities require three components for their specification, so they are tensors of the first rank. The number of components required to specify a quantity is 3^n, where n is the rank of the tensor.[1] The elastic constant that relates stress with strain in an elastic solid is a fourth-rank tensor with 81 components in the general case.

Example The displacements of points in a deformed elastic solid (u) are related to the coordinates of the points (x) by a vector relationship $u_i = e_{ij}x_j$. Expand this tensor expression.

Since j is the dummy suffix, summation will take place over $j = 1, 2, 3$.

$$u_1 = \sum e_{1j}x_j = e_{11}x_1 + e_{12}x_2 + e_{13}x_3$$
$$u_2 = \sum e_{2j}x_j = e_{21}x_1 + e_{22}x_2 + e_{23}x_3$$
$$u_3 = \sum e_{3j}x_j = e_{31}x_1 + e_{32}x_2 + e_{33}x_3$$

The coefficients in these equations are the components of the strain tensor.

The product of two vectors **A** and **B** having components (A_1, A_2, A_3) and (B_1, B_2, B_3) results in a second-rank tensor T_{ij}. The components of this tensor can be displayed as a 3×3 matrix.

$$T_{ij} = \begin{vmatrix} T_{11} & T_{12} & T_{13} \\ T_{21} & T_{22} & T_{23} \\ T_{31} & T_{32} & T_{33} \end{vmatrix} = \begin{vmatrix} A_1B_1 & A_1B_2 & A_1B_3 \\ A_2B_1 & A_2B_2 & A_2B_3 \\ A_3B_1 & A_3B_2 & A_3B_3 \end{vmatrix}$$

On transformation of axes the vector components become (A_1', A_2', A_3') and (B_1', B_2', B_3'). We wish to find the relationship between the nine components of T_{ij} and the nine components of T_{ij}' after the transformation of axes.

$$A_i' = a_{ij}A_j \qquad B_k' = a_{kl}B_l$$

or

$$A_i'B_k' = (a_{ij}A_j)(a_{kl}B_l) \tag{2-25}$$
$$T_{ik}' = a_{ij}a_{kl}T_{jl}$$

Since stress is a second-rank tensor, the components of the stress tensor can be written as

$$\sigma_{ij} = \begin{vmatrix} \sigma_{11} & \sigma_{12} & \sigma_{13} \\ \sigma_{21} & \sigma_{22} & \sigma_{23} \\ \sigma_{31} & \sigma_{32} & \sigma_{33} \end{vmatrix} = \begin{vmatrix} \sigma_x & \tau_{xy} & \tau_{xz} \\ \tau_{yx} & \sigma_y & \tau_{yz} \\ \tau_{zx} & \tau_{zy} & \sigma_z \end{vmatrix}$$

[1] A more precise relationship is $N = k^n$, where N is the number of components required for the description of a tensor of the nth rank in a k-dimensional space. For a two dimensional space only four components are required to describe a second-rank tensor.

The transformation of the stress tensor σ_{ij} from the x_1, x_2, x_3 system of axes to the x'_1, x'_2, x'_3 axes is given by

$$\sigma_{kl} = a_{ki}a_{lj}\sigma_{ij} \tag{2-26}$$

where i and j are dummy suffixes and k and l are free suffixes. To expand the tensor equation, we first sum over $j = 1, 2, 3$.

$$\sigma_{kl} = a_{ki}a_{l1}\sigma_{i1} + a_{ki}a_{l2}\sigma_{i2} + a_{ki}a_{l3}\sigma_{i3}$$

Now summing over $i = 1, 2, 3$

$$\begin{aligned}\sigma_{kl} = {} & a_{k1}a_{l1}\sigma_{11} + a_{k1}a_{l2}\sigma_{12} + a_{k1}a_{l3}\sigma_{13} \\ & + a_{k2}a_{l1}\sigma_{21} + a_{k2}a_{l2}\sigma_{22} + a_{k2}a_{l3}\sigma_{23} \\ & + a_{k3}a_{l1}\sigma_{31} + a_{k3}a_{l2}\sigma_{32} + a_{k3}a_{l3}\sigma_{33}\end{aligned} \tag{2-27}$$

For each value of k and l there will be an equation similar to (2-27). Thus, to find the equation for the normal stress in the x'_1 direction, let $k = 1$ and $l = 1$

$$\begin{aligned}\sigma_{11} = {} & a_{11}a_{11}\sigma_{11} + a_{11}a_{12}\sigma_{12} + a_{11}a_{13}\sigma_{13} \\ & + a_{12}a_{11}\sigma_{21} + a_{12}a_{12}\sigma_{22} + a_{12}a_{13}\sigma_{23} \\ & + a_{13}a_{11}\sigma_{31} + a_{13}a_{12}\sigma_{32} + a_{13}a_{13}\sigma_{33}\end{aligned}$$

The reader should verify that this reduces to Eq. (2-18) when recast in the symbolism of Sec. 2-5.

Similarly, if we want to determine the shear stress on the x' plane in the z' direction, that is, $\tau_{x'z'}$, let $k = 1$ and $l = 3$

$$\begin{aligned}\sigma_{13} = {} & a_{11}a_{31}\sigma_{11} + a_{11}a_{32}\sigma_{12} + a_{11}a_{33}\sigma_{13} \\ & + a_{12}a_{31}\sigma_{21} + a_{12}a_{32}\sigma_{22} + a_{12}a_{33}\sigma_{23} \\ & + a_{13}a_{31}\sigma_{31} + a_{13}a_{32}\sigma_{32} + a_{13}a_{33}\sigma_{33}\end{aligned}$$

It is perhaps worth emphasizing again that it is immaterial what letters are used for subscripts in tensor notation. Thus, the transformation of a second-rank tensor could just as well be written as $T'_{st} = a_{sp}a_{tq}T_{pq}$, where T_{pq} are the components in the original unprimed axes and T'_{st} are the components referred to the new primed axes. The transformation law for a third-rank tensor is written

$$T'_{stv} = a_{sp}a_{tq}a_{vr}T_{pqr}$$

The material presented so far in this section is really little more than tensor notation. However, even with the minimal topics that have been discussed we have gained a powerful shorthand method for writing the often unwieldy equations of continuum mechanics. (The student will find that this will greatly ease the problem of remembering equations.) We have also gained a useful technique for transforming a tensor quantity from one set of axes to another. There are only a few additional facts about tensors that we need to consider. The student interested in pursuing this topic further is referred to a number of applications-oriented texts on cartesian tensors.[1]

[1] L. G. Jaeger, "Cartesian Tensors in Engineering Science," Pergamon Press, New York, 1966.

A useful quantity in tensor theory is the Kronecker delta δ_{ij}. The Kronecker delta is a second-rank unit isotropic tensor, that is, it has identical components in any coordinate system.

$$\delta_{ij} = \begin{vmatrix} 1 & 0 & 0 \\ 0 & 1 & 0 \\ 0 & 0 & 1 \end{vmatrix} = \begin{cases} 1 & i = j \\ 0 & i \neq j \end{cases} \tag{2-28}$$

Multiplication of a tensor or products of tensors by δ_{ij} result in a reduction of two in the rank of the tensor. This is called *contraction* of the tensor. The rule is stated here without proof but examples are given so we can make use of this operation in subsequent discussions. Consider the product of two second-rank tensors $A_{pq}B_{vw}$. This multiplication would produce a fourth-rank tensor, nine equations each with nine terms. If we multiply the product by δ_{qw}, it is reduced to a second-rank tensor.

$$A_{pq}B_{vw}\delta_{qw} = A_{pq}B_{vq}$$

The "rule" is, replace w by q and drop δ_{qw}. The process of contraction can be repeated several times. Thus, $A_{pq}B_{vw}\delta_{qw}\delta_{pv}$ reduces to $A_{pq}B_{vq}\delta_{pv}$ on the first contraction, and then to $A_{pq}B_{pq}$, which is a zero-rank tensor (scalar).

If we apply contraction to the second-rank stress tensor

$$\sigma_{ij}\delta_{ij} = \sigma_{ii} = \sigma_{11} + \sigma_{22} + \sigma_{33} = I_1$$

we obtain the first invariant of the tensor (a scalar).

The invariants of the stress tensor may be determined readily from the matrix of its components. Since $\sigma_{12} = \sigma_{21}$, etc., the stress tensor is a *symmetric tensor.*

$$\sigma_{ij} = \begin{vmatrix} \sigma_{11} & \sigma_{12} & \sigma_{13} \\ \sigma_{12} & \sigma_{22} & \sigma_{23} \\ \sigma_{13} & \sigma_{23} & \sigma_{33} \end{vmatrix}$$

The first invariant is the trace of the matrix, i.e., the sum of the main diagonal terms

$$I_1 = \sigma_{11} + \sigma_{22} + \sigma_{33}$$

The second invariant is the sum of the principal minors. A minor of an element of a matrix is the determinant of the next lower order which remains when the row and column in which the element stands are suppressed. Thus, taking each of the principal (main diagonal) terms in order and suppressing that row and column we have

$$I_2 = \begin{vmatrix} \sigma_{22} & \sigma_{23} \\ \sigma_{23} & \sigma_{33} \end{vmatrix} + \begin{vmatrix} \sigma_{11} & \sigma_{13} \\ \sigma_{13} & \sigma_{33} \end{vmatrix} + \begin{vmatrix} \sigma_{11} & \sigma_{12} \\ \sigma_{12} & \sigma_{22} \end{vmatrix}$$

Finally, the third invariant is the determinant of the entire matrix of the components of the stress tensor.

As an example of the advantages of the concepts that are provided by tensor notation we shall derive again the equations for principal stress that were developed in Sec. 2-5. The reader is warned that it is easy to lose the physical significance in the mathematical manipulation. It is a basic theorem of tensor theory that there is some orientation of the coordinate axes such that the components of a symmetric tensor of rank 2 will all be equal to zero for $i \neq j$. This is equivalent to stating that the concepts of principal stress and principal axes are inherent in the tensor character of stress.

The three force summation equations, Eqs. (2-17), can be written as

$$\sigma_{nj} = a_{ni}\sigma_{ij} \tag{2-29}$$

where the suffix n is used to denote that we are dealing with the angles to the normal of an oblique plane. If we let the oblique plane be a principal plane and let the normal stress on it be σ_p, then we can write

$$\sigma_{nj} = a_{pj}\sigma_p \tag{2-30}$$

Combining Eqs. (2-29) and (2-30)

$$(a_{ni}\sigma_{ij} - a_{pj}\sigma_p) = 0 \tag{2-31}$$

But, $a_{pj} = a_{pi}\delta_{ji}$ (replace i by j and drop δ_{ji})

$$a_{ni}\sigma_{ij} - \sigma_p a_{pi}\delta_{ji} = 0$$

However, $a_{ni} = a_{pi}$, since the principal stress lies in the direction of the normal to the oblique plane, so

$$(\sigma_{ij} - \sigma_p\delta_{ji})a_{pi} = 0 \tag{2-32}$$

Expanding Eq. (2-32) will give the three equations (2-13), since $a_{p1} = l$, $a_{p2} = m$, etc., and $\delta_{ji} = 0$ when $j \neq i$. For Eq. (2-32) to have a nontrivial solution in a_{pi} the determinant of the coefficients must vanish, resulting in

$$|\sigma_{ij} - \sigma_p\delta_{ji}| = \begin{vmatrix} \sigma_x - \sigma_p & \tau_{xy} & \tau_{xz} \\ \tau_{yx} & \sigma_y - \sigma_p & \tau_{yz} \\ \tau_{zx} & \tau_{zy} & \sigma_z - \sigma_p \end{vmatrix} = 0$$

which yields the cubic equation Eq. (2-14). The coefficients of this equation in tensor notation are

$$I_1 = \sigma_{ii}$$

$$I_2 = \tfrac{1}{2}(\sigma_{ik}\sigma_{ki} - \sigma_{ii}\sigma_{kk})$$

$$I_3 = \tfrac{1}{6}(2\sigma_{ij}\sigma_{jk}\sigma_{ki} - 3\sigma_{ij}\sigma_{ji}\sigma_{kk} + \sigma_{ii}\sigma_{jj}\sigma_{kk})$$

The fact that only dummy subscripts appear in these equations indicates the scalar nature of the invariants of the stress tensor.

2-7 MOHR'S CIRCLE—THREE DIMENSIONS

The discussion given in Sec. 2-4 of the representation of a two-dimensional state of stress by means of Mohr's circle can be extended to three dimensions. Figure 2-10 shows how a triaxial state of stress, defined by the three principal stresses,

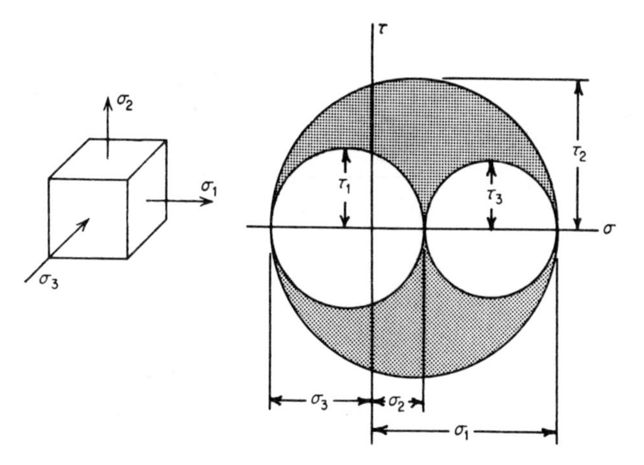

Figure 2-10 Mohr's circle representation of a three-dimensional state of stress.

can be represented by three Mohr's circles. It can be shown[1] that all possible stress conditions within the body fall within the shaded area between the circles in Fig. 2-10.

While the only physical significance of Mohr's circle is that it gives a geometrical representation of the equations that express the transformation of stress components to different sets of axes, it is a very convenient way of visualizing the state of stress. Figure 2-11 shows Mohr's circle for a number of common states of stress. Note that the application of a tensile stress σ_2 at right angles to an existing tensile stress σ_1 (Fig. 2-11*c*) results in a decrease in the principal shear stress on two of the three sets of planes on which a principal shear stress acts. However, the maximum shear stress is not decreased from what it would be for uniaxial tension, although if only the two-dimensional Mohr's circle had been used, this would not have been apparent. If a tensile stress is applied in the third principal direction (Fig. 2-11*d*), the maximum shear stress is reduced appreciably. For the limiting case of equal triaxial tension (hydrostatic tension) Mohr's circle reduces to a point, and there are no shear stresses acting on any plane in the body. The effectiveness of biaxial- and triaxial-tension stresses in reducing the shear stresses results in a considerable decrease in the ductility of the material, because plastic deformation is produced by shear stresses. Thus, brittle fracture is invariably associated with triaxial stresses developed at a notch or stress raiser. However, Fig. 2-11*e* shows that, if compressive stresses are applied lateral to a tensile stress, the maximum shear stress is larger than for the case of either uniaxial tension or compression. Because of the high value of shear stress relative to the applied tensile stress the material has an excellent opportunity to

[1] A. Nadai, "Theory of Flow and Fracture of Solids," 2d ed., pp. 96–98, McGraw-Hill Book Company, New York, 1950.

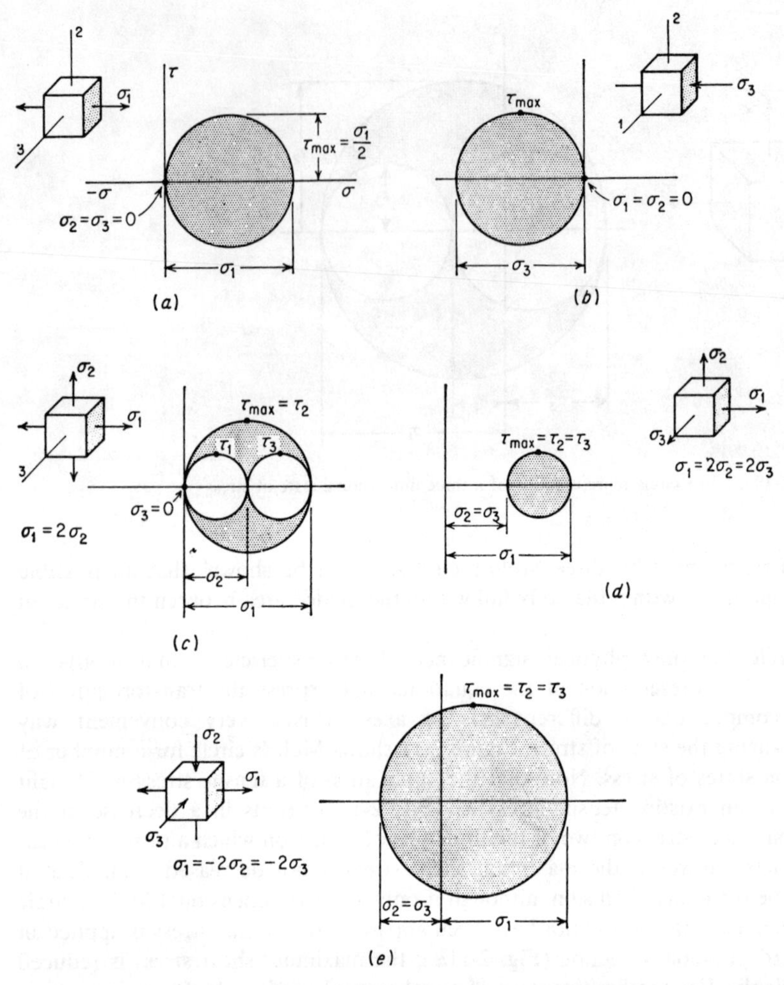

Figure 2-11 Mohr's circles (three-dimensional) for various states of stress. (*a*) Uniaxial tension; (*b*) uniaxial compression; (*c*) biaxial tension; (*d*) triaxial tension (unequal); (*e*) uniaxial tension plus biaxial compression.

deform plastically without fracturing under this state of stress. Important use is made of this fact in the plastic working of metals. For example, greater ductility is obtained in extrusion through a die than in simple uniaxial tension because the reaction of the metal with the die will produce lateral compressive stresses.

2-8 DESCRIPTION OF STRAIN AT A POINT

The displacement of points in a continuum may result from rigid-body translation, rotation, and deformation. The deformation of a solid may be made up of

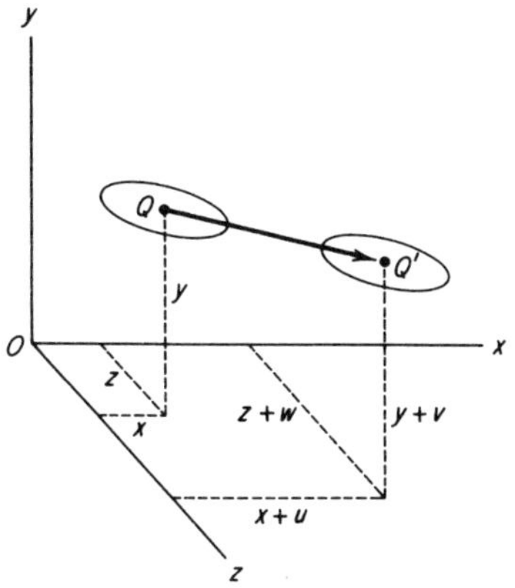

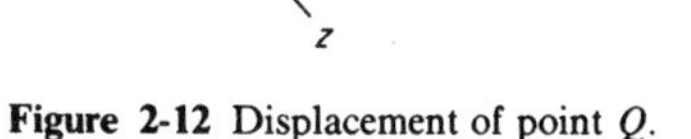

Figure 2-12 Displacement of point Q.

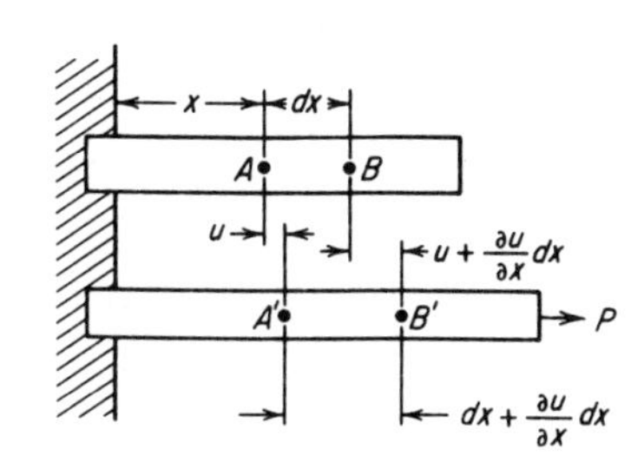

Fig. 2-13 One-dimensional strain.

dilatation, change in volume, or *distortion*, change in shape. Situations involving translation and rotation are usually treated in the branch of mechanics called *dynamics*. Small deformations are the province of elasticity theory, while larger deformations are treated in the disciplines of plasticity and hydrodynamics. The equations developed in this section are basically geometrical, so that they apply to all types of continuous media.

Consider a solid body in fixed coordinates, x, y, z (Fig. 2-12). Let a combination of deformation and movement displace point Q to Q' with new coordinates $x + u$, $y + v$, $z + w$. The components of the displacement are u, v, w. The displacement of Q is the vector $\mathbf{u}_Q = f(u, v, w)$. If the displacement vector is constant for all particles in the body then there is no strain. However, in general, u_i is different from particle to particle so that displacement is a function of distance, $u_i = f(x_i)$. For elastic solids and small displacements, u_i is a linear function of x_i, homogeneous displacements, and the displacement equations are linear. However, for other materials the displacement may not be linear with distance, which leads to cumbersome mathematical relationships.

To start our discussion of strain, consider a simple one-dimensional case (Fig. 2-13). In the undeformed state points A and B are separated by a distance dx. When a force is applied in the x direction A moves to A' and B moves to B'. Since displacement u, in this one-dimensional case, is a function of x, B is displaced slightly more than A since it is further from the fixed end. The normal strain is given by

$$e_x = \frac{\Delta L}{L} = \frac{A'B' - AB}{AB} = \frac{dx + \dfrac{\partial u}{\partial x}\,dx - dx}{dx} = \frac{\partial u}{\partial x} \tag{2-33}$$

For this one-dimensional case, the displacement is given by $u = e_x x$. To generalize this to three dimensions, each of the components of the displacement

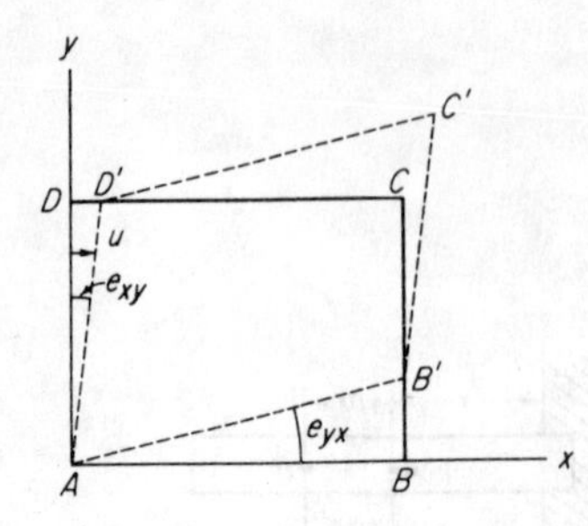

Figure 2-14 Angular distortion of an element.

will be linearly related to each of the three initial coordinates of the point.

$$\begin{aligned} u &= e_{xx}x + e_{xy}y + e_{xz}z \\ v &= e_{yx}x + e_{yy}y + e_{yz}z \\ w &= e_{zx}x + e_{zy}y + e_{zz}z \end{aligned} \tag{2-34}$$

or

$$u_i = e_{ij}x_j \tag{2-35}$$

The coefficients relating displacement with the coordinates of the point in the body are the components of the relative displacement tensor. Three of these terms can be identified readily as the normal strains.

$$e_{xx} = \frac{\partial u}{\partial x} \qquad e_{yy} = \frac{\partial v}{\partial y} \qquad e_{zz} = \frac{\partial w}{\partial z} \tag{2-36}$$

However, the other six coefficients require further scrutiny.

Consider an element in the xy plane which has been distorted by shearing stresses (Fig. 2-14). The element has undergone angular distortion. The displacement of points along the line AD is parallel to the x axis, but this component of displacement increases in proportion to the distance out along the y axis. Thus, referring to Eq. (2-34)

$$e_{xy} = \frac{DD'}{DA} = \frac{\partial u}{\partial y} \tag{2-37}$$

Similarly, for the angular distortion of the x axis

$$e_{yx} = \frac{BB'}{AB} = \frac{\partial v}{\partial x} \tag{2-38}$$

These shear displacements are positive when they rotate a line from one positive axis towards another positive axis. By similar methods the rest of the components of the displacement tensor can be seen to be

$$e_{ij} = \begin{vmatrix} e_{xx} & e_{xy} & e_{xz} \\ e_{yx} & e_{yy} & e_{yz} \\ e_{zx} & e_{zy} & e_{zz} \end{vmatrix} = \begin{vmatrix} \dfrac{\partial u}{\partial x} & \dfrac{\partial u}{\partial y} & \dfrac{\partial u}{\partial z} \\ \dfrac{\partial v}{\partial x} & \dfrac{\partial v}{\partial y} & \dfrac{\partial v}{\partial z} \\ \dfrac{\partial w}{\partial x} & \dfrac{\partial w}{\partial y} & \dfrac{\partial w}{\partial z} \end{vmatrix} \tag{2-39}$$

In general, displacement components such as e_{xy}, e_{yx}, etc., produce both *shear strain* and *rigid-body rotation*. Figure 2-15 illustrates several cases. Since we need to identify that part of the displacement that results in strain, it is important to break the displacement tensor into a strain contribution and a rotational contribution. Fortunately, a basic postulate of tensor theory states that any second-rank tensor can be decomposed into a symmetric tensor and an antisymmetric (skew-symmetric) tensor.

$$e_{ij} = \tfrac{1}{2}(e_{ij} + e_{ji}) + \tfrac{1}{2}(e_{ij} - e_{ji}) \tag{2-40}$$

or

$$e_{ij} = \varepsilon_{ij} + \omega_{ij} \tag{2-41}$$

where $\varepsilon_{ij} = \dfrac{1}{2}\left(\dfrac{\partial u_i}{\partial x_j} + \dfrac{\partial u_j}{\partial x_i}\right)$ and is called the *strain tensor*

$\omega_{ij} = \dfrac{1}{2}\left(\dfrac{\partial u_i}{\partial x_j} - \dfrac{\partial u_j}{\partial x_i}\right)$ and is called the *rotation tensor*

$$\varepsilon_{ij} = \begin{vmatrix} \varepsilon_{xx} & \varepsilon_{xy} & \varepsilon_{xz} \\ \varepsilon_{yx} & \varepsilon_{yy} & \varepsilon_{yz} \\ \varepsilon_{zx} & \varepsilon_{zy} & \varepsilon_{zz} \end{vmatrix} = \begin{vmatrix} \dfrac{\partial u}{\partial x} & \dfrac{1}{2}\left(\dfrac{\partial u}{\partial y} + \dfrac{\partial v}{\partial x}\right) & \dfrac{1}{2}\left(\dfrac{\partial u}{\partial z} + \dfrac{\partial w}{\partial x}\right) \\ \dfrac{1}{2}\left(\dfrac{\partial u}{\partial y} + \dfrac{\partial v}{\partial x}\right) & \dfrac{\partial v}{\partial y} & \dfrac{1}{2}\left(\dfrac{\partial v}{\partial z} + \dfrac{\partial w}{\partial y}\right) \\ \dfrac{1}{2}\left(\dfrac{\partial u}{\partial z} + \dfrac{\partial w}{\partial x}\right) & \dfrac{1}{2}\left(\dfrac{\partial v}{\partial z} + \dfrac{\partial w}{\partial y}\right) & \dfrac{\partial w}{\partial z} \end{vmatrix} \tag{2-42}$$

$$\omega_{ij} = \begin{vmatrix} \omega_{xx} & \omega_{xy} & \omega_{xz} \\ \omega_{yx} & \omega_{yy} & \omega_{yz} \\ \omega_{zx} & \omega_{zy} & \omega_{zz} \end{vmatrix} = \begin{vmatrix} 0 & \dfrac{1}{2}\left(\dfrac{\partial u}{\partial y} - \dfrac{\partial v}{\partial x}\right) & \dfrac{1}{2}\left(\dfrac{\partial u}{\partial z} - \dfrac{\partial w}{\partial x}\right) \\ \dfrac{1}{2}\left(\dfrac{\partial v}{\partial x} - \dfrac{\partial u}{\partial y}\right) & 0 & \dfrac{1}{2}\left(\dfrac{\partial v}{\partial z} - \dfrac{\partial w}{\partial y}\right) \\ \dfrac{1}{2}\left(\dfrac{\partial w}{\partial x} - \dfrac{\partial u}{\partial z}\right) & \dfrac{1}{2}\left(\dfrac{\partial w}{\partial y} - \dfrac{\partial v}{\partial z}\right) & 0 \end{vmatrix} \tag{2-43}$$

Note that ε_{ij} is a symmetric tensor since $\varepsilon_{ij} = \varepsilon_{ji}$, that is, $\varepsilon_{xy} = \varepsilon_{xz}$, etc. ω_{ij} is an antisymmetric tensor since $\omega_{ij} = -\omega_{ji}$, that is, $\omega_{xy} = -\omega_{yx}$. If $\omega_{ij} = 0$, the deformation is said to be irrotational.

By substituting Eq. (2-41) into Eq. (2-35), we get the general displacement equations

$$u_i = \varepsilon_{ij} x_j + \omega_{ij} x_j \tag{2-44}$$

Earlier in Sec. 1-9 the shear strain γ was defined as the total angular change from a right angle. Referring to Fig. 2-15*a*, $\gamma = e_{xy} + e_{yx} = \varepsilon_{xy} + \varepsilon_{yx} = 2\varepsilon_{xy}$.

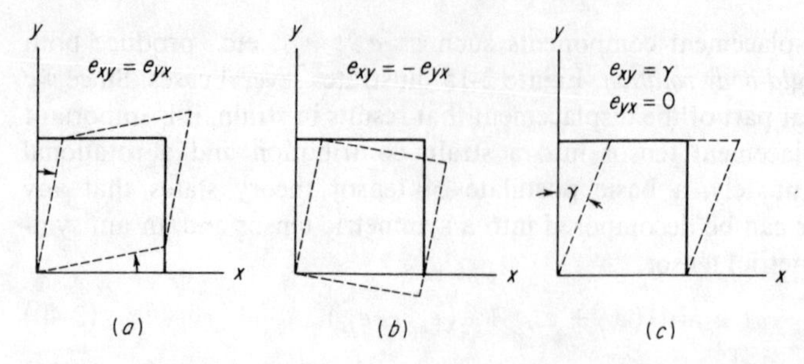

Figure 2-15 Some examples of displacement with shear and rotation. (*a*) Pure shear without rotation; (*b*) pure rotation without shear; (*c*) simple shear. Simple shear involves a shape change produced by displacements along a single set of parallel planes. Pure shear involves a shape change produced by equal shear displacements on two sets of perpendicular planes.

This definition of shear strain, $\gamma_{ij} = 2\varepsilon_{ij}$, is called the *engineering shear strain*.

$$\begin{aligned} \gamma_{xy} &= \frac{\partial u}{\partial y} + \frac{\partial v}{\partial x} \\ \gamma_{xz} &= \frac{\partial w}{\partial x} + \frac{\partial u}{\partial z} \\ \gamma_{yz} &= \frac{\partial w}{\partial y} + \frac{\partial v}{\partial z} \end{aligned} \tag{2-45}$$

This definition of shear strain commonly is used in engineering elasticity. However, the shear strain defined according to Eq. (2-45) *is not a tensor quantity.*

Because of the obvious advantages in the transformation of tensors by the methods discussed in Sec. 2-6, it is profitable to use the strain tensor as defined by Eq. (2-42). Since the strain tensor is a second-rank tensor, it has all of the properties that have been described earlier for stress. Thus, the strain tensor may be transformed from one set of coordinate axes to a new system of axes by

$$\varepsilon_{kl} = a_{ki}a_{lj}\varepsilon_{ij} \tag{2-46}$$

For simplicity, equations for strain analogous with those for stress can be written directly by substituting ε for σ and $\gamma/2$ for τ. Thus, the normal strain on an oblique plane is given by

$$\varepsilon = \varepsilon_x l^2 + \varepsilon_y m^2 + \varepsilon_z n^2 + \gamma_{xy} lm + \gamma_{yz} mn + \gamma_{xz} ln$$

[Compare the above with Eq. (2-18).]

In complete analogy with stress, it is possible to define a system of coordinate axes along which there are no shear strains. These axes are the principal strain axes. For an isotropic body the direction of principal strains coincide with principal stress directions.[1] An element oriented along one of the principal strain axes will undergo pure extension or contraction without any rotation or shear strain. The three principal strains are the roots of the cubic equation

$$\varepsilon^3 - I_1\varepsilon^2 + I_2\varepsilon - I_3 = 0 \tag{2-47}$$

where

$$I_1 = \varepsilon_x + \varepsilon_y + \varepsilon_z$$

$$I_2 = \varepsilon_x\varepsilon_y + \varepsilon_y\varepsilon_z + \varepsilon_z\varepsilon_x - \tfrac{1}{4}\left(\gamma_{xy}^2 + \gamma_{zx}^2 + \gamma_{yz}^2\right)$$

$$I_3 = \varepsilon_x\varepsilon_y\varepsilon_z + \tfrac{1}{4}\gamma_{yx}\gamma_{zx}\gamma_{yz} - \tfrac{1}{4}\left(\varepsilon_x\gamma_{yz}^2 + \varepsilon_y\gamma_{zx}^2 + \varepsilon_z\gamma_{xy}^2\right)$$

The directions of the principal strains are obtained from the three equations analogous to Eqs. (2-13)

$$2l(\varepsilon_x - \varepsilon) + m\gamma_{xy} + n\gamma_{xz} = 0$$

$$l\gamma_{xy} + 2m(\varepsilon_y - \varepsilon) + n\gamma_{yz} = 0$$

$$l\gamma_{xz} + m\gamma_{yz} + 2n(\varepsilon_z - \varepsilon) = 0$$

Continuing the analogy between stress and strain equations, the equation for the *principal shearing strains* can be obtained from Eq. (2-20).

$$\gamma_1 = \varepsilon_2 - \varepsilon_3$$

$$\gamma_{max} = \gamma_2 = \varepsilon_1 - \varepsilon_3 \tag{2-48}$$

$$\gamma_3 = \varepsilon_1 - \varepsilon_2$$

In general, the deformation of a solid involves a combination of volume change and change in shape. Therefore, we need a way to determine how much of the deformation is due to these contributions. The *volume strain*, or cubical dilatation, is the change in volume per unit volume. Consider a rectangular parallelepiped with edges dx, dy, dz. The volume in the strained condition is $(1 + \varepsilon_x)(1 + \varepsilon_y)(1 + \varepsilon_z)\,dx\,dy\,dz$, since only normal strains result in volume change. The volume strain Δ is

$$\Delta = \frac{(1 + \varepsilon_x)(1 + \varepsilon_y)(1 + \varepsilon_z)\,dx\,dy\,dz - dx\,dy\,dz}{dx\,dy\,dz}$$

$$= (1 + \varepsilon_x)(1 + \varepsilon_y)(1 + \varepsilon_z) - 1$$

which for small strains, after neglecting the products of strains, becomes

$$\Delta = \varepsilon_x + \varepsilon_y + \varepsilon_z \tag{2-49}$$

Note that the volume strain is equal to the first invariant of the strain tensor, $\Delta = \varepsilon_x + \varepsilon_y + \varepsilon_z = \varepsilon_1 + \varepsilon_2 + \varepsilon_2$. We can also define $(\varepsilon_x + \varepsilon_y + \varepsilon_z)/3$ as the *mean strain* or the hydrostatic (spherical) component of strain.

$$\varepsilon_m = \frac{\varepsilon_x + \varepsilon_y + \varepsilon_z}{3} = \frac{\varepsilon_{kk}}{3} = \frac{\Delta}{3} \tag{2-50}$$

[1] For a derivation of this point see C. T. Wang, "Applied Elasticity," pp. 26–27, McGraw-Hill Book Company, New York, 1953.

That part of the strain tensor which is involved in shape change rather than volume change is called the *strain deviator* ε'_{ij}. To obtain the deviatoric strains, we simply subtract ε_m from each of the normal strain components. Thus,

$$\varepsilon'_{ij} = \begin{vmatrix} \varepsilon_x - \varepsilon_m & \varepsilon_{xy} & \varepsilon_{xz} \\ \varepsilon_{yx} & \varepsilon_y - \varepsilon_m & \varepsilon_{yz} \\ \varepsilon_{zx} & \varepsilon_{zy} & \varepsilon_z - \varepsilon_m \end{vmatrix}$$

$$= \begin{vmatrix} \dfrac{2\varepsilon_x - \varepsilon_y - \varepsilon_z}{3} & \varepsilon_{xy} & \varepsilon_{xz} \\ \varepsilon_{yx} & \dfrac{2\varepsilon_y - \varepsilon_z - \varepsilon_x}{3} & \varepsilon_{yz} \\ \varepsilon_{zx} & \varepsilon_{zy} & \dfrac{2\varepsilon_z - \varepsilon_x - \varepsilon_y}{3} \end{vmatrix} \tag{2-51}$$

The division of the total strain tensor into deviatoric and dilatational strains is given in tensor notation by

$$\varepsilon_{ij} = \varepsilon'_{ij} + \varepsilon_m = \left(\varepsilon_{ij} - \frac{\Delta}{3}\delta_{ij}\right) + \frac{\Delta}{3}\delta_{ij} \tag{2-52}$$

For example, when ε_{ij} are the principal strains, $(i = j)$, the strain deviators are $\varepsilon'_{11} = \varepsilon_{11} - \varepsilon_m$, $\varepsilon'_{22} = \varepsilon_{22} - \varepsilon_m$, $\varepsilon'_{33} = \varepsilon_{33} - \varepsilon_m$. These strains represent elongations or contractions along the principal axes that change the shape of the body at constant volume.

2-9 MOHR'S CIRCLE OF STRAIN

Except in a few cases involving contact stresses, it is not possible to measure stress directly. Therefore, experimental measurements of stress are actually based on measured strains and are converted to stresses by means of Hooke's law and the more general relationships which are given in Sec. 2-11. The most universal strain-measuring device is the bonded-wire resistance gage, frequently called the SR-4 strain gage.[1] These gages are made up of several loops of fine wire or foil of special composition, which are bonded to the surface of the body to be studied. When the body is deformed, the wires in the gage are strained and their electrical resistance is altered. The change in resistance, which is proportional to strain, can be accurately determined with a simple Wheatstone-bridge circuit. The high sensitivity, stability, comparative ruggedness, and ease of application make resistance strain gages a very powerful tool for strain determination.

[1] For a treatment of strain gages and other techniques of experimental stress analysis see J. W. Dally, and W. F. Riley, "Experimental Stress Analysis," 2d ed., McGraw-Hill Book Company, New York, 1978.

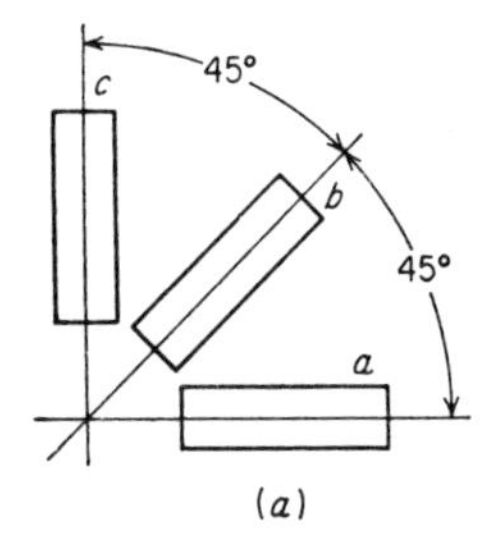

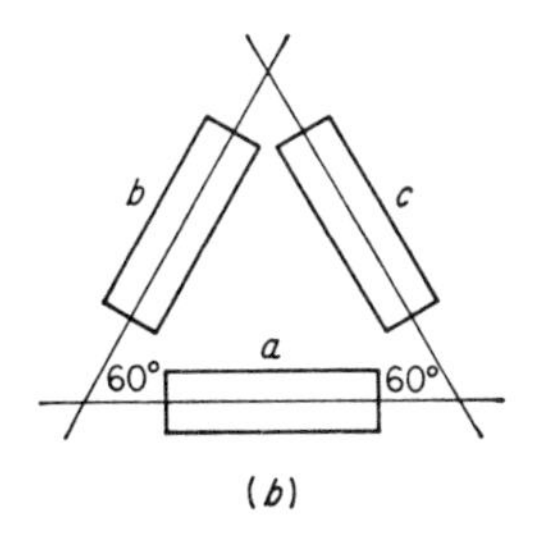

Figure 2-16 Typical strain-gage rosettes. (*a*) Rectangular; (*b*) delta.

For practical problems of experimental stress analysis if is often important to determine the principal stresses. If the principal directions are known, gages can be oriented in these directions and the principal stresses determined quite readily. In the general case the direction of the principal strains will not be known, so that it will be necessary to determine the orientation and magnitude of the principal strains from the measured strains in arbitrary directions. Because no stress can act perpendicular to a free surface, strain-gage measurements involve a two-dimensional state of stress. The state of strain is completely determined if ε_x, ε_y, and γ_{xy} can be measured. However, strain gages can make only direct readings of linear strain, while shear strains must be determined indirectly. Therefore, it is the usual practice to use three strain gages separated at fixed angles in the form of a "rosette," as in Fig. 2-16. Strain-gage readings at three values of θ will give three simultaneous equations similar to Eq. (2-53) which can be solved for ε_x, ε_y, and γ_{xy}. The two-dimensional version of Eq. (2-47) can then be used to determine the principal strains.

$$\varepsilon_\theta = \varepsilon_x \cos^2 \theta + \varepsilon_y \sin^2 \theta + \gamma_{xy} \sin \theta \cos \theta \tag{2-53}$$

A more convenient method of determining the principal strains from strain-gage readings than the solution of three simultaneous equations in three unknowns is the use of Mohr's circle. In constructing a Mohr's circle representation of strain, values of linear normal strain ε are plotted along the x axis, and the shear strain divided by 2 is plotted along the y axis. Figure 2-17 shows the Mohr's circle construction[1] for the generalized strain-gage rosette illustrated at the top of the figure. Strain-gage readings ε_a, ε_b, and ε_c are available for three gages situated at arbitrary angles α and β. The objective is to determine the magnitude and orientation of the principal strains ε_1 and ε_2.

1. Along an arbitrary axis $X'X'$ lay off vertical lines *aa*, *bb*, and *cc* corresponding to the strains ε_a, ε_b, and ε_c.
2. From any point on the line *bb* (middle strain gage) draw a line *DA* at an angle α with *bb* and intersecting *aa* at point *A*. In the same way, lay off *DC* intersecting *cc* at point *C*.

[1] G. Murphy, *J. Appl. Mech.*, vol. 12, p. A209, 1945; F. A. McClintock, *Proc. Soc. Exp. Stress Anal.*, vol. 9, p. 209, 1951.

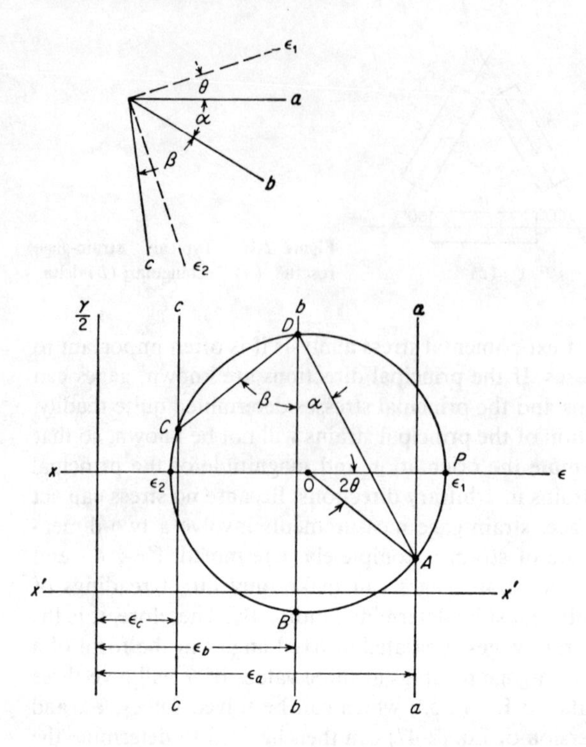

Figure 2-17 Mohr's circle for determination of principal strains.

3. Construct a circle through A, C, and D. The center of this circle is at O, determined by the intersection of the perpendicular bisectors to CD and AD.
4. Points A, B, and C on the circle give the values of ε and $\gamma/2$ (measured from the new x axis through O) for the three gages.
5. Values of the principal strains are determined by the intersection of the circle with the new x axis through O. The angular relationship of ε_1 to the gage a is one-half the angle AOP on the Mohr's circle ($AOP = 2\theta$).

2-10 HYDROSTATIC AND DEVIATOR COMPONENTS OF STRESS

Having introduced the concept that the strain tensor can be divided into a hydrostatic or mean strain and a strain deviator, it is important to consider the physical significance of a similar operation on the stress tensor. The total stress tensor can be divided into a *hydrostatic* or *mean stress tensor* σ_m, which involves only pure tension or compression, and a *deviator stress tensor* σ'_{ij}, which represents the shear stresses in the total state of stress (Fig. 2-18). In direct analogy with the situation for strain, the hydrostatic component of the stress tensor produces only elastic volume changes and does not cause plastic deformation. Experiment shows that the yield stress of metals is independent of hydrostatic

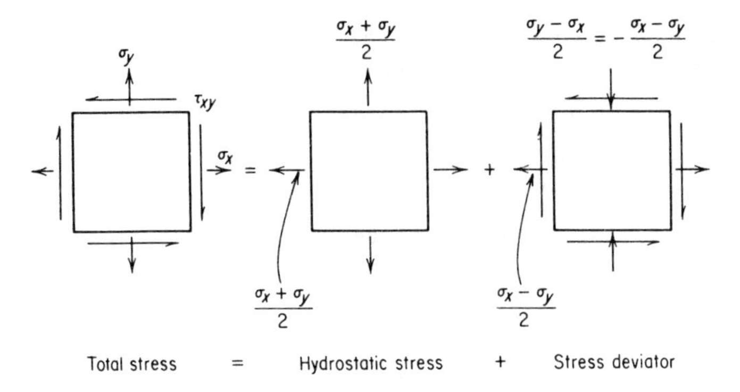

Figure 2-18 Resolution of total stress into hydrostatic stress and stress deviator.

stress, although the fracture strain is strongly influenced by hydrostatic stress. Because the stress deviator involves the shearing stresses, it is important in causing plastic deformation. In Chap. 3 we shall see that the stress deviator is useful in formulating theories of yielding.

The hydrostatic or mean stress is given by

$$\sigma_m = \frac{\sigma_{kk}}{3} = \frac{\sigma_x + \sigma_y + \sigma_z}{3} = \frac{\sigma_1 + \sigma_2 + \sigma_3}{3} \tag{2-54}$$

The decomposition of the stress tensor is given by

$$\sigma_{ij} = \sigma'_{ij} + \tfrac{1}{3}\delta_{ij}\sigma_{kk} \tag{2-55}$$

Therefore,

$$\sigma'_{ij} = \sigma_{ij} - \sigma_m \delta_{ij} \tag{2-56}$$

$$\sigma'_{ij} = \begin{vmatrix} \dfrac{2\sigma_x - \sigma_y - \sigma_z}{3} & \tau_{xy} & \tau_{xz} \\ \tau_{yx} & \dfrac{2\sigma_y - \sigma_z - \sigma_x}{3} & \tau_{yz} \\ \tau_{zx} & \tau_{zy} & \dfrac{2\sigma_z - \sigma_x - \sigma_y}{3} \end{vmatrix} \tag{2-57}$$

It can be seen readily that the stress deviator involves shear stresses. For example, referring σ'_{ij} to a system of principal axes,

$$\sigma'_1 = \frac{2\sigma_1 - \sigma_2 - \sigma_3}{3} = \frac{(\sigma_1 - \sigma_2) + (\sigma_1 - \sigma_3)}{3}$$

$$\sigma'_1 = \frac{2}{3}\left(\frac{\sigma_1 - \sigma_2}{2} + \frac{\sigma_1 - \sigma_3}{2}\right) = \frac{2}{3}(\tau_3 + \tau_2) \tag{2-58}$$

where τ_3 and τ_2 are principal shearing stresses.

Since σ'_{ij} is a second-rank tensor, it has principal axes. The principal values of the stress deviator are the roots of the cubic equation[1]

$$(\sigma')^3 - J_1(\sigma')^2 - J_2\sigma' - J_3 = 0 \tag{2-59}$$

where J_1, J_2, J_3 are the invariants of the deviator stress tensor. J_1 is the sum of the principal terms in the diagonal of the matrix of components of σ'_{ij}.

$$J_1 = (\sigma_x - \sigma_m) + (\sigma_y - \sigma_m) + (\sigma_z - \sigma_m) = 0 \tag{2-60}$$

J_2 can be obtained from the sum of the principal minors of σ'_{ij}.

$$\begin{aligned} J_2 &= \tau_{xy}^2 + \tau_{yz}^2 + \tau_{xz}^2 - \sigma'_x\sigma'_y - \sigma'_y\sigma'_z - \sigma'_z\sigma'_x \\ &= \tfrac{1}{6}\left[(\sigma_x - \sigma_y)^2 + (\sigma_y - \sigma_z)^2 + (\sigma_z - \sigma_x)^2 + 6\left(\tau_{xy}^2 + \tau_{yz}^2 + \tau_{xz}^2\right)\right] \end{aligned} \tag{2-61}$$

The third invariant J_3 is the determinant of Eq. (2-57).

2-11 ELASTIC STRESS-STRAIN RELATIONS

Up till now our discussion of stress and strain has been perfectly general and applicable to any continuum. Now, if we want to relate the stress tensor with the strain tensor, we must introduce the properties of the material. Equations of this nature are called *constitutive equations*. In this chapter we shall consider only constitutive equations for elastic solids. Moreover, initially we shall only consider isotropic elastic solids.

In Chap. 1 we saw that elastic stress is linearly related to elastic strain by means of the modulus of elasticity (Hooke's law).

$$\sigma_x = E\varepsilon_x \tag{2-62}$$

where E is the modulus of elasticity in tension or compression. While a tensile force in the x direction produces an extension along that axis, it also produces a contraction in the transverse y and z directions. The transverse strain has been found by experience to be a constant fraction of the strain in the longitudinal direction. This is known as *Poisson's ratio*, denoted by the symbol ν.

$$\varepsilon_y = \varepsilon_z = -\nu\varepsilon_x = -\frac{\nu\sigma_x}{E} \tag{2-63}$$

Only the absolute value of ν is used in calculations. Poisson's ratio is 0.25 for a perfectly isotropic elastic material, but for most metals the values[2] of ν are closer to 0.33.

To develop the stress-strain relations for a three-dimensional state of stress, consider a unit cube subjected to normal stresses σ_x, σ_y, σ_z and shearing stresses τ_{xy}, τ_{yz}, τ_{zx}. Because the elastic stresses are small and the material is isotropic, we can assume that normal stress σ_x does not produce shear strain on the x, y, or z planes and that a shear stress τ_{xy} does not produce normal strains on the x, y, or

[1] Note that we use a negative sign for the coefficient of σ'. Compare with Eq. (2-14).

[2] W. Koster and H. Franz, *Metall. Rev.*, vol. 6, pp. 1–55, 1961.

z planes. We can then apply the principle of superposition[1] to determine the strain produced by more than one stress component. For example, the stress σ_x produces a normal strain ε_x and two transverse strains $\varepsilon_y = -\nu\varepsilon_x$ and $\varepsilon_z = -\nu\varepsilon_x$. Thus,

Stress	Strain in the x direction	Strain in the y direction	Strain in the z direction
σ_x	$\varepsilon_x = \dfrac{\sigma_x}{E}$	$\varepsilon_y = -\dfrac{\nu\sigma_x}{E}$	$\varepsilon_z = -\dfrac{\nu\sigma_x}{E}$
σ_y	$\varepsilon_x = -\dfrac{\nu\sigma_y}{E}$	$\varepsilon_y = \dfrac{\sigma_y}{E}$	$\varepsilon_z = -\dfrac{\nu\sigma_y}{E}$
σ_z	$\varepsilon_x = -\dfrac{\nu\sigma_z}{E}$	$\varepsilon_y = -\dfrac{\nu\sigma_z}{E}$	$\varepsilon_z = \dfrac{\sigma_z}{E}$

By superposition of the components of strain in the x, y, and z directions

$$\begin{aligned}
\varepsilon_x &= \frac{1}{E}\left[\sigma_x - \nu(\sigma_y + \sigma_z)\right] \\
\varepsilon_y &= \frac{1}{E}\left[\sigma_y - \nu(\sigma_z + \sigma_x)\right] \\
\varepsilon_z &= \frac{1}{E}\left[\sigma_z - \nu(\sigma_x + \sigma_y)\right]
\end{aligned} \tag{2-64}$$

The shearing stresses acting on the unit cube produce shearing strains.

$$\tau_{xy} = G\gamma_{xy} \qquad \tau_{yz} = G\gamma_{yz} \qquad \tau_{xz} = G\gamma_{xz} \tag{2-65}$$

The proportionality constant G is the *modulus of elasticity in shear*, or the *modulus of rigidity*. Values of G are usually determined from a torsion test.

We have seen that the stress-strain equations for an *isotropic* elastic solid involve three constants, E, G, and ν. Typical values of these constants for a number of metals are given in Table 2-1.

Still another elastic constant is the *bulk modulus* or the *volumetric modulus of elasticity* K. The bulk modulus is the ratio of the hydrostatic pressure to the dilatation that it produces

$$K = \frac{\sigma_m}{\Delta} = \frac{-p}{\Delta} = \frac{1}{\beta} \tag{2-66}$$

where $-p$ is the hydrostatic pressure and β is the compressibility.

Many useful relationships may be derived between the elastic constants E, G, ν, K. For example, if we add up the three equations (2-64),

$$\varepsilon_x + \varepsilon_y + \varepsilon_z = \frac{1 - 2\nu}{E}(\sigma_x + \sigma_y + \sigma_z)$$

[1] The principle of superposition states that two strains may be combined by direct superposition. The order of application has no effect on the final strain of the body.

Table 2-1 Typical room-temperature values of elastic constants for isotropic materials

Material	Modulus of elasticity, 10^{-6} psi (GPa)	Shear modulus, 10^{-6} psi (GPa)	Poisson's ratio
Aluminum alloys	10.5 (72.4)	4.0 (27.5)	0.31
Copper	16.0 (110)	6.0 (41.4)	0.33
Steel (plain carbon and low-alloy)	29.0 (200)	11.0 (75.8)	0.33
Stainless steel (18-8)	28.0 (193)	9.5 (65.6)	0.28
Titanium	17.0 (117)	6.5 (44.8)	0.31
Tungsten	58.0 (400)	22.8 (157)	0.27

The term on the left is the volume strain Δ, and the term on the right is $3\sigma_m$.

$$\Delta = \frac{1-2\nu}{E} 3\sigma_m$$

or

$$K = \frac{\sigma_m}{\Delta} = \frac{E}{3(1-2\nu)} \tag{2-67}$$

Another important relationship is the expression relating E, G, and ν. This equation is usually developed in a first course in strength of materials.[1]

$$G = \frac{E}{2(1+\nu)} \tag{2-68}$$

Many other relationships can be developed between these four isotropic elastic constants. For example,

$$E = \frac{9K}{1+3K/G} \qquad \nu = \frac{1-2G/3K}{2+2G/3K}$$

$$G = \frac{3(1-2\nu)K}{2(1+\nu)} \qquad K = \frac{E}{9-3E/G}$$

Equations (2-64) and (2-65) may be expressed succinctly in tensor notation

$$\varepsilon_{ij} = \frac{1+\nu}{E}\sigma_{ij} - \frac{\nu}{E}\sigma_{kk}\delta_{ij} \tag{2-69}$$

[1] For a geometric development see D. C. Drucker, "Introduction to Mechanics of Deformable Solids," pp. 64–65, McGraw-Hill Book Company, New York, 1967. For a derivation based on isotropy and transformation of axes see Chou and Pagano, op. cit., pp. 58–59.

For example, if $i = j = x$,

$$\varepsilon_{xx} = \frac{1+\nu}{E}\sigma_{xx} - \frac{\nu}{E}(\sigma_{xx} + \sigma_{yy} + \sigma_{zz})(1)$$

$$= \frac{1}{E}\left[\sigma_{xx} - \nu(\sigma_{yy} + \sigma_{zz})\right]$$

If $i = x$ and $j = y$,

$$\varepsilon_{xy} = \frac{\gamma_{xy}}{2} = \frac{1+\nu}{E}\tau_{xy} - \frac{\nu}{E}\sigma_{kk}(0)$$

where

$$\frac{1+\nu}{E} = \frac{1}{2G} \quad \text{and} \quad \gamma_{xy} = \frac{1}{G}\tau_{xy}$$

2-12 CALCULATION OF STRESSES FROM ELASTIC STRAINS

Since for small elastic strains there is no coupling between the expressions for normal stress and strain and the equations for shear stress and shear strain, it is possible to invert Eqs. (2-64) and (2-65) to solve for stress in terms of strain. From Eq. (2-64),

$$\sigma_x + \sigma_y + \sigma_z = \frac{E}{1-2\nu}(\varepsilon_x + \varepsilon_y + \varepsilon_z) \tag{2-70}$$

$$\varepsilon_x = \frac{1+\nu}{E}\sigma_x - \frac{\nu}{E}(\sigma_x + \sigma_y + \sigma_z) \tag{2-71}$$

Substitution of Eq. (2-70) into Eq. (2-71) gives

$$\sigma_x = \frac{E}{1+\nu}\varepsilon_x + \frac{\nu E}{(1+\nu)(1-2\nu)}(\varepsilon_x + \varepsilon_y + \varepsilon_z) \tag{2-72}$$

or in tensor notation

$$\sigma_{ij} = \frac{E}{1+\nu}\varepsilon_{ij} + \frac{\nu E}{(1+\nu)(1-2\nu)}\varepsilon_{kk}\delta_{ij} \tag{2-73}$$

Upon expansion, Eq. (2-73) gives three equations for normal stress and six equations for shear stress. Equation (2-72) is often written in a briefer form by letting

$$\frac{\nu E}{(1+\nu)(1-2\nu)} = \lambda \quad \text{Lamé's constant}$$

and noting that $\Delta = \varepsilon_x + \varepsilon_y + \varepsilon_z$.

$$\sigma_x = 2G\varepsilon_x + \lambda\Delta \tag{2-74}$$

The stresses and the strains can be broken into deviator and hydrostatic components. The deviatoric response (distortion) is related to the stress deviator by

$$\sigma'_{ij} = \frac{E}{1+\nu}\varepsilon'_{ij} = 2G\varepsilon'_{ij} \tag{2-75}$$

while the relationship between hydrostatic stress and mean strain is

$$\sigma_{ii} = \frac{E}{1 - 2\nu}\varepsilon_{kk} = 3K\varepsilon_{kk} \tag{2-76}$$

For a case of *plane stress* ($\sigma_3 = 0$), two simple and useful equations relating stress to strain may be obtained by solving simultaneously two of the equations of (2-64).

$$\begin{aligned} \sigma_1 &= \frac{E}{1 - \nu^2}(\varepsilon_1 + \nu\varepsilon_2) \\ \sigma_2 &= \frac{E}{1 - \nu^2}(\varepsilon_2 + \nu\varepsilon_1) \end{aligned} \tag{2-77}$$

A situation of plane stress exists typically in a thin sheet loaded in the plane of the sheet or a thin-wall tube loaded by internal pressure where there is no stress normal to a free surface.

Another important situation is *plane strain* ($\varepsilon_3 = 0$), which occurs typically when one dimension is much greater than the other two, as in a long rod or a cylinder with restrained ends. Some type of physical restraint exists to limit the strain in one direction, so

$$\varepsilon_3 = \frac{1}{E}[\sigma_3 - \nu(\sigma_1 + \sigma_2)] = 0$$

but

$$\sigma_3 = \nu(\sigma_1 + \sigma_2)$$

Therefore, a stress exists even though the strain is zero. Substituting this value into Eq. (2-64), we get

$$\begin{aligned} \varepsilon_1 &= \frac{1}{E}[(1 - \nu^2)\sigma_1 - \nu(1 + \nu)\sigma_2] \\ \varepsilon_2 &= \frac{1}{E}[(1 - \nu^2)\sigma_2 - \nu(1 + \nu)\sigma_1] \\ \varepsilon_3 &= 0 \end{aligned} \tag{2-78}$$

Example Strain-gage measurements made on the free surface of a steel plate indicate that the principal strains are 0.004 and 0.001 in/in. What are the principal stresses?

Since this is a condition of plane stress, Eqs. (2-77) apply. From Table 2-1, $E = 200$ GPa and $\nu = 0.33$.

$$\sigma_1 = \frac{E}{1 - \nu^2}(\varepsilon_1 + \nu\varepsilon_2) = \frac{200}{1 - 0.109}\{0.004 + 0.33(0.001)\}$$

$$= \frac{200}{0.891}(0.004 + 0.0003) = 0.965 \text{ GPa} = 965 \text{ MPa} = 140{,}000 \text{ psi}$$

$$\sigma_2 = \frac{E}{1 - \nu^2}(\varepsilon_2 + \nu\varepsilon_1) = \frac{200}{0.891}(0.001 + 0.0013) = 0.516 \text{ GPa} = 516 \text{ MPa}$$

Note the error that would result if the principal stresses were computed by simply multiplying Young's modulus by the strain.

$$\sigma_1 = E\varepsilon_1 = 200(0.004) = 800 \text{ MPa} \quad \text{incorrect}$$

$$\sigma_2 = E\varepsilon_1 = 200(0.001) = 200 \text{ MPa} \quad \text{incorrect}$$

2-13 STRAIN ENERGY

The *elastic strain energy* U is the energy expended by the action of external forces in deforming an elastic body. Essentially all the work performed during elastic deformation is stored as elastic energy, and this energy is recovered on the release of the applied forces. Energy (or work) is equal to a force multiplied by the distance over which it acts. In the deformation of an elastic body, the force and deformation increase linearly from initial values of zero so that the average energy is equal to one-half of their product. This is also equal to the area under the load-deformation curve.

$$U = \tfrac{1}{2}P\delta$$

For an elemental cube that is subjected to only a tensile stress along the x axis, the elastic strain energy is given by

$$\begin{aligned} dU &= \tfrac{1}{2}P\,du = \tfrac{1}{2}(\sigma_x A)(\varepsilon_x\,dx) \\ &= \tfrac{1}{2}(\sigma_x\varepsilon_x)(A\,dx) \end{aligned} \tag{2-79}$$

Equation (2-79) describes the total elastic energy absorbed by the element. Since $A\,dx$ is the volume of the element, the *strain energy per unit volume* or strain energy density U_0 is given by

$$U_0 = \frac{1}{2}\sigma_x\varepsilon_x = \frac{1}{2}\frac{\sigma_x^2}{E} = \frac{1}{2}\varepsilon_x^2 E \tag{2-80}$$

Note that the lateral strains which accompany deformation in simple tension do not enter into the expression for strain energy because forces do not exist in the direction of the lateral strains.

By the same type of reasoning, the strain energy per unit volume of an element subjected to *pure shear* is given by

$$U_0 = \frac{1}{2}\tau_{xy}\gamma_{xy} = \frac{1}{2}\frac{\tau_{xy}^2}{G} = \frac{1}{2}\gamma_{xy}^2 G \tag{2-81}$$

The elastic strain energy for a general three-dimensional stress distribution may be obtained by superposition.

$$U_0 = \tfrac{1}{2}(\sigma_x\varepsilon_x + \sigma_y\varepsilon_y + \sigma_z\varepsilon_z + \tau_{xy}\gamma_{xy} + \tau_{xz}\gamma_{xz} + \tau_{yz}\gamma_{yz}) \tag{2-82}$$

or in tensor notation

$$U_0 = \tfrac{1}{2}\sigma_{ij}\varepsilon_{ij} \tag{2-83}$$

Substituting the equations of Hooke's law [Eqs. (2-64) and (2-65)] for the strains in Eq. (2-82) results in an expression for strain energy per unit volume expressed solely in terms of the stress and the elastic constants

$$U_0 = \frac{1}{2E}\left(\sigma_x^2 + \sigma_y^2 + \sigma_z^2\right) - \frac{\nu}{E}\left(\sigma_x\sigma_y + \sigma_y\sigma_z + \sigma_x\sigma_z\right) + \frac{1}{2G}\left(\tau_{xy}^2 + \tau_{xz}^2 + \tau_{yz}^2\right) \tag{2-84}$$

Also, by substituting Eqs. (2-74) into Eq. (2-82), the stresses are eliminated, and the strain energy is expressed in terms of strains and the elastic constants

$$U_0 = \tfrac{1}{2}\lambda\Delta^2 + G\left(\varepsilon_x^2 + \varepsilon_y^2 + \varepsilon_z^2\right) + \tfrac{1}{2}G\left(\gamma_{xy}^2 + \gamma_{xz}^2 + \gamma_{yz}^2\right) \tag{2-85}$$

It is interesting to note that the derivative of U_0 with respect to any strain component gives the corresponding stress component. For example,

$$\frac{\partial U_0}{\partial \varepsilon_x} = \lambda\Delta + 2G\varepsilon_x = \sigma_x \tag{2-86}$$

In the same way, $\partial U_0/\partial \sigma_x = \varepsilon_x$. Methods of calculation using strain energy to arrive at stresses and strains are powerful tools in elasticity analysis. Some of the better known techniques are Castigliano's theorem, the theorem of least work, and the principal of virtual work.

2-14 ANISOTROPY OF ELASTIC BEHAVIOR

Up to this point we have considered elastic behavior from a simple phenomenological point of view, i.e., Hooke's law was presented as a well-established empirical law and our attention was directed at developing useful relationships between stress and strain in an isotropic elastic solid. In this section we consider the fact that the elastic constants of a crystal vary markedly with orientation. However, first it is important to discuss briefly the nature of the elastic forces between atoms.

When a force is applied to a crystalline solid, it either pulls the atoms apart or pushes them together. The applied force is resisted by the forces of attraction or repulsion between the atoms. A convenient way to look at this is with an energy-distance diagram (Fig. 2-19), which represents the interaction energy (potential energy) between two atoms as they are separated by a distance a. When the external force is zero, the atoms are separated by a distance equal to the equilibrium spacing $a = a_0$. For small applied forces, the atoms will find a new equilibrium spacing a at which the external and internal forces are balanced. The displacement of the atom is $u = a - a_0$. Since force is the derivative of potential energy with distance [compare Eq. (2-86)], the force to produce a given equilibrium displacement is

$$P = \frac{d\phi(u)}{du} \tag{2-87}$$

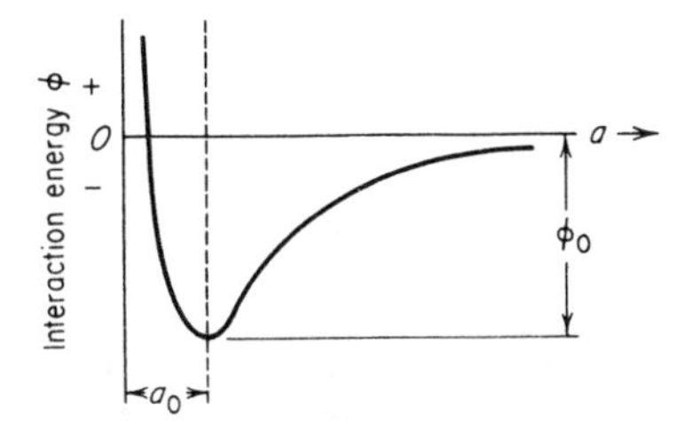

Figure 2-19 Interaction energy vs. separation between atoms

where $\phi(u)$ is the interaction bond energy at a displacement u. Thus, the force on a bond is a function of displacement u. For each displacement there is a characteristic value of force $P(u)$. Moreover, the deformation of the bonds between atoms is reversible. When the displacement returns to some initial value u_1 after being extended to u_2 the force returns to its previous value $P(u_1)$.

In an elastic solid the bond energy is a continuous function of displacement.[1] Thus, we can express $\phi(u)$ as a Taylor series

$$\phi(u) = \phi_0 + \left(\frac{d\phi}{du}\right)_0 u + \frac{1}{2}\left(\frac{d^2\phi}{du^2}\right)_0 u^2 + \cdots \tag{2-88}$$

where ϕ_0 is the energy at $u = 0$ and the differential coefficients are measured at $u = 0$. Since the force is zero when $a = a_0$, $d\phi/du = 0$

$$\phi(u) = \phi_0 + \frac{1}{2}\left(\frac{d^2\phi}{du^2}\right)_0 u^2$$

$$P = \frac{d\phi(u)}{du} = \left(\frac{d^2\phi}{du^2}\right)_0 u \tag{2-89}$$

The coefficient $(d^2\phi/du^2)_0$ is the curvature of the energy-distance curve at $u = a_0$. Since it is independent of u, the coefficient is a constant, and Eq. (2-89) is equivalent to $P = ku$, which is Hooke's law in its original form. When Eq. (2-89) is expressed in terms of stress and strain, the coefficient is directly proportional to the elastic constant of the material. It has the same value for both tension and compression since it is independent of the sign of u. Thus, we have shown that the elastic constant is determined by the sharpness of curvature of the minimum in the energy-distance curve. It is therefore a basic property of the material, not readily changed by heat treatment or defect structure, although it would be expected to decrease with increasing temperature. Moreover, since the binding forces will be strongly affected by distance between atoms, the elastic constants will vary with direction in the crystal lattice.

[1] This development follows that given by A. H. Cottrell, "The Mechanical Properties of Matter," pp. 84–85, John Wiley & Sons, Inc., New York, 1964.

In the generalized case[1] Hooke's law may be expressed as

$$\varepsilon_{ij} = S_{ijkl}\sigma_{kl} \tag{2-90}$$

and

$$\sigma_{ij} = C_{ijkl}\varepsilon_{kl} \tag{2-91}$$

where S_{ijkl} is the *compliance tensor* and C_{ijkl} is the *elastic stiffness* (often called just the elastic constants). Both S_{ijkl} and C_{ijkl} are fourth-rank tensor quantities. If we expanded Eq. (2-90) or (2-91), we would get nine equations, each with nine terms, 81 constants in all. However, we know that both ε_{ij} and σ_{ij} are symmetric tensors, that is, $\sigma_{ij} = \sigma_{ji}$, which immediately leads to appreciable simplification. Thus, we can write

$$\varepsilon_{ij} = S_{ijkl}\sigma_{kl} \quad \text{or} \quad \varepsilon_{ij} = S_{ijlk}\sigma_{lk}$$

and since

$$S_{ijkl}\sigma_{kl} = S_{ijlk}\sigma_{lk}$$

$$\sigma_{kl} = \sigma_{lk} \quad \text{and} \quad S_{ijkl} = S_{ijlk}$$

Also, we could write

$$\varepsilon_{ij} = S_{ijkl}\sigma_{kl} = \varepsilon_{ji} = S_{jikl}\sigma_{kl}$$

$$S_{ijkl} = S_{jikl}$$

Therefore, because of the symmetry of the stress and strain tensors, only 36 of the components of the compliance tensor are independent and distinct terms. The same is true of the elastic stiffness tensor.

Expanding Eq. (2-91) and taking into account the above relationships gives equations like

$$\begin{aligned}
\sigma_{11} &= C_{1111}\varepsilon_{11} + C_{1122}\varepsilon_{22} + C_{1133}\varepsilon_{33} + C_{1123}(2\varepsilon_{23}) + C_{1113}(2\varepsilon_{13}) + C_{1112}(2\varepsilon_{12}) \\
&\cdots\cdots\cdots\cdots\cdots\cdots\cdots\cdots\cdots\cdots \\
&\cdots\cdots\cdots\cdots\cdots\cdots\cdots\cdots\cdots\cdots \\
\sigma_{23} &= C_{2311}\varepsilon_{11} + C_{2322}\varepsilon_{22} + C_{2333}\varepsilon_{33} + C_{2323}(2\varepsilon_{23}) + C_{2313}(2\varepsilon_{13}) + C_{2312}(2\varepsilon_{12}) \\
&\cdots\cdots\cdots\cdots\cdots\cdots\cdots\cdots\cdots\cdots \\
&\cdots\cdots\cdots\cdots\cdots\cdots\cdots\cdots\cdots\cdots
\end{aligned} \tag{2-92}$$

These equations show that, in contrast to the situation for an isotropic elastic solid, Eq. (2-72), for an anisotropic elastic solid both normal strains and shear strains are capable of contributing to a normal stress.

[1] An excellent text that deals with the anisotropic properties of crystals in tensor notation is J. F. Nye, "Physical Properties of Crystals," Oxford University Press, London, 1957. For a treatment of anisotropic elasticity see R. F. S. Hearmon, "An Introduction to Applied Anisotropic Elasticity," Oxford University Press, London, 1961. A fairly concise but complete discussion of crystal elasticity is given by S. M. Edelglass, "Engineering Materials Science," pp. 277–301. The Ronald Press Company, New York, 1966.

In expanding Eq. (2-90), we express the shearing strains by the more conventional engineering shear strain $\gamma = 2\varepsilon$.

$$
\begin{aligned}
\varepsilon_{11} &= S_{1111}\sigma_{11} + S_{1122}\sigma_{22} + S_{1133}\sigma_{33} + 2S_{1123}\sigma_{23} + 2S_{1113}\sigma_{13} + 2S_{1112}\sigma_{12} \\
&\cdots\cdots\cdots\cdots\cdots\cdots\cdots\cdots\cdots\cdots \\
&\cdots\cdots\cdots\cdots\cdots\cdots\cdots\cdots\cdots\cdots \\
\gamma_{23} &= 2\varepsilon_{23} = 2S_{2311}\sigma_{11} + 2S_{2322}\sigma_{22} + 2S_{2333}\sigma_{33} + 4S_{2323}\sigma_{23} \\
&\qquad + 4S_{2313}\sigma_{13} + 4S_{2312}\sigma_{12} \\
&\cdots\cdots\cdots\cdots\cdots\cdots\cdots\cdots\cdots\cdots \\
&\cdots\cdots\cdots\cdots\cdots\cdots\cdots\cdots\cdots\cdots
\end{aligned}
\tag{2-93}
$$

The usual convention for designating components of elastic compliance and elastic stiffness uses only two subscripts instead of four. This is called the *contracted notation*. The subscripts simply denote the row and column in the matrix of components in which they fall.

$$
\begin{aligned}
\sigma_{11} &= C_{11}\varepsilon_{11} + C_{12}\varepsilon_{22} + C_{13}\varepsilon_{33} + C_{14}\gamma_{23} + C_{15}\gamma_{13} + C_{16}\gamma_{12} \\
&\cdots\cdots\cdots\cdots\cdots\cdots\cdots\cdots\cdots\cdots \\
&\cdots\cdots\cdots\cdots\cdots\cdots\cdots\cdots\cdots\cdots \\
\sigma_{23} &= C_{41}\varepsilon_{11} + C_{42}\varepsilon_{22} + C_{43}\varepsilon_{33} + C_{44}\gamma_{23} + C_{45}\gamma_{13} + C_{46}\gamma_{12} \\
&\cdots\cdots\cdots\cdots\cdots\cdots\cdots\cdots\cdots\cdots \\
&\cdots\cdots\cdots\cdots\cdots\cdots\cdots\cdots\cdots\cdots
\end{aligned}
\tag{2-94}
$$

and

$$
\begin{aligned}
\varepsilon_{11} &= S_{11}\sigma_{11} + S_{12}\sigma_{22} + S_{13}\sigma_{33} + S_{14}\sigma_{23} + S_{15}\sigma_{13} + S_{16}\sigma_{12} \\
&\cdots\cdots\cdots\cdots\cdots\cdots\cdots\cdots\cdots\cdots \\
&\cdots\cdots\cdots\cdots\cdots\cdots\cdots\cdots\cdots\cdots \\
\sigma_{23} &= S_{41}\sigma_{11} + S_{42}\sigma_{22} + S_{43}\sigma_{33} + S_{44}\sigma_{23} + S_{45}\sigma_{13} + S_{46}\sigma_{12} \\
&\cdots\cdots\cdots\cdots\cdots\cdots\cdots\cdots\cdots\cdots \\
&\cdots\cdots\cdots\cdots\cdots\cdots\cdots\cdots\cdots\cdots
\end{aligned}
\tag{2-95}
$$

By comparing coefficients in Eqs. (2-92) and (2-94) and Eqs. (2-93) and (2-95) we note, for example, that

$$C_{2322} = C_{42} \qquad C_{1122} = C_{12}$$

$$S_{1122} = C_{12} \qquad 2S_{2311} = C_{41} \qquad 4S_{2323} = S_{44}$$

The elastic stiffness constants are defined by equations like

$$C_{11} = \frac{\Delta\sigma_{11}}{\Delta\varepsilon_{11}} \qquad \text{all } \varepsilon_{ij} \text{ constant except } \varepsilon_{11}$$

Unfortunately, a measurement such as this is difficult to do experimentally since the specimen must be constrained mechanically to prevent strains such as ε_{23}. It is much easier to experimentally determine the coefficients of the elastic compliance from equations of the type

$$S_{11} = \frac{\Delta\varepsilon_{11}}{\Delta\sigma_{11}} \qquad \text{all } \sigma_{ij} \text{ constant except } \sigma_{11}$$

If the components of S_{ij} have been determined experimentally, then the components of C_{ij} can be determined by matrix inversion.

At this stage we have 36 independent constants, but further reduction in the number of independent constants is possible. By using the relationship given in Eq. (2-86), we can show that the constants are symmetrical, that is, $C_{ij} = C_{ji}$. For example,

$$\frac{\partial U}{\partial \varepsilon_{11}} = \sigma_{11} = C_{11}\varepsilon_{11} + C_{12}\varepsilon_{22} + C_{13}\varepsilon_{33} + C_{14}\gamma_{23} + C_{15}\gamma_{13} + C_{16}\gamma_{12}$$

$$\frac{\partial^2 U}{\partial \varepsilon_{11}\,\partial \varepsilon_{22}} = C_{12}$$

$$\frac{\partial U}{\partial \varepsilon_{22}} = \sigma_{22} = C_{21}\varepsilon_{11} + C_{22}\varepsilon_{22} + C_{23}\varepsilon_{33} + C_{24}\gamma_{23} + C_{25}\gamma_{13} + C_{26}\gamma_{12}$$

$$\frac{\partial^2 U}{\partial \varepsilon_{22}\,\partial \varepsilon_{11}} = C_{21}$$

$$\therefore \quad \frac{\partial^2 U}{\partial \varepsilon_{11}\,\partial \varepsilon_{22}} = \frac{\partial^2 U}{\partial \varepsilon_{22}\,\partial \varepsilon_{11}} = C_{12} = C_{21}$$

In general, $C_{ij} = C_{ji}$ and $S_{ij} = S_{ji}$. Now, we start with 36 constants C_{ij}, but of these there are six consants where $i = j$. This leaves 30 constants where $i \neq j$, but only one-half of these are independent constants since $C_{ij} = C_{ji}$. Therefore, for the general anisotropic linear elastic solid there are $30/2 + 6 = 21$ independent elastic constants.

As a result of symmetry conditions found in different crystal structures the number of independent elastic constants can be reduced still further.

Crystal structure	Rotational symmetry	Number of independent elastic constants
Triclinic	None	21
Monoclinic	1 twofold rotation	13
Orthorhombic	2 perpendicular twofold rotations	9
Tetragonal	1 fourfold rotation	6
Hexagonal	1 sixfold rotation	5
Cubic	4 threefold rotations	3
Isotropic		2

Table 2-2 Stiffness and compliance constants for cubic crystals

Metal	C_{11}	C_{12}	C_{44}	S_{11}	S_{12}	S_{44}
Aluminum	10.82	6.13	2.85	1.57	−0.57	3.51
Copper	16.84	12.14	7.54	1.49	−0.62	1.33
Iron	23.70	14.10	11.60	0.80	−0.28	0.86
Tungsten	50.10	19.80	15.14	0.26	−0.07	0.66

Stiffness constants in units of 10^{10} Pa.
Compliance in units of 10^{-11} Pa.

For a cubic crystal structure

$$C_{11} = \frac{S_{11} + S_{12}}{(S_{11} - S_{12})(S_{11} + 2S_{12})}$$

$$C_{12} = \frac{-S_{12}}{(S_{11} - S_{12})(S_{11} + 2S_{12})} \tag{2-96}$$

$$C_{44} = \frac{1}{S_{44}}$$

The modulus of elasticity in any direction of a cubic crystal (described by the direction cosines l, m, n) is given by

$$\frac{1}{E} = S_{11} - 2\left[(S_{11} - S_{12}) - \frac{1}{2}S_{44}\right](l^2m^2 + m^2n^2 + l^2n^2) \tag{2-97}$$

Typical values of elastic constants for cubic metals are given in Table 2-2.

By comparing the generalized Hooke's law Eqs. (2-95) with the equations using the common technical moduli Eq. (2-64) we can conclude that the elastic constants for an isotropic material are given by

$$S_{11} = \frac{1}{E} \qquad S_{12} = -\frac{\nu}{E} \qquad S_{44} = \frac{1}{G}$$

Since S_{11} and S_{12} are the independent constants, their relationship to S_{44} can be obtained from Eq. (2-68)

$$G = \frac{E}{2(1 + \nu)} = \frac{1}{2(1/E + \nu/E)}$$

$$G = \frac{1}{S_{44}} = \frac{1}{2(S_{11} - S_{12})}$$

or

$$S_{44} = 2(S_{11} - S_{12}) \tag{2-98}$$

Comparable equations relating the elastic stiffness constants can be developed from Eqs. (2-95) and (2-74).

$$C_{12} = \lambda \quad \text{Lamé's constant}$$

$$C_{11} = 2G + \lambda \tag{2-99}$$

$$C_{44} = \tfrac{1}{2}(C_{11} - C_{12})$$

The technical elastic moduli E, ν, and G are usually measured by direct static measurements in the tension or torsion tests. However, where more precise measurements are required or where measurements are required in small single-crystal specimens cut along specified directions, dynamic techniques using measurement of frequency or elapsed time are frequently employed. Dynamic measurements involve very small atomic displacements and low stresses compared with static modulus measurements. The velocity of propagation of a displacement down a cylindrical-crystal specimen is given by

$$v_x = \frac{\omega\lambda}{2\pi}\sqrt{\frac{E_x}{\rho}} \tag{2-100}$$

where ω is the natural frequency of vibration of a stress pulse of wavelength λ in a crystal of density ρ. Dynamic techniques consist of measuring either the natural frequency of vibration or the elapsed time for an ultrasonic pulse to travel down the specimen and return. Because the strain cycles produced in dynamic testing occur at high rates, there is very little time for heat transfer to take place. Thus, dynamic measurements of elastic constants are obtained under adiabatic conditions, while static elastic measurements are obtained under essentially isothermal conditions. There is a small difference between adiabatic and isothermal elastic moduli.[1]

$$E_{adi} = \frac{E_{iso}}{1 - \dfrac{E_{iso}T\alpha^2}{9c}} \tag{2-101}$$

where α is the volume coefficient of thermal expansion and c is the specific heat. Since the specific heat of a solid is large compared to a gas, the difference between adiabatic and isothermal moduli is not great and can be ignored for practical purposes.

Example Determine the modulus of elasticity for tungsten and iron in the $\langle 111 \rangle$ and $\langle 100 \rangle$ directions. What conclusions can be drawn about their elastic anisotropy? From Table 2-2

	S_{11}	S_{12}	S_{44}
Fe:	0.80	−0.28	0.86
W:	0.26	−0.07	0.66

[1] For a derivation of Eq. (2-101) see S. M. Edelgass, op. cit., pp. 294–297.

The direction cosines for the chief directions in a cubic lattice are:

Directions	l	m	n
$\langle 100 \rangle$	1	0	0
$\langle 110 \rangle$	$1/\sqrt{2}$	$1/\sqrt{2}$	0
$\langle 111 \rangle$	$1/\sqrt{3}$	$1/\sqrt{3}$	$1/\sqrt{3}$

For iron:

$$\frac{1}{E_{111}} = 0.80 - 2\{(0.80 + 0.28) - 0.86/2\}\left(\frac{1}{9} + \frac{1}{9} + \frac{1}{9}\right)$$

$$\frac{1}{E_{111}} = 0.80 - 2(1.08 - 0.43)\left(\frac{1}{3}\right) = 0.80 - 1.30\left(\frac{1}{3}\right)$$

$$= 0.80 - 0.43$$

$$E_{111} = \frac{1}{0.37} = 2.70 \times 10^{11}\ \text{Pa}$$

$$\frac{1}{E_{100}} = 0.80 - 1.30(0) = 0.80 \qquad E_{100} = 1.25 \times 10^{11}\ \text{Pa}$$

For tungsten:

$$\frac{1}{E_{111}} = 0.26 - 2\left\{(0.26 + 0.07) - \frac{0.66}{2}\right\}\left(\frac{1}{3}\right)$$

$$\frac{1}{E_{111}} = 0.26 - 2\{0.33 - 0.33\}\left(\frac{1}{3}\right) = 0.26$$

$$E_{111} = \frac{1}{0.26} = 3.85 \times 10^{11}\ \text{Pa} = 385\ \text{GPa}$$

$$\frac{1}{E_{100}} = 0.26 - 2\left\{(0.26 + 0.07) - \frac{0.66}{2}\right\}(0) = 0.26$$

$$E_{100} = \frac{1}{0.26} = 3.85 \times 10^{11}\ \text{Pa}$$

Therefore, we see that tungsten is elastically isotropic while iron is elastically anisotropic.

2-15 STRESS CONCENTRATION

A geometrical discontinuity in a body, such as a hole or a notch, results in a nonuniform stress distribution at the vicinity of the discontinuity. At some region near the discontinuity the stress will be higher than the average stress at distances removed from the discontinuity. Thus, a *stress concentration* occurs at the

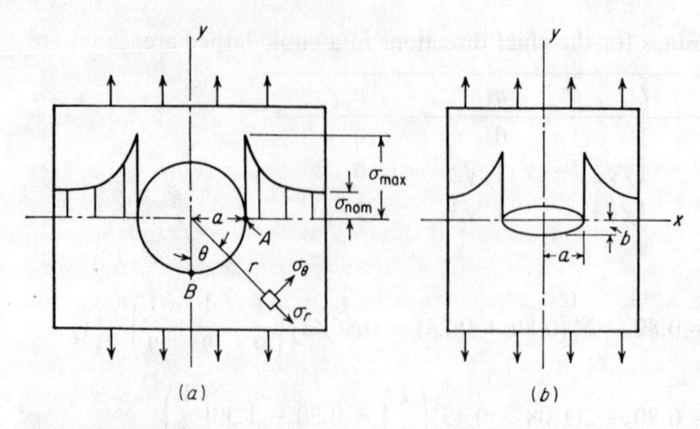

Figure 2-20 Stress distributions due to (a) circular hole and (b) elliptical hole.

discontinuity, or *stress raiser*. Figure 2-20a shows a plate containing a circular hole which is subjected to a uniaxial load. If the hole were not present, the stress would be uniformly distributed over the cross section of the plate and it would be equal to the load divided by the cross-sectional area of the plate. With the hole present, the distribution is such that the axial stress reaches a high value at the edges of the hole and drops off rapidly with distance away from the hole.

The stress concentration is expressed by a theoretical stress-concentration factor K_t. Generally K_t is described as the ratio of the maximum stress to the nominal stress based on the net section, although some workers use a value of nominal stress based on the entire cross section of the member in a region where there is no stress concentrator.

$$K_t = \frac{\sigma_{\text{max}}}{\sigma_{\text{nominal}}} \tag{2-102}$$

In addition to producing a stress oncentration, a notch also creates a localized condition of biaxial or triaxial stress. For example, for the circular hole in a plate subjected to an axial load, a radial stress is produced as well as a longitudinal stress. From elastic analysis,[1] the stresses produced in an infinitely wide plate containing a circular hole and axially loaded can be expressed as

$$\sigma_r = \frac{\sigma}{2}\left(1 - \frac{a^2}{r^2}\right) + \frac{\sigma}{2}\left(1 + 3\frac{a^4}{r^4} - 4\frac{a^2}{r^2}\right)\cos 2\theta$$

$$\sigma_\theta = \frac{\sigma}{2}\left(1 + \frac{a^2}{r^2}\right) - \frac{\sigma}{2}\left(1 + 3\frac{a^4}{r^4}\right)\cos 2\theta \tag{2-103}$$

$$\tau = -\frac{\sigma}{2}\left(1 - 3\frac{a^4}{r^4} + 2\frac{a^2}{r^2}\right)\sin 2\theta$$

[1] Timoshenko and Goodier, *op. cit.*, pp. 78–81.

Examination of these equations shows that the maximum stress occurs at point A when $\theta = \pi/2$ and $r = a$. For this case

$$\sigma_\theta = 3\sigma = \sigma_{\max} \tag{2-104}$$

where σ is the uniform tensile stress applied at the ends of the plate. The theoretical stress-concentration factor for a plate with a circular hole is therefore equal to 3. Further study of these equations shows that $\sigma_\theta = -\sigma$ for $r = a$ and $\theta = 0$. Therefore, when a tensile stress is applied to the plate, a compressive stress of equal magnitude exists at the edge of the hole at point B in a direction perpendicular to the axis of loading in the plane of the plate.

Another interesting case for which an analytical solution for the stress concentration is available[1] is the case of a small elliptical hole in a plate. Figure 2-20b shows the geometry of the hole. The maximum stress at the ends of the hole is given by the equation

$$\sigma_{\max} = \sigma\left(1 + 2\frac{a}{b}\right) \tag{2-105}$$

Note that, for a circular hole ($a = b$), the above equation reduces to Eq. (2-104). Equation (2-105) shows that the stress increases with the ratio a/b. Therefore, a very narrow hole, such as a crack, normal to the tensile direction will result in a very high stress concentration.

Mathematical complexities prevent the calculation of elastic stress-concentration factors in all but the simplest geometrical cases. Much of this work has been compiled by Neuber,[2] who has made calculations for various types of notches. Stress-concentration factors for practical problems are usually determined by experimental methods.[3] Photoelastic analysis[4] of models is the most widely used technique. This method is especially applicable to plane-stress problems, although it is possible to make three-dimensional photoelastic analyses. Figure 2-21 shows typical curves for the theoretical stress-concentration factor of certain machine elements that were obtained by photoelastic methods. Much of the information on stress concentrations in machine parts has been collected by Peterson.[5]

The effect of a stress raiser is much more pronounced in a brittle material than in a ductile material. In a ductile material, plastic deformation occurs when the yield stress is exceeded at the point of maximum stress. Further increase in load produces a local increase in strain at the critically stressed region with little increase in stress. Because of strain hardening, the stress increases in regions adjacent to the stress raiser, until, if the material is sufficiently ductile, the stress distribution becomes essentially uniform. Thus, a ductile metal loaded statically

[1] C. E. Inglis, *Trans. Inst. Nav. Archit.*, pt. 1, pp. 219–230, 1913.

[2] H. Neuber, "Theory of Notch Stresses," English translation, J. W. Edwards, Publisher, Incorporated, Ann Arbor, Mich., 1946.

[3] M. Hetenyi, "Handbook on Experimental Stress Analysis," John Wiley & Sons, Inc., New York, 1950.

[4] M. M. Frocht, "Photoelasticity," John Wiley & Sons, Inc., New York, 1955.

[5] R. E. Peterson, "Stress-Concentration Factors," John Wiley & Sons, Inc., New York, 1974.

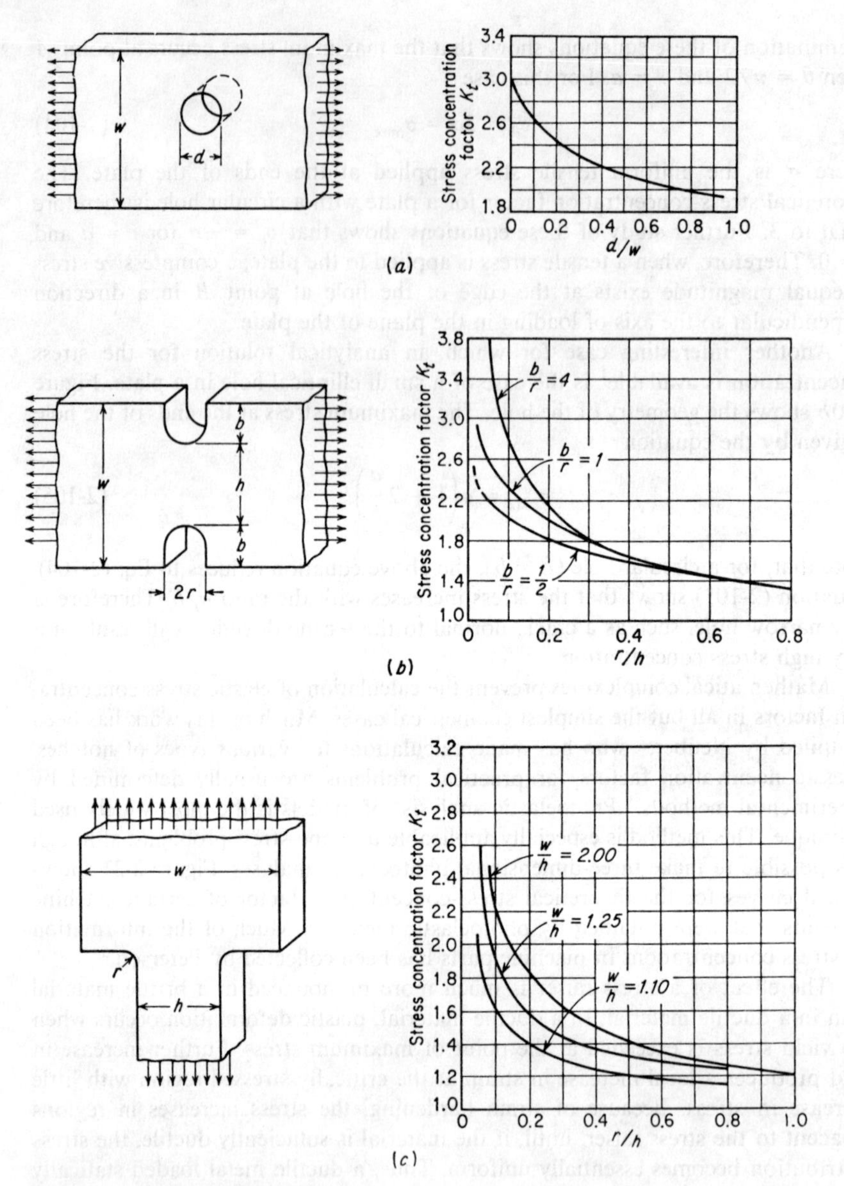

Figure 2-21 Theoretical stress-concentration factors for different geometrical shapes. (*After G. H. Neugebauer, Prod. Eng. (NY), vol. 14, pp. 82–87, 1943.*)

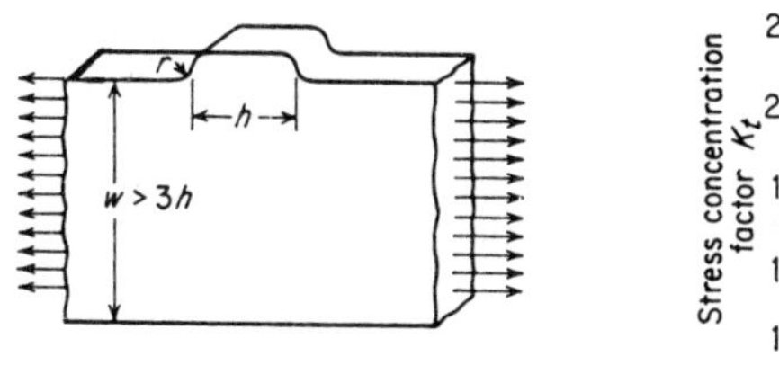

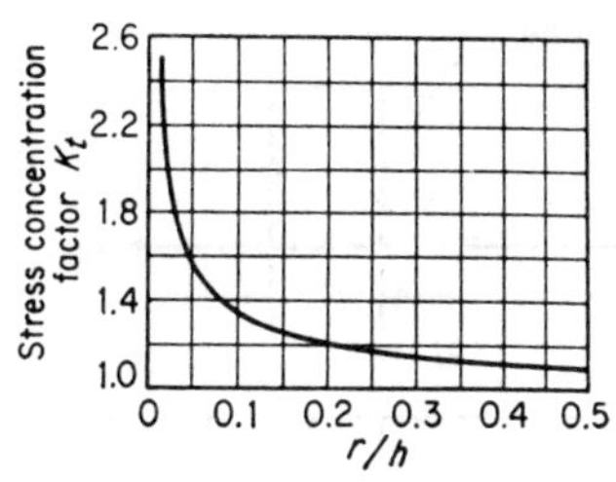

(d)

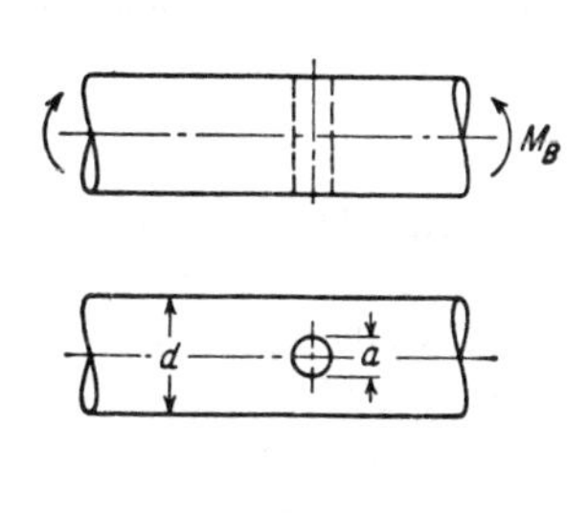

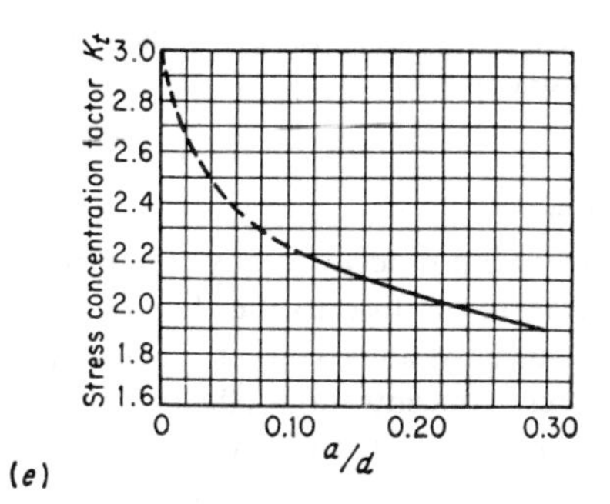

(e)

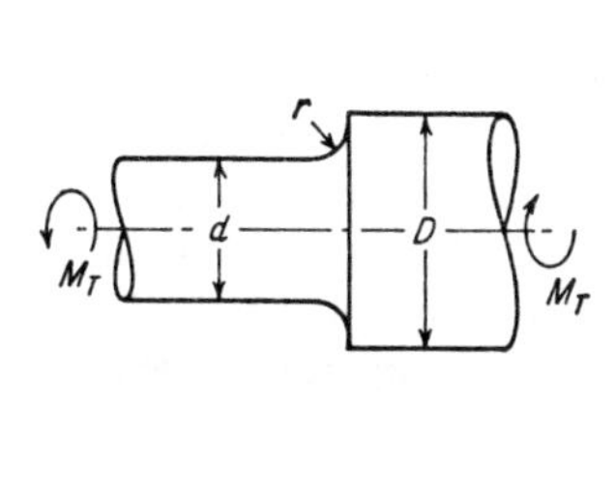

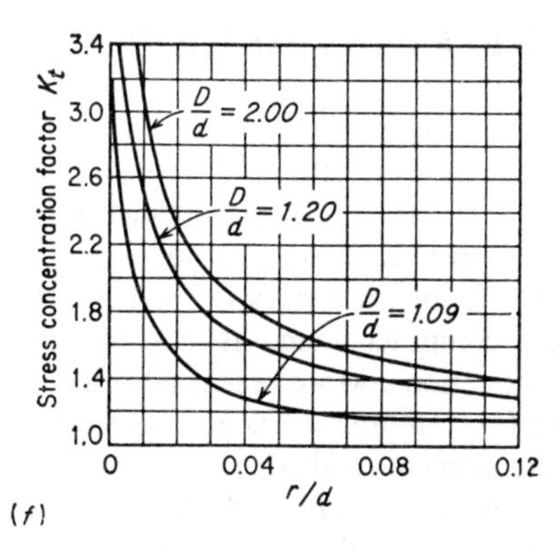

(f)

Figure 2-21 (*Continued*)

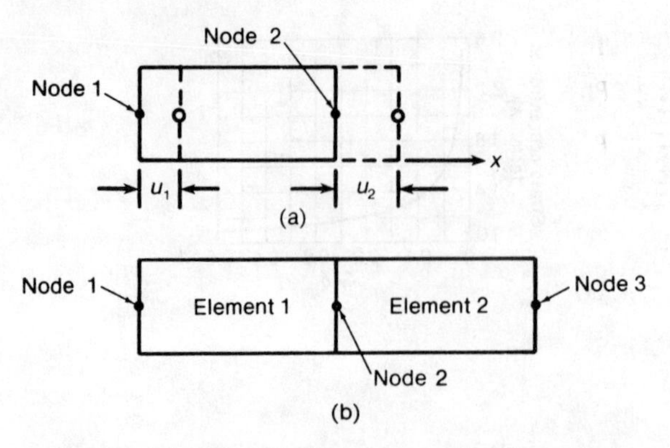

Figure 2-22 (*a*) Simple rectangular element to illustrate finite element analysis; (*b*) two elements joined to model a structure.

will not develop the full theoretical stress-concentration factor. However, redistribution of stress will not occur to any extent in a brittle material, and therefore a stress concentration of close to the theoretical value will result. Although stress raisers are not usually dangerous in ductile materials subjected to static loads, appreciable stress-concentration effects will occur in ductile materials under fatigue conditions of alternating stresses. Stress raisers are very important in the fatigue failure of metals and will be discussed further in Chap. 12.

2-16 FINITE ELEMENT METHOD

The finite element method (FEM) is a very powerful technique for determining stresses and deflections in structures too complex to analyze by strictly analytical methods. With this method the structure is divided into a network of small elements connected to each other at node points. Finite element analysis grew out of matrix methods for the analysis of structures when the widespread availability of the digital computer made it possible to solve systems of hundreds of simultaneous equations. More recent advances in computer graphics and availability of powerful computer work stations have given even greater emphasis to the spread of finite element methods throughout engineering practice.

A very simplified concept of the finite element method[1] is given in Fig. 2-22. A simple one-dimensional two-node element is shown in Fig. 2-22*a*. For this element, each node has one degree of freedom as it is displaced by u_1 and u_2. The

[1] O. Zienkiewicz, "The Finite Element Method," 3d ed., McGraw-Hill, New York, 1977; L. J. Segerlind, "Applied Finite Element Analysis," John Wiley & Sons, New York, 1976; K. H. Heubner, and E. A. Thornton, "The Finite Element Method for Engineers," John Wiley & Sons, New York, 1982.

equations relating the forces applied to the nodes to their displacements are

$$\begin{aligned} P_1 &= k_{11}u_1 + k_{12}u_2 \\ P_2 &= k_{21}u_1 + k_{22}u_2 \end{aligned} \tag{2-106}$$

The stiffness coefficients k_{ij} are calculated by the computer program based on the elastic properties of the material and the geometry of the finite element. The stiffness equations above are manipulated in the computer in matrix form.

$$\begin{Bmatrix} P_1 \\ P_2 \end{Bmatrix} = \begin{bmatrix} k_{11}k_{12} \\ k_{21}k_{22} \end{bmatrix} \begin{Bmatrix} u_1 \\ u_2 \end{Bmatrix} \tag{2-107}$$

When a second element is added to the first, Fig. 2-22*b*, a new set of matrix equations is generated.

$$\begin{Bmatrix} P_2 \\ P_3 \end{Bmatrix} = \begin{bmatrix} k_{22}k_{23} \\ k_{32}k_{33} \end{bmatrix} \begin{Bmatrix} u_2 \\ u_3 \end{Bmatrix} \tag{2-108}$$

When these two elements are combined into a structure we can use the principle of superposition to arrive at the stiffness for the two-element structure.

$$\begin{Bmatrix} P_1 \\ P_2 \\ P_3 \end{Bmatrix} = \begin{bmatrix} k_{11} & k_{12} & 0 \\ k_{21} & k_{22} + k_{22} & k_{23} \\ 0 & k_{32} & k_{33} \end{bmatrix} \begin{Bmatrix} u_1 \\ u_2 \\ u_3 \end{Bmatrix} \tag{2-109}$$

Finite element analysis was originally developed for two-dimensional (plane-stress) situations. A three-dimensional structure causes orders of magnitude increase in the number of simultaneous equations; but by using higher order elements and faster computers, these problems are being handled by the FEM. Figure 2-23 shows a few of the elements available for FEM analysis. Figure 2-23*a* is the basic triangular element. It is the simplest two-dimensional element, and it is also the element most often used. An assemblage of triangles can always represent a two-dimensional domain of any shape. The six-node triangle (*b*) increases the degrees of freedom available in modeling. The quadrilateral element (*c*) is a combination of two basic triangles. Its use reduces the number of elements necessary to model some situations. Elements (*d*) and (*e*) are three-dimensional but require only two independent variables for their description. These elements are used for problems that possess axial symmetry in cylindrical coordinates Figure 2-23*d* is a one-dimensional ring element and (*e*) is a two-dimensional triangular element. Three-dimensional FEM models are best constructed from isoparametric elements with curved sides. Figure 2-23*f* is an isoparametric triangle; (*g*) is a tetrahedron; and (*h*) is a hexahedron. These elements are most useful when it is desirable to approximate curved boundaries with a minimum number of elements.

A finite element solution involves calculating the stiffness matrices for every element in the structure. These elements are then assembled into an overall

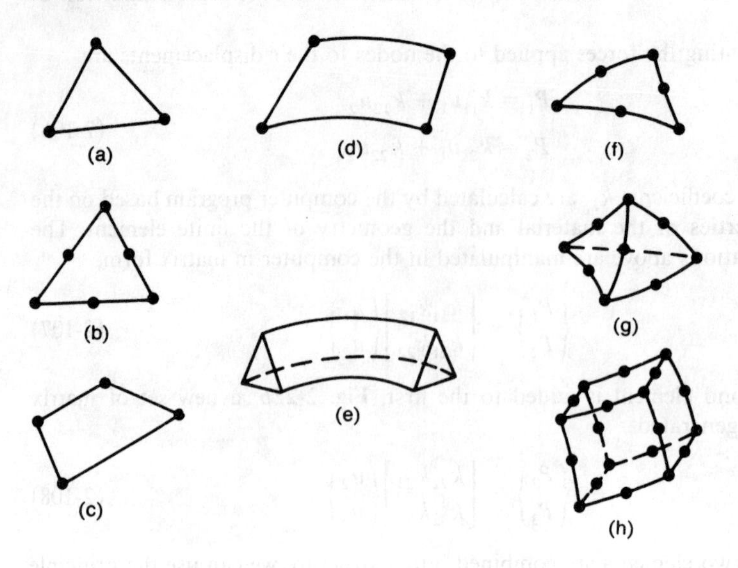

Figure 2-23 Some common elements used in FEM analysis.

stiffness matrix $[K]$ for the complete structure.

$$\{P\} = [K]\{u\} \tag{2-110}$$

The force matrix is known because it consists of numerical values of loads and reactions computed prior to the start of the finite element analysis. The displacements $\{u\}$ are the unknowns and they are solved for in Eq. (2-110) to give the displacements of all the nodes. When this is multiplied by the matrix of coordinate positions of the nodes $[B]$ and the matrix of elastic constants $[D]$ the stress is known at every nodal point.

$$\{\sigma\} = [D][B]\{u\} \tag{2-111}$$

A cumbersome part of the finite element solution is the preparation of the input data. The topology of the element mesh must be described in the computer program with the node numbers and the coordinates of the node points, along with the element numbers and the node numbers associated with each element. Tabulating all of this information is an extremely tedious bookkeeping task which is very error prone for a structure containing hundreds of nodes. Fortunately, modern technology has eliminated these problems and in the process has greatly expanded the utilization of finite element methods. Preprocessors allow the finite element mesh to be positioned on a drawing of the structure and the nodal coordinates and element connectivity to be automatically input. Postprocessing routines display the output of the finite element analysis in graphic form, allowing the user to quickly evaluate the information instead of wading through reams of numerical printouts.

BIBLIOGRAPHY

Jaeger, J. C.: "Elasticity, Fracture, and Flow," 3d ed., Methuen & Co., Ltd., London, 1971.
Love, A. E. H.: "A Treatise on the Mathematical Theory of Elasticity," 4th ed., Dover Publications, Inc., New York, 1949.
Reid, C. N.: "Deformation Geometry for Materials Scientists," Pergamon Press, New York, 1973.
Timoshenko, S. P., and J. N. Goodier: "Theory of Elasticity," 3d ed., McGraw-Hill Book Company, New York, 1961.
Wang, C. T.: "Applied Elasticity," McGraw-Hill Book Company, New York, 1953.
Urgarl, A. C., and S. K. Fenster: "Advanced Strength of Materials and Applied Elasticity," Elsevier, 1975.

CHAPTER

3

ELEMENTS OF THE THEORY OF PLASTICITY

3-1 INTRODUCTION

The theory of plasticity deals with the behavior of materials at strains where Hooke's law is no longer valid. A number of aspects of plastic deformation make the mathematical formulation of a theory of plasticity more difficult than the description of the behavior of an elastic solid. For example, plastic deformation is not a reversible process like elastic deformation. Elastic deformation depends only on the initial and final states of stress and strain, while the plastic strain depends on the loading path by which the final state is achieved. Moreover, in plastic deformation there is no easily measured constant relating stress to strain as with Young's modulus for elastic deformation. The phenomenon of strain hardening is difficult to accommodate within the theory of plasticity without introducing considerable mathematical complexity. Also, several aspects of real material behavior, such as plastic anisotropy, elastic hysteresis, and the Bauschinger effect (see Sec. 3-2) cannot be treated easily by plasticity theory. Nevertheless, the theory of plasticity has been one of the most active areas of continuum mechanics, and considerable progress has been made in developing a theory which can solve important engineering problems.

The theory of plasticity is concerned with a number of different types of problems. From the viewpoint of design, plasticity is concerned with predicting the maximum load which can be applied to a body without causing excessive yielding. The yield criterion must be expressed in terms of stress in such a way that it is valid for all states of stress. The designer is also concerned with plastic deformation in problems where the body is purposely stressed beyond the yield stress into the plastic region. For example, plasticity must be considered in designing for processes such as autofrettage, shrink fitting, and the overspeeding

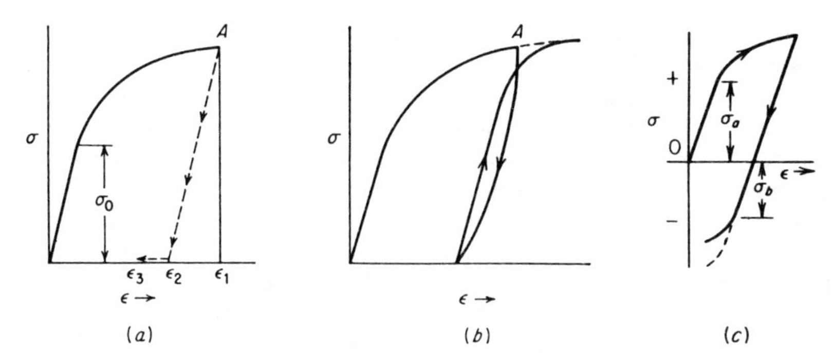

Figure 3-1 Typical true stress-strain curves for a ductile metal.

of rotor disks. The consideration of small plastic strains allows economies in building construction through the use of the theory of limit design.

The analysis of large plastic strains is required in the mathematical treatment of the plastic forming of metals. This aspect of plasticity will be considered in Part Four. It is very difficult to describe, in a rigorous analytical way, the behavior of a metal under these conditions. Therefore, certain simplifying assumptions are usually necessary to obtain a tractable mathematical solution.

Another aspect of plasticity is concerned with acquiring a better understanding of the mechanism of the plastic deformation of metals. Interest in this field is centered on the imperfections in crystalline solids. The effect of metallurgical variables, crystal structure, and lattice imperfections on the deformation behavior are of chief concern. This aspect of plasticity is considered in Part Two.

3-2 THE FLOW CURVE

The stress-strain curve obtained by uniaxial loading, as in the ordinary tension test, is of fundamental interest in plasticity when the curve is plotted in terms of true stress σ and true strain ε. True stress and true strain are discussed in the next section. The purpose of this section[1] is to describe typical stress-strain curves for real metals and to compare them with the theoretical flow curves for ideal materials.

The true stress-strain curve for a typical ductile metal, such as aluminum, is illustrated in Fig. 3-1*a*. Hooke's law is followed up to some yield stress σ_0. (The value of σ_0 will depend upon the accuracy with which strain is measured.) Beyond σ_0, the metal deforms plastically. Most metals strain-harden in this region, so that increases in strain require higher values of stress than the initial yield stress σ_0. However, unlike the situation in the elastic region, the stress and strain are not related by any simple constant of proportionality. If the metal is strained to point

[1] See Chap. 8 for a more complete discussion of the mathematics of the true stress-strain curve.

A, when the load is released the total strain will immediately decrease from ε_1 to ε_2 by an amount σ/E. The strain decrease $\varepsilon_1 - \varepsilon_2$ is the *recoverable elastic strain*. However, the strain remaining is not all permanent plastic strain. Depending upon the metal and the temperature, a small amount of the plastic strain $\varepsilon_2 - \varepsilon_3$ will disappear with time. This is known as *anelastic behavior*. Generally the anelastic strain is neglected in mathematical theories of plasticity.

Usually the stress-strain curve on unloading from a plastic strain will not be exactly linear and parallel to the elastic portion of the curve (Fig. 3-1*b*). Moreover, on reloading the curve will generally bend over as the stress approaches the original value of stress from which it was unloaded. With a little additional plastic strain the stress-strain curve becomes a continuation of what it would have been had no unloading taken place. The hysteresis behavior resulting from unloading and loading from a plastic strain is generally neglected in plasticity theories.

If a specimen is deformed plastically beyond the yield stress in one direction, e.g., in tension, and then after unloading to zero stress it is reloaded in the opposite direction, e.g., in compression, it is found that the yield stress on reloading is less than the original yield stress. Referring to Fig. 3-1*c*, $\sigma_b < \sigma_a$. This dependence of the yield stress on loading path and direction is called the *Bauschinger effect*. The Bauschinger effect is commonly ignored in plasticity theory, and it is usual to assume that the yield stress in tension and compression are the same.

A true stress-strain curve is frequently called a *flow curve* because it gives the stress required to cause the metal to flow plastically to any given strain. Many attempts have been made to fit mathematical equations to this curve. The most common is a power expression of the form

$$\sigma = K\varepsilon^n \tag{3-1}$$

where K is the stress at $\varepsilon = 1.0$ and n, the strain-hardening coefficient, is the slope of a log-log plot of Eq. (3-1). This equation can be valid only from the beginning of plastic flow to the maximum load at which the specimen begins to neck down.

Even the simple mathematical expression for the flow curve that is given by Eq. (3-1) can result in considerable mathematical complexity when it is used with the equations of the theory of plasticity. Therefore, in this field it is common practice to devise idealized flow curves which simplify the mathematics without deviating too far from physical reality. Figure 3-2*a* shows the flow curve for a *rigid, perfectly plastic* material. For this idealized material, a tensile specimen is completely rigid (zero elastic strain) until the axial stress equals σ_0, whereupon the material flows plastically at a constant flow stress (zero strain hardening). This type of behavior is approached by a ductile metal which is in a highly cold worked condition. Figure 3-2*b* illustrates the flow curve for a perfectly plastic material with an elastic region. This behavior is approached by a material such as plain carbon steel which has a pronounced yield-point elongation (see Sec. 6-5). A more realistic approach is to approximate the flow curve by two straight lines

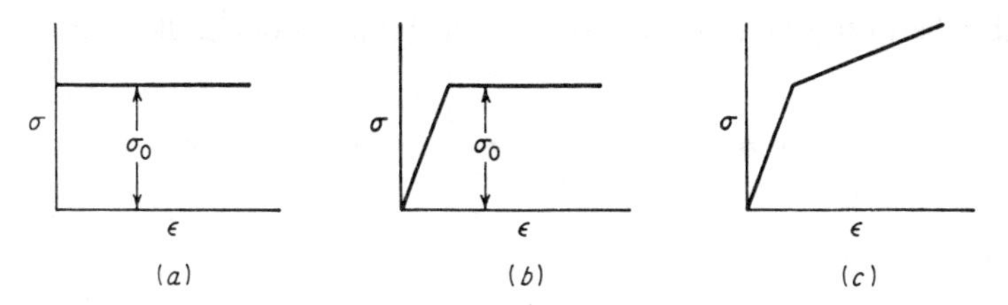

Figure 3-2 Idealized flow curves. (*a*) Rigid ideal plastic material; (*b*) ideal plastic material with elastic region; (*c*) piecewise linear (strain-hardening) material.

corresponding to the elastic and plastic regions (Fig. 3-2*c*). This type of curve results in somewhat more complicated mathematics.

3-3 TRUE STRESS AND TRUE STRAIN

The engineering stress-strain curve does not give a true indication of the deformation characteristics of a material because it is based entirely on the original dimensions of the specimen, and these dimensions change continuously during the test. Also in metalworking processes, such as wiredrawing, the workpiece undergoes appreciable change in cross-sectional area. Thus, measures of stress and strain which are based on the instantaneous dimensions are needed. Since dimensional changes are small in elastic deformation, it was not necessary to make this distinction in the previous chapter.

Equation (1-1) describes the conventional concept of unit linear strain, namely, the change in length referred to the original unit length.

$$e = \frac{\Delta L}{L_0} = \frac{1}{L_0}\int_{L_0}^{L} dL$$

This definition of strain is satisfactory for elastic strains where ΔL is very small. However, in plastic deformation the strains are frequently large, and during the extension the gage length changes considerably. Ludwik[1] first proposed the definition of true strain, or natural strain, ε, which obviates this difficulty. In this definition of strain the change in length is referred to the instantaneous gage length, rather than to the original gage length.

$$\varepsilon = \sum \frac{L_1 - L_0}{L_0} + \frac{L_2 - L_1}{L_1} + \frac{L_3 - L_2}{L_2} + \cdots \tag{3-2}$$

or

$$\varepsilon = \int_{L_0}^{L} \frac{dL}{L} = \ln \frac{L}{L_0} \tag{3-3}$$

[1] P. Ludwik, "Elemente der technologischen Mechanik," Springer-Verlag OHG, Berlin, 1909.

The relationship between true strain and conventional linear strain[1] follows from Eq. (1-1).

$$e = \frac{\Delta L}{L_0} = \frac{L - L_0}{L_0} = \frac{L}{L_0} - 1$$

$$e + 1 = \frac{L}{L_0}$$

$$\varepsilon = \ln \frac{L}{L_0} = \ln (e + 1) \tag{3-4}$$

Values of true strain and conventional linear strain are given for comparison:

True strain ε	0.01	0.10	0.20	0.50	1.0	4.0
Conventional strain e	0.01	0.105	0.22	0.65	1.72	53.6

Thus, the two measures of strain give nearly identical results up to a strain of 0.1.

The advantage of using true strain should be apparent from the following example: Consider a uniform cylinder which is extended to twice its original length. The linear strain is then $e = (2L_0 - L_0)/L_0 = 1.0$, or a strain of 100 percent. To achieve the same amount of negative linear strain in compression, the cylinder would have to be squeezed to zero thickness. However, intuitively we should expect that the strain produced in compressing a cylinder to half its original length would be the same as, although opposite in sign to, the strain produced by extending the cylinder to twice its length. If true strain is used, equivalence is obtained for the two cases. For extension to twice the original length, $\varepsilon = \ln(2L_0/L_0) = \ln 2$. For compression to half the original length, $\varepsilon = \ln[(L_0/2)/L_0] = \ln \frac{1}{2} = -\ln 2$.

Another advantage of working with true strain is that the total true strain is equal to the sum of the incremental true strains. This can be seen from the following example. Consider a rod initially 2.0 in long that is elongated in three increments, each increment being a conventional strain of $e = 0.1$.

Increment	Length of rod	
0	2.00	
1	2.20	$e_{0-1} = 0.2/2.0 = 0.1$
2	2.42	$e_{1-2} = 0.22/2.20 = 0.1$
3	2.662	$e_{2-3} = 0.242/2.42 = 0.1$

[1] The reader is warned against confusion in notation for various measures of strain. We in general shall use ε for strain. The true strain will be implied unless otherwise specified. For elastic strains ($\varepsilon < 0.01$) the numerical values of ε and e are identical and on occasion we may use e for linear strain, especially where we want to designate a small elastic strain. In other texts true strain is sometimes denoted by δ or $\bar{\varepsilon}$.

We note that the total conventional strain $e_{0-3} = 0.662/2.0 = 0.331$ is not equal to $e_{0-1} + e_{1-2} + e_{2-3}$. However, if we use true strain, the sum of the increments equals the total strain.

$$\varepsilon_{0-1} + \varepsilon_{1-2} + \varepsilon_{2-3} = \ln\frac{2.2}{2.0} + \ln\frac{2.42}{2.2} + \ln\frac{2.662}{2.42} = \ln\frac{2.662}{2.0} = \varepsilon_{0-3}$$

One of the basic characteristics of plastic deformation is that a metal is essentially incompressible. The density changes measured on metals after large plastic strain are less than 0.1 percent. Therefore, as a good engineering approximation we can consider that the *volume* of a solid *remains constant* during plastic deformation.

In Sec. 2-8 we determined the volume strain by considering a cube of initial volume $dx\,dy\,dz$ which when deformed had a volume $dx\,(1 + e_x)\,dy\,(1 + e_y)\,dz\,(1 + e_z)$. The volume strain Δ is given by

$$\Delta = \frac{\Delta V}{V} = \frac{(1 + e_x)(1 + e_y)(1 + e_z)\,dx\,dy\,dz - dx\,dy\,dz}{dx\,dy\,dz}$$

or

$$\Delta = (1 + e_x)(1 + e_y)(1 + e_z) - 1$$

When we previously determined Δ for small elastic strains, it was permissible to neglect products of strains compared to the strain itself, but this is no longer possible when the larger plastic strains are being considered. Since the volume change is zero for plastic deformation,

$$\Delta + 1 = 0 + 1 = (1 + e_x)(1 + e_y)(1 + e_z)$$

or

$$\ln 1 = 0 = \ln(1 + e_x) + \ln(1 + e_y) + \ln(1 + e_z)$$

But $\varepsilon_x = \ln(1 + e_x)$, etc.

$$\therefore\quad \varepsilon_x + \varepsilon_y + \varepsilon_z = \varepsilon_1 + \varepsilon_2 + \varepsilon_3 = 0 \tag{3-5}$$

Equation (3-5) represents the first invariant of the strain tensor when strain is expressed as true strain. It is a very useful relationship in plasticity problems. Note particularly that Eq. (3-5) is not valid for elastic strains since there is an appreciable volume change relative to the magnitude of the elastic strains. Thus, if we add up the three equations for Hooke's law (2-64)

$$\Delta = e_x + e_y + e_z = \frac{1 - 2\nu}{E}(\sigma_x + \sigma_y + \sigma_z)$$

we see that Δ can be zero only if $\nu = \frac{1}{2}$. This result can be interpreted that Poisson's ratio is equal to $\frac{1}{2}$ for a plastic material for which $\Delta = 0$.

Because of constancy of volume $A_0 L_0 = AL$, and Eq. (3-3) can be written in terms of either length or area.

$$\varepsilon = \ln\frac{L}{L_0} = \ln\frac{A_0}{A} \tag{3-6}$$

True stress is the load at any instant divided by the cross-sectional area over which it acts. The *engineering stress*, or conventional stress, is the load divided by the original area. In considering elastic behavior it was not necessary to make this distinction, but in certain problems in plasticity, particularly when dealing with the mathematics of the tension test (Chap. 8), it is important to distinguish between these two definitions of stress. True stress will be denoted by the familiar symbol σ, while engineering stress will be denoted by s.

True stress $$\sigma = \frac{P}{A} \tag{3-7}$$

Engineering stress $$s = \frac{P}{A_0} \tag{3-8}$$

The true stress may be determined from the engineering stress as follows:

$$\sigma = \frac{P}{A} = \frac{P}{A_0}\frac{A_0}{A}$$

But, by the constancy-of-volume relationships

$$\frac{A_0}{A} = \frac{L}{L_0} = e + 1$$

$$\therefore \quad \sigma = \frac{P}{A_0}(e+1) = s(e+1) \tag{3-9}$$

Example A tensile specimen with a 0.505-in initial diameter and 2.0-in gage length reaches maximum load at 20,000 lb and fractures at 16,000 lb. The minimum diameter at fracture is 0.425 in. Determine the engineering stress at maximum load (the ultimate tensile strength) and the true fracture stress.

$$\text{Engineering stress at maximum load} = \frac{P_{max}}{A_{max}} = \frac{20{,}000}{\pi/4(0.505)^2} = \frac{20{,}000}{0.20} = 100{,}000 \text{ psi}$$

$$\text{True fracture stress} = \frac{P_f}{A_f} = \frac{16{,}000}{\pi/4(0.425)^2} = \frac{16{,}000}{0.142} = 112{,}600 \text{ psi}$$

Determine the true strain at fracture

$$\varepsilon_f = \ln\frac{A_0}{A_f} = \ln\left(\frac{0.505}{0.425}\right)^2 = 2\ln 1.188 = 2(0.172) = 0.344$$

What is the engineering strain at fracture?

$$\varepsilon = \ln(1+e); \qquad \exp(\varepsilon) = (1+e); \qquad \exp(0.344) = 1 + e_f$$

$$e_f = 1.410 - 1.000 = 0.410$$

3.4 YIELDING CRITERIA FOR DUCTILE METALS

The problem of deducing mathematical relationships for predicting the conditions at which plastic yielding begins when a material is subjected to any possible combination of stresses is an important consideration in the field of plasticity. In uniaxial loading, as in a tension test, macroscopic plastic flow begins at the yield stress σ_0. It is expected that yielding under a situation of combined stresses can be related to some particular combination of principal stresses. There is at present no theoretical way of calculating the relationship between the stress components to correlate yielding for a three-dimensional state of stress with yielding in the uniaxial tension test.

The yielding criteria are essentially empirical relationships. However, a yield criterion must be consistent with a number of experimental observations, the chief of which is that pure hydrostatic pressure does not cause yielding in a continuous solid.[1] As a result of this, the hydrostatic component of a complex state of stress does not influence the stress at which yielding occurs. Therefore, we look for the stress deviator to be involved with yielding. Moreover, for an isotropic material, the yield criterion must be independent of the choice of axes, i.e., it must be an invariant function. These considerations lead to the conclusion that the yield criteria must be some function of the invariants of the stress deviator. At present there are two generally accepted criteria for predicting the onset of yielding in ductile metals.

Von Mises' or Distortion-Energy Criterion

Von Mises (1913) proposed that yielding would occur when the second invariant of the stress deviator J_2 exceeded some critical value.

$$J_2 = k^2 \tag{3-10}$$

where $J_2 = \frac{1}{6}[(\sigma_1 - \sigma_2)^2 + (\sigma_2 - \sigma_3)^2 + (\sigma_3 - \sigma_1)^2]$.

To evaluate the constant k and relate it to yielding in the tension test, we realize that at yielding in uniaxial tension $\sigma_1 = \sigma_0$, $\sigma_2 = \sigma_3 = 0$

$$\sigma_0^2 + \sigma_0^2 = 6k^2$$

$$\sigma_0 = \sqrt{3}\,k \tag{3-11}$$

Substituting Eq. (3-11) in Eq. (3-10) results in the usual form of the von Mises' yield criterion

$$\sigma_0 = \frac{1}{\sqrt{2}}\left[(\sigma_1 - \sigma_2)^2 + (\sigma_2 - \sigma_3)^2 + (\sigma_3 - \sigma_1)^2\right]^{1/2} \tag{3-12}$$

[1] A significant influence of hydrostatic or mean stress of modest values on yielding has been observed in glassy polymers such as PMMA. S. S. Sternstein and L. Ongchin, *Polym. Prepr. Am. Chem. Soc. Div. Polym. Chem.*, September 1969.

or from Eq. (2-61)

$$\sigma_0 = \frac{1}{\sqrt{2}}\left[(\sigma_x - \sigma_y)^2 + (\sigma_y - \sigma_z)^2 + (\sigma_z - \sigma_x)^2 + 6(\tau_{xy}^2 + \tau_{yz}^2 + \tau_{xz}^2)\right]^{1/2} \quad (3\text{-}13)$$

Equation (3-12) or (3-13) predicts that yielding will occur when the differences of stresses on the right side of the equation exceed the yield stress in uniaxial tension σ_0.

Example Stress analysis of a spacecraft structural member gives the state of stress shown below. If the part is made from 7075-T6 aluminum alloy with $\sigma_0 = 500$ MPa, will it exhibit yielding? If not, what is the safety factor?

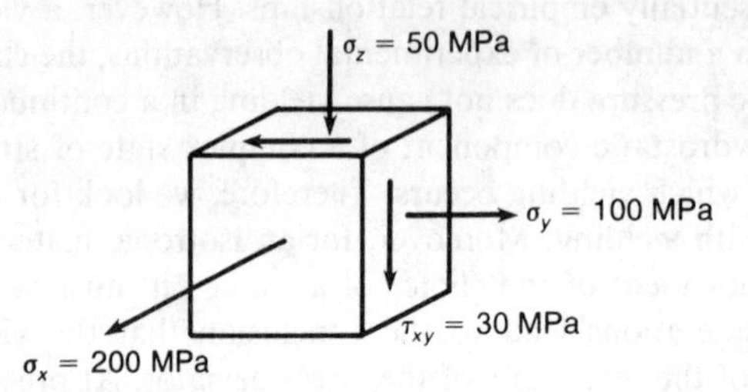

From Eq. (3-13)

$$\sigma_0 = \frac{1}{\sqrt{2}}\left[(200 - 100)^2 + (100 - (-50))^2 + (-50 - 200)^2 + 6(30)^2\right]^{1/2}$$

$$\sigma_0 = \frac{1}{\sqrt{2}}(100{,}400)^{1/2} = \frac{316.859}{\sqrt{2}} = 224 \text{ MPa}$$

Since the value of σ_0 calculated from the yield criterion is less than the yield strength of the aluminum alloy, yielding will not occur. The safety factor is $500/224 = 2.2$.

To identify the constant k in Eq. (3-10), consider the state of stress in pure shear, as is produced in a torsion test.

$$\sigma_1 = -\sigma_3 = \tau \qquad \sigma_2 = 0$$

$$\text{at yielding} \qquad \sigma_1^2 + \sigma_1^2 + 4\sigma_1^2 = 6k^2$$

$$\therefore \quad \sigma_1 = k$$

so that k represents the yield stress in pure shear (torsion). Therefore, the von Mises' criterion predicts that the yield stress in torsion will be less than in uniaxial tension according to

$$k = \frac{1}{\sqrt{3}}\sigma_0 = 0.577\sigma_0 \quad (3\text{-}14)$$

To summarize, note that the von Mises' yield criterion implies that yielding is not dependent on any particular normal stress or shear stress, but instead, yielding depends on a function of all three values of principal shearing stress. Since the yield criterion is based on differences of normal stresses, $\sigma_1 - \sigma_2$, etc., the criterion is independent of the component of hydrostatic stress. Since the von Mises' yield criterion involves squared terms, the result is independent of the sign of the individual stresses. This is an important advantage since it is not necessary to know which are the largest and smallest principal stresses in order to use this yield criterion.

Von Mises originally proposed this criterion because of its mathematical simplicity. Subsequently, other workers have attempted to give it physical meaning. Hencky (1924) showed that Eq. (3-12) was equivalent to assuming that yielding occurs when the *distortion energy* reaches a critical value. The distortion energy is that part of the total strain energy per unit volume that is involved in change of shape as opposed to a change in volume.

Example The fact that the total strain energy can be split into a term depending on change of volume and a term depending on distortion can be seen by expressing Eq. (2-84) in terms of principal stresses.

$$U_0 = \frac{1}{2E}\left[\sigma_1^2 + \sigma_2^2 + \sigma_3^2 - 2\nu(\sigma_1\sigma_2 + \sigma_2\sigma_3 + \sigma_1\sigma_3)\right] \qquad (3\text{-}15)$$

or expressing in terms of the invariants of the stress tensor

$$U_0 = \frac{1}{2E}\left[I_1^2 - 2I_2(1 + \nu)\right] \qquad (3\text{-}16)$$

This equation is more meaningful if we express it in terms of the bulk modulus (volume change) and the shear modulus (distortion). From Sec. 2-11,

$$E = \frac{9GK}{3K + G} \qquad \nu = \frac{3K - 2G}{6K + 2G}$$

Substituting into Eq. (3-16)

$$U_0 = \frac{I_1^2}{18K} + \frac{1}{6G}\left(I_1^2 - 3I_2\right) \qquad (3\text{-}17)$$

Equation (3-17) is important because it shows that the total strain energy can be split into a term depending on change of volume and a term depending on distortion.

$$(U_0)_{\text{distortion}} = \frac{1}{6G}\left(\sigma_1^2 + \sigma_2^2 + \sigma_3^2 - \sigma_1\sigma_2 - \sigma_2\sigma_3 - \sigma_1\sigma_3\right)$$

or

$$(U_0)_{\text{distortion}} = \frac{1}{12G}\left[(\sigma_1 - \sigma_2)^2 + (\sigma_2 - \sigma_3)^2 + (\sigma_3 - \sigma_1)^2\right] \qquad (3\text{-}18)$$

For a uniaxial state of stress, $\sigma_1 = \sigma_0$, $\sigma_2 = \sigma_3 = 0$

$$(U_0)_{\text{distortion}} = \frac{1}{12G}2\sigma_0^2$$

or

$$\sigma_0 = \frac{1}{\sqrt{2}}\left[(\sigma_1 - \sigma_2)^2 + (\sigma_2 - \sigma_3)^2 + (\sigma_3 - \sigma_1)^2\right]^{1/2} \tag{3-19}$$

Another physical interpretation given to the von Mises' yield criterion is that it represents the critical value of the octahedral shear stress (see Sec. 3-9). This is the shear stress on the octahedral planes which make equal angles with the principal axes. Still another interpretation is that it represents the mean square of the shear stress averaged over all orientations in the solid.[1]

Maximum-Shear-Stress or Tresca Criterion

This yield criterion assumes that yielding occurs when the maximum shear stress reaches the value of the shear stress in the uniaxial-tension test. From Eq. (2-21), the maximum shear stress is given by

$$\tau_{max} = \frac{\sigma_1 - \sigma_3}{2} \tag{3-20}$$

where σ_1 is the algebraically largest and σ_3 is the algebraically smallest principal stress.

For uniaxial tension, $\sigma_1 = \sigma_0$, $\sigma_2 = \sigma_3 = 0$, and the shearing yield stress τ_0 is equal to $\sigma_0/2$. Substituting in Eq. (3-20),

$$\tau_{max} = \frac{\sigma_1 - \sigma_3}{2} = \tau_0 = \frac{\sigma_0}{2}$$

Therefore, the maximum-shear-stress criterion is given by

$$\sigma_1 - \sigma_3 = \sigma_0 \tag{3-21}$$

For a state of pure shear, $\sigma_1 = -\sigma_3 = k$, $\sigma_2 = 0$, the maximum-shear-stress criterion predicts that yielding will occur when

$$\sigma_1 - \sigma_3 = 2k = \sigma_0$$

or

$$k = \frac{\sigma_0}{2}$$

so that the maximum-shear-stress criterion may be written

$$\sigma_1 - \sigma_3 = \sigma_1' - \sigma_3' = 2k \tag{3-22}$$

We note that the maximum-shear-stress criterion is less complicated mathematically than the von Mises' criterion, and for this reason it is often used in engineering design. However, the maximum-shear criterion does not take into consideration the intermediate principal stress. It suffers from the major difficulty that it is necessary to known in advance which are the maximum and minimum principal stresses. Moreover, the general form of the maximum-shear-stress criterion, Eq. (3-23), is far more complicated than the von Mises' criterion, Eq.

[1] See G. Sines, "Elasticity and Strength," pp. 54–56, Allyn and Bacon, Inc., Boston, 1969.

(3-10), and for this reason the von Mises' criterion is preferred in most theoretical work.

$$4J_2^3 - 27J_3^2 - 36k^2J_2^2 + 96k^4J_2 - 64k^6 = 0 \tag{3-23}$$

Example Use the maximum-shear-stress criterion to establish whether yielding will occur for the stress state shown in the previous example.

$$\tau_{max} = \frac{\sigma_x - \sigma_z}{2} = \frac{\sigma_0}{2}$$

$$200 - (-50) = \sigma_0$$

$$\sigma_0 = 250 \text{ MPa}$$

Again, the calculated value of σ_0 is less than the yield strength of the material.

3-5 COMBINED STRESS TESTS

The conditions for yielding under states of stress other than uniaxial and torsion loading can be studied conveniently with thin-wall tubes. Axial stress can be combined with torsion to produce various combinations of shear stress to normal stress intermediate between the values obtained separately in tension and torsion. Alternatively, a hydrostatic pressure may be introduced to produce a circumferential hoop stress in the tube.[1]

For the stresses shown in Fig. 3-3, from Eq. (2-9) the principal stresses are

$$\sigma_1 = \frac{\sigma_x}{2} + \left(\frac{\sigma_x^2}{4} + \tau_{xy}^2\right)^{1/2}$$

$$\sigma_2 = 0 \tag{3-24}$$

$$\sigma_3 = \frac{\sigma_x}{2} - \left(\frac{\sigma_x^2}{4} + \tau_{xy}^2\right)^{1/2}$$

Therefore, the maximum-shear-stress criterion of yielding is given by

$$\left(\frac{\sigma_x}{\sigma_0}\right)^2 + 4\left(\frac{\tau_{xy}}{\sigma_0}\right)^2 = 1 \tag{3-25}$$

and the distortion-energy theory of yielding is expressed by

$$\left(\frac{\sigma_x}{\sigma_0}\right)^2 + 3\left(\frac{\tau_{xy}}{\sigma_0}\right)^2 = 1 \tag{3-26}$$

[1] See for example S. S. Hecker, *Metall. Trans.*, vol. 2, pp. 2077–2086, 1971. A unique method for determining the yield locus of a flat sheet has been presented by D. Lee and W. A. Backofen, *Trans. Metall. Soc. AIME*, vol. 236, pp. 1077–1084, 1966. This method is well suited for studying the anisotropy of rolled sheet.

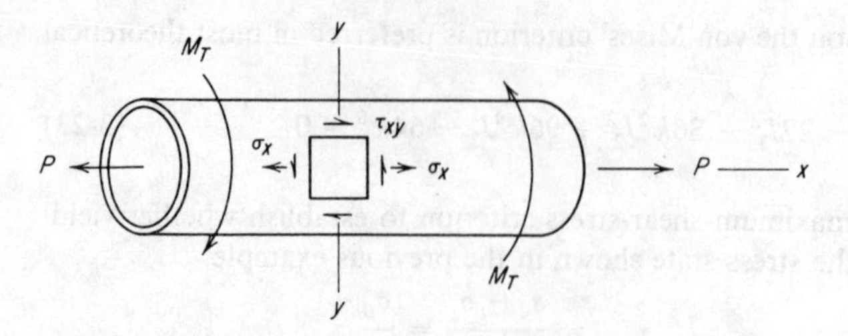

Figure 3-3 Combined tension and torsion in a thin-walled tube.

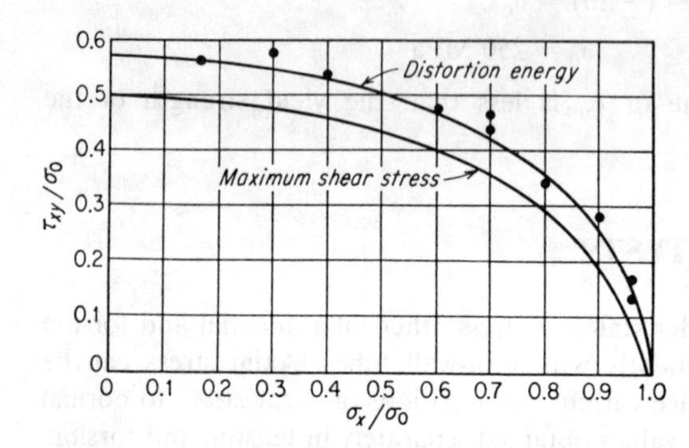

Figure 3-4 Comparison between maximum-shear-stress theory and distortion-energy (von Mises') theory.

Both equations define an ellipse. Figure 3-4 shows that the experimental results[1] agree best with the distortion-energy theory.

3-6 THE YIELD LOCUS

For a biaxial plane-stress condition ($\sigma_2 = 0$) the von Mises' yield criterion can be expressed mathematically as

$$\sigma_1^2 + \sigma_3^2 - \sigma_1\sigma_3 = \sigma_0^2 \tag{3-27}$$

This is the equation of an ellipse whose major semiaxis is $\sqrt{2}\,\sigma_0$ and whose minor semiaxis is $\sqrt{\frac{2}{3}}\,\sigma_0$. The plot of Eq. (3-27) is called a *yield locus* (Fig. 3-5). Several important points on the yield ellipse corresponding to particular stress-ratio loading paths are noted on the figure.

The yield locus for the maximum-shear-stress criterion falls inside of the von Mises' yield ellipse. Note that the two yielding criteria predict the same yield stress for conditions of uniaxial stress and balanced biaxial stress ($\sigma_1 = \sigma_3$). The greatest divergence between the two criteria occurs for pure shear ($\sigma_1 = -\sigma_3$). The yield stress predicted by the von Mises' criterion is 15.5 percent greater than the yield stress predicted by the maximum-shear-stress criterion.

[1] G. I. Taylor and H. Quinney, *Proc. R. Soc. London Ser. A.*, vol. 230A, pp. 323–362, 1931.

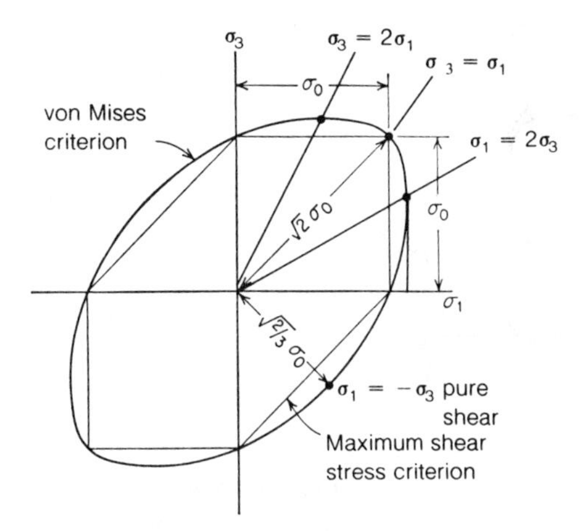

Figure 3-5 Comparison of yield criteria for plane stress.

3-7 ANISOTROPY IN YIELDING

The yielding criteria considered so far assume that the material is isotropic. While this may be the case at the start of plastic deformation, it certainly is no longer a valid assumption after the metal has undergone appreciable plastic deformation. Moreover, most fabricated metal shapes have anisotropic properties, so that it is likely that the tubular specimens used for basic studies of yield criteria incorporate some degree of anisotropy. Certainly the von Mises' criterion as formulated in Eq. (3-12) would not be valid for a highly oriented cold-rolled sheet or a fiber-reinforced composite material.

Hill[1] has formulated the von Mises' yield criterion for an anisotropic material having orthotropic symmetry.

$$F(\sigma_y - \sigma_z)^2 + G(\sigma_z - \sigma_x)^2 + H(\sigma_x - \sigma_y)^2 + 2L\tau_{yz}^2 + 2M\tau_{zx}^2 + 2N\tau_{xy}^2 = 1$$

where $F, G, \ldots, N$ are constants defining the degree of anisotropy. For principal axes of orthotropic symmetry

$$F(\sigma_2 - \sigma_3)^2 + G(\sigma_3 - \sigma_1)^2 + H(\sigma_1 - \sigma_2)^2 = 1 \tag{3-28}$$

If X is the yield stress in the 1 direction, Y is the yield stress in the 2 direction, Z is the yield stress in the 3 direction, then by substituting into Eq. (3-28) we can evaluate the constants by

$$G + H = \frac{1}{X^2} \qquad H + F = \frac{1}{Y^2} \qquad F + G = \frac{1}{Z^2}$$

Lubahn and Felgar[2] give detailed plasticity calculations for anisotropic behavior.

[1] R. Hill, *Proc. R. Soc. London, Ser. B*, vol. 193, pp. 281–297, 1948.

[2] J. D. Lubahn and R. P. Felgar, "Plasticity and Creep of Metals," chap. 13, John Wiley & Sons, Inc., New York, 1961.

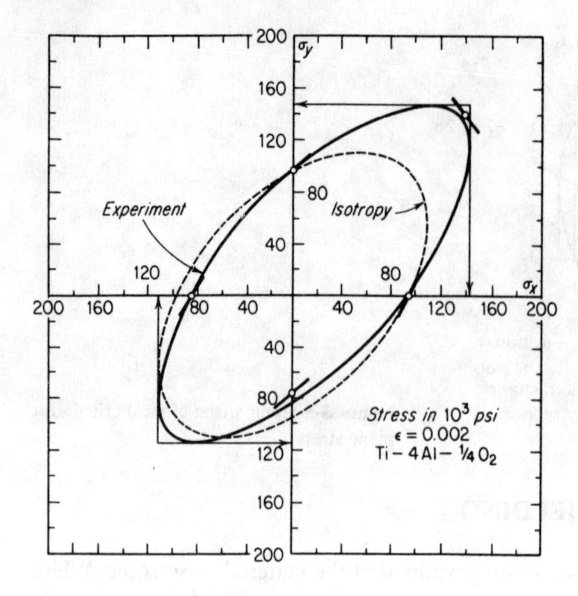

Figure 3-6 Yield locus for textured titanium-alloy sheet. *(From D. Lee and W. A. Backofen, Trans. Metall. Soc. AIME, vol. 236, p. 1083, 1966. By permission of the publishers.)*

On a plane-stress yield locus, such as Fig. 3-5, anisotropic yielding results in distortion of the yield ellipse. Figure 3-6 shows the yield locus for highly textured titanium alloy sheet.[1] Note that the experimentally determined curve is nonsymmetric when compared with the ideal isotropic curve.

An important aspect of yield anisotropy is *texture hardening*.[2] Consider a highly textured sheet that is fabricated into a thin-wall pressure vessel, so that the thickness stress σ_3 is negligible. From Eq. (3-28)

$$F\sigma_2^2 + G\sigma_1^2 + H(\sigma_1^2 - 2\sigma_1\sigma_2 + \sigma_2^2) = 1$$

$$(G + H)\sigma_1^2 + (F + H)\sigma_2^2 - 2H\sigma_1\sigma_2 = 1$$

or

$$\left(\frac{\sigma_1}{X}\right)^2 + \left(\frac{\sigma_2}{Y}\right)^2 - 2HXY\left(\frac{\sigma_1}{X}\frac{\sigma_2}{Y}\right) = 1 \qquad (3\text{-}29)$$

For simplicity, we shall assume that the yield stresses in the plane of the sheet are equal, that is, $X = Y$. Thus,

$$\sigma_1^2 + \sigma_2^2 - 2HY^2\sigma_1\sigma_2 = Y^2$$

and

$$G = F = \frac{1}{2Z^2} \qquad HY^2 = 1 - \frac{1}{2}\left(\frac{Y}{Z}\right)^2$$

However, the yield stress in the thickness direction of the sheet, Z, is a difficult property to measure. This problem can be circumvented by measuring the R

[1] These curves were obtained with the method of D. Lee and W. A. Backofen, op. cit.

[2] W. A. Backofen, W. F. Hosford, Jr., and J. J. Burke, *ASM Trans Q.*, vol. 55, p. 264, 1962.

value, the ratio of the width strain to the thickness strain

$$R = \frac{\ln(w_0/w)}{\ln(t_0/t)} \tag{3-30}$$

Since $(Z/Y)^2 = \frac{1}{2}(1 + R)$, the equation or the yield locus can be written as

$$\sigma_1^2 + \sigma_2^2 - \frac{2R}{1+R}\sigma_1\sigma_2 = Y^2 \tag{3-31}$$

High through-thickness yield stress Z results in low-thickness strain and a high value of R. The extent of strengthening from the texture effect can be seen from Fig. 3-6. For a spherical pressure vessel $\sigma_1 = \sigma_2$. Thus, by moving out a 45° line on Fig. 3-6, we see that the resistance to yielding increases markedly with increased R.

3-8 YIELD SURFACE AND NORMALITY

The relationships that have been developed for yield criteria, Eqs. (3-12) and (3-21), can be represented geometrically by a cylinder oriented at equal angles to the $\sigma_1, \sigma_2, \sigma_3$ axes (Fig. 3-7). A state of stress which gives a point inside of the cylinder represents elastic behavior. Yielding begins when the state of stress reaches the surface of the cylinder, which is called the *yield surface*. The radius of the cylinder MN is the stress deviator. Since the axis of the cylinder OM makes equal angles with the principal stress axes, $l = m = n = 1/\sqrt{3}$, and from Eq. (2-18), $\sigma = (\sigma_1 + \sigma_2 + \sigma_3)/3 = \sigma_m$. Therefore, the axis of the cylinder is the hydrostatic component of stress. Since plastic deformation is not influenced by hydrostatic stress, the generator of the yield surface is a straight line parallel to OM, so that the radius of the cylinder is constant. As plastic deformation occurs we can consider that the yield surface expands outward, maintaining its same geometric shape.

The yield surface shown in Fig. 3-7 is a circular cylinder if its represents the von Mises' yield criterion. If a plane is passed through this surface parallel to the σ_2 axis, it intersects on the $\sigma_1\sigma_3$ plane as an ellipse (see Fig. 3-5). The yield surface

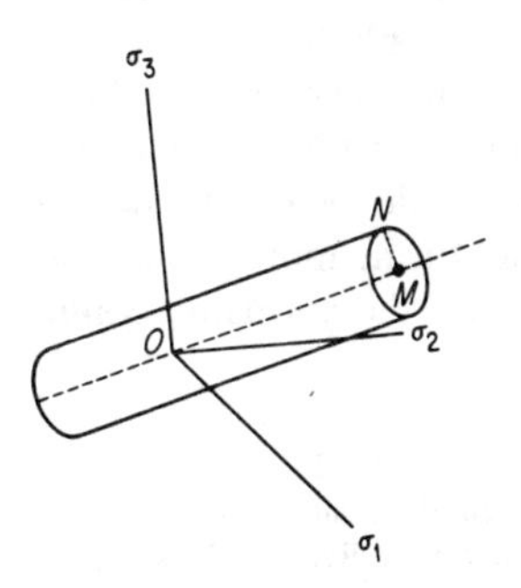

Figure 3-7 Yield surface for von Mises' criterion.

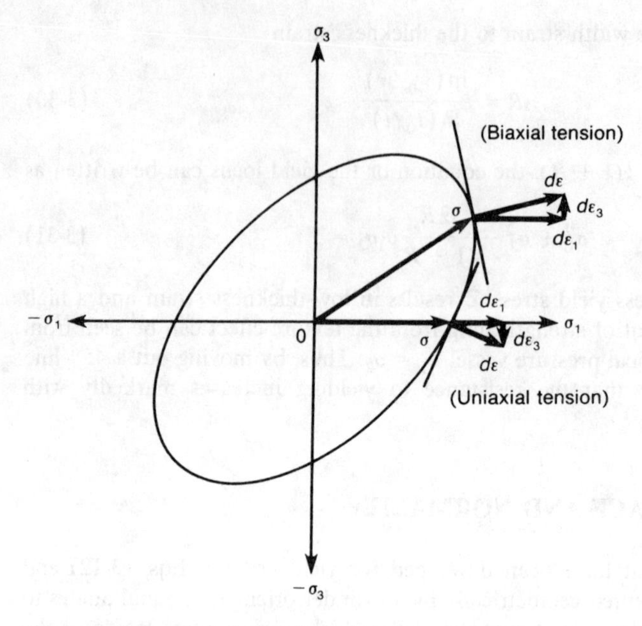

Figure 3-8 Example of the usefulness of the normality rule in working with the yield locus. Note the total strain vector $d\varepsilon$ is normal to the yield locus.

for the maximum-shear-stress criterion is a hexagonal cylinder. It should be noted that although the yield surface is an important concept in plasticity theory, there is no extensive body of experimental data on the shape of the surface. There is some work[1] which indicates that the yield surface is not a cylinder of uniform radius.

Drucker[2] has shown that the total plastic strain vector must be normal to the yield surface. As a consequence, any acceptable yield surface must be convex about its origin. Because of normality there is no component of the total strain vector that acts in the direction of σ_m. Therefore, the hydrostatic component of stress does not act to expand the yield surface. Because the deviatoric component of stress acts in the same direction as the total strain vector their dot product causes the plastic work as the yield surface is expanded by plastic deformation.

The normality rule also is useful in constructing experimental yield loci.[3] Figure 3-8 shows that the total strain vector $d\varepsilon$ is normal to the yield locus. We are looking at the projection of $d\varepsilon$ on the 1-3 plane. If the yield locus is known we can establish the ratio $d\varepsilon_1 : d\varepsilon_3$ from the normality rule. In the more usual case, $d\varepsilon_1/d\varepsilon_3$ is known experimentally and when combined with the normality condition they establish part of the yield locus.

[1] L. W. Hu, J. Markowitz, and T. A. Bartush, *Exp. Mech.* vol. 6, pp. 58–65, 1966.
[2] D. C. Drucker, Proceedings 1st U.S. National Congress of Applied Mechanics, p. 487, 1951.
[3] W. A. Backofen, "Deformation Processing," pp. 58–72, Addison-Wesley, Reading, Mass., 1972.

3-9 OCTAHEDRAL SHEAR STRESS AND SHEAR STRAIN

The octahedral stresses are a particular set of stress functions which are important in the theory of plasticity. They are the stresses acting on the faces of a three-dimensional octahedron which has the geometric property that the faces of the planes make equal angles with each of the three principal directions of stress. For such a geometric body, the angle between the normal to one of the faces and the nearest principal axis is 54°44′, and the cosine of this angle is $1/\sqrt{3}$. This is equivalent to {111} plane in an fcc crystal lattice.

The stress acting on each face of the octahedron can be resolved[1] into a normal octahedral stress σ_{oct} and an octahedral shear stress lying in the octahedral plane, τ_{oct}. The normal octahedral stress is equal to the hydrostatic component of the total stress.

$$\sigma_{oct} = \frac{\sigma_1 + \sigma_2 + \sigma_3}{3} = \sigma_m \tag{3-32}$$

The octahedral shear stress τ_{oct} is given by

$$\tau_{oct} = \tfrac{1}{3}\left[(\sigma_1 - \sigma_2)^2 + (\sigma_2 - \sigma_3)^2 + (\sigma_3 - \sigma_1)^2\right]^{1/2} \tag{3-33}$$

Since the normal octahedral stress is a hydrostatic stress, it cannot produce yielding in solid materials. Therefore, the octahedral shear stress is the component of stress responsible for plastic deformation. In this respect, it is analogous to the stress deviator.

If it is assumed that a critical octahedral shear stress determines yielding, the failure criterion can be written as

$$\tau_{oct} = \tfrac{1}{3}\left[(\sigma_1 - \sigma_2)^2 + (\sigma_2 - \sigma_3)^2 + (\sigma_3 - \sigma_1)^2\right]^{1/2} = \frac{\sqrt{2}}{3}\sigma_0$$

or

$$\sigma_0 = \frac{1}{\sqrt{2}}\left[(\sigma_1 - \sigma_2)^2 + (\sigma_2 - \sigma_3)^2 + (\sigma_3 - \sigma_1)^2\right]^{1/2} \tag{3-34}$$

Since Eq. (3-34) is identical with the equation already derived for the distortion-energy theory, the two yielding theories give the same results. In a sense, the octahedral theory can be considered the *stress equivalent* of the distortion-energy theory. According to this theory, the octahedral shear stress corresponding to yielding in uniaxial stress is given by

$$\tau_{oct} = \frac{\sqrt{2}}{3}\sigma_0 = 0.471\sigma_0 \tag{3-35}$$

Octahedral strains are referred to the same three-dimensional octahedron as the octahedral stresses. The octahedral linear strain is given by

$$\varepsilon_{oct} = \frac{\varepsilon_1 + \varepsilon_2 + \varepsilon_3}{3} \tag{3-36}$$

[1] A. Nadai, "Theory of Flow and Fracture of Solids," 2d Ed., vol. I, pp. 99–105, McGraw-Hill Book Co., New York, 1950.

Octahedral shear strain is given by

$$\gamma_{oct} = \tfrac{2}{3}\left[(\varepsilon_1 - \varepsilon_2)^2 + (\varepsilon_2 - \varepsilon_3)^2 + (\varepsilon_3 - \varepsilon_1)^2\right]^{1/2} \tag{3-37}$$

3-10 INVARIANTS OF STRESS AND STRAIN

It is frequently useful to simplify the representation of a complex state of stress or strain by means of invariant functions of stress and strain. If the plastic stress-strain curve (the flow curve) is plotted in terms of invariants of stress and strain, approximately the same curve will be obtained regardless of the state of stress. For example, the flow curves obtained in a uniaxial-tension test and a biaxial-torsion test of a thin tube with internal pressure will coincide when the curves are plotted in terms of invariant stress and strain functions.

Nadai[1] has shown that the octahedral shear stress and shear strain are invariant functions which describe the flow curve independent of the type of test. However, the most frequently used invariant function to describe plastic deformation is *effective stress* $\bar{\sigma}$ or *effective strain* $\bar{\varepsilon}$.

$$\bar{\sigma} = \frac{\sqrt{2}}{2}\left[(\sigma_1 - \sigma_2)^2 + (\sigma_2 - \sigma_3)^2 + (\sigma_3 - \sigma_1)^2\right]^{1/2} \tag{3-38}$$

$$d\bar{\varepsilon} = \frac{\sqrt{2}}{3}\left[(d\varepsilon_1 - d\varepsilon_2)^2 + (d\varepsilon_2 - d\varepsilon_3)^2 + (d\varepsilon_3 - d\varepsilon_1)^2\right]^{1/2} \tag{3-39}$$

The above equation for effective strain can be simplified as[2]

$$d\bar{\varepsilon} = \left[\tfrac{2}{3}\left(d\varepsilon_1^2 + d\varepsilon_2^2 + d\varepsilon_3^2\right)\right]^{1/2} \tag{3-40}$$

or in terms of total plastic strain

$$\bar{\varepsilon} = \left[\tfrac{2}{3}\left(\varepsilon_1^2 + \varepsilon_2^2 + \varepsilon_3^2\right)\right]^{1/2} \tag{3-41}$$

The strains used in Eqs. (3-39), (3-40), and (3-41) should be the plastic portion of the total strain. Frequently this is indicated by the notation ε_i^p, where $\varepsilon_i^p = \varepsilon_i(\text{total}) - \varepsilon_i(\text{elastic})$. In dealing with problems in metalworking the elastic strain is negligible, but in plasticity problems involving strains at a notch, overstressing of pressure vessels, etc., the elastic strains usually cannot be ignored.

Example Show that the equations for significant stress and strain reduce to the values for a tensile test.

[1] A. Nadai, *J. Appl. Phys.*, vol. 8, p. 205, 1937.

[2] W. E. Hosford and R. M. Caddell, "Metal Forming: Mechanics and Metallurgy," pp. 44–46, Prentice-Hall, Inc., Englewood Cliffs, N.J., 1983.

For a tensile test $\sigma_1 \neq 0$; $\sigma_2 = \sigma_3 = 0$, so from Eq. (3-38)

$$\bar{\sigma} = \frac{\sqrt{2}}{2}\left[\sigma_1^2 + \sigma_1^2\right] = \frac{\sqrt{2}\sqrt{2}}{2}\sigma_1 = \sigma_1$$

The strains in the tensile test are ε_1; $\varepsilon_2 = \varepsilon_3 \neq \varepsilon_1$ but from $\varepsilon_1 + \varepsilon_2 + \varepsilon_3 = 0$

$$\varepsilon_1 + 2\varepsilon_2 = 0 \quad \text{and} \quad d\varepsilon_1 = -2\,d\varepsilon_2 = -2\,d\varepsilon_3$$

$$d\bar{\varepsilon} = \left[\frac{2}{3}\left(d\varepsilon_1^2 + d\varepsilon_2^2 + d\varepsilon_3^2\right)\right]^{1/2} = \left[\frac{2}{3}\left(d\varepsilon_1^2 + \frac{d\varepsilon_1^2}{4} + \frac{d\varepsilon_1^2}{4}\right)\right]^{1/2}$$

$$d\bar{\varepsilon} = \left[\tfrac{2}{3}\left(\tfrac{6}{4}\right)d\varepsilon_1^2\right]^{1/2} = d\varepsilon_1$$

Thus the power law expression for the flow curve, Eq. (3-1) may be used as a first approximation to predict the plastic stress-strain behavior in other than tensile forms of loading.

$$\bar{\sigma} = K\bar{\varepsilon}^n \tag{3-42}$$

3-11 PLASTIC STRESS-STRAIN RELATIONS

Having discussed the relationships between stress state and plastic yielding, it is now necessary to consider the relations between stress and strain in plastic deformation. In the elastic region the strains are uniquely determined by the stresses through Hooke's law without regard to how the stress state was achieved. This is not the case for plastic deformation. In the plastic region the strains in general are not uniquely determined by the stresses but depend on the entire history of loading. Therefore, in plasticity it is necessary to determine the differentials or *increments of plastic strain* throughout the loading path and then obtain the total strain by integration or summation. As a simple example, consider a rod 1 in long extended in tension to $1\frac{1}{4}$ in and then compressed to the original 1 in length.

On the basis of total deformation

$$\varepsilon = \int_1^{1\frac{1}{4}} \frac{dL}{L} + \int_{1\frac{1}{4}}^{1} \frac{dL}{L} = 0$$

However, on an incremental basis

$$\varepsilon = \int_1^{1\frac{1}{4}} \frac{dL}{L} + \int_{1\frac{1}{4}}^{1} - \frac{dL}{L} = 2\ln 1\tfrac{1}{4} = 0.445$$

For the particular class of loading paths in which all the stresses increase in the same ratio, *proportional loading*, i.e.,

$$\frac{d\sigma_1}{\sigma_1} = \frac{d\sigma_2}{\sigma_2} = \frac{d\sigma_3}{\sigma_3}$$

the plastic strains are independent of the loading path and depend only on the final state of stress.

There are two general categories of plastic stress-strain relationships. *Incremental* or *flow theories* relate the stresses to the plastic strain increments. *Deformation* or *total strain theories* relate the stresses to the total plastic strain. Deformation theory simplifies the solution of plasticity problems, but the plastic strains in general cannot be considered independent of loading path.

Levy-Mises Equations (Ideal Plastic Solid)

The relationship between stress and strain for an ideal plastic solid, where the elastic strains are negligible, are called *flow rules* or the Levy-Mises equations. If we consider yielding under uniaxial tension, then $\sigma_1 \neq 0$, $\sigma_2 = \sigma_3 = 0$, and $\sigma_m = \sigma_1/3$. Since only the deviatoric stresses cause yielding

$$\sigma_1' = \sigma_1 - \sigma_m = \frac{2\sigma_1}{3}; \qquad \sigma_2' = \sigma_3' = \frac{-\sigma_1}{3}$$

from which we find

$$\sigma_1' = -2\sigma_2' = -2\sigma_3' \tag{3-43}$$

From the condition of constancy of volume in plastic deformation

$$d\varepsilon_1 = -2\,d\varepsilon_2 = -2\,d\varepsilon_3 \tag{3-44}$$

so that

$$\frac{d\varepsilon_1}{d\varepsilon_2} = -2 = \frac{\sigma_1'}{\sigma_2'} \tag{3-45}$$

This can be generalized to the Levy-Mises equation

$$\frac{d\varepsilon_1}{\sigma_1'} = \frac{d\varepsilon_2}{\sigma_2'} = \frac{d\varepsilon_3}{\sigma_3'} = d\lambda \tag{3-46}$$

These equations express the fact that at any instant of deformation the ratio of the plastic strain increments to the current deviatoric stresses is constant.

By using Eqs. (2-57) the above equations can be written in terms of the actual stresses.

$$d\varepsilon_1 = \tfrac{2}{3}\,d\lambda\left[\sigma_1 - \tfrac{1}{2}(\sigma_2 + \sigma_3)\right], \quad \text{etc.}$$

To evaluate $d\lambda$ we utilize the effective strain, Eq. (3-39), which yields $d\bar{\varepsilon} = \frac{2}{3}\,d\lambda\,\bar{\sigma}$.

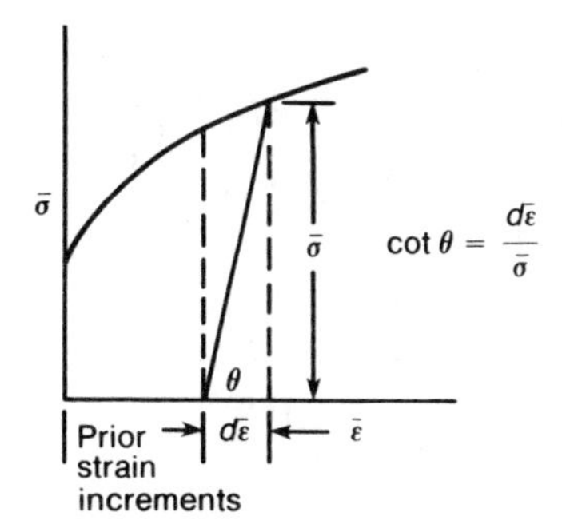

Figure 3-9 Method of establishing $d\bar{\varepsilon}/\bar{\sigma}$ in Eq. (3-47).

The Levy-Mises equations then become

$$
\begin{aligned}
d\varepsilon_1 &= \frac{d\bar{\varepsilon}}{\bar{\sigma}}\left[\sigma_1 - \frac{1}{2}(\sigma_2 + \sigma_3)\right] \\
d\varepsilon_2 &= \frac{d\bar{\varepsilon}}{\bar{\sigma}}\left[\sigma_2 - \frac{1}{2}(\sigma_3 + \sigma_1)\right] \qquad (3\text{-}47) \\
d\varepsilon_3 &= \frac{d\bar{\varepsilon}}{\bar{\sigma}}\left[\sigma_3 - \frac{1}{2}(\sigma_1 + \sigma_2)\right]
\end{aligned}
$$

The similarity with Eqs. (2-64) for the elastic solid should be noted. In place of $1/E$ the flow rules have a ratio $d\bar{\varepsilon}/\bar{\sigma}$ which changes throughout the course of the deformation. In place of ν they have the value $\frac{1}{2}$. The proportionality constant $d\bar{\varepsilon}/\bar{\sigma}$ is evaluated from an effective stress-effective strain curve for an increment of plastic strain $d\bar{\varepsilon}$ in the manner shown in Fig. 3-9.

Example An aluminum thin-walled tube (radius/thickness = 20) is closed at each end and pressurized to 1,000 psi to cause plastic deformation. Neglect the elastic strain and find the plastic strain in the circumferential (hoop) direction of the tube. The plastic stress-strain curve is given by $\bar{\sigma} = 25(\bar{\varepsilon})^{0.25}$, where stress is in ksi.

From the strength of materials equations for thin-walled pressure vessels, the stresses on the outside of the tube are:

$$\sigma_\theta = \sigma_1 = \frac{pr}{t} \quad \text{(circumferential direction)}$$

$$\sigma_l = \sigma_2 = \frac{pr}{2t} = \frac{\sigma_1}{2} \quad \text{(longitudinal direction)}$$

$$\sigma_r = \sigma_3 = 0 \quad \text{(radial direction)}$$

From the Levy-Mises equations

$$d\varepsilon_1 = \frac{d\bar{\varepsilon}}{\bar{\sigma}}\left[\sigma_1 - \frac{1}{2}(\sigma_2 + \sigma_3)\right] = \frac{d\bar{\varepsilon}}{\bar{\sigma}}\left[\sigma_1 - \frac{\sigma_1}{4}\right] = \frac{d\bar{\varepsilon}}{\bar{\sigma}}\left(\frac{3\sigma_1}{4}\right)$$

$$d\varepsilon_3 = \frac{d\bar{\varepsilon}}{\bar{\sigma}}\left[\sigma_3 - \frac{1}{2}(\sigma_1 + \sigma_2)\right] = \frac{d\bar{\varepsilon}}{\bar{\sigma}}\left[0 - \frac{3\sigma_1}{4}\right]$$

$$\therefore \quad d\varepsilon_1 = -d\varepsilon_3 \quad \text{and from} \quad d\varepsilon_1 + d\varepsilon_2 + d\varepsilon_3 = 0 \qquad d\varepsilon_2 = 0$$

$$\bar{\sigma} = \frac{1}{\sqrt{2}}\left[\left(\sigma_1 - \frac{\sigma_1}{2}\right)^2 + \left(\frac{\sigma_1}{2} - 0\right)^2 + (0 - \sigma_1)^2\right]^{1/2}$$

$$= \frac{1}{\sqrt{2}}\left[\frac{6}{4}\sigma_1\right]^2 = \frac{\sqrt{3}}{2}\sigma_1$$

$$\sigma_1 = \frac{pr}{t} = 1000(20) = 20 \text{ ksi} \qquad \bar{\sigma} = \frac{\sqrt{3}}{2}(20) = 17.3 \text{ ksi}$$

$$\bar{\sigma} = 25(\bar{\varepsilon})^{0.25} \qquad \bar{\varepsilon} = \left(\frac{17.3}{25}\right)^{1/0.25} = (0.693)^4 = 0.230$$

$$d\bar{\varepsilon} = \frac{\sqrt{2}}{3}\left[(d\varepsilon_1 - 0)^2 + (0 - (-d\varepsilon_1))^2 + (-d\varepsilon_1 - d\varepsilon_1)^2\right]^{1/2}$$

$$d\bar{\varepsilon} = \frac{\sqrt{2}}{3}\sqrt{6}\, d\varepsilon_1 = \frac{2}{\sqrt{3}}\, d\varepsilon_1$$

$$d\varepsilon_1 = \frac{\sqrt{3}}{2}\, d\bar{\varepsilon} \qquad \varepsilon_1 = \frac{\sqrt{3}}{2}\int_0^{\bar{\varepsilon}} d\bar{\varepsilon} = \frac{\sqrt{3}}{2}(\bar{\varepsilon}) = \frac{\sqrt{3}}{2}(0.230) = 0.199$$

Prandtl-Reuss Equations (Elastic-Plastic Solid)

The Levy-Mises equations can only be applied to problems of large plastic deformation because they neglect elastic strains. To treat the important, but more difficult problems in the elastic-plastic region it is necessary to consider both elastic and plastic components of strain. These equations were proposed by Prandtl (1925) and Reuss (1930).

The total strain increment is the sum of an elastic strain increment de^E and a plastic strain increment $d\varepsilon^P$.

$$d\varepsilon_{ij} = de_{ij}^E + d\varepsilon_{ij}^P \tag{3-48}$$

From Eqs. (2-52) and (2-69), the elastic strain increment is given by

$$de_{ij}^E = \left(de_{ij} - \frac{de_{kk}}{3}\delta_{ij}\right) + \frac{de_{kk}}{3}\delta_{ij} = \frac{1+\nu}{E}\, d\sigma_{ij} - \frac{\nu}{E}\sigma_{kk}\delta_{ij}$$

or

$$de_{ij}^E = \frac{1+\nu}{E}\, d\sigma'_{ij} + \frac{1-2\nu}{E}\frac{d\sigma_{kk}}{3}\delta_{ij} \tag{3-49}$$

The plastic strain increment is given by the Levy-Mises equations, which can be written as

$$d\varepsilon_{ij}^{P} = \frac{3}{2}\frac{d\bar{\varepsilon}}{\bar{\sigma}}\sigma'_{ij} \tag{3-50}$$

Thus, the stress, strain relations for an elastic-plastic solid are given by

$$d\varepsilon_{ij} = \frac{1+\nu}{E}\, d\sigma'_{ij} + \frac{1+2\nu}{E}\frac{d\sigma_{kk}}{3}\delta_{ij} + \frac{3}{2}\frac{d\bar{\varepsilon}}{\bar{\sigma}}\sigma'_{ij} \tag{3-51}$$

Solution of Plasticity Problems

The Levy-Mises and Prandtl-Reuss equations provide relations between the increments of plastic strain and the stresses. The basic problem is to calculate the next increment of plastic strain for a given state of stress when the loads are increased incrementally. If all of the increments of strain are known, then the total plastic strain is simply determined by summation. To do this we have available a set of plastic stress-strain relationships, either Eqs. (3-47) or (3-51), a yield criterion, and a basic relationship for the flow behavior of the material in terms of a curve of $\bar{\sigma}$ vs. $\bar{\varepsilon}$. In addition, a complete solution also must satisfy the equations of equilibrium, the strain-displacement relations, and the boundary conditions. The reader is referred to the several excellent texts on plasticity listed at the end of this chapter for examples of detailed solutions.[1] Although the incremental nature of plasticity solutions in the past has resulted in much labor and infrequent application of the available techniques, the current widespread use of digital computers and finite element analysis should make plasticity analysis of engineering problems more commonplace.

3-12 TWO-DIMENSIONAL PLASTIC FLOW—SLIP-LINE FIELD THEORY

In many practical problems, such as rolling and strip drawing, all displacements can be considered to be limited to the *xy* plane, so that strains in the *z* direction can be neglected in the analysis. This is known as a condition of *plane strain*. When a problem is too difficult to an exact three-dimensional solution, a good indication of the stresses often can be obtained by consideration of the analogous plane-strain problem.

Since a plastic material tends to deform in all directions, to develop a plane-strain condition it is necessary to constrain flow in one direction. Constraint can be produced by an external lubricated barrier, such as a die wall (Fig. 3-10*a*), or it can arise from a situation where only part of the material is deformed and the rigid (elastic) material outside the plastic region prevents the spread of deformation (Fig. 3-10*b*).

[1] A number of plasticity problems are worked out in great detail in Lubahn and Felgar op. cit., Chaps 8 and 9.

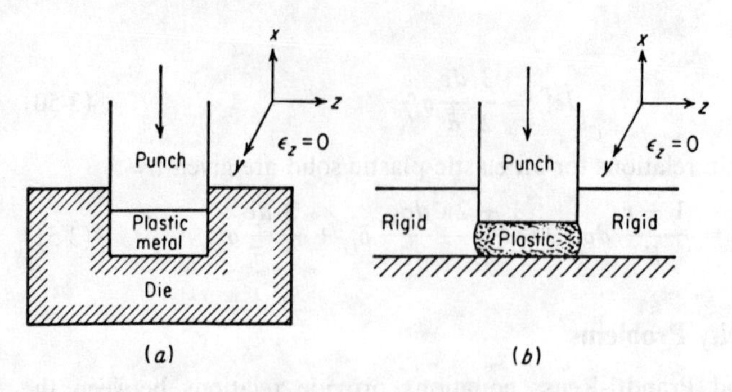

Figure 3-10 Methods of developing plastic constraint.

If the plane-strain deformation occurs on planes parallel to the *xy* plane, then

$$\varepsilon_z = \varepsilon_{xz} = \varepsilon_{yz} = 0 \quad \text{and} \quad \tau_{xz} = \tau_{yz} = 0$$

Since $\tau_{xz} = \tau_{yz} = 0$, it follows that σ_z is a principal stress. From the Levy-Mises equations, Eq. (3-47)

$$d\varepsilon_z = 0 = \frac{d\bar{\varepsilon}}{\bar{\sigma}}\left[\sigma_z - \frac{1}{2}(\sigma_x + \sigma_y)\right]$$

and
$$\sigma_z = \frac{\sigma_x + \sigma_y}{2} \tag{3-52}$$

Note that although the strain is zero in the *z* direction, a restraining stress acts in this direction.

Equation (3-52) could just as well have been written in terms of the principal stresses $\sigma_3 = (\sigma_1 + \sigma_2)/2$.

This principal stress will be intermediate between σ_1 and σ_2, so that the maximum-shear-stress yield criterion is given by

$$\sigma_1 - \sigma_2 = \sigma_0 = 2k \tag{3-53}$$

where *k* is the yield stress in pure shear.

If the value for the intermediate principal stress σ_3 is substituted into the von Mises' yield criterion, Eq. (3-12) it reduces to

$$\sigma_1 - \sigma_2 = \frac{2}{\sqrt{3}}\sigma_0 \tag{3-54}$$

However, for the von Mises' yield criterion $\sigma_0 = \sqrt{3}\,k$ so that Eq. (3-54) becomes

$$\sigma_1 - \sigma_2 = 2k \tag{3-55}$$

Thus, for a state of plane strain the maximum-shear stress and von Mises' yield criteria are equivalent. It can be considered that two-dimensional plastic flow will begin when the shear stress reaches a critical value of *k*.

Slip-line field theory is based on the fact that any general state of stress in plane strain consists of *pure shear* plus a *hydrostatic pressure*. We could show this

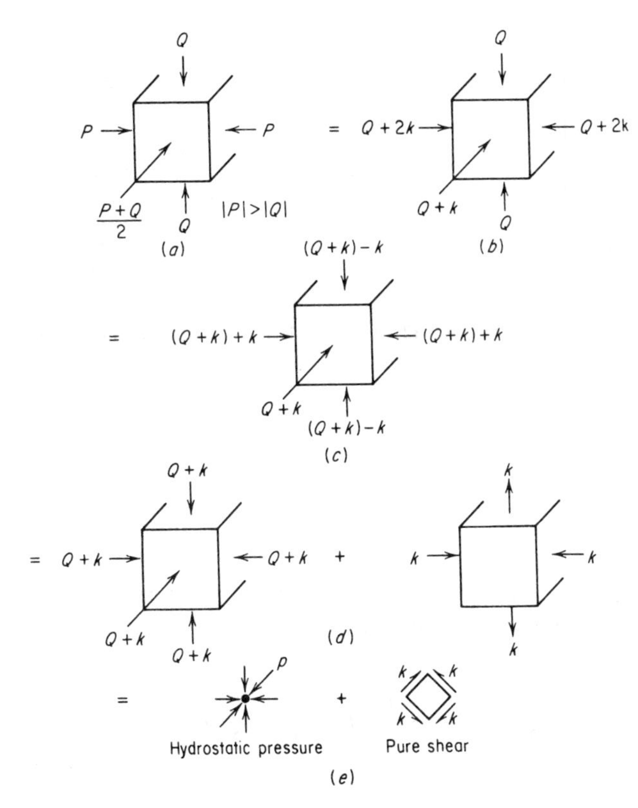

Figure 3-11 Demonstration that a state of stress in plane strain may be expressed as the sum of a hydrostatic stress and pure shear.

by applying the equations for transformation of stress from one set of axes to another, Eqs. (2-5) to (2-7), but it perhaps is more instructive to see this diagrammatically. In Fig. 3-11, let the state of stress consist of $\sigma_1 = -Q$, $\sigma_3 = -P$, and $\sigma_2 = (-P - Q)/2$. The maximum shear stress is given by

$$\tau_{max} = \sigma_1 - \sigma_3 = 2k$$

$$-Q + P = 2k$$

or in Fig. 3-11*b*,

$$P = Q + 2k$$

But we can write the state of stress in Fig. 3-11*b* as in Fig. 3-11*c*, which in turn can be written as the sum of a hydrostatic pressure and a biaxial state of stress Fig. 3-11*d*. The latter is the stress state in pure torsion, which for planes rotated by 45° consists of pure shear stresses. Thus, a general state of stress in plane strain can be decomposed into a hydrostatic state of stress p (in this case compression) and a state of pure shear k. The components of the stress tensor for

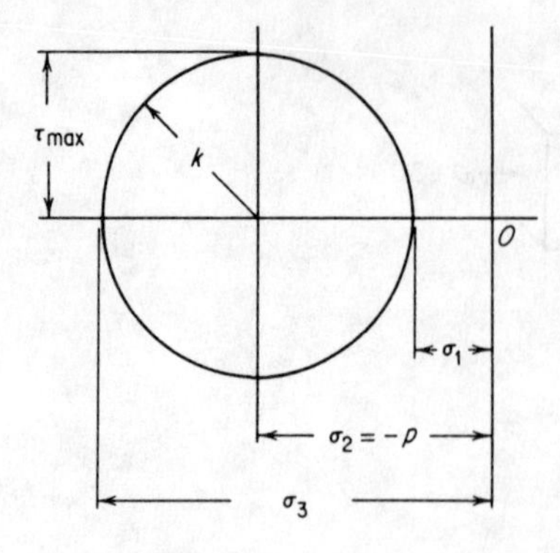

Figure 3-12 Mohr's circle representation of stresses in Fig. 3-9*a*.

plane strain are

$$\sigma_{ij} = \begin{vmatrix} p & k & 0 \\ k & p & 0 \\ 0 & 0 & p \end{vmatrix}$$

Mohr's circle representation for the state of stress given in Fig. 3-11 is shown in Fig. 3-12. If $\sigma_1 = -Q$ and $\sigma_3 = -P$, then $\sigma_2 = (-Q - P)/2 = -p$. This follows because

$$p = \sigma_m = \frac{\sigma_1 + \sigma_2 + \sigma_3}{3} = -\frac{1}{3}\left(Q + \frac{Q}{2} + \frac{P}{2} + P\right)$$

$$\therefore \quad p = \frac{Q + P}{2} = -\sigma_2$$

Also, the radius of Mohr's circle is $\tau_{max} = k$, where k is the yield stress in pure shear. Thus, using Fig. 3-12, we can express the principal stresses

$$\sigma_1 = -p + k$$
$$\sigma_2 = -p$$
$$\sigma_3 = -p - k$$

The slip-line field theory for plane strain allows the determination of stresses in a plastically deformed body when the deformation is not uniform throughout the body. In addition to requiring plane-strain conditions, the theory assumes an isotropic, homogeneous, rigid ideal plastic material. For such a non-strain-hardening material k is everywhere constant but p may vary from point to point. The state of stress at any point can be determined if we can find the magnitude of p and the direction of k. The lines of maximum shear stress occur in two orthogonal directions α and β. These lines of maximum shear stress are called *slip lines* and have the property that shear strain is a maximum and linear strain is zero tangent to their directions. The slip lines give the direction of p at any point and the changes in magnitude of p are deduced from the rotation of the slip line

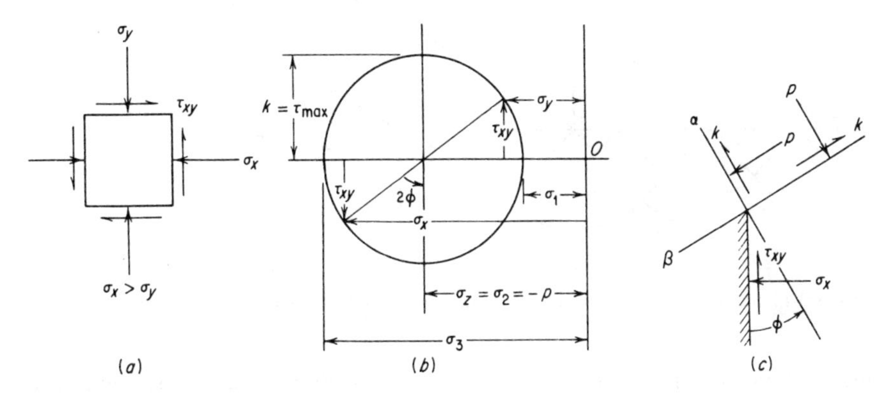

Figure 3-13 (*a*) Stress state on physical body; (*b*) Mohr's circle for (*a*); (*c*) relationship of physical body and α and β slip lines.

between one point and another in the field. It should be noted that the slip lines referred to in this section are geometric constructions which define the characteristic directions of the hyperbolic partial differential equations for the stress under plane-strain conditions. These slip lines bear no relationship to the slip lines observed under the microscope on the surface of a plastically deformed metal.

To arrive at the equations for calculating stress through the use of slip-line fields, we must now relate the stresses on a physical body in the xy coordinate system to p and k. Figure 3-13*b* shows the Mohr's circle representation of the stress state given in Fig. 3-13*a*. The stresses may be expressed as

$$\sigma_x = -p - k \sin 2\phi$$

$$\sigma_y = -p - (-k \sin 2\phi) = -p + k \sin 2\phi$$

$$\sigma_z = -p$$

$$\tau_{xy} = k \cos 2\phi$$

where 2ϕ is a counterclockwise angle on Mohr's circle from the *physical* x plane to the first plane of maximum shear stress. This plane of maximum shear stress is known as an α slip line. The relationship between the stress state on the physical body and the α and β slip lines is given in Fig. 3-13*c*.

The variation of hydrostatic pressure p with change in direction of the slip lines is given by the *Hencky equations*

$$p + 2k\phi = \text{constant along an } \alpha \text{ line} \qquad (3\text{-}56)$$

$$p - 2k\phi = \text{constant along a } \beta \text{ line}$$

These equations are developed[1] from the equilibrium equations in plane strain. The use of the Hencky equations will be illustrated with the example of the

[1] See for example W. Johnson and P. B. Mellor, "Plasticity for Mechanical Engineers," pp. 263–265, D. Van Nostrand Company, Inc., Princeton, N.J., 1962.

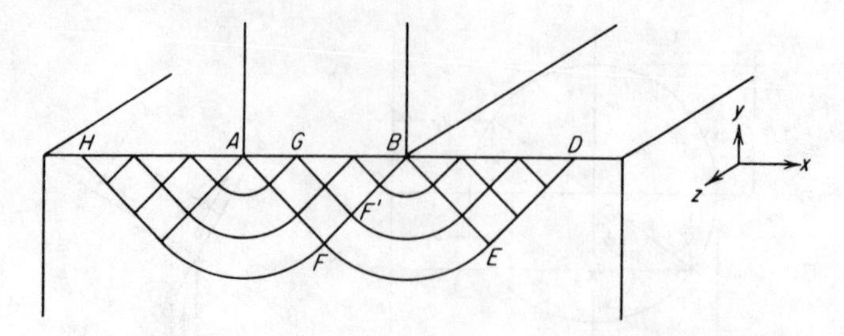

Figure 3-14 Slip-line field for frictionless indentation with a flat punch.

indentation of a thick block with a flat frictionless punch. The slip-line field shown in Fig. 3-14 was first suggested by Prandtl[1] in 1920. At the free surface on the frictionless interface between the punch and the block the slip lines meet the surface at 45° (see Prob. 3-15). We could construct the slip-line field by starting with triangle AFB, but we would soon see that if all plastic deformation were restricted to this region, the metal could not move because it would be surrounded by rigid (elastic) material. Therefore, the plastic zone described by the slip-line field must be extended along the free surface to AH and BD.

To determine the stresses from the slip-line field, we start with a simple point such as D. Since D is on a free surface, there is no stress normal to this surface.

$$\sigma_y = 0 = -p + k \sin 2\phi$$

and
$$\sigma_x = -p - k \sin 2\phi = -p - p = -2p$$

The stresses at point D are shown in Fig. 3-15. From the Mohr's circle we learn that $p = k$. In order to use the Hencky equations we need to know whether the slip line through D is an α or β line. This is done most simply from the following sign convention:

> For a counterclockwise rotation about the point of intersection of two slip lines, starting from an α-line the direction of the algebraically highest principal stress σ_1 is crossed before a β line is crossed.

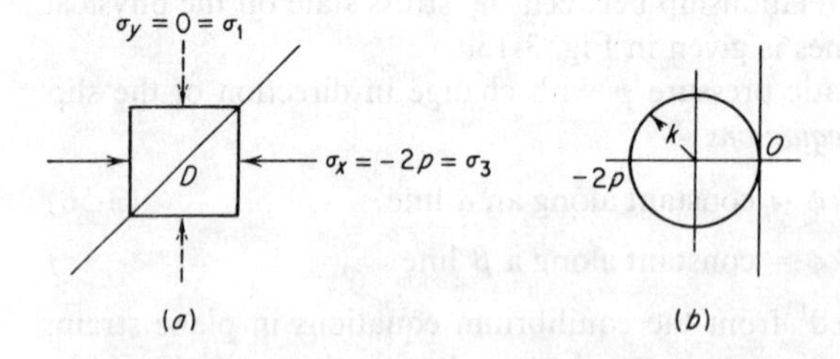

Figure 3-15 (a) Stresses at point D; (b) Mohr's circle.

[1] A different slip-line field was later suggested by R. Hill. Although the slip field is different, it leads to the same value of indentation pressure. This illustrates the fact that slip-line field solutions are not necessarily unique.

Applying this convention, we see that the slip line from D to E is an α line. Thus, the first Hencky equation applies,

$$p + 2k\phi = C_1$$

and if we use DE as the reference direction so $\phi = 0$,

$$p = C_1 = k$$

Because DE is straight p is constant from D to E and

$$p_D = p_E = k$$

Between E and F the tangent to the α slip line rotates through $\pi/2$ rad. Since the tangent to the α line rotates *clockwise*, $d\phi = -\pi/2$. If we write the Hencky equation in differential form, for clarity

$$dp + 2k\,d\phi = 0$$

or

$$(p_F - p_E) + 2k(\phi_F - \phi_E) = 0$$

$$p_F - k + 2k\left(-\frac{\pi}{2} - 0\right) = 0$$

$$p_F = k(\pi + 1)$$

Note that the pressure at F' is the same as at F because the slip line is straight and that the value of p under the punch face at G is also the same. (We stayed away from A and B at the punch edges because these are points of pressure discontinuity.) To find the punch pressure required to indent the block, it is necessary to convert the hydrostatic pressure at the punch interface into the vertical stress σ_y.

$$p_F = p_{F'} = p_G = k(\pi + 1)$$

$$\sigma_y = -p_G + k \sin 2\phi$$

From Fig. 3-13c, recall that the angle ϕ is measured by the counterclockwise angle from the physical x axis to the α line.

$$\sigma_y = -k(\pi + 1) + k \sin 2\left(\frac{3\pi}{4}\right)$$

$$\sigma_y = -k\pi - k - k = -2k\left(1 + \frac{\pi}{2}\right) \tag{3-57}$$

If we trace out other slip lines, we shall find in the same way that the normal compressive stress under the punch is $2k(1 + \pi/2)$, and the pressure is uniform.

Since $k = \sigma_0/\sqrt{3}$,

$$\sigma_y = \frac{2\sigma_0}{\sqrt{3}}\left(1 + \frac{\pi}{2}\right) \approx 3\sigma_0 \tag{3-58}$$

This shows that the yield pressure for the indentation of a thick block with a narrow punch is nearly three times the stress required for the yielding of a cylinder in frictionless compression. This increase in flow stress is a *geometrical constraint* resulting from the localized deformation under the narrow punch.

The example described above is one of the simplest situations that involves slip-line fields. In the general case the slip-line field selection must also satisfy certain velocity conditions to assure equilibrium. Prager[1] and Thomsen[2] have given general procedures for constructing slip-line fields. However, there is no easy method of checking the validity of a solution. Partial experimental verification of theoretically determined slip-line fields has been obtained for mild steel by etching techniques[3] which delineate the plastically deformed regions. Highly localized plastic regions can be delineated by an etching technique in Fe-3% Si steel.[4]

BIBLIOGRAPHY

Calladine, C. R.: "Engineering Plasticity," Pergamon Press Inc., New York, 1969.

Hill, R.: "The Mathematical Theory of Plasticity," Oxford University Press, New York, 1950.

Johnson, W., and P. B. Mellor: "Engineering Plasticity," Van Nostrand Reinhold Company, New York, 1973.

Johnson, W., R. Sowerby, and J. B. Haddow: "Plane-Strain Slip-Line Fields," Pergamon Press, New York, 1981.

Mendelson, A.: "Plasticity: Theory and Application," The Macmillan Company, New York, 1968.

Nadai, A.: "Theory of Flow and Fracture of Solids," 2d ed., vol. I, McGraw-Hill Book Company, New York, 1950; vol. II, 1963.

Prager, W., and P. G. Hodge: "Theory of Perfectly Plastic Solids," John Wiley & Sons, New York, 1951.

Slater, R. A. C.: "Engineering Plasticity—Theory and Application to Metal Forming Processes," John Wiley & Sons, New York, 1977.

[1] W. Prager, *Trans. R. Inst. Technol. Stockholm*, no. 65, 1953.

[2] E. G. Thomsen, *J. Appl. Mech.*, vol 24, pp 81–84, 1957.

[3] B. Hundy, *Metallurgia*, vol 49, no. 293, pp 109–118, 1954.

[4] G. T. Hahn, P. N. Mincer and A. R. Rosenfield, *Exp. Mech.*, vol. 11, pp. 248–253, 1971.

PART TWO

METALLURGICAL FUNDAMENTALS

CHAPTER

FOUR

PLASTIC DEFORMATION OF SINGLE CRYSTALS

4-1 INTRODUCTION

The previous three chapters have been concerned with the phenomenological description of the elastic and plastic behavior of metals. It has been shown that formal mathematical theories have been developed for describing the mechanical behavior of metals based upon the simplifying assumptions that metals are homogeneous and isotropic. That this is not true should be obvious to anyone who has examined the structure of metals under a microscope. However, for fine-grained metals subjected to static loads within the elastic range the theories are perfectly adequate for design. Within the plastic range the theories describe the observed behavior, although not with the precision which is frequently desired. For conditions of dynamic and impact loading we are forced, in general, to rely heavily on experimentally determined data. As the assumption that we are dealing with an isotropic homogeneous medium becomes less tenable, our ability to predict the behavior of metals under stress by means of the theories of elasticity and plasticity decreases.

Following the discovery of the diffraction of x-rays by metallic crystals by Von Laue in 1912 and the realization that metals were fundamentally composed of atoms arranged in specific geometric lattices there have been a great many investigations of the relationships between atomic structure and the plastic behavior of metals. Much of the fundamental work on the plastic deformation of metals has been performed with single-crystal specimens, so as to eliminate the complicating effects of grain boundaries and the restraints imposed by neighboring grains and second-phase particles. Techniques for preparing single crystals have been described in a number of sources.[1-4]

The basic mechanisms of plastic deformation in single crystals will be discussed in this chapter. The dislocation theory, which plays such an important part in modern concepts of plastic deformation, will be introduced in this chapter to the extent needed to provide a qualitative understanding. A more detailed consideration of dislocation theory will be found in Chap. 5. Using dislocation theory as the main tool, consideration will be given to the strengthening mechanisms in polycrystalline solids in Chap. 6. Primary consideration will be given to tensile deformation. The fundamental deformation behavior in creep and fatigue will be covered in chapters in Part Three specifically devoted to these subjects. This part closes with a chapter on the fundamental aspects of fracture (Chap. 7).

4-2 CONCEPTS OF CRYSTAL GEOMETRY

X-ray diffraction analysis shows that the atoms in a metal crystal are arranged in a regular, repeated three-dimensional pattern. The atom arrangement of metals is most simply portrayed by a crystal lattice in which the atoms are visualized as hard balls located at particular locations in a geometrical arrangements.

The most elementary crystal structure is the simple cubic lattice (Fig. 4-1). This is the type of structure cell found for ionic crystals, such as NaCl and LiF, but not for any of the metals. Three mutually perpendicular axes are arbitrarily placed through one of the corners of the cell. Crystallographic planes and directions will be specified with respect to these axes in terms of *Miller indices*. A crystallographic plane is specified in terms of the length of its intercepts on the

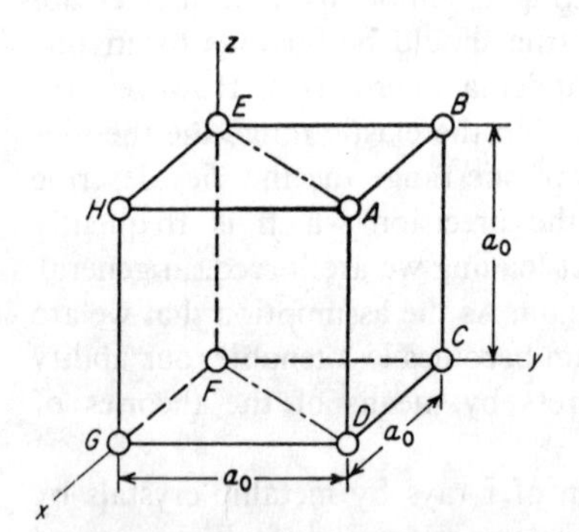

Figure 4-1 Simple cubic structure.

[1] R. W. K. Honeycombe, *Metall. Rev.*, vol. 4, no. 13, pp. 1–47, 1969.

[2] A. N. Holden, *Trans. Am. Soc. Met.*, vol. 42, pp. 319–346, 1950.

[3] W. D. Lawson and S. Nielsen, "Preparation of Single Crystals," Academic Press, Inc., New York, 1958.

[4] J. J. Gilman (ed.), "The Art and Science of Growing Crystals," John Wiley & Sons, Inc., New York, 1963.

three axes, measured from the origin of the coordinate axes. To simplify the crystallographic formulas, the reciprocals of these intercepts are used. They are reduced to a lowest common denominator to give the Miller indices (*hkl*) of the plane. For example, the plane *ABCD* in Fig. 4-1 is parallel to the *x* and *z* axes and intersects the *y* axis at one interatomic distance a_0. Therefore, the indices of the plane are $1/\infty$, $1/1$, $1/\infty$, or $(hkl) = (010)$. Plane *EBCF* would be designated as the $(\bar{1}00)$ plane, since the origin of the coordinate system can be moved to *G* because every point in a space lattice has the same arrangement of points as every other point. The bar over one of the integers indicates that the plane intersects one of the axes in a negative direction. There are six crystallographically equivalent planes of the type (100), any one of which can have the indices (100), (010), (001), $(\bar{1}00)$, $(0\bar{1}0)$, $(00\bar{1})$ depending upon the choice of axes. The notation $\{100\}$ is used when they are to be considered as a group, or *family of planes*.

Crystallographic directions are indicated by integers in brackets: [*uvw*]. Reciprocals are not used in determining directions. As an example, the direction of the line *FD* is obtained by moving out from the origin a distance a_0 along the *x* axis and moving an equal distance in the positive *y* direction. The indices of this direction are then [110]. A family of crystallographically equivalent directions would be designated $\langle uvw \rangle$. For the cubic lattice only, a direction is always perpendicular to the plane having the same indices.

Many of the common metals have either a body-centered cubic (bcc) or face-centered cubic (fcc) crystal structure. Figure 4-2*a* shows a body-centered cubic structure cell with an atom at each corner and another atom at the body center of the cube. Each corner atom is surrounded by eight adjacent atoms, as is the atom located at the center of the cell. Therefore, there are two atoms *per structure cell* for the body-centered cubic structure ($\frac{8}{8} + 1$). Typical metals which have this crystal structure are alpha iron, columbium, tantalum, chromium, molybenum, and tungsten. Figure 4-2*b* shows the structure cell for a face-centered cubic crystal structure. In addition to an atom at each corner, there is an atom at

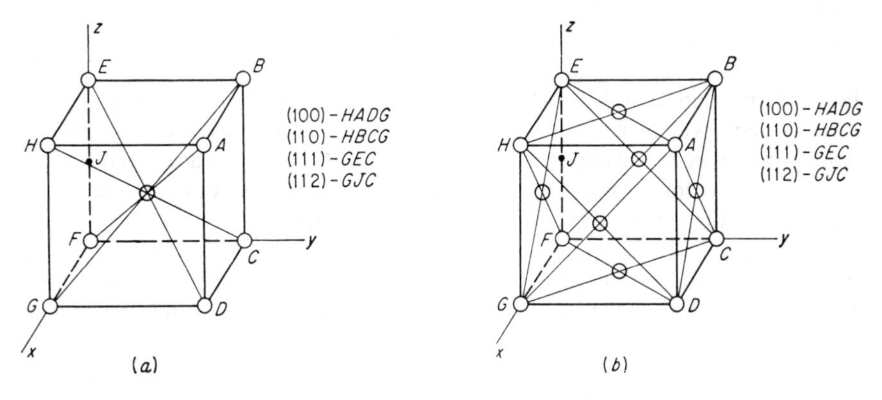

Figure 4-2 (*a*) Body-centered cubic structure; (*b*) face-centered cubic structure.

the center of each of the cube faces. Since these latter atoms belong to two unit cells, there are four atoms per structure cell in the face-centered cubic structure ($\frac{8}{8} + \frac{6}{2}$). Aluminum, copper, gold, lead, silver, and nickel are common face-centered cubic metals.

For cubic systems there is a set of simple relationships between a direction $[uvw]$ and a plane (hkl) which are very useful.

1. $[uvw]$ is normal to (hkl) when $u = h$; $v = k$; $w = l$. [111] is normal to (111).
2. $[uvw]$ is parallel to (hkl), i.e., $[uvw]$ lies in (hkl), when $hu + kv + lw = 0$. $[11\bar{2}]$ is a direction in (111).
3. Two planes $(h_1k_1l_1)$ and $(h_2k_2l_2)$ are normal if $h_1h_2 + k_1k_2 + l_1l_2 = 0$. (001) is perpendicular to (100) and (010). (110) is perpendicular to $(1\bar{1}0)$.
4. Two directions $u_1v_1w_1$ and $u_2v_2w_2$ are normal if $u_1u_2 + v_1v_2 + w_1w_2 = 0$. [100] is perpendicular to [001]. [111] is perpendicular to $[11\bar{2}]$.
5. Angles between planes $(h_1k_1l_1)$ and $(h_2k_2l_2)$ are given by

$$\cos\theta = \frac{h_1h_2 + k_1k_2 + l_1l_2}{\left(h_1^2 + k_1^2 + l_1^2\right)^{1/2}\left(h_2^2 + k_2^2 + l_2^2\right)^{1/2}}$$

The third common metallic crystal structure is the hexagonal close-packed (hcp) structure[1] (Fig. 4-3). In order to specify planes and directions in the hcp structure, it is convenient to use the Miller-Bravais system with four indices of the type $(hkil)$. These indices are based on four axes; the three axes a_1, a_2, a_3 are 120° apart in the basal plane, and the vertical c axis is normal to the basal plane. These axes and typical planes in the hcp crystal structure are given in Fig. 4-3. The third index is related to the first two by the relation $i = -(h + k)$.

The face-centered cubic and hexagonal close-packed structures can both be built up from a stacking of close-packed planes of spheres. Figure 4-4 shows that there are two ways in which the spheres can be stacked. The first layer of spheres is arranged so that each sphere is surrounded by and just touching six other spheres. This corresponds to the solid circles in Fig. 4-4. A second layer of close-packed spheres can be placed over the bottom layer so that the centers of the atoms in the second plane cover one-half the number of valleys in the bottom layer (dashed circles in Fig. 4-4). There are two ways of adding spheres to give a third close-packed plane. Although the spheres in the third layer must fit into the valleys in the second plane, they may lie either over the valleys not covered in the first plane (the dots in Fig. 4-4) or directly above the atoms in the first plane (the crosses in Fig. 4-4). The first possibility results in a stacking sequence *ABCABC*..., which is found for the {111} planes of an fcc structure. The other possibility results in the stacking sequence *ABAB*..., which is found for the (0001) basal plane of the hcp structure. For the ideal hcp packing, the ratio of

[1] A detailed review of the crystallography and deformation in hcp metals is given by P. G. Partridge, *Metall. Rev.*, no. 118 and *Met. Mater.* vol. 1, no. 11, pp. 169–194, 1967.

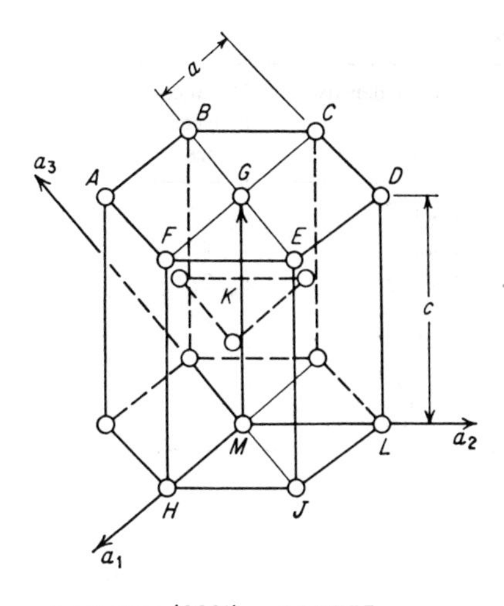

Figure 4-3 Hexagonal close-packed structure.

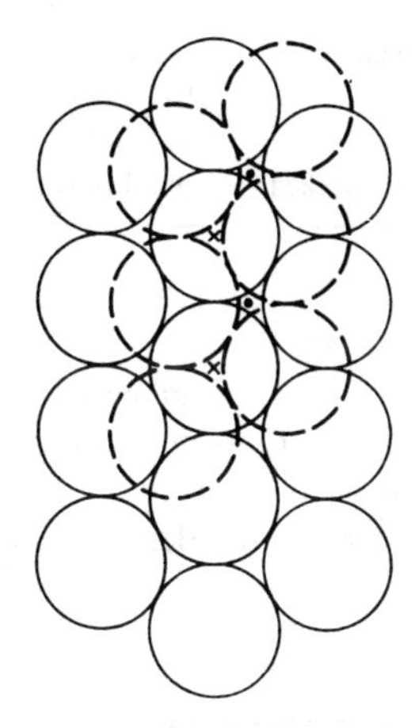

Figure 4-4 Stacking of close-packed spheres.

c/a is $\sqrt{\frac{8}{3}}$, or 1.633. Table 4-1 shows that actual hcp metals deviate from the ideal c/a ratio.

The fcc and hcp structures are both close-packed structures. Seventy-four percent of the volume of the unit cell is occupied by atoms, on a hard sphere model, in the fcc and hcp structures. This is contrasted with 68 percent packing for a bcc unit cell and 52 percent of the volume occupied by atoms in the simple cubic unit cell.

Table 4-1 Axial ratios of some hexagonal metals

Metal	c/a
Be	1.567
Ti	1.587
Mg	1.623
Ideal hcp	1.633
Zn	1.856
Cd	1.886

Table 4-2 Atomic density of low-index planes

Crystal structure	Plane	Atomic density, atoms per unit area	Distance between planes
Face-centered cubic	Octahedral {111}	$4/\sqrt{3}\,a_0^2$	$a_0/\sqrt{3}$
	Cube {100}	$2/a_0^2$	$a_0/2$
	Dodecahedral {110}	$2/\sqrt{2}\,a_0^2$	$a_0/2\sqrt{2}$
Body-centered cubic	Dodecahedral {110}	$2/\sqrt{2}\,a_0^2$	$a_0/\sqrt{2}$
	Cube {100}	$1/a_0^2$	$a_0/2$
	Octahedral {111}	$1/\sqrt{3}\,a_0^2$	$a_0/2\sqrt{3}$
Hexagonal close-packed	Basal {0001}	$2/\sqrt{3}\,a_0^2$	c

Plastic deformation is generally confined to the low-index planes, which have a higher density of atoms per unit area than the high-index planes. Table 4-2 lists the atomic density per unit area for the common low-index planes. Note that the planes of greatest atomic density also are the most widely spaced planes for the crystal structure.

4-3 LATTICE DEFECTS

Real crystals deviate from the perfect periodicity that was assumed in the previous section in a number of important ways. While the concept of the perfect lattice is adequate for explaining the *structure-insensitive* properties of metals, for a better understanding of the *structure-sensitive* properties it has been necessary to consider a number of types of lattice defects. The description of the structure-sensitive properties then reduces itself largely to describing the behavior of these defects.

Structure-insensitive	Structure-sensitive
Elastic constants	Electrical Conductivity
Melting point	Semiconductor properties
Density	Yield stress
Specific heat	Fracture strength
Coefficient of thermal expansion	Creep strength

As is suggested by the above brief tabulation, practically all the mechanical properties are structure-sensitive properties. Only since the realization of this fact, in relatively recent times, have really important advances been made in understanding the mechanical behavior of materials.

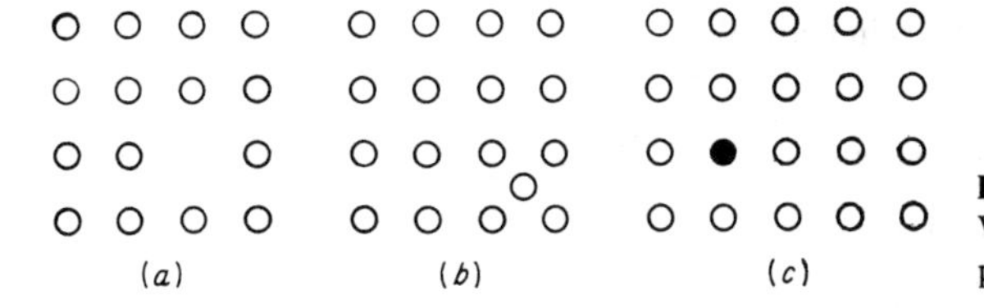

Figure 4-5 Point defects. (*a*) Vacancy; (*b*) interstitial; (*c*) impurity atom.

The term *defect*, or *imperfection*, is generally used to describe any deviation from an orderly array of lattice points. When the deviation from the periodic arrangement of the lattice is localized to the vicinity of only a few atoms it is called a *point defect*, or *point imperfection*. However, if the defect extends through microscopic regions of the crystal, it is called a *lattice imperfection*. Lattice imperfections may be divided into *line defects* and *surface*, or *plane*, *defects*. Line defects obtain their name because they propagate as lines or as a two-dimensional net in the crystal. The edge and screw dislocations that are discussed in this section are the common line defects encountered in metals. Surface defects arise from the clustering of line defects into a plane. Low-angle boundaries and grain boundaries are surface defects (see Chap. 5). The *stacking fault* between two close-packed regions of the crystal that have alternate stacking sequences (Sec. 4-11) and twinned region of a crystal (Sec. 4-10) are other examples of surface defects. It is important to note that even at the places where the long-range periodicity of the crystal structure breaks down, as at dislocations and stacking faults, it does so only in certain well-defined ways. Thus, the defects in crystals have regular and reproducible structures and properties.

Point Defects

Figure 4-5 illustrates three types of point defects. A *vacancy*, or vacant lattice site,[1] exists when an atom is missing from a normal lattice position (Fig. 4-5*a*). In pure metals, small numbers of vacancies are created by thermal excitation, and these are thermodynamically stable at temperatures greater than absolute zero. At equilibrium, the fraction of lattices that are vacant at a given temperature is given approximately by the equation

$$\frac{n}{N} = e^{-E_s/kT} \tag{4-1}$$

where n is the number of vacant sites in N sites and E_s is the energy required to move an atom from the interior of a crystal to its surface. Table 4-3 illustrates how the fraction of vacant lattice sites in a metal increases rapidly with temperature. By rapid quenching from close to the melting point, it is possible to trap in a greater than equilibrium number of vacancies at room temperature. Higher than

[1] A. C. Damask and G. J. Dienes, "Point Defects in Metals," Gordon and Breach, Science Publishers, Inc., New York, 1963.; C. P. Flynn, "Point Defects and Diffusion," Clarendon Press, Oxford, 1972.

Table 4-3 Equilibrium vacancies in a metal

Temperature, °C	Approximate fraction of vacant lattice sites
500	1×10^{-10}
1000	1×10^{-5}
1500	5×10^{-4}
2000	3×10^{-3}
	$E_s \approx 1$ ev

equilibrium concentrations of vacancies can also be produced by extensive plastic deformation (cold-work) or as the result of bombardment with high-energy nuclear particles. When the density of vacancies becomes relatively large, it is possible for them to cluster together to form voids.

An atom that is trapped inside the crystal at a point intermediate between normal lattice positions is called an *interstitial atom*, or interstitialcy (Fig. 4-5*b*). The interstitial defect occurs in pure metals as a result of bombardment with high-energy nuclear particles (radiation damage), but it does not occur frequently as a result of thermal activation.

The presence of an *impurity atom* at a lattice position (Fig. 4-5*c*) or at an interstitial position results in a local disturbance of the periodicity of the lattice, the same as for vacancies and interstitials.

It is important to realize that no material is completely pure. Most commercially "pure" materials contain usually 0.01 to 1 percent impurities, while ultrapurity materials, such as germanium and silicon crystals for transistors, contain purposely introduced foreign atoms on the order of one part in 10^{10}. In alloys, foreign atoms are added usually in the range 1 to 50 percent to impart special properties.

Line Defects-Dislocations

The most important two-dimensional, or line, defect is the *dislocation*. The dislocation is the defect responsible for the phenomenon of slip, by which most metals deform plastically. Therefore, one way of thinking about a dislocation is to consider that it is the region of localized lattice disturbance separating the slipped and unslipped regions of a crystal. In Fig. 4-6, *AB* represents a dislocation lying in the slip plane, which is the plane of the paper. It is assumed that slip is advancing to the right. All the atoms above area *C* have been displaced one atomic distance in the slip direction; the atoms above *D* have not yet slipped. *AB* is then the boundary between the slipped and unslipped regions. It is shown shaded to indicate that for a few atomic distances on each side of the dislocation line there is a region of atomic disorder in which the slip distance is between zero and one atomic spacing. As the dislocation moves, slip occurs in the area over

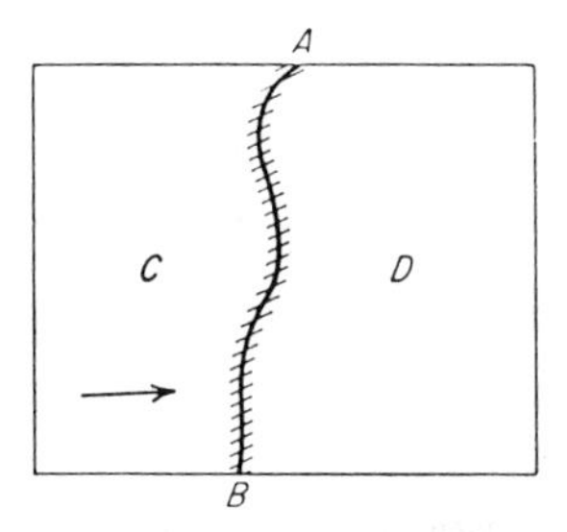

Figure 4-6 A dislocation in a slip plane.

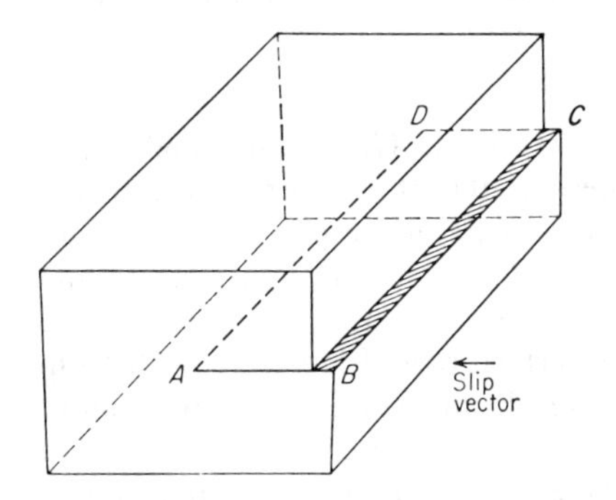

Figure 4-7 Edge dislocation produced by slip in a simple cubic lattice. Dislocation lies along *AD*, perpendicular to slip direction. Slip has occurred over area *ABCD*. (*From W. T. Read, Jr., "Dislocations in Crystals," p. 2, McGraw-Hill Book Company, New York, 1953.*)

which it moves. In the absence of obstacles, a dislocation can move easily on the application of only a small force; this helps explain why real crystals deform much more readily than would be expected for a crystal with a perfect lattice. Not only are dislocations important for explaining the slip of crystals, but they are also intimately connected with nearly all other mechanical phenomena such as strain hardening, the yield point, creep, fatigue, and brittle fracture.

The two basic types of dislocations are the edge dislocation and the screw dislocation. The simplest type of dislocation, which was originally suggested by Orowan, Polanyi, and Taylor, is called the *edge dislocation*, or Taylor-Orowan dislocation. Figure 4-7 shows the slip that produces an edge dislocation for an element of crystal having a simple cubic lattice. Slip has occurred in the direction of the slip vector over the area *ABCD*. The boundary between the right-hand slipped part of the crystal and the left-hand part which has not yet slipped is the line *AD*, the edge dislocation. Note that the parts of the crystal above the slip plane have been displaced, in the direction of slip, with respect to the part of the crystal below the slip plane by an amount indicated by the shaded area in Fig. 4-7. All points in the crystal which were originally coincident across the slip plane have been displaced relative to each other by this same amount. The amount of displacement is equal to the *Burgers vector* **b** of the dislocation. A defining characteristic of an edge dislocation is that its Burgers vector is always perpendicular to the dislocation line.

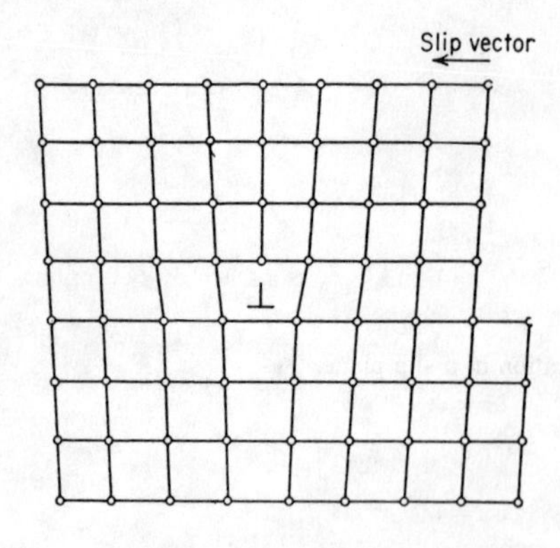

Figure 4-8 Atomic arrangement in a plane normal to an edge dislocation. *(From W. T. Read, Jr., "Dislocations in Crystals," p. 3, McGraw-Hill Book Company, New York, 1953.)*

Although the exact arrangement of atoms along *AD* is not known, it is generally agreed that Fig. 4-8 closely represents the atomic arrangement in a plane normal to the edge dislocation *AD*. The plane of the paper in this figure corresponds to a (100) plane in a simple cubic lattice and is equivalent to any plane parallel to the front face of Fig. 4-7. Note that the lattice is distorted in the region of the dislocation. There is one more vertical row of atoms above the slip plane than below it. The atomic arrangement results in a compressive stress above the slip plane and a tensile stress below the slip plane. An edge dislocation with the extra plane of atoms above the slip plane, as in Fig. 4-8, by convention is called a *positive edge dislocation* and is frequently indicated by the symbol ⊥. If the extra plane of atoms lies below the slip plane, the dislocation is a negative edge dislocation, ⊤.

A pure edge dislocation can glide or slip in a direction perpendicular to its length. However, it may move vertically by a process known as *climb*, if diffusion of atoms or vacancies can take place at an appreciable rate. Consider Fig. 4-8. For the edge dislocation to move upward (positive direction of climb), it is necessary to remove the extra atom directly over the symbol ⊥ or to add a vacancy to this spot. One such atom would have to be removed for every atomic spacing that the dislocation climbs. Conversely, if the dislocation moved down, atoms would have to be added. Atoms could be removed from the extra plane of atoms by the extra atom interacting with a lattice vacancy. Atoms are added to the extra plane for negative climb by the diffusion of an atom from the surrounding crystal, creating a vacancy. Since movement by climb is diffusion-controlled, motion is much slower than in glide and less likely except at high temperatures.

The second basic type of dislocation is the *screw*, or Burgers, dislocation. Figure 4-9 shows a simple example of a screw dislocation. The upper part of the crystal to the right of *AD* has moved relative to the lower part in the direction of the slip vector. No slip has taken place to the left of *AD*, and therefore *AD* is a dislocation line. Thus, the dislocation line is parallel to its Burgers vector, or slip vector, and by definition this must be a screw dislocation. Consider the trace of a

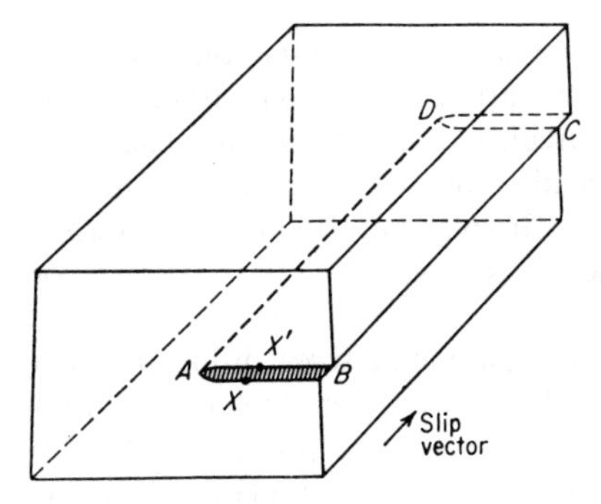

Figure 4-9 Slip that produces a screw dislocation in a simple cubic lattice. Dislocation lies along AD, parallel to slip direction. Slip has occurred over the area $ABCD$. *(From W. T. Read, Jr., "Dislocations in Crystals," p. 15, McGraw-Hill Book Company, New York, 1953.)*

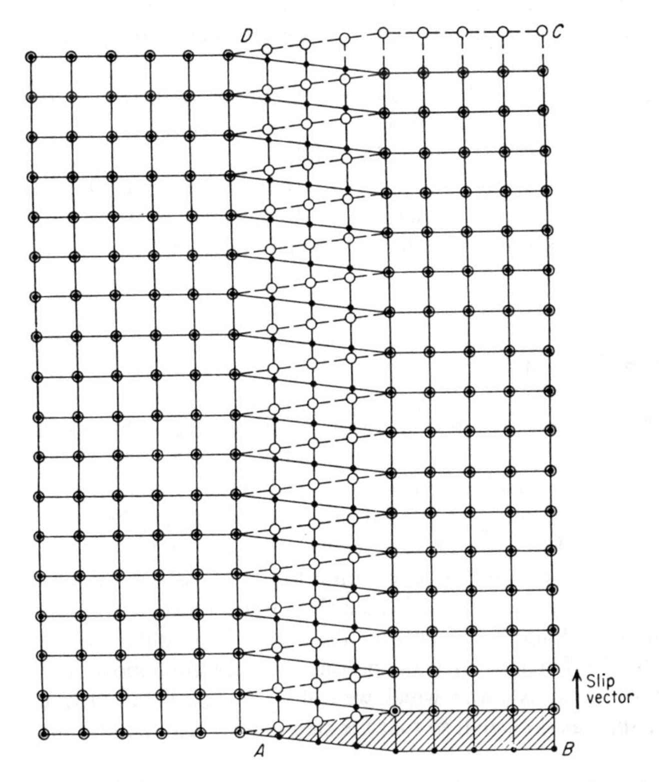

Figure 4-10 Atomic arrangement around the screw dislocation shown in Fig. 4-9. The plane of the figure is parallel to the slip plane. $ABCD$ is the slipped area, and AD is the screw dislocation. Open circles represent atoms in the atomic plane just above the slip plane, and the solid circles are atoms in the plane just below the slip plane. *(From W. T. Read, Jr., "Dislocations in Crystals," p. 17, McGraw-Hill Book Company, New York, 1953.)*

circuit around the dislocation line, on the front face of the crystal. Starting at X and completing a counterclockwise circuit, we arrive at X', one atomic plane behind that containing X. In making this circuit we have traced the path of a right-handed screw. Every time a circuit is made around the dislocation line, the end point is displaced one plane parallel to the slip plane in the lattice. Therefore, the atomic planes are arranged around the dislocation in a spiral staircase or screw.

The arrangement of atoms (in two dimension) around a screw dislocation in a simple cubic lattice is shown in Fig. 4-10. In this figure we are looking down on the slip plane in Fig. 4-9. The open circles represent atoms just above the slip plane, and the solid circles are atoms just below the slip plane. A screw dislocation does not have a preferred slip plane, as an edge dislocation has, and therefore the motion of a screw dislocation is less restricted than the motion of an edge dislocation. However, movement by climb is not possible with a screw dislocation.

For the present, the discussion of dislocations will be limited to the geometrical concepts presented in this section. After a more complete discussion of the plastic deformation of single crystals and polycrystalline specimens, we shall return to a detailed discussion of dislocation theory in Chap. 5. Among the topics covered will be the effect of crystal structure on dislocation geometry, the experimental evidence for dislocations, and the interaction between dislocations.

4-4 DEFORMATION BY SLIP

The usual method of plastic deformation in metals is by the sliding of blocks of the crystal over one another along definite crystallographic planes, called *slip planes*. As a very crude approximation, the slip, or glide of a crystal can be considered analogous to the distortion produced in a deck of cards when it is pushed from one end. Figure 4-11 illustrates this classical picture of slip. In Fig. 4-11*a*, a shear stress is applied to a metal cube with a top polished surface. Slip occurs when the shear stress exceeds a critical value. The atoms move an integral number of atomic distances along the slip plane, and a step is produced in the polished surface (Fig. 4-11*b*). When we view the polished surface from above with a microscope, the step shows up as a line, which we call a *slip line*. If the surface is then repolished after slip has occurred, so that the step is removed, the slip line will disappear (Fig. 4-11*c*).

Because of the translational symmetry of a crystal lattice, the crystal structure is perfectly restored after slip has taken place provided that the deformation was uniform. Each atom in the slipped part of the crystal moves forward the same integral number of lattice spacings. Note that slip lines are due to changes in surface elevation and that the surface must be suitably prepared for microscopic observation prior to deformation if the slip lines are to be observed. Figure 4-12 shows straight slip lines in copper.

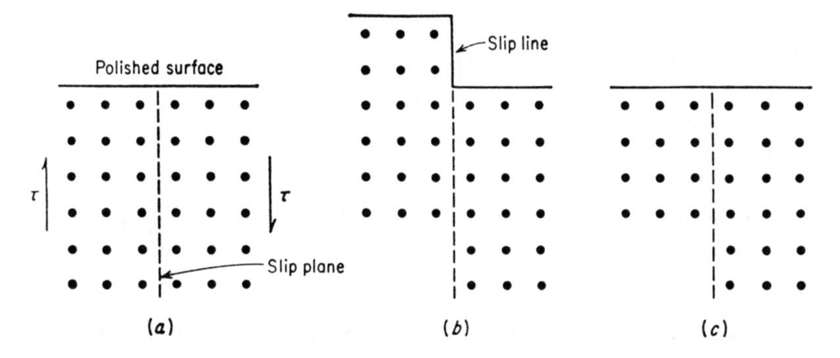

Figure 4-11 Schematic drawing of classical idea of slip.

Figure 4-12 Straight slip lines in copper (500 ×). *(Courtesy W. L. Phillips.)*

The fine structure of slip lines has been studied at high magnification by means of the electron microscope. What appears as a line, or at best a narrow band at 1,500 diameters' magnification in the optical microscope can be resolved by the electron microscope as discrete slip lamellae at 20,000 diameters, shown schematically in Fig. 4-13.

The fact that a single crystal remains a single crystal after homogeneous plastic deformation imposes limitations on the way in which plastic deformation may occur. Slip occurs most readily in specific directions on certain crystallographic planes. Generally the slip plane is the plane of greatest atomic density (Table 4-2) and the slip direction is the closest-packed direction within the slip plane. Since the planes of greatest atomic density are also the most widely spaced planes in the crystal structure, the resistance to slip is generally less for these

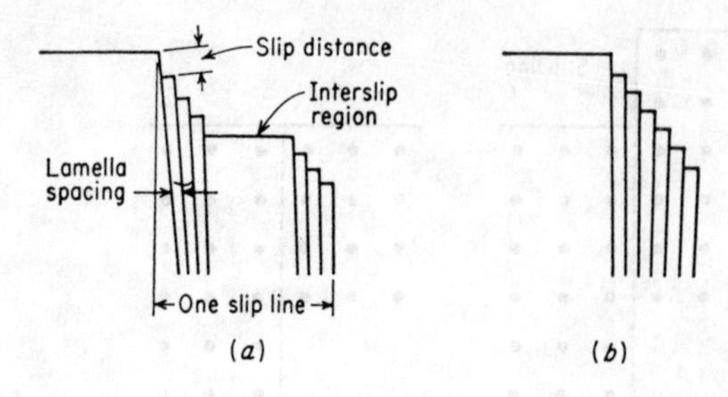

Figure 4-13 Schematic drawing of the fine structure of a slip band. (*a*) Small deformation; (*b*) large deformation.

planes than for any other set of planes. The slip plane together with the slip direction establishes the *slip system*.

In the hexagonal close-packed metals, the only plane with high atomic density is the basal plane (0001). The axes $\langle 1120 \rangle$ are the close-packed directions. For zinc, cadmium, magnesium, and cobalt slip occurs on the (0001) plane in the $\langle 11\bar{2}0 \rangle$ directions.[1] Since there is only one basal plane per unit cell and three $\langle 11\bar{2}0 \rangle$ directions, the hcp structure possesses three slip systems. The limited number of slip systems is the reason for the extreme orientation dependence and low ductility in hcp crystals.

In the face-centered cubic structure, the $\{111\}$ octahedral planes and the $\langle 110 \rangle$ directions are the close-packed systems. There are eight $\{111\}$ planes in the fcc unit cell. However, the planes at opposite faces of the octahedron are parallel to each other, so that there are only four *sets* of octahedral planes. Each $\{111\}$ plane contains three $\langle 110 \rangle$ directions (the reverse directions being neglected). Therefore, the fcc lattice has 12 possible slip systems.

Example Determine the slip systems for slip on a (111) plane in a fcc crystal and sketch the result.

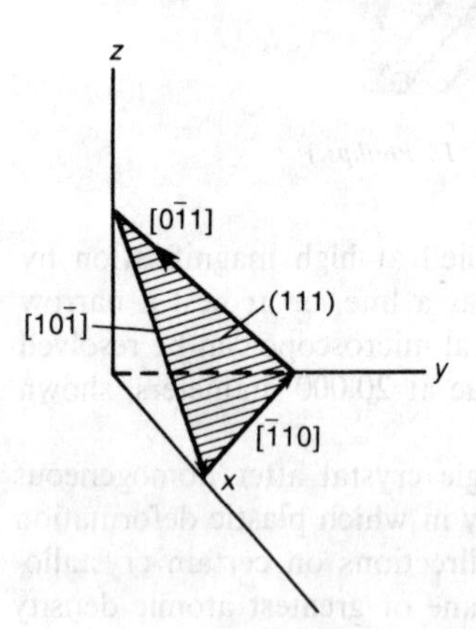

[1] Zirconium and titanium, which have low c/a ratios, slip primarily on the prism and pyramidal planes in the $\langle 11\bar{2}0 \rangle$ direction.

Slip direction in fcc is $\langle 110 \rangle$ type direction. Slip directions are most easily established from a sketch of the (111) plane. To prove that these slip directions lie in the slip plane $hu + kv + lw = 0$

$$(1)(1) + (1)(0) + (1)(-1) = 0$$

$$(1)(-1) + (1)(1) + (1)(0) = 0$$

$$(1)(0) + (1)(-1) + (1)(1) = 0$$

The bcc structure is not a close-packed structure like the fcc or hcp structures. Accordingly, there is no one plane of predominant atomic density, as (111) in the fcc structure and (0001) in the hcp structure. The $\{110\}$ planes have the highest atomic density in the bcc structure, but they are not greatly superior in this respect to several other planes. However, in the bcc structure the $\langle 111 \rangle$ direction is just as close-packed as the $\langle 110 \rangle$ and $\langle 11\bar{2}0 \rangle$ directions in the fcc and hcp structures. Therefore, the bcc metals obey the general rule that the slip direction is the close-packed direction, but they differ from most other metals by not having a definite single slip plane. Slip in bcc metals is found to occur on the $\{110\}$, $\{112\}$, and $\{123\}$ planes, while the slip direction is always the [111] direction. There are 48 possible slip systems, but since the planes are not so close-packed as in the fcc structure, higher shearing stresses are usually required to cause slip.

Slip lines in bcc metals have a wavy appearance. This is due to the fact that slip occurs on several planes, $\{110\}, \{112\}, \{123\}$ but always in the close-packed $\langle 111 \rangle$ direction which is common to each of these planes. Dislocations can readily move from one type of plane to another by cross sip, giving rise to the irregular wavy slip bands.

Certain metals show additional slip systems with increased temperature. Aluminum deforms on the $\{110\}$ plane at elevated temperature, while in magnesium the $\{10\bar{1}1\}$ pyramidal plane plays an important role in deformation by slip above 225°C. In all cases the slip direction remains the same when the slip plane changes with temperature.

4-5 SLIP IN A PERFECT LATTICE

If slip is assumed to occur by the translation of one plane of atoms over another, it is possible to make a reasonable estimate[1] of the shear stress required for such a movement in a perfect lattice. Consider two planes of atoms subjected to a

[1] J. Frenkel, *Z. Phys.*, vol. 37, p. 572, 1926.

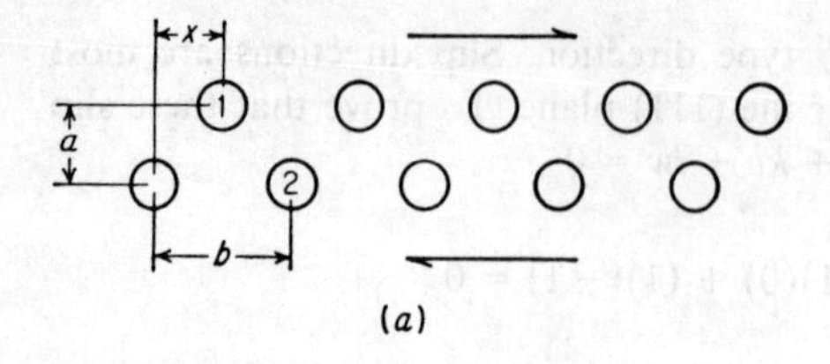

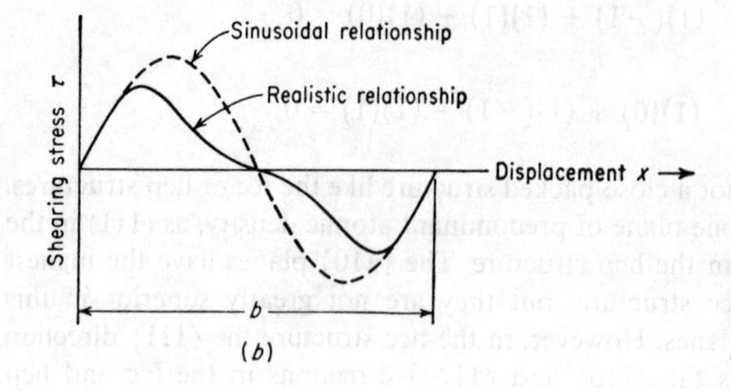

Figure 4-14 (*a*) Shear displacement of one plane of atoms over another atomic plane; (*b*) variation of shearing stress with displacement in slip direction.

homogeneous shear stress (Fig. 4-14). The shear stress is assumed to act in the slip plane along the slip direction. The distance between atoms in the slip directions is b, and the spacing between adjacent lattice planes is a. The shear stress causes a displacement x in the slip direction between the pair of adjacent lattice planes. The shearing stress is initially zero when the two planes are in coincidence, and it is also zero when the two planes have moved one identity distance b, so that point 1 in the top plane is over point 2 on the bottom plane. The shearing stress is also zero when the atoms of the top plane are midway between those of the bottom plane, since this is a symmetry position. Between these positions each atom is attracted toward the nearest atom of the other row, so that the shearing stress is a periodic function of the displacement.

As a first approximation, the relationship between shear stress and displacement can be expressed by a sine function

$$\tau = \tau_m \sin \frac{2\pi x}{b} \tag{4-2}$$

where τ_m is the amplitude of the sine wave and b is the period. At small values of displacement, Hooke's law should apply.

$$\tau = G\gamma = \frac{Gx}{a} \tag{4-3}$$

For small values of x/b, Eq. (4-2) can be written

$$\tau \approx \tau_m \frac{2\pi x}{b} \tag{4-4}$$

Combining Eqs. (4-3) and (4-4) provides an expression for the maximum shear stress at which slip should occur.

$$\tau_m = \frac{G}{2\pi}\frac{b}{a} \tag{4-5}$$

As a rough approximation, b can be taken equal to a, with the result that the theoretical shear strength of a perfect crystal is approximately equal to the shear modulus divided by 2π.

$$\tau_m = \frac{G}{2\pi} \tag{4-6}$$

The shear modulus for metals is in the range 20 to 150 GPa. Therefore, Eq. (4-6) predicts that the theoretical shear stress will be in the range (3 to 30 GPa), while actual values of the shear stress required to produce plastic deformation in metal single crystals are in the range 0.5 to 10 MPa. Even if more refined calculations are used to correct the sine-wave assumption, the value of τ_m cannot be made equal to the observed shear stress. Tyson,[1] using a computer solution of the interatomic force equations, predicted $\tau_m = G/16$ for an fcc metal, $G/8$ for an NaCl structure, and $G/4$ for a covalently bonded diamond structure. Since the theoretical shear strength of metal crystals is at least 100 times greater than the observed shear strength, it must be concluded that a mechanism other than bodily shearing of planes of atoms is responsible for slip. In the next section it is shown that dislocations provide such a mechanism.

4-6 SLIP BY DISLOCATION MOVEMENT

The concept of the dislocation was first introduced to explain the discrepancy between the observed and theoretical shear strengths of metals. For the dislocation concept to be valid it is necessary to show (1) that the motion of a dislocation through a crystal lattice requires a stress far smaller than the theoretical shear stress, and (2) that the movement of the dislocation produces a step, or slip band, at the free surface.

In a perfect lattice all atoms above and below the slip plane are in minimum energy positions. When a shear stress is applied to the crystal, the same force opposing the movement acts on all the atoms. This is the model for slip presented in Fig. 4-14. When there is a dislocation in the crystal, the atoms well away from the dislocation are still in the minimum energy positions but at the dislocation only a small movement of the atoms is required. Referring to Fig. 4-15*a*, the extra plane of atoms at the edge dislocation initially is at 4. Under the action of the shear stress, a very small movement of atoms to the right will allow this half plane to line up with the half plane 5′, at the same time cutting the half plane 5 from its neighbors below the slip plane. By this process the edge dislocation line has moved from its initial position between planes 4′ and 5′ to a new position

[1] W. R. Tyson, *Philos. Mag.*, vol. 14, pp. 925–936, 1966.

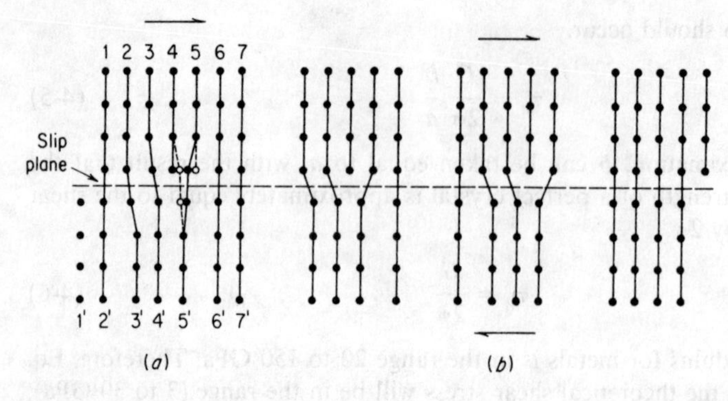

Figure 4-15 (*a*) Atom movements near dislocation in slip; (*b*) movement of an edge dislocation.

between planes 5′ and 6′. Since the atoms around the dislocations are symmetrically placed on opposite sides of the extra half plane, equal and opposite forces oppose and assist the motion. Thus, in a first approximation there is no net force on the dislocation and the stress required to move the dislocation is zero. The continuation of this process under the stresses shown in Fig. 4-15 moves the dislocation to the right. When the extra half plane of atoms reaches a free surface (Fig. 4-15*b*), it results in a slip step of one Burgers vector, or one atomic distance for the simple cubic lattice.

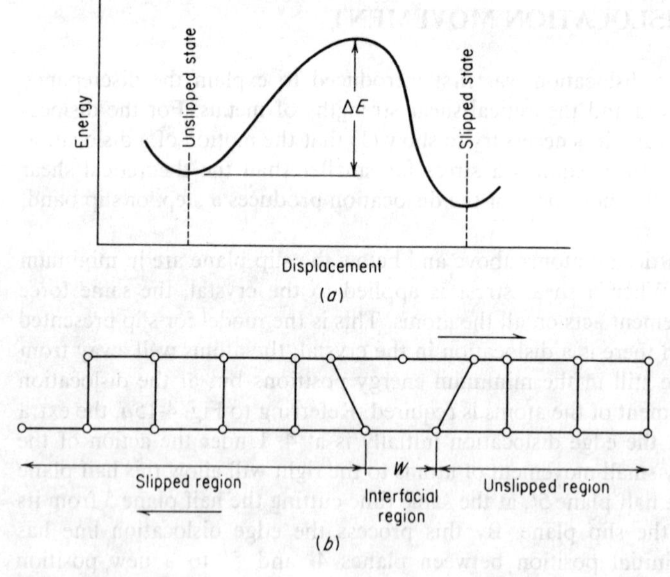

Figure 4-16 (*a*) Energy change from unslipped to slipped state; (*b*) stages in growth of slipped region.

A particularly instructive way of looking at slip by dislocation motion has been proposed by Cottrell.[1] Consider that plastic deformation is the transition from an unslipped to a slipped state (Fig. 4-16*a*). Since the process is opposed by an energy barrier ΔE, in order to facilitate the process it is logical to assume that the material will not all make the transition simultaneously. To minimize the energetics of the process, the slipped material will grow at the expense of the unslipped region by the advance of an interfacial region (Fig. 4-16*b*). The interfacial region is a *dislocation*. To minimize the energy for the transition, we expect the interface thickness w to be narrow. The distance w is the *width* of the dislocation. The smaller the width of the dislocation, the lower is the interfacial energy, but the wider the dislocation, the lower is the elastic energy of the crystal because then the atomic spacing in the slip direction is closer to its equilibrium spacing. Thus, the equilibrium width of the dislocation is determined by a balance between these two opposing energy changes.

The dislocation width is important because it determines the force required to move a dislocation through the crystal lattice. This force is called the *Peierls-Nabarro force*. The Peierls stress is the shear stress required to move a dislocation through a crystal lattice in a particular direction

$$\tau_p \approx \frac{2G}{1-\nu} e^{-2\pi w/b} \approx \frac{2G}{1-\nu} e^{-[2\pi a/(1-\nu)b]} \tag{4-7}$$

where a is the distance between slip planes and b is the distance between atoms in the slip direction. Note that the dislocation width appears in the exponential term in Eq. (4-7), so that the Peierls stress will be very sensitive to the atomic position at the core of a dislocation. These are not known with any high degree of accuracy and, since Eq. (4-7) was derived for the sinusoidal force-distance law that has only limited validity, the equation cannot be used for precise calculations. However, it is accurate enough to show that the stress needed to move a dislocation in a metal is quite low.

In spite of these limitations, the Peierls equation has important conceptual value. It shows that materials with wide dislocations will require a low stress to move the dislocations. Physically, this is so because when the dislocation is wide, the highly distorted region at the core of the dislocation is not localized on any particular atom in the crystal lattice. In ductile metals the dislocation width is of the order of 10 atomic spacings. However, in ceramic materials with directional covalent bonds, the interfacial energy is high and the dislocations are narrow and relatively immobile. This fact combined with the restriction on slip systems imposed by the requirements of electrostatic forces results in the low degree of plasticity of ceramic materials. Ceramics become more ductile at high temperatures because thermal activation helps the dislocations overcome the energy barrier.

[1] A. H. Cottrell, "An Introduction to Metallurgy," pp. 266–269, Edward Arnold (Publishers) Ltd. London, 1967.

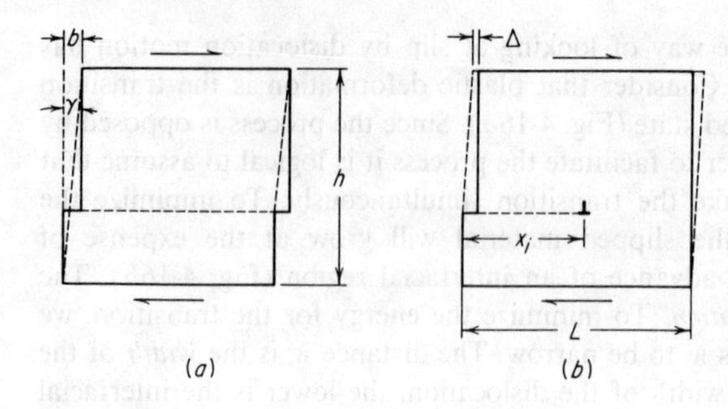

Figure 4-17 (*a*) Shear strain associated with passage of a single dislocation through a crystal; (*b*) shear strain due to motion of dislocation part way through a crystal.

The fact that slip occurs in close-packed directions means that b is minimized, and from Eq. (4-7) the Peierls stress will be lower. If $a < b$, as would occur for closely spaced but loosely packed planes, the Peierls stress would be high. Thus, Eq. (4-7) provides a basis for the observation that slip occurs most readily on close-packed planes in the close-packed directions. When the crystal structure is complex, without highly close-packed planes and directions, dislocations tend to be immobile. This causes the brittleness and high hardness of intermetallic compounds.

It is instructive to calculate the shear strain that results from the motion of a dislocation through a single crystal. The dislocation produces a slip offset of length b, so that the shear strain (γ from Fig. 4-17*a*) is $\gamma = b/h$. If the crystal is 1 cm high, the passage of a single dislocation produces a shear strain of only 3×10^{-8}. It is obvious that the motion of many dislocations is involved in producing strains of engineering significance. The total macroscopic plastic strain is the sum of all the small strains due to a very large number of individual dislocations. Thus if three dislocations on three parallel planes moved through the crystal, the shear strain would be $\gamma = 3b/h$.

Now consider the case where a dislocation has moved part way through the crystal along the slip plane (Fig. 4-17*b*). Since b is very small compared with L or h, the displacement δ_i for a dislocation at an intermediate position between $x_i = 0$ and $x_i = L$ would be proportional to the fractional displacement x_i/L.

$$\delta_i = \frac{x_i b}{L}$$

The total displacement of the top of the crystal relative to the bottom for many dislocations on many slip planes will be

$$\Delta = \sum \delta_i = \frac{b}{L} \sum_i^N x_i \tag{4-8}$$

where N is the total number of dislocations that have moved in the volume of the crystal.

The macroscopic shear strain is

$$\gamma = \frac{\Delta}{h} = \frac{b}{hL}\sum_i^N x_i \tag{4-9}$$

If the average distance that dislocations have moved is $\bar{x}$, where

$$\bar{x} = \frac{\sum_i^N x_i}{N}$$

then

$$\gamma = \frac{bN\bar{x}}{hL} \tag{4-10}$$

This equation can best be written in terms of the *dislocation density* ρ

$$\gamma = b\rho\bar{x} \tag{4-11}$$

where $\rho = N/hL$. The dislocation density is the total length of dislocation line per unit volume or, alternatively, the number of dislocation times that cut through a unit cross-sectional area. Either way, ρ has units of number of dislocations per square centimeter. The argument used in deriving Eq. (4-11) is applicable to both edge and screw dislocations. However, such highly simplified motion by parallel dislocations rarely occurs in real metals, so that Eq. (4-11) needs to be modified to account for complexities in geometry and dislocation configuration.

It is often useful to express Eq. (4-11) in terms of shear-strain rate.

$$\dot{\gamma} = \frac{d\gamma}{dt} = b\rho\frac{d\bar{x}}{dt} = b\rho\bar{v} \tag{4-12}$$

where $\bar{v}$, the average dislocation velocity, is a quantity which can be measured experimentally in many systems. From Eq. (4-12) we see that if we want to describe macroscopic plastic deformation in terms of dislocation behavior we need to know (1) the crystal structure in order to evaluate b, (2) the number of mobile dislocations ρ, and (3) the average dislocation velocity $\bar{v}$. The quantities ρ and $\bar{v}$ will depend on stress, time, temperature, and prior thermomechanical history.

Measurements of the velocity of dislocation motion[1] in a number of ionic crystals and metals have shown that the velocity is a very strong function of the shear stress in the slip plane. This is given by the equation $v = A\tau^{m'}$, where m' is a constant varying from about 1.5 to 40 for different materials. There is a critical stress required to start dislocations moving, with small increases in stress leading to large increases in dislocation velocity. Above about 10 cm/s the velocity increases less rapidly and tends toward an upper limit close to the shear wave velocity of the material.

[1] W. G. Johnston and J. H. Gilman, *J. Appl. Phys.*, vol. 30, p. 129, 1959.

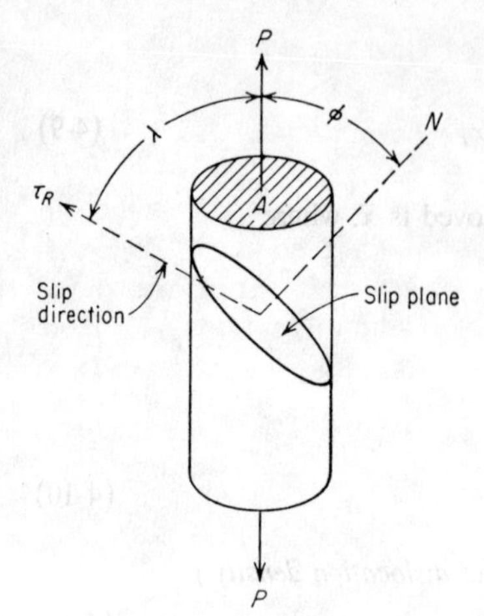

Figure 4-18 Diagram for calculating critical resolved shear stress.

4-7 CRITICAL RESOLVED SHEAR STRESS FOR SLIP

The extent of slip in a single crystal depends on the magnitude of the shearing stress produced by external loads, the geometry of the crystal structure, and the orientation of the active slip planes with respect to the shearing stresses. Slip begins when the shearing stress on the slip plane in the slip direction reaches a threshold value called the *critical resolved shear stress*. This value[1] is really the single-crystal equivalent of the yield stress of an ordinary stress-strain curve. The value of the critical resolved shear stress depends chiefly on composition and temperature.

The fact that different tensile loads are required to produce slip in single crystals of different orientation can be rationalized by a critical resolved shear stress; this was first recognized by Schmid.[2] To calculate the critical resolved shear stress from a single crystal tested in tension, it is necessary to know, from x-ray diffraction, the orientation with respect to the tensile axis of the plane on which slip first appears and the slip direction. Consider a cylindrical single crystal with cross-sectional area A (Fig. 4-18). The angle[3] between the *normal* to the slip plane and the tensile axis is ϕ, and the angle which the slip direction makes with the tensile axis is λ. The area of the slip plane inclined at the angle ϕ will be $A/\cos\phi$, and the component of the axial load acting in the slip plane in the slip

[1] In practice it is very difficult to determine the stress at which the first slip bands are produced. In most cases, the critical shear stress is obtained by the intersection of the extrapolated elastic and plastic region of the stress-strain curve.

[2] E. Schmid, *Z. Elektrochem.*, vol. 37, 447, 1931.

[3] Note that ϕ and λ are complementary angles only in the special case when the slip direction is in the plane defined by the stress axis and the normal to the slip plane.

direction is $P \cos \lambda$. Therefore, the critical resolved shear stress is given by

$$\tau_R = \frac{P \cos \lambda}{A/\cos \phi} = \frac{P}{A} \cos \phi \cos \lambda \tag{4-13}$$

Equation (4-13) gives the shear stress resolved on the slip plane in the slip direction. This shear stress is a maximum when $\phi = \lambda = 45°$, so that $\tau_R = \frac{1}{2}P/A$. If the tension axis is normal to the slip plane ($\lambda = 90°$) or if it is parallel to the slip plane ($\phi = 90°$), the resolved shear stress is zero. Slip will not occur for these extreme orientations since there is no shear stress on the slip plane. Crystals close to these orientations tend to fracture rather than slip.

Example Determine the tensile stress that is applied along the $[1\bar{1}0]$ axis of a silver crystal to cause slip on the $(1\bar{1}\bar{1})[0\bar{1}1]$ system. The critical resolved shear stress is 6 MPa.

The angle between tensile axis $[1\bar{1}0]$ and normal to $(1\bar{1}\bar{1})$ is

$$\cos \phi = \frac{(1)(1) + (-1)(-1) + (0)(-1)}{\sqrt{(1)^2 + (-1)^2 + (0)^2}\sqrt{(1)^2 + (-1)^2 + (-1)^2}} = \frac{2}{\sqrt{2}\sqrt{3}} = \frac{2}{\sqrt{6}}$$

The angle between tensile axis $[1\bar{1}0]$ and slip direction $[0\bar{1}\bar{1}]$ is

$$\cos \lambda = \frac{(1)(0) + (-1)(-1) + (0)(-1)}{\sqrt{2}\sqrt{(0)^2 + (-1)^2 + (-1)^2}} = \frac{1}{\sqrt{2}\sqrt{2}} = \frac{1}{2}$$

From Eq. (4-13)

$$\sigma = \frac{P}{A} = \frac{\tau_R}{\cos \phi \cos \lambda} = \frac{6}{2/\sqrt{6} \times \frac{1}{2}} = 6\sqrt{6} = 14.7 \text{ MPa}$$

Table 4-4 gives values of critical resolved shear stress for a number of metals. The importance of small amounts of impurities in increasing the critical resolved shear stress is shown by the data for silver and copper. Alloying-element additions have even a greater effect, as shown by the data for gold-silver alloys in Fig. 4-19. Note that a large increase in the resistance to slip is produced by alloying gold and silver even though these atoms are very much alike in size and electronegativity, and hence they form a solid solution over the complete range of composition. In solid solutions, where the solute atoms differ considerably in size from the solvent atoms, an even greater increase in critical resolved shear stress would be observed.

The magnitude of the critical resolved shear stress of a crystal is determined by the interaction of its population of dislocations with each other and with defects such as vacancies, interstitials, and impurity atoms. This stress is, of course, greater than the stress required to move a single dislocation, but it is appreciably lower than the stress required to produce slip in a perfect lattice. On the basis of this reasoning, the critical resolved shear stress should decrease as the

Table 4-4 Room-temperature slip systems and critical resolved shear stress for metal single crystals

Metal	Crystal structure	Purity, %	Slip plane	Slip direction	Critical shear stress, MPa	Ref.
Zn	hcp	99.999	(0001)	[11$\bar{2}$0]	0.18	a
Mg	hcp	99.996	(0001)	[1120]	0.77	b
Cd	hcp	99.996	(0001)	[11$\bar{2}$0]	0.58	c
Ti	hcp	99.99	(1010)	[11$\bar{2}$0]	13.7	d
		99.9	(1010)	[11$\bar{2}$0]	90.1	d
Ag	fcc	99.99	(111)	[110]	0.48	e
		99.97	(111)	[110]	0.73	e
		99.93	(111)	[110]	1.3	e
Cu	fcc	99.999	(111)	[110]	0.65	e
		99.98	(111)	[110]	0.94	e
Ni	fcc	99.8	(111)	[110]	5.7	e
Fe	bcc	99.96	(110)	[111]	27.5	f
			(112)			
			(123)			
Mo	bcc	. . .	(110)	[111]	49.0	g

[a] D. C. Jillson, *Trans. AIME*, vol. 188, p. 1129, 1950.
[b] E. C. Burke and W. R. Hibbard, Jr., *Trans. AIME*, vol. 194, p. 295, 1952.
[c] E. Schmid, "International Conference on Physics," vol. 2, Physical Society, London. 1935.
[d] A. T. Churchman, *Proc. R. Soc. London Ser. A*, vol. 226A, p. 216, 1954.
[e] F. D. Rosi, *TRans., AIME*, vol. 200, p. 1009, 1954.
[f] J. J. Cox, R. F. Mehl, and G. T. Horne, *Trans. Am. Soc. Met.*, vol. 49, p. 118, 1957.
[g] R. Maddin and N. K. Chen, *Trans. AIME*, vol. 191, p. 937, 1951.

density of defects decreases, provided that the total number of imperfections is not zero. When the last dislocation is eliminated, the critical resolved shear stress should rise abruptly to the high value predicted for the shear strength of a perfect crystal. Experimental evidence for the effect of decreasing defect density is shown by the fact that the critical resolved shear stress of soft metals can be reduced to less than one-third by increasing the purity. At the other extreme, micron-diame-

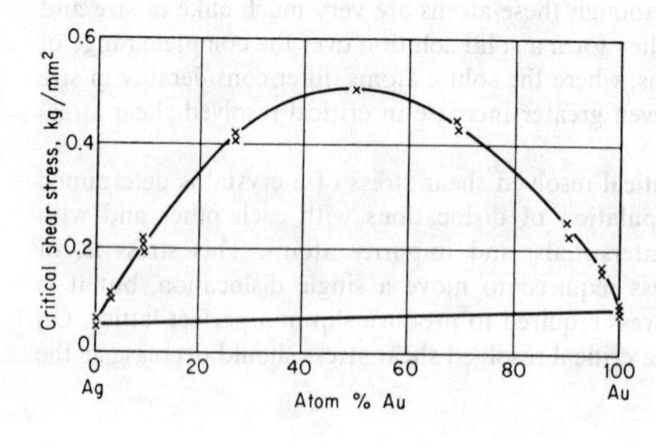

Figure 4-19 Variation of critical resolved shear stress with composition in silver-gold-alloy single crystals. *(From G. Sachs and J. Weerts, Z. Phys., vol. 62, p.473, 1930.)*

ter single crystal filaments, or whiskers, can be grown essentially dislocation-free. Tensile tests[1] on these filaments have given strengths which are approximately equal to the calculated strength of a perfect crystal.

The ratio of the resolved shear stress to the axial stress is called the *Schmid factor m*. For a single crystal loaded in tension or compression along its axis, $m = \cos\phi \cos\lambda$. Hartley and Hirth[2] have presented graphical methods of determining m for any crystal orientation and slip system.

It is observed experimentally that a single crystal will slip when the resolved shear stress on the slip plane reaches a critical value. This behavior, known as Schmid's law, is best demonstrated with hcp metals where the limited number of slip systems allows large differences in orientation between the slip plane and the tensile axis (see Prob. 4-8). In fcc metals the high symmetry results in so many equivalent slip systems that it is possible to get a variation in the yield stress of only about a factor of 2 because of differences in the orientation of the slip plane with the tensile axis.

4-8 DEFORMATION OF SINGLE CRYSTALS

Most studies of the mechanical properties of single crystals are made by subjecting the crystal to simple uniaxial tension. In the ordinary tension test, the movement of the crosshead of the testing machine constrains the specimen at the grips since the grips must remain in line. Therefore, the specimen is not permitted to deform freely by uniform glide on every slip plane along the gage length of the specimen, as is pictured in Fig. 4-20*a*. Instead, the slip planes rotate toward the tensile axis since the tensile axis of the specimen remains fixed, as in Fig. 4-20*b*.

Since plastic flow occurs by slip on certain planes in particular directions, the measured increase in length of the specimen for a given amount of slip will depend on the orientations of the slip plane and direction with the specimen axis.[3] The fundamental measure of plastic strain in a single crystal is the crystallographic glide strain γ. Glide strain is the relative displacement of two parallel slip planes separated at a unit distance. The equations relating glide strain with specimen extension can be derived from Fig. 4-21. As the single crystal elongates, the slip direction rotates toward the tensile axis. For simplicity in Fig. 4-21, the glide elements are kept fixed and the tensile axis is rotated as the crystal elongates from L_0 to L_1. The two cases are equivalent geometrically. Moreover, for simplicity the orientation of the slip plane is given by the angle χ between the axis of the glide ellipse and the tensile axis rather than the angle ϕ between the normal to the glide ellipse (slip plane) and the tensile axis. With this selection of angles, $\tau_R = P/A \sin\chi \cos\lambda$. From triangle ABB', using the law of sines, we can

[1] S. S. Brenner, *J. Appl. Phys.*, vol. 27, pp. 1484–1491, 1956.

[2] C. S. Hartley and J. P. Hirth, *Trans. Metall. Soc. AIME*, vol. 223, pp. 1415–1419, 1965.

[3] E. Schmid and W. Boas, "Plasticity of Crystals," English translation, pp. 58–60, F. A. Hughes & Co., London, 1950.

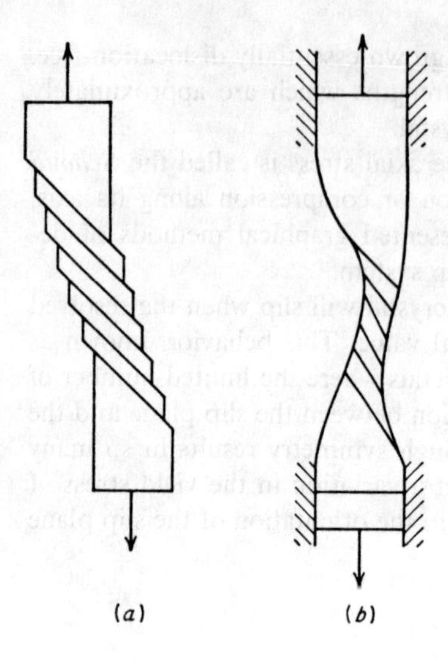

Figure 4-20 (*a*) Tensile deformation of single crystal without constraint; (*b*) rotation of slip planes due to constraint.

see that

$$\frac{L_0}{\sin \lambda_1} = \frac{L_1}{\sin (180 - \lambda_0)} = \frac{L_1}{\sin \lambda_0}$$

From triangles *ABC* and *AB′C*

$$AC = L_0 \sin \chi_0 = L_1 \sin \chi_1$$

The *glide strain* is defined as the total amount of slip divided by the thickness of the glide packet

$$\lambda = \frac{BB'}{AC}$$

Again, from the law of sines

$$BB' = \frac{L_1 \sin (\lambda_0 - \lambda_1)}{\sin \lambda_0}$$

Substitution in the expression for glide strain, and after considerable trigonometric manipulation, results in

$$\gamma = \frac{1}{\sin \chi_0} \left\{ \left[\left(\frac{L_1}{L_0} \right)^2 - \sin^2 \lambda_0 \right]^{1/2} - \cos \lambda_0 \right\} \qquad (4\text{-}14)$$

Thus, the glide shear strain may be determined from the initial orientation of the slip plane and slip direction (χ_0 and λ_0) and the extension of the specimen L_1/L_0. This analysis assumes that slip occurs on only a single slip system.

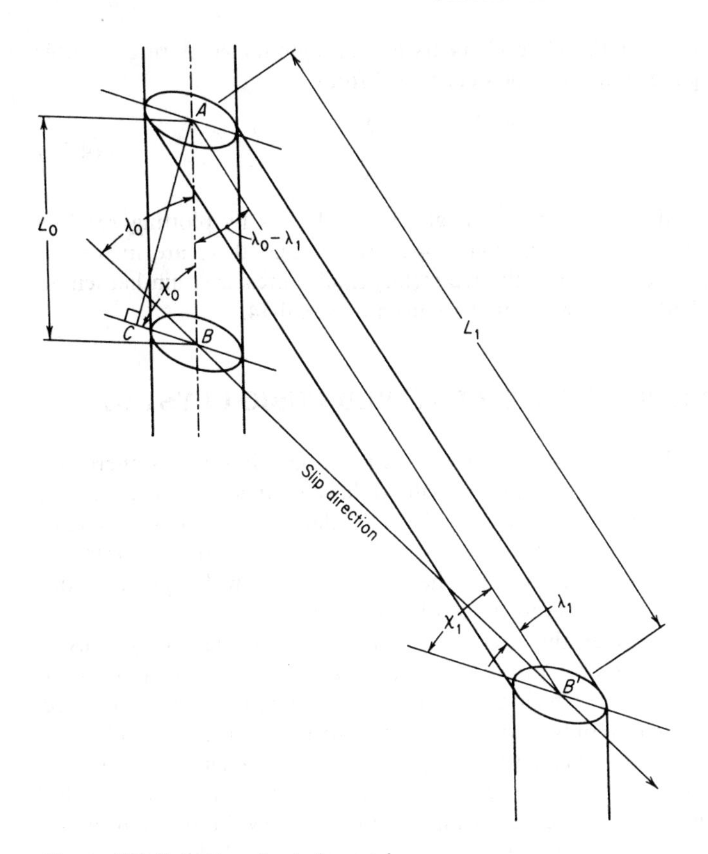

Figure 4-21 Extension of a single crystal.

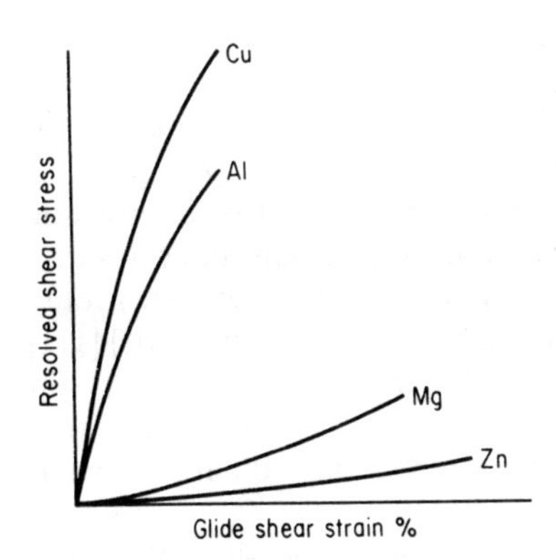

Figure 4-22 Typical single-crystal stress-strain curves.

If the orientation of the glide elements can be determined during or after deformation, the glide strain may be determined from

$$\gamma = \frac{\cos \lambda_1}{\sin \chi_1} - \frac{\cos \lambda_0}{\sin \chi_0} \tag{4-15}$$

The fundamental way to present single-crystal data is by plotting *resolved shear stress vs. glide shear strain*. Figure 4-22 shows that there are important differences between metals. Typically, fcc metals exhibit greater strain hardening than hcp metals. Additional details will be given in Sec. 4-14.

4-9 DEFORMATION OF FACE-CENTERED CUBIC CRYSTALS

Because fcc crystals have high symmetry and 12 potential slip systems, there is a wide choice of slip systems. The slip plane will not have to undergo much rotation before the resolved shear stress becomes high on another $\{111\}\langle 110\rangle$ slip system. The initial operative slip system, the primary slip system, will be the one with the highest Schmid factor, $m = \sin \chi \cos \lambda$. The primary system will depend on the orientation of the crystal relative to the tensile stress axis.

The relationship between the stress axis and the 12 possible slip systems is best shown on a stereographic projection[1] (Fig. 4-23), where each of the unit triangles defines a region in which a particular slip system operates. There are four $\langle 111\rangle$ poles *ABCD* representing the normals to the octahedral $\{111\}$ slip planes. Slip directions are indicated I through IV. For a specimen axis at *P*, the slip system *B*IV will be operative. ϕ_0 and λ_0 are given by the great circles through *B*-*P*-IV. We can use the stereographic plot to follow the rotation of the slip system toward the tensile axis. As the specimen elongates, λ decreases and ϕ increases. However it is more convenient to consider that the slip system remains fixed and the specimen axis rotates.

As the specimen elongates, the specimen axis eventually reaches the $[001]$-$[\bar{1}11]$ boundary at P'. Now the resolved shear stress is equal on the *primary slip system* and the *conjugate slip system* $(\bar{1}\bar{1}1)[011]$. At this point deformation proceeds on the two slip systems simultaneously to produce *duplex slip* or *multiple slip*.

Under the microscope conjugate slip appears as another set of intersecting slip lines. The fact that slip can occur equally on both slip systems indicates that latent strain hardening must have occurred on the conjugate system when only the primary system was acting. The specimen axis rotates along the $[001]$-$[\bar{1}11]$ boundary to the $[\bar{1}12]$ pole, which is midway between the two operative slip directions $[\bar{1}01]$ and $[011]$. When the specimen axis reaches $[\bar{1}12]$ it stays at that orientation until the specimen necks down and fractures. Thus, the introduction

[1] For a description of stereographic projection see C. S. Barrett and T. B. Massalski, "Structure of Metals," 3rd ed., chap. 2, McGraw-Hill Book Company, New York, 1966; A. Kelly and G. W. Groves, "Crystallography and Crystal Defects, chap. 2, Addison-Wesley Publishing Company, Inc., Reading, Mass., 1970.

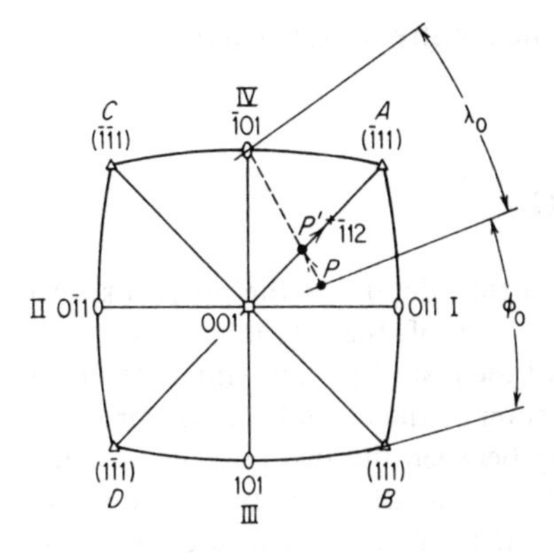

Figure 4-23 Standard (001) stereographic projection or a cubic crystal.

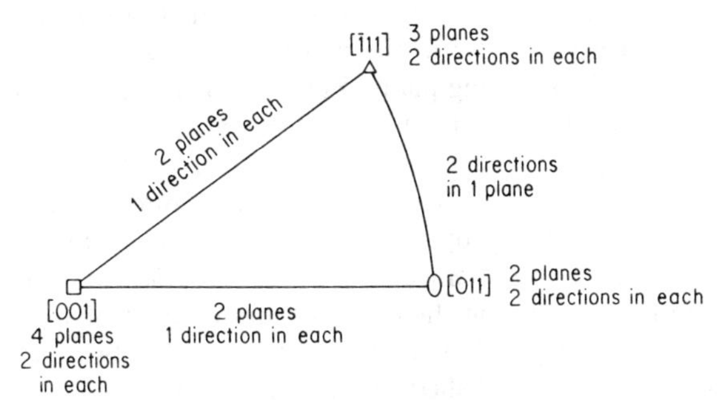

Figure 4-24 Operative slip systems along boundaries of stereographic triangle.

of duplex slip interrupts the free rotation of the glide system and leads to lower ductility at fracture[1] than in the case of hcp crystals where easy glide on only a single slip system occurs out to fracture. For duplex slip on primary and conjugate slip systems, the glide shear strain is given by[2]

$$\gamma = \sqrt{6} \ln \frac{1 + \sqrt{2} \cot \beta}{1 + \sqrt{2} \cot \beta_0} \tag{4-16}$$

where β is the angle between the stress axis and the [001]-[$\bar{1}$11] boundary.

Crystals whose axes lie at orientations along the boundaries of the stereographic triangle represent a special situation because the critical resolved shear stress will be the same on more than one slip system. Therefore, plastic deformation will begin on more than one slip plane and they will initially deform by duplex slip. Figure 4-24 shows the number of operative slip systems in a cubic crystal at these orientations. Deformation by duplex slip results in a high degree of strain hardening because of interaction between dislocations on two intersecting slip systems. This is shown in Fig. 4-22 where Mg and Zn deform on a single

[1] Duplex slip results in high ductility in polycrystalline material for reasons discussed in Chap. 5.
[2] D. K. Bowen and J. W. Christian, *Philos. Mag.*, vol. 12, pp. 369–378, 1965.

slip system (because of the hcp geometry) while the stress-strain curves for Al and Cu are for crystals oriented for duplex slip.

4-10 DEFORMATION BY TWINNING

The second important mechanism by which metals deform is the process known as twinning.[1] Twinning results when a portion of the crystal takes up an orientation that is related to the orientation of the rest of the untwinned lattice in a definite, symmetrical way. The twinned portion of the crystal is a mirror image of the parent crystal. The plane of symmetry between the two portions is called the *twinning plane*. Figure 4-25 illustrates the classical atomic picture of twinning. Figure 4-25*a* represents a section perpendicular to the surface in a cubic lattice with a low-index plane parallel to the paper and oriented at an angle to the plane of polish. The twinning plane is perpendicular to the paper. If a shear stress is applied, the crystal will twin about the twinning plane (Fig. 4-25*b*). The region to the right of the twinning plane is undeformed. To the left of this plane, the planes of atoms have sheared in such a way as to make the lattice a mirror image across the twin plane. In a simple lattice such as this, each atom in the twinned region moves by a homogeneous shear a distance proportional to its distance from the twin plane. In Fig. 4-25*b*, open circles represent atoms which have not moved, dashed circles indicate the original positions in the lattice of atoms which change position, and solid circles are the final positions of these atoms in the twinned region. Note that the twin is visible on the polished surface because of the change in elevation produced by the deformation and because of the difference in crystallographic orientation between the deformed and undeformed regions. If the surface were polished down to section *AA*, the difference in elevation would be eliminated but the twin would still be visible after etching because it possesses a different orientation from the untwinned region.

It should be noted that twinning differs from slip in several specific respects. In slip, the orientation of the crystal above and below the slip plane is the same after deformation as before, while twinning results in an orientation difference across the twin plane. Slip is usually considered to occur in discrete multiples of the atomic spacing, while in twinning the atom movements are much less than an atomic distance. Slip occurs on relatively widely spread planes, but in the twinned region of a crystal every atomic plane is involved in the deformation.

Twins may be produced by mechanical deformation or as the result of annealing following plastic deformation. The first type are known as *mechanical*

[1] For complete reviews on this subject see E. O. Hall, "Twinning and Diffusionless Transformations in Metals," Butterworth & Co. (Publishers), Ltd., London, 1954, R. W. Cahn, *Adv. Phys.*, vol. 3, pp. 363–445, 1954, and S. Mahagan and D. F. Williams, *Int. Metall. Rev.*, vol. 18, pp. 43–61, June 1973. A good presentation of the crystallography of twinning is given by R. E. Reed-Hill, "Physical Metallurgy Principles," chap. 15, D. Van Nostrand Company, Inc., Princeton, N.J., 1964, and R. W. Hertzberg, "Deformation an Fracture Mechanics of Engineering Materials," 2d ed., chap. 4, John Wiley & Sons, New York, 1983.

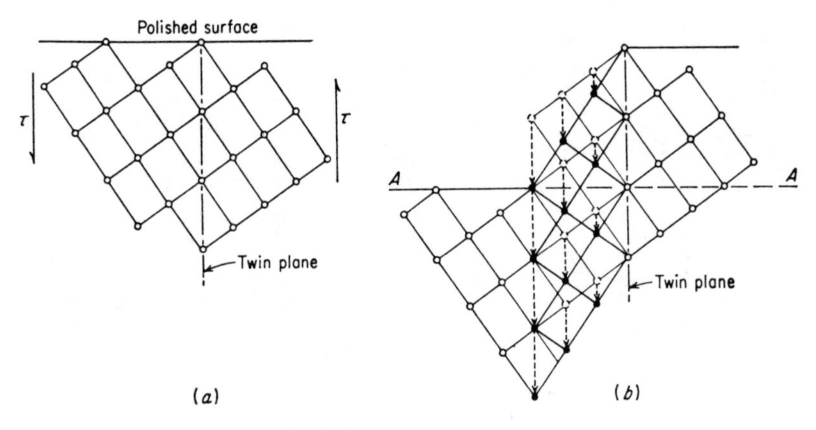

Figure 4-25 Classical picture of twinning.

twins; the latter are called *annealing twins*. Mechanical twins are produced in bcc or hcp metals under conditions of rapid rate of loading (shock loading) and decreased temperature. Face-centered cubic metals are not ordinarily considered to deform by mechanical twinning, although gold-silver alloys twin fairly readily when deformed at low temperature, and mechanical twins have been produced in copper by tensile deformation at 4°K and by shock loading. Twins can form in a time as short as a few microseconds, while for slip there is a delay time of several milliseconds before a slip band is formed. Under certain conditions, twins can be heard to form with a click or loud report [tin cry]. If twinning occurs during a tensile test, it produces serrations in the stress-strain curve.

Twinning occurs in a definite direction on a specific crystallographic plane for each crystal structure. Table 4-5 lists the common twin planes and twin directions. It is not known whether or not there is a critical resolved shear stress for twinning. However, twinning is not a dominant deformation mechanism in metals which possess many possible slip systems. Twinning generally occurs when the slip systems are restricted or when something increases the critical resolved shear stress so that the twinning stress is lower than the stress for slip. This explains the occurrence of twinning at low temperatures or high strain rates in bcc and fcc metals or in hcp meals at orientations which are unfavorable for basal slip.

Table 4-5 Twin planes and twin directions

Crystal structure	Typical examples	Twin plane	Twin direction
bcc	α-Fe, Ta	(112)	[111]
hcp	Zn, Cd, Mg, Ti	$(10\bar{1}2)$	$[\bar{1}011]$
fcc	Ag, Au, Cu	(111)	[112]

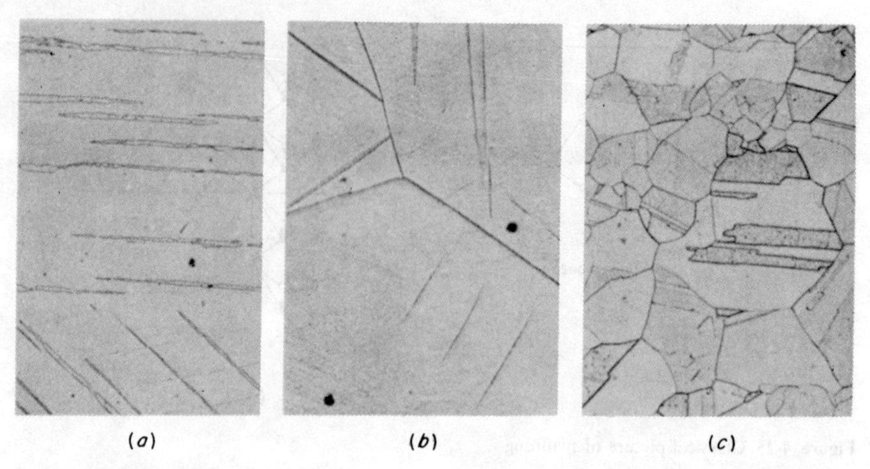

Figure 4-26 Microstructures of twins. (*a*) Neumann bands in iron; (*b*) mechanical twins produced in zinc by polishing; (*c*) annealing twins in gold-silver alloy.

The lattice strains needed to produce a twin configuration in a crystal are small, so that the amount of gross deformation that can be produced by twinning is small. For example,[1] the maximum extension which it is possible to produce in a zinc crystal when the entire crystal is converted into a twin on the {1012} plane is only 7.39 percent. The important role of twinning in plastic deformation comes not from the strain produced by the twinning process but from the fact that orientation changes resulting from twinning may place new slip systems in a favorable orientation with respect to the stress axis so that additional slip can take place. Thus, twinning is important in the overall deformation of metals with a low number of slip systems, such as the hcp meals. However, it should be understood that only a relatively small fraction of the total volume of a crystal is reoriented by twinning, and therefore hcp metals will, in general, possess less ductility than metals with a greater number of slip systems.

Figure 4-26 shows some metallographic features of twins in several different systems. Figure 4-26*a* is an example of mechanical twins in iron (Neumann bands). Note that the width of the twins can be readily resolved at rather low magnification. The boundaries of the twins etch at about the same rate as grain boundaries, indicating that they are rather high-energy boundaries. Figure 4.26*b* shows the broad, lens-shaped twins commonly found in hcp metals. Note that twins do not extend beyond a grain boundary. Figure 4-26*c* shows annealing twins in an fcc gold-silver alloy. Annealing twins are usually broader and with straighter sides than mechanical twins. The energy of annealing twin boundaries is about 5 percent of the average grain-boundary energy. Most fcc metals form annealing twins. Their presence in the microstructure is a good indication that the

[1] Barrett, *op. cit.*, p. 384.

metal has been given mechanical deformation prior to annealing, since it is likely that they grow from twin nuclei produced during deformation.

A process closely related to twinning is the formation of a *martensite* region by a diffusionless shear transformation. Although both processes produce a local region of new lattice orientation, the basic difference is that in a martensite plate the crystal structure is different from the parent crystal. The driving force for twinning is the applied shear stress, while in the martensite transformation, the driving force is the free energy difference between the parent crystal and the martensitic phase. This thermodynamic driving force may be assisted by the applied shear stress.

4-11 STACKING FAULTS

In an earlier section, it was shown that the atomic arrangement on the {111} plane of an fcc structure and the {0001} plane of an hcp structure could be obtained by the stacking of close-packed planes of spheres. For the fcc structure, the stacking sequence of the planes of atoms is given by *ABC ABC ABC*. For the hcp structure, the stacking sequence is given by *AB AB AB*.

Errors, or faults, in the stacking sequence can be produced in most metals by plastic deformation.[1] Slip on the {111} plane in an fcc lattice produces a deformation stacking fault by the process shown in Fig. 4-27*b*. Slip has occurred between an *A* and a *B* layer. The stacking sequence then becomes *ABC AC AB*. Comparison of this faulted stacking sequence (Fig. 4-27*b*) with the stacking sequence for an hcp structure without faults *CACA* (Fig. 4-27*d*) shows that the deformation stacking fault contains four layers of an hcp sequence. Therefore, the formation of a stacking fault in an fcc metal is equivalent to the formation of a thin hcp region. Another way in which a stacking fault could occur in an fcc metal is by the sequence[2] shown in Fig. 4-27*c*. The stacking sequence *ABC ACB CA* is called an extrinsic, or twin, stacking fault. The three layers *ACB* constitute the twin. Thus, stacking faults in fcc metals can also be considered as submicroscopic twins of nearly atomic thickness. The reason why mechanical twins of microscopically resolvable width are not formed readily when fcc metals are deformed is that the formation of stacking faults is so energetically favorable.

The differences in the deformation behavior of fcc metals are due to the differences in stacking-fault behavior. The creation of a region with hcp stacking *CACA* introduces a region with higher free energy than the fcc structure. A stacking fault in a fcc metal, when viewed from dislocation theory, is an *extended dislocation* consisting of a thin hcp region bounded by *partial dislocations* (Fig. 4-28). The nearly parallel partial dislocations tend to repel each other, but this is counterbalanced by the surface tension of the stacking fault pulling them to-

[1] Very precise x-ray diffraction measurements are needed to detect the presence of stacking faults. For example, see B. E. Warren and E. P. Warekois, *Acta Metall.* vol. 3 p. 473, 1955.

[2] C. N. J. Wagner, *Acta Metall.*, vol. 5, pp. 427–434, 1957.

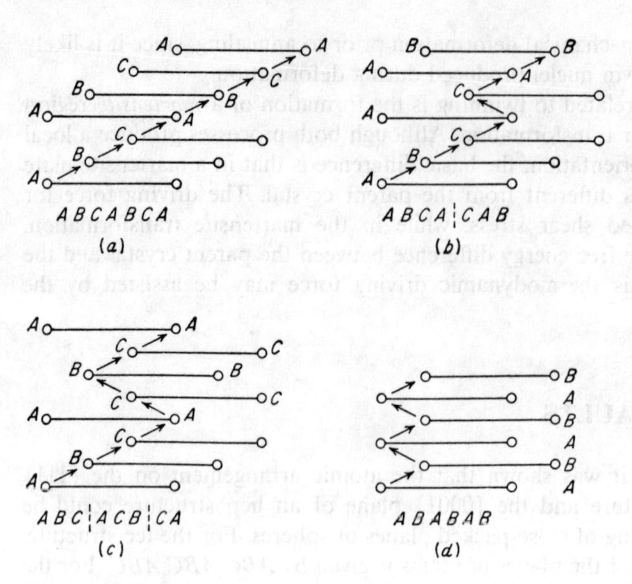

Figure 4-27 Faulted structures. (*a*) fcc packing; (*b*) deformation fault in fcc; (*c*) twin fault in fcc; (*d*) hcp packing.

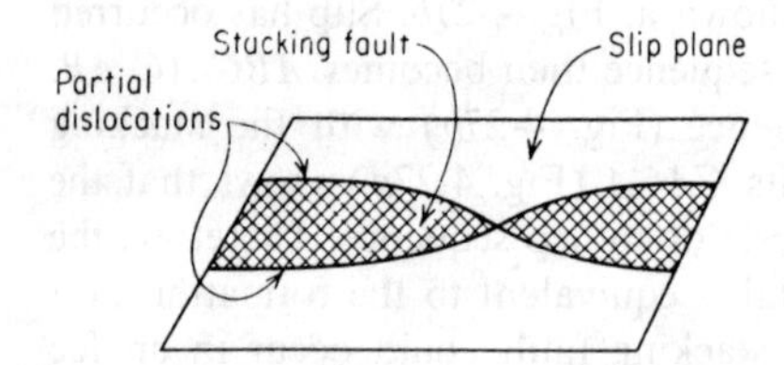

Figure 4-28 Schematic model of a stacking fault.

gether. The lower the stacking-fault energy the greater the separation between the partial dislocations and the wider the stacking fault.[1] Typical values for stacking-fault energy (SFE) are given in Table 4-6. The data for the stainless steels illustrate that SFE is very sensitive to chemical composition.

Stacking faults influence the plastic deformation in a number of ways. Metals with wide stacking faults (low SFE) strain-harden more rapidly, twin easily on annealing, and show a different temperature dependence of flow stress than metals with narrow stacking faults. Metals with high SFE have a deformation substructure of dislocation tangles and cells, while low-SFE metals show a deformation substructure of banded, linear arrays of dislocations. The behavior of stacking faults is considered in greater detail in Chap. 5.

[1] L. E. Murr, "Interfacial Phenomena in Metals and Alloys," Addison-Wesley, Reading, Mass., 1975.

Table 4-6 Typical values of stacking-fault energy

Metal	Stacking-fault energy, $mJ/m^2 = erg/cm^2$
Brass	< 10
303 stainless steel	8
304 stainless steel	20
310 stainless steel	45
Silver	~ 25
Gold	~ 50
Copper	~ 80
Nickel	~ 150
Aluminum	~ 200

4-12 DEFORMATION BANDS AND KINK BANDS

Inhomogeneous deformation of a crystal results in regions of different orientation called *deformation bands*. When slip occurs without restraint in a perfectly homogeneous fashion, the slip lines are removed by subsequent polishing of the surface. Deformation bands, however, can be observed even after repeated polishing and etching because they represent a region of different crystallographic orientation. In single crystals, deformation bands several millimeters wide may occur, while in polycrystalline specimens microscopic observation is needed to see them. The tendency for the formation of deformation bands is greater in polycrystalline specimens because the restraints imposed by the grain boundaries make it easy for orientation differences to arise in a grain during deformation. Deformation bands generally appear irregular in shape but are elongated in the direction of principal strain. The outline of the bands is generally indistinct and poorly defined, indicating a general fading out of the orientation difference. Deformation bands have been observed in both fcc and bcc metals, but not in hcp metals.

Consideration of the equation for critical resolved shear stress shows that it will be difficult to deform a hexagonal crystal when the basal plane is nearly parallel to the crystal axis. Orowan[1] found that if a cadmium crystal of this orientation were loaded in compression, it would deform by a localized region of the crystal suddenly snapping into a tilted position with a sudden shortening of the crystal. The buckling, or *kinking*, behavior is illustrated in Fig. 4-29. The horizontal lines represent basal planes, and the planes designated p are the kink planes at which the orientation suddenly changes. Distortion of the crystal is essentially confined to the kink band. Further study of kink bands by Hess and Barrett[2] showed that they can be considered to be a simple type of deformation

[1] E. Orowan, *Nature*, vol. 149, p. 643, 1942.

[2] J. A. Hess and C. S. Barrett, *Trans. Metall. Soc. AIME*, vol. 185, p. 599–1949.

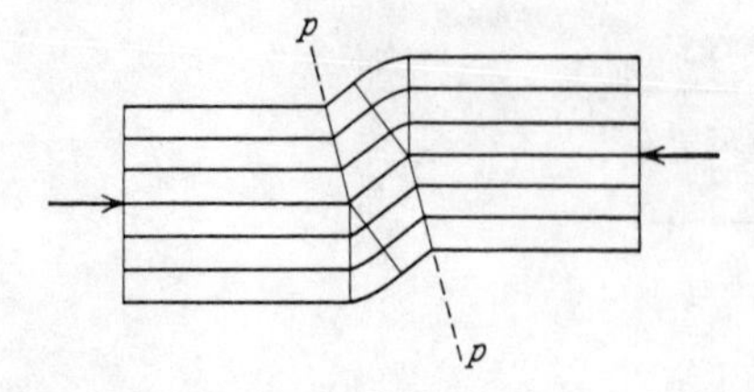

Figure 4-29 Kink band.

band. Kink bands have also been observed in zinc crystals tested in tension, where a nonuniform distribution of slip can produce a bending moment which can cause kink formation.

4-13 MICROSTRAIN BEHAVIOR

The ordinary engineering yield strength determined at an offset strain of 0.002 represents a stress at which a very great number of dislocations have already moved. From a dislocation viewpoint this is a very macroscopic measurement of initial yielding. Therefore, considerable experimentation has gone into developing reliable techniques for work in the preyield microstrain region at plastic strains less than 0.001. Mechanical property studies on single crystals at plastic strains of the order of 10^{-5} are important in providing data for studying basic dislocation behavior. The key to these studies[1] is a technique for loading with near-perfect alignment and a transducer, such as a linear variable differential transformer (LVDT), which provides reproducible strain measurements at a level of 10^{-6}.

The shape of the stress-strain curve on repeated loading to higher stresses in the microplasticity regime is shown in Fig. 4-30. At very low stresses a load-unload cycle produces a single straight line with the material exhibiting ideal elastic behavior (Fig. 4-30*a*). Deviation from ideal elastic behavior occurs at some higher stress τ_E, the *true elastic limit* (Fig. 4-30*b*). Above τ_E the load-unload cycle produces a closed mechanical hysteresis loop which closely approximates a parallelogram. The initial high elastic modulus corresponds well with dynamic measurements of modulus by high-frequency methods. The lower modulus corresponds to the relaxed modulus characteristic of usual static tension tests. Stress cycling above some value τ_A causes permanent plastic deformation since the hysteresis loop does not close on unloading (Fig. 4-30*d*). This stress is generally known as the anelastic limit τ_A. Finally, at much higher stresses gross yielding occurs. Typical values for molybdenum are: $\tau_E = 0.5$ MPa, $\tau_A = 5$ MPa, τ_0 (macroyield) = 50 MPa.

[1] A. Lawley and J. D. Meakin, "Microplasticity," C. J. McMahon (ed.), John Wiley & Sons, Inc., New York, 1968.

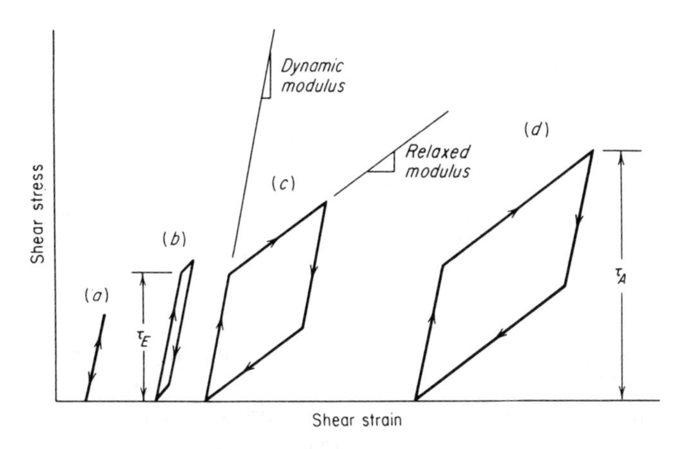

Figure 4-30 Types of stress-strain curves found in microplasticity regime.

4-14 STRAIN HARDENING OF SINGLE CRYSTALS

One of the chief characteristics of the plastic deformation of metals is the fact that the shear stress required to produce slip continuously increases with increasing shear strain. The increase in the stress required to cause slip because of previous plastic deformation is known as *strain hardening*, or *work hardening*. An increase in flow stress of over 100 percent from strain hardening is not unusual in single crystals of ductile metals.

Strain hardening is caused by dislocations interacting with each other and with barriers which impede their motion through the crystal lattice. Hardening due to dislocation interaction is a complicated problem because it involves large groups of dislocations, and it is difficult to specify group behavior in a simple mathematical way. It is known that the number of dislocations in a crystal increases with strain over the number present in the annealed crystal. The dislocation density of a good annealed crystal is 10^5 to 10^6 cm^{-2}, while the observed dislocation density in cold-worked metal is 10^{10} to 10^{12} cm^{-2}. Dislocation multiplication can arise from condensation of vacancies, by regeneration under applied stress from existing dislocations by either the Frank-Read mechanism or a multiple cross-slip mechanism (see Chap. 5) or by emission of dislocations from a high-angle grain boundary.

One of the earliest dislocation concepts to explain strain hardening was the idea that dislocations pile up on slip planes at barriers in the crystal. The pile-ups produce a *back stress* which opposes the applied stress on the slip plane. The existence of a back stress was demonstrated experimentally by shear tests on zinc single crystals.[1] Zinc crystals are ideal for crystal-plasticity experiments because

[1] E. H. Edwards, J. Washburn, and E. R. Parker, *Trans. AIME*, vol. 197, p. 1525, 1953.

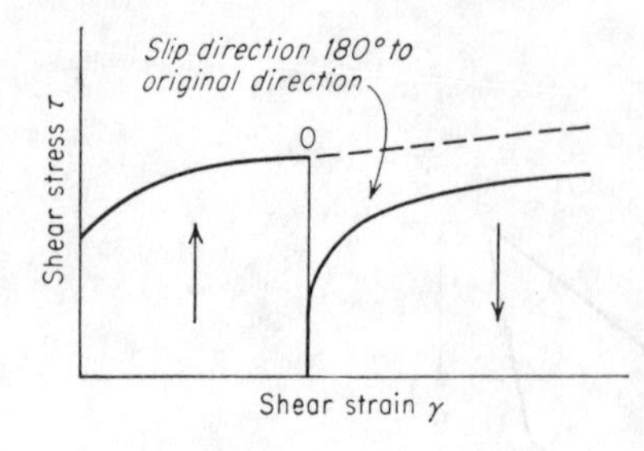

Figure 4-31 Effect of complete reversal of slip direction on stress-strain curve. *(From E. H. Edwards, J. Washburn, and E. R. Parker Trans. AIME, vol. 197, p. 1526, 1953.)*

they slip only on the basal plane, and hence complications due to duplex slip are easily avoided. In Fig. 4-31, the crystal is strained to point *O*, unloaded, and then reloaded in the direction opposite to the original slip direction. Note that on reloading the crystal yields at a lower shear stress than when it was first loaded. This is because the back stress developed as a result of dislocations piling up at barriers during the first loading cycle is aiding dislocation movement when the direction of slip is reversed. Furthermore, when the slip direction is reversed, dislocations of opposite sign could be created at the same sources that produced the dislocations responsible for strain in the first slip direction. Since dislocations of opposite sign attract and annihilate each other, the net effect would be a further softening of the lattice. This explains the fact that the flow curve in the reverse direction lies below the curve for continued flow in the original direction. The lowering of the yield stress when deformation in one direction is followed by deformation in the opposite direction is called the *Bauschinger effect*.[1] While all metals exhibit a Bauschinger effect, it may not always be of the magnitude shown here for zinc crystals. Moreover, the flow curve after reversal of direction does not fall below the original flow curve for all metals.

The existence of back stress and its importance to strain hardening in metals having been established, the next step is to identify the barriers to dislocation motion in single crystals. Microscopic precipitate particles and foreign atoms can serve as barriers, but other barriers which are effective in pure single crystals must be found. Such barriers arise from the fact that glide dislocations on intersecting slip planes may combine with one another to produce a new dislocation that is not in a slip direction. The dislocation of low mobility that is produced by a dislocation reaction is called a *sessile dislocation*. Since sessile dislocations do not lie on the slip plane of low shear stress, they act as a barrier to dislocation motion until the stress is increased to a high enough level to break down the barrier. The most important dislocation reaction, which leads to the formation of sessile dislocations, is the formation of Lomer-Cottrell barriers in fcc metals by slip on intersecting {111} planes.

Another mechanism of strain hardening, in addition to that due to the back stress resulting from dislocation pile-ups at barriers, is believed to occur when dislocations moving in the slip plane cut through other dislocations intersecting

[1] J. Bauschinger, *Zivilingur.*, vol. 27, pp. 289–347, 1881.

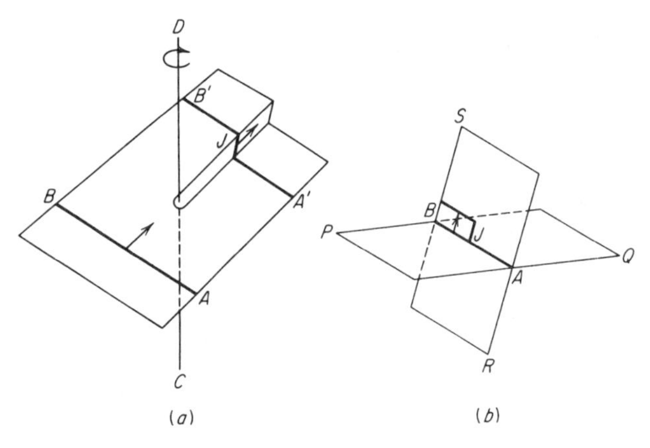

Figure 4-32 Formation of a jog J (a) by a dislocation cutting through a screw dislocation as it glides from AB to $A'B'$; (b) by part of a screw dislocation line AB cross slipping from the primary slip plane PQ into the plane RS. *(From A. H. Cottrell, "The Mechanical Properties of Matter," John Wiley & Sons, Inc., New York, 1964. By permission of the publishers.)*

the active slip plane. The dislocations threading through the active slip plane are often called a *dislocation forest*, and this strain-hardening process is referred to as the *intersection* of a forest of dislocations.

Figure 4-32a shows that dislocation intersection results in a small step or jog in the dislocation line. Jogs on a dislocation restrict its motion so that they contribute to strain hardening. Jogs are also formed by a screw dislocation *cross slipping* from the primary slip plane to another plane which contains the common slip direction (Fig. 4-32b).

The phenomenon of *cross slip* is restricted to screw dislocations. Since the line of a screw dislocation and its Burgers vector are parallel, this does not define a specific plane as with an edge dislocation (where **b** is perpendicular to the dislocation line). To a screw dislocation, all directions around its axis look the same, and it can glide on any plane as long as it moves parallel to its original orientation. The slip plane of a screw dislocation can be any plane containing the dislocation, and it can cross slip from one plane to another just so long as both planes contain a common slip direction. This permits screw dislocations (or screw components of mixed dislocations) to detour around obstacles and barriers. If cross slip of screw dislocations could not occur, dislocation motion would be impeded early in the deformation process and the rate of strain hardening (slope of stress-strain curve) would be very high and fracture would occur at low values of strain. The formation of jogs in screw dislocations impedes their motion and may even lead to the formation of vacancies and interstitials if the jogs are forced to move nonconservatively. Jogs in edge dislocations do not impede their motion. All these processes require an increased expenditure of energy, and therefore they contribute to hardening.

Strain hardening due to a dislocation cutting process arises from short-range forces occurring over distances less than 5 to 10 interatomic distances. This

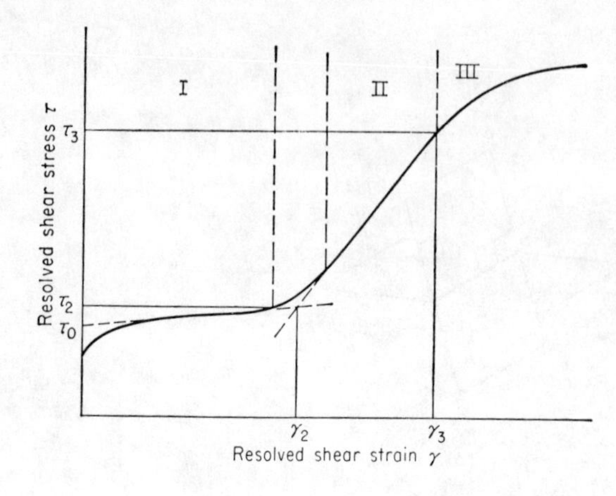

Figure 4-33 Generalized flow curve for fcc single crystals.

hardening can be overcome at finite temperatures with the help of thermal fluctuations, and therefore it is temperature- and strain-rate-dependent. On the other hand, strain hardening arising from dislocation pile-up at barriers occurs over longer distances, and therefore it is relatively independent of temperature and strain rate. Accordingly, data on the temperature and strain-rate dependence of strain hardening can be used[1] to determine the relative contribution of the two mechanisms.

When the stress-strain curves for single crystals are plotted as resolved shear stress vs. shear strain, certain generalizations can be made for all fcc metals. Following the notation proposed by Seeger,[2] the flow curve for pure-metal single crystals can be divided into three stages (Fig. 4-33).

Stage I, the region of *easy glide*, is a stage in which the crystal undergoes little strain hardening. During easy glide, the dislocations are able to move over relatively large distances without encountering barriers. The low strain hardening produced during this stage implies that most of the dislocations escape from the crystal at the surface. During easy glide, slip always occurs on only one slip system. For this reason, stage I slip is sometimes called *laminar flow*.

Stage II is a nearly linear part of the flow curve where strain hardening increases rapidly. In this stage, slip occurs on more than one set of planes. The length of the active slip lines decreases with increasing strain, which is consistent with the formation of a greater number of Lomer-Cottrell barriers with increasing strain. During stage II, the ratio of the strain-hardening coefficient (the slope of the curve) to the shear modulus is nearly independent of stress and temperature, and approximately independent of crystal orientation and purity. The fact that the slope of the flow curve in sage II is nearly independent of temperature agrees

[1] Z. S. Basinski, *Philos. Mag.*, vol. 4, ser. 8, pp. 393–432, 1959. For an extensive review see H. Conrad, *J. Met.*, pp. 582–588, July 1964.

[2] A. Seeger, in "Dislocations and Mechanical Properties of Crystals," John Wiley & Sons, Inc., New York, 1957.

with the theory that assumes the chief strain-hardening mechanism to be piled-up groups of dislocations.

As a result of slip on several slip systems, lattice irregularities are formed. Dislocation tangles begin to develop and these eventually result in the formation of a dislocation cell structure consisting of regions almost free of dislocations surrounded by material of high dislocation density (about five times the average dislocation density). Although the heterogeneity of dislocation distribution makes precise measurements difficult, measurements over a wide range of systems show that the average dislocation density in stage II correlates with resolved shear stress according to

$$\tau = \tau_0 + \alpha G b \rho^{1/2} \tag{4-17}$$

where τ_0 is the shear stress needed to move a dislocation in the absence of other dislocations and α is a numerical constant which varies from 0.3 to 0.6 for different fcc and bcc metals.

Stage III is a region of decreasing rate of strain hardening. The processes occurring during this stage are often called *dynamical recovery*. In this region of the flow curve, the stresses are high enough so that dislocations can take part in processes that are suppressed at lower stresses. Cross slip is believed to be the main process by which dislocations, piled up at obstacles during stage II, can escape and reduce the internal-strain field. The stress at which stage III begins, τ_3, is strongly temperature-dependent. Also, the flow stress of a crystal strained into stage III is more temperature-dependent than if it had been strained only into stage II. This temperature dependence suggests that the intersection of forests of dislocations is the chief strain-hardening mechanism in stage III.

The curve shown in Fig. 4-33 represents a general behavior for fcc metals. Certain deviations from a three-stage flow curve have been observed. For example, metals with a high stacking-fault energy, like aluminum, usually show only a very small stage II region at room temperature because they can deform so easily by cross slip. The shape and magnitude of a single-crystal flow curve, particularly during the early stages, depends upon the purity of the metal, the orientation of the crystal, the temperature at which it is tested, and the rate at which it is

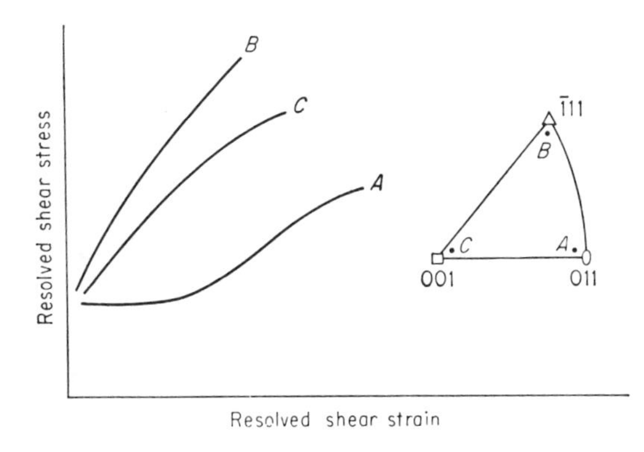

Figure 4-34 Effect of specimen orientation on the shape of the flow curve for fcc single crystals.

strained. The easy-glide region is much more prominent in hcp crystals than in fcc metals. A region of easy glide in the flow curve is favored by slip on a single system, high purity, low temperature, absence of surface oxide films, an orientation favorable for simple slip, and a method of testing which minimizes extraneous bending stresses. Figure 4-34 shows that crystal orientation can have a very strong effect on the flow curve of fcc single crystals. When the tensile axis is parallel to a ⟨011⟩ direction, one slip system is carrying appreciably more shear stress than any other and the flow curve shows a relatively large region of easy glide. When the tensile axis is close to a ⟨100⟩ or ⟨111⟩ direction, the stress on several slip systems is not very different and the flow curves show rapid rates of strain hardening.

The value of the resolved shear stress at a given shear strain decreases with increasing temperature. With increasing temperature both stage I and stage II decrease in extent until at high temperature the stress-strain curve shows entirely parabolic stage III behavior.

BIBLIOGRAPHY

Barrett, C. S., and T. B. Massalski, "Structure of Metals," 3rd ed., McGraw-Hill Book Company, New York, 1966.

Clarebrough, L. M. and M. E. Hargreaves: Work Hardening of Metals, in "Progress in Metal Physics," vol. 8, Pergamon Press, Ltd., London, 1959.

Honeycombe, R. W. K.: "The Plastic Deformation of Metals," 2d ed., Edward Arnold, Baltimore, Md., 1984.

Maddin, R. and N. K. Chen: Geometric Aspects of the Plastic Deformation of Metal Single Crystals, in "Progress in Metal Physics," vol. 5, Pergamon Press, Ltd., London, 1954.

Nabarro, F. R. N., Z. S. Basinski, and D. B. Holt: Plasticity of Pure Single Crystals, *Adv. Phys.*, vol. 13, pp. 193–323, 1964.

Reid, C. N.: "Deformation Geometry for Materials Scientists," Pergamon Press, New York, 1973.

Schmid, E. and W. Boas: "Plasticity of Crystals," English translation, F. A. Hughes & Co., London, 1950.

CHAPTER

FIVE

DISLOCATION THEORY

5-1 INTRODUCTION

A dislocation is the linear lattice defect that is responsible for nearly all aspects of the plastic deformation of metals. This concept was introduced in Chap. 4, where the geometry of edge and screw dislocations was presented for the case of a simple cubic lattice. It was shown that the existence of a dislocationlike defect is necessary to explain the low values of yield stress observed in real crystals. A general picture of the factors which impede dislocation motion and lead to strain hardening was also given.

This chapter is intended to present a more complete treatment of dislocation theory. Techniques for observing dislocations in metals are discussed. The effect on dislocation behavior of considering real fcc, bcc, or hcp crystal structures are considered. The origin of dislocations and the mechanisms for their multiplication are discussed. Interaction of dislocations with other dislocations, vacancies, and foreign atoms is discussed in some detail. The object of this chapter is the presentation of the basic geometric and mathematical relationships which describe dislocation behavior. These relationships will be used to explain mechanical behavior and strengthening mechanisms in subsequent chapters of this book.

5-2 OBSERVATION OF DISLOCATIONS

The concept of the dislocation was proposed independently by Taylor, Orowan, and Polanyi[1] in 1934, but the idea lay relatively undeveloped until the end of World War II. There followed a period of approximately 10 years in which the

[1] G. I. Taylor, *Proc. R. Soc. London*, vol. 145A, p. 362, 1934; E. Orowan, *Z. Phys.*, vol. 89, pp. 605, 614, 634, 1934; M. Polanyi, *Z. Phys.*, vol. 89, p. 660, 1934.

theory of dislocation behavior was developed extensively and applied to practically every aspect of the plastic deformation of metals. Because there were no really reliable methods for detecting dislocations in real materials, it was necessary to build much of this theory on the basis of indirect observations of dislocation behavior. However, in the last 20 years extensive research has developed a variety of techniques for observing and studying dislocations in real materials. These studies leave no doubt that dislocations exist, and more importantly, they have provided experimental verification for most of the theoretical concepts of dislocation theory.

The resolving power of the best electron microscope would have to be improved by a factor of 5 to 10 in order to observe directly the distortion of the individual lattice planes around a dislocation in a metal crystal.[1] Practically all the experimental techniques for detecting dislocations utilize the strain field around a dislocation to increase its effective size. These experimental techniques can be roughly classified into two categories, those involving chemical reactions with the dislocation, and those utilizing the physical changes at the site of a dislocation.[2] Chemical methods include etch-pit techniques and precipitation techniques. Methods based on the physical structure at a dislocation site include transmission electron microscopy of thin films and x-ray diffraction techniques.

The simplest chemical technique is the use of an etchant which forms a pit at the point where a dislocation intersects the surface. Etch pits are formed at dislocation sites because the strain field surrounding the dislocation causes preferential chemical attack. A great deal of information about dislocation behavior in the ionic crystal LiF has been obtained in this way by Gilman and Johnson.[3] Important information about dislocations in metals has also been obtained with etch-pit techniques. Figure 5-1 shows the excellent resolution obtainable from etch-pit studies on alpha brass.[4] Pits only 500 Å apart have been resolved. In the region of heavy slip shown in this electron micrograph the dislocation density is 10^{10} cm^{-2}.

In metals, etch-pit formation at dislocations appears to be dependent on purity.[5] Because of solute segregation to the dislocation, the region around the

[1] It has been possible by means of an electron microscope to observe this lattice distortion in an organic crystal of platinum phthalocyanine, which has a very large lattice spacing (12 A) [J. W. Menter, *Proc. R. Soc. London Ser. A*; vol. 236A, p. 119, 1956]. An indication of the lattice distortion at a dislocation in metals has been obtained by making use of the magnification resulting from moiré patterns produced by electron transmission through two thin overlapping crystals with slightly different orientations or lattice spacings. See G. A. Bassett, J. W. Menter, and D. W. Pashley, *Proc. R. Soc. London Ser. A*, vol. 246A, p. 345, 1958.

[2] Several excellent reviews of experimental techniques have been published. See P. B. Hirsch., *Metall. Rev.*, vol. 4, no, 14, pp. 101–140, 1959; J. Nutting, Seeing Dislocations, in "The Structure of Metals," Institution of Metallurgists, Interscience Publishers, Inc., New York, 1959; S. Amelinckx, The Direct Observation of Dislocations, *Solid State Phys. Suppl.* 6, 1964.

[3] J. J. Gilman and W. G. Johnston, in "Dislocations and Mechanical Properties of Crystals," John Wiley & Sons, Inc., New York, 1957.

[4] J. D. Meakin and H. G. F. Wilsdorf, *Trans. Metall. Soc. AIME*, vol. 218, pp. 737–745, 1960.

[5] A summary of etch-pit techniques in metals is given by L. C. Lowell, F. L. Vogel, and J. H. Wernick, *Met. Prog.*, vol. 75, pp. 96–96D, 1959.

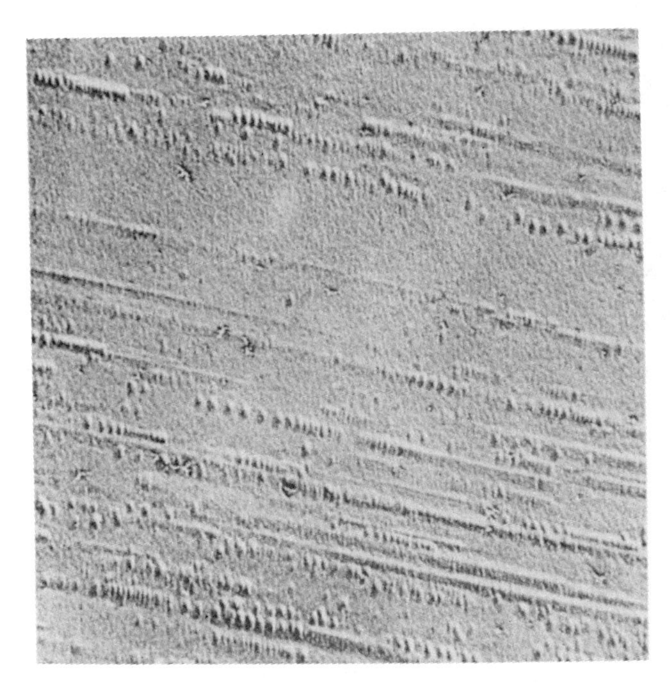

Figure 5-1 Etch pits on slip bands in alpha brass crystals (5,000 ×). *(From J. D. Meakin and H. G. F. Wilsdorf, Trans. Metall. Soc. AIME, vol. 218, p. 740, 1960.)*

dislocation becomes anodic to the surrounding metal, and consequently preferential etching occurs at the dislocation. Figure 6-4 shows an etch-pit structure in an iron-silicon alloy which was made visible by diffusion of carbon atoms to the dislocations. Etch-pit techniques are useful because they can be used with bulk samples.

In certain systems it may be possible to distinguish between edge and screw dislocations and between positive and negative edge dislocations. The technique can also be used to study the movement of dislocations. However, care must be taken to ensure that pits are formed only at dislocation sites and that all dislocations intersecting the surface are revealed. Because the etch pits have a finite size and are difficult to resolve when they overlap, the etch-pit technique is limited generally to crystals with a low dislocation density of about 10^6 cm^{-2}.

A similar method of detecting dislocations is to form a visible precipitate along the dislocation lines. Usually a small amount of impurity is added to form the precipitate after suitable heat treatment. The procedure is called "decoration" of dislocations. This technique was first used by Hedges and Mitchell[1] to decorate dislocations in AgBr with photolytic silver. It has since been used with many

[1] J. M. Hedges and J. W. Mitchell, *Philos. Mag.*, vol. 44, p. 223, 1953.

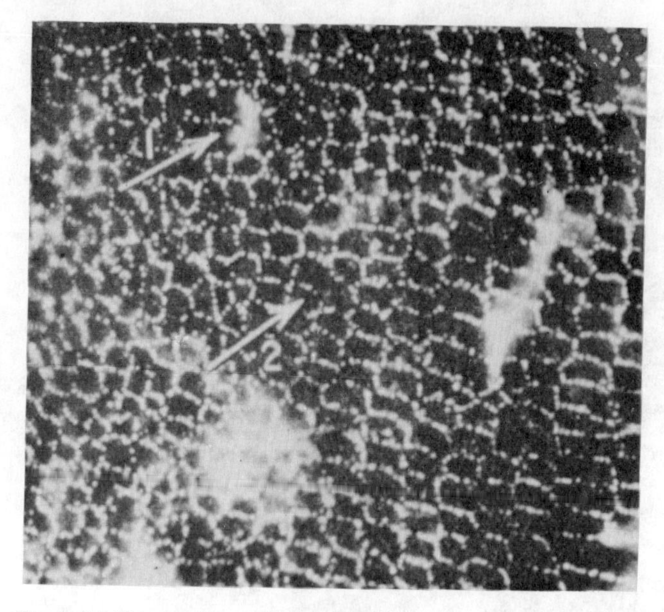

Figure 5-2 Hexagonal network of dislocations in NaCl detected by a decoration technique. *(From S. Amelinckx, in: "Dislocations and Mechanical Properties of Crystals," John Wiley & Sons, Inc., New York, 1957. By permission of the publishers.)*

other ionic crystals,[1] such as AgCl, NaCl, KCl, and CaF_2. With these optically transparent crystals this technique has the advantage that it shows the internal structure of the dislocation lines. Figure 5-2 shows a hexagonal network of dislocations in a NaCl crystal which was made visible by decoration. Although dislocation decoration has not been used extensively with metals, some work has been done along these lines with the Al-Cu precipitation-hardening system and with silicon crystals.

The most powerful method available today for the detection of dislocations in metals is transmission electron microscopy of thin foils.[2] Thin sheet, less than 1 mm thick, is thinned after deformation by electropolishing to a thickness of about 1,000 Å. At this thickness the specimen is transparent to electrons in the electron microscope. Although the crystal lattice cannot be resolved, individual dislocation lines can be observed because the intensity of the diffracted electron beam is altered by the strain field of the dislocation. The width of the diffraction image of a dislocation in a thin foil is about 100 Å, so that this technique can be used at dislocation densities up to about 10^{11} cm^{-2}. By means of this technique it has

[1] S. Amelinckx, in "Dislocations and Mechanical Properties of Crystals," John Wiley & Sons, Inc., New York, 1957.

[2] G. Thomas and M. J. Goringe, "Transmission Electron Microscopy," Wiley-Interscience, New York, 1979.

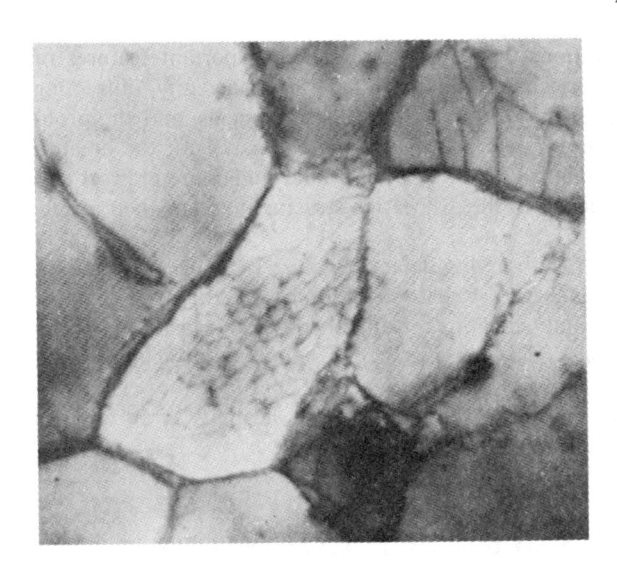

Figure 5-3 Dislocation network in cold-worked aluminum (32,500 ×). *(From P. B. Hirsch, R. W. Horne, and M. J. Whelan, Philos. Mag., ser. 8, vol. 1, p. 677, 1956.)*

been possible to observe dislocation networks (Fig. 5-3), stacking faults, dislocation pile-up at grain boundaries (Fig. 6-1), Lomer-Cottrell barriers, and many other structural features of dislocation theory. Dislocation movement has been observed by generating thermal stresses in the thin film with the electron beam or by using special straining fixtures in the electron microscope.

Transmission electron microscopy is the most powerful and universally applicable technique available for studying dislocations in solids. By application of the kinematic[1] and dynamic[2] theories of electron diffraction, it is possible to make detailed analysis of the images to determine the number of dislocations, their Burgers vectors, and the slip planes on which they lie. However, this technique is not without disadvantages. Since only a miniscule volume of material is examined with thin films, great care must be exerted to obtain a representative sample. It is possible to alter the defect structure during sectioning and polishing to a thin film, and dislocation structures may relax in a very thin foil. The greatest defect of transmission electron microscopy is that it is not very effective in detecting long-range stresses, nor does it give very much information about slip-line lengths or surface step heights.

The dislocation structure of a crystal can be detected by x-ray microscopy. The most common techniques are the Berg-Barrett reflection method[3] and the Lang topography method.[4] Unfortunately, the limiting resolution of these tech-

[1] P. B. Hirsch, A. Howie, and M. J. Whelan, *Philos. Trans. R. Soc. London Ser. A*, vol. A252, pp. 499–529, 1960.

[2] A. Howie and M. J. Whelan, *Proc. R. Soc. London Ser. A*, vol. A263, pp. 217–237, 1961; *Proc. R. Soc. London Ser. A*, vol. A267, pp. 206–230, 1962.

[3] C. S. Barrett, *Trans. Metall. Soc. AIME*, vol. 161, pp. 15–64, 1945.

[4] A. R. Lang, *J. Appl. Phys.*, vol. 30, pp. 1748–1755, 1959.

niques is about 10^5 dislocations per square centimeter. Important features of plastically deformed metal such as the mean size of coherently diffracting domains (crystallite size), the microstrains within these domains, and the probability of lattice faulting can be determined by detailed analysis of the breadth and shape of x-ray diffraction peaks.[1] X-ray peak profile studies are indirect and statistical, and depend on models of the dislocation structure to relate their results to dislocation behavior.

The ultimate technique for observing defect structure is the field-ion microscope.[2] Since its resolution is 2 to 3 Å, individual atoms may be distinguished. Thus, it is the only experimental technique for directly observing vacancies. At its present state of development, field-ion microscopy is limited to metals with a strong binding force, such as W, Mo, or Pt. Since the specimen must be a fine wire bent into a sharp tip, the method is limited in applicability and examines only a very small area of the surface.

5-3 BURGERS VECTOR AND THE DISLOCATION LOOP

The Burgers vector **b** is the vector which defines the magnitude and direction of slip. Therefore, it is the most characteristic feature of a dislocation. It has already been shown that for a *pure* edge dislocation the Burgers vector is perpendicular to the dislocation line, while for a *pure* screw dislocation the Burgers vector is parallel to the dislocation line. The macroscopic slip produced by the motion of an edge dislocation is shown in Fig. 5-4*a* and by a screw dislocation in Fig. 5-4*b*. Both the shear stress and final deformation are identical for both situations, but for an edge dislocation the dislocation line moves parallel to the slip direction while the screw dislocation moves at right angles to it. The geometric properties of dislocations are summarized in Table 5-1.

Actually, dislocations in real crystals are rarely straight lines and rarely lie in a single plane. In general, a dislocation will be partly edge and partly screw in character. As shown by Figs. 5-2 and 5-3, dislocations will ordinarily take the form of curves or loops, which in three dimensions form an interlocking dislocation network. In considering a dislocation loop in a slip plane any small segment of the dislocation line can be resolved into edge and screw components. For example, in Fig. 5-5, the dislocation loop is pure screw at point *A* and pure edge at point *B*, while along most of its length it has mixed edge and screw components. Note, however, that the Burgers vector is the same along the entire dislocation loop. If this were not so, part of the crystal above the slipped region would have to slip by a different amount relative to another part of the crystal and this would mean that another dislocation line would run across the slipped region.

[1] B. E. Warren, X-ray Studies of Deformed Metals, *Prog. Met. Phys.*, vol. 8, pp. 147–202, 1958.

[2] E. W. Muller, "Direct Observation of Imperfections in Crystals," Interscience Publishers, Inc., New York, 1962.

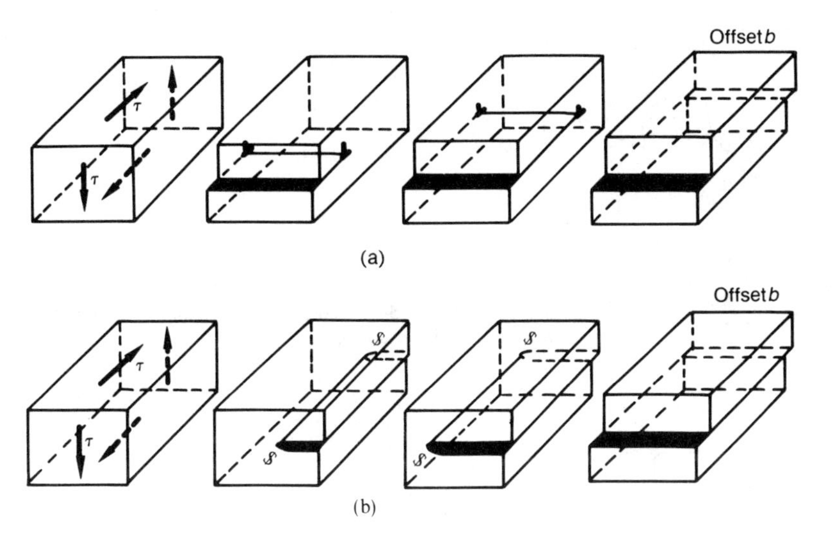

Figure 5-4 (*a*) Macroscopic deformation of a cube produced by glide of an edge dislocation. (*b*) Macroscopic deformation of a cube produced by glide of a screw dislocation. Note that the end result is identical for both situations.

Table 5-1 Geometric properties of dislocations

Dislocation property	Type of dislocation	
	Edge	Screw
Relationship between dislocation line and b	perpendicular	parallel
Slip direction	parallel to b	parallel to b
Direction of dislocation line movement relative to b (slip direction)	parallel	perpendicular
Process by which dislocation may leave slip plane	climb	cross-slip

A convenient way of defining the Burgers vector of a dislocation is with a *Burgers circuit*. Consider the positive edge dislocation shown in Fig. 5-6*a*. If we start at a lattice point and imagine a clockwise path traced from atom to atom an equal distance in each direction, we find that at the finish of the path the circuit does not close. The closure failure from finish to start is the Burgers vector **b** of the dislocation. (If we had made the Burgers circuit around the dislocation in the anticlockwise direction, the direction of the Burgers vector would have been in the opposite sense.) Moreover, if we traverse a Burgers circuit about the screw dislocation shown in Fig. 5-6*b*, we would find the closure error pointing out of the front face of the crystal. This is a right-handed screw dislocation since in

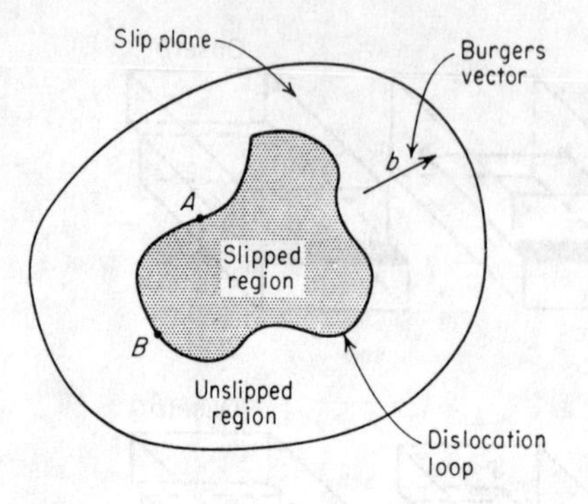

Figure 5-5 Dislocation loop lying in a slip plane (schematic).

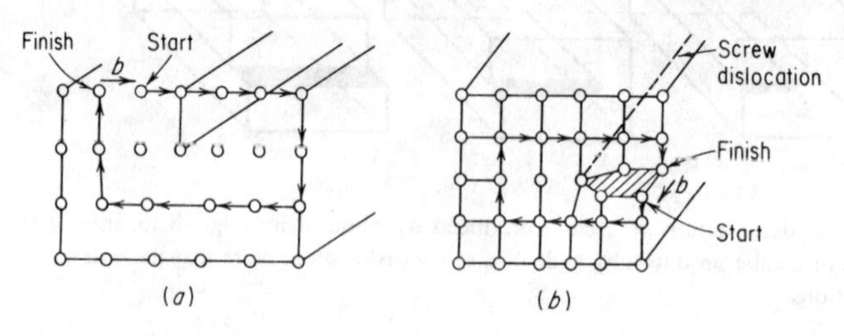

Figure 5-6 Burgers circuits. (*a*) Around positive edge dislocation; (*b*) around a right-handed screw dislocation.

traversing the circuit around the dislocation line, we advance the helix one atomic plane into the crystal.

The process of *cross slip* illustrated in Fig. 5-7, will serve as an example of dislocation loops. In Fig. 5-7*a* a small loop of dislocation line with $\mathbf{b} = a_0/2[\bar{1}01]$ is moving on a (111) plane in an fcc crystal. The dislocation loop is pure positive edge at w and pure negative edge at y. At x the dislocation is a right-handed

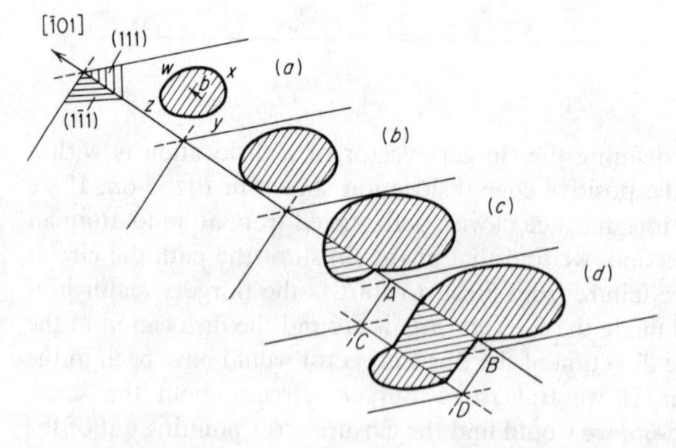

Figure 5-7 Cross slip in a face-centered cubic crystal. *(From D. Hull, "Introduction to Dislocations," p. 56, Pergamon Press, New York, 1965. By permission of the publishers.)*

screw while at z the dislocation loop is a pure left-handed screw dislocation. At some stage (Fig. 5-7*b*), the shear stress causing expansion of the loop tends to move the dislocation on the intersecting $(1\bar{1}1)$ plane. Since the dislocation is pure screw at z, it is free to move on this plane. In Fig. 5-7*c* the loop has expanded on the second plane, while in Fig. 5-7*d* *double cross slip* has taken place as the loop glides back onto the original (111) plane. Note that during the glide of the dislocation on the cross-slip plane only the screw component of the loop has moved.

Because a dislocation represents the boundary between the slipped and unslipped region of a crystal, topographic considerations require that it either must be a closed loop or else must end at the free surface of a crystal or at a grain boundary. In general, a dislocation line cannot end inside of a crystal. The exception is at a *node*, where three or four dislocation lines meet. At a node two dislocations with Burgers vectors $\mathbf{b}_1$ and $\mathbf{b}_2$ combine to produce a resultant dislocation $\mathbf{b}_3$. The vector $\mathbf{b}_3$ is given by the vector sum of $\mathbf{b}_1$ and $\mathbf{b}_2$.

Since the periodic force field of the crystal lattice requires that atoms must move from one equilibrium position to another, it follows that the Burgers vector must always connect one equilibrium lattice position with another. Therefore, the crystal structure will determine the possible Burgers vectors. A dislocation with a Burgers vector equal to one lattice spacing is said to be a *dislocation of unit strength*. Because of energy considerations dislocations with strengths larger than unity are generally unstable and dissociate into two or more dislocations of lower strength. The criterion for deciding whether or not dissociation will occur is based on the fact that the strain energy of a dislocation is proportional to the square of its Burgers vector. Therefore, the dissociation reaction $\mathbf{b}_1 \rightarrow \mathbf{b}_2 + \mathbf{b}_3$ will occur when $b_1^2 > b_2^2 + b_3^2$, but not if $b_1^2 < b_2^2 + b_3^2$.

Dislocations with strengths less than unity are possible in close-packed lattices where the equilibrium positions are not the edges of the structure cell. A Burgers vector is specified by giving its components along the axes of the crystallographic structure cell. Thus, the Burgers vector for slip in a cubic lattice from a cube corner to the center of one face has the components $a_0/2, a_0/2, 0$. The Burgers vector is $[a_0/2\ a_0/2\ 0]$, or as generally written $\mathbf{b} = (a_0/2)[110]$. The strength of a dislocation with Burgers vector $a_0[uvw]$ is $|b| = a_0[u^2 + v^2 + w^2]^{1/2}$. For example, the magnitude of the Burgers vector given above is $|b| = a_0/\sqrt{2}$.

In adding Burgers vectors, each of the corresponding components are added separately. Thus $b_1 + b_2 = a_0[110] + a_0[211] = a_0[321]$. In adding or subtracting components common unit vectors must be used. Thus $a_0/3[112] + a_0/6[11\bar{1}]$ must be expressed as $a_0/6[224] + a_0/6[11\bar{1}] = a_0/6[333] = a_0/2[111]$.

Example Determine whether the dislocation dissociation reaction is feasible.

$$b_1 = b_2 + b_3$$

$$\frac{a}{2}[0\bar{1}1] = \frac{a}{6}[1\bar{2}1] + \frac{a}{6}[\bar{1}\bar{1}2]$$

Since this is a vector equation the x, y, and z components of the right-hand side of the equation must equal the x, y, and z components of the left side (original dislocation).

$$x \text{ components:} \quad 0 = \tfrac{1}{6} - \tfrac{1}{6}$$

$$y \text{ components:} \quad -\tfrac{1}{2} = -\tfrac{2}{6} - \tfrac{1}{6} = -\tfrac{1}{2}$$

$$z \text{ components:} \quad \tfrac{1}{2} = \tfrac{1}{6} + \tfrac{2}{6} = \tfrac{1}{2}$$

For the dissociation to be energetically favorable $b_1^2 > b_2^2 + b_3^2$

$$b_1 = \frac{a}{2}\left[0 + (-1)^2 + (1)^2\right]^{1/2} = \frac{\sqrt{2}\,a}{2} \qquad b_1^2 = \frac{a^2}{2}$$

$$b_2 = \frac{a}{6}\left[(1)^2 + (-2)^2 + (1)^2\right]^{1/2} = \frac{\sqrt{6}\,a}{6} \qquad b_2^2 = \frac{a^2}{6}$$

$$b_3 = \frac{a}{6}\left[(-1)^2 + (-1)^2 + (2)^2\right]^{1/2} = \frac{\sqrt{6}\,a}{6} \qquad b_3^2 = \frac{a^2}{6}$$

$\therefore\ b_1^2 > b_2^2 + b_3^2$ and the dislocation reaction is feasible.

A dislocation of unit strength, or *unit dislocation*, has a minimum energy when its Burgers vector is parallel to a direction of closest atomic packing in the lattice. This agrees with the experimental observation that crystals always slip in the close-packed directions. A unit dislocation of this type is also said to be a *perfect dislocation* because translation equal to one Burgers vector produces an identity translation. For a perfect dislocation there is perfect alignment of atom planes above and below the slip plane within the dislocation loop. A unit dislocation parallel to the slip direction cannot dissociate further unless it becomes an *imperfect dislocation*, where a translation of one Burgers vector does not result in an identity translation. A stacking fault is produced by the dissociation of a unit dislocation into two imperfect dislocations. For a stacking fault to be stable, the decrease in energy due to dissociation must be greater than the increase in interfacial energy of the faulted region.

5-4 DISLOCATIONS IN THE FACE-CENTERED CUBIC LATTICE

Slip occurs in the fcc lattice on the {111} plane in the ⟨110⟩ direction. The shortest lattice vector is $(a_0/2)[110]$, which connects an atom at a cube corner with a neighboring atom at the center of a cube face. The Burgers vector is therefore $(a_0/2)[110]$.

However, consideration of the atomic arrangement on the {111} slip plane shows that slip will not take place so simply. Figure 5-8 represents the atomic

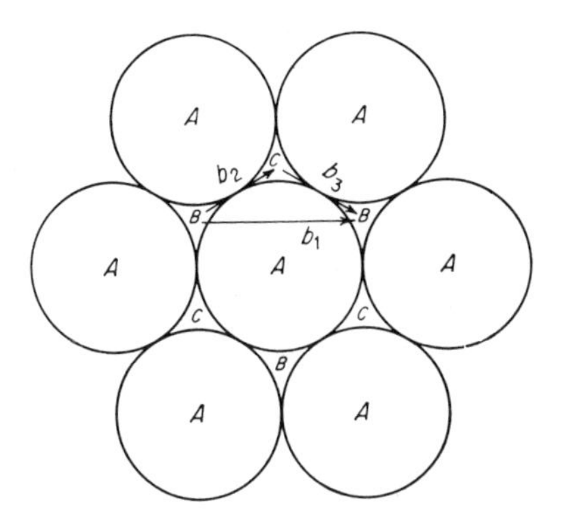

Figure 5-8 Slip in a close-packed (111) plane in an fcc lattice. *(After A. H. Cottrell, "Dislocations and Plastic Flow in Crystals," p. 73, Oxford University Press, New York, 1953. By permission of the publishers.)*

packing on a close-packed (111) plane. It has already been shown (Fig. 4-4) that the {111} planes are stacked on a sequence *ABCABC*.... The vector $b = (a_0/2)[10\bar{1}]$ defines one of the observed slip directions. The same shear displacement as produced by b_1 can be accomplished by the two-step path $b_2 + b_3$. The latter displacement is more energetically favorable but it causes the perfect dislocation to decompose into two partial dislocations.

$$\mathbf{b}_1 \rightarrow \mathbf{b}_2 + \mathbf{b}_3$$

$$\frac{a_0}{2}[10\bar{1}] \rightarrow \frac{a_0}{6}[2\bar{1}\bar{1}] + \frac{a_0}{6}[11\bar{2}]$$

The above reaction is energetically favorable since there is a decrease in strain energy proportional to the change $a_0^2/2 \rightarrow a_0^2/3$.

Original dislocation	Product of reaction	
$\lvert b_1\rvert = a_0[\frac{1}{4} + 0 + \frac{1}{4}]^{1/2}$	$\lvert b_2\rvert = a_0[\frac{4}{36} + \frac{1}{36} + \frac{1}{36}]^{1/2}$	
$\lvert b_1\rvert = \frac{\sqrt{2}}{2} a_0$	$\lvert b_2\rvert = \frac{a_0}{\sqrt{6}}$	$\lvert b_3\rvert = \frac{a_0}{\sqrt{6}}$
$b_1^2 = \frac{a_0^2}{2}$	$> b_2^2 = \frac{a_0^2}{6}$	$b_3^2 = \frac{a_0^2}{6}$
b_1^2	$> b_2^2 + b_3^2$	

Slip by this two-stage process creates a stacking fault *ABCAC¦ABC* in the stacking sequence. As Fig. 5-9 shows, the dislocation with Burgers vector $\mathbf{b}_1$ has been dissociated into two *partial dislocations* $\mathbf{b}_2$ and $\mathbf{b}_3$. This dislocation reaction was suggested by Heidenreich and Shockley,[1] and therefore this dislocation arrangement is often known as *Shockley partials*, since the dislocations are imperfect ones which do not produce complete lattice translations. Figure 5-9 represents the situation looking down on (111) along $[11\bar{1}]$. *AB* represents the perfect dislocation line having the full slip vector $\mathbf{b}_1$. This dissociates according to

[1] R. D. Heidenreich and W. Shockley, "Report on Strength of Solids," p. 37, Physical Society, London, 1948.

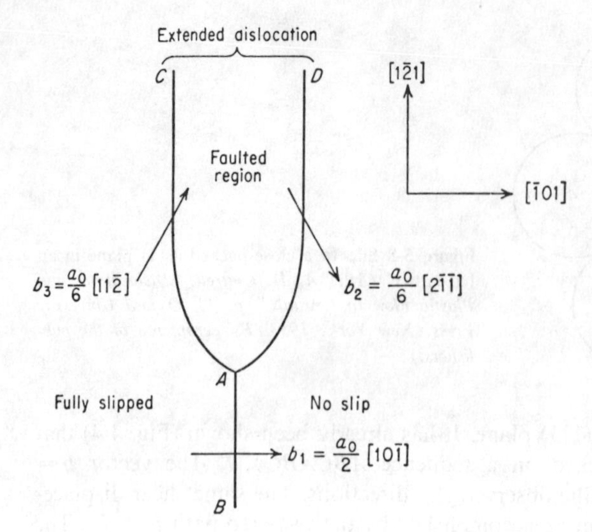

Figure 5-9 Dissociation of a dislocation into two partial dislocations.

the above reaction into partial dislocations with Burgers vectors $\mathbf{b}_2$ and $\mathbf{b}_3$. The combination of the two partials AC and AD is known as an *extended dislocation*. The region between them is a stacking fault representing a part of the crystal which has undergone slip intermediate between full slip and no slip. Because $\mathbf{b}_2$ and $\mathbf{b}_3$ are at a 60° angle, there will be a repulsive force between them (Sec. 5-9). However, the surface tension of the stacking fault tends to pull them together. The partial dislocations will settle at an equilibrium separation determined primarily by the stacking-fault energy. As was discussed in Sec. 4-11, the stacking-fault energy can vary considerably for different fcc metals and alloys and this in turn can have an important influence on their deformation behavior.

Dissociation of unit dislocations is independent of the character (edge, screw, or mixed) of the dislocation. However, unlike the unextended screw dislocation, the extended screw dislocation defines a specific slip plane, the {111} plane of the fault, and it will be constrained to move in this plane. The partial dislocations move as a unit maintaining the equilibrium width of the faulted region. Because of this restriction to a specific slip plane, an extended screw dislocation cannot cross slip unless the partial dislocations recombine into a perfect dislocation. Constrictions in the stacking fault ribbon which permit cross slip are possible (Fig. 4-28), but this requires energy. The greater the width of the stacking fault (or the lower the stacking-fault energy) the more difficult it is to produce constrictions in the stacking faults. This explains why cross slip is quite prevalent in aluminum, which has a very narrow stacking-fault ribbon, while it is not observed usually in copper, which has a wide stacking-fault ribbon.

Extended dislocations are readily detected by transmission electron microscopy. Figure 5-10 shows the characteristic fringe pattern of the stacking fault between the extended dislocations.

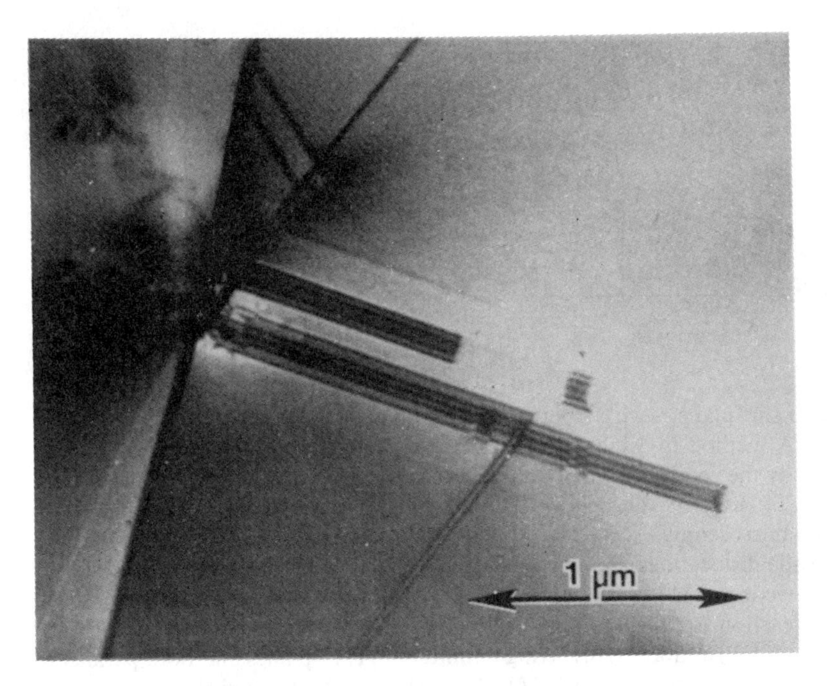

Figure 5-10 Group of stacking faults in 302 stainless steel stopped at boundary on left-hand side. *(Courtesy of Prof. H. G. F. Wilsdorf, University of Virginia.)*

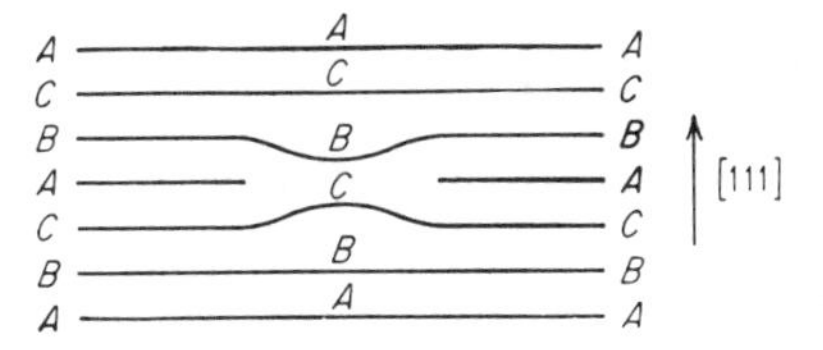

Figure 5-11 A Frank partial dislocation or sessile dislocation. *(After A. H. Cottrell, "Dislocations and Plastic Flow in Crystals," p. 75, Oxford University Press, New York, 1953. By permission of the publishers.)*

Frank[1] pointed out that another type of partial dislocation can exist in the fcc lattice. Figure 5-11 illustrates a set of (111) planes viewed from the edge. The center part of the middle *A* plane is missing. An edge dislocation is formed in this region with a Burgers vector $(a_0/3)[111]$. This is called a *Frank partial dislocation*. Its Burgers vector is perpendicular to the central stacking fault. Since glide must be restricted to the plane of the stacking fault and the Burgers vector is normal to this plane, the Frank partial dislocation cannot move by glide. For this reason it is called a *sessile dislocation*. A sessile dislocation can move only by the diffusion of atoms or vacancies to or from the fault, i.e., by the process of climb. Because climb is not a likely process at ordinary temperatures, sessile dislocations provide obstacles to the movement of other dislocations. Dislocations which glide freely

[1] F. C. Frank, *Proc. Phys. Soc. London*, vol. 62A, p. 202, 1949.

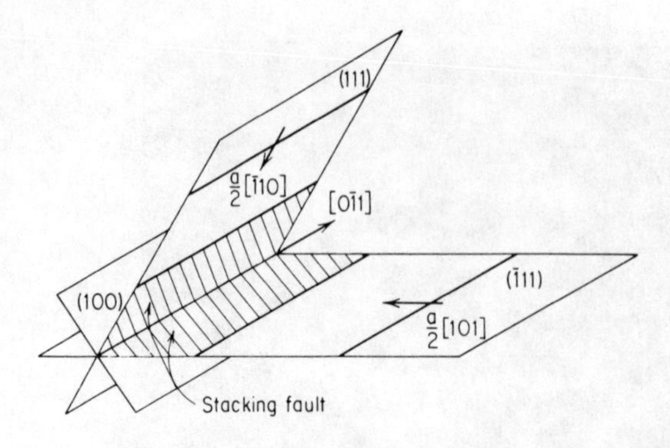

Figure 5-12 Lomer-Cottrell barrier.

over the slip plane, such as perfect dislocations or Shockley partials, are called *glissile*. A method by which a missing row of atoms can be created in the (111) plane is by the condensation of a disk of vacancies on that plane. Evidence for the collapse of disks of vacancies in aluminum has been obtained by transmission electron microscopy.[1]

Sessile dislocations are produced in the fcc lattice by the glide of dislocations on intersecting {111} planes during duplex slip. The sessile dislocation produced by the reaction is called a *Lomer-Cottrell barrier*. Consider two perfect dislocations $a_0/2[\bar{1}10]$ and $a_0/2[101]$ lying in different {111} planes and both parallel to the line of intersection of the {111} planes (Fig. 5-12). These dislocations attract each other and move toward their line of intersection. Lomer[2] suggested that they react according to

$$\frac{a_0}{2}[101] + \frac{a_0}{2}[\bar{1}10] \rightarrow \frac{a_0}{2}[011]$$

to produce a new dislocation of reduced energy. This new dislocation lies parallel to the line of intersection of the initial slip planes in the (100) plane bisecting the slip planes. Its Burgers vector lying in the (100) plane is normal to the line of intersection so it is a pure edge dislocation. Since (100) is not a close-packed slip plane in the fcc lattice, this dislocation will not glide freely. However, it is not a true sessile dislocation in the sense of the Frank partial because it is not an imperfect dislocation.

Cottrell[3] showed that the product of Lomer's reaction could be made strictly immobile if it is considered that dislocations on the {111} planes of an fcc metal are normally dissociated into partials. The leading partial dislocations on each slip plane will interact with each other in a reaction of the type

$$\frac{a_0}{6}[\bar{1}2\bar{1}] + \frac{a_0}{6}[1\bar{1}2] \rightarrow \frac{a_0}{6}[011]$$

[1] P. B. Hirsch, J. Silcox, R. E. Smallman, and K. H. Westmacott, *Philos. Mag.*, vol. 3, p. 897, 1958.

[2] W. M. Lomer, *Philos. Mag.*, vol. 42, p. 1327, 1951.

[3] A. H. Cottrell, *Philos. Mag.*, vol. 43, p. 645, 1952.

Like before, the new dislocation $a_0/6[011]$ lies parallel to the line of intersection of the slip plane and has a pure edge character in the (100) plane. The dislocation is sessile because its Burgers vector does not lie in either of the planes of its stacking faults.

Lomer-Cottrell barriers can be overcome at high stresses and/or temperatures. A mathematical analysis of the stress required to break down a barrier either by slip on the (100) plane or by reaction back into the dislocations from which the barrier formed has been given by Stroh.[1] However, it has been shown[2] that for the important case of screw dislocations piled up at Lomer-Cottrell barriers, the screw dislocations can escape the pile-up by cross slip before the stress is high enough to collapse the barrier. While the formation of Lomer-Cottrell barriers is an important mechanism in the strain hardening of fcc metals, they do not constitute the chief contribution to strain hardening.

Because of the multiplicity of slip systems in the fcc lattice, a large number of dislocation reactions of the types discussed above are possible. These have been worked out in detail by Hirth.[3] Also, the Thompson tetrahedron[4] is a useful geometrical method for visualizing the geometry of these reactions.

5-5 DISLOCATIONS IN THE HEXAGONAL CLOSE-PACKED LATTICE

The basal plane of the hcp lattice is a close-packed plane with the stacking sequence *ABABAB*.... Slip occurs on the basal plane (0001) in the $\langle 11\bar{2}0\rangle$ direction (Fig. 4-3). The smallest unit vector for the hcp structure has a length a_0 and lies in the close-packed $\langle 11\bar{2}0\rangle$ direction. Therefore, the Burgers vector is $a_0[11\bar{2}0]$. Dislocations in the basal plane can reduce their energy by dissociating into Shockley partials according to the reaction

$$a_0[11\bar{2}0] \rightarrow a_0[10\bar{1}0] + a_0[01\bar{1}0]$$

The stacking fault produced by this reaction lies in the basal plane, and the extended dislocation which forms it is confined to glide in this plane.

5-6 DISLOCATIONS IN THE BODY-CENTERED CUBIC LATTICE

Slip occurs in the $\langle 111\rangle$ direction in the bcc lattice. The shortest lattice vector extends from an atom corner to the atom at the center of the unit cube. Therefore, the Burgers vector is $(a_0/2)[111]$. It will be recalled that slip lines in iron have been found to occur on $\{110\}$, $\{112\}$, and $\{123\}$, although in other bcc

[1] A. N. Stroh, *Philos. Mag.*, vol. 1, sec, 8, p. 489, 1956.
[2] A. Seeger, J. Diehl, S. Mader, and R. Rebstook, *Philos. Mag.*, vol. 2, p. 323, 1957.
[3] J. P. Hirth, *J. Appl. Phys.*, vol. 32, pp. 700–706, 1961.
[4] N. Thompson, *Proc. Phys. Soc. London Ser. B*, vol. B66, p. 481, 1953.

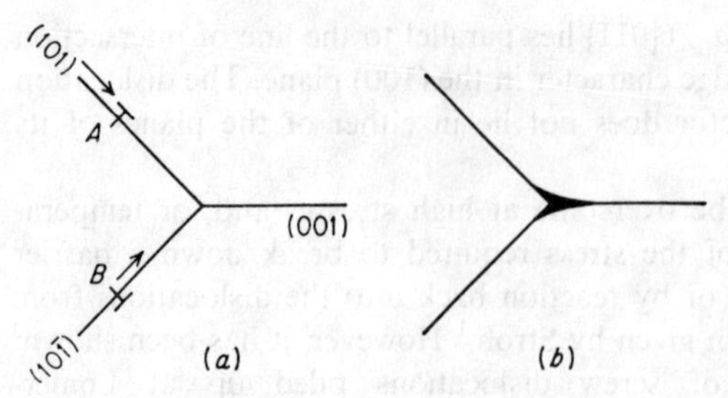

Figure 5-13 Slip on intersecting (110) planes. (*From A. H. Cottrell, Trans. AIME, vol. 212, p. 196, 1958.*)

metals slip appears to occur predominantly on the {110} planes. While the slip plane is normally (110), it is important to realize that three {110}-type planes intersect in a [111] direction. Thus, screw dislocations with a $(a_0/2)[111]$ Burgers vector may move at random on the {111} planes with a high resolved shear stress. This is the origin of the wavy and poorly defined slip lines in iron.

Extended dislocations are not commonly observed in bcc metals as they are in fcc and hcp metals. Although dislocation reactions involving partials have been suggested,[1] there are no well-established reactions that have been substantiated by extensive experimentation.

However, Cottrell[2] has suggested a dislocation reaction which appears to lead to the formation of immobile dislocations in the bcc lattice. This reaction has been shown to be one of the mechanisms for producing a crack nucleus for brittle fracture. It is also a mechanism for producing $a_0[001]$ dislocation networks that are observed in iron. In Fig. 5-13, dislocation A with Burgers vector $(a_0/2)[\bar{1}\bar{1}1]$ is gliding on the (101) plane, while dislocation B with $(a_0/2)[111]$ is gliding on the intersecting $(10\bar{1})$ slip plane. The two dislocations come together and react to lower their strain energy by providing a pure edge dislocation which lies on the (001) plane.

$$\frac{a_0}{2}[\bar{1}\bar{1}1] + \frac{a_0}{2}[111] \rightarrow a_0[001]$$

Since the (001) is not a close-packed slip plane in the bcc lattice, the dislocation is immobile. Moreover, the (001) plane is the cleavage plane along which brittle fracture occurs.

5-7 STRESS FIELDS AND ENERGIES OF DISLOCATIONS

A dislocation is surrounded by an elastic stress field that produces forces on other dislocations and results in interaction between dislocations and solute atoms. For the case of a perfect dislocation a good approximation of the stress field can be

[1] For a summary see J. P. Hirth and J. Lothe, "Theory of Dislocations," pp. 344–353, McGraw-Hill Book Company, New York, 1968.

[2] A. H. Cottrell, *Trans. Metall. Soc. AIME*, vol. 212, p. 192, 1958.

obtained from the mathematical theory of elasticity for continuous media. However the equations obtained are not valid close to the core of the dislocation line. The equations given below apply to straight edge and screw dislocations in an isotropic crystal.[1] The stress around a straight dislocation will be a good approximation to that around a curved dislocation at distances that are small compared with the radius of curvature. Appreciably greater complexity results from the consideration of a crystal with anisotropic elastic constants.[2]

Figure 5-14 represents the cross section of a cylindrical piece of elastic material containing an edge dislocation running through point O parallel to the z axis (normal to the plane of the figure). The original undistorted cylinder without a dislocation is shown by the dashed line. The dislocation was produced by making a radial cut along the plane $y = 0$ (line OA), sliding the cut surfaces along each other the distance AA', and joining them back together again. This sequence of operations[3] produces a positive edge dislocation running along the z axis with a strain field identical with that around a dislocation model such as that of Fig. 4-8. Since the dislocation line is parallel to the z axis, strains in that direction are zero and the problem can be treated as one in plane strain.

For the case of a straight edge dislocation in an elastically isotropic material the stresses, in terms of three orthogonal coordinate axes, are given by the following equations. The notation is the same as that used in Chaps. 1 and 2.

$$\sigma_x = -\tau_0 \frac{by(3x^2 + y^2)}{x^2 + y^2} \tag{5-1}$$

$$\sigma_y = \tau_0 \frac{by(x^2 - y^2)}{(x^2 + y^2)^2} \tag{5-2}$$

$$\sigma_z = \nu(\sigma_x + \sigma_y) = \frac{2\tau_0 \nu y}{x^2 + y^2} \tag{5-3}$$

where

$$\tau_0 = \frac{G}{2\pi(1 - \nu)}$$

$$\tau_{xy} = \tau_0 \frac{bx(x^2 - y^2)}{(x^2 + y^2)^2} \tag{5-4}$$

$$\tau_{xz} = \tau_{yz} = 0 \tag{5-5}$$

[1] For derivations see F. R. N. Nabarro, *Adv. Phys.*, vol. 1, no. 3, pp. 271–395, 1952; W. T. Read, Jr., "Dislocations in Crystals," pp. 114–123, McGraw-Hill Book Company, New York, 1953; J. D. Eshelby, *Brit J. Appl. Phys.*, vol. 17, pp. 1131–1135, 1966; R. M. Caddell, "Deformation and Fracture of Solids," pp. 152–160, Prentice-Hall Inc., Englewood Cliffs, N.J., 1980.

[2] J. D. Eshelby, W. T. Read, and W. Shockley, *Acta Metall.*, vol. 1, pp. 351–359, 1953.

[3] It is interesting that this problem was analyzed by Volterra in 1907, long before the concept of dislocations was originated. The mathematical details may be found in A. E. H. Love, "A Treatise on the Mathematical Theory of Elasticity," pp. 221–228, Cambridge University Press, New York, 1934.

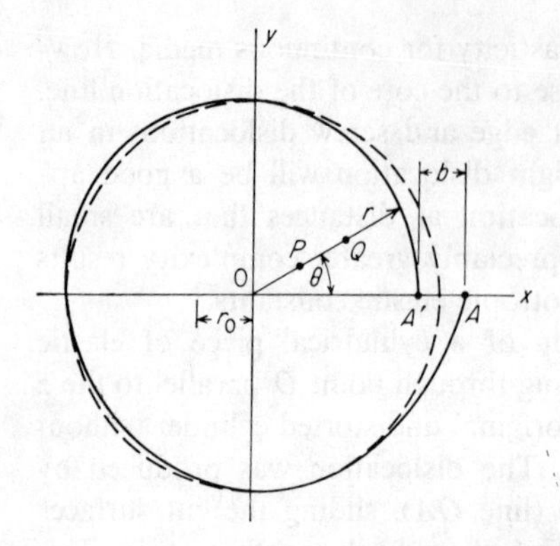

Figure 5-14 Deformation of a circle containing an edge dislocation. The unstrained circle is shown by a dashed lin. The solid line represents the circle after the dislocation has been introduced.

The largest normal stress σ_x is along the x axis, and is compressive above the slip plane and tensile below the slip plane. The shear stress τ_{xy} is a maximum in the slip plane, i.e., when $y = 0$.

For polar coordinates, the equations are

$$\sigma_r = \sigma_\theta = \frac{-\tau_0 b \sin\theta}{r} \tag{5-6}$$

$$\tau_{r\theta} = \tau_{\theta r} = \tau_0 \frac{b \cos\theta}{r} \tag{5-7}$$

σ_r acts in the radial direction, while σ_θ acts in a plane perpendicular to r. Note that the stresses vary inversely with distance from the dislocation line. Since the stress becomes infinite at $r = 0$, a small cylindrical region $r = r_0$ around the dislocation line must be excluded from the analysis. Estimates of r_0 indicate that it is on the order of 5 to 10 Å.

A straight screw dislocation in an isotropic medium has complete cylindrical symmetry. For a rectangular-coordinate system only two components of stress are not equal to zero.

$$\tau_{xz} = -\frac{Gb}{2\pi}\frac{y}{x^2 + y^2} \tag{5-8}$$

$$\tau_{yz} = \frac{Gb}{x}\frac{2\pi}{x^2 + y^2} \tag{5-9}$$

Since there is no extra half plane of atoms in a screw dislocation, there are no tensile or compressive normal stresses. The stress field is simply one of shear. The radial symmetry of this stress field is apparent when the shear stress is expressed in a polar-coordinate system.

$$\tau_{\theta z} = \frac{Gb}{2\pi r} \tag{5-10}$$

The strain energy involved in the formation of an edge dislocation can be estimated from the work involved in displacing the cut OA in Fig. 5-14 a distance b along the slip plane.

$$U = \frac{1}{2}\int_{r_0}^{r_1} \tau_{r\theta} b\, dr = \frac{1}{2}\int_{r_0}^{r_1} \tau_0 b^2 \cos\theta \frac{dr}{r} \tag{5-11}$$

But $\cos\theta = 1$ along the slip plane $y = 0$, so that the strain energy is given by

$$U = \frac{Gb^2}{4\pi(1-\nu)} \ln \frac{r_1}{r_0} \tag{5-12}$$

In the same way, the strain energy of a screw dislocation is given by

$$U = \frac{1}{2}\int_{r_0}^{r_1} \tau_{\theta z} b\, dr = \frac{Gb^2}{4\pi} \ln \frac{r_1}{r_0} \tag{5-13}$$

Note that, in accordance with our assumption up to this point, the strain energy per unit length of dislocation is proportional to Gb^2.

The total strain energy of a dislocation is the sum of the elastic strain energy [Eq. (5-12) or (5-13)] plus the energy of the core of the dislocation. Although estimates of the core energy are quite approximate, quantum-mechanical calculations indicate that the core energy is about one-fifteenth of the total energy. As a good approximation, the core energy is added to the elastic strain energy by taking $r_0 \approx b$, so that the total energy per unit length of screw dislocation is given by

$$U_t = \frac{Gb^2}{4\pi} \ln \frac{r_1}{b} \tag{5-14}$$

Typical values for an annealed crystal are $r_1 = 10^{-5}$ cm and $b = 2 \times 10^{-8}$ cm. Since the natural logarithm of a large number changes slowly, for this range of values $\ln(r_1/b) \approx 2\pi$, and the dislocation energy per unit length (ignoring small differences in Eqs. (5-12) and (5-13) simplifies to

$$U = \frac{Gb^2}{2} \tag{5-15}$$

The strain energy of a dislocation is about 8 eV for each atom plane threaded by the dislocation (see Prob. 5-8) while the core energy is the order of 0.5 eV per atom plane. This large, positive strain energy means that the free energy of a crystal is increased by the introduction of a dislocation. Since nature tries to minimize the free energy of a system, a crystal with dislocations is thermodynamically unstable and will try to lower its free energy by the elimination of dislocations in a process such as annealing. This situation should be contrasted with that of point defects (vacancies), which are thermodynamically stable crystal defects. At a given temperature there is an equilibrium concentration of vacancies given by Eq. (4-1).

5-8 FORCES ON DISLOCATIONS

When an external force of sufficient magnitude is applied to a crystal, the dislocations move and produce slip. Thus, there is a force acting on a dislocation line which tends to drive it forward. Figure 5-15 shows a dislocation line moving in the direction of its Burgers vector under the influence of a uniform shear stress τ. An element of the dislocation line ds is displaced in the direction of slip normal to ds by an amount dl. The area swept out by the line element is then $ds\,dl$. This corresponds to an average displacement of the crystal above the slip plane to the crystal below the slip plane of an amount $(ds\,dl/A)b$, where A is the area of the slip plane. The applied force creating the shear stress is τA. The work done when the increment of slip occurs is

$$dW = \tau A\left(\frac{ds\,dl}{A}\right)b$$

The force on a dislocation is always defined as a force F *per unit length* of dislocation line. Since $F = dW/dl$ and remembering that this is a force per unit length (ds), we have

$$F = \frac{dW}{dl\,ds} = \tau b \tag{5-16}$$

This force is normal to the dislocation line at every point along its length and is directed toward the unslipped part of the glide plane. Since the Burgers vector is constant along a curved dislocation line, if τ is constant, the value of F will be the same at any point along the dislocation line but its direction always will be normal to the dislocation line. As a result, the force on a dislocation is not necessarily in the same direction as the applied stress.

All of the equations for the strain energy of a dislocation in Sec. 5-7 were expressed *per unit length of dislocation*. Because the strain energy of a dislocation line is proportional to its length, work must be performed to increase its length. Therefore, a dislocation possesses a *line tension* which attempts to minimize its energy by shortening its length. For a curved dislocation line the line tension produces a restoring force which tends to straighten it out. The line tension has

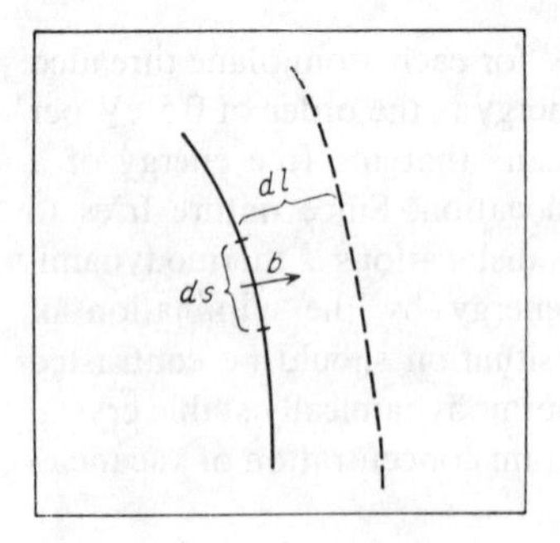

Figure 5-15 Force acting on a dislocation line.

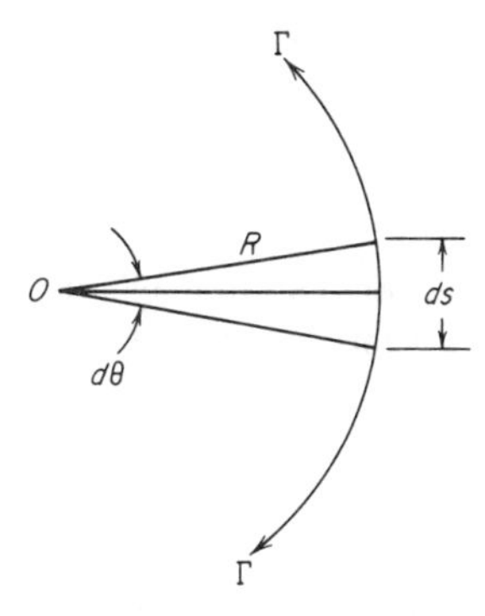

Figure 5-16 Forces on a curved dislocation line.

units of energy per unit length and is analogous to the surface energy of a soap bubble or liquid.

Consider the curved dislocation line in Fig. 5-16. Thc line tension Γ will produce a force tending to straighten the line and the line only will remain curved if there is a shear stress which produces a force on the dislocation to resist the line tension. We wish to determine the value of the shear stress τ that is required to maintain the dislocation line at a radius of curvature R. The angle subtended by an element of arc ds is $d\theta = ds/R$. The outward force on the dislocation line is $\tau b\,ds$. The opposing inward force due to the line tension is $2\Gamma \sin(d\theta/2)$, which simplifies to $\Gamma\,d\theta$ for small values of $d\theta$. For the dislocation line to remain in the curved configuration

$$\Gamma\,d\theta = \tau b\,ds$$

or

$$\tau = \frac{\Gamma}{bR} \tag{5-17}$$

But, Γ is an energy per unit length, and we have seen in Sec. 5-7 that a good approximation for the energy per unit length of dislocation is $Gb^2/2 \approx \Gamma$. Thus, the shear stress required to bend a dislocation to a radius R is

$$\tau \approx \frac{Gb}{2R} \tag{5-18}$$

5-9 FORCES BETWEEN DISLOCATIONS

Dislocations of opposite sign on the same slip plane will attract each other, run together, and annihilate each other. This can be seen readily for the case of an edge dislocation (Fig. 4-8) where the superposition of a positive and negative dislocation on the same slip plane would eliminate the extra plane of atoms and therefore the dislocation would disappear. Conversely, dislocations of like sign on the same slip plane will repel each other. We can understand this by considering the energy changes. For two dislocations separated at a large distance, the elastic strain energy for the combined situations will be

$$2 \cdot \frac{Gb^2}{4\pi(1-\nu)} \ln \frac{r_1}{r_0}$$

When the two dislocations are very close together the configuration can be approximated by a single dislocation of strength $2b$. For this case, the elastic strain energy will be

$$\frac{G(2b)^2}{4\pi(1-\nu)} \ln\left(\frac{r_1}{r_0}\right)$$

Since this is twice the energy of the dislocations when they are separated by a large distance, the dislocations will tend to repel each other to reduce the total elastic strain energy. When unlike dislocations are on closely spaced neighboring slip planes, complete annihilation cannot occur. In this situation, they combine to form a row of vacancies (for the case of $\frac{\perp}{\top}$) or an interstitial atom (for the case $\perp\,\top$).

The simplest situation to consider is the force between two parallel screw dislocations. Since the stress field of a screw dislocation is radially symmetrical, the force between the dislocations is a radial force which depends only on the distance of separation r.

$$F_r = \tau_{\theta z} b = \frac{Gb^2}{2\pi r} \tag{5-19}$$

The force is attractive for dislocations of opposite sign (antiparallel screws) and repulsive for dislocations of the same sign (parallel screws).

Example Two parallel and straight screw dislocations lie in a copper grain with a grain size about 0.04 mm. If the dislocations are separated by a distance of 1,200 Å, determine the total force on each dislocation. The dislocations are of opposite sign.

$F = Gb^2/2\pi r$. This will be an attractive force because the dislocations are of opposite sign.

For copper $G = 40$ GPa; $b = 2.5$ Å.

$$F = \frac{(40 \times 10^9 \text{ N/m}^2)(2.5 \times 10^{-10} \text{ m})^2}{2\pi(1200 \times 10^{-10} \text{ m})} = 3.3 \times 10^{-3} \text{ N/m}$$

The total force on the dislocation is approximately

$$F_T = Fl = (3.3 \times 10^{-3} \text{ N/m}) \times (4 \times 10^{-2} \times 10^{-3}) \text{ m} = 1.3 \times 10^{-7} \text{ N}$$

Consider now the forces between two parallel edge dislocations with the same Burgers vectors. Referring to Fig. 5-14, the edge dislocations are at P and Q, parallel to the z axis, with their Burgers vectors along the x axis. The force between them is not a central force, and so it is necessary to consider both a

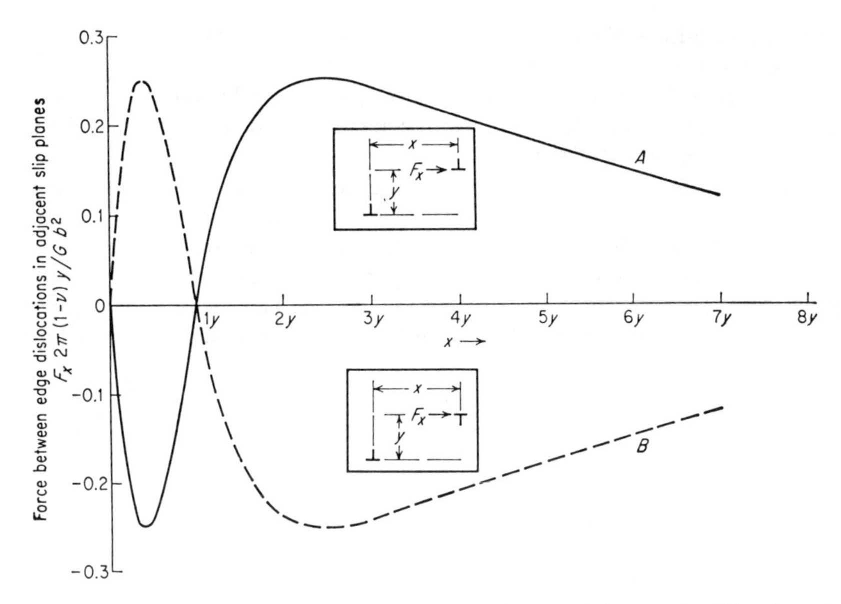

Figure 5-17 Graphical representation of Eq. (5-21). Solid curve A is for two edge dislocations of same sign. Dashed curve B is for two unlike edge dislocations. *(After A. H. Cottrell, "Dislocations and Plastic Flow in Crystals," p. 48, Oxford University Press, New York, 1953. By permission of the publishers.)*

radial and a tangential component. The force per unit length is given by[1]

$$F_r = \frac{Gb^2}{2\pi(1-\nu)}\frac{1}{r} \qquad F_\theta = \frac{Gb^2}{2\pi(1-\nu)}\frac{\sin 2\theta}{r} \tag{5-20}$$

Because edge dislocations are mainly confined to the slip plane, the force component along the x direction, which is the slip direction, is of most interest.

$$\begin{aligned} F_x &= F_r \cos\theta - F_\theta \sin\theta \\ &= \frac{Gb^2 x(x^2 - y^2)}{2\pi(1-\nu)(x^2+y^2)^2} \end{aligned} \tag{5-21}$$

Figure 5-17 is a plot of the variation of F_x with distance x, where x is expressed in units of y. Curve A is for dislocations of the same sign; curve B is for dislocations of opposite sign. Note that dislocations of the same sign repel each other when $x > y$ ($\theta < 45°$) and attract each other when $x < y$ ($\theta > 45°$). The reverse is true for dislocations of opposite sign. F_x is zero at $x = 0$ and $x = y$. The situation $x = 0$, where the edge dislocations lie vertically above one

[1] A. H. Cottrell, "Dislocations and Plastic Flow in Crystals," p. 46, Oxford University Press, New York, 1953.

another, is a condition of equilibrium. Thus, theory predicts that a vertical array of edge dislocations of the same sign is in stable equilibrium. This is the arrangement of dislocations that exists in a low-angle grain boundary of the tilt variety. The calculation of forces for more complex situations are discussed by Read[1] and Weertman and Weertman[2].

In Sec. 5-4 we saw that an extended dislocation can glide only when the Burgers vector of both partial dislocations lies in the plane of the stacking fault. The force of repulsion between two edge dislocations on the same slip plane is given by Eq. (5-20). This force applies to the partial dislocations (Fig. 5-9) but it is opposed by the surface energy of the faulted region. The equilibrium spacing between the partial dislocations, d, is given by setting $r = d$ and $F = \gamma$ in Eq. (5-20)

$$d = \frac{Gb^2}{2\pi(1 - \nu)\gamma} \tag{5-22}$$

where γ is the stacking-fault energy per unit area, J/m^2. We note from Eq. (5-22) that metals with a high stacking-fault energy, like aluminum, result in a narrow faulted region.

A free surface exerts a force of attraction on a dislocation, since escape from the crystal at the surface would reduce its strain energy. Koehler[3] has shown that this force is approximately equal to the force which would be exerted in an infinite solid between the dislocation and one of opposite sign located at the position of its image on the other side of the surface. This *image force* is equal to

$$F = \frac{Gb^2}{4\pi(1 - \nu)} \frac{1}{r}$$

for an edge dislocation. However, it should be noted that metal surfaces are often covered with thin oxide films. A dislocation approaching a surface with a coating of an elastically harder material will encounter a repulsive rather than an attractive image force.

5-10 DISLOCATION CLIMB

An edge dislocation can glide only in the slip plane containing the dislocation line and its Burgers vector. However, under certain conditions an edge dislocation can move out of the slip plane onto a parallel plane directly above or below the slip plane. This is the process of *dislocation climb*. This type of movement is termed nonconservative, as compared with conservative movement when a dislocation glides in its slip plane.

[1] Read, *op. cit.*, p. 131.

[2] J. Weertman and J. R. Weertman, "Elementary Dislocation Theory," pp. 65–72, The Macmillan Company, New York, 1964.

[3] J. S. Koehler, *Phys. Rev.*, vol. 60, p. 397, 1941.

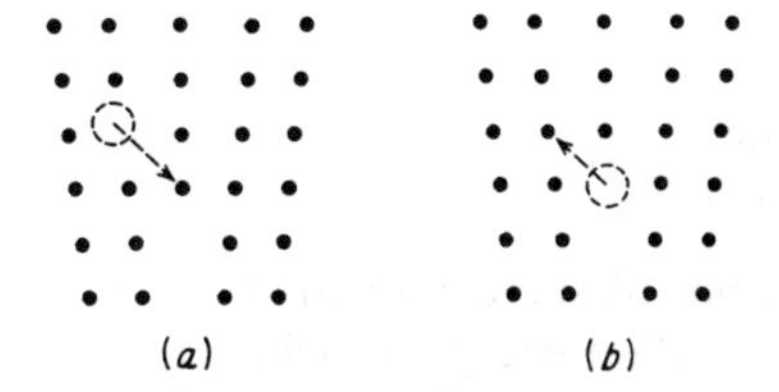

Figure 5-18 (*a*) Diffusion of vacancy to edge dislocation; (*b*) dislocation climbs up one lattice spacing.

Dislocation climb occurs by the diffusion of vacancies or interstitials to or away from the site of the dislocation. Since climb is diffusion-controlled, it is thermally activated and occurs more readily at elevated temperature. In *positive climb* atoms are removed from the extra half plane of atoms at a positive edge dislocation so that this extra half plane moves up one atom spacing. In *negative climb* a row of atoms is added below the extra half plane so that the dislocation line moves down one atom spacing. The usual mechanism for positive climb is by a vacancy diffusing to the dislocation and the extra atom moving into the vacant lattice site (Fig. 5-18). It is also possible, but less energetically favorable, for the atom to break loose from the extra half plane and become an interstitial atom. To produce negative climb, atoms must be added to the extra half plane of atoms. This can occur by atoms from the surrounding lattice joining the extra half plane, which creates a flux of vacancies away from the dislocation, or less probably, by interstitial atoms diffusing to the dislocation. The accumulation of vacancies at the dislocation core causes a free-energy change which results in an *osmotic force* on the dislocation.[1]

The existence of a compressive stress in the slip direction causes a force in the upward direction of climb. Similarly, a tensile stress normal to the extra half plane causes a force in the direction of negative climb. The superposition of stress on the high temperature needed for diffusion results in an increased rate of climb.

It is highly unlikely that complete rows of atoms are removed or added at the extra half plane in the climb process. In actual fact individual vacancies or small clusters of vacancies diffuse to the dislocation and climb occurs over a short segment of the dislocation line. This results in the formation of short steps or *jogs* along the dislocation line. Climb proceeds by the nucleation and motion of jogs. There will be a thermodynamic equilibrium number of jogs per unit length of dislocation, given by

$$n_j = n_0 e^{-U_j/kT} \tag{5-23}$$

where n_0 is the number of atom sites per unit length of dislocation and U_j is the activation energy to nucleate a jog. The activation energy of a jog is about 1 eV. The activation energy for climb is given by

$$U_c = U_j + U_v + U_m = U_j + U_d \tag{5-24}$$

[1] J. P. Hirth and J. Lothe, "Theory of Dislocations," pp. 506–509, McGraw-Hill Book Company, New York, 1968.

where U_v = the energy of formation of a vacancy

U_m = the energy of movement of a vacancy
U_d = the activation energy for self-diffusion

If the dislocations are heavily jogged due to prior plastic deformation (see Sec. 5-11), then nucleation is negligible and the activation energy for climb will be determined only by the activation energy for self-diffusion. Dislocation climb is an important mechanism in the creep of metals, where the activation energy for steady-state creep is equal to the activation energy for self-diffusion.

Climb is not possible with screw dislocations since in this case there is no extra half plane of atoms. Because the Burgers vector of a screw dislocation is parallel to the dislocation, it is free to slip on any plane which contains the dislocation line and the Burgers vector. No diffusion of atoms is needed to allow the screw dislocation to move on to another slip plane. However, a higher stress or activation energy may be needed since the resolved shear stress may not be as high as on the original slip plane.

5-11 INTERSECTION OF DISLOCATIONS

Since even well-annealed crystals contain many dislocations, it is not infrequent that a dislocation moving in its slip plane will intersect other dislocations crossing the slip plane. We have already seen (Sec. 4-14) that dislocation intersection mechanisms play an important role in the strain-hardening process.

The intersection of two dislocations produces a sharp break, a few atom spacings in length, in the dislocation line. These breaks can be of two types.

A *jog* is a sharp break in the dislocation moving it out of the slip plane.
A *kink* is a sharp break in the dislocation line which remains in the slip plane.

The intersection of two edge dislocations with Burgers vectors at right angles to each other is illustrated in Fig. 5-19. An edge dislocation XY with Burgers vector $\mathbf{b}_1$ is moving on plane P_{XY}. It cuts through dislocation AD, with Burgers vector $\mathbf{b}_2$, lying on plane P_{AD}. The intersection produces a *jog* PP' in dislocation AD. The resulting jog is parallel to $\mathbf{b}_1$, but it has a Burgers vector $\mathbf{b}_2$ since it is part of the dislocation line $APP'D$. The length of the jog is equal to $\mathbf{b}_1$. It can be seen that the jog resulting from this intersection of two edge dislocations has an edge orientation, and therefore, it readily can glide with the rest of the dislocation. To determine which dislocation forms the jog, we note that a jog forms when the Burgers vector of the intersecting dislocation is normal to the other dislocation line ($\mathbf{b}_1$ is normal to AD and jogs AD, while $\mathbf{b}_2$ is parallel to XY and no jog is formed). The direction of the jog is parallel to the Burgers vector of the intersecting dislocation.

The intersection of two orthogonal edge dislocations with parallel Burgers vectors is shown in Fig. 5-20. In this case both dislocations are jogged. The length

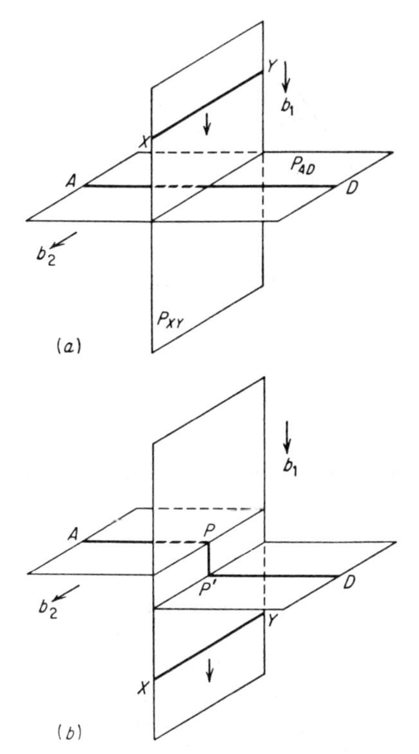

Figure 5-19 Intersection of two edge dislocations. *(From W. T. Read, Jr., "Dislocations in Crystals," McGraw-Hill Book Company, New York, 1953.)*

of jog PP' is $\mathbf{b}_1$ and the length of jog QQ' is $\mathbf{b}_2$. In this case the jogs both have a screw orientation and lie in the original slip planes of the dislocations rather than on a neighboring slip plane as in the previous case. Jogs of this type which lie in the slip plane instead of normal to it are usually called *kinks*. Kinks in dislocations are unstable since during glide they can line up and annihilate the offset.

The intersection of a screw dislocation and an edge dislocation is shown in Fig. 5-21*a*. Intersection produces a jog with an edge orientation on the edge dislocation and a kink with an edge orientation on the screw dislocation. The intersection of two screw dislocations (Fig. 5-21*b*) produces jogs of edge orientation in both screw dislocations. This is the most important type of intersection from the viewpoint of plastic deformation.

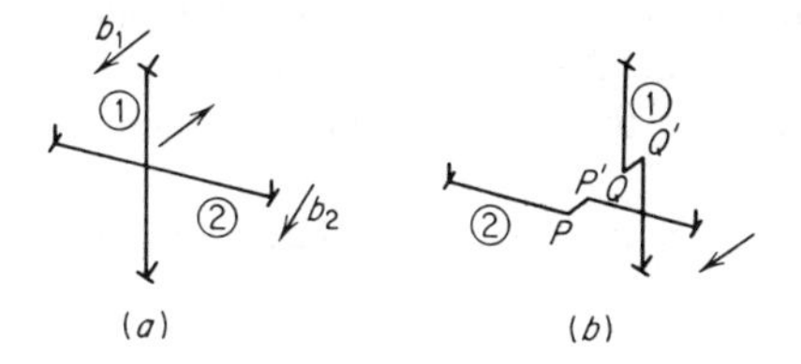

Figure 5-20 Intersection of edge dislocations with parallel Burgers vectors. (*a*) Before intersection; (*b*) after intersection.

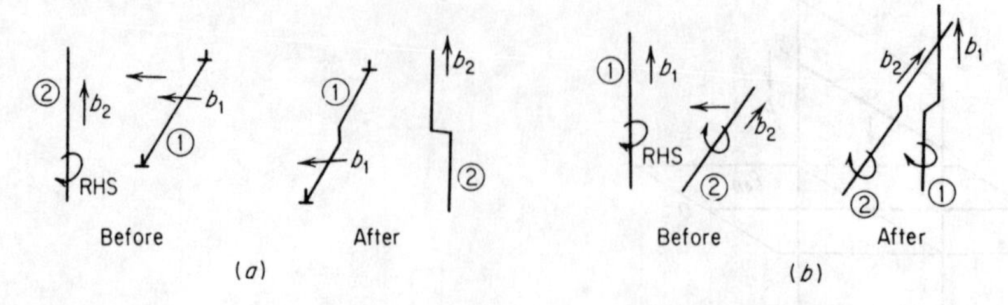

Figure 5-21 (*a*) Intersection of edge and screw dislocation; (*b*) intersection of two screw dislocations.

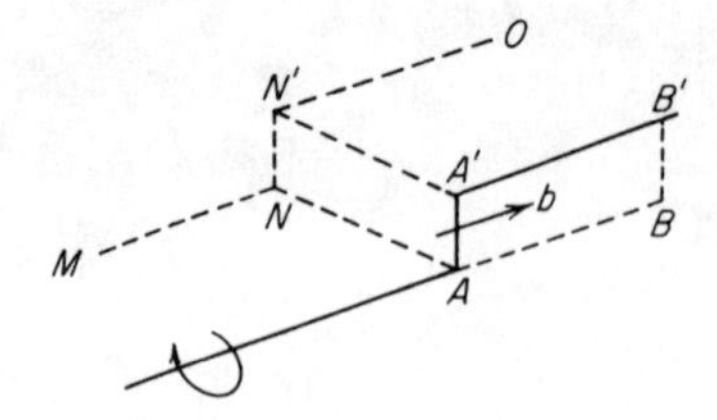

Figure 5-22 Movement of jog on screw dislocation. The jog is constrained to move along the dislocation in plane *AA′BB′*.

The jogs produced by the intersection of two edge dislocations (of either orientation) are able to glide readily because they lie in the slip planes of the original dislocations. The only difference between the motion of the jogged dislocation and an ordinary edge dislocation is that instead of gliding along a single plane, it glides over a stepped surface. Thus, jogs in pure edge dislocations do not affect the motion of the dislocations on which they lie. However, for the screw dislocations the jogs all have an edge orientation. Since an edge dislocation can glide freely only in a plane containing its line and its Burgers vector, the only way the jog can move by slip (conservative motion) is along the axis of the screw dislocation (Fig. 5-22). The only way that the screw dislocation can slip to a new position such as MNN′O and take its jog with it is by a nonconservative process such as climb. Dislocation climb is a thermally activated process and therefore the movement of jogged screw dislocations will be temperature-dependent. At temperatures where climb cannot occur, the motion of screw dislocations will be impeded by jogs. This is consistent with the experimental observation[1] that screw dislocations move more slowly through a crystal than edge dislocations.

5-12 JOGS

The creation of jogged segments on dislocation lines has many important implications in the theory of the plastic deformation of metals. A stable jog represents an increase in the length of the dislocation line, and therefore produces an increase

[1] N. K. Chen and R. B. Pond, *Trans. AIME*, vol. 194, p. 1085, 1952; W. G. Johnston and J. J. Gilman, *J. Appl. Phys.*, vol. 30, p. 121, 1957.

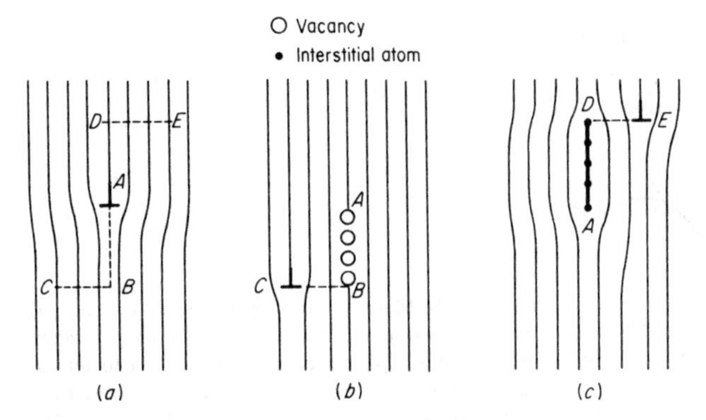

Figure 5-23 Creation of point defects by nonconservative motion of jogs of edge orientation. *(From D. Hull, "Introduction to Dislocations," p. 134, Pergamon Press, New York, 1965. By permission of the publishers.)*

in the energy of the crystal. The energy of a jog will be somewhat less than the energy of a dislocation per atom plane since its entire length lies in the distorted material of the core of the parent dislocation. Since the energy of a dislocation is $\alpha G b^2$ (where α is about 0.5 to 1.0), the energy of a jog of length b_2 in a dislocation of Burgers vector $\mathbf{b}_1$ is

$$U_j = \alpha G b_1^2 b_2 \tag{5-25}$$

where $\alpha \approx 0.2$. The energy of a jog is about 0.5 to 1.0 eV in metals.

We have seen that the most significant dislocation intersection is the case of two screw dislocations where nonconservative jogs are produced. For all other intersections the jogs are able to move readily with their dislocations. The nonconservative motion of jogs on screw dislocations needs further discussion. Figure 5-23 shows a section of an edge jog in a screw dislocation which is gliding on a plane parallel to the paper in the direction *DB* or *BD*. For the edge jog to move nonconservatively, it must either eliminate matter (create vacancies) or create matter (interstitials). If the jog moves nonconservatively from *A* to *B* and then glides along its slip plane to *C*, a row of vacancies will be created as shown in Fig. 5-23*b*. These vacancies will have a line tension and create a drag on the jog. However, if there is sufficient thermal activation, the vacancies will diffuse into the lattice and the drag will be removed. If the jog moves nonconservatively from *A* to *D* and then glides to *E* (Fig. 5-23*c*), a row of interstitials will be produced along *AD*.

As a screw dislocation moves through its slip plane it will, in general, encounter a forest of screw dislocations and many intersections will occur. Some of these will produce vacancy jogs and others will be interstitial jogs. These will tend to glide along the length of the screw dislocation and annihilate each other, leaving a net concentration of jogs all of the same sign. Then, because of their mutual repulsion, they will spread out along the dislocation line at approximately

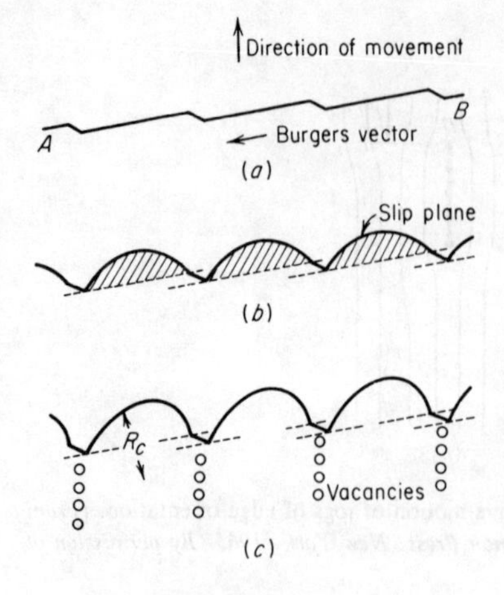

Figure 5-24 Movement of jogged screw dislocation. (*a*) Straight dislocation under zero stress; (*b*) dislocation bowed out in slip plane between the jogs due to applied shear stress; (*c*) movement of dislocation leaving trails of vacancies behind the jogs. *(From D. Hull, "Introduction to Dislocations," p. 136, Pergamon Press, New York, 1965. By permission of the publishers.)*

evenly spaced intervals. Figure 5-24 illustrates this situation. Under an applied shear stress τ acting in the slip direction, the jogs will act as pinning points. The dislocation will bow out between the jogs (Fig. 5-24*b*) with a radius of curvature given by Eq. (5-18). At some critical radius R_c the shear stress required to further decrease R is greater than the stress needed to nonconservative climb. Then the dislocation will move forward leaving a trail of vacancies (or interstitials) behind each jog. Estimates[1] of the energy required to form a vacancy or an interstitial atom at a jog in an fcc metal are 4.8 eV for an interstitial atom and 0.7 eV for a vacancy.

$$U_1 = \alpha_1 G b^3 \tag{5-26}$$

where $\alpha_1 \approx 1.0$ for an interstitial atom and 0.2 for a vacancy. The work done by the applied stress in moving the jogs forward one atomic spacing by the formation of vacancies or interstitials is

$$W = (\tau b l) b = \tau b^2 l \tag{5-27}$$

where l is the spacing between jogs. By equating Eqs. (5-26) and (5-27), we obtain the shear stress required to generate a defect and move the dislocation in the absence of thermal activation.

$$\tau = \alpha_1 \frac{Gb}{l} \tag{5-28}$$

At elevated temperatures thermal activation assists in the formation of vacancies. The activation energy for the movement of the jog or the formation of defects at the jog will be

$$U = \alpha_1 G b^3 - \tau b^2 l \tag{5-29}$$

[1] A. Seeger, "Defects in Crystalline Solids," p. 381, Physical Society, London, 1955.

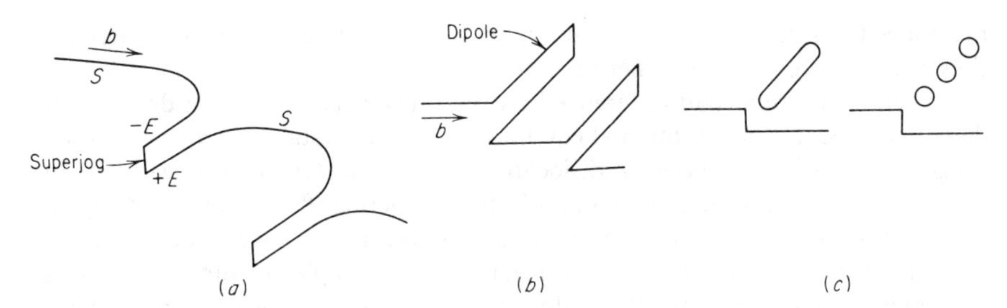

Figure 5-25 (*a*) Bowing out of dislocation between superjogs to produce (*b*) dislocation dipoles; (*c*) formation of dislocation loops from dipole.

Since α_1 is much greater for interstitial atoms than for vacancies, vacancy formation wll occur more readily than interstitial atom formation. Good agreement between the kinetics of annealing vacancies in quenched wires and in the annealing of cold-worked wires provides strong experimental evidence that vacancies are formed by plastic deformation.

Under certain conditions where the stress is high enough, the jogs strung out along the dislocation line can be forced together to form a *superjog* where the step height is of the order of 5 to 30b. As the shear stress increases, the dislocation bows out between the superjogs (Fig. 5-25a), generating long segments of edge orientation. As this continues, it produces elongated dislocation loops or dislocation dipoles (Fig. 5-25b). The dipole, in turn, can lower its elastic energy by forming a prismatic loop which later breaks into isolated loops (Fig. 5-25c). Loops can also form at high temperature by the collection of vacancies. The appearance of debris in the form of loops in transmission electron micrographs is a direct result of the formation of edge jogs on screw dislocations.

For very large superjogs, when the step height is greater than about 200 Å, the distance between the two dislocation segments is large enough to prevent mutual interaction. In this case the dislocations behave as separate single-ended sources (see Sec. 5-14).

Jog formation in extended dislocations is a complicated phenomena.[1] Intersection can only occur in extended dislocations by first forming a constriction in the region of the intersection.

5-13 DISLOCATION SOURCES

The low yield strength of pure crystals leads to the conclusion that dislocations must exist in completely annealed crystals and in crystals carefully solidified from the melt. The high strain energy of a dislocation, about 8 eV per atom plane,

[1] M. J. Whelan, *Proc. R. Soc. London Ser. A*, vol. 249A, p. 114, 1959; P. B. Hirsch, *Philos. Mag.*, vol. 7, p. 67, 1962.

precludes the generation of dislocations by thermal activation and indicates that they must be produced by other processes.

There is an important difference between line defects and point defects. The density of dislocations in thermal equilibrium with a crystal is very low. There is no general relationship between dislocation density and temperature as exists for vacancies. Since dislocations are not affected by thermal fluctuations at temperatures below which recrystallization occurs, a metal can have widely different dislocation densities depending upon prior history. Completely annealed material will contain about 10^6 to 10^8 dislocation lines per square centimeter, while heavily cold-worked metal will have a dislocation density of 10^{10} to 10^{12} dislocation lines per square centimeter.

All metals, with the exception of tiny whiskers, initially contain an appreciable number of dislocations produced as the result of the growth of the crystal from the melt or vapor phase. Gradients of temperature and composition may produce misalignments between neighboring dendrite arms growing from the same nucleus. These result in dislocations arranged in networks (Fig. 5-2) or in grain boundaries.

There is a theoretical[1] and experimental basis for believing that irregularities at grain boundaries (grain boundary ledges and steps) are responsible for emitting dislocations. It is thought that emission from grain boundaries is an important source of dislocations in the early stages of plastic deformation. In single crystals small surface steps which serve as stress raisers can act as dislocation sources. The dislocation density in single crystals is appreciably higher at the surface than at the interior[2] because of this effect. However, it is not significant in polycrystalline material because the majority of the grains are not in contact with the surface.

Another way that dislocations can form is by the aggregation and collapse of vacancies to form a disk or prismatic loop. In fcc crystals these loops form on {111} planes where they are called Frank partials (Fig. 5-11).

The stress to homogeneously nucleate a dislocation by rupturing atomic bonds in a perfect lattice is quite high, of the order $G/15$ to $G/30$. However, heterogeneous nucleation of dislocations is possible from high local stresses at second-phase particles or as a result of phase transformations.

5-14 MULTIPLICATION OF DISLOCATIONS

One of the original stumbling blocks in the development of dislocation theory was the formulation of a reasonable mechanism by which sources originally present in the metal could produce new dislocations by the process of slip. Such a mechanism is required when it is realized that the surface displacement at a slip band is due to the movement of about 1,000 dislocations over the slip plane. Thus, the number of dislocation sources initially present in a metal could not account for

[1] J. C. M. Li, *Trans. Metall. Soc. AIME*, vol. 227, p. 239, 1963.

[2] P. N. Pangborn, S. Weissmann, and I. R. Kramer, *Met. Trans.*, vol. 12A, p. 109, 1981.

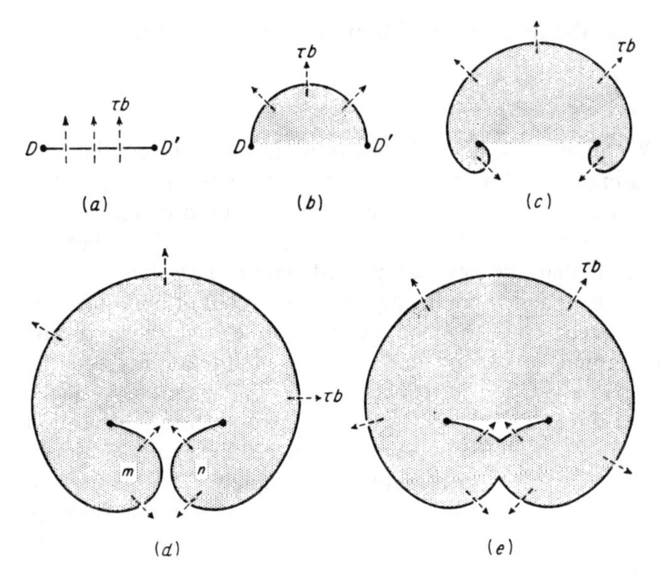

Figure 5-26 Schematic representation of the operation of a Frank-Read source. *(From W. T. Read, Jr., "Dislocations in Crystals," McGraw-Hill Book Company, New York, 1953.)*

the observed slip-band spacing and displacement unless there were some way in which each source could produce large amounts of slip before it became immobilized. Moreover, if there were no source generating dislocations, cold-work should decrease, rather than increase, the density of dislocations in a single crystal. Thus, there must be a method of generating dislocations or of multiplying the number initially present to produce the high dislocation density found in cold-worked metal. The scheme by which dislocations could be generated from existing dislocations was proposed by Frank and Read[1] and is commonly called a *Frank-Read source*.

Consider a dislocation line DD' lying in a slip plane (Fig. 5-26*a*). The plane of the figure is the slip plane. The dislocation line leaves the slip plane at points D and D', so that it is immobilized at these points. This could occur if D and D' were nodes where the dislocation in the plane of the paper intersects dislocations in other slip planes, or the anchoring could be caused by impurity atoms. If a shear stress τ acts in the slip plane, the dislocation line bulges out and produces slip. For a given stress the dislocation line will assume a certain radius of curvature given by Eq. (5-18). The maximum value of shear stress is required when the dislocation bulge becomes a semicircle so that R has the minimum value $l/2$ (Fig. 5-26*b*). From the approximation that $\Gamma \approx 0.5Gb^2$ and Eq. (5-18),

[1] F. C. Frank and W. T. Read, *Phys. Rev.*, vol. 79, pp. 722–723, 1950.

it readily can be seen that the stress required to produce this configuration is

$$\tau \approx \frac{Gb}{l} \tag{5-30}$$

Beyond this point R will increase and the dislocation loop will continue to expand under a decreasing stress (Fig. 5-26*c*). When the loop reaches Fig. 5-26*d*, the segments at m and n will meet and annihilate each other to form a large loop and a new dislocation (Fig. 5-26*e*). The stage shown in Fig. 5-26*d* can best be understood if we assume that the original pinned length DD' has a screw orientation. Then segments m and n are in edge orientation but of opposite sign, so that annihilation will occur. Once the loop moves into the stage shown in Fig. 5-26*c*, the loop can continue to expand under increased shear stress and the pinned segment DD' is in a position to repeat the process. This process can be repeated over and over again at a single source, each time producing a dislocation loop which produces slip of one Burgers vector along the slip plane. However, once the source is initiated it does not continue indefinitely. The back stress produced by the dislocations piling up along the slip plane (see Sec. 5-16) opposes the applied stress and when this equals the critical stress given by Eq. (5-30), the source ceases to operate.

Although double-ended Frank-Read sources have been observed experimentally, their occurrence is not highly frequent. Several other multiplication mechanisms have been observed. A single-ended source can arise where part of a screw dislocation lies in the slip plane while part is immobilized by lying out of the slip plane. The immobilization could also arise from a superjog on a screw dislocation. Rotation around the immobilized segment of the dislocation by the dislocation segment in the slip plane results in a spiral slip step, with the height of the step proportional to the number of revolutions in the slip plane. This spiraling around the immobilized dislocation segment also results in an increase in the total length of dislocation line.

The double-ended Frank-Read source produces a large slip step in the slip plane, but it is restricted to operation in a single plane. This cannot explain the observation that slip bands widen with increased strain as well as increasing the slip offset. A *multiple cross-slip* mechanism provides a ready explanation for this behavior. In Fig. 5-8 a screw dislocation on AB can cross slip into position CD. The edge segments AC and BD on the cross-slip plane can be considered as superjogs which are relatively immobile and pin the dislocations on the AB and CD planes. The dislocation segments lying on these planes can expand as a Frank-Read source. When cross slip can occur readily the Frank-Read sources may not complete a cycle and there will be one continuous dislocation line lying on each of many parallel glide planes connected by jogs. This produces a wide slip band.

At elevated temperature double-pinned edge segments can bow out under the influence of an osmotic driving force due to vacancy flux in the manner described for the Frank-Read source. The supersaturation of vacancies needed to operate this source is as small as 2 percent at high temperature.[1] This multiplication mechanism is known as a *Bardeen-Herring source*.[2]

5-15 DISLOCATION-POINT DEFECT INTERACTIONS

Isolated solute atoms and vacancies are centers of elastic distortion just as are dislocations. Therefore, point defects and dislocations will interact elastically and exert forces on each other. To a good approximation the strains around a point defect distort the lattice spherically, just as though an elastic sphere of radius a' had been forced into a spherical hole of radius a in an elastic continuum. The resulting strain is $\varepsilon = (a' - a)/a$. If the point defect is a vacancy, the radius a is the radius of the atom normally at the lattice site, while if the defect is an interstitial atom, a corresponds to the average radius of an empty interstitial site. The volume change produced by the point effect is given by

$$\Delta V = 4\pi a^3 \varepsilon \tag{5-31}$$

Because we are concerned only with spherical distortions, the interaction only occurs with the hydrostatic component of the dislocation stress field. The elastic interaction energy between the dislocation and the point defect is

$$U_i = \sigma_m \Delta V \tag{5-32}$$

The hydrostatic stress of a positive edge dislocation at r, θ (Fig. 5-14) from the dislocation is

$$\sigma_m = \frac{(1 + \nu) Gb \sin \theta}{3\pi (1 - \nu) r} \tag{5-33}$$

so that the interaction energy is

$$U_i = \frac{4(1 + \nu) Gba^3 \varepsilon \sin \theta}{3(1 - \nu) r} \tag{5-34}$$

However, this expression includes only the energy external to the point defect. When the strain energy due to elastic distortion of the solute atom is considered, the complete expression for interaction energy is

$$U_i = 4Gba^3 \varepsilon \frac{\sin \theta}{r} = A \frac{\sin \theta}{r} \tag{5-35}$$

Since Eq. (5-35) is derived from elasticity theory,[3] it is not strictly correct near the

[1] Hirth and Lothe, op. cit., pp. 565–568.

[2] J. Bardeen and C. Herring, in "Imperfections in Nearly Perfect Crystals," p. 261, John Wiley & Sons, Inc., New York, 1952.

[3] J. Weertman and J. R. Weertman, "Elementary Dislocation Theory," The Macmillan Company, New York, pp. 173–177; B. A. Bilby, *Proc. Phys. Soc., London*, vol. A63, p. 191, 1950; D. M. Barrett, G. Wong, and W. D. Nix, *Acta Met.*, vol. 30, pp. 2035–2041, 1982.

core of the dislocation where linear elasticity theory is no longer applicable. Since the maximum interaction energy occurs in this location, Eq. (5-35) provides only an estimate.

A negative value of interaction energy indicates attraction between the point defect and the dislocation, while a positive value denotes repulsion. A solute atom larger than the solvent atom ($\varepsilon > 1$) will be repelled from the compression side of a positive edge dislocation ($0 < \theta < \pi$) and will be attracted to the tension side ($\pi < \theta < 2\pi$). An atom smaller than the solvent atom ($\varepsilon < 1$) will be attracted to a site on the compression side of a positive edge dislocation. Similarly vacancies[1] will be attracted to regions of compression and interstitials will collect at regions of tension.

Since the strain fields around a point defect are spherically symmetrical, such a defect will produce no net force on a screw dislocation because its stress field is pure shear. Thus there is no hydrostatic stress field around a screw dislocation which can be relaxed by the presence of a point defect. However, some point defects (such as interstitial carbon atoms in a bcc lattice) produce a nonspherical distortion[2] so that there will be an interaction energy between a screw dislocation and the defect.

Even if the point defect has the same volume as the atoms in the lattice, an interaction with a dislocation can occur if the defect has different elastic constants than the matrix.[3] There will be an attraction if the point defect is elastically softer than the matrix and a repulsion if the point defect is elastically harder. For this type of interaction a binding force will exist between a screw dislocation and a spherically symmetrical point defect.

Since the energy associated with a point defect is affected by its proximity to a dislocation, Eq. (5-35), we should expect the concentration of defects to be different in the vicinity of a dislocation line. Assuming a Boltzmann distribution for the concentration of point defects, the concentration in the vicinity of a dislocation C is related to the average concentration C_0 by the relationship

$$C = C_0 e^{-U_i/kT} \tag{5-36}$$

The concentration of point defects around a dislocation exceeds the average value when U_i is negative and is less than the average when U_i is positive. An increased concentration of solute atoms around a dislocation is called an *impurity atmosphere* or an *impurity cloud*. The concentration C cannot exceed one solute atom per lattice site or per interstitial site. At temperatures so low that $C_0 e^{-U_i/kT}$ exceeds 1, the sites near the dislocation core have been saturated with solute atoms. Under these conditions the impurity atmosphere is said to have "condensed" on the dislocation lines.

[1] R. Bullough and R. C. Newman, *Philos. Mag.*, vol. 7, 529, 1962.
[2] A. W. Cochardt, G. Schoek, and H. Wiedersich, *Acta Metall.*, vol. 3, pp. 533–537, 1955.
[3] R. L. Fleischer, *Acta Metall.*, vol. 11, p. 203, 1963.

Interaction of dislocations with solute atoms is important for the explanation of yield-point behavior, strain aging, and solid-solution strengthening. These topics will be considered more fully in Chap. 6.

5-16 DISLOCATION PILE-UPS

Dislocations frequently pile up on slip planes at barriers such as grain boundaries, second phases, or sessile dislocations. The leading dislocation in the pile-up is acted on not only by the applied shear stress but also by the interaction force with the other dislocations in the pile-up. This leads to a high concentration of stress on the leading dislocation in the pile-up. When many dislocations are contained in the pile-up, the stress on the dislocation at the head of a pile-up can approach the theoretical shear stress of the crystal. This high stress either can initiate yielding on the other side of the barrier, or in other instances, it can nucleate a crack at the barrier. Dislocations piled up against a barrier produce a *back stress* acting to oppose the motion of additional dislocations along the slip plane in the slip direction.

The dislocations in the pile-up will be tightly packed together near the head of the array and more widely spaced toward the source (Fig. 5-27). The distribution of dislocations of like sign in a pile-up along a single slip plane has been studied by Eshelby, Frank, and Nabarro.[1] The number of dislocations that can occupy a distance L along the slip plane between the source and the obstacle is

$$n = \frac{k\pi\tau_s L}{Gb} \tag{5-37}$$

where τ_s is the average resolved shear stress in the slip plane and k is a factor close to unity. For an edge dislocation $k = 1 - \nu$, while for a screw dislocation $k = 1$. When the source is located at the center of a grain of diameter D, the number of dislocations in the pile-up is given by

$$n = \frac{k\pi\tau_s D}{4Gb} \tag{5-38}$$

The factor 4 is used instead of the expected factor of 2 because the back stress on the source arises from dislocations piled up on both sides of the source.

A piled-up array of n dislocations can be considered for many purposes to be a giant dislocation with Burgers vector nb. At large distances from the array the stress due to the dislocations can be considered to be due to a dislocation of strength nb located at the center of gravity three-quarters of the distance from the

[1] J. D. Eshelby, F. C. Frank, and F. R. N. Nabarro, *Philos. Mag.*, vol. 42, p. 351, 1951; calculations for more complicated types of pile-ups have been given by A. K. Head, *Philos. Mag.*, vol. 4, pp. 295–302, 1959; experimental confirmation of theory has been obtained by Meakin and Wilsdorf, *op. cit.*, pp. 745–752.

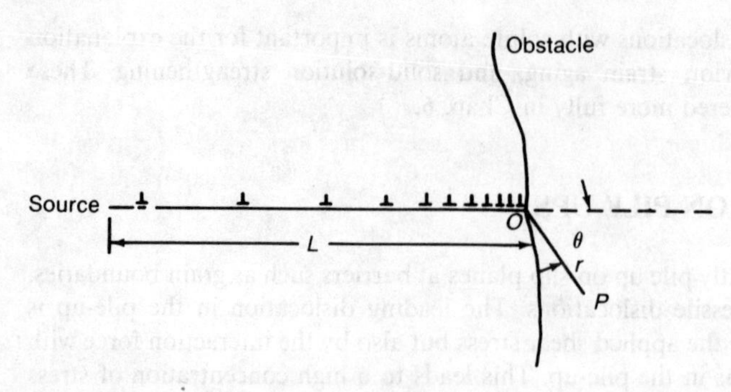

Figure 5-27 Dislocation pile-up at an obstacle.

source to the head of the pile-up. The total slip produced by a pile-up can be considered that due to a single dislocation *nb* moving a distance $3L/4$. Very high forces act on the dislocations at the head of the pile-up. This force is equal to $nb\tau_s$, where τ_s is the average resolved shear stress on the slip plane. Koehler[1] has pointed out that large tensile stresses of the order of $n\tau$ will be produced at the head of a pile-up. Stroh[2] has made a somewhat more detailed analysis of the stress distribution at the head of a dislocation pile-up. Using the coordinate system given in Fig. 5-27, he showed that the tensile stress normal to a line *OP* is given by

$$\sigma = \frac{3}{2}\left(\frac{L}{r}\right)^{1/2} \tau_s \sin\theta \cos\frac{\theta}{2} \tag{5-39}$$

The maximum value of σ occurs at $\cos\theta = \frac{1}{3}$ or $\theta = 70.5°$. For this situation

$$\sigma_{\max} = \frac{2}{\sqrt{3}}\left(\frac{L}{r}\right)^{1/2} \tau_s \tag{5-40}$$

The shear stress acting in the plane *OP* is given by

$$\tau = \beta\tau_s\left(\frac{L}{r}\right)^{1/2} \tag{5-41}$$

where β is an orientation-dependent factor which is close to unity.

The number of dislocations which can be supported by an obstacle will depend on the type of barrier, the orientation relationship between the slip plane and the structural features at the barrier, the material, and the temperature. Breakdown of a barrier can occur by slip on a new plane, by climb of dislocations around the barrier, or by the generation of high enough tensile stresses to produce a crack.

[1] J. S. Koehler. *Phys. Rev.*, vol. 85, p. 480, 1952.
[2] A. N. Stroh, *Proc. Roy. Soc.* (*London*), vol. 223, pp. 404–414, 1954.

The equations describing dislocation pile-ups are obtained using the concept of continuous dislocations. In this method of calculation[1] discrete dislocations with finite Burgers vectors are replaced by continuously distributed dislocations with infinitesimal Burgers vectors. This concept provides a link between the discrete nature of dislocations and continuum theory. It is particularly useful in dealing with cracks and fracture.

BIBLIOGRAPHY

Cottrell, A. H.: "Dislocations and Plastic Flow in Crystals," Oxford University Press, New York, 1953.

Friedel, J.: "Dislocations," Addison-Wesley Publishing Company, Inc., Reading, Mass., 1964.

Hirth, J. P., and J. Lothe: "Theory of Dislocations," 2d ed. Wiley-Interscience, New York, 1982.

Hull, D., and D. J. Bacon: "Introduction to Dislocations," 3d ed., Pergamon Press, New York, 1984.

Nabarro, F. R. N. (ed): "Dislocations in Solids," North-Holland Publishing Company, 1983.

Nabarro, F. R. N.: "Theory of Crystal Dislocations," Oxford University Press, New York, 1967.

Read, W. T., Jr.: "Dislocations in Crystals," McGraw-Hill Book Company, New York, 1953.

Van Bueren, H. G.: "Imperfections in Crystals," Interscience Publishers, Inc., New York, 1960.

Weertman, J., and J. R. Weertman: "Elementary Dislocation Theory," The Macmillan Company, New York, 1964.

[1] R. Bullough, *Philos. Mag.*, vol. 9, p. 917, 1964.

CHAPTER
SIX
STRENGTHENING MECHANISMS

6-1 INTRODUCTION

Chapter 4 considered the plastic deformation of single crystals in terms of the movement of dislocations and the basic deformation mechanisms of slip and twinning. Single-crystal specimens represent the most ideal condition for study. The simplifications which result from the single-crystal condition materially assist in describing the deformation behavior in terms of crystallography and defect structure. However, with the exception of solid-state electronic devices, single crystals are rarely used for engineering applications because of limitations involving their strength, size, and production.[1] Commercial metal products invariably are made up of a tremendous number of individual crystals or grains. The individual grains of the polycrystalline aggregate do not deform in accordance with the relatively simple laws which describe plastic deformation in single crystals because of the restraining effect of the surrounding grains.

Chapter 5 was concerned with the basic relationships that govern dislocation behavior. It should be clear from this that strength is inversely related to dislocation mobility and that even in high-purity single crystals there are a number of possible factors that can affect the strength and mechanical behavior. Thus, the crystal structure determines the number and type of slip systems, fixes the Burgers vector, and determines the lattice friction stress (Peierls stress) which sets the base strength level and temperature dependence of strength. In close-packed structures the stacking-fault energy determines the extent of dislocation dissociation, which influences the ease of cross slip and the subsequent strain-

[1] An outstanding exception is the use of single-crystal turbine blades for jet engines where the elimination of grain boundaries greatly increases the resistance to thermal shock and fatigue. See F. L. Versnyder and M. E. Shank, *Mater. Sci. Eng.*, vol. 6, pp. 213–243, 1970.

hardening rate. The purity and method of preparation determine the initial dislocation density and substructure. These limited variables introduce enough complexity that mechanical behavior cannot in general be predicted with high precision as a function of strain, strain rate, temperature, and stress rate.

However, the introduction of still greater complexity is needed to produce materials of highest strength and usefulness. Thus, fine grain size is often desired for high strength, large additions of solute atoms are added to increase strength and bring about new phase relationships, fine particles may be added to increase strength and phase transformations may be utilized to increase strength. These various strengthening mechanisms and other aspects of the deformation of polycrystalline materials are considered in this chapter.

6-2 GRAIN BOUNDARIES AND DEFORMATION

The boundaries between grains in a polycrystalline aggregate are a region of disturbed lattice only a few atomic diameters wide. In the general case, the crystallographic orientation changes abruptly in passing from one grain to the next across the grain boundary. The ordinary high-angle grain boundary represents a region of random misfit between the adjoining crystal lattices.[1] As the difference in orientation between the grains on each side of the boundary decreases, the state of order in the boundary increases. For the limiting case of a low-angle boundary where the orientation difference across the boundary may be less than 1° (see Sec. 6-4), the boundary is composed of a regular array of dislocations.

Figure 6-1*a* schematically illustrates the structure at a high-angle grain boundary. Note the unorganized structure, with a few atoms belonging to both grains, while most belong to neither. Those atoms that belong to both grains are called coincidence sites. This grain boundary structure contains *grain-boundary dislocations* (Fig. 6-1*b*). These are not mobile dislocations producing extensive slip; rather, their chief role is that they group together within the boundary to form a step or *grain-boundary ledge*. As the misorientation angle of the grain boundary increases the density of the ledges increases. Grain-boundary ledges are effective sources of dislocations.

High-angle grain boundaries are boundaries of rather high surface energy. For example, a grain boundary in copper has an interfacial surface energy of about 600 mJ/m^2, while the energy of a twin boundary is only about 25 mJ/m^2. Because of their high energy, grain boundaries serve as preferential sites for solid-state reactions such as diffusion, phase transformations, and precipitation reactions. The high energy of a grain boundary usually results in a higher concentration of solute atoms at the boundary than in the interior of the grains.

[1] For a review of the proposed models of grain boundaries see H. Hu (ed.), "The Nature and Behavior of Grain Boundaries," Plenum Press, New York, 1972; H. Gleiter, *Mater. Sci. Eng.*, vol. 52, p. 91, 1982.

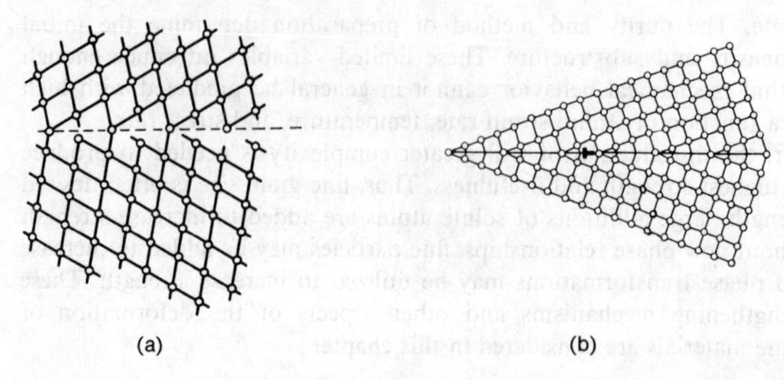

Figure 6-1 (*a*) Schematic atomic model of a grain boundary; (*b*) dislocation model of a grain boundary.

This makes it difficult to separate the pure mechanical effect of grain boundaries on properties from an effect due to impurity segregation.

When a single crystal is deformed in tension, it is usually free to deform on a single slip system for a good part of the deformation and to change its orientation by lattice rotation as extension takes place. However, individual grains in a polycrystalline specimen are not subjected to a single uniaxial stress system when the specimen is deformed in tension. In a polycrystal, continuity must be maintained, so that the boundaries between the deforming crystals remain intact. Although each grain tries to deform homogeneously in conformity with the deformation of the specimen as a whole, the constraints imposed by continuity cause considerable differences in the deformation between neighboring grains and within each grain. Studies of the deformation in coarse-grain aluminum[1] showed that the strain in the vicinity of a grain boundary usually differs markedly from the strain at the center of the grain. Although the strain is continuous across boundaries, there may be a steep strain gradient in this region. As the grain size decreases and strain increases, the deformation becomes more homogeneous. Because of the constraints imposed by the grain boundaries slip occurs on several systems, even at low strains. Also, this causes slip to occur on non-close-packed planes in regions near the grain boundary. Slip in polycrystalline aluminum has been observed on the {100}, {110}, and {113} planes. The fact that different slip systems can operate in adjacent regions of the same grain results in complex lattice rotations which result in the formation of deformation bands. Since more slip systems are usually operative near the grain boundary, the hardness usually will be higher near the boundary than in the center of a grain. As the grain diameter is reduced more of the effects of grain boundaries will be felt at the grain center. Thus, the strain hardening of a fine grain size metal will be greater than in a coarse-grain polycrystalline aggregate.

[1] W. Boas and M. E. Hargreaves, *Proc. R. Soc. London Ser. A*, vol. A193, p. 89, 1948; V. M. Urie and H. L. Wain, *J. Inst. Met.*, vol. 81, p. 153, 1952.

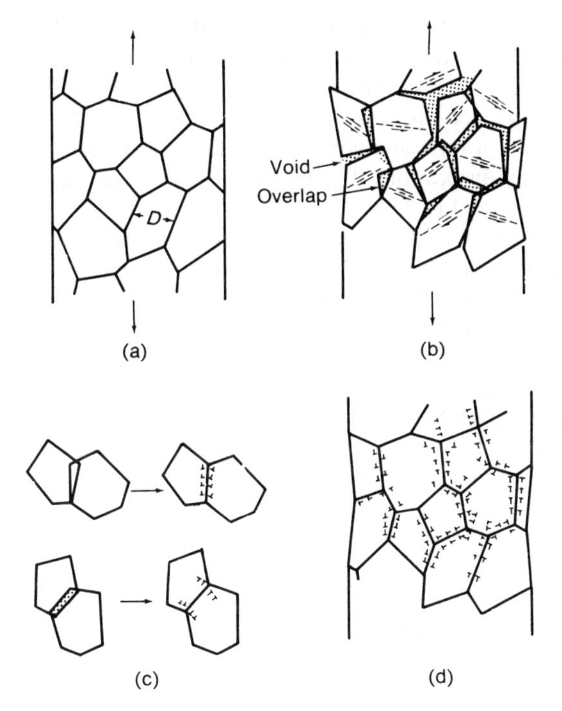

Figure 6-2 Ashby's model of deformation of a polycrystal (*a*). Polycrystal deforms in macroscopic uniform way, produces overlap, and voids at boundaries (*b*). These can be corrected by introducing geometrically necessary dislocations at (*c*) and (*d*). Note that statistical dislocations are not shown. *(From M. F. Ashby, Phil. Mag., vol. 21, p. 413, 1970.)*

Von Mises[1] showed that for a crystal to undergo a general change of shape by slip requires the operation of five independent slip systems. This arises from the fact that an arbitrary deformation is specified by the six components of the strain tensor, but because of the requirement of constant volume ($\Delta V = 0 = \varepsilon_{11} + \varepsilon_{22} + \varepsilon_{33}$), there are only five independent strain components. Crystals which do not possess five independent slip systems are never ductile in polycrystalline form, although small plastic elongation may be obtained if there is twinning or a favorable preferred orientation. Cubic metals easily satisfy this requirement, which accounts for their general high ductility. Hexagonal close-packed and other low-symmetry metals do not satisfy this requirement and have low ductility at room temperature in polycrystalline form. Polycrystalline Zn and Mg become ductile at some elevated temperature at which nonbasal slip can become operative and increase the number of slip systems to at least five.

The concerns for continuity, which lead to more complex deformation modes in polycrystals than in single crystals, led Ashby[2] to suggest a dislocation model for polycrystal deformation (Fig. 6-2). The distinction is made between *statistically stored dislocations*, those which encounter and trap one another randomly, as in a single crystal, and *geometrically necessary dislocations*, which are generated as

[1] R. Von Mises, *Z. Angew. Math. Mech.*, vol. 8, p. 161, 1928.
[2] M. F. Ashby, *Phil. Mag.*, ser. 8, vol. 21, pp. 399–424, 1970.

a result of nonuniform strain in the crystal. In Ashby's model, the polycrystal is deformed by disassembling it into constituent grains and allowing each to slip according to Schmid's law (Fig. 6-2*b*). This process generates statistically stored dislocations. However, this generates overlaps and voids between the grains. Now, each of these discrepancies is taken in turn, and corrected by the introduction of appropriate geometrically necessary dislocations (Fig. 6-2*c*) until the grains again fit together. The reassembled polycrystal is shown in Fig. 6-2*d*.

At temperatures above about one-half of the melting point, deformation can occur by sliding along the grain boundaries. *Grain-boundary sliding* becomes more prominent with increased temperature and decreasing strain rate, as in creep. The restriction of deformation to the grain-boundary region is one of the primary sources of high-temperature fracture. Because impurities tend to segregate to grain boundaries, intergranular fracture is strongly influenced by composition. A rough way of distinguishing when grain-boundary sliding becomes prominent is with the *equicohesive temperature*. Above this temperature the grain boundary region is weaker than the grain interior and strength increases with increasing grain size. Below the equicohesive temperature the grain-boundary region is stronger than the grain interior and strength increases with decreasing grain size (increasing grain-boundary area).

The strengthening mechanisms discussed in this chapter are those which impede the conservative motion of dislocations. Generally speaking they are operative at temperatures below about $0.5T_m$, where T_m is the melting temperature in degrees Kelvin. High temperature deformation of metals is considered in Chap. 13.

6-3 STRENGTHENING FROM GRAIN BOUNDARIES

Direct evidence for the mechanical strengthening of grain boundaries is provided by experiments[1] on bicrystals in which the orientation difference between a longitudinal grain boundary was varied in a systematic manner. The yield stress of the bicrystals increased linearly with increasing misorientation across the grain boundary, and extrapolation to zero misorientation angle gave a value close to that of the yield stress of a single crystal. These results imply that a simple grain boundary has little inherent strength and that the strengthening due to grain boundaries results from mutual interference to slip within the grains.

Several attempts have been made to calculate the stress-strain curve for a polycrystal from stress-strain curves for single crystals. In Chap. 4 we saw that the resolved shear stress in a single crystal was given by

$$\tau = \sigma \sin \chi \cos \lambda = \frac{\sigma}{M} \tag{6-1}$$

where M is an orientation factor (the reciprocal of the Schmid factor). For a polycrystal the orientation factor M varies from grain to grain and it is necessary

[1] B. Chalmers, *Proc. R. Soc. London Ser. A*, vol. A193, p. 89, 1948; R. Clark and B. Chalmers, *Acta. Metall.*, vol. 2, p. 80, 1954.

to determine some average orientation factor $\bar{M}$. The best estimate for an fcc lattice is $\bar{M} = 3.1$, obtained by G. I. Taylor[1] based on the use of the von Mises compatibility condition and assuming that all grains undergo the same deformation as the overall deformation. The energy expended in deforming a polycrystal must be equal to the sum of the increments of work performed on each of the n slip systems.

$$\sigma \, d\varepsilon = \sum_{i=1}^{n} \tau_i \, d\gamma_i \tag{6-2}$$

If we assume that the critical shear stress is the same on each slip system, then

$$\frac{\sigma}{\tau} = \frac{\sum_{i=1}^{n} |d\gamma_i|}{d\varepsilon} = \bar{M} \tag{6-3}$$

The value of $\bar{M}$ was determined[2] by finding the combination of slip systems that minimized the value of $\Sigma|d\gamma_i|$ but still satisfied the required continuity at the grain boundaries. To get reasonable agreement, single-crystal curves must be used which involve the same slip mechanisms as the polycrystalline specimen. Since polycrystals involve multiple slip, the single-crystal curves for fcc metals should have the $\langle 111 \rangle$ or $\langle 100 \rangle$ orientation for which easy glide is a minimum. The polycrystalline curve is calculated from the reations

$$\sigma = \bar{M}\tau \quad \text{and} \quad \varepsilon = \frac{\gamma}{\bar{M}} \tag{6-4}$$

Quite reasonable agreement has been found.[3] By combining these equations we find that the strain-hardening rate for an fcc polycrystal should be about 9.5 times that for a single crystal.

$$\frac{d\sigma}{d\varepsilon} = \bar{M}^2 \frac{d\tau}{d\gamma} \tag{6-5}$$

Hall-Petch Relation

A general relationship between yield stress (and other mechanical properties) and grain size was proposed by Hall[4] and greatly extended by Petch.[5]

$$\sigma_0 = \sigma_i + kD^{-1/2} \tag{6-6}$$

[1] G. J. Taylor, *J. Inst. Met.*, vol. 62, p. 307, 1938.

[2] Essentially the same value of $\bar{M} = 3.1$ was found by a more rigorous treatment. J. F. W. Bishop and R. Hill, *Philos. Mag.*, vol. 42, pp. 414–427, 1298–1307, 1951. The Taylor and Bishop-Hill approaches have been shown to be equivalent and have been generalized by G. Y. Chin and W. L. Mammel, *Trans. Metall. Soc. AIME*, vol. 245, pp. 1211–1214, 1969.

[3] U. F. Kocks, *Acta Metall.*, vol. 6, p. 85, 1958; for a detailed analysis see U. F. Kocks, *Metall. Trans.*, vol. 1, pp. 1121–1143, 1970.

[4] E. O. Hall, *Proc. Phys. Soc. London*, vol. 643, p. 747, 1951.

[5] N. J. Petch, *J. Iron Steel Inst. London*, vol. 173, p. 25, 1953; N. Hansen and B. Ralph, *Acta Metall.*, vol. 30, pp. 411–417, 1982.

where σ_0 = the yield stress
σ_i = the "friction stress," representing the overall resistance of the crystal lattice to dislocation movement
k = the "locking parameter," which measures the relative hardening contribution of the grain boundaries
D = grain diameter

The Hall-Petch equation was originally based on yield-point measurements in low-carbon steel. It has been found to express the grain-size dependence of the flow stress at any plastic strain out to ductile fracture and also to express the variation of brittle fracture stress with grain size and the dependence of fatigue strength on grain size.[1] The Hall-Petch equation also has been found to apply not only to grain boundaries but to other kinds of boundaries such as ferrite-cementite in pearlite, mechanical twins, and martensite plates.

The original dislocation model for the Hall-Petch equation was based on the concept that grain boundaries act as barriers to dislocation motion. Consider a dislocation source at the center of a grain D which sends out dislocations to pile-up at the grain boundary. The stress at the tip of this pile-up must exceed some critical shear stress τ_c to continue slip past the grain-boundary barrier. From Eq. (5-38) we have

$$\tau_c = n\tau_s = \frac{\pi\tau_s^2 D}{4Gb} \tag{6-7}$$

The resolved shear stress τ_s required to overcome the barrier can be taken equal to the applied stress less the friction stress to overcome lattice resistance to dislocation motion τ_i.

$$\tau_s = \tau - \tau_i \tag{6-8}$$

Therefore,

$$\tau_c = \frac{\pi(\tau - \tau_i)^2 D}{4Gb}$$

and

$$\tau = \tau_i + \left(\frac{\tau_c 4Gb}{\pi D}\right)^{1/2} = \tau_i + k'D^{-1/2} \tag{6-9}$$

Expressing (6-9) in terms of normal stresses, results in Eq. (6-6).

The factor k is the slope of the straight line that is obtained when σ_0 is plotted against $D^{-1/2}$. Many physical interpretations have been ascribed to the term, which is roughly independent of temperature. The term σ_i is the intercept[2] along the ordinate in a plot of σ_0 vs $D^{-1/2}$. It is interpreted as the friction stress needed to move unlocked dislocations along the slip plane. This term depends strongly on temperature, strain, and alloy (impurity) content.

[1] R. W. Armstrong, *Metall. Trans.*, vol. 1, pp. 1169–1176, 1970.

[2] The various techniques used to determine σ_i are discussed by R. Phillips and J. F. Chapman, *J. Iron Steel Inst. London*, vol. 203, pp. 511–513, 1965.

While the Hall-Petch equation is a very general relationship, it must be used with some caution. For example, if Eq. (6-6) were extrapolated to the smallest grain size imaginable (approximately 40 Å), it would predict strength levels close to the theoretical shear strength. Such an extrapolation is in error because the equations for the stresses in a pile-up on which Eq. (6-6) is based were derived for large pile-ups containing more than 50 dislocations. For small pile-ups other equations must be considered.[1]

The growing realization of the importance of grain boundaries as dislocation sources has cast considerable doubt on the dislocation pile-up model for the Hall-Petch equation. A more general model proposed by Li[2] avoids the description of the stresses at grain boundaries and instead concentrates on the influence of grain size on the dislocation density, and hence, on the yield or flow stress. The flow stress is given in terms of dislocation density by

$$\sigma_0 = \sigma_i + \alpha G b \rho^{1/2} \tag{6-10}$$

where σ_i has the same meaning as in Eq. (6-6), α is a numerical constant generally between 0.3 and 0.6, and ρ is the dislocation density. The justification for this equation was given in Sec. 4-14. The tie-in with grain size is based on the experimental observation that ρ is an inverse function of the grain size. Thus, $\rho = 1/D$

$$\sigma_0 = \sigma_i + \alpha G b D^{-1/2} = \sigma_i + k' D^{-1/2} \tag{6-11}$$

Grain-Size Measurement

Grain size is measured with a light microscope by counting the number of grains within a given area, by determining the number of grains (or grain boundaries) that intersect a given length of random line, or by comparing with standard-grain-size charts. Most grain-size measurements invoke assumptions relative to the shape and size distribution of the grains and, therefore, must be interpreted with some degree of caution. As pointed out by DeHoff and Rhines,[3] the most applicable technique is that which provides structural information which may be correlated with property data and which may be accomplished by relatively simple measurements on the polished surface.

Most grain-size measurements seek to correlate the interaction of grain boundaries with a specific mechanical property. Thus, a measurement of the grain-boundary area per unit volume S_v is a useful parameter. Smith and Guttman[4] have shown that S_v can be calculated without assumptions concerning

[1] R. W. Armstrong, Y. T. Chou, R. A. Fisher, and N. Louat, *Philos. Mag.*, vol. 14, p. 943, 1966.
[2] J. C. M. Li, *Trans. Metall. Soc. AIME*, vol. 227, pp. 239–247, 1963.
[3] R. T. DeHoff and F. N. Rhines, "Quantitative Microscopy," pp. 201–266, McGraw-Hill Book Company, 1968.
[4] C. S. Smith and L. Guttman, *Trans. AIME*, vol. 197, p. 81, 1953.

grain shape and size distribution from measurements of the mean number of intercepts of random test lines with grain boundaries per unit length of test line N_L.

$$S_v = 2N_L \tag{6-12}$$

If a mean grain diameter D is required from S_v, this may be obtained by assuming constant-size spherical grains and noting that each boundary is shared by two adjacent grains

$$2S_v = \frac{4\pi(D/2)^2}{4\pi/3(D/2)^3}$$

or

$$D = \frac{3}{S_v} = \frac{3}{2N_L} \tag{6-13}$$

An average grain size may also be obtained from measurements of the number of grains per unit area on a polished surface N_A. Fullman[1] has shown that the mean area on a plane of polish passing through spheres of constant size is

$$A = \frac{2}{3}\pi\left(\frac{D}{2}\right)^2 = \frac{\pi}{6}D^2 \tag{6-14}$$

Thus, measurements of N_A, the reciprocal of A, give a grain size defined as

$$D = \sqrt{\frac{6}{\pi N_A}} \tag{6-15}$$

Many studies have employed the mean intercept length of random test lines as a measure of grain size. This determination is made by dividing the total length of test line by the number of grains traversed. By comparison with Eq. (6-13), it can be seen that the "mean intercept length" grain size will be somewhat smaller than the actual grain size.

A very common method of measuring grain size in the United States is to compare the grains at a fixed magnification with the American Society for Testing and Materials (ASTM) grain-size charts. The ASTM grain-size number n is related to N^*, the number of grains per square inch at a magnification of 100× by the relationship

$$N^* = 2^{n-1} \tag{6-16}$$

Table 6-1 compares the ASTM grain-size numbers with several other useful measures of grain size.

[1] R. L. Fullman, *Trans. AIME*, vol. 197, pp. 447 and 1267, 1953.

Table 6-1 Comparison of grain-size measuring systems†

ASTM no.	Grains/in^2 at 100×	Grains/mm^2	Grains/mm^3	Average grain diameter, mm
−3	0.06	1	0.7	1.00
−2	0.12	2	2	0.75
−1	0.25	4	5.6	0.50
0	0.5	8	16	0.35
1	1	16	45	0.25
2	2	32	128	0.18
3	4	64	360	0.125
4	8	128	1,020	0.091
5	16	256	2,900	0.062
6	32	512	8,200	0.044
7	64	1,024	23,000	0.032
8	128	2,048	65,000	0.022
9	256	4,096	185,000	0.016
10	512	8,200	520,000	0.011
11	1,024	16,400	1,500,000	0.008
12	2,048	32,800	4,200,000	0.006

† "Metals Handbook," American Society for Metals, Metals Park, Ohio, 1948.

Example If a steel has a value of $\sigma_i = 150\ \text{MN/m}^2$ and $k = 0.70\ \text{MN/m}^{3/2}$, what is the value of yield stress if the grain size is ASTM no. 6?

$$N^* = 2^{6-1} = 32\ \text{grains/in}^2 \quad \text{at } 100\times$$

$$N = \frac{32}{(0.01)(0.01)} = 320{,}000\ \text{grains/in}^2 \quad \text{at } 1\times$$

$$N = 320{,}000\left(\frac{\text{grain}}{\text{in}^2}\right) \times \frac{1}{(25.4)^2}\left(\frac{\text{in}^2}{\text{mm}^2}\right) \times \left(\frac{10^{+3}}{1}\right)^2 \frac{\text{mm}^2}{\text{m}^2}$$

$$= 496 \times 10^6\ \text{grain/m}^2$$

From Eq. (6-15) $D \approx \sqrt{1/N}$ or $D^2 \approx 1/N$

$$D^2 \approx 20 \times 10^{-10}\ \text{m}^2 \qquad D \approx 44.7 \times 10^{-6}\ \text{m} \qquad \frac{1}{\sqrt{D}} = 149\ \text{m}^{-1/2}$$

$$\sigma_0 = \sigma_i + kD^{-1/2} = 150 + 0.70(149) = 150 + 104.3 = 254.3\ \text{MPa}$$

6-4 LOW-ANGLE GRAIN BOUNDARIES

A definite substructure can exist within the grains surrounded by high-energy grain boundaries. The subgrains are low-angle boundaries in which the difference in orientation across the boundary may be only a few minutes of arc or, at most, a few degrees. Because of this small orientation difference, special x-ray techniques are required to detect the existence of a substructure network. Subgrain boundaries are lower-energy boundaries than grain boundaries, and therefore they etch less

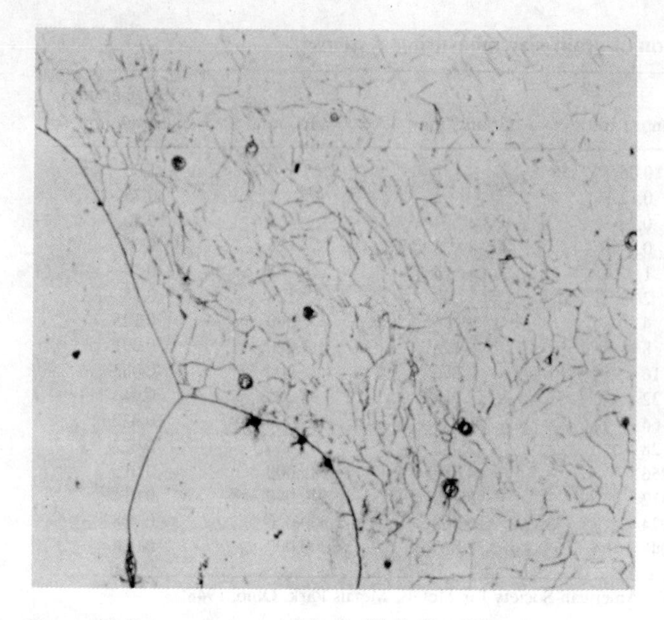

Figure 6-3 Substructure network in Fe-3% Si alloy (250×).

readily than grain boundaries. However, in many metals they can be detected in the microstructure by metallographic procedures (Fig. 6-3).

A low-angle boundary contains a relatively simple arrangement of dislocations. The simplest situation is the case of a tilt boundary. Figure 6-4*a* illustrates two cubic crystals with a common [001] axis. The slight difference in orientation between the grains is indicated by the angle θ. In Fig. 6-4*b* the two crystals have been joined to form a bicrystal containing a low-angle boundary. Along the boundary the atoms adjust their position by localized deformation to produce a smooth transition from one grain to the other. However, elastic deformation cannot accommodate all the misfit, so that some of the atom planes must end on the grain boundary. Where the atom planes end, there is an edge dislocation. Therefore, low-angle tilt boundaries can be considered to be an array of edge dislocations. From the geometry of Fig. 6-4*b* the relationship between θ and the spacing between dislocations is given by

$$\theta = 2\tan^{-1}\frac{b}{2D} \approx \frac{b}{D} \tag{6-17}$$

where b is the magnitude of the Burgers vector of the lattice.

The validity of the dislocation model of the low-angle boundary is found in the fact it is possible to calculate the grain-boundary energy as a function of the difference in orientation between the two grains. So long as the angle does not become greater than about 20°, good agreement is obtained between the mea-

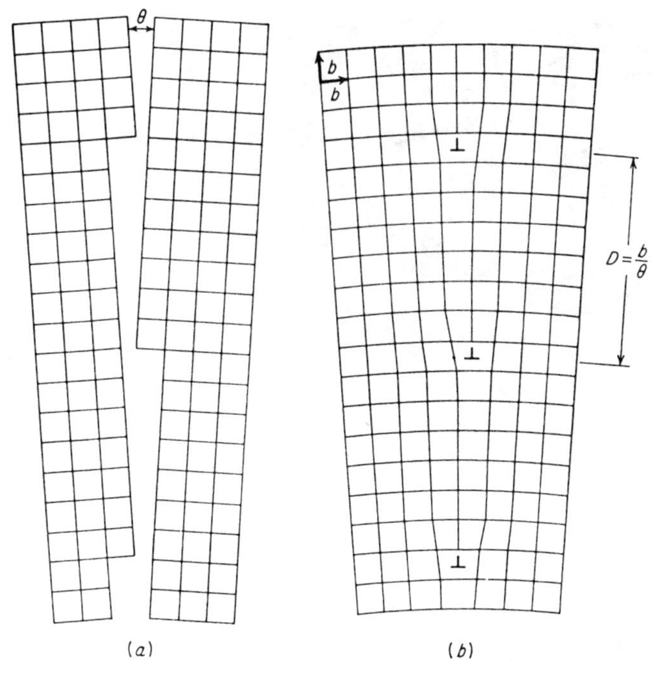

Figure 6-4 Diagram of low-angle grain boundary. (*a*) Two grains having a common [001] axis and angular difference in orientation of θ; (*b*) two grains joined together to form a low-angle grain boundary made up of an array of edge dislocations. *(From W. T. Read, Jr., "Dislocations in Crystals," p. 157, McGraw-Hill Book Company, New York, 1953.)*

sured values of grain-boundary energy and the values calculated on the basis of the dislocation model. Other evidence for the dislocation nature of low-angle boundaries comes from metallographic observations. If the angle is low, so that the spacing between dislocations is large, it is often possible to observe that the boundary is composed of a row of etch pits, with each pit corresponding to the site of an edge dislocation (Fig. 6-5).

Subboundaries or low-angle boundaries can be produced in a number of ways. They may be produced during crystal growth, during high-temperature creep deformation, or as the result of a phase transformation. The veining in ferrite grains is a well-known example of a substructure resulting from the internal stresses accompanying a phase transformation. Perhaps the most general method of producing a substructure network is by introducing a small amount of deformation (from about 1 to 10 percent prestrain) and following this with an annealing treatment to rearrange the dislocations into subgrain boundaries. The amount of deformation and temperature must be low enough to prevent the formation of new grains by recrystallization. This process has been called recrystallization *in situ*, or polygonization.

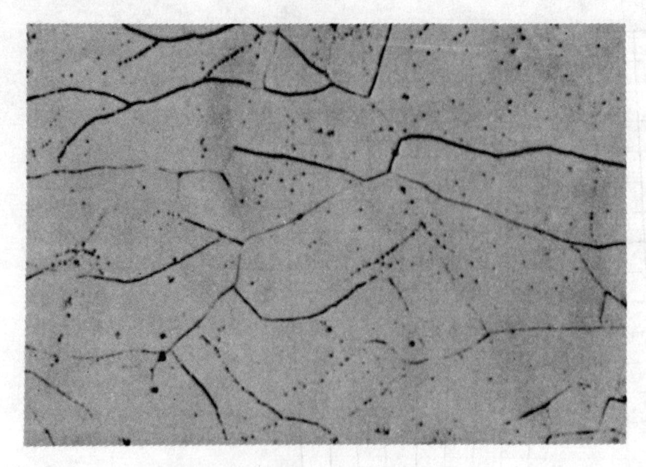

Figure 6-5 Etch-pit structures along low-angle grain boundaries in Fe-Si alloy (1,000×).

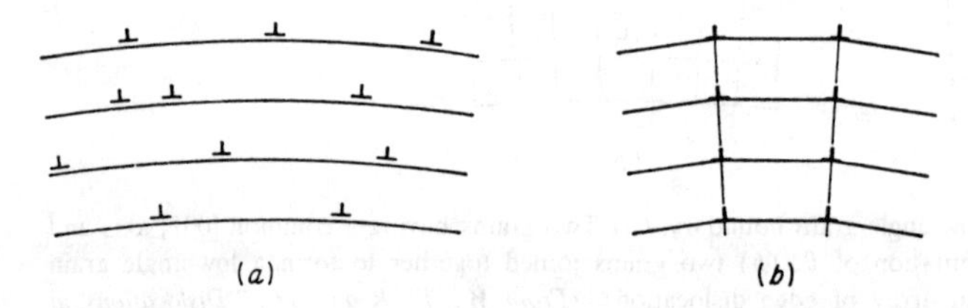

Figure 6-6 Movement of dislocations to produce polygonization (schematic).

The term *polygonization* was used originally to describe the situation that occurs when a single crystal is bent to a relatively small radius of curvature and then annealed. Bending results in the introduction of an excess number of dislocations of one sign. These dislocations are distributed along the bent-glide planes as shown in Fig. 6-6*a*. When the crystal is heated, the dislocations group themselves into the lower-energy configuration of a low-angle boundary by dislocation climb. The resulting structure is a polygonlike network of low-angle grain boundaries (Fig. 6-6*b*).

Since low-angle boundaries consist of simple dislocation arrays, a study of their properties should provide valuable information on dislocation behavior. Parker and Washburn[1] demonstrated that a low-angle boundary moves as a unit when subjected to a shear stress, in complete agreement with what would be expected for a linear dislocation array. It has also been found that the boundary angle decreases with increasing distance of shear. This means that the boundary loses dislocations as it moves, a fact which would be expected if dislocations were

[1] E. R. Parker and J. Washburn, *Trans. Metall. Soc. AIME*, vol. 194, pp. 1076–1078, 1952.

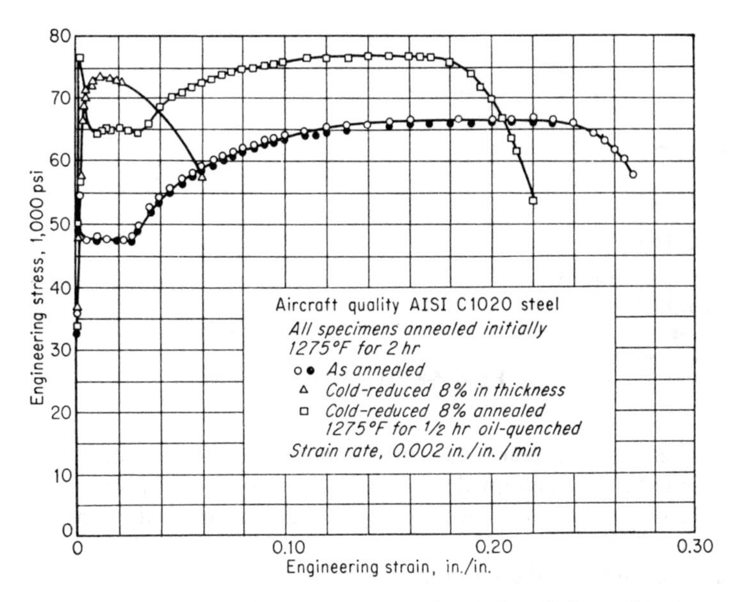

Figure 6-7 Effect of a substructure of low-angle grain boundaries on the stress-strain curve of SAE 1020 steel. *(From E. R. Parker and J. Washburn, "Impurities and Imperfections," p. 155, American Society for Metals, Metals Park, Ohio, 1955. By permission of the publishers.)*

held up at imperfections such as foreign atoms, precipitated particles, and other dislocations.

The effect of a substructure of low-angle grain boundaries on the stress-strain curve of 1020 steel is shown in Fig. 6-7. Note that the material that was cold-reduced and annealed, so as to produce a substructure, has a higher yield point and tensile strength than both the annealed material and the material which as only cold-reduced. Moreover, the ductility of the material containing a substructure is almost as good as the ductility of the annealed steel.

6-5 YIELD-POINT PHENOMENON

Many metals, particularly low-carbon steel, show a localized, heterogeneous type of transition from elastic to plastic deformation which produces a yield point in the stress-strain curve. Rather than having a flow curve with a gradual transition from elastic to plastic behavior, such as was shown in Fig. 3-1, metals with a yield point have a flow curve or, what is equivalent, a load-elongation diagram similar to Fig. 6-8. The load increases steadily with elastic strain, drops suddenly, fluctuates about some approximately constant value of load, and then rises with further strain. The load at which the sudden drop occurs is called the *upper yield point*. The constant load is called the *lower yield point*, and the elongation which occurs at constant load is called the *yield-point elongation*. The deformation

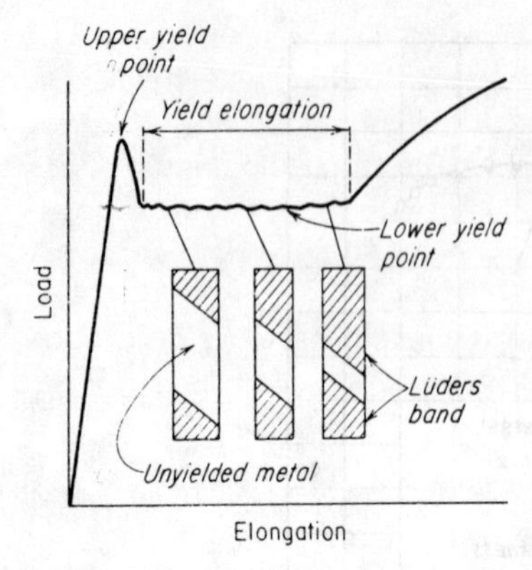

Figure 6-8 Typical yield-point behavior.

occurring throughout the yield-point elongation is heterogeneous. At the upper yield point a discrete band of deformed metal, often readily visible with the eye, appears at a stress concentration such as a fillet, and coincident with the formation of the band the load drops to the lower yield point. The band then propagates along the length of the specimen, causing the yield-point elongation. In the usual case several bands will form at several points of stress concentration. These bands are generally at approximately 45° to the tensile axis. They are usually called *Lüders bands*, Hartmann lines, or stretcher strains, and this type of deformation is sometimes referred to as the Piobert effect. When several Lüders bands are formed, the flow curve during the yield-point elongation will be irregular, each jog corresponding to the formation of a new Lüders band. After the Lüders bands have propagated to cover the entire length of the specimen test section, the flow will increase with strain in the usual manner. This marks the end of the yield-point elongation.

The yield-point phenomenon was found originally in low-carbon steel. A pronounced upper and lower yield point and a yield-point elongation of over 10 percent can be obtained with this material under proper conditions. More recently the yield point has come to be accepted as a general phenomenon, since it has been observed in a number of other metals and alloys. In addition to iron and steel, yield points have been observed in polycrystalline molybdenum, titanium, and aluminum alloys and in single crystals of iron, cadmium, zinc, alpha and beta brass, and aluminum. Usually the yield point can be associated with small amounts of interstitial or substitutional impurities. For example, it has been shown[1] that almost complete removal of carbon and nitrogen from low-carbon steel by wet-hydrogen treatment will remove the yield point. However, only about 0.001 percent of either of these elements is required for a reappearance of the yield point.

A number of experimental factors affect the attainment of a sharp upper yield point. A sharp upper yield point is promoted by the use of an elastically rigid

[1] J. R. Low and M. Gensamer, *Trans. AIME*, vol. 158, p. 207, 1944.

(hard) testing machine, very careful axial alignment of the specimen, the use of specimens free from stress concentrations, high rate of loading, and, frequently, testing at subambient temperatures. If, through careful avoidance of stress concentrations, the first Lüders band can be made to form at the middle of the test specimen, the upper yield point can be roughly twice the lower yield point. However, it is more usual to obtain an upper yield point 10 to 20 percent greater than the lower yield point.

The onset of general yielding occurs at a stress where the average dislocation sources can create slip bands through a good volume of the material. Thus, the general yield stress can be expressed as

$$\sigma_0 = \sigma_s + \sigma_i \tag{6-18}$$

where σ_s is the stress to operate the dislocation sources and σ_i is the friction stress representing the combining effect of all the obstacles to the motion of dislocations arising from the sources.

If the stress to operate the sources is high, then the initial yield stress is high. The explanation of the yield-point phenomenon in terms of dislocation behavior arose originally from the idea that the dislocation sources were locked or pinned by solute atom interactions (Sec. 5-15). The explanation[1] of this behavior was one of the early triumphs of dislocation theory. Carbon or nitrogen atoms in iron readily diffuse to the position of minimum energy just below the extra plane of atoms in a positive edge dislocation. The elastic interaction is so strong that the impurity atmosphere becomes completely saturated and condenses into a row of atoms along the core of the dislocation. The breakaway stress required to pull a dislocation line away from a line of solute atoms is

$$\sigma \approx \frac{A}{b^2 r_0^2} \tag{6-19}$$

where A is given by Eq. (5-35) and $r_0 \approx 2 \times 10^{-8}$ cm is the distance from the dislocation core to the line of solute atoms. When the dislocation line is pulled free from the influence of the solute atoms, slip can occur at a lower stress. Alternatively, where dislocations are strongly pinned, such as by carbon and nitrogen in iron, new dislocations must be generated to allow the flow stress to drop. This explains the origin of the upper yield stress (the drop in load after yielding has begun). The dislocations released into the slip plane pile-up at grain boundaries. As discussed in Sec. 6-3, the pile-up produces a stress concentration at its tip which combines with the applied stress in the next grain to unlock sources (or create new dislocations) and in this way a Lüders band propagates over the specimen. The magnitude of the yield-point effect will depend on the interaction energy. Eq. (5-35), and the concentration of solute atoms at the dislocations, Eq. (5-36).

[1] A. H. Cottrell and B. A. Bilby, *Proc. Phys. Soc. London*, vol. 62A, pp. 49–62, 1949; see also E. O. Hall, "Yield Point Phenomena in Metals, Plenum Publishing Company, New York, 1970.

Although dislocation locking by interstitial atoms originally was advanced as a mechanism of the yield point, subsequent research showed that a yield-point phenomenon was a very general behavior that was found in such diverse materials as LiF and Ge crystals and copper whiskers. In these materials the dislocation density is quite low and impurity pinning cannot explain the effect. A more general theory has been advanced[1] for all materials which exhibit a *yield drop*, i.e., where the stress falls rapidly once yielding begins. Impurity locking thus becomes a special case of yield-point behavior.

The relation between the strain rate imposed on the material by the test and the dislocation motion is given by

$$\dot{\varepsilon} = b\rho\bar{v} \tag{6-20}$$

where ρ is the density of mobile dislocations and $\bar{v}$ is the average dislocation velocity. The dislocation density increases with strain and $\bar{v}$ is a very strong function of stress

$$\bar{v} = \left(\frac{\tau}{\tau_0}\right)^{m'} \tag{6-21}$$

where τ_0 is the resolved shear stress corresponding to unit velocity. For materials with a low initial dislocation density (or with strongly pinned dislocations, as in iron) the only way that $b\rho\bar{v}$ can match the imposed strain rate is for $\bar{v}$ to be high. But, according to Eq. (6-21), this can only be achieved at a high stress. However, once some dislocations begin to move they begin to multiply and ρ increases rapidly. Although this introduces some strain hardening, it is more than compensated for by the fact that $\bar{v}$ can drop and with it the stress needed to move the dislocations can drop. Thus, the stress required to deform the specimen decreases once yielding begins (yield drop). Finally, the increasing dislocation density produces increased strain hardening through dislocation interactions and the stress begins to increase with further strain.

According to this model the controlling parameters are the mobile dislocation density and the exponent describing the stress dependence of dislocation velocity, m'. From Eqs. (6-20) and (6-21), we can express conditions at the upper and lower yield points by

$$\frac{\tau_u}{\tau_L} = \left(\frac{\rho_L}{\rho_u}\right)^{1/m'} \tag{6-22}$$

For small values of m' ($m' < 15$), the ratio τ_u/τ_L will be large and there will be a strong yield drop. For iron ($m' \approx 35$), the yield drop will only be substantial if ρ_u is less than about 10^3 cm^{-2}. Since the dislocation density of annealed iron is at least 10^6 cm^{-2}, this requires that most of the dislocations must be pinned. Pinning can arise from the solute-dislocation interaction or by precipitation of

[1] W. G. Johnston and J. J. Gilman, *J. Appl. Phys.*, vol. 30, p. 129, 1959. The theory has been extended to cover yielding in iron and other bcc metals by G. T. Hahn, *Acta Metall.*, vol. 10, pp. 727–738, 1962.

fine carbides or nitrides along the dislocation. The yield point occurs as a result of unlocking the dislocations by a high stress, or for case of strong pinning, by creating new dislocations at the points of stress concentration.

6-6 STRAIN AGING

Strain aging is a type of behavior, usually associated with the yield-point phenomenon, in which the strength of a metal is increased and the ductility is decreased on heating at a relatively low temperature after cold-working. This behavior can best be illustrated by considering Fig. 6-9, which schematically describes the effect of strain aging on the flow curve of a low-carbon steel. Region *A* of Fig. 6-9 shows the stress-strain curve for a low-carbon steel strained plastically through the yield-point elongation to a strain corresponding to point *X*. The specimen is then unloaded and retested without appreciable delay or any heat treatment (region *B*). Note that on reloading the yield point does not occur, since the dislocations have been torn away from the atmosphere of carbon and nitrogen atoms. Consider now that the specimen is strained to point *Y* and unloaded. If it is reloaded after aging for several days at room temperature or several hours at an aging temperature like 300°F, the yield point will reappear. Moreover, the yield point will be increased by the aging treatment from *Y* to *Z*. The reappearance of the yield point is due to the diffusion of carbon and nitrogen atoms to the dislocations during the aging period to form new atmospheres of interstitials anchoring the dislocations. Support for this mechanism is found in the fact that the activation energy for the return of the yield point on aging is in good agreement with the activation energy for the diffusion of carbon in alpha iron.

Nitrogen plays a more important role in the strain aging of iron than carbon because it has a higher solubility and diffusion coefficient and produces less complete precipitation during slow cooling. From a practical standpoint it is important to eliminate strain aging in deep-drawing steel because the reap-

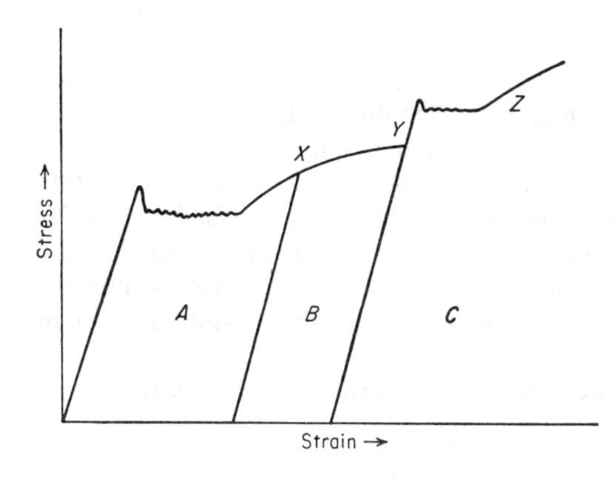

Figure 6-9 Stress-strain curves for low-carbon steel showing strain aging. Region *A*, original material strained through yield point. Region *B*, immediately retested after reaching point *X*. Region *C*, reappearance and increase in yield point after aging at 300°F.

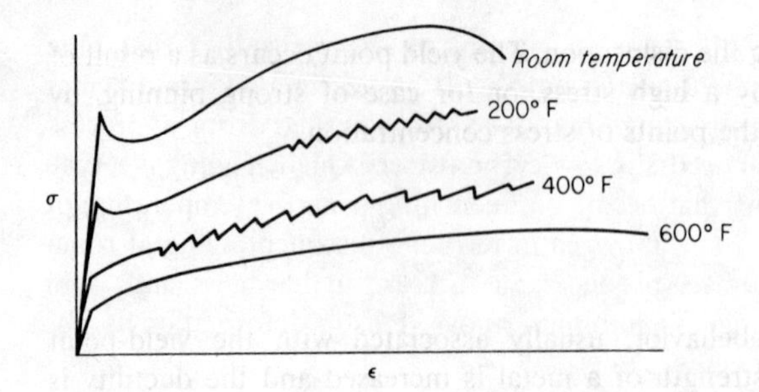

Figure 6-10 Portevin-LeChatelier effect in iron (schematic).

pearance of the yield point can lead to difficulties with surface markings or "stretcher strains" due to the localized heterogeneous deformation. To control strain aging, it is usually desirable to lower the amount of carbon and nitrogen in solution by adding elements which will take part of the interstitials out of solution by forming stable carbides or nitrides. Aluminum, vanadium, titaniun, columbium, and boron have been added for this purpose. While a certain amount of control over strain aging can be achieved, there is no commercial low-carbon steel which is completely non-strain aging. The usual industrial solution to this problem is to deform the metal to point *X* by roller leveling or a skin-pass rolling operation and use it immediately before it can age. The local plastic deformation by rolling produces sufficient fresh dislocations so that subsequent plastic flow can occur without a yield point.

The occurrence of strain aging is a fairly general phenomenon in metals. In addition to the return of the yield point and an increase in the yield stress after aging, strain aging also produces a decrease in ductility and a low value of strain-rate sensitivity.[1] Strain aging also is associated with the occurrence of serrations in the stress-strain curve[2] (discontinuous or repeated yielding). This dynamic strain-aging behavior (Fig. 6-10) is called the *Portevin-LeChatelier effect*. The solute atoms are able to diffuse in the specimen at a rate faster than the speed of the dislocations so as to catch and lock them. Therefore, the load must increase and when the dislocations are torn away from the solute atoms there is a load drop. This process occurs many times, causing the serrations in the stress-strain-curve.

Dynamic strain aging is not the only phenomena that can cause a serrated stress-strain curve. Mechanical twinning occurring during deformation or stress-assisted martensitic transformations will produce the same effect.

For plain carbon steel discontinuous yielding occurs in the temperature region of 450 to 700°F. This temperature region is known as the *blue brittle region* because steel heated in this temperature region shows a decreased tensile ductility and decreased notched-impact resistance. This temperature range is also the region in which steels show a minimum in strain-rate sensitivity and a maximum

[1] The strain-rate sensitivity is the change in stress required to produce a certain change in strain rate at constant temperature (see Chap. 8).

[2] Observations and theories of serrated yielding and the static yield point are reviewed by B. J. Brindley and P. J. Worthington, *Met. Rev.*, no. 145, *Met. Mater.*, vol. 4, no. 8, pp. 101–114, 1970.

in the rate of strain aging. All these facts point to the realization that blue brittleness is not a separate phenomenon but is just an accelerated strain aging.

The phenomenon of strain aging should be distinguished from a process known as *quench aging*, which occurs in low-carbon steels. Quench aging is a type of true precipitation hardening that occurs on quenching from the temperature of maximum solubility of carbon and nitrogen in ferrite. Subsequent aging at room temperature, or somewhat above, produces an increase in hardness and yield stress, as in the age hardening of aluminum alloys. Plastic deformation is not necessary to produce quench aging.

6-7 SOLID-SOLUTION STRENGTHENING

The introduction of solute atoms into solid solution in the solvent-atom lattice invariably produces an alloy which is stronger than the pure metal. There are two types of solid solutions. If the solute and solvent atoms are roughly similar in size, the solute atoms will occupy lattice points in the crystal lattice of the solvent atoms. This is called *substitutional solid solution*. If the solute atoms are much smaller than the solvent atoms, they occupy interstitial positions in the solvent lattice. Carbon, nitrogen, oxygen, hydrogen, and boron are the elements which commonly form *interstitial solid solutions*.

The factors which control the tendency for the formation of substitutional solid solutions have been uncovered chiefly through the work of Hume-Rothery. If the sizes of the two atoms, as approximately indicated by the lattice parameter, differ by less than 15 percent, the size factor is favorable for solid-solution formation. When the size factor is greater than 15 percent, the extent of solid solubility is usually restricted to less than 1 percent. Metals which do not have a strong chemical affinity for each other tend to form solid solutions, while metals which are far apart on the electromotive series tend to form intermetallic compounds. The relative valence of the solute and solvent also is important. The solubility of a metal with higher valence in a solvent of lower valence is more extensive than for the reverse situation. For example, zinc is much more soluble in copper than is copper in zinc. This relative-valence effect can be rationalized to a certain extent in terms of the electron-atom ratio.[1] For certain solvent metals, the limit of solubility occurs at approximately the same value of electron-atom ratio for solute atoms of different valence. Finally, for complete solid solubility over the entire range of composition the solute and solvent atoms must have the same crystal structure.

The acquisition of fundamental information about the causes of solid-solution hardening has been a slow process. Early studies[2] of the increase in hardness

[1] For example, an alloy of 30 atomic percent Zn in Cu has an electron-atom ratio of 1.3 $(3 \times 2) + (7 \times 1) = 13$ valence electrons per $3 + 7 = 10$ atoms.

[2] A. L. Norbury, *Trans. Faraday Soc.*, vol. 19, pp. 506–600, 1924; R. M. Brick, D. L. Martin, and R. P. Angier, *Trans. Am. Soc. Met.*, vol. 31, pp. 675–698, 1943; J. H. Frye and W. Hume-Rothery, *Proc. R. Soc. London*, vol. 181, pp. 1–14, 1942.

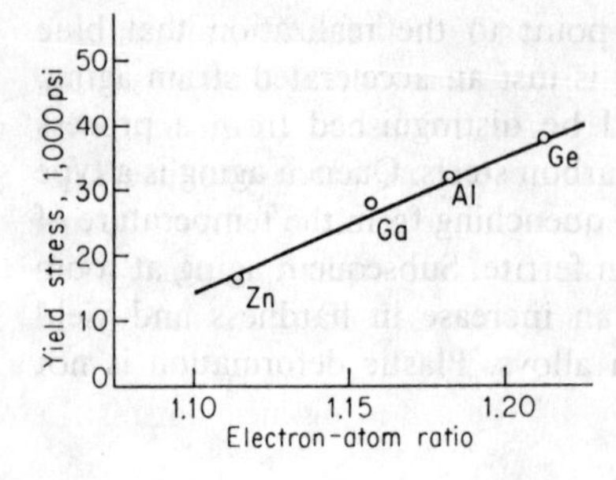

Figure 6-11 Effect of electron-atom ratio on the yield stress of copper solid-solution alloys. *(From W. R. Hibbard, Jr., Trans. Metall. Soc. AIME, vol. 212, p. 3, 1958)*

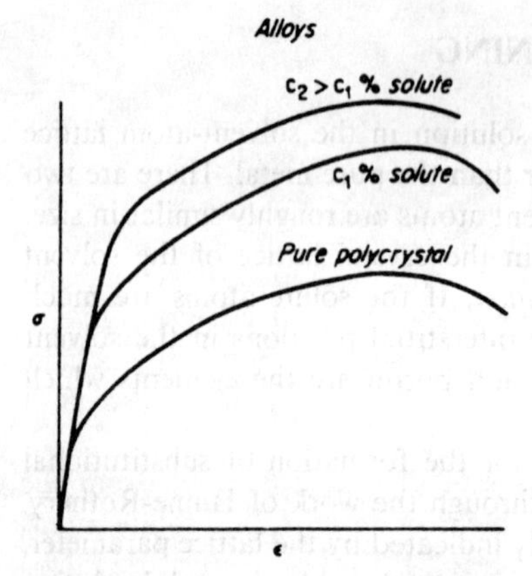

Figure 6-12 Effect of solute alloy additions on stress-strain curve.

resulting from solid-solution additions showed that the hardness increase varies directly with the difference in the size of the solute and solvent atoms, or with the change in lattice parameter resulting from the solute addition. However, it is apparent that size factor alone cannot explain solid-solution hardening. An improvement in correlation of data[1] results when the relative valence of the solute and solvent are considered in addition to the lattice-parameter distortion. The importance of valence is shown in Fig. 6-11, where the yield stress of copper alloys of constant lattice parameter is plotted against the electron-atom ratio.[2] Further results[3] show that alloys with equal grain size, lattice parameter, and electron-atom ratio have the same initial yield stress, but the flow curves differ at larger strains.

The usual result of solute additions is to raise the yield stress and the level of the stress-strain curve as a whole (Fig. 6-12). As was seen in Sec. 6-5, solute atoms also frequently produce a yield-point effect. Since solid-solution alloy additions

[1] J. E. Dorn, P. Pietrokowsky, and T. E. Tietz, *Trans. AIME*, vol. 188, pp. 933–943, 1950.
[2] W. R. Hibbard, Jr., *Trans. Metall. Soc. AIME*, vol. 212, pp. 1–5, 1958.
[3] N. G. Ainslie, R. W. Guard, and W. R. Hibbard, *Trans. Metall. Soc. AIME*, vol. 215, pp. 42–48, 1959.

affect the entire stress-strain curve, we are led to conclude that solute atoms have more influence on the frictional resistance to dislocation motion σ_i than on the static locking of dislocations. Solute atoms fall into two broad categories[1] with respect to their relative strengthening effect. Those atoms which produce non-spherical distortions, such as most interstitial atoms, have a relative strengthening effect per unit concentration of about three times their shear modulus, while solute atoms which produce spherical distortion, such as substitutional atoms, have a relative strengthening of about $G/10$.

Solute atoms can interact with dislocations by the following mechanisms:

elastic interaction
modulus interaction
stacking-fault interaction
electrical interaction
short-range order interaction
long-range order interaction

Of these various mechanisms, the elastic interaction, modulus interaction, and long-range order interaction are long range, i.e., they are relatively insensitive to temperature and continue to act to about $0.6T_m$. The other three interactions constitute short-range barriers and only contribute strongly to the flow stress at lower temperatures.

Elastic interaction between solute atoms and dislocations arises from the mutual interaction of elastic stress fields which surround misfitting solute atoms and the core of edge dislocations [according to Eq. (5-35)]. The relative size factor is $\varepsilon_a = 1/a(da/dc)$ where a is the lattice parameter and c is the atomic concentration of the solute. The strengthening due to elastic interaction is directly proportional to the misfit of the solute. Substitutional solutes only impede the motion of edge dislocations. However, interstitial solutes have both dilatation and shear components, and can interact with both edge and screw dislocations.

A *modulus interaction* occurs if the presence of a solute atom locally alters the modulus of the crystal. If the solute has a smaller shear modulus than the matrix the energy of the strain field of the dislocation will be reduced and there will be an attraction between solute and matrix. The modulus interaction is similar to the elastic interaction but, because a change in shear modulus is accompanied by a local change in bulk modulus, both edge and screw dislocations will be subject to this interaction.

Stacking-fault interactions arise because solute atoms preferentially segregate to the stacking faults contained in extended dislocations. (This was first pointed out by Suzuki, so this also is called Suzuki or chemical interaction.) For this to happen the solute must have a preferential solubility in the hcp structure of the

[1] R. L. Fleischer, Solid Solution Hardening in D. Peckner (ed.), "The Strengthening of Metals," Reinhold Publishing Corporation, New York, 1964; C. G. Schmidt and A. K. Miller, *Acta. Metall.*, vol. 30, pp. 615–625, 1982.

stacking fault. As the concentration of the solute within the stacking fault increases it lowers the stacking-fault energy and increases the separation of the partial dislocations. Therefore, the motion of the extended dislocations is made more difficult and additional work must be done to constrict the pair of partial dislocations.

Electrical interaction arises from the fact that some of the charge associated with solute atoms of dissimilar valence remains localized around the solute atom. The solute atoms become charge centers and can interact with dislocations which have electrical dipoles.[1] The electrical interaction of solute atoms is much weaker than the elastic and modulus interaction, and becomes important only when there is a large valence difference between solute and matrix and the elastic misfit is small.

Short-range order interaction arises from the tendency for solute atoms to arrange themselves so that they have more than the equilibrium number of disimilar neighbors. The opposite of short-range order is *clustering*, where like solute atoms tend to group together in regions of the lattice. Strengthening occurs because the movement of a dislocation through a region of short-range order or clustering reduces the degree of local order. This process of disordering will cause an increase in the energy of the alloy and, to sustain the energetically unfavorable dislocation motion, extra work (represented by the interaction energy) must be provided.

Long-range order interaction arises in alloys which form superlattices. In a superlattice there is a long-range periodic arrangement of dissimilar atoms. The movement of a dislocation through a superlattice creates regions of disorder called *anti-phase boundaries* (APB) because the atoms across the slip plane have become "out of phase" with respect to the energetically preferred superlattice structure. The dislocation dissociates into two pairs of ordinary dislocations separated by the APB. The details of the geometry and the dislocation reactions are quite complex.[2] The width of the APB is the result of a balance between the elastic repulsion of the two dislocations of the same sign and the energy of the APB. The stress required to move a dislocation through a long-range region is

$$\tau_0 = \gamma/t \tag{6-23}$$

where γ is the energy of the APB and t is the spacing of the APBs. Because more APBs are produced as slip proceeds, the rate of strain hardening is higher in the ordered condition than the disordered state. Ordered alloys with a fine domain size (~ 50 Å) are stronger than the disordered state but ordered alloys with a large domain size generally have a yield stress lower than the disordered state.

The resistance to dislocation motion that constitutes solid-solution strengthening can come from one or more of these factors. At first it might appear

[1] A. H. Cottrell, S. C. Hunter and F. R. N. Nabarro, *Phil. Mag.*, vol. 44, p. 1,064, 1953.
[2] N. S. Stoloff and R. G. Davies, *Prog. Mater. Sci.*, vol. 13, p. 1, 1966.

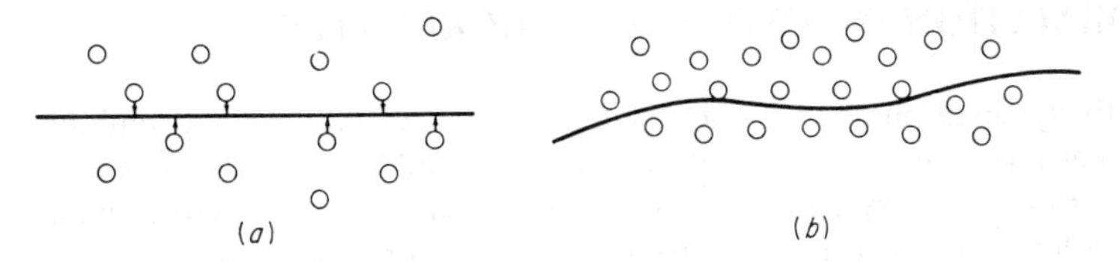

Figure 6-13 (*a*) Straight dislocation line in random solid solution; (*b*) flexible dislocation line.

that dislocations would not be impeded by the interactions from solute atoms since on the average as many interactions will tend to promote motion as retard it (Fig. 6-13*a*). In a random solid solution, provided the dislocation remains straight, there will be no net force on the dislocation since the algebraic sum of all interaction energies will be zero. Mott and Nabarro[1] provided the key theory by pointing out that dislocation lines generally are not straight. Rather, dislocation lines are *flexible* so that the entire line does not have to move simultaneously and can take up lower energy positions by bending around regions of high interaction energy. As was shown in Sec. 5-8, the smallest radius of curvature that the dislocation line can accommodate to under a local stress τ_i at a solute atom is

$$R \approx \frac{Gb}{2\tau_i} \tag{6-24}$$

The degree of interaction that the dislocation will have will depend on the average spacing λ of the barriers. For individual solute atoms distributed through the crystal lattice, λ is very small and is given by

$$\lambda = \frac{a}{c^{1/3}} \tag{6-25}$$

where a is the interatomic spacing and c is the atomic concentration of solute. The spacing λ of the local solute stress fields will be much smaller than the radius of curvature to which the dislocation can be bent by the local stresses. Thus, $\lambda \ll R$ and the dislocation will move in lengths much greater than λ (Fig. 6-13*b*).

The central problem that must be solved before these interaction mechanisms can be used to calculate the strength of an alloy is to develop a method for averaging the interactions between gliding dislocations and solute atoms which lie at various distances from the dislocation line. A statistical treatment of the problem which allows for considering contributions from barriers of different strengths has been advanced by Labusch.[2]

[1] N. F. Mott and F. R. N. Nabarro, "Report on Conference on the Strength of Solids," p. 1, *Phys. Soc., London*, 1948. This theory has been extended by T. Stefansky and J. E. Dorn, *Trans. Metall. Soc. AIME*, vol. 245, pp. 1869–1876, 1969.

[2] R. Labusch, *Phys. Stat. Sol.*, vol. 41, p. 659, 1970.

6-8 DEFORMATION OF TWO-PHASE AGGREGATES

Only a relatively small number of alloy systems permit extensive solid solubility between two or more elements, and only a relatively small hardening effect can be produced in most alloy systems by solid-solution additions. Therefore, many commercial alloys contain a heterogeneous microstructure consisting of two or more metallurgical phases. A number of different microstructures may be encountered,[1] but in general they fall into the two classes illustrated in Fig. 6-14. Figure 6-14*a* represents the aggregated type of two-phase structure in which the size of the second-phase particles is of the order of the grain size of the matrix. This is typified by beta brass particles in an alpha brass matrix or by pearlite colonies in a ferrite matrix in annealed steel (Fig. 15-20). The other general type of structure is the dispersed two-phase structure (Fig. 6-14*b*) in which each particle is completely surrounded by a matrix of a single orientation (grain). Generally the particle size of the second phase is much finer for the dispersed structure and may extend down to submicroscopic dimensions in the early stages of precipitation. The theories of strengthening in dispersion-hardened systems have been studied extensively and will be considered in Sec. 6-9.

The strengthening produced by second-phase particles is usually additive to the solid-solution strengthening produced in the matrix. For two-phase alloys produced by equilibrium methods, the existence of a second phase ensures maximum solid-solution hardening because its presence resulted from supersaturation of the matrix phase. Moreover, the presence of second-phase particles in the continuous matrix phase results in localized internal stresses which modify the plastic properties of the continuous phase. Many factors must be considered for a complete understanding of strengthening from second-phase particles. These factors include the size, shape, number, and distribution of the second-phase particles, the strength, ductility, and strain-hardening behavior of the matrix and second phase, the crystallographic fit between the phases, and the interfacial energy and interfacial bonding between the phases. It is almost impossible to vary these factors independently in experiments, and it is very difficult to measure many of these quantities with any degree of precision.

In a multiphase alloy, each phase contributes certain things to the overall properties of the aggregate. If the contributions from each phase are independent, then the properties of the multiphase alloy will be a weighted average of the properties of the individual phases. For example, the density of a two-phase alloy will be equal to the sum of the volume fraction of each phase times its density. However, for the structure-sensitive mechanical properties the properties of the aggregate are generally influenced by interaction between the two phases. Two simple hypotheses may be used to calculate the properties of a two-phase alloy from the properties of the individual ductile phases. If it is assumed that the

[1] For a well-illustrated discussion of the relationship between phase diagrams, microstructure, and properties see R. M. Brick, R. B. Gordon, and A. Phillips, "Structure and Properties of Alloys," 3rd ed., McGraw-Hill Book Company, New York, 1965.

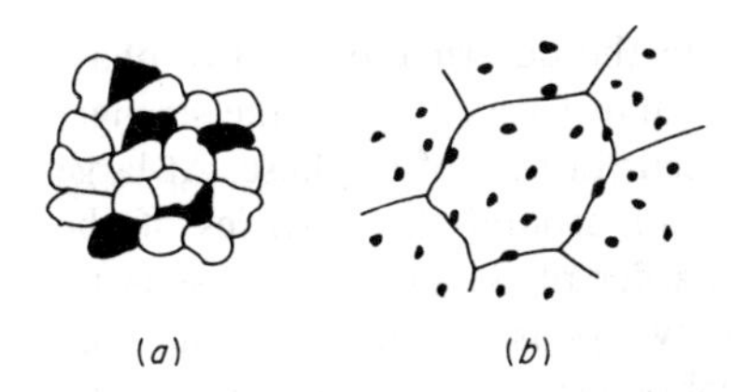

Figure 6-14 Types of two-phase microstructures. (*a*) Aggregated structure; (*b*) dispersed structure.

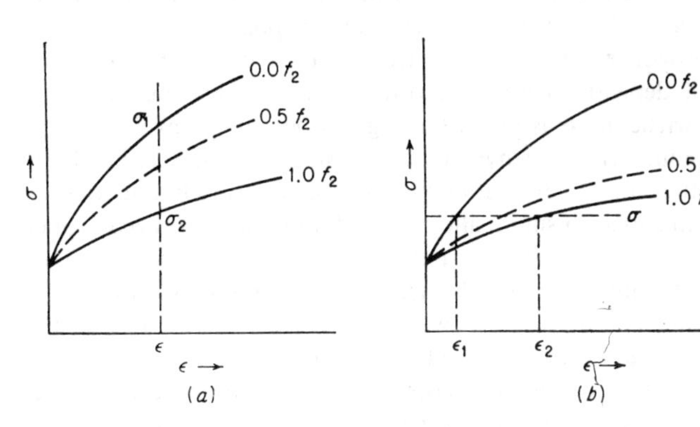

Figure 6-15 Estimate of flow stress of two-phase alloy. (*a*) Equal strain; (*b*) equal stress. *(From J. E. Dorn and C. D. Starr, "Relation of Properties to Microstructure," pp. 77–78, American Society for Metals, Metals Park, Ohio, 1954. By permission of the publishers.)*

strain in each phase is equal, the average stress in the alloy for a given strain will increase linearly with the volume fraction of the strong phase.

$$\sigma_{avg} = f_1\sigma_1 + f_2\sigma_2 \tag{6-26}$$

The volume fraction of phase 1 is f_1, and $f_1 + f_2 = 1$. Figure 6-15*a* shows the calculation of the flow curve for an alloy with 0.5 volume fraction of phase 2 on the basis of the equal-strain hypothesis. An alternative hypothesis is to assume that the two phases are subjected to equal stresses. The average strain in the alloy at a given stress is then given by

$$\varepsilon_{avg} = f_1\varepsilon_1 + f_2\varepsilon_2 \tag{6-27}$$

Figure 6-15*b* shows the flow curve for a 0.5-volume-fraction alloy on the basis of equal-stress hypothesis. Both these hypotheses are simple approximations, and the strengths of alloys containing two ductile phases usually lie somewhere between the values predicted by the two models.

The deformation of an alloy consisting of two ductile phases depends on the volume fraction of the two phases and the total deformation. Experiments have shown[1] that not all second-phase particles produce strengthening. In order for particle strengthening to occur there must be a strong particle-matrix bond.

[1] B. I. Edelson and W. M. Baldwin, Jr., *Trans. Am. Soc. Met.*, vol. 55, p. 230, 1962.

Limited experimental information is available on the deformation of two-phase alloys.[1] Slip will occur first in the weaker phase, and if very little of the stronger phase is present, most of the deformation wil occur in the softer phase. At large deformations flow of the softer matrix will occur around the particles of the harder phase. At about 30 vol percent of the harder phase the soft phase is no longer the completely continuous phase and the two phases tend to deform more or less with equal strains. At above about 70 vol percent of the harder phase the deformation is largely controlled by the properties of this phase

The mechanical properties of an alloy consisting of a ductile phase and a hard brittle phase will depend on how the brittle phase is distributed in the microstructure. If the brittle phase is present as a grain-boundary envelope, as in oxygen-free copper-bismuth alloys or hypereutectoid steel, the alloy is brittle. If the brittle phase is in the form of discontinuous particles at grain boundaries, as when oxygen is added to copper-bismuth alloys or with internally oxidized copper or nickel, the brittleness of the alloy is reduced somewhat. A condition of optimum strength and ductility is obtained when the brittle phase is present as a fine dispersion uniformly distributed throughout the softer matrix. This is the situation in heat-treated steel with a tempered martensitic structure.

One of the first correlations between microstructures of two-phase alloys and yield stress was made by Gensamer *et al.*[2] for the aggregates of cementite (iron carbide) and ferrite found in annealed, normalized, and spheroidized steel. In these coarse-dispersed second-phase aggregates they found that the flow stress at a true strain of 0.2 was inversely proportional to the logarithm of the mean interparticle spacing[3] (mean free ferrite path) (Fig. 6-16). This relationship also has been found to hold for the finer cementite particles in a tempered martensite,[4] the coarse particles in overaged Al-Cu alloys[5] and spheroidized steel,[6] and Co-WC sintered carbides.[7] The degree of strengthening produced by dispersed particles is illustrated by Fig. 6-17. The bottom curve represents the saturated solid solution of Al-Cu. The top curve is for the solid solution containing 5 vol percent of fine $CuAl_2$ particles, while the middle curve is for a coarse dispersion of the same volume fraction.

A strong theoretical basis for the strength of coarse two-phase aggregates has not been developed. In general terms second phases block slip so that plastic deformation is not uniform in the matrix. Strain in the matrix is localized and increased over the average strain of the specimen. This leads to strengthening

[1] H. Unkel, *J. Inst. Met.*, vol. 61, p. 171, 1937; L. M. Clarebrough and G. Perger, *Aust. J. Sci. Res. Ser. A*, vol. A5, p. 114, 1952.

[2] M. Gensamer, E. B. Pearsall, W. S. Pellini, and J. R. Low, Jr., *Trans. Am. Soc. Met*, vol. 30, pp. 983–1020, 1942.

[3] The mean free ferrite path p is given by $p = (1 - f_c)/N_L$, where f_c is the volume fraction of cementite and N_L is the number of carbide-particles intersected per unit length of random line through the microstructure.

[4] A. M. Turkalo and J. R. Low, Jr., *Trans. Metall. Soc. AIME*, vol. 212, pp. 750–758, 1958.

[5] C. D. Starr, R. B. Shaw, and J. E. Dorn, *Trans. Am. Soc. Met.*, vol. 46, pp. 1075–1088, 1954.

[6] C. T. Liu and J. Gurland, *Trans. Metall. Soc. AIME*, vol. 242, pp. 1535–1542, 1968.

[7] J. Gurland and P. Bardzil, *Trans. AIME*, vol. 203, p. 311, 1955.

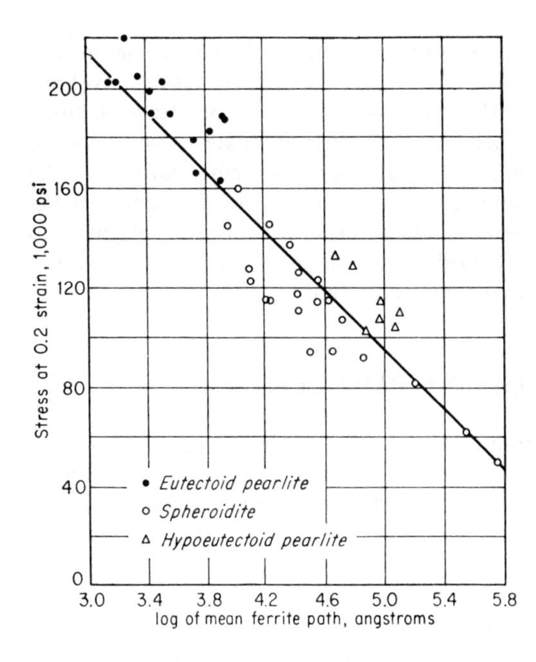

Figure 6-16 Flow stress vs. logarithm of mean free ferrite path in steels with pearlitic and spheroidal distribution of carbides. *(From M. Gensamer, E. S. Pearsall, W. S. Pellini, and J. R. Low, Trans. ASM, vol. 30, p. 1003, 1942.)*

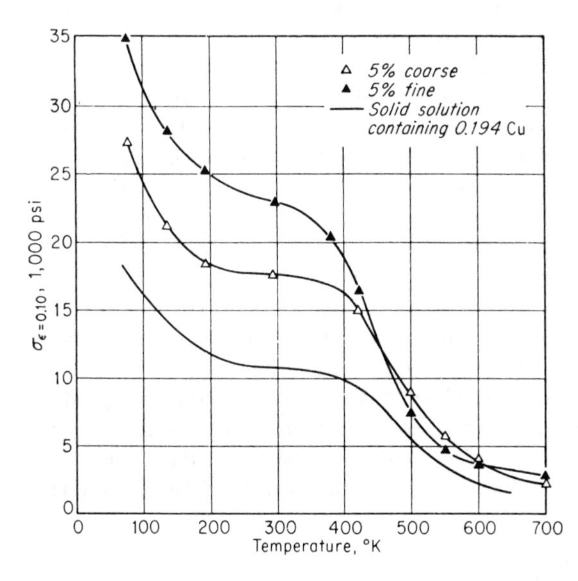

Figure 6-17 Variation of flow stress with temperature for Al-Cu alloy containing 5 volume percent fine and coarse second-phase particles. *(From C. D. Starr, R. B. Shaw, and J. E. Dorn, Trans. ASM, vol. 46, p. 1085, 1954.)*

from local plastic constraint. A good example can be found in the relative strengthening from pearlite and spheroidite in steel (Fig. 8-25). At equal volume fraction of cementite phase, the pearlitic structure will have a greater yield stress because the ferrite matrix sandwiched between cementite platelets will be more constrained than will the ferrite surrounding the more spheroidal particles of the spheroidized microstructure. An interesting approach to this problem, just beginning, is the application of the limit design theories of continuum plasticity to the prediction of the strength of model (idealized) microstructures[1] and the use of the finite element method to model the microstructure.[2]

6-9 STRENGTHENING FROM FINE PARTICLES

Small second-phase particles distributed in a ductile matrix are a common source of alloy strengthening. In *dispersion hardening*[3] the hard particles are mixed with matrix powder and consolidated and processed by powder metallurgy techniques. However, very many alloy systems can be strengthened by precipitation reactions in the solid state.

Precipitation hardening, or *age hardening*,[4] is produced by solution treating and quenching an alloy in which a second phase is in solid solution at the elevated temperature but precipitates upon quenching and aging at a lower temperature. The age-hardening aluminum alloys and copper-beryllium alloys are common examples. For precipitation hardening to occur, the second phase must be soluble at an elevated temperature but must exhibit decreasing solubility with decreasing temperature. By contrast, the second phase in dispersion-hardening systems has very little solubility in the matrix, even at elevated temperatures. Usually there is atomic matching, or *coherency*, between the lattices of the precipitate and the matrix, while in dispersion-hardened systems there generally is no coherency between the second-phase particles and the matrix. The requirement of a decreasing solubility with temperature places a limitation on the number of useful precipitation-hardening alloy systems. On the other hand, it is at least theoretically possible to produce an almost infinite number of dispersion-hardened systems by mixing finely divided metallic powders and second-phase particles (oxides, carbides, nitrides, borides, etc.) and consolidating them by powder metallurgy techniques. Advantage has been taken of this method to produce dispersion-hardened systems which are thermally stable at very high temperatures. Because of the finely dispersed second-phase particles, these alloys are much more resistant to recrystallization and grain growth than single-phase

[1] D. C. Drucker, *J. Mater.*, vol. 1, p. 873–910, 1966; T. W. Butler and D. C. Drucker, *J. Appl. Mech.*, vol. 95, pp. 780–784, 1973.

[2] R. E. Smelser and J. L. Swedlow, *Mater. Sci. Eng.*, vol. 46, pp. 175–190, 1980; S. Ankem and H. Margolin, *Metall. Trans.*, vol. 13A, pp. 595–609, 1982.

[3] R. F. Decker, *Metall. Trans.*, vol. 4, pp. 2495–2518, 1973.

[4] J. W. Martin, "Precipitation Hardening," Pergamon Press, New York, 1968.

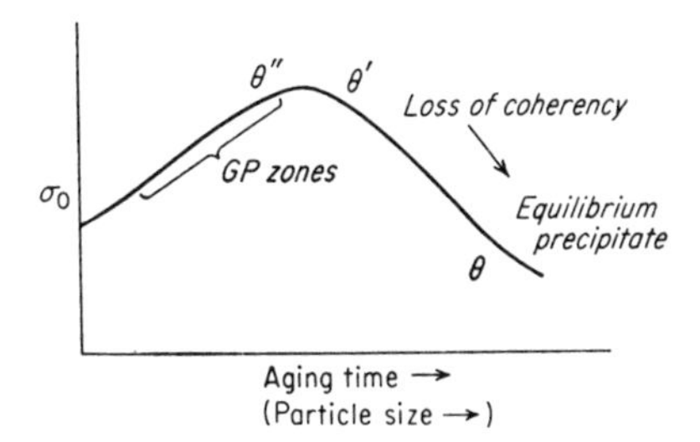

Figure 6-18 Variation of yield stress with aging time (schematic).

alloys. Because there is very little solubility of the second-phase constituent in the matrix, the particles resist growth or overaging to a much greater extent than the second-phase particles in a precipitation-hardening system.

The formation of a coherent precipitate in a precipitation-hardening system, such as Al-Cu, occurs in a number of steps. After quenching from solid solution the alloy contains regions of solute segregation, or clustering. Guiner and Preston first detected this local clustering with special x-ray techniques, and therefore this structure is known as a GP zone. The clustering may produce local strain, so that the hardness of GP[1] is higher than for the solid solution. With additional aging the hardness is increased further by the ordering of larger clumps of copper atoms on the {100} planes of the matrix. This structure is known as GP[2], or θ''. Next, definite precipitate platelets of $CuAl_2$, or θ', which are coherent with the matrix, form on the {100} planes of the matrix. The coherent precipitate produces an increased strain field in the matrix and a further increase in hardness. With still further aging the equilibrium phase $CuAl_2$, or θ, is formed from the transition lattice θ'. These particles are no longer coherent with the matrix, and therefore the hardness is lower than at the stage when coherent θ' was present. For most precipitation hardening alloys the resolution with the light microscope of the first precipitate occurs after the particles are no longer coherent with the matrix. Continued aging beyond this stage produces particle growth and further decrease in hardness. Figure 6-18, illustrates the way in which strength varies with aging time or particle size. The sequence of events in the Al-Cu system is particularly complicated. Although other precipitation-hardening systems may not have so many stages, it is quite common for a coherent precipitate to form and then lose coherency when the particle grows to a critical size.

The deformation of alloys with fine particle strengthening is well illustrated by the stress-strain curves in Fig. 6-19. Careful studies on A1-4.5% Cu single crystals[1] have correlated observations of slip lines with the deformation behavior. When the crystals are solution treated and quenched so that the alloy contains all of the copper in supersaturated solid solution, the yield stress is raised significantly over that for pure aluminum. The rate of strain hardening (slope of stress-strain curve) is low and characteristic of easy glide. Slip bands are broad

[1] G. Greetham and R. W. K. Honeycombe, *J. Inst. Met.*, vol. 89, p. 13, 1960–1961; J. G. Byrne, M. E. Fine, and A. Kelly, *Philos. Mag.*, vol. 6, pp. 1119–1143, 1961.

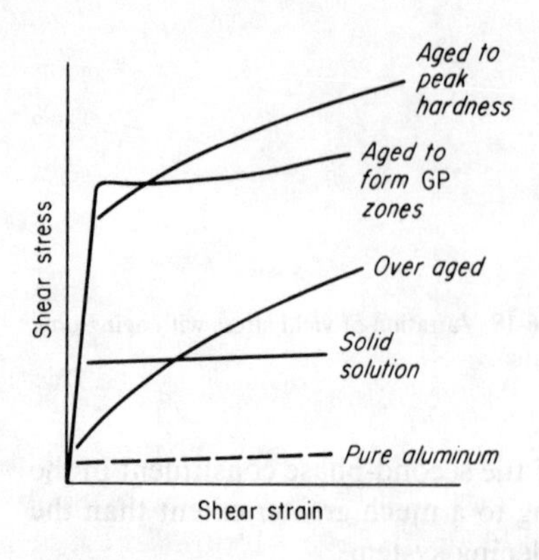

Figure 6-19 Stress-strain curves for Al-Cu single crystals in various conditions (schematic).

and widely spaced. When the crystal is aged to form coherent GP zones, the yield stress is raised significantly and a yield drop occurs. The rate of strain hardening is still low and slip lines can be distinguished although they are finer and more closely spaced. The yield drop and low rate of strain hardening suggest that dislocations cut through the zones once the stress reaches a high enough value. Although crystals aged to peak hardness show a slight decrease in yield strength, the rate of strain hardening is significantly increased. In this condition slip lines are very short or indistinct. This suggests that dislocations are no longer cutting through particles to form well-defined slip bands but are moving around particles so as to bypass them. In the overaged conditions where particles are noncoherent and relatively coarse, the yield strength is low but the rate of strain hardening is very high. Dislocations accumulate in tangles around the particles in the process of passing between them. This promotes slip on secondary slip systems and promotes strain hardening of the matrix. The high stresses from the dislocation loops around the particles tend to make the particles conform, by elastic deformation, to the plastic deformation of the matrix. Very high elastic strains are developed in the particles and the particles support a large fraction of the total load much as in fiber-reinforced materials (Sec. 6-10). The limit of strength in overaged alloys or dispersion-strengthened alloys is the yielding, or fracture, of the particles, or the tearing of the matrix away from the particles. However, since the particles are usually very fine intermetallic compounds, they possess high strength.

The degree of strengthening resulting from second-phase particles depends on the distribution of particles in the ductile matrix. In addition to shape, the second-phase dispersion can be described by specifying the volume fraction, average particle diameter, and mean interparticle spacing. These factors are all interrelated so that one factor cannot be changed without affecting the others (see Prob. 6-9). For example, for a given volume fraction of second phase, reducing the particle size decreases the average distance between particles. For a given size particle, the distance between particles decreases with an increase in the volume fraction of second phase.

The strongest alloys are produced by combining the effects of precipitation and strain hardening. If plastic deformation precedes the aging treatment, a finer

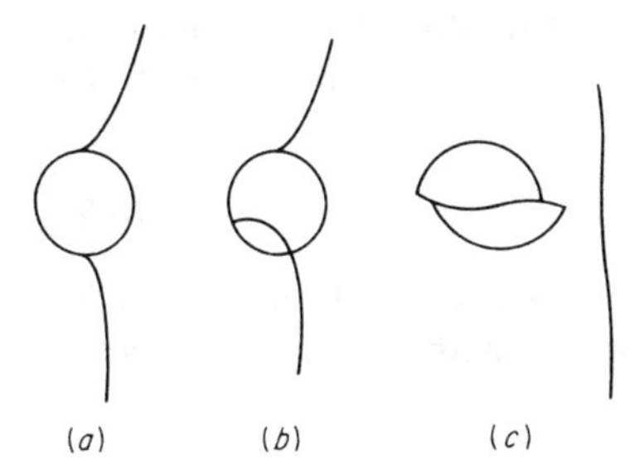

Figure 6-20 A dislocation cutting a particle.

dispersion is produced when particles nucleate on the dislocations in the matrix. The strongest alloys seem to be those in which the particles are formed in dense dislocation cell structures in the deformed matrix. Extensive plastic deformation of alloys containing fine, strong, dispersed particles can result in very high strength, as in cold-drawn steel wire.

We have seen that there are several ways in which fine particles can act as barriers to dislocations. They can act as strong impenetrable particles through which the dislocations can move only by sharp changes in curvature of the dislocation line. On the other hand, they can act as coherent particles through which dislocations can pass, but only at stress levels much above those required to move dislocations through the matrix phase. Thus, second-phase particles act in two distinct ways to retard the motion of dislocations. The particles either may be cut by the dislocations or the particles resist cutting and the dislocations are forced to bypass them. A critical parameter of the dispersion of particles is the interparticle spacing λ. Interparticle spacing has been subject to many interpretations and expressed by many parameters.[1] A simple expression for the linear mean free path is

$$\lambda = \frac{4(1 - f)r}{3f} \tag{6-28}$$

where f is the volume fraction of spherical particles of radius r.

When the particles are small and/or soft, dislocations can cut and deform the particles as shown in Fig. 6-20. There are six properties of the particles which affect the ease with which they can be sheared. These strengthening mechanisms are:

coherency strains
stacking-fault energy
ordered structure
modulus effect
interfacial energy and morphology
lattice friction stress

[1] C. W. Corti, P. Cotterill and G. Fitzpatrick, *Int. Met. Rev.*, vol. 19, pp. 77–88, June 1974.

Mott and Nabarro[1] recognized that the strain field resulting from the mismatch between a particle and the matrix would be a source of strengthening. The increase in yield stress is given by

$$\Delta\sigma \approx 2G\varepsilon f \tag{6-29}$$

where f is the volume fraction of the dispersed phase and ε is the measure of the strain field. A more sophisticated estimation[2] of the strengthening due to coherency strains is

$$\Delta\sigma \approx 6G\left(\frac{r}{b}\right)^{1/2} f^{1/2}\varepsilon^{3/2} \tag{6-30}$$

where r is the particle radius.

For precipitates which have stacking-fault energies significantly different from the matrix the interaction betwen the dislocations and particles can be dominated by the local variation of fault width when glide dislocations enter the particles. Hirsch and Kelly[3] have shown that the increase in flow stress is proportional to the difference in stacking fault energy between the particle γ_p and the matrix γ_m.

$$\Delta\sigma \approx C\left(\frac{\gamma_m - \gamma_p}{b}\right)\left\{\frac{3k(\alpha)\ln(\gamma_m/\gamma_p)}{E}\right\}F_1 f^{2/3} \tag{6-31}$$

where $k(\alpha)$ is the partial dislocation separating force times the separation distance and F_1 is a complex function of stacking-fault width and particle size. For this mechanism to operate the particle must have a structure which gives rise to extended dislocations.

Strengthening due to ordered particles is responsible for the good high-temperature strength of many superalloys. If the particles have an ordered structure then anti-phase boundaries are introduced when they are sheared. The increment of hardening is given by[4]

$$\Delta\sigma \approx \frac{2}{\sqrt{\pi E}}\left(\frac{\gamma_{apb}}{b}\right)^{3/2} r^{1/2} f^{1/2} \tag{6-32}$$

The model of strengthening from ordered particles depends on the details of the size and spacing of the particles.

Because the energy of a dislocation depends linearly on the local modulus, particles which have a modulus which differs significantly from the matrix will raise or lower the energy of a dislocation as it passes through them. This

[1] N. F. Mott and F. R. N. Nabarro, *op. cit.*

[2] V. Gerold and H. Haberkorn, *Phys. Status. Solidi*, vol. 16, p. 675, 1966.

[3] P. B. Hirsch and A. Kelly, *Phil. Mag.*, vol. 12, p. 881, 1965.

[4] L. M. Brown and R. K. Hamm, "Strengthening Methods in Crystals," A. Kelly and R. B. Nicholson (eds.), p. 9, Elsevier, London, 1971.

strengthening effect is given by

$$\Delta\sigma \approx \frac{\Delta G}{2\pi^2}\left[\frac{3|\Delta G|}{Gb}\right]^{1/2}\left[0.8 - 0.143\ln\left(\frac{r}{b}\right)\right]^{3/2} r^{1/2} f^{1/2} \tag{6-33}$$

In most alloys there is not enough of a modulus difference between the matrix and the particle to produce a strong strengthening effect. However, a large ΔG will be produced when voids are present and these can contribute to the strengthening.

When a particle is sheared by a dislocation a step which is one Burgers vector high is produced at the particle-matrix interface. Since this process increases the surface area of the particle there is an associated increase in surface energy which must be supplied by the external stress. Kelly and Nicholson[1] have shown that the increment in strengthening is given by

$$\Delta\sigma \approx \frac{2\sqrt{6}}{\pi}\gamma_s \frac{f}{r} \tag{6-34}$$

where γ_s is the particle-matrix surface energy. This equation is derived for the case of spherical particles but the greatest strengthening due to surface effects arises from thin plate-shaped precipitates, like the GP zones or θ'' precipitates in the Al-Cu system, where the surface-to-volume ratio of the particle is high.

Finally, there will be a strengthening increment due to the Peierls stress in the particle and the matrix.[2]

$$\Delta\sigma \approx \frac{5.2 f^{1/3} r^{1/2}}{G^{1/2} b^2}(\sigma_p - \sigma_m) \tag{6-35}$$

where σ_p and σ_m are the strength of the particle and matrix, respectively.

Deformation which occurs by the shearing of particles produces little strain hardening. The slip bands are planar and coarse. While there are good theoretical models for describing the strength increase from particular mechanisms of dislocation-particle interactions there is no clear understanding of how the contributions from two or more mechanisms should be combined. The summation of these mechanisms leads to an increase in strength with particle size (see left-hand side of Fig. 6-18). Eventually a point is reached where the cutting of particles becomes very difficult, and instead the dislocations find ways of moving around the particles.

For the case of overaged noncoherent precipitates Orowan[3] proposed the mechanism illustrated in Fig. 6-21. The yield stress is determined by the shear

[1] A. Kelly and R. B. Nicholson, *Prog. Mater. Sci.*, vol. 10, no. 3, p. 151, 1963.

[2] H. Gleiter and E. Hornbogen, *Mater. Sci. Eng.*, vol. 2, pp. 285–302, 1967.

[3] E. Orowan, discussion in "Symposium on Internal Stresses," p. 451, Institute of Metals, London, 1947.

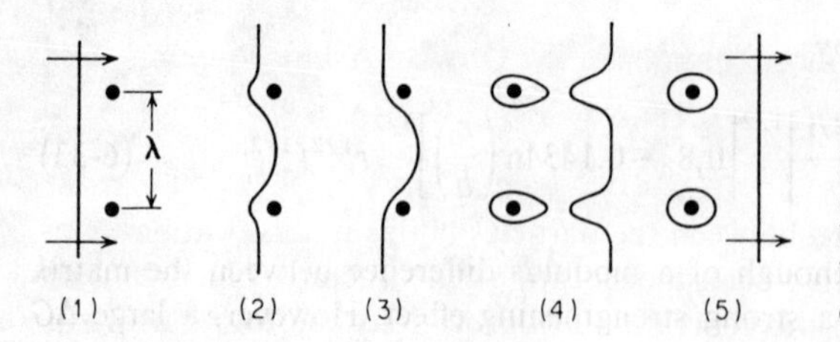

Figure 6-21 Schematic drawing of stages in passage of a dislocation between widely separated obstacles, based on Orowan's mechanism of dispersion hardening.

stress required to bow a dislocation line between two particles separated by a distance λ, where $\lambda > R$. In Fig. 6-21, stage 1 shows a straight dislocation line approaching two particles. At stage 2 the line is beginning to bend, and at stage 3 it has reached the critical curvature. The dislocation can then move forward without further decreasing its radius of curvature. From Eq. (5-18), $R = Gb/2\tau_0$ and $\lambda = 2R$, so that the stress required to force the dislocation between the obstacles is

$$\tau_0 = \frac{Gb}{\lambda} \tag{6-36}$$

Since the segments of dislocation that meet on the other side of the particle are of opposite sign, they can annihilate each other over part of their length, leaving a dislocation loop around each particle (stage 4). The original dislocation is then free to move on (stage 5). Every dislocation gliding over the slip plane adds one loop around the particle. These loops exert a back stress on dislocation sources which must be overcome for additional slip to take place. This requires an increase in shear stress, with the result that dispersed noncoherent particles cause the matrix to strain-harden rapidly. The strain hardening due to short-range stress can be calculated from Hart's model,[1] while that due to the average internal stress may also be determined.[2] The strain-hardening rate due to average internal stresses is

$$\frac{d\sigma}{d\varepsilon} = \frac{(7 - 5\nu)}{10(1 - \nu^2)} \frac{fE}{1 - f} \tag{6-37}$$

The basic Orowan equation has been modified by introducing more refined estimates of the dislocation line tension,[3] by using the planar spacing λ_p, for the mean free path,[4] and by adding a correction for the interaction between dislocation segments on either side of the particle.[5] These lead to a number of versions

[1] E. W. Hart, *Acta Metall.*, vol. 20, p. 272, 1972.
[2] T. Mori and K. Tanaka, *Acta Metall.*, vol. 21, p. 571, 1973.
[3] A. Kelly and R. B. Nicholson, *op. cit.*
[4] U. F. Kochs, *Philos. Mag.*, vol. 13, p. 54, 1966.
[5] M. F. Ashby, *Acta Metall.*, vol. 14, p. 679, 1966.

of the equation, of which the most common is the Orowan-Ashby equation[1]

$$\Delta\sigma = \frac{0.13Gb}{\lambda} \ln \frac{r}{b} \tag{6-38}$$

The bowing of dislocations between the particles builds up dislocation loops around the particles, but it also creates a dislocation cell structure (Fig. 6-28*b*) on the particles. This arises from the generation of dislocations due to the necessity for retaining continuity between the nondeforming particles and the matrix. If the substructure is strong, it can result in Hall-Petch type of strengthening rather than Orowan strengthening.

$$\Delta\sigma = \sigma_0 + k\lambda^{-1/2} \tag{6-39}$$

Most theory of strengthening with second-phase particles is based on idealized spherical particles. However, particle shape can be important, primarily by changing λ. At equal volume fraction, rods and plates strengthen about twice as much as spherical particles.[2]

Role of Slip Character

Both solid solution and particle strengthening have a significant influence on the deformation characteristics of the alloy. The character of the slip process can have an important influence on the tensile ductility, strain-hardening rate, fatigue crack initiation, and fatigue crack growth rate. It must be realized that increase in strength that is not accompanied by improvements in those other very structure-sensitive properties may not be useful in an engineering situation.

The slip character can be characterized as planar or wavy slip and as coarse or fine slip.[3] Fine wavy slip produces the most homogeneous deformation and generally leads to the best ductility at a given strength level. As a general rule, solid solution additions tend to promote either fine or coarse planar slip, because most alloy additions reduce the stacking-fault energy, thereby making cross slip difficult. Solutes which promote short-range order cause coarse planar slip.

Strengthening with fine particles can produce a planar-to-wavy or wavy-to-planar slip mode transition depending on the nature of the dislocation-particle interactions. Particles which are sheared by dislocations tend to produce slip which is coarse and planar. Particles which are bypassed by dislocations lead to fine wavy slip.

Example An aluminum-4% copper alloy has a yield stress of 600 MPa. Estimate the particle spacing and particle size in this alloy.

At this strength level we are dealing with a precipitation-hardening alloy that has been aged beyond the maximum strength. The strengthening mechanism is dislocation bypassing of particles. From Eq. (6-36) $\lambda = Gb/\tau_0$. From

[1] M. F. Ashby, in *Proc. Second Bolton Landing Conf. on Oxide Dispersion Strengthening*, Gordon and Breach, Science Publishers, Inc., New York, 1968.

[2] P. M. Kelly, *Scr. Metall.*, vol. 6, pp. 647–656, 1972.

[3] J. C. Williams, A. W. Thompson, and R. G. Baggerly, *Scr. Metall.*, vol. 8, p. 625, 1974.

Table 2-1 $G \approx 27.6$ GPa; $b \approx 2.5 \times 10^{-8}$ cm $= 2.5 \times 10^{10}$ m; $\tau_0 = 600/2 =$ 300 MPa

$$\lambda = \text{interparticle spacing} = \frac{27.6 \times 10^9\ \text{Pa}(2.5 \times 10^{-10}\ \text{m})}{3 \times 10^8\ \text{Pa}}$$

$$= 2.3 \times 10^{-8}\ \text{m} = 0.023\ \mu\text{m}$$

To estimate the particle size we first need to determine the volume fraction of precipitate particles. The maximum volume of precipitate would be obtained on slow cooling from the melt. We can estimate this quantity from the phase diagram for the aluminum-copper system.

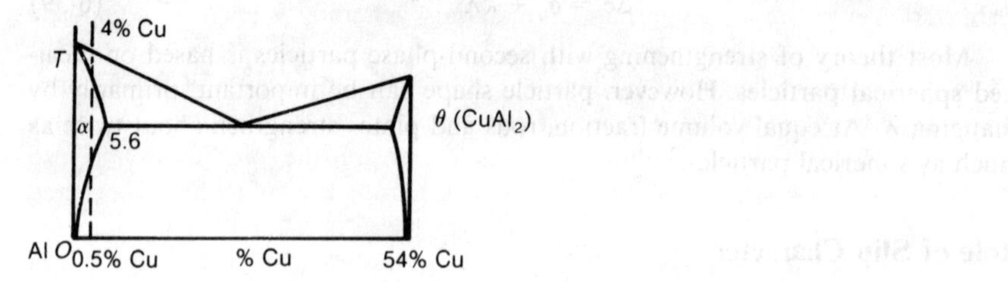

$$\text{wt\% } \alpha\text{-phase (Al)} = \frac{54 - 4}{54 - 0.5} = 93.5$$

$$\text{wt\% } \theta\text{-phase } (CuAl_2) = \frac{4 - 0.5}{54 - 0.5} = 6.5$$

$$\text{volume of } \alpha = \frac{93.5\ \text{g}}{2.70\ \text{g/cm}^3} = 34.6\ \text{cm}^3$$

$$\text{volume of } \theta = \frac{6.5\ \text{g}}{4.43\ \text{g/cm}^3} = 1.5\ \text{cm}^3$$

volume fraction of $\alpha = 0.96$

volume fraction of $\theta = 0.04$

Assuming that the particles are spherical, we can use Eq. (6-28)

$$r = \frac{3f\lambda}{4(1-f)} = \frac{3(0.04)(0.023)}{4(1-0.04)} = 0.0007\ \mu\text{m}$$

6-10 FIBER STRENGTHENING

Materials of high strength, and especially high *strength-to-weight ratio*, can be produced by incorporating fine fibers in a ductile matrix. The fibers must have high strength and high elastic modulus while the matrix must be ductile and nonreactive with the fibers. Because of their very high strength, whiskers of materials such as Al_2O_3 have been used with good results, but most fiber-strengthened materials use fibers of boron or graphite or metal wires such as

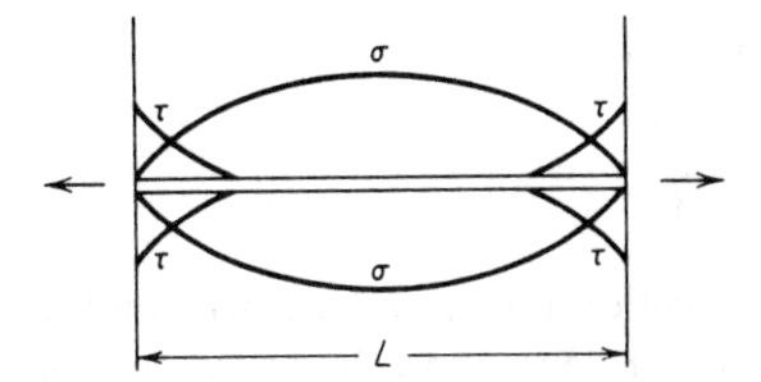

Figure 6-22 The variation of stresses along a fiber.

tungsten. The fibers may be long and continuous, or they may be discontinuous. Metals and polymers have been used as matrix materials. Glass-fiber-reinforced polymers are the most common fiber-strengthened materials. Fiber-reinforced materials are an important group of materials generally known as composite materials.[1]

An important distinction between fiber-strengthened and dispersion-strengthened metals is that in fiber strengthening the high modulus fibers carry essentially all of the load. The matrix serves to transmit the load to the fibers, to protect fibers from surface damage, and to separate the individual fibers and blunt cracks which arise from fiber breakage. There is a fair analogy between fiber-strengthened materials and steel reinforcing bars in concrete. The analysis of the strength of fiber-strengthened materials involves the direct application of the continuum principles of Chap. 2 to the microscopic level.[2] There is no need to invoke dislocation theory since the material behavior is essentially elastic.

Because the fibers and the matrix have quite different elastic moduli a complex stress distribution will be developed when a composite body is loaded uniaxially in the direction of the fibers. A fairly rigorous analysis[3] shows that shear stresses develop at the fiber-matrix interface. The distribution of this shear stress τ and the axial tensile stress in the fiber σ along the length of the fiber is given in Fig. 6-22.

Strength and Moduli of Composites

To a reasonable approximation the modulus and strength of a fiber-reinforced composite are given by a *rule of mixtures*. Using the model shown in Fig. 6-22, if we apply a tensile force P in the direction of the fiber, we can assume that the fiber and the matrix will strain equally, i.e., $e_f = e_m = e_c$.

$$P = \sigma_f A_f + \sigma_m A_m$$

where A_f and A_m are the cross-sectional areas of fiber and matrix. The average

[1] L. J. Broutman and R. H. Krock, "Modern Composite Materials," Addison-Wesley Publishing Co., Reading, Mass., 1967; A. Kelly and G. J. Davies, *Metall. Rev.*, vol. 10, p. 1, 1965; A. Kelly, *Metall. Trans.*, vol. 3, pp. 2313–2325, 1972; M. R. Piggott, "Load Bearing Fibre Composites," Pergamon Press, Oxford, 1980.

[2] This micromechanics approach to composite materials is well described by H. T. Corten, Micromechanics and Fracture Behavior of Composites, in L. J. Broutman and R. H. Krock, "Modern Composite Materials," see chap. 2.

[3] H. L. Cox, *Br. J. Appl. Phys.*, vol. 3, p. 72, 1952.

composite strength is $\sigma_c = P/A_c$ where $A_c = A_f + A_m$.

$$\sigma_c = \frac{P}{A_c} = \frac{\sigma_f A_f}{A_c} + \frac{\sigma_m A_m}{A_c}$$

But A_f/A_c represents the fraction of total cross section taken up by fibers, and if we multiply by the total length of composite this represents the volume fraction of fibers, f_f. In a similar way A_m/A_c represents f_m, and $f_f + f_m = 1$. Therefore

$$\sigma_c = \sigma_f f_f + \sigma_m f_m = \sigma_f f_f + (1 - f_f)\sigma_m \tag{6-40}$$

Since we are dealing with elastic behavior, $\sigma_f = E_f e_f$ and $\sigma_m = E_m e_m$. Therefore

$$P_c = E_f e_f A_f + E_m e_m A_m$$

and $P_c = E_c e_c A_c$. Assuming that $e_c = e_f = e_m$, and transforming from area ratio to volume fraction, we arrive at a rule of mixtures relation for the modulus of the composite.

$$E_c = E_f f_f + E_m f_m = E_f f_f + (1 - f_f) E_m \tag{6-41}$$

Example Boron fibers, $E_f = 380$ GPa, are made into a unidirectional composite with an aluminum matrix, $E_m = 60$ GPa. What is the modulus parallel to the fibers for 10 and 60 vol%.

$$E_c = E_f f_f + (1 - f_f) E_m$$

$$f_f = 0.10 \qquad E_c = 380(0.10) + 0.90(60) = 92 \text{ GPa}$$

$$f_f = 0.60 \qquad E_c = 380(0.60) + 0.40(60) = 252 \text{ GPa}$$

The uniaxial tensile stress-strain curves for unidirectional continuous fibers show four stages (Fig. 6-23). In stage 1 both fibers and matrix undergo elastic deformation. Young's modulus for the composite E_c can be determined from a simple "rule of mixtures" addition of the elastic modulus of the matrix and the fiber. In stage 2 the matrix deforms plastically and the fiber still is elastic. This stage begins at approximately the yield strain of the matrix material without fibers. To calculate E_c in stage 2, E_m must be replaced by the slope of the matrix

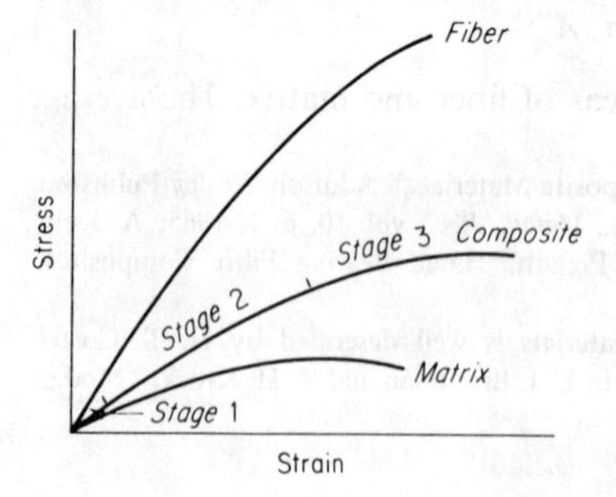

Figure 6-23 Stages in stress-strain curves of the fiber, matrix, and fiber-reinforced composite material.

stress-strain curve.

$$E_c = E_f f_f + \left(\frac{d\sigma}{d\varepsilon}\right)_m f_m \tag{6-42}$$

Since the slope of the plastic part of the matrix stress-strain curve is less than E_m, the last term in Eq. (6-42) is small and we can express the composite modulus as

$$E_c \approx E_f f_f \tag{6-43}$$

The composite responds in a quasi-elastic manner in stage 2. When the composite is unloaded the fibers return to their original length but the matrix is deformed into compression.[1] Stage 3 occurs when both the fibers and the matrix undergo plastic deformation. Since many of the high strength–high modulus fibers like boron are brittle, they fracture on entering stage 3, but metal wire fibers show this region. Finally, in stage 4 the fibers fracture and the composite as a whole soon fractures.

Assuming that the fibers all have the same strength (which usually is not true), the ultimate tensile strength of the composite is given by

$$\sigma_{cu} = \sigma_{fu} f_f + \sigma'_m (1 - f_f) \tag{6-44}$$

where σ_{fu} is the ultimate tensile strength of the fiber and σ'_m is the flow stress in the matrix at a strain equal to the fiber breaking stress.

To obtain any benefit from the presence of fibers, the strength of the composite must be greater than the strength of the strain-hardened matrix, i.e., $\sigma_{cu} \geq \sigma_{mu}$. Thus,

$$\sigma_{fu} f_f + \sigma'_m (1 - f_f) \geq \sigma_{mu}$$

which leads to a *critical fiber volume* which must be exceeded for fiber strengthening to occur.

$$f_{\text{crit}} = \frac{\sigma_{mu} - \sigma'_m}{\sigma_{fu} - \sigma'_m} \tag{6-45}$$

For small values of f_f the strength of the composite may not follow Eq. (6-44) because there are an insufficient number of fibers to effectively restrain the elongation of the matrix. As a result, the fibers are rapidly stressed to fracture. Under these conditions the metal matrix will carry part of the load by strain hardening. If we assume that when this occurs all fibers fracture ($\sigma_f = 0$), then from Eq. (6-44),

$$\sigma_{cu} \geq \sigma_{mu} (1 - f_f) \tag{6-46}$$

This expression serves to define the minimum volume fraction of fiber which must be exceeded to have real reinforcement.

$$f_{\min} = \frac{\sigma_{mu} - \sigma'_m}{\sigma_{fu} + \sigma_{mu} - \sigma'_m} \tag{6-47}$$

[1] H. P. Cheskis and R. W. Heckel, *Metall. Trans.*, vol. 1, pp. 1931–1942, 1970.

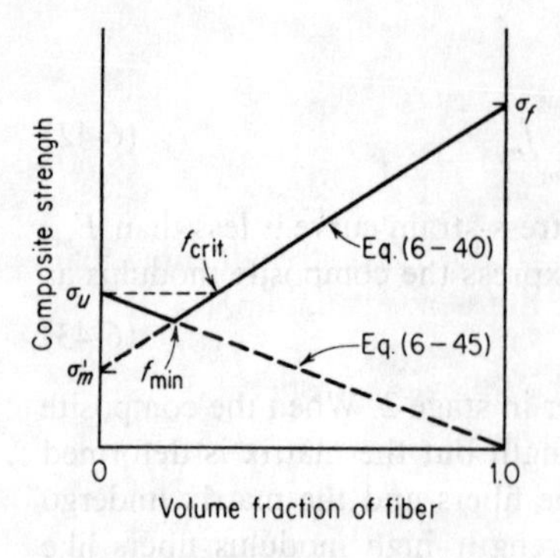

Figure 6-24 Theoretical variation of composite strength with volume fraction of fiber (for reinforcement with continuous fibers).

The relationships between composite strength and volume fraction of fibers and the positions of $f_{\min}$ and f_{crit} are shown in Fig. 6-24.

Load Transfer

When a fiber-reinforced composite is loaded the matrix acts to transfer the load to the fibers. The load transfer agent is the bond between the matrix and the fiber, which is represented by the interfacial shear stress τ_i. The existence of the bond strength builds up stresses in the fibers. The high shear stresses at the ends of the fiber (see Fig. 6-22) mean that the metal matrix will flow plastically. In order to fully utilize the high strength of the fiber it is necessary that the plastic zone in the matrix not extend from the fiber ends to its midlength before the strain in the fiber reaches its failure strain.

If we neglect strain hardening, then the shear stress on the fiber surface will not exceed the yield stress τ_0. Taking an equilibrium of forces between the axial stress in the fiber, σ_f, and the shear stress acting on a length $L/2$, because the fiber is being loaded from both ends, we get

$$\sigma_f \frac{\pi d^2}{4} = \tau_0 \pi d \frac{L}{2}$$

and

$$L_c = \frac{\sigma_f d}{2\tau_0} \tag{6-48}$$

When the fiber length exceeds L_c it is possible to load it to breaking stress by means of load transfer through the matrix deforming plastically around it.

The average stress in the fiber is less than the maximum fiber stress at the midlength of the fiber, even when $L > L_c$.

$$\bar{\sigma}_f = \sigma_f \left(1 - \frac{1 - \beta}{L/L_c} \right) \tag{6-49}$$

where β is a load transfer function. For an ideally plastic matrix the buildup of stress from the end to a distance $L_c/2$ from the end is linear, and $\beta = 0.5$.

Therefore, Eq. (6-44) should be written

$$\sigma_{cu} = \sigma_{fu} f_f \left(1 - \frac{1-\beta}{L/L_c}\right) + \sigma_m'(1 - f_f) \tag{6-50}$$

This shows that discontinuous fibers will always produce less strengthening than continuous fibers. However, if L/L_c is large, the difference is unimportant.

Example A fiber-reinforced composite has the following properties:

$$\sigma_{fu} = 5 \text{ GPa} \qquad f_f = 0.50$$
$$\sigma_m' = 100 \text{ MPa} \qquad d = 100\ \mu\text{m}$$
$$\tau_0 = 80 \text{ MPa}$$

(*a*) Determine the composite strength if the fibers have a length of 10 cm. Assume $\beta = 0.5$

$$L_c = \frac{\sigma_f d}{2\tau_0} = \frac{5 \times 10^9 \times 100 \times 10^{-6}}{2 \times 80 \times 10^6} = 3.12 \times 10^{-3} \text{ m}$$

$$\sigma_{cu} = \sigma_{fu} f_f \left(1 - \frac{L_c}{2L}\right) + \sigma_m'(1 - f_f)$$

$$= (5 \times 10^9)(0.5)\left(1 - \frac{3.12 \times 10^{-3}}{2 \times 10^{-1} \text{ m}}\right) + 100 \times 10^6(0.50)$$

$$= \left[(5 \times 10^9)(0.9844) + 100 \times 10^6\right]0.50$$

$$= (4.922 \times 10^9 + 0.1 \times 10^9)0.5$$

$$= 5.022 \times 10^9/2 = 2.51 \text{ GPa}$$

(*b*) Determine the composite strength if $L = 2$ mm $= 2 \times 10^{-3}$ m

$$\sigma_{cu} = (5 \times 10^9)(0.5)\left(1 - \frac{3.12 \times 10^{-3}}{2 \times 2 \times 10^{-3}}\right) + 100 \times 10^6(0.50)$$

$$= (5 \times 10^9)(0.50)(1 - 0.78) + 100 \times 10^6(0.50)$$

$$= \left[(5 \times 10^9)(0.22) + 100 \times 10^6\right]0.50$$

$$= (1.1 \times 10^9 + 0.1 \times 10^9)0.50$$

$$= 0.6 \text{ GPa} = 600 \text{ MPa}$$

Anisotropy

A unidirectional array of fibers in a matrix is a highly anisotropic material. When such a composite is loaded at an angle to the fibers (Fig. 6-25), three strength parameters must be considered. The stress required to produce failure by flow parallel to the fibers is σ_c, given by Eq. (6-44). The shear stress required to produce failure by shear in the matrix or at the fiber-matrix interface is τ_s, while σ_s is the tensile stress required to produce failure of the composite in a direction

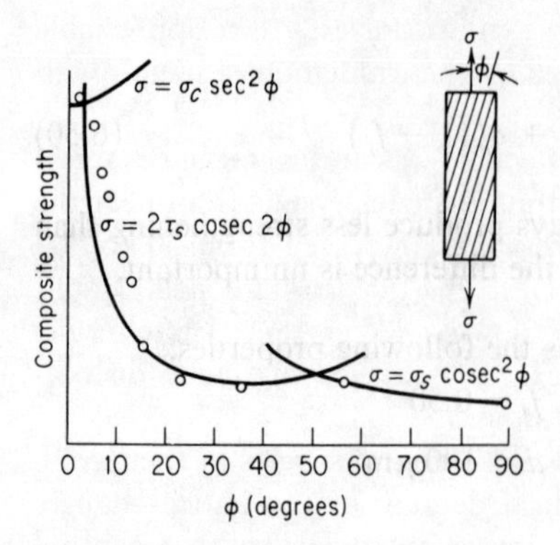

Figure 6-25 Variation of composite strength with angle between fibers and tensile axis.

normal to the fibers. The tensile stress to produce failure of the composite by fracture of the fibers is

$$\sigma = \sigma_c \sec^2 \phi \tag{6-51}$$

If failure occurs by shear in the direction of the fibers on a plane parallel to the fibers, the failure stress is

$$\sigma = 2\tau_s \operatorname{cosec} 2\phi \tag{6-52}$$

Failure by flow of the matrix transverse to the fibers or tensile failure of the interface requires a stress

$$\sigma = \sigma_s \operatorname{cosec}^2 \phi \tag{6-53}$$

These failure criteria are plotted in Fig. 6-25, where it is seen that the strength of a unidirectional composite falls off significantly at small departures from the fiber orientation.

One of the consequences of the anisotropy of fiber composites is that they display *shear coupling*. This means that an axial stress produces shear strains and a shear stress produces axial strains. In an isotropic material a uniaxially applied load produces only axial and transverse normal strains. But, in a fiber-reinforced material, in addition to those normal strains resulting from a uniaxial load there is a shear strain. This is compensated for in practice by using a cross-ply laminate in which the fibers have a different orientation in each layer. These effects obviously complicate the design with composite materials.[1]

6-11 STRENGTHENING DUE TO POINT DEFECTS

In Sec. 5-12 it was shown that the movement of jogs produced by dislocation intersections can lead to the formation of point defects, either vacant lattice sites or interstitial atoms. Previously, it had been shown that quenching from tempera-

[1] J. E. Ashton, J. C. Halpin, and P. H. Pettit, "Primer on Composite Materials Analysis," Technomic Publishing Company, Stamford, Conn., 1969; R. M. Jones, "Mechanics of Composite Materials," McGraw-Hill, New York, 1975; J. R. Vinson and T. W. Chou, "Composite Materials and Their Use in Structures," Halsted Press, New York, 1975.

tures near the melting point retains an excess of vacancies. Also, appreciable concentrations of point defects can be produced by the irradiation of metals with high-energy atomic particles.

The most basic experiments[1] on the effect of vacancies on mechanical properties were performed by quenching aluminum single crystals from near the melting point. The critical resolved shear stress was increased from 50 to 500 g/cm^2 by the presence of quenched-in vacancies. The quench-hardened crystals showed coarse slip bands compared to the soft slow-cooled crystals. These results can be explained by the assumption that the excess vacancies migrate to dislocations and pin them in a way similar to solute atoms.

Fast-moving atomic particles create interstitials and vacancies in their collision with a solid metal. Ignoring the structural details[2] of the lattice changes brought about by high-energy radiation, it is important to realize that neutron irradiation can have marked effects on the mechanical properties of metals. In the tensile stress-strain curve the yield stress is increased by a factor of 2 to 4 for an annealed metal. Face-centered cubic metals, such as aluminum and copper, develop a sharp yield point after irradiation, but in bcc metals, like steel and molybdenum, the yield point is frequently eliminated. From an engineering point of view the most serious consequence of neutron radiation is a dramatic increase in the ductile-to-brittle transition temperature (see Sec. 14-4) in structural steels.

6-12 MARTENSITE STRENGTHENING

The transformation of austenite to martensite by a diffusionless shear-type transformation in quenching of steel is one of the most common strengthening processes used in engineering materials. Although martensitic transformations occur in a number of metallurgical systems,[3] only the alloys based on iron and carbon show such a pronounced strengthening effect. Figure 6-26 shows how the hardness of martensite varies with carbon content and compares this degree of strengthening with that achieved in dispersed aggregates of iron and cementite.

The high strength of martensite implies that there are many strong barriers to dislocation motion in this structure. The complexity of the system allows for considerable controversy and hardening mechanisms abound, but it appears that there are two main contributions[4] to the high strength of martensite. Kelly and Nutting[5] have identified two structures in quenched iron-carbon alloys with the

[1] R. Maddin and A. H. Cottrell, *Philos. Mag.*, vol. 46, p. 735, 1955.

[2] For a review see G. H. Vineyard, Radiation Hardening, in "Strengthening Mechanisms in Solids," American Society for Metals, Metals Park, Ohio, 1962.

[3] C. S. Barrett and T. B. Massalski, "Structure of Metals," 3rd ed., pp. 517–531, McGraw-Hill Book Company, New York, 1966.

[4] M. Cohen, *Trans. Metall. Soc. AIME*, vol. 224, p. 638, 1962; W. Leslie and R. Sober, *Trans. Am. Soc. Metall.*, vol. 60, p. 459, 1967.

[5] P. M. Kelley and J. Nutting, *J. Iron Steel Inst.*, vol. 197, p. 199, 1961; Also J. Nutting in "Strengthening Mechanisms in Solids," chap. 4, American Society for Metals, Metals Park, Ohio, 1962.

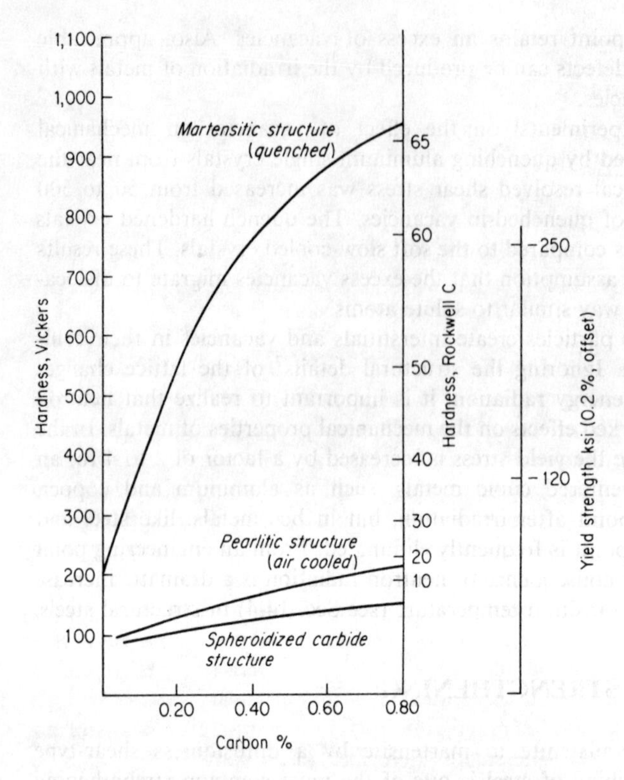

Figure 6-26 Hardness of various transformation products in steel. *(From E. Bain and H. W. Paxton, "Alloying Elements in Steel," 2nd ed., p. 37, American Society for Metals, Metals Park, Ohio, 1961. Copyright, American Society of Metals, 1961.)*

aid of transmission electron microscopy. The conventional martensite has a plate structure with a unique habit plane and an internal structure of parallel twins each about 0.1 μm thick within the plates. The other type of martensite structure is a block martensite containing a high dislocation density of 10^{11} to 10^{12} dislocations per square centimeter, comparable to that in a highly deformed metal. Thus, part of the high strength of martensite arises from the effective barriers to slip provided by the fine twin structure or the high dislocation density.

The second important contribution to the strength of martensite comes from the carbon atoms. Figure 6-26 shows that the hardness of martensite is very sensitive to carbon content below 0.4 percent. On rapidly transforming from austenite to ferrite in the quench, the solubility of carbon in iron is greatly reduced. The carbon atoms strain the ferrite lattice and this strain can be relieved by redistribution of carbon atoms by diffusion at room temperature. One result is that a strong binding is set up between dislocations and the carbon atoms. We have already seen that this restricts the motion of dislocations. Another result is

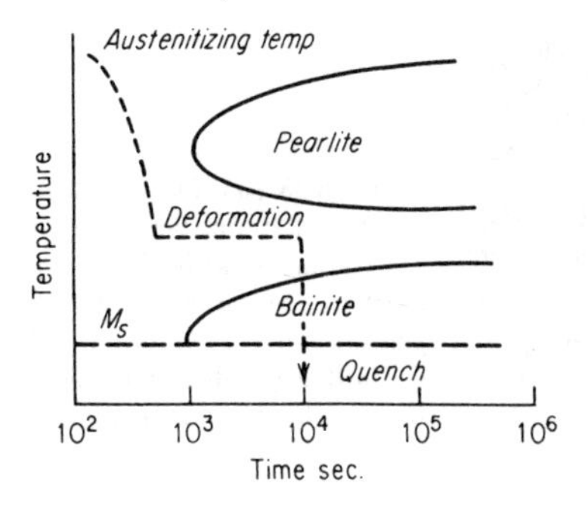

Figure 6-27 Time-temperature-transformation diagram showing steps in ausforming process.

the formation of carbon atom clusters on {100} planes. This is very similar to the GP zones discussed earlier in conjunction with the age hardening of aluminum alloys. The contribution to strength from the barriers in the martensite structure is essentially independent of carbon content, while the strengthening due to carbon atom clustering and dislocation interaction increases approximately linearly with carbon content.

An area of considerable interest has been the development of thermal-mechanical processes in which martensite is formed from an austenitic matrix which previously had been strengthened by plastic deformation.[1] This process is called *ausforming*. The plastic deformation of austenite must be accomplished without transformation to pearlite or bainite. Thus, it is necessary to work with an alloy steel which has a stable austenite region in its time-temperature-transformation (TTT) curve (Fig. 6-27). The steel is deformed in amounts in excess of 50 percent, usually by rolling, and then quenched to below the M_s to form martensite. For a given alloy, the temperature of deformation and the amount of deformation are the principal variables. Highest strengths are achieved by the greatest possible deformation at the lowest temperature at which transformation does not occur. The dislocation density of ausformed martensite is very high (10^{13} cm^{-2}) and the dislocations are usually uniformly distributed. Precipitation is more important than in ordinary quenched martensite, with the precipitates providing sites for dislocation multiplication and pinning.[2] As a result of these strengthening mechanisms ausformed steels can reach very high yield strengths of 300,000 to 400,000 psi at reductions of area ranging from 40 to 20 percent.

6-13 COLD-WORKED STRUCTURE

In Chap. 4 strain hardening was attributed to the interaction of dislocations with other dislocations and with other barriers to their motion through the lattice. So long as slip takes place on only a single set of parallel planes, as with single

[1] S. V. Radcliffe and E. B. Kula, in "Fundamentals of Deformation Processing," Syracuse University Press, 1964.

[2] O. Johari and G. Thomas, *Trans. Am. Soc. Met.*, vol. 58, pp. 563–578, 1965.

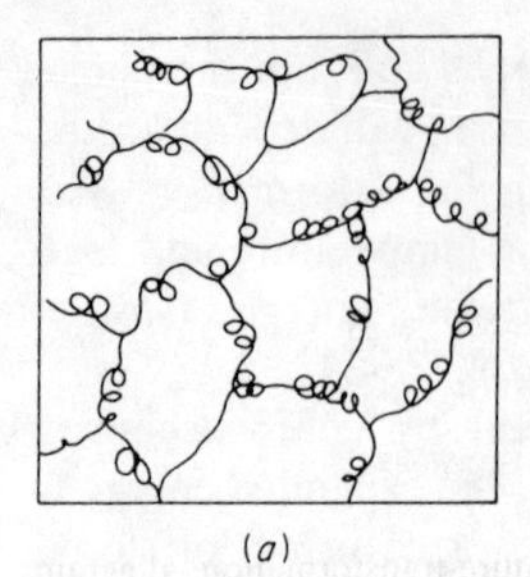

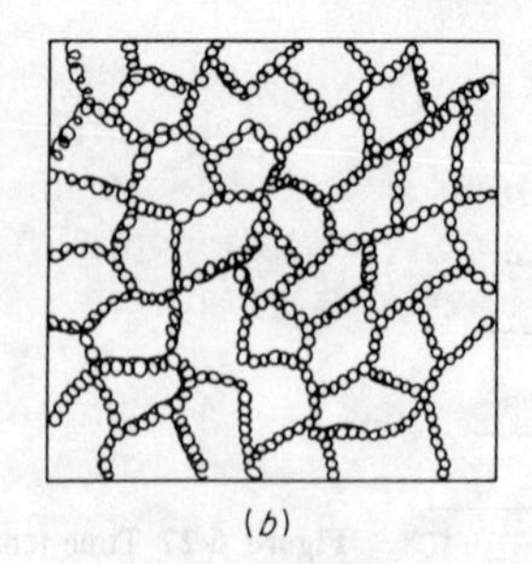

Figure 6-28 (*a*) Deformed to 10 percent strain. Beginning of cell formation with dislocation tangles; (*b*) deformed to 50 percent strain. Equilibrium cell size with heavy dislocation density in cell walls (schematic).

crystals of hcp metals, only a small amount of strain hardening occurs. However, even with single crystals extensive easy glide is not a general phenomenon, and with polycrystalline specimens it is not observed. Because of the mutual interference of adjacent grains in a polycrystalline specimen multiple slip occurs readily, and there is appreciable strain hardening. Plastic deformation which is carried out in a temperature region and over a time interval such that the strain hardening is not relieved is called *cold-work*.

Plastic deformation produces an increase in the number of dislocations, which by virtue of their interaction results in a higher state of internal stress. An annealed metal contains about 10^6 to 10^8 dislocations per square centimeter, while a severely plastically deformed metal contains about 10^{12} dislocations per square centimeter. Strain hardening or cold-work can be readily detected by x-ray diffraction, although detailed analysis of the x-ray patterns in terms of the structure of the cold-worked state is not usually possible. In Laue patterns cold-work produces a blurring, or *asterism*, of the spots. For Debye-Scherrer patterns the lines are broadened by cold-work. X-ray line broadening can be due to both a decrease in size of the diffraction unit, as would occur if the grains were fragmented by cold-work, and an increase in lattice strain due to dislocation interaction. Techniques for analyzing the entire peak profile of x-ray lines and separating out the contribution due to lattice strain and particle size have been developed.[1]

Considerable detailed knowledge on the structure of the cold-worked state has been obtained from thin-film electron microscopy. In the early stages of plastic deformation slip is essentially on primary glide planes and the dislocations form coplanar arrays. As deformation proceeds, cross slip takes place and multiplication processes operate. The cold-worked structure forms high-dislocation-density regions or *tangles*, which soon develop into tangled networks. Thus, the characteristic structure of the cold-worked state is a *cellular substructure* in which high-density-dislocation tangles form the cell walls (Fig. 6-28). The cell structure is usually well developed at strains of around 10 percent. The cell size decreases with strain at low deformation but soon reaches a fixed size, indicating

[1] B. E. Warren, in "Progress in Metal Physics," vol. 8, pp. 147–202, Pergamon Press, Ltd., London, 1959.

that as strain proceeds the dislocations sweep across the cells and join the tangle in the cell walls. The exact nature of the cold-worked structure will depend on the material, the strain, the strain rate, and the temperature of deformation. The development of a cell structure is less pronounced for low temperature and high strain-rate deformation and in materials with low stacking-fault energies (so that cross slip is difficult).

Most of the energy expended in deforming a metal by cold-working is converted into heat. However, roughly about 10 percent of the expended energy is stored in the lattice as an increase in internal energy. Reported values of stored energy[1] range from about 0.01 to 1.0 cal/g of metal. The magnitude of the stored energy increases with the melting point of the metal and with solute additions. For a given metal the amount of stored energy depends on the type of deformation process, e.g., wire drawing vs. tension. The stored energy increases with strain up to a limiting value corresponding to saturation. It increases with decreasing temperature of deformation. Very careful calorimeter measurements are required to measure the small amounts of energy stored by cold-working.

The major part of the stored energy is due to the generation and interaction of dislocations during cold-working. Vacancies account for part of the stored energy for metals deformed at very low temperature. However, vacancies are so much more mobile than dislocations that they readily escape from most metals deformed at room temperature. Stacking faults and twin faults are probably responsible for a small fraction of the stored energy. A reduction in short-range order during the deformation of solid solutions may also contribute to stored energy. Elastic strain energy accounts for only a minor part of the measured stored energy.

6-14 STRAIN HARDENING

Strain hardening or cold-working is an important industrial process that is used to harden metals or alloys that do not respond to heat treatment. The rate of strain hardening can be gaged from the slope of the flow curve. Generally, the rate of strain hardening is lower for hcp metals than for cubic metals. Increasing temperature also lowers the rate of strain hardening. For alloys strengthened by solid-solution additions the rate of strain hardening may be either increased or decreased compared with the behavior for the pure metal. However, the final strength of a cold-worked solid-solution alloy is almost always greater than that of the pure metal cold-worked to the same extent.

Figure 6-29 shows the typical variation of strength and ductility parameters with increasing amount of cold-work. Since in most cold-working processes one or two dimensions of the metal are reduced at the expense of an increase in the other dimensions, cold-work produces elongation of the grains in the principal direction of working. Severe deformation produces a reorientation of the grains

[1] For a comprehensive review of the stored energy of cold-work see M. B. Bever, D. L. Holt, and A. L. Titchener, "Progress in Materials Science," vol. 17, Pergamon Press, Ltd., London, 1973.

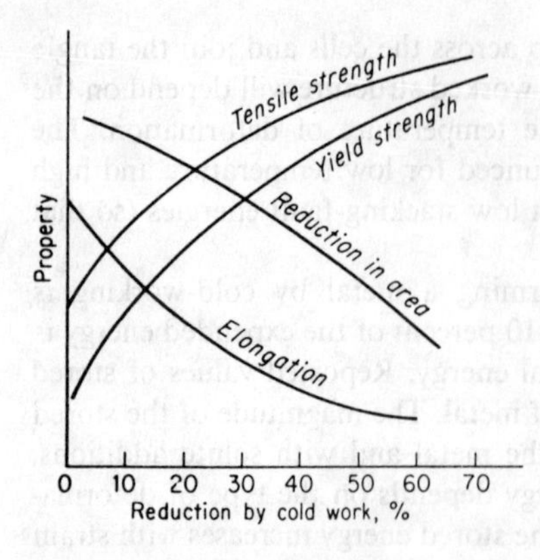

Figure 6-29 Variation of tensile properties with amount of cold-work.

into a preferred orientation (Sec. 6-17). In addition to the changes in tensile properties shown in Fig. 6-29, cold-working produces changes in other physical properties. There is usually a small decrease in density of the order of a few tenths of a percent, an appreciable decrease in electrical conductivity due to an increased number of scattering centers, and a small increase in the thermal coefficient of expansion. Because of the increased internal energy of the cold-worked state, chemical reactivity is increased. This leads to a general decrease in corrosion resistance and in certain alloys introduces the possibility of stress-corrosion cracking.

A high rate of strain hardening implies mutual obstruction of dislocations gliding on intersecting systems. This can come about (1) through interaction of the stress fields of the dislocations, (2) through interactions which produce sessile locks, and (3) through the interpenetration of one slip system by another (like cutting trees in a forest) which results in the formation of dislocation jogs.

The basic equation relating flow stress (strain hardening) to structure is

$$\sigma_0 = \sigma_i + \alpha G b \rho^{1/2} \tag{6-54}$$

Much attention has been given to developing theories of strain hardening based on dislocation models. Theories based on each of the three processes listed above result[1] in equations of the form of Eq. (6-54).

Thin-film electron micrographs perhaps impart a wrong impression of the environment in which dislocations move in metals. McLean[2] gives a graphic description of the situation involving elastic interactions between dislocations and between dislocations and second-phase particles. A metal which has been plastically deformed a few percent in strain contains 50,000 km or more of dislocation line in *each* cubic centimeter. Moreover, if this cubic centimeter were enlarged to the size of a large auditorium, these dislocations would be seen to be arranged as

[1] For a review of theories of strain hardening see D. McLean, "Mechanical Properties of Metals," pp. 153–161, John Wiley & Sons, Inc., New York, 1962; also M. A. Meyers and K. K. Chawla, "Mechanical Metallurgy," chap. 9, Prentice-Hall, Inc., Englewood Cliffs, N.J., 1984.

[2] D. McLean, *Trans. Metall. Soc. AIME*, vol. 242, pp. 1193–1203, 1968.

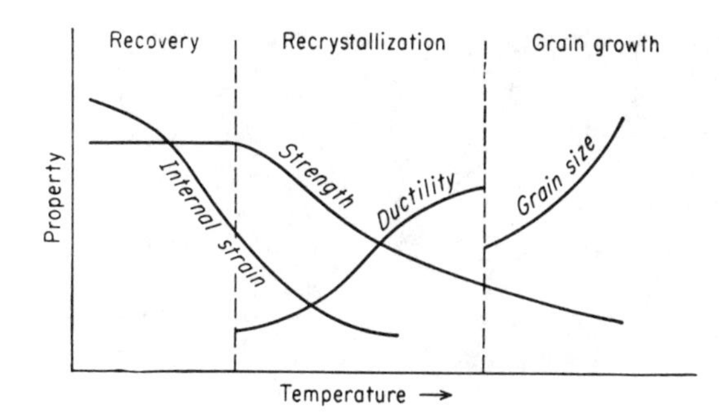

Figure 6-30 Schematic drawing indicating recovery, recrystallization, and grain growth and chief property changes in each region.

an extremely fine irregular three-dimensional spider's web, with a mesh spacing varying from 0.1 to 1.0 mm. With this type of structure a moving dislocation can hardly avoid intersecting other dislocations and traversing the stress field of other dislocations. Because thin foils for electron microscopy sample such a small amount of material they tend to miss most of the dislocation nodes and give the impression that the dislocation network is less tightly linked than it really is.

6-15 ANNEALING OF COLD-WORKED METAL

The cold-worked state is a condition of higher internal energy than the undeformed metal. Although the cold-worked dislocation cell structure is mechanically stable, it is not thermodynamically stable. With increasing temperature the cold-worked state becomes more and more unstable. Eventually the metal softens and reverts to a strain-free condition. The overall process by which this occurs is known as annealing.[1] Annealing is very important commercially because it restores the ductility to a metal that has been severely strain-hardened. Therefore, by interposing annealing operations after severe deformation it is possible to deform most metals to a very great extent.

The process of annealing can be divided into three fairly distinct processes: recovery, recrystallization, and grain growth. Figure 6-30 will help to distinguish between these processes. *Recovery* is usually defined as the restoration of the physical properties of the cold-worked metal without any observable change in microstructure. Electrical conductivity increases rapidly toward the annealed value during recovery, and lattice strain, as measured with x-rays, is appreciably reduced. The properties that are most affected by recovery are those which are sensitive to point defects. The strength properties, which are controlled by dislocations, are not affected at recovery temperatures. An exception to this is single crystals of hcp metals which have deformed on only one set of planes (easy

[1] For detailed reviews of annealing, see P. A. Beck, *Adv. Phys.*, vol. 3, pp. 245–324, 1954; J. E. Burke and D. Turnbull, in "Progress in Metal Physics," vol. 3, Interscience Publishers, Inc., New York, 1952; J. D. Verhoeven, "Fundamentals of Physical Metallurgy," chap. 10, John Wiley and Sons, Inc., New York, 1975.

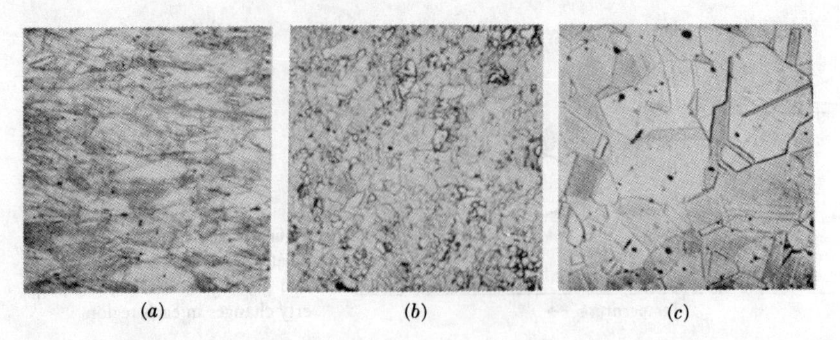

(a) (b) (c)

Figure 6-31 Changes in microstructure of cold-worked 70-30 brass with annealing. (*a*) Cold-worked 40 percent; (*b*) 440°C, 15 min; (*c*) 575°C, 15 min (150×). *(Courtesy L. A. Monson.)*

glide). For this situation it is possible to recover completely the yield stress of a strain-hardened crystal without producing recrystallization. *Recrystallization* is the replacement of the cold-worked structure by a new set of strain-free grains. Recrystallization is readily detected by metallographic methods and is evidenced by a decrease in hardness or strength and an increase in ductility. The density of dislocations decreases considerably on recrystallization, and all effects of strain hardening are eliminated. The stored energy of cold-work is the driving force for both recovery and recrystallization. If the new strain-free grains are heated at a temperature greater than that required to cause recrystallization, there will be a progressive increase in grain size. The driving force for grain growth is the decrease in free energy resulting from a decreased grain-boundary area due to an increase in grain size. Figure 6-31 shows the progression from a cold-worked microstructure to a fine recrystallized grain structure, and finally to a larger grain size by grain growth.

The recrystallization process[1] consists of the nucleation of a strain-free region whose boundary can transform the strained matrix into strain-free material as it moves. In the growth of the boundary out from the nucleus the dislocations are annihilated in the region swept through. This requires that the moving boundary be a high-angle boundary so that there is a high degree of "misfit" to accommodate the dislocations. At least two distinct nucleation mechanisms have been identified for recrystallization. The first is called *strain-induced boundary migration*, where a strain-free nucleus is formed when one of the existing grain boundaries moves into its neighbor, leaving a strain-free recrystallized region behind. The boundary moves into the grain which contains the higher dislocation density in the local region. In the second nucleation mechanism new grain boundaries are formed in regions of sharp lattice curvature through subgrain

[1] For a discussion of the mechanisms and kinetics of recovery, recrystallization, and grain growth see P. G. Shewmon, "Transformations in Metals," chap. 3, McGraw-Hill Book Company, New York, 1969; also "Recrystallization, Grain Growth, and Textures," American Society for Metals, Metals Park, Ohio, 1966.

growth. This mechanism seems to predominate at high strains, with nuclei appearing at grain boundaries, twin boundaries, or at inclusions or second-phase particles. The nuclei form only in regions which through inhomogeneous deformation have been rotated into an orientation appreciably different from that of the matrix.

Six main variables influence recrystallization behavior. They are (1) amount of prior deformation, (2) temperature, (3) time, (4) initial grain size, (5) composition, and (6) amount of recovery or polygonization prior to the start of recrystallization. Because the temperature at which recrystallization occurs depends on the above variables, it is not a fixed temperature in the sense of a melting temperature. For practical considerations a recrystallization temperature can be defined as the temperature at which a given alloy in a highly cold-worked state completely recrystallizes in 1 h. The relationship of the above variables to the recrystallization process can be summarized[1] as follows:

1. A minimum amount of deformation is needed to cause recrystallization.
2. The smaller the degree of deformation, the higher the temperature required to cause recrystallization.
3. Increasing the annealing time decreases the recrystallization temperature. However, temperature is far more important than time. Doubling the annealing time is approximately equivalent to increasing the annealing temperature 10°C.
4. The final grain size depends chiefly on the degree of deformation and to a lesser extent on the annealing temperature. The greater the degree of deformation and the lower the annealing temperature, the smaller the recrystallized grain size.
5. The larger the original grain size, the greater the amount of cold-work required to produce an equivalent recrystallization temperature.
6. The recrystallization temperature decreases with increasing purity of the metal. Solid-solution alloying additions always raise the recrystallization temperature.
7. The amount of deformation required to produce equivalent recrystallization behavior increases with increased temperature of working.
8. For a given reduction in cross section, different metalworking processes, such as rolling, drawing, etc., produce somewhat different effective deformations. Therefore, identical recrystallization behavior may not be obtained.

Because the driving force for grain growth is appreciably lower than the driving force for recrystallization, at a temperature at which recrystallization occurs readily grain growth will occur slowly. However, grain growth is strongly temperature-dependent, and a grain-coarsening region will soon be reached in which the grains increase in size very rapidly. Grain growth is inhibited consider-

[1] R. F. Mehl, Recrystallization, in "Metals Handbook," pp. 259–268, American Society for Metals, Metals Park, Ohio, 1948.

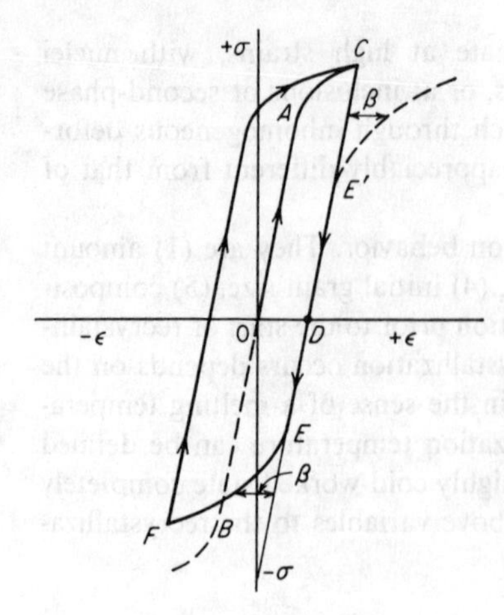

Figure 6-32 Bauschinger effect and hysteresis loop.

ably by the presence of a fine dispersion of second-phase particles, which restricts grain-boundary movement.

Under certain conditions, some of the grains of a fine-grained recrystallized metal will begin to grow rapidly at the expense of the other grains when heated at a higher temperature. This phenomenon is known as *exaggerated*, or *abnormal, grain growth*. The driving force for exaggerated grain growth is the decrease in surface energy, not stored energy, but because the phenomenon shows kinetics similar to those of recrystallization it is often called *secondary recrystallization*.

6-16 BAUSCHINGER EFFECT

In an earlier discussion on the strain hardening of single crystals it was shown that generally a lower stress is required to reverse the direction of slip on a certain slip plane than to continue slip in the original direction. The directionality of strain hardening is called the Bauschinger effect. Figure 6-32 is an example of the type of stress-strain curve that is obtained when the Bauschinger effect is considered. The Bauschinger effect is a general phenomenon in polycrystalline metals.

The initial yield stress of the material in tension is A. If the same ductile material were tested in compression, the yield strength would be approximately the same, point B on the dashed curve. Now, consider that a new specimen is loaded in tension past the tensile yield stress to C along the path O-A-C. If the specimen is then unloaded, it will follow the path C-D, small elastic-hysteresis effects being neglected. If now a compressive stress is applied, plastic flow will begin at the stress corresponding to point E, which is appreciably lower than the original compressive yield stress of the material. While the yield stress in tension was increased by strain hardening from A to C, the yield stress in compression was decreased. This is the *Bauschinger effect*. The phenomenon is reversible, for had the specimen originally been stressed plastically in compression, the yield stress in tension would have been decreased. One way of describing the amount of Bauschinger effect is by the Bauschinger strain β (Fig. 6-32). This is the difference in strain between the tension and compression curves at a given stress.

If the loading cycle in Fig. 6-32 is completed by loading further in compression to point F, then unloading, and reloading in tension, a mechanical-hysteresis loop is obtained. The area under the loop will depend on the initial overstrain beyond the yield stress and the number of times the cycle is repeated. If the cycle is repeated many times, failure by fatigue is likely to occur.

The Bauschinger effect can have important consequences in metal-forming applications. For example, it can be important in the bending of steel plates[1] and results in work-softening[2] when severely cold-worked metals are subjected to stresses of reversed sign. The best example of this is the straightening of drawn bars or rolled sheet by passing through rollers which subject the material to alternating bending stresses. Such roller-leveling operations can reduce the yield strength and increase the elongation from its cold-worked value.

The mechanism of the Bauschinger effect lies in the structure of the cold-worked state. Orowan[3] has pointed out that during plastic deformation dislocations will accumulate at barriers in tangles, and eventually form cells. Now, when the load is removed, the dislocation lines will not move appreciably because the structure is mechanically stable. However, when the direction of loading is reversed, some dislocation lines can move an appreciable distance at a low shear stress because the barriers to the rear of the dislocations are not likely to be so strong and closely spaced as those immediately in front. This gives rise to initial yielding at a lower stress level when the loading direction is reversed.

6-17 PREFERRED ORIENTATION (TEXTURE)

A metal which has undergone a severe amount of deformation, as in rolling or wire drawing, will develop a *preferred orientation*, or texture, in which certain crystallographic planes tend to orient themselves in a preferred manner with respect to the direction of maximum strain. The tendency for the slip planes in a single crystal to rotate parallel to the axis of principal strain was considered previously. The same situation exists in a polycrystalline aggregate, but the complex interactions between the multiple slip systems makes analysis of the polycrystalline situation much more difficult. Since the individual grains in a polycrystalline aggregate cannot rotate freely, lattice bending and fragmentation will occur.

[1] S. T. Rolfe, R. P. Haak, and J. H. Gross, *Trans. Amer. Soc. Mech. Eng., J. Basic Eng.*, vol. 90, pp. 403–408, 1968.

[2] N. H. Polakowski, *Am. Soc. Test. Mater. Proc.*, vol. 63, p. 535, 1963.

[3] E. Orowan, Causes and Effects of Internal Stresses, in "Internal Stresses and Fatigue in Metals," Elsevier Publishing Company, New York, 1959.

Preferred orientations are determined by x-ray methods. The x-ray pattern of a fine-grained randomly oriented metal will show rings corresponding to different planes where the angles satisfy the condition for Bragg reflection. If the grains are randomly oriented, the intensity of the rings will be uniform for all angles, but if a preferred orientation exists, the rings will be broken up into short arcs, or spots. The dense areas of the x-ray photograph indicate the orientation of the poles of the planes corresponding to the diffraction ring in question. The orientation of the grains of a particular crystallographic orientation with respect to the principal directions of working is best shown by means of a *pole figure*. For a description of the methods of determining pole figures and a compilation of pole figures describing the deformation textures in many metals, the reader is referred to Barrett.[1]

A preferred orientation can be detected with x-rays after about a 20 to 30 percent reduction in cross-sectional area by cold-working. At this stage of reduction there is appreciable scatter in the orientation of individual crystals about the ideal orientation. The scatter decreases with increasing reduction, until at about 80 to 90 percent reduction the preferred orientation is essentially complete. The type of preferred orientation, or deformation texture, which is developed depends primarily on the number and type of slip systems available and on the principal strains. Other factors which may be important are the temperature of deformation and the type of texture present prior to deformation.

The simplest deformation texture is produced by the drawing or rolling of a wire or rod. This is often referred to as a *fiber texture* because of its similarity to the arrangement in naturally fibrous materials. It is important to note that a distinction should be made between the *crystallographic* fibering produced by crystallographic reorientation of the grains during deformation and *mechanical fibering*, which is brought about by the alignment of inclusions, cavities, and second-phase constituents in the main direction of mechanical working. Mechanical and crystallographic fibering are important factors in producing directional mechanical properties of plastically worked metal shapes such as sheet and rods.[2] This will be discussed further in Chap. 8.

In an ideal wire texture a definite crystallographic direction lies parallel to the wire axis, and the texture is symmetrical around the wire or fiber axis. Body-centered cubic metals have a fiber texture with the $\langle 110 \rangle$ direction parallel to the wire axis. Face-centered cubic metals can have a double fiber texture with both $\langle 111 \rangle$ and $\langle 100 \rangle$ directions parallel to the wire axis. The $\langle 111 \rangle$ texture is favored by easy cross slip and predominates in high stacking-fault metals such as aluminum. Silver and brass with low stacking-fault energies have a predominantly $\langle 100 \rangle$ texture. In hcp metals the basal planes rotate so that the $\langle 10\bar{1}0 \rangle$ direction coincides with the wire axis (for magnesium).

[1] C. S. Barrett and T. B. Massalski, "The Structure of Metals," 3rd ed., chaps. 20 and 21, McGraw-Hill Book Company, New York, 1966; also H. J. Bunge, "Texture Analysis in Materials Science," Butterworths, London, 1982.

[2] D. V. Wilson, *Met. Technol.*, vol. 2, pp. 8–20, 1975.

The deformation texture of a sheet produced by rolling is described by the crystallographic planes parallel to the surface of the sheet as well as the crystallographic directions in that plane which are parallel to the direction of rolling. Two rolling textures[1] predominate in face-centered cubic metals and alloys. During initial deformation a $\{110\}\langle 112 \rangle$ texture (α-brass-type texture) is developed, but if extensive cross slip occurs, this changes to a $\{112\}\langle 111 \rangle$ texture (copper-type texture) with more extensive plastic deformation. There is good correlation[2] between stacking-fault energy (relative ease of cross slip) and the type of texture. High stacking-fault energy and high temperature deformation favor the copper-type structure $\{112\}\langle 111 \rangle$. In body-centered cubic metals the predominant rolling texture consists of $\{100\}$ planes oriented parallel to the plane of the sheet with the $\langle 110 \rangle$ directions in the rolling direction, but other texture elements such as $\{112\}\langle 110 \rangle$, and $\{111\}\langle 112 \rangle$ may be found. For hcp metals the basal plane tends to be parallel with the rolling plane with $\langle 2\bar{1}\bar{1}0 \rangle$ aligned in the rolling direction.

The preferred orientation resulting from plastic deformation is strongly dependent on the slip and twinning systems available for deformation, but it is not generally affected by processing variables such as die angle, roll diameter, roll speed, and reduction per pass. The most important mechanical variables are the geometry of flow and the amount of deformation (reduction). Thus, the same deformation texture is produced whether a rod is made by rolling or drawing.

The recrystallization of a cold-worked metal generally produces a preferred orientation which is different from and stronger than that existing in the deformed metal. This is called an *annealing texture*, or *recrystallization texture*. An outstanding example is the cube texture in copper, where the $\{110\}$ plane lies parallel to the rolling plane with a $\langle 001 \rangle$ direction parallel to the direction of rolling. Since the existence of a recrystallization texture depends on a preferential orientation of the nuclei of the recrystallized grains, the resulting texture is strongly dependent on the texture produced by the deformation. Other important variables which affect the annealing texture are the composition, the initial grain size and grain orientation of the alloy, and the annealing temperature and time. Generally the factors which favor the formation of a fine recrystallized grain size also favor the formation of an essentially random orientation of recrystallized grains. Moderate cold reductions and low annealing temperatures are beneficial.

Sometimes the formation of a strong texture in a finished sheet is beneficial. One of the best examples is cube-oriented silicon-iron transformer sheet, where the energy losses are minimized by orienting the grains in the easy direction of magnetization. The use of texture to resist yielding in titanium plate was discussed in Chap. 3, and in Chap. 20 we shall consider in detail how proper texture can greatly increase the deep-drawing quality of low-carbon steel. On the other hand, a strong preferred orientation will result in an anisotropy in mechanical properties in the plane of the sheet. This can result in uneven response of the

[1] R. E. Smallman, *J. Inst. Met.*, vol. 84, pp. 10–18, 1955–1956.

[2] I. S. Dilamore and W. T. Roberts, *Metall. Rev.*, vol. 10, no. 39, 1965.

material during forming and fabrication operations, and it must be recognized as a factor in design.

BIBLIOGRAPHY

Cottrell, A. H.: "An Introduction to Metallurgy," Edward Arnold (Publishers) Ltd., London, 1967.

Felbeck, D. K., and A. G. Atkins: "Strength and Fracture of Engineering Solids," Prentice-Hall, Inc., Englewood Cliffs, N.J., 1984.

Honeycombe, R. W. K.: "The Plastic Deformation of Metals," Edward Arnold (Publishers) Ltd., London, 2d ed., 1984.

Kelly, A.: "Strong Solids," 2nd ed., Oxford University Press, London, 1974.

Kelly, A., and R. B. Nicholson (eds.): "Strengthening Methods in Crystals," Halsted Press, New York, 1971.

McLean, D.: "Mechanical Properties of Metals," John Wiley & Sons, Inc., New York, 1962.

Reed-Hill, R. E.: "Physical Metallurgy Principles," 2d ed., D. Van Nostrand Company, Inc., New York, 1973.

Tien, J. K., and G. S. Ansell (eds.): "Alloy and Microstructural Design," Academic Press, New York, 1976.

Williams, J. C., and A. W. Thompson: Strengthening of Metals and Alloys, in "Metallurgical Treatises," J. K. Tien and J. F. Elliott (eds.), Met. Soc. of AIME, Warrendale, Pa., 1981.

Wyatt, O. H. and D. Dew-Hughes: "Metals, Ceramics and Polymers," Cambridge University Press, London, 1974.

Zackay, V. F. (ed): "High Strength Materials," John Wiley & Sons, Inc., New York, 1965.

CHAPTER
SEVEN
FRACTURE

7-1 INTRODUCTION

Fracture is the separation, or fragmentation, of a solid body into two or more parts under the action of stress. The process of fracture can be considered to be made up of two components, crack initiation and crack propagation. Fractures can be classified into two general categories, ductile fracture and brittle fracture. A ductile fracture is characterized by appreciable plastic deformation prior to and during the propagation of the crack. An appreciable amount of gross deformation is usually present at the fracture surfaces. Brittle fracture in metals is characterized by a rapid rate of crack propagation, with no gross deformation and very little microdeformation. It is akin to cleavage in ionic crystals. The tendency for brittle fracture is increased with decreasing temperature, increasing strain rate, and triaxial stress conditions (usually produced by a notch). Brittle fracture is to be avoided at all cost, because it occurs without warning and usually produces disastrous consequences.

This chapter will present a broad picture of the fundamentals of the fracture of metals. Since most of the research has been concentrated on the problem of brittle fracture, this topic will be given considerable prominence. The engineering aspects of brittle fracture will be considered in greater detail in Chap. 11. Fracture occurs in characteristic ways, depending on the state of stress, the rate of application of stress, and the temperature. Unless otherwise stated, it will be assumed in this chapter that fracture is produced by a single application of a uniaxial tensile stress. Fracture under more complex conditions will be considered in later chapters. Typical examples are fracture due to torsion (Chap. 10), fatigue (Chap. 12), and creep (Chap. 13), and low-temperature brittle fracture, temper embrittlement, or hydrogen embrittlement (Chap. 14).

7-2 TYPES OF FRACTURE IN METALS

Metals can exhibit many different types of fracture, depending on the material, temperature, state of stress, and rate of loading. The two broad categories of ductile and brittle fracture have already been considered. Figure 7-1 schematically illustrates some of the types of tensile fractures which can occur in metals. A brittle fracture (Fig. 7-1*a*) is characterized by separation normal to the tensile stress. Outwardly there is no evidence of deformation, although with x-ray diffraction analysis it is possible to detect a thin layer of deformed metal at the fracture surface. Brittle fractures have been observed in bcc and hcp metals, but not in fcc metals unless there are factors contributing to grain-boundary embrittlement.

Ductile fractures can take several forms. Single crystals of hcp metals may slip on successive basal planes until finally the crystal separates by shear (Fig. 7-1*b*). Polycrystalline specimens of very ductile metals, like gold or lead, may actually be drawn down to a point before they rupture (Fig. 7-1*c*). In the tensile fracture of moderately ductile metals the plastic deformation eventually produces a necked region (Fig. 7-1*d*). Fracture begins at the center of the specimen and then extends by a shear separation along the dashed lines in Fig. 7-1*d*. This results in the familiar "cup-and-cone" fracture.

Fractures are classified with respect to several characteristics, such as strain to fracture, crystallographic mode of fracture, and the appearance of the fracture. Gansamer[1] has summarized the terms commonly used to describe fractures as follows:

Behavior described	Terms used	
Crystallographic mode	Shear	Cleavage
Appearance of fracture	Fibrous	Granular
Strain to fracture	Ductile	Brittle

A shear fracture occurs as the result of extensive slip on the active slip plane. This type of fracture is promoted by shear stresses. The cleavage mode of fracture is controlled by tensile stresses acting normal to a crystallographic cleavage plane. A fracture surface which is caused by shear appears at low magnification to be gray and fibrous, while a cleavage fracture appears bright or granular, owing to reflection of light from the flat cleavage surfaces. Fracture surfaces frequently consist of a mixture of fibrous and granular fracture, and it is customary to report the percentage of the surface area represented by one of these categories. Based on metallographic examination, fractures in polycrystalline samples are classified as either *transgranular* (the crack propagates through the grains) or *intergranular*

[1] M. Gensamer, General Survey of the Problem of Fatigue and Fracture, in "Fatigue, and Fracture of Metals," John Wiley & Sons, Inc., New York, 1952.

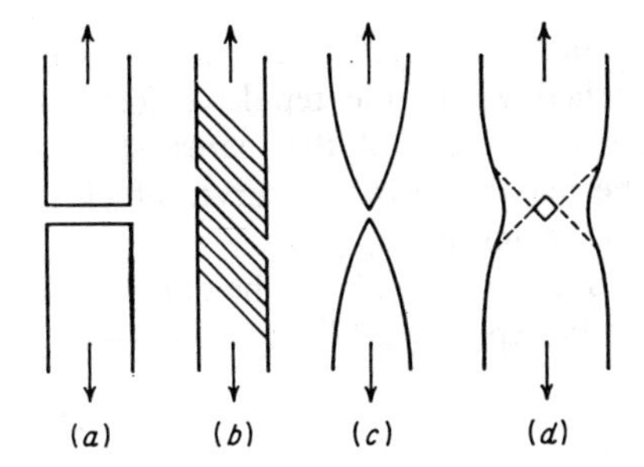

Figure 7-1 Types of fractures observed in metals subjected to uniaxial tension. (*a*) Brittle fracture of single crystals and polycrystals; (*b*) shearing fracture in ductile single crystals; (*c*) completely ductile fracture in polycrystals; (*d*) ductile fracture in polycrystals.

(the crack propagates along the grain boundaries). A ductile fracture is one which exhibits a considerable degree of deformation. The boundary between a ductile and brittle fracture is arbitrary and depends on the situation being considered. For example, nodular cast iron is ductile when compared with ordinary gray iron; yet it would be considered brittle when compared with mild steel. As a further example, a deeply notched tensile specimen will exhibit little gross deformation; yet the fracture could occur by a shear mode.

7-3 THEORETICAL COHESIVE STRENGTH OF METALS

Metals are of great technological value, primarily because of their high strength combined with a certain measure of plasticity. In the most basic terms the strength is due to the cohesive forces between atoms. In general, high cohesive forces are related to large elastic constants, high melting points, and small coefficients of thermal expansion. Figure 7-2 shows the variation of the cohesive force between two atoms as a function of the separation between these atoms. This curve is the resultant of the attractive and repulsive forces between the atoms. The interatomic spacing of the atoms in the unstrained condition is indicated by a_0. If the crystal is subjected to a tensile load, the separation between atoms will be increased. The repulsive force decreases more rapidly with increased separation than the attractive force, so that a net force between atoms

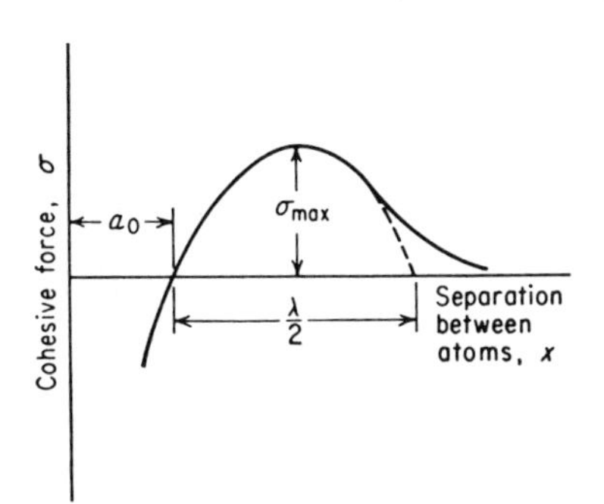

Figure 7-2 Cohesive force as a function of the separation between atoms.

balances the tensile load. As the tensile load is increased still further, the repulsive force continues to decrease. A point is reached where the repulsive force is negligible and the attractive force is decreasing because of the increased separation of the atoms. This corresponds to the maximum in the curve, which is equal to the theoretical cohesive strength of the material.

A good approximation to the theoretical cohesive strength can be obtained if it is assumed that the cohesive force curve can be represented by a sine curve.

$$\sigma = \sigma_{max} \sin \frac{2\pi x}{\lambda} \tag{7-1}$$

where σ_{max} is the theoretical cohesive strength and $x = a - a_0$ is the displacement in atomic spacing in a lattice with wave length λ. For small displacements, $\sin x \approx x$, and

$$\sigma = \sigma_{max} \frac{2\pi x}{\lambda} \tag{7-2}$$

Also, if we restrict consideration to a brittle elastic solid, then from Hooke's law

$$\sigma = Ee = \frac{Ex}{a_0} \tag{7-3}$$

Eliminating x from Eqs. (7-2) and (7-3), we have

$$\sigma_{max} = \frac{\lambda}{2\pi} \frac{E}{a_0} \tag{7-4}$$

If we make the reasonable assumption that $a_0 \approx \lambda/2$, then

$$\sigma_{max} = E/\pi \tag{7-5}$$

Therefore, the potential exists for high values of cohesive strength.

When fracture occurs in a brittle solid all of the work expended in producing the fracture goes into the creation of two new surfaces. Each of these surfaces has a *surface energy* of γ_s J/m^2. The work done per unit area of surface in creating the fracture is the area under the stress-displacement curve.

$$U_0 = \int_0^{\lambda/2} \sigma_{max} \sin \frac{2\pi x}{\lambda} dx = \frac{\lambda \sigma_{max}}{\pi} \tag{7-6}$$

But this energy is equal to the energy required to create the two new fracture surfaces.

$$\frac{\lambda \sigma_{max}}{\pi} = 2\gamma_s$$

or

$$\lambda = \frac{2\pi \gamma_s}{\sigma_{max}} \tag{7-7}$$

and substituting into Eq. (7-4) gives

$$\sigma_{max} = \left(\frac{E\gamma_s}{a_0} \right)^{1/2} \tag{7-8}$$

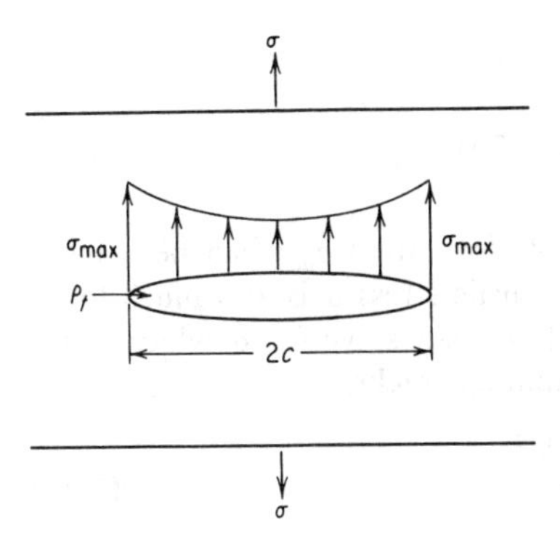

Figure 7-3 Elliptical crack model.

Using expressions for the force-displacement curve which are more complicated than the sine-wave approximation results in estimates of σ_{max} from $E/4$ to $E/15$. A convenient choice is to say that $\sigma_{max} \approx E/10$.

Example Determine the cohesive strength of a silica fiber, if $E = 95$ GPa, $\gamma_s = 1{,}000$ erg/cm² and $a_0 = 1.6$ Å.

$$\gamma_s = 10^3 \times 10^{-3} = 1 \text{ J/m}^2 \qquad a_0 = 1.6 \times 10^{-10} \text{ m}$$

$$\sigma_{max} = \left(\frac{E\gamma_s}{a_0}\right)^{1/2} = \left(\frac{95 \times 10^9 \times 1}{1.6 \times 10^{-10}}\right)^{1/2} = (5.937 \times 10^{20})^{1/2} = 24.4 \text{ GPa}$$

$$\text{units:} \quad \left(\frac{\text{N}}{\text{m}^2} \times \frac{\text{J}}{\text{m}^2} \times \frac{1}{\text{m}}\right)^{1/2} = \left(\frac{\text{N}}{\text{m}^2} \times \frac{\text{N}-\text{m}}{\text{m}^2} \times \frac{1}{\text{m}}\right)^{1/2} = \frac{\text{N}}{\text{m}^2}$$

Experience with high-strength steels shows that a fracture strength in excess of 300,000 psi (2 GPa) is exceptional. Engineering materials typically have fracture stresses that are 10 to 1000 times lower than the theoretical value. The only materials that approach the theoretical value are tiny, defect-free metallic whiskers and very-fine-diameter silica fibers. This leads to the conclusions that flaws or cracks are responsible for the lower-than-ideal fracture strength of engineering materials.

Postponing until later the question of where the cracks come from, it is a logical extension of the idea of stress concentration (Sec. 2-15) to show how the presence of cracks[1] will result in a reduced fracture strength. Figure 7-3 shows a thin elliptical crack in an infinitely wide plate. The crack has a length $2c$ and a radius of curvature at its tip of ρ_t. The maximum stress at the tip of the crack σ_{max}

[1] E. Orowan, *Welding J.*, vol. 34, pp. 157s–160s, 1955.

is given by[1]

$$\sigma_{\max} = \sigma\left[1 + 2\left(\frac{c}{\rho_t}\right)^{1/2}\right] \approx 2\sigma\left(\frac{c}{\rho_t}\right)^{1/2} \tag{7-9}$$

This approach assumes that the theoretical cohesive stress $\sigma_{\max}$ can be reached locally at the tip of a crack while the average tensile stress σ is at a much lower value. Therefore, equating Eq. (7-8) and (7-9), we can solve for σ which is the nominal fracture stress σ_f of the material containing cracks.

$$\sigma_f \approx \left(\frac{E\gamma_s\rho_t}{4a_0c}\right)^{1/2} \tag{7-10}$$

The sharpest possible crack would be one where $\rho_t = a_0$, so that

$$\sigma_f \approx \left(\frac{E\gamma_s}{4c}\right)^{1/2} \tag{7-11}$$

Example Calculate the fracture stress for a brittle material with the following properties:

$$E = 100 \text{ GPa} \qquad \gamma_s = 1 \text{ J/m}^2 \qquad a_0 = 2.5 \times 10^{-10} \text{ m}$$

The crack length is $c = 10^4 a_0 = 2.5\ \mu\text{m}$

$$\sigma_f = \left(\frac{E\gamma_s}{4c}\right)^{1/2} = \left(\frac{100 \times 10^9 \times 1}{4 \times 2.5 \times 10^{-6}}\right)^{1/2} = (10^{16})^{1/2} = 10^8 \text{ Pa} = 100 \text{ MPa}$$

Note that the fracture stress is $E/1000$ while the theoretical cohesive strength is $E/5$. Thus, we see that a very small crack produces a very great decrease in the stress for fracture.

7-4 GRIFFITH THEORY OF BRITTLE FRACTURE

The first explanation of the discrepancy between the observed fracture strength of crystals and the theoretical cohesive strength was proposed by Griffith.[2] Griffith's theory in its original form is applicable only to a perfectly brittle material such as glass. However, while it cannot be applied directly to metals, Griffith's ideas have had great influence on the thinking about the fracture of metals.

Griffith proposed that a brittle material contains a population of fine cracks which produce a stress concentration of sufficient magnitude so that the theoreti-

[1] C. E. Inglis, *Trans. Inst. Nav. Archit.*, vol. 55, pt. I, pp. 219–230, 1913. Equation (7-9) is equivalent to Eq. (2-105) since for an ellipse $\rho = b^2/a$ and $a = c$ in Fig. 7-3.

[2] A. A. Griffith, *Philos. Trans. R. Soc. London*, vol. 221A, pp. 163–198, 1920; *First Int. Congr. Appl. Mech.*, Delft, 1924, p. 55; this paper has been reprinted with annotations in *Trans. Am. Soc. Met.*, vol. 61, pp. 871–906, 1968.

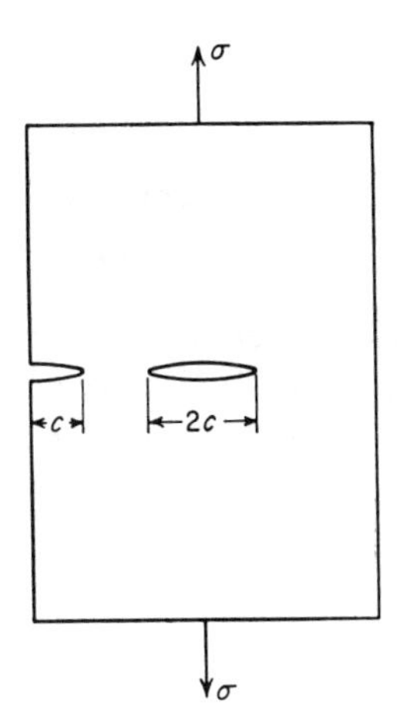

Figure 7-4 Griffith crack model.

cal cohesive strength is reached in localized regions at a nominal stress which is well below the theoretical value. When one of the cracks spreads into a brittle fracture, it produces an increase in the surface area of the sides of the crack. This requires energy to overcome the cohesive force of the atoms, or, expressed in another way, it requires an increase in surface energy. The source of the increased surface energy is the elastic strain energy which is released as the crack spreads. Griffith established the following criterion for the propagation of a crack: *A crack will propagate when the decrease in elastic strain energy is at least equal to the energy required to create the new crack surface*. This criterion can be used to determine the magnitude of the tensile stress which will just cause a crack of a certain size to propagate as a brittle fracture.

Consider the crack model shown in Fig. 7-4. The thickness of the plate is negligible, and so the problem can be treated as one in plane stress. The cracks are assumed to have an elliptical shape. For a crack at the interior the length is $2c$, while for an edge crack it is c. The effect of both types of crack on the fracture behavior is the same. The stress distribution for an elliptical crack was determined by Inglis.[1] A decrease in strain energy results from the formation of a crack. The elastic strain energy per unit of plate thickness is equal to

$$U_E = -\frac{\pi c^2 \sigma^2}{E} \tag{7-12}$$

where σ is the tensile stress acting normal to the crack of length $2c$. A negative sign is used because growth of the crack releases elastic strain energy. The surface

[1] C. E. Inglis, *op. cit.*; Eq. (7-12) can be understood if we consider that the strain energy resides in a circular region of radius c around the crack. The strain energy per unit volume is $\sigma^2/2E$, so that U_E per unit thickness is $\sigma^2(\pi c^2)/2E$. The factor $\frac{1}{2}$ drops out in the more rigorous analysis.

energy due to the presence of the crack is

$$U_s = 4c\gamma_s \tag{7-13}$$

The total change in potential energy resulting from the creation of the crack is

$$\Delta U = U_s + U_E \tag{7-14}$$

According to Griffith's criterion, the crack will propagate under a constant applied stress σ if an incremental increase in crack length produces no change in the total energy of the system; i.e., the increased surface energy is compensated by a decrease in elastic strain energy.

$$\frac{d\Delta U}{dc} = 0 = \frac{d}{dc}\left(4c\gamma_s - \frac{\pi c^2\sigma^2}{E}\right)$$

$$4\gamma_s - \frac{2\pi c\sigma^2}{E} = 0$$

$$\sigma = \left(\frac{2E\gamma_s}{\pi c}\right)^{1/2} \tag{7-15}$$

Equation (7-15) gives the stress required to propagate a crack in a brittle material as a function of the size of the microcrack. Note that this equation indicates that the fracture stress is inversely proportional to the square root of the crack length. Thus, increasing the crack length by a factor of 4 reduces the fracture stress by one-half.

For a plate which is thick compared with the length of the crack (plane strain) the Griffith equation is given by

$$\sigma = \left[\frac{2E\gamma_s}{(1-\nu^2)\pi c}\right]^{1/2} \tag{7-16}$$

Analysis of the three-dimensional case, where the crack is a very flat oblate spheroid,[1] results only in a modification to the constant in Griffith's equation. Therefore, the simplification of considering only the two-dimensional case introduces no large error.

Let us look briefly at the equation for fracture stress derived from a stress concentration point of view, Eq. (7-10), and the Griffith equation, Eq. (7-15). Equation (7-10) can be written as

$$\sigma_f = \left(\frac{2E\gamma_s}{\pi c}\frac{\pi\rho_t}{8a_0}\right)^{1/2} \quad \text{or} \quad \sigma_f \approx \left(\frac{2E\gamma_s}{\pi c}\frac{\rho_t}{3a_0}\right)^{1/2}$$

When $\rho_t = 3a_0$ this equation reduces to the Griffith equation. Thus, $\rho_t = 3a_0$ is the lower limit of the effective radius of an elastic crack. In other words, σ_f cannot approach zero as ρ_t approaches zero. When $\rho_t < 3a_0$ the stress to produce brittle

[1] R. A. Sack, *Proc. Phys. Soc. London*, vol. 58, p. 729, 1946.

fracture is given by Eq. (7-15), but when $\rho_t > 3a_0$ the fracture stress is given by Eq. (7-10).

The Griffith's equation shows a strong dependence of fracture strength on crack length. Griffith's theory satisfactorily predicts the fracture strength of a completely brittle material such as glass. In glass, reasonable values of crack length of about 1 μm are calculated from Eq. (7-15). For zinc crystals Griffith's theory predicts a critical crack length of several millimeters. This average crack length could easily be greater than the thickness of the specimen, and therefore the theory does not apply.

The importance of the surface energy term can be demonstrated by carrying out the fracture in solutions of surface active chemicals. The fracture stress of ice when tested in bending in air is about 150 psi. If the bend specimen is sprayed with methyl chloride, to lower γ_s, the fracture stress is reduced to about 75 psi. The sensitivity of the fracture of brittle solids to surface conditions has been termed the *Joffe Effect*.[1] Use is made of surface active agents to make rock drilling more easy. In metallurgical systems the surface energy can be reduced by surface adsorption of an element in solid solution. For example, the addition of 0.5% Sb to Cu reduces the surface energy from about 1,800 to 1,000 erg/cm^2. Since solute concentration often builds up in grain boundaries, this can lead to intergranular embrittlement.

It is well established that even metals which fail in a completely brittle manner have undergone some plastic deformation prior to fracture. This is substantiated by x-ray diffraction studies of fracture surfaces[2] and by metallographic studies of fracture (see Sec. 7-6). Therefore, Griffith's equation for the fracture stress does not apply for metals. One way of realizing that the fracture stress of a material which undergoes plastic deformation before fracture is greater than that of a truly brittle (elastic) material is to consider Eq. (7-10). Plastic deformation at the root of the crack would be expected to blunt the tip of the crack and increase ρ_t, thus increasing the fracture stress.

Orowan[3] suggested that the Griffith equation would be made more compatible with brittle fracture in metals by the inclusion of a term γ_p expressing the plastic work required to extend the crack wall.

$$\sigma_f = \left[\frac{2E(\gamma_s + \gamma_p)}{\pi c}\right]^{1/2} \approx \left(\frac{E\gamma_p}{c}\right)^{1/2} \tag{7-17}$$

The surface-energy term can be neglected since estimates of the plastic-work term are about 10^2 to 10^3 J/m^2 compared with values of γ_s of about 1 to 2 J/m^2.

[1] A. F. Joffe, "The Physics of Crystals," McGraw-Hill Book Company, New York, 1928.

[2] E. P. Klier, *Trans. Am. Soc. Met.*, vol. 43, pp. 935–957, 1951; L. C. Chang, *J. Mech. Phys. Solids*, vol. 3, pp. 212–217, 1955; D. K. Felbeck and E. Orowan, *Welding J.*, vol. 34, pp. 570s–757s, 1955.

[3] E. Orowan, in "Fatigue and Fracture of Metals," Symposium at Massachusetts Institute of Technology, John Wiley & Sons, Inc., New York, 1952.

7-5 FRACTURE OF SINGLE CRYSTALS

The brittle fracture of single crystals is considered to be related to the resolved normal stress on the cleavage plane. Sohncke's law states that fracture occurs when the resolved normal stress reaches a critical value. Considering the situation used to develop the resolved shear stress for slip (Fig. 4-18), the component of the tensile force which acts normal to the cleavage plane is $P\cos\phi$, where ϕ is the angle between the tensile axis and the normal to the plane. The area of the cleavage plane is $A/(\cos\phi)$. Therefore, the critical normal stress for brittle fracture is

$$\sigma_c = \frac{P\cos\phi}{A/\cos\phi} = \frac{P}{A}\cos^2\phi \tag{7-18}$$

The cleavage planes for certain metals and values of the critical normal stress are given in Table 7-1.

Although Sohncke's law has been accepted for over 40 years, it is not based on very extensive experimental evidence. Doubt was cast on its reliability by fracture studies[1] on zinc single crystals at −77 and −196°C. The resolved normal cleavage stress was found to vary by over a factor of 10 for a large difference in orientation of the crystals. This variation from the normal-stress law may be due to plastic strain prior to fracture, although it is doubtful that this could account for the observed discrepancy.

Several modes of ductile fracture in single crystals are shown in Fig. 7-1. Under certain conditions hcp metals tested at room temperature or above will

[1] A. Deruyttere and G. B. Greenough, *J. Inst. Met.*, vol. 84, pp. 337–345, 1955–1956.

Table 7-1 Critical normal stress for cleavage of single crystals†

Metal	Crystal lattice	Cleavage plane	Temperature, °C	Critical normal stress, kg/mm^2
Iron	bcc	(100)	−100	26
			−185	27.5
Zinc (0.03% Cd)	hcp	(0001)	−185	0.19
Zinc (0.13% Cd)	hcp	(0001)	−185	0.30
Zinc (0.53% Cd)	hcp	(0001)	−185	1.20
Magnesium	hcp	(0001), $(10\bar{1}1)$		
		$(10\bar{1}2)$, $(10\bar{1}0)$		
Tellurium	Hexagonal	$(10\bar{1}0)$	20	0.43
Bismuth	Rhombohedral	(111)	20	0.32
Antimony	Rhombohedral	$(11\bar{1})$	20	0.66

† Data from C. S. Barrett, "Structure of Metals," 2d ed., McGraw-Hill Book Company, New York, 1952; N. J. Petch, The Fracture of Metals, in "Progress in Metal Physics," vol. 5, Pergamon Press, Ltd., London, 1954.

shear only on a restricted number of basal planes. Fracture will then occur by "shearing off" (Fig. 7-1*b*). More usually, slip will occur on systems other than the basal plane, so that the crystal necks down and draws down almost to a point before rupture occurs. The usual mode of fracture in fcc crystals is the formation of a necked region due to multiple slip, followed by slip on one set of planes until fracture occurs. The crystal can draw down to a chisel edge or a point (if multiple slip continues to fracture). The best stress criterion for ductile fracture in fcc metals appears to be the resolved shear stress on the fracture plane (which is usually the slip plane).

The mode of fracture in bcc iron crystals is strongly dependent on temperature, purity, heat treatment, and crystal orientation.[1] Crystals located near the [001] corner of the stereographic triangle show no measurable ductility when tested in tension at $-196°C$, while crystals closer to $[\bar{1}11]$ and [011] orientations may rupture by drawing down to a chisel edge when tested at the same temperature. An interesting point is that the change from brittle to ductile fracture is very sharp, occurring over a change in orientation of only about 2°.

7-6 METALLOGRAPHIC ASPECTS OF FRACTURE

Because of the prominence of the Griffith theory, it has been natural for metallurgists to use their microscopes in a search for Griffith cracks in metals. However, based on observations up to the magnifications available with the electron microscope, there is no reliable evidence that Griffith cracks exist in metals in the unstressed condition. There is, however, a considerable amount of experimental evidence to show that microcracks can be produced by plastic deformation.

Metallographic evidence of the formation of microcracks at nonmetallic inclusions in steel as a result of plastic deformation has existed for a number of years. These microcracks do not necessarily produce brittle fracture. However, they do contribute to the observed anisotropy in the ductile-fracture strength. The fact that vacuum-melted steel, which is very low in inclusions, shows a reduction in the fracture anisotropy supports the idea of microcracks being formed at second-phase particles.

An excellent correlation between plastic deformation, microcracks, and brittle fracture was made by Low.[2] He showed that for mild steel of a given grain size tested at $-196°C$ brittle fracture occurs in tension at the same value of stress that is required to produce yielding in compression. Microcracks only one or two grains long were observed. More detailed studies of the conditions for microcrack

[1] N. P. Allen, B. E. Hopkins, and J. E. McLennan, *Proc. R. Soc. London*, vol. 234A, p. 221, 1956.

[2] J. R. Low, I.U.T.A.M., Madrid Colloquium, "Deformation and Flow of Solids," p. 60, Springer-Verlag OHG, Berlin, 1956.

Figure 7-5 Microcracks produced in iron by tensile deformation at $-140°C$ ($250\times$). *(Courtesy G. T. Hahn.)*

formation have been made[1] with tensile tests on mild steel at carefully controlled subzero temperatures. Figure 7-5 illustrates a microcrack found in a specimen before it fractured.

Detailed experiments demonstrate that the cracks responsible for brittle-cleavage-type fracture are not initially present in the material but are produced by the deformation process. The fact that at appropriate temperatures appreciable numbers of microcracks are present shows that the conditions for the initiation of a crack are not necessarily the same as conditions for the propagation of a crack. The process of cleavage fracture should be considered to be made up of three steps. (1) plastic deformation to produce dislocation pile-ups, (2) crack initiation, and (3) crack propagation.

The initiation of microcracks can be greatly influenced by the presence and nature of second-phase particles.[2] A common situation is for the particle to crack during deformation. Resistance to cracking improves if the particle is well bonded to the matrix. Small particles ($r < 1\ \mu m$) and spherical particles are more resistant to cracking. If the dispersion of second-phase particles is readily cut by the dislocations, then there will be planar slip and relatively large dislocation pile-ups will occur. This will lead to high stresses, easy initiation of microcracks, and brittle behavior. However, if the second-phase consists of a dispersion of fine impenetrable particles, the slip distance is greatly reduced and, correspondingly, the number of dislocations that can be sustained in a pile-up is reduced. Also, once cracks are formed they will be forced to bow between the particles, increasing the effective interfacial energy. Thus, fine dispersions of particles can lead to increased toughness under the proper circumstances. A soft, ductile phase

[1] G. T. Hahn, W. S. Owen, B. L. Averbach, and M. Cohen, *Welding J.*, vol. 38, pp. 367s–376s, 1959.

[2] R. F. Decker, *Metall. Trans.*, vol. 4, pp. 2508–2611, 1973.

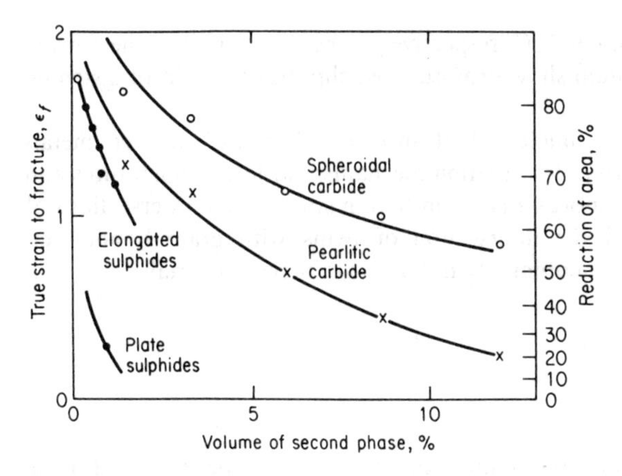

Figure 7-6 Effect of second-phase particles on tensile ductility. *(From T. Gladman, B. Holmes, and L. D. McIvor, "Effect of Second-Phase Particles on the Mechanical Properties of Steel," p. 78, Iron and Steel Institute, London, 1971.)*

can also impart ductility to a brittle matrix. The ductile phase must be thick enough to yield before large dislocation pile-ups are created against it.

In the brittle fracture of mild steel the large microcracks observed in the ferrite grains are invariably associated with a fractured carbide particles located somewhere in the grain or on the surrounding grain boundary.[1] Fracture of the carbide particle by the stress field of a pile-up is an essential intermediate event between the formation of a dislocation pile-up and cleavage of the ferrite. The formation of a crack in a carbide can initiate cleavage fracture in the adjacent ferrite if the local stress is sufficiently high.

Most brittle fractures occur in a transgranular manner. However, if the grain boundaries contain a film of brittle constituent, as in sensitized austenitic stainless steel or molybdenum alloys containing oxygen, nitrogen, or carbon, the fracture will occur in an intergranular manner. Intergranular failure can also occur without the presence of a microscopically visible precipitate at the grain boundaries. Apparently, segregation at the grain boundaries can lower the surface energy sufficiently to cause intergranular failure. The embrittlement produced by the addition of antimony to copper and oxygen to iron and the temper embrittlement of alloy steels are good examples.

Ductile fracture (Sec. 7-10) starts with the initiation of *voids*, most commonly at second-phase particles. The particle geometry, size, and bonding with the matrix are important parameters. Figure 7-6 shows the influence of common second phases in steel on the tensile ductility.

The character of the slip band can also influence the fracture behavior. Cross slip of screw dislocations broadens the slip band and makes it more difficult to crack a particle when the slip band impinges upon this obstacle. For example,[2] the ductile-to-brittle transition temperatures for the bcc metals columbium and

[1] C. J. McMahon, Jr., and M. Cohen, *Acta Metall.*, vol. 13, p. 591, 1965.

[2] A. R. Rosenfield, E. Votava, and G. T. Hahn, in "Ductility," pp. 75–76, American Society for Metals, Metals Park, Ohio, 1968.

tungsten are about −120 and 300°C, respectively, when measured in the tension test. Dislocations in columbium show profuse cross slip, but those in tungsten do not cross slip at all.

Cleavage cracks may be nucleated at mechanical twins.[1] In bcc metals twinning becomes the preferred deformation mechanism at low temperatures and high rates of deformation. Important crack initiation sites are the intersections of twins with other twins and the intersection of twins with grain boundaries. Cleavage of twins occurs most commonly in larger grain size material.

7-7 FRACTOGRAPHY

Important information about the nature of fracture can be obtained from microscopic examination of the fracture surface. This study is usually called *fractography*. Fractography is most commonly done using the scanning election microscope (SEM). The large depth of focus (important in examining rough fracture surfaces) and the fact that the actual surface can be examined make the SEM an important tool for research and for failure analysis.[2]

On a microscopic scale the commonly observed fracture modes are cleavage, quasi-cleavage, and dimpled rupture. Other fracture modes such as fatigue, intergranular fracture, and creep-rupture will be discussed in chapters dealing with those specific mechanical behaviors.

Cleavage fracture represents brittle fracture occurring along crystallographic planes. The characteristic feature of cleavage fracture is flat facets which generally are about the size of the ferrite grain (in steel). Usually, the flat facets exhibit "river marking" (Fig. 7-7). The river markings are caused by the crack moving through the crystal along a number of parallel planes which form a series of plateaus and connecting ledges. These are indications of the absorption of energy by local deformation. The direction of the "river pattern" represents the direction of crack propagation.

Quasi-cleavage fracture is related but distinct to cleavage fracture. It is observed chiefly in low-temperature fracture of quenched and tempered steels. The term quasi-cleavage is used because the facets on the fracture surface are not true cleavage planes (Fig. 7-8). These facets usually correspond in size with the prior austenite grain size. Quasi-cleavage fractures often exhibit dimples and tear ridges around the periphery of the facets.

Dimpled rupture (Fig. 7-9) is characterized by cup-like depressions that may be equiaxial, parabolic, or elliptical, depending on the stress state.[3] This type of fracture surface denotes a ductile fracture. Microvoids are initiated at second-phase

[1] D. Hull, in D. C. Drucker and J. J. Gilman (eds.), "Fracture in Solids," pp. 417–453. Interscience Publishers, Inc., New York, 1963.; D. Hull, *Acta Met.*, vol. 9, p. 191, 1961.

[2] M. Coster and J. L. Chesmont, *Int. Met. Rev.*, vol. 28, no. 4, pp. 228–250, 1983.

[3] C. D. Beachem, *Metall. Trans.*, vol. 6A, pp. 377–383, 1975.

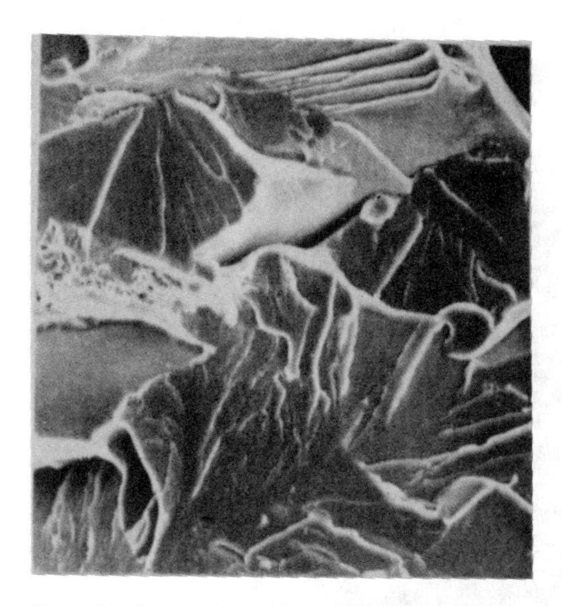

Figure 7-7 Cleavage in ferrite caused by overload fracture of hot extruded carbon steel (2,000×). *(Courtesy of D. A. Meyn, Naval Research Laboratory.)*

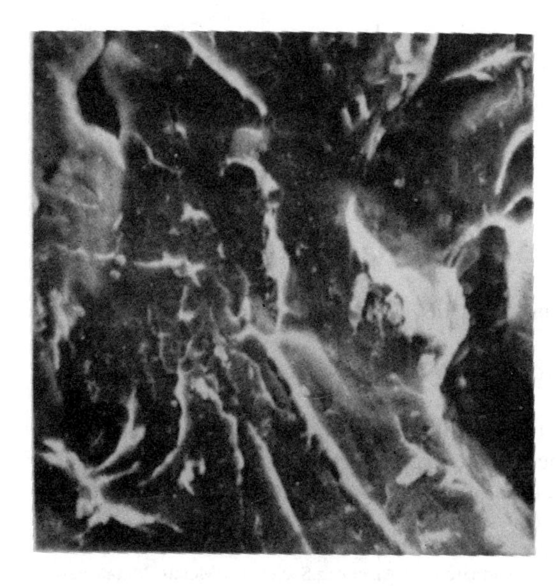

Figure 7-8 Quasi-cleavage in tempered martensite of 17-4 PH stainless steel (4,600×). *(Courtesy of D. A. Meyn, Naval Research Laboratory.)*

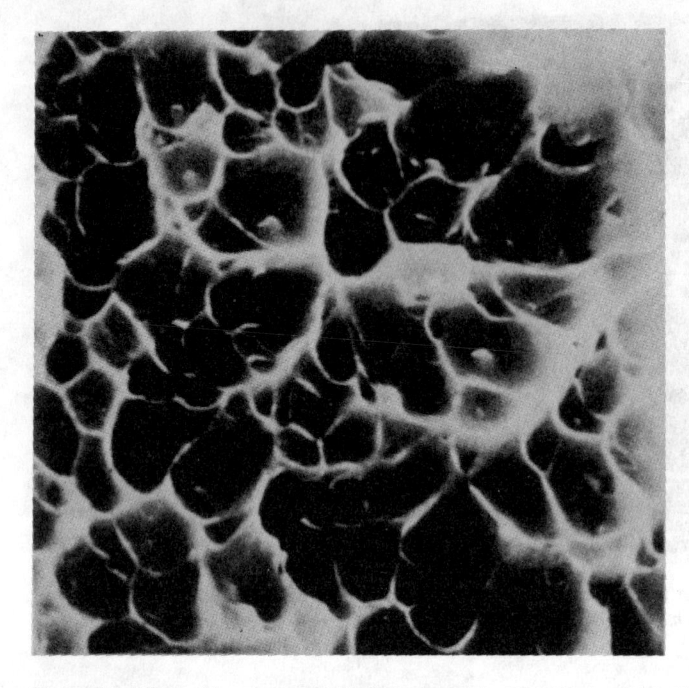

Figure 7-9 Microvoid coalescence (dimpled rupture) in 2024-T4 aluminum (12,000 ×). *(Courtesy of D. A. Meyn, Naval Research Laboratory.)*

particles, the voids grow, and eventually the ligaments between the microvoids fracture.

Identification of the cause of fracture through fractography has become a standard investigative technique. Compilations of fractographs are available.[1]

7-8 DISLOCATION THEORIES OF BRITTLE FRACTURE

The process of brittle fracture consists of three stages.

1. Plastic deformation which involves the pile-up of dislocations along their slip planes at an obstacle.
2. The buildup of shear stress at the head of the pile-up to nucleate a microcrack.
3. In some cases the stored elastic strain energy drives the microcrack to complete fracture without further dislocation movement in the pile-up. More typically in

[1] D. Bhattacharya (ed.), "IITRI Fracture Handbook," IITRI, Chicago, 1978; "Metals Handbook," 8th ed., vol. 9, "Fractography and Atlas of Fractographs," American Society for Metals, Metals Park, OH, 1974; "SEM/TEM Electron Fractography Handbook," Metals and Ceramics Information Center, Battelle Columbus Labs, 1975.

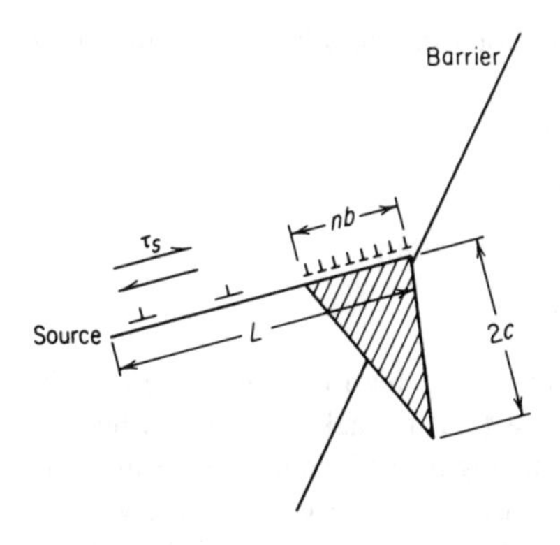

Figure 7-10 Model of microcrack formation at a pile-up of edge dislocations.

metals, a distinct growth stage is observed in which an increased stress is required to propagate the microcrack. Thus, the fracture stress, σ_f, is the stress required to propagate the microcracks.

The idea that the high stresses produced at the head of a dislocation pile-up could produce fracture was first advanced by Zener.[1] The model is shown in Fig. 7-10. The shear stress acting on the slip plane squeezes the dislocations together. At some critical value of stress the dislocations at the head of the pile-up are pushed so close together that they coalesce into a wedge crack or cavity dislocation of height nb and length $2c$. Stroh[2] has shown that provided the stress concentration at the head of the pile-up is not relieved by plastic deformation, then the tensile stress at the pile-up is given by Eq. (5-41), which can be equated to the theoretical cohesive stress, Eq. (7-8). Thus,

$$(\tau_s - \tau_i)\left(\frac{L}{r}\right)^{1/2} = \left(\frac{E\gamma_s}{a_0}\right)^{1/2}$$

and microcrack nucleation occurs at

$$\tau_s = \tau_i + \left(\frac{Er\gamma_s}{La_0}\right)^{1/2} \tag{7-19}$$

where L is the length of the blocked slip band and r is the distance from the tip of the pile-up to the point where the crack is forming. If we let $r \approx a_0$ and $E \approx 2G$, then Eq. (7-19) becomes

$$\tau_s = \tau_i + \left(\frac{2G\gamma_s}{L}\right)^{1/2} \tag{7-20}$$

[1] C. Zener, The Micro-mechanism of Fracture, in "Fracturing of Metals," American Society for Metals, Metals Park, Ohio, 1948.

[2] A. N. Stroh, *Adv. Phys.*, vol. 6, p. 418, 1957; the development given here follows that of A. S. Tetelman and A. J. McEvily, Jr., "Fracture of Structural Materials," chap. 6, John Wiley & Sons, Inc., New York, 1967.

But, from Eq. (5-37), the number of dislocations in the slip band can be expressed as

$$nb \approx L\frac{\tau_s - \tau_i}{G} \tag{7-21}$$

On eliminating L from Eqs. (7-20) and (7-21) we obtain

$$(\tau_s - \tau_i)nb \approx 2\gamma_s \tag{7-22}$$

This form of the equation for microcrack nucleation was proposed by Cottrell.[1] It has the direct physical significance that a crack will form when the work done by the applied shear stress in producing a displacement nb equals the work done in moving the dislocations against the friction stress plus the work in producing the new fracture surfaces. It is interesting that Eq. (7-22) does not contain a crack-length term $2c$. Thus, the crack grows by plastic deformation so long as the dislocation source continues to force dislocations into the pile-up. It should be realized that only shearing stresses are involved in forcing the dislocations together. Tensile stresses are not involved in the microcrack nucleation process, and indeed, cleavage cracks can be formed in compression.[2] However, a tensile stress is needed to make the microcracks propagate. The fact that normal stresses (hydrostatic stress state) are not involved in microcrack nucleation leads to the conclusion that the crack-propagation stage is ordinarily more difficult than crack initiation in metals, since experience shows that fracture is strongly influenced by the hydrostatic component of stress (Sec. 7-13). Also, the occurrence of nonpropagating microcracks supports this point of view. Modification[3] of Stroh's equations to allow for a nonuniform stress field on the pile-up and dislocations of both signs in the pile-up has shown that the stress required to nucleate a microcrack is significantly lower than originally thought.

There is strong evidence that in most engineering materials the most difficult step is the propagation of deformation-produced microcracks through a strong barrier such as a grain boundary. Thus, the grain size will have a strong influence on brittle-fracture behavior. Petch[4] found that the grain-size dependence of brittle fracture in iron and steel could be expressed by

$$\sigma_f = \sigma_i + k_f D^{-1/2} \tag{7-23}$$

This is analogous to Eq. (6-6) for the grain-size dependence of the yield point and the flow stress. To develop the dislocation model for Eq. (7-23), we express Eq. (7-22) in terms of normal stress

$$\sigma nb \approx 4\gamma_s \tag{7-24}$$

and adopt a model in which the dislocation source is at the center of a grain of

[1] A. H. Cottrell, *Trans. Metall. Soc. AIME*, vol. 212, pp. 192–203, 1958.
[2] A. Gilbert et al., *Acta Metall.*, vol. 12, p. 754, 1964.
[3] E. Smith and J. T. Barnby, *Met. Sci. J.*, vol. 1, pp. 56–64, 1967.
[4] N. J. Petch, *J. Iron Steel Inst. London*, vol. 174, p. 25, 1953.

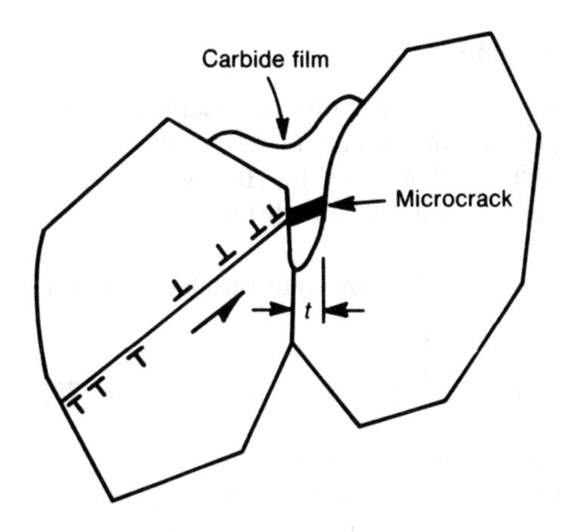

Figure 7-11 Smith's model of microcrack formation in grain boundary carbide film.

diameter D, so that $L = D/2$. Thus, substituting Eq. (7-21) into Eq. (7-22)

$$\sigma(\tau_s - \tau_i)D = 8G\gamma_s$$

But, experience shows that microcracks form when the shear stress equals the yield stress so that, from Eq. (6-6)

$$\tau_s - \tau_i = \tau_0 - \tau_i = 2k'D^{-1/2}$$

and

$$\sigma_f = \frac{4G\gamma_s}{k'D^{1/2}} \tag{7-25}$$

This equation represents the stress required to propagate a microcrack of length D in brittle fracture. In Eq. (7-25) k' is a parameter related to release of dislocations from a pile-up. It may be appropriate to think of k' as a microstructural stress intensity factor.

We have seen how brittle carbides play a critical role in producing brittle fracture in mild steel. Smith[1] has proposed a model shown in Fig. 7-11. Deformation on a slip band within the ferrite grain provides a stress which may initiate a crack in a brittle-grain boundary carbide of thickness t. The surface energy of the carbide is lower than that of the ferrite, $\gamma_c < \gamma_f$. Therefore, whether the crack will propagate through the ferrite will depend on the increase in the energy of the system with increasing crack length c. The criterion for fracture in microstructure of grain size D and carbide thickness t is

$$\left(\frac{t}{D}\right)\sigma_f^2 + \tau_{\text{eff}}\left[1 + \left(\frac{4}{\pi}\right)\left(\frac{t}{D}\right)^{1/2}\frac{\tau_i}{\tau_{\text{eff}}}\right]^2 > \frac{4E\gamma_p}{\pi(1-\nu^2)D} \tag{7-26}$$

where σ_f = the fracture stress required to propagate the crack nucleus

$\tau_{\text{eff}} = \tau_{\text{app}} - \tau_i$

γ_p = plastic work involved in propagating the crack nucleus into the ferrite.

[1] E. Smith, "Proc. Conf. on Physical Basis of Yield and Fracture," Inst. of Physics, Oxford, pp. 36–46, 1966.

A Theory of the Ductile-to-Brittle Transition

An important engineering phenomenon is the transition in ductile-to-brittle fracture behavior with decreasing temperature that occurs in steel and other bcc materials. This is discussed in considerable detail in Chap. 14. Here we consider a theory of the ductile-to-brittle transition based on the dislocation concepts presented earlier in this section.

Cottrell[1] has reformulated Eq. (7-22) so that the important variables in brittle fracture are easily shown. This equation is

$$\left(\tau_i D^{1/2} + k'\right)k' = G\gamma_s\beta \tag{7-27}$$

where τ_i = the resistance of the lattice to dislocation movement
k' = a parameter related to the release of dislocations from a pile-up
γ_s = the effective surface energy and includes the energy of plastic deformation
β = a term which expresses the ratio of shear stress to normal stress. For torsion $\beta = 1$; for tension $\beta = \frac{1}{2}$; for a notch $\beta \approx \frac{1}{3}$.

Equation (7-27) expresses the limiting condition for the formation of a propagating crack from a pile-up of glide dislocations. If the left side of the equation is smaller than the right side, a microcrack can form but it cannot grow. This is the case of nonpropagating microcracks. When the left side of the equation is greater than the right, a propagating brittle fracture can be produced at a shear stress equal to the yield stress. Thus, this equation describes a ductile-to-brittle transition. Since many of the metallurgical parameters change with temperature, there would be a *transition temperature* at which the fracture would change from ductile to brittle.

The parameter k' is important since it determines the number of dislocations that are released into a pile-up when a source is unlocked. Materials with a high value of k' (e.g., iron and molybdenum) are more prone to brittle fracture than materials with lower values (e.g., columbium and tantalum). Strengthening mechanisms which depend on dislocation locking are likely to result in brittleness. The importance of grain size as a parameter in brittle fracture is shown in Fig. 7-12, where below a certain grain size there is measurable ductility at fracture. Actually, the grain size term in Eq. (7-27) should be interpreted more properly as the slip-band length. Generaly this is controlled by the grain size, but in a alloy containing a fine precipitate the particle spacing will determine the slip distance. The very fine carbide precipitates in quenched and tempered steel which cause short slip distances result in the low transition temperature for this material.

A high value of frictional resistance leads to brittle fracture, since high stresses must be reached before yielding occurs. The directional bonding in ceramics results in high τ_i and high inherent hardness and brittleness. In bcc

[1] Cottrell, op. cit.

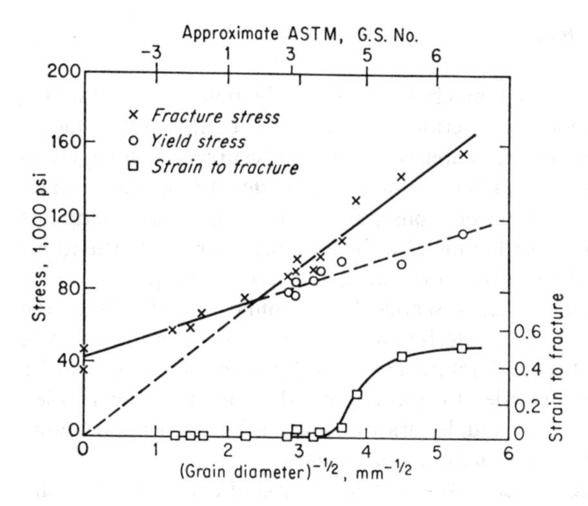

Figure 7-12 Effect of grain size on the yield and fracture stresses for a low-carbon steel tested in tension at $-196°C$. *(From J. R. Low, in: "Relation of Properties to Microstructure," American Society for Metals, Metals Park, OH, 1954. By permission of the publishers.)*

metals the frictional resistance increases rapidly as the temperature falls below room temperature[1] and thus leads to a ductile-to-brittle transition. The τ_i term enters into Eq. (7-27) as the product of $D^{1/2}$, so that a fine-grain metal can withstand higher values of τ_i (lower temperatures) before becoming brittle. Many of the effects of the composition of steel on the ductile-to-brittle transition are due to changes in D, k', or τ_i. For example, manganese decreases the grain size and reduces k', where silicon produces large grain size and increases τ_i.

If the effective surface energy is large at a given temperature, then brittle fracture is suppressed. The contribution from plastic deformation will depend on the number of available slip systems and the number of mobile dislocations at the tip of the crack. Thus, zinc is brittle in large grain sizes because of its limited slip systems, while bcc metals can be brittle because impurities lock most of the mobile dislocations. Neither of these conditions exist in fcc metals and they do not ordinarily fail in brittle fracture. Various environmental factors such as corrosion or hydrogen penetration may lower the surface energy.

It is well known that the presence of a notch greatly increases the tendency for brittle fracture. The complicated effects of a notch will be considered in Sec. 7-11. The effect of a notch in decreasing the ratio of shear stress to tensile stress is covered in Eq. (7-27) by the constant β. Strain rate or rate of loading is not an explicit term in Eq. (7-27). However, it is well known that increasing strain rate raises both τ_i and τ_0. Strain rate interacts with the notch effect in an important way. Because the deformation is localized in the vicinity of the notch, there is a strain-concentration effect and the local strain is much higher than the average value.

[1] From Eq. (6-6) it follows that $\tau_0 D^{1/2} = \tau_i D^{1/2} + k'$, so that Eq. (7-27) can be written $\tau_0 k' D^{1/2} = G\gamma_s \beta$. The yield stress τ_0 for bcc metals increases rapidly with decreasing temperature.

7-9 DUCTILE FRACTURE

Ductile fracture has been studied much less extensively than brittle fracture, probably because it is a much less serious problem. Up to this point ductile fracture has been defined rather ambiguously as fracture occurring with appreciable gross plastic deformation. Another important characteristic of ductile fracture, which should be apparent from previous considerations of brittle fracture, is that it occurs by a slow tearing of the metal with the expenditure of considerable energy. Many varieties of ductile fractures can occur during the processing of metals and their use in different types of service. For simplification, the discussion in this section will be limited to ductile fracture of metals produced in uniaxial tension. Other aspects of tensile fracture are considered in Chap. 8. Ductile fracture in tension is usually preceded by a localized reduction in diameter called *necking*. Very ductile metals may actually draw down to a line or a point before separation. This kind of failure is usually called *rupture*.

The stages in the development of a ductile "cup-and-cone" fracture are illustrated in Fig. 7-13. Necking begins at the point of plastic instability where the increase in strength due to strain hardening fails to compensate for the decrease in cross-sectional area (Fig. 7-13*a*). This occurs at maximum load or at a true

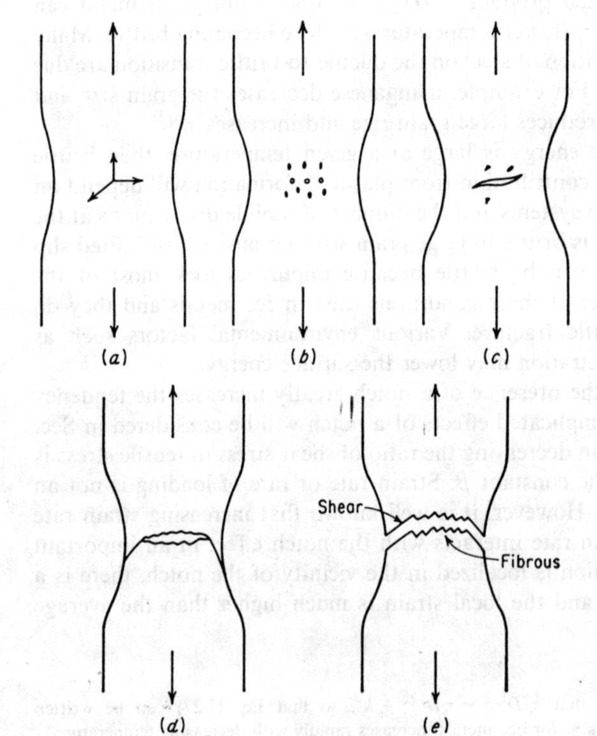

Figure 7-13 Stages in the formation of a cup-and-cone fracture.

strain equal to the strain-hardening coefficient (see Sec. 8-3). The formation of a neck introduces a triaxial state of stress in the region. A hydrostatic component of tension acts along the axis of the specimen at the center of the necked region. Many fine cavities form in this region (Fig. 7-13*b*), and under continued straining these grow and coalesce into a central crack (Fig. 7-13*c*). This crack grows in a direction perpendicular to the axis of the specimen until it approaches the surface of the specimen. It then propagates along localized shear planes at roughly 45° to the axis to form the "cone" part of the fracture (Fig. 7-13*d*).

Detailed study[1] of the ductile fracture process shows that the central crack which forms early tends to concentrate the deformation at its tip in narrow bands of high shear strain. These shear bands are at angles of 50 to 60° to the transverse direction. Sheets of voids are nucleated in these bands, and the voids grow and coalesce into local fracture of the *void sheet*. While the average direction of crack growth is radially outward in the direction transverse to the tensile axis, on a finer scale the crack zig-zags back and forth across the transverse plane by void-sheet formation. Thus, crack growth in ductile fracture is essentially by a process of *void coalescence*. Coalescence occurs by elongation of the voids and elongation of the bridges of material between the voids. This leads to the formation of a fracture surface[2] consisting of elongated "dimples," as if it had formed from numerous holes which were separated by thin walls until it fractures.

The voids which are the basic source of ductile fracture are nucleated heterogeneously at sites where compatibility of deformation is difficult. The preferred sites for void formation are inclusions, second-phase particles, or fine oxide particles, while in high-purity metals voids can form at grain-boundary triple points. Particles as small as 50 Å have been found to nucleate voids so that the absence of voids on metallographic examination may not be a reliable indicator of whether void formation has occurred. In the tension test voids form prior to necking but after a neck is formed and hydrostatic tensile stresses develop, the void formation becomes much more prominent. The frequency of occurrence of nucleating particles should have a strong influence on ductile fracture.[3] It has been shown[4] that the true strain to fracture decreases rapidly with increasing volume fraction of second phase particles (Fig. 7-6). Careful metallographic study[5] of ductile fracture in carbon steels containing pearlite has shown that a combination of applied tensile stress and concentrated shear zone are required to initiate voids. The suggested mechanism is shown in Fig. 7-14. Carbides that are parallel to the applied tensile stress crack first (Fig. 7-14*a*). A concentrated shear zone at about 50° to the tensile axis causes cracking of adjacent carbide plates (Fig. 7-14*b*). The voids grow (Fig. 7-14*c*) and coalesce to

[1] H. C. Rogers, *Trans. Metall. Soc. AIME*, vol. 218, p. 298, 1966; H. C. Rogers, in "Ductility," chap. 2, American Society for Metals, Metals Park, Ohio, 1968.

[2] C. D. Beachem, *Trans. Am. Soc. Met.*, vol. 50, pp. 318–326, 1963.

[3] For a detailed review see A. R. Rosenfield, *Metall. Rev.* 121, *Met. Mater.*, April 1968 and J. L. Mogford, *Metall. Rev.*, vol. 12, pp. 49–60, 1967.

[4] B. Edelson and W. Baldwin, *Trans. Am. Soc. Met.*, vol. 55, pp. 230–250, 1962.

[3] L. E. Miller and G. C. Smith, *J. Iron Steel Inst. London*, vol. 208, pp. 998–1005, 1970.

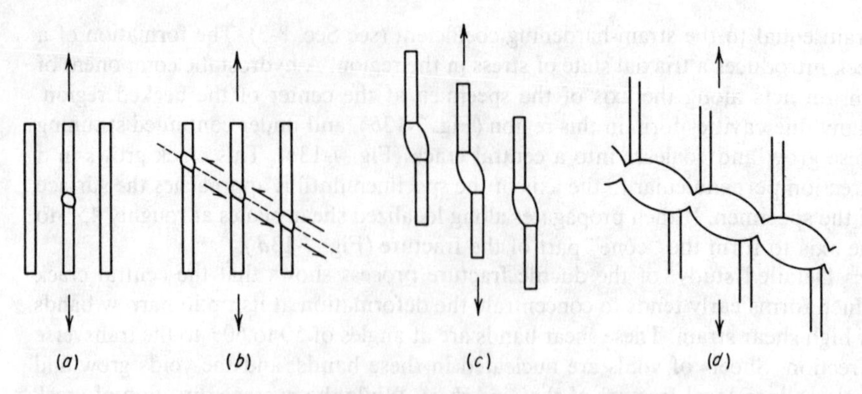

Figure 7-14 Mechanism of ductile fracture in a pearlitic steel. *(After Miller and Smith.)*

form the ductile fracture (Fig. 7-14*d*). Particle shape can have an important influence on ductile fracture. When the particles are more spherical than platelike as in spheroidized pearlite, cracking of the carbides is much more difficult and the ductility is increased. Cracking of spheroidized carbides is more difficult because dislocations in the ferrite matrix can cross slip around them more easily than for platelike carbides, and thus avoid the buildup of high stresses at pile-ups. Also, because of the smaller contact area between spheroidized carbides and the matrix, the tensile stresses generated in the particles will be less than for lamellar carbides. The very fine rounded carbides in quenched and tempered steels are very resistant to void formation and this accounts for the good ductility of this structure at high-strength levels. Finally since the second-phase particles invariably will be distorted in shape by plastic deformation processes, like rolling, it is common to find that the resistance to ductile fracture (ductility) varies greatly with orientation in a rolled sheet or plate.

An important start toward developing an analytical treatment of ductile fracture has been made by McClintock[1] using a model consisting of cylindrical holes initially of radius b_0 and average spacing l_0. The strain to fracture is given by

$$\varepsilon_f = \frac{(1-n)\ln(l_0/2b_0)}{\sinh\left[(1-n)(\sigma_a+\sigma_b)/(2\bar{\sigma}/\sqrt{3})\right]} \tag{7-28}$$

for a material with a stress-strain curve given by $\sigma = K\varepsilon^n$. In this equation σ_a and σ_b are the stresses parallel and perpendicular to the axis of the cylindrical holes, respectively, and $\bar{\sigma}$ is the true flow stress. While Eq. (7-28) does not give close agreement with the limited data available it certainly predicts the proper variation of fracture strain with the important variables. Equation (7-28) indicates that ductility decreases as the void fraction increases, as the strain-hardening exponent

[1] F. A. McClintock, in "Ductility," chap. 9, American Society for Metals, Metals Park, Ohio, 1968; also F. A. McClintock, *J. Appl. Mech.*, vol. 90, pp. 363–371, 1968.

n decreases, and as the stress state changes from uniaxial tension to triaxial tension. A somewhat different analysis[1] of ductile fracture by the internal necking of cavities leads to the same general conclusions.

7-10 NOTCH EFFECTS

The changes produced by the introduction of a notch have important consequences in the fracture process. For example, the presence of a notch will increase appreciably the ductile-brittle transition temperature of a steel. From a discussion of elastic stress concentration in Sec. 2-15, we expect that a notch creates a local stress peak at the root of the notch. Plastic flow begins at the notch root when this local stress reaches the yield strength of the material. The plastic flow relieves the high elastic stress and limits the peak stress to the yield stress of the material. However, the chief effect of the notch is not in introducing a stress concentration but in producing a triaxial state of stress at the notch.

The elastic stress distribution is shown in Fig. 7-15*a* for a notch in a thin plate. When the plate is loaded in the y direction to a stress less than the elastic limit, the stress distribution of σ_y is shown. At the same time a transverse elastic stress σ_x is produced by the geometry of the notch. This can be understood physically if we imagine a series of small tensile specimens at the tip of the notch. The distribution of σ_y is given by Fig. 7-15*a*. If each tensile specimen were able to deform freely, it would undergo a tensile strain ε_y in response to the local value of σ_y. Because σ_y falls off rapidly in going away from the notch root, there will be an elastic strain gradient below the notch. However, each tensile strain ε_y will also have associated with it a lateral contraction ε_x due to Poisson's ratio. If each tensile element is free to deform independently of its neighbors, this is given by $\varepsilon_x = -\nu\varepsilon_y$. Thus, all of the interfaces between the tensile specimens would pull apart. In order to maintain continuity a tensile stress σ_x must exist across each interface. At the free surface of the notch ($x = 0$) the tensile element can undergo lateral contraction without any restraint from one side and $\sigma_x = 0$. The necessary lateral stress to maintain continuity increases with distance in from the root of the notch, but it decreases at large values of x because the difference in longitudinal strain between adjacent elements becomes small as the σ_y distribution flattens out at large values of x. Therefore, σ_x will rise fairly steeply with x and then fall more slowly as shown in Fig. 7-15*a*.

For the plane-stress condition of tensile loading of a thin plate the stress in the thickness direction is small and can be ignored. However, this definitely is not the case for plane-strain deformation where the thickness B in the z direction is large relative to the notch or crack depth. For plane-strain deformation $\varepsilon_z = 0$ and $\sigma_z = \nu(\sigma_y + \sigma_x)$. The elastic stress distribution along the x axis for a thick notched plate loaded uniaxially in the y direction is shown in Fig. 7-15*b*. The value of σ_z falls to zero at both surfaces of the plate ($z = \pm B/2$) but rises rapidly

[1] P. F. Thomason, *J. Inst. Met.*, vol. 96, pp. 360–365, 1968.

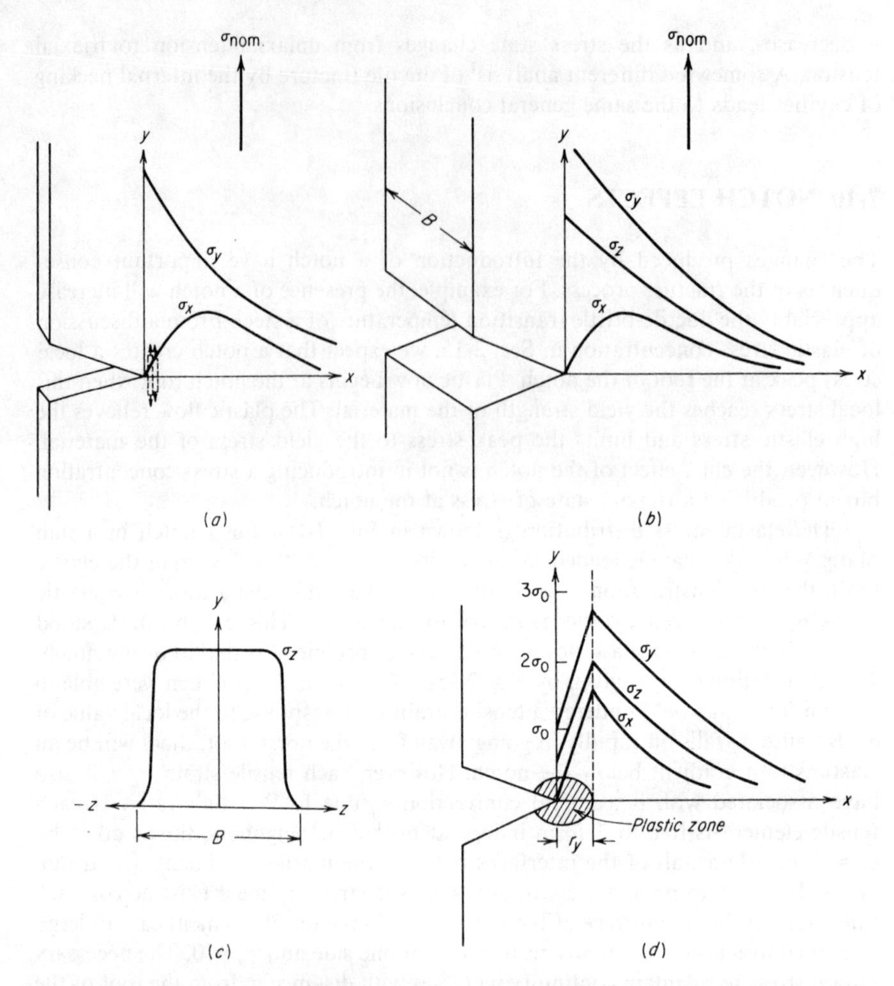

Figure 7-15 (*a*) Elastic stresses beneath a notch in a thin plate (planes stress); (*b*) elastic stresses beneath a notch in plane strain; (*c*) distribution of σ_z with z at $x = 0$ (plane strain); (*d*) distribution of stresses during local yielding (plane strain).

with distance in from the free surfaces. The distribution of σ_z with z at the notch root ($x = 0$) is shown in Fig. 7-15*c*. The values of σ_y and σ_x are nearly independent of z. Figure 7-15*b* shows that stressing a thick plate provides a high degree of elastic triaxiality. As the thickness B decreases, the values of σ_y and σ_x fall by less than 10 percent but the peak value of σ_z is strongly dependent on B. In plane stress, when B is very small, $\sigma_z = 0$.

The existence of transverse stresses raises the average value of longitudinal stress at which yielding occurs. For a Tresca yield criterion, $\sigma_0 = \sigma_1 - \sigma_3$. In an unnotched tension specimen the longitudinal stress alone measures yielding since

$\sigma_0 = \sigma_y - 0$. In plane strain, yielding starts at the root of the notch because $\sigma_x = 0$ at this free surface (Fig. 7-15). However, just below the notch, $\sigma_0 = \sigma_y - \sigma_x$. Since the basic material yield strength σ_0 is the same, whether notched or unnotched, it takes a higher value of longitudinal stress σ_y to produce yielding in a notched specimen. The distribution of the principal stresses with distance in from the root of the notch is sketched in Fig. 7-15*d*. When local yielding occurs, the value of σ_y drops from its high elastic value (see Fig. 7-15*b*) to the value of σ_0. Once the first tensile element at the notch root has yielded, it deforms plastically at constant volume with $\nu = 0.5$ instead of the elastic value of $\nu = 0.3$. Therefore, the transverse strain σ_x is larger than for the elastic case and a larger value of σ_x must be applied to maintain cohesion at the element interfaces. Therefore, as the plastic zone spreads from the notch root the value of σ_x increases much more steeply with distance than in the elastic case. The maximum value of σ_x is found at the elastic-plastic interface. Within the plastically yielding region the value of σ_y is given by the yield criterion, $\sigma_y = \sigma_0 + \sigma_x$ and σ_z is given by $\sigma_z = 0.5(\sigma_y + \sigma_x)$. With increasing stress the plastic zone moves inward until at some point the entire region beneath the notch becomes plastic.

As a result of the triaxial stress state produced by the notch the general yield stress of a notched specimen is greater than the uniaxial yield stress σ_0 because it is more difficult to spread the yielded zone in the presence of triaxial stresses. The ratio of notched-to-unnotched flow stress is referred to as the *plastic-constraint factor q*. Unlike elastic stress concentration, which can reach values in excess of $K_t = 10$ as the notch is made sharper and deeper, Orowan[1] has shown that the plastic-constraint factor cannot exceed a value of 2.57. Thus, the triaxial stress state of a notch results in "notch-strengthening" in a ductile metal, but in a material prone to brittle fracture the increased tensile stresses from the plastic constraint can exceed the critical value for fracture before the material undergoes general plastic yielding.

The steep gradients of stress which exist at a notch imply that there also are sharp gradients of strain. Although there are no exact methods for determining the local strain distribution in a strain-hardening material, the Neuber[2] approximation is useful for determining the magnitude of the strain concentration K_ε.

$$K_\varepsilon(K_\sigma) = K_t^2 \tag{7-29}$$

where K_ε = the plastic-strain-concentration factor
K_σ = the plastic-stress-concentration factor
K_t = the elastic-stress-concentration factor.

[1] E. Orowan, *Trans. Inst. Eng. Schipbuild. Scot.*, vol. 89, p. 165, 1945; measurements of plastic constraint for low-temperature cleavage fracture of mild steel have been made by G. T. Hahn and A. R. Rosenfield, *ASM Trans. Q*, vol. 59, pp. 909–199, 1966.

[2] H. Neuber, *J. Appl. Mech.*, vol. 28, p. 544, 1961.

Therefore, another effect of the notch is to produce high, locally concentrated strain.[1] The accompanying strain hardening can lead to ductile void formation that can become converted into brittle cracks. because the plastically strain-hardened volume beneath the notch is small, the concentration of strain leads to cracking without the expenditure of much plastic work. Still another important consequence of the plastic strain concentration at a notch is that the *local* strain rate will be much higher than the average strain rate. Since brittle fracture depends strongly on strain rate, this can be an important, but easily overlooked factor.

In summary, a notch increases the tendency for brittle fracture in four important ways:

By producing high local stresses
By introducing a triaxial tensile state of stress
By producing high local strain hardening and cracking
By producing a local magnification to the strain rate

7-11 CONCEPT OF THE FRACTURE CURVE

The true-stress-true-strain curve, or flow curve, represents the stress required to cause plastic flow at any particular value of plastic strain. Plastic flow is terminated by fracture when strain hardening, triaxial stress, or high strain rate inhibit the plastic deformation sufficiently to cause a stress that is high enough to break the material. It was proposed by Ludwik[2] that a metal has a *fracture stress curve* in addition to a flow curve and that fracture occurs when the flow curve intersects the fracture curve (Fig. 7-16).

In principle, a point on the fracture curve is obtained by plastically straining a specimen to a given value of strain and then straining without gross plastic deformation until fracture occurs. In practice, a notch and/or low temperature have been used to prevent further plastic deformation before fracture. However, the fact that the embrittling effect of a notch is limited to a plastic-constraint factor of about 2.5 and the realization that even brittle fracture of metals at low temperature is preceded by plastic deformation have made it clear that good experimental measurements of the fracture stress curve cannot be made for metals.[3] However, this does not prohibit using the concept of the fracture stress

[1] For an experimental illustration see R. Taggart, D. H. Polonis, and L. A. James, *Exp. Mech.*, vol. 7, pp. 1–6, 1967.

[2] P. Ludwick, *Z. Ver. deut. Ing.*, vol. 71, pp. 1532–1538, 1927.

[3] Drucker has presented theoretical arguments that the fracture stress may in certain cases decrease with increasing strain; D. C. Drucker, in "Fracture of Solids," chap. 1, Interscience Publishers, Inc., New York, 1963.

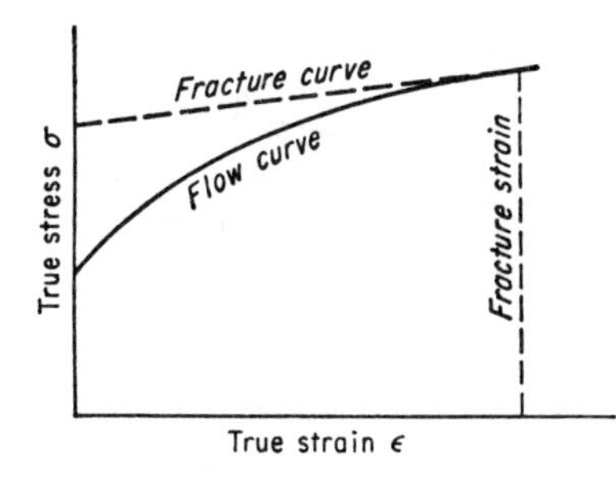

Figure 7-16 Schematic drawing of intersection of flow curve and fracture curve according to the Ludwik theory.

curve, in a qualitative sense, where it is useful for thinking about fracture problems.

The chief factors that influence the ductile-to-brittle transition in bcc metals are stress state, temperature, and strain rate. The transition behavior can be explained in terms of the relative resistance to shear and cleavage and how they change with temperature. Figure 7-17 shows the variation of shear resistance σ_0 and cleavage resistance σ_f with temperature. In agreement with available data σ_f is a less sensitive function of temperature than σ_0. For unnotched specimens the flow stress is lower than the fracture stress at all temperatures above the transition temperature. Thus, the material deforms plastically well before fracture takes place, and the material is ductile. Below the transition temperature $\sigma_f < \sigma_0$, and the material fractures before it can flow plastically.

Now, if a notch is machined in the specimen, the flow stress is raised by about a factor of 2.5 due to plastic constraint, while the fracture stress essentially is unchanged. Figure 7-17 shows how this produces a large increase in the transition temperature. Increasing the strain rate raises the flow stress in the same way as plastic constraint.

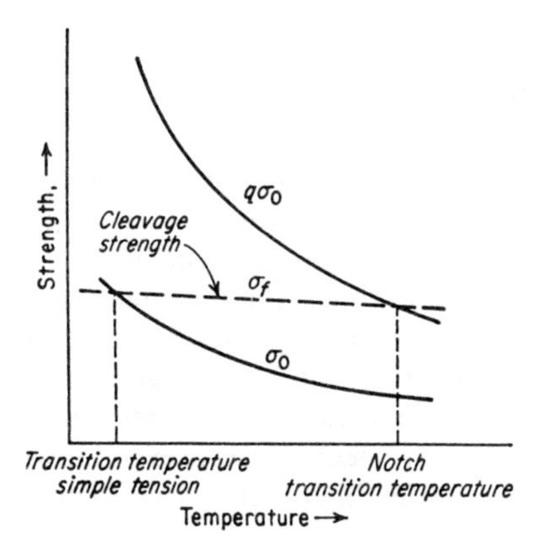

Figure 7-17 Schematic description of transition temperature.

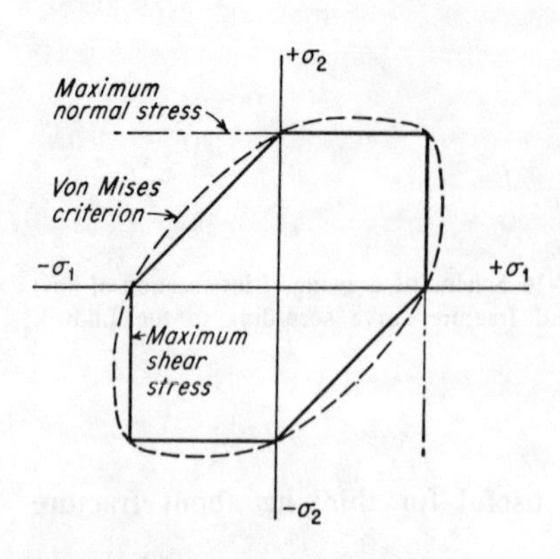

Figure 7-18 Proposed fracture criteria for biaxial state of stress in ductile metals.

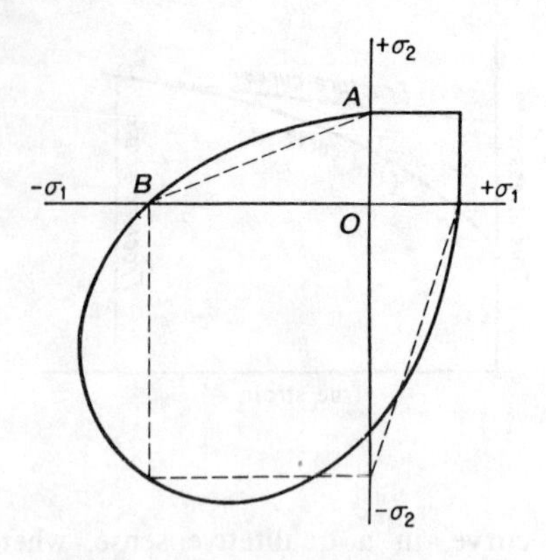

Figure 7-19 Biaxial fracture criterion for brittle cast iron —. Shaw's criterion for brittle materials --.

7-12 FRACTURE UNDER COMBINED STRESSES

The phenomenological approach to fracture is concerned with uncovering the general macroscopic laws which describe the fracture of metals under all possible states of stress. This same approach was discussed in Chap. 3 with regard to the prediction of yielding under complex states of stress. The problem of determining general laws for the fracture strength of metals is quite difficult because fracture is so sensitive to prior plastic straining and temperature. In principle we can conceive of a three-dimensional fracture surface in terms of the three principal stresses σ_1, σ_2, and σ_3. For any combination of principal stresses the metal will fracture when the limiting surface is reached.

Most experimentation in this field has been with biaxial states of stress where one of the principal stresses is zero. Tubular specimens in which an axial tensile or compressive load is superimposed on the circumferential stress produced by internal pressure are ordinarily used for this type of work. For accurate results bulging or necking during the later stages of the test must be avoided. This makes it difficult to obtain good data for very ductile metals.

Figure 7-18 illustrates the fracture criteria which have been most frequently proposed for fracture under a biaxial state of stress. The maximum-shear-stress criterion and the von Mises', or distortion-energy, criterion have already been considered previously in the discussion of yielding criteria. The maximum-normal-stress criterion proposes that fracture is controlled only by the magnitude of the greatest principal stress. Available data on ductile metals such as aluminum and magnesium alloys[1] and steel[2] indicate that the maximum-shear stress criterion for fracture results in the best agreement. Agreement between experiment and theory is not nearly so good as for the case of yielding criteria. The fracture

[1] J. E. Dorn, "Fracturing of Metals," American Society for Metals, Metals Park, Ohio, 1948.
[2] E. A. Davis, *J. Appl. Mech.*, vol. 12, pp. A13–A24, 1945.

criterion for a brittle cast iron[1] is shown in Fig. 7-19. Note that the normal stress criterion is followed in the tension-tension region and that the fracture strength increases significantly as one of the principal stresses becomes compressive. Two theories[2,3] which consider the stress concentration of graphite flakes in cast iron are in good agreement with the fracture data. Shaw[4] has shown a fracture criterion for brittle materials which is in substantial agreement. In this criterion, $OA = 3OB$.

7-13 EFFECT OF HIGH HYDROSTATIC PRESSURE ON FRACTURE

The importance of triaxial tensile stresses in initiating brittle fracture was emphasized in Sec. 7-10. It then should not be surprising to learn that triaxial compressive stress (hydrostatic pressure) resists fracture and increases the ductility. Hydrostatic pressure is utilized in many metalworking operations, like wiredrawing and extrusion, to produce large plastic deformation which would not be possible in the absence of a high component of hydrostatic pressure.

The pioneering work in this field was done by Bridgman,[5] who subjected metals to tension tests with superimposed hydrostatic pressures up to 450,000 psi. More recently, mechanical tests at high pressure[6] and metalworking studies of hydrostatic extrusion have been active areas of research. The effect of superimposed hydrostatic pressure on the ductility in tension is shown in Fig. 7-20. Metals that are normally ductile at atmospheric pressure show a behavior similar to curve 1. For example copper and aluminum reach reductions of area very close to 100 percent at a pressure of about 80 ksi. Curve 2 would be typical of brittle materials such as cast iron or zinc, while curve 3 would be more representative of very brittle materials such as tungsten or marble.

Since a hydrostatic component of stress exerts no shear stress, it cannot increase the number of dislocations in a pile-up or squeeze them closer together. This is the reason for the statement given earlier that hydrostatic stress influences crack propagation but not crack initiation. Compressive hydrostatic stresses act to close up small pores or separations at phase interfaces and generally make the fracture-propagation process more difficult. Studies of the fracture of iron-carbon alloys[7] have shown that high hydrostatic pressure does not suppress the fracture of carbides, but it does reduce void growth in the ferrite matrix. Hydrostatic

[1] W. R. Clough and M. E. Shank, *Trans. Am. Soc. Met.*, vol. 49, pp. 241–262, 1957.

[2] L. F. Coffin, Jr., *J. Appl. Mech.*, vol. 17, p. 233, 1950.

[3] J. C. Fisher, *ASTM Bull.*, 181, p. 74, April, 1952.

[4] J. Takagi and M. C. Shaw, *Trans. ASME J. Eng. Ind.*, vol. 105, pp. 143–148, 1983.

[5] P. W. Bridgman, "Studies in Large Plastic Flow and Fracture," McGraw-Hill Book Company, New York, 1952.

[6] H. L. D. Pugh (ed.), "The Mechanical Behavior of Materials Under Pressure," Elsevier Publishing Company, New York, 1970.

[7] T. E. Davidson and G. S. Ansell, *Trans. Metall. Soc. AIME*, vol. 245, pp. 2383–2390, 1969.

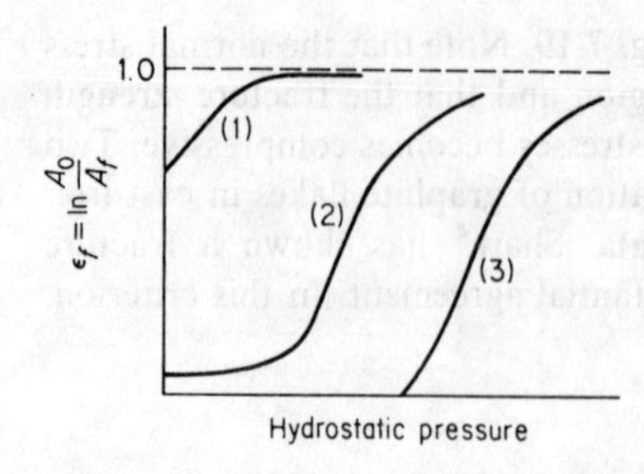

Figure 7-20 Effect of hydrostatic pressure on ductility at fracture in tension. Case 1 is a normally ductile material. Case 2 is a low-ductility material. Case 3 is a brittle material.

pressure at elevated temperature, hot isostatic pressing (HIP), is used commercially to close porosity in castings and powder metallurgy parts and improve the ductility and toughness.

BIBLIOGRAPHY

Averbach, B. L., D. K. Felback, G. T. Hahn, and D. A. Thomas (eds.): "Fracture," The Technology Press of The Massachusetts Institute of Technology and John Wiley & Sons, Inc., New York, 1959.

Drucker, D. C., and J. J. Gilman (eds.): "Fracture of Solids," *Metall. Soc. Conf.*, vol. 20, Interscience Publishers, Inc., New York, 1963.

"Fracture of Engineering Materials," American Society for Metals, Metals Park, Ohio, 1964.

Knott, J. F.: "Fundamentals of Fracture Mechanics," Butterworths, London, 1973.

Lawn, B. R. and T. R. Wilshaw: "Fracture of Brittle Solids," Cambridge University Press, Cambridge, 1975.

Liebowitz, H. (ed.): "Fracture, An Advanced Treatise," Academic Press, Inc., New York, 1969.

Vol. I, "Microscopic and Macroscopic Fundamentals"
Vol. II, "Mathematical Fundamentals"
Vol. III, "Engineering Fundamentals and Environmental Effects"
Vol. IV, "Engineering Fracture Design"
Vol. V, "Fracture Design of Structures"
Vol. VI, "Fracture of Metals"
Vol. VII, "Fracture of Nonmetals and Composites"

Pugh, S. F.: *Br. J. Appl. Phys.*, vol. 18, pp. 129–162, 1967.

Tetleman, A. S. and A. J. McEvily: "Fracture of Structural Materials," John Wiley & Sons, Inc., New York, 1967.

PART THREE

APPLICATIONS TO MATERIALS TESTING

CHAPTER

EIGHT

THE TENSION TEST

8-1 ENGINEERING STRESS-STRAIN CURVE

The engineering tension test is widely used to provide basic design information on the strength of materials and as an acceptance test for the specification of materials. In the tension test[1] a specimen is subjected to a continually increasing uniaxial tensile force while simultaneous observations are made of the elongation of the specimen. An engineering stress-strain curve is constructed from the load-elongation measurements (Fig. 8-1). The significant points on the engineering stress-strain curve have already been considered in Sec. 1-5, while the appearance of a yield point in the stress-strain curve was covered in Sec. 6-5. The stress used in this stress-strain curve is the *average* longitudinal stress in the tensile specimen.[2] It is obtained by dividing the load by the *original area* of the cross section of the specimen.

$$s = \frac{P}{A_0} \tag{8-1}$$

[1] Standard Methods of Tension Testing of Metallic Materials, ASTM Designation E8-69, "Annual Book of ASTM Standards," American Society for Testing and Materials, Philadelphia.

[2] In this chapter we use s to designate engineering stress and σ to designate true stress. Elsewhere in this text we have used σ more generally to indicate stress.

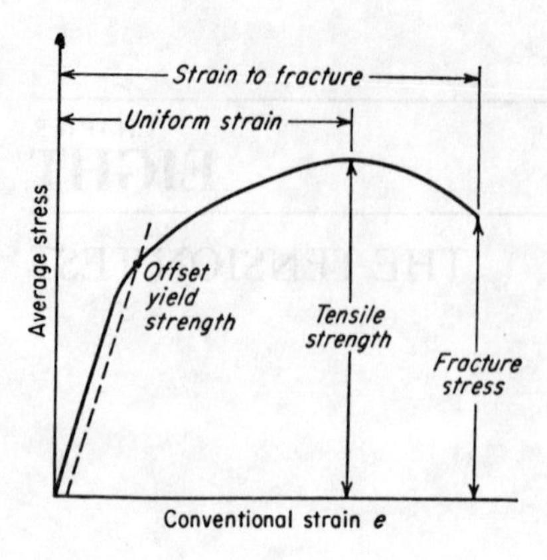

Figure 8-1 The engineering stress-strain curve.

The strain used for the engineering stress-strain curve is the *average* linear strain, which is obtained by dividing the elongation of the gage length of the specimen, δ, by its original length.

$$e = \frac{\delta}{L_0} = \frac{\Delta L}{L} = \frac{L - L_0}{L_0} \tag{8-2}$$

Since both the stress and the strain are obtained by dividing the load and elongation by constant factors, the load-elongation curve will have the same shape as the engineering stress-strain curve. The two curves are frequently used interchangeably.

The shape and magnitude of the stress-strain curve of a metal will depend on its composition, heat treatment, prior history of plastic deformation, and the strain rate, temperature, and state of stress imposed during the testing. The parameters which are used to describe the stress-strain curve of a metal are the *tensile strength*, *yield strength* or *yield point*, *percent elongation*, and *reduction of area*. The first two are strength parameters; the last two indicate ductility.

The general shape of the engineering stress-strain curve (Fig. 8-1) requires further explanation. In the elastic region stress is linearly proportional to strain. When the load exceeds a value corresponding to the yield strength, the specimen undergoes gross plastic deformation. It is permanently deformed if the load is released to zero. The stress to produce continued plastic deformation increases with increasing plastic strain, i.e., the metal strain-hardens. The volume of the specimen remains constant during plastic deformation, $AL = A_0 L_0$, and as the specimen elongates, it decreases uniformly along the gage length in cross-sectional area. Initially the strain hardening more than compensates for this decrease in area and the engineering stress (proportional to load P) continues to rise with increasing strain. Eventually a point is reached where the decrease in specimen cross-sectional area is greater than the increase in deformation load arising from strain hardening. This condition will be reached first at some point in the

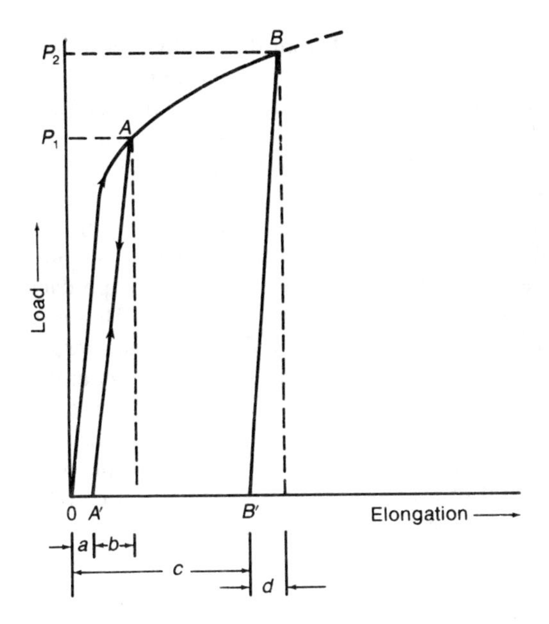

Figure 8-2 Loading and unloading curves showing elastic recoverable strain and plastic deformation.

specimen that is slightly weaker than the rest. All further plastic deformation is concentrated in this region, and the specimen begins to neck or thin down locally. Because the cross-sectional area now is decreasing far more rapidly than the deformation load is increased by strain hardening, the actual *load* required to deform the specimen falls off and the engineering stress by Eq. (8-1) likewise continues to decrease until fracture occurs.

Consider a tensile specimen that has been loaded to a value in excess of the yield stress and then the load is removed (Fig. 8-2). The loading follows the path O-A-A'. Note that the slope of the unloading curve A-A' is parallel to the elastic modulus on loading. The recoverable elastic strain on unloading is $b = \sigma_1/E = (P_1/A_0)/E$. The permanent plastic deformation is the offset a in Fig. 8-2. Note that elastic deformation is always present in the tension specimen when it is under load. If the specimen were loaded and unloaded along the path O-A-B-B', the elastic strain would be greater than on loading to P_1, since $P_2 > P_1$, but the elastic deformation (d) would be less than the plastic deformation (c).

Tensile Strength

The tensile strength, or ultimate tensile strength (UTS), is the maximum load divided by the original cross-sectional area of the specimen.

$$s_u = \frac{P_{max}}{A_0} \tag{8-3}$$

The tensile strength is the value most often quoted from the results of a tension test; yet in reality it is a value of little fundamental significance with regard to the strength of a metal. For ductile metals the tensile strength should be regarded as a measure of the maximum load which a metal can withstand under the very restrictive conditions of uniaxial loading. It will be shown that this value bears little relation to the useful strength of the metal under the more complex conditions of stress which are usually encountered. For many years it was customary to base the strength of members on the tensile strength, suitably reduced by a factor of safety. The current trend is to the more rational approach of basing the static design of ductile metals on the yield strength. However, because of the long practice of using the tensile strength to determine the strength of materials, it has become a very familiar property, and as such it is a very useful identification of a material in the same sense that the chemical composition serves to identify a metal or alloy. Further, because the tensile strength is easy to determine and is a quite reproducible property, it is useful for the purposes of specifications and for quality control of a product. Extensive empirical correlations between tensile strength and properties such as hardness and fatigue strength are often quite useful. For brittle materials, the tensile strength is a valid criterion for design.

Measures of Yielding

The stress at which plastic deformation or yielding is observed to begin depends on the sensitivity of the strain measurements. With most materials there is a gradual transition from elastic to plastic behavior, and the point at which plastic deformation begins is hard to define with precision. Various criteria for the initiation of yielding are used depending on the sensitivity of the strain measurements and the intended use of the data.

1. *True elastic limit* based on microstrain measurements at strains on order of 2×10^{-6} in/in (see Sec. 4-13). This elastic limit is a very low value and is related to the motion of a few hundred dislocations.
2. *Proportional limit* is the highest stress at which stress is directly proportional to strain. It is obtained by observing the deviation from the straight-line portion of the stress-strain curve.
3. *Elastic limit* is the greatest stress the material can withstand without any measurable permanent strain remaining on the complete release of load. With increasing sensitivity of strain measurement, the value of the elastic limit is decreased until at the limit it equals the true elastic limit determined from microstrain measurements. With the sensitivity of strain usually employed in engineering studies (10^{-4} in/in), the elastic limit is greater than the proportional limit. Determination of the elastic limit requires a tedious incremental loading-unloading test procedure.
4. The *yield strength* is the stress required to produce a small specified amount of plastic deformation. The usual definition of this property is the *offset yield*

strength determined by the stress corresponding to the intersection of the stress-strain curve and a line parallel to the elastic part of the curve offset by a specified strain (Fig. 8-1). In the United States the offset is usually specified as a strain of 0.2 or 0.1 percent (e = 0.002 or 0.001).

$$s_0 = \frac{P_{(\text{strain offset}=0.002)}}{A_0} \tag{8-4}$$

A good way of looking at offset yield strength is that after a specimen has been loaded to its 0.2 percent offset yield strength and then unloaded it will be 0.2 percent longer than before the test. The offset yield strength is often referred to in Great Britain as the *proof stress*, where offset values are either 0.1 or 0.5 percent. The yield strength obtained by an offset method is commonly used for design and specification purposes because it avoids the practical difficulties of measuring the elastic limit or proportional limit.

Some materials have essentially no linear portion to their stress-strain curve, for example, soft copper or gray cast iron. For these materials the offset method cannot be used and the usual practice is to define the yield strength as the stress to produce some total strain, for example, $\varepsilon = 0.005$.

Measures of Ductility

At our present degree of understanding, ductility is a qualitative, subjective property of a material. In general, measurements of ductility are of interest in three ways[1]:

1. To indicate the extent to which a metal can be deformed without fracture in metalworking operations such as rolling and extrusion.
2. To indicate to the designer, in a general way, the ability of the metal to flow plastically before fracture. A high ductility indicates that the material is "forgiving" and likely to deform locally without fracture should the designer err in the stress calculation or the prediction of severe loads.
3. To serve as an indicator of changes in impurity level or processing conditions. Ductility measurements may be specified to assess material "quality" even though no direct relationship exists between the ductility measurement and performance in service

The conventional measures of ductility that are obtained from the tension test are the engineering strain at fracture e_f (usually called the *elongation*) and the *reduction of area* at fracture q. Both of these properties are obtained after fracture

[1] G. E. Dieter, Introduction to Ductility, in: "Ductility," American Society for Metals, Metals Park, Ohio, 1968.

by putting the specimen back together and taking measurements of L_f and A_f.

$$e_f = \frac{L_f - L_0}{L_0} \tag{8-5}$$

$$q = \frac{A_0 - A_f}{A_0} \tag{8-6}$$

Both elongation and reduction of area usually are expressed as a percentage.

Because an appreciable fraction of the plastic deformation will be concentrated in the necked region of the tension specimen, the value of e_f will depend on the gage length L_0 over which the measurement was taken (see Sec. 8-5). The smaller the gage length the greater will be the contribution to the overall elongation from the necked region and the higher will be the value of e_f. Therefore, when reporting values of *percentage elongation*, the gage length L_0 always should be given.

The reduction of area does not suffer from this difficulty. Reduction of area values can be converted into an equivalent *zero-gage-length elongation* e_0. From the constancy of volume relationship for plastic deformation $AL = A_0L_0$, we obtain

$$\frac{L}{L_0} = \frac{A_0}{A} = \frac{1}{1-q}$$

$$e_0 = \frac{L - L_0}{L_0} = \frac{A_0}{A} - 1 = \frac{1}{1-q} - 1 = \frac{q}{1-q} \tag{8-7}$$

This represents the elongation based on a very short gage length near the fracture.

Another way to avoid the complication from necking is to base the percentage elongation on the uniform strain out to the point at which necking begins. The uniform elongation e_u correlates well with stretch-forming operation. Since the engineering stress-strain curve often is quite flat in the vicinity of necking, it may be difficult to establish the strain at maximum load without ambiguity. In this case the method suggested by Nelson and Winlock[1] is useful.

Modulus of Elasticity

The slope of the initial linear portion of the stress-strain curve is the modulus of elasticity, or Young's modulus. The modulus of elasticity is a measure of the stiffness of the material. The greater the modulus, the smaller the elastic strain resulting from the application of a given stress. Since the modulus of elasticity is needed for computing deflections of beams and other members, it is an important design value.

The modulus of elasticity is determined by the binding forces between atoms. Since these forces cannot be changed without changing the basic nature of the material, it follows that the modulus of elasticity is one of the most structure-insensitive of the mechanical properties. It is only slightly affected by alloying

[1] P. G. Nelson and J. Winlock, *ASTM Bull.*, vol. 156, p. 53, January 1949.

Table 8-1 Typical values of modulus of elasticity at different temperatures

	Modulus of elasticity, psi $\times 10^{-6}$				
Material	Room temp.	400°F	800°F	1000°F	1200°F
Carbon steel	30.0	27.0	22.5	19.5	18.0
Austenitic stainless steel	28.0	25.5	23.0	22.5	21.0
Titanium alloys	16.5	14.0	10.7	10.1	
Aluminium alloys	10.5	9.5	7.8		

additions, heat treatment, or cold-work.[1] However, increasing the temperature decreases the modulus of elasticity. The modulus is usually measured at elevated temperatures by a dynamic method.[2] Typical values[3] of the modulus of elasticity for common engineering metals at different temperatures are given in Table 8-1.

Example A standard 0.505-in-diameter tensile specimen has a 2.00 in gage length. The load corresponding to the 0.2 percent offset is 15,000 lb and the maximum load is 18,500 lb. Fracture occurs at 16,200 lb. The diameter after fracture is 0.315 in and the gage length at fracture is 2.53 in. Calculate the standard properties of the material from the tension test.

$$A_0 = \frac{\pi}{4}(0.505)^2 = 0.200 \text{ in}^2,$$

$$A_f = \frac{\pi}{4}(0.315)^2 = 0.078 \text{ in}^2$$

ultimate tensile strength: $$s_u = \frac{P_{max}}{A_0} = \frac{18{,}500}{0.200} = 92{,}500 \text{ psi}$$

0.2 percent offset yield strength: $$s_0 = \frac{P_y}{A_0} = \frac{15{,}000}{0.200} = 60{,}000 \text{ psi}$$

breaking stress: $$s_f = \frac{P_f}{A_0} = \frac{16{,}200}{0.200} = 81{,}000 \text{ psi}$$

elongation: $$e_f = \frac{L_f - Lo}{Lo} = \frac{2.53 - 2.00}{2.00} = 26.5 \text{ percent}$$

reduction of area: $$q = \frac{A_0 - A_f}{A_0} = \frac{0.200 - 0.078}{0.200} = 61 \text{ percent}$$

[1] D. J. Mack, *Trans. AIME*, vol. 166, pp. 68–85, 1946.

[2] P. E. Armstrong, Measurement of Elastic Constants, in R. F. Bunshaw (ed.), "Techniques of Metals Research," vol. V, pt. 2, chap. 9, Interscience Publishers, Inc., New York, 1971.

[3] Standard Method of Test for Young's Modulus at Room Temperature, ASTM E111-61, op. cit., pp. 409–413.

If $E = 30 \times 10^6$ psi the elastic recoverable strain at maximum load is

$$e_E = \frac{P_{max}/A_0}{E} = \frac{9.25 \times 10^4}{30 \times 10^6} = 0.0031$$

If the elongation at maximum load (the uniform elongation) is 22 percent, what is the plastic strain at maximum load?

$$e_p = e_{total} - e_E = 0.2200 - 0.0031 = 0.2169$$

Resilience

The ability of a material to absorb energy when deformed elastically and to return it when unloaded is called *resilience*. This is usually measured by the *modulus of resilience*, which is the strain energy per unit volume required to stress the material from zero stress to the yield stress σ_0. Referring to Eq. (2-80), the strain energy per unit volume for uniaxial tension is

$$U_0 = \tfrac{1}{2}\sigma_x e_x$$

From the above definition the modulus of resilience is

$$U_R = \frac{1}{2} s_0 e_0 = \frac{1}{2} s_0 \frac{s_0}{E} = \frac{s_0^2}{2E} \tag{8-8}$$

This equation indicates that the ideal material for resisting energy loads in applications where the material must not undergo permanent distortion, such as mechanical springs, is one having a high yield stress and a low modulus of elasticity. Table 8-2 gives some values of modulus of resilience for different materials.

Toughness

The toughness of a material is its ability to absorb energy in the plastic range. The ability to withstand occasional stresses above the yield stress without fracturing is particularly desirable in parts such as freight-car couplings, gears, chains, and crane hooks. Toughness is a commonly used concept which is difficult to pin down and define. One way of looking at toughness is to consider that it is the

Table 8-2 Modulus of resilience for various materials

Material	E, psi	s_0, psi	Modulus of resilience, U_R
Medium-carbon steel	30×10^6	45,000	33.7
High-carbon spring steel	30×10^6	140,000	320
Duraluminum	10.5×10^6	18,000	17
Copper	16×10^6	4,000	5.3
Rubber	150	300	300
Acrylic polymer	0.5×10^6	2,000	4.0

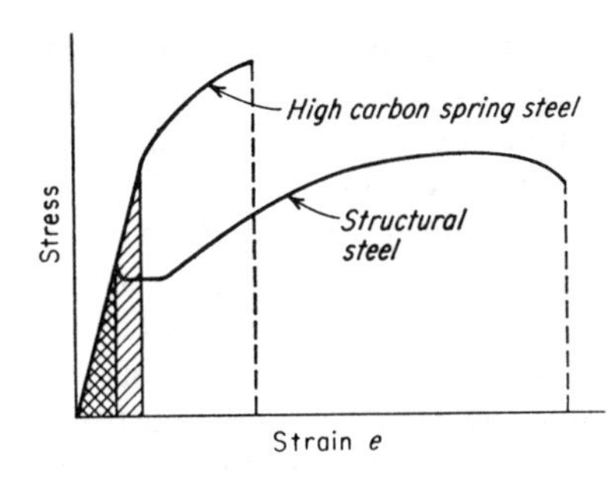

Figure 8-3 Comparison of stress-strain curves for high- and low-toughness materials.

total area under the stress-strain curve. This area is an indication of the amount of work per unit volume which can be done on the material without causing it to rupture. Figure 8-3 shows the stress-strain curves for high- and low-toughness materials. The high-carbon spring steel has a higher yield strength and tensile strength than the medium-carbon structural steel. However, the structural steel is more ductile and has a greater total elongation. The total area under the stress-strain curve is greater for the structural steel, and therefore it is a tougher material. This illustrates that toughness is a parameter which comprises *both* strength and ductility. The crosshatched regions in Fig. 8-3 indicate the modulus of resilience for each steel. Because of its higher yield strength, the spring steel has the greater resilience.

Several mathematical approximations for the area under the stress-strain curve have been suggested. For ductile metals which have a stress-strain curve like that of the structural steel, the area under the curve can be approximated by either of the following equations:

$$U_T \approx s_u e_f \tag{8-9}$$

or

$$U_T \approx \frac{s_0 + s_u}{2} e_f \tag{8-10}$$

For brittle materials the stress-strain curve is sometimes assumed to be a parabola, and the area under the curve is given by

$$U_T \approx \tfrac{2}{3} s_u e_f \tag{8-11}$$

All these relations are only approximations to the area under the stress-strain curves. Further, the curves do not represent the true behavior in the plastic range, since they are all based on the original area of the specimen.

8-2 TRUE-STRESS–TRUE-STRAIN CURVE

The engineering stress-strain curve does not give a true indication of the deformation characteristics of a metal because it is based entirely on the original dimensions of the specimen, and these dimensions change continuously during the test. Also, ductile metal which is pulled in tension becomes unstable and necks down during the course of the test. Because the cross-sectional area of the specimen is decreasing rapidly at this stage in the test, the load required to continue deformation falls off. The average stress based on original area likewise decreases, and this produces the fall-off in the stress-strain curve beyond the point of maximum load. Actually, the metal continues to strain-harden all the way up to fracture, so that the stress required to produce further deformation should also increase. If the *true stress*, based on the actual cross-sectional area of the specimen, is used, it is found that the stress-strain curve increases continuously up to fracture. If the strain measurement is also based on instantaneous measurements, the curve which is obtained is known as a *true-stress–true-strain* curve. This is also known as a *flow curve* (Sec. 3-2) since it represents the basic plastic-flow characteristics of the material. Any point on the flow curve can be considered the yield stress for a metal strained in tension by the amount shown on the curve. Thus, if the load is removed at this point and then reapplied, the material will behave elastically throughout the entire range of reloading.

The definitions of true stress and true strain were given in Sec. 3-3. The true stress σ is expressed in terms of engineering stress s by

$$\sigma = \frac{P}{A_0}(e+1) = s(e+1) \tag{8-12}$$

The derivation of Eq. (8-12) assumes both constancy of volume and a homogeneous distribution of strain along the gage length of the tension specimen. Thus, Eq. (8-12) should only be used until the onset of necking. Beyond maximum load the true stress should be determined from actual measurements of load and cross-sectional area.

$$\sigma = P/A \tag{8-13}$$

The true strain ε may be determined from the engineering or conventional strain e by

$$\varepsilon = \ln(e+1) \tag{8-14}$$

This equation is applicable only to the onset of necking for the reasons discussed above. Beyond maximum load the true strain should be based on actual area or diameter measurements.

$$\varepsilon = \ln\frac{A_0}{A} = \ln\frac{(\pi/4)D_0^2}{(\pi/4)D^2} = 2\ln\frac{D_0}{D} \tag{8-15}$$

Figure 8-4 compares the true-stress–true-strain curve with its corresponding engineering stress-strain curve. Note that because of the relatively large plastic strains, the elastic region has been compressed into the y axis. In agreement with Eqs. (8-12) and (8-14), the true-stress–true-strain curve is always to the left of the engineering curve until the maximum load is reached. However, beyond maximum load the high localized strains in the necked region that are used in Eq. (8-15) far exceed the engineering strain calculated from Eq. (8-2). Frequently the flow curve is linear from maximum load to fracture, while in other cases its slope

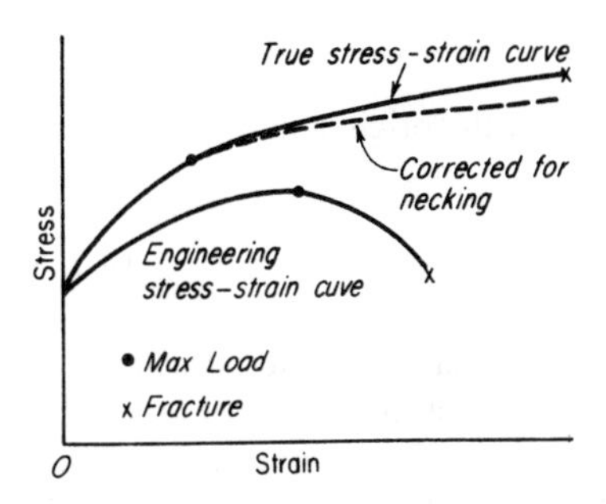

Figure 8-4 Comparison of engineering and true stress-strain curves.

continuously decreases up to fracture. The formation of a necked region or mild notch introduces triaxial stresses which make it difficult to determine accurately the longitudinal tensile stress on out to fracture (see Sec. 8-3).

The following parameters usually are determined from the true-stress–true-strain curve.

True Stress at Maximum Load

The true stress at maximum load corresponds to the true tensile strength. For most materials necking begins at maximum load at a value of strain where the true stress equals the slope of the flow curve (see Sec. 8-3). Let σ_u and ε_u denote the true stress and true strain at maximum load when the cross-sectional area of the specimen is A_u. The ultimate tensile strength is given by

$$s_u = \frac{P_{\max}}{A_0}$$

and

$$\sigma_u = \frac{P_{\max}}{A_u} \qquad \varepsilon_u = \ln \frac{A_0}{A_u}$$

Eliminating $P_{\max}$ yields

$$\sigma_u = s_u \frac{A_0}{A_u}$$

and

$$\sigma_u = s_u e^{\varepsilon_u} \tag{8-16}$$

True Fracture Stress

The true fracture stress is the load at fracture divided by the cross-sectional area at fracture. This stress should be corrected for the triaxial state of stress existing in the tensile specimen at fracture. Since the data required for this correction are often not available, true-fracture-stress values are frequently in error.

True Fracture Strain

The true fracture strain ε_f is the true strain based on the original area A_0 and the area after fracture A_f.

$$\varepsilon_f = \ln \frac{A_0}{A_f} \tag{8-17}$$

This parameter represents the maximum true strain that the material can withstand before fracture and is analogous to the total strain to fracture of the engineering stress-strain curve. Since Eq. (8-14) is not valid beyond the onset of necking, it is not possible to calculate ε_f from measured values of e_f. However, for cylindrical tensile specimens the reduction of area q is related to the true fracture strain by the relationship

$$\varepsilon_f = \ln \frac{1}{1-q} \tag{8-18}$$

True Uniform Strain

The true uniform strain ε_u is the true strain based only on the strain up to maximum load. It may be calculated from either the specimen cross-sectional area A_u or the gage length L_u at maximum load. Equation (8-14) may be used to convert conventional uniform strain to true uniform strain. The uniform strain is often useful in estimating the formability of metals from the results of a tension test.

$$\varepsilon_u = \ln \frac{A_0}{A_u} \tag{8-19}$$

True Local Necking Strain

The local necking strain ε_n is the strain required to deform the specimen from maximum load to fracture.

$$\varepsilon_n = \ln \frac{A_u}{A_f} \tag{8-20}$$

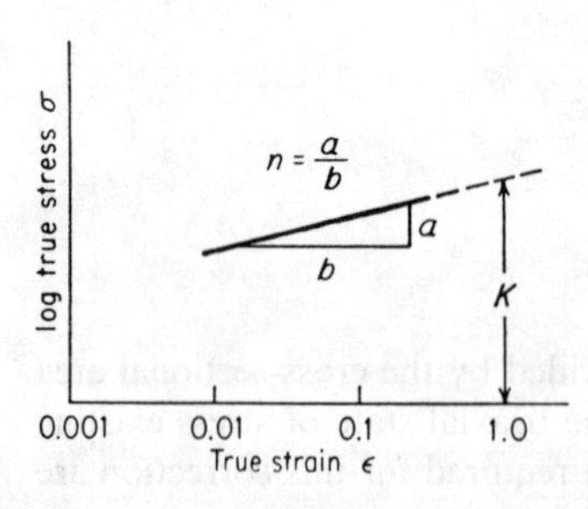

Figure 8-5 Log-log plot of true stress-strain curve.

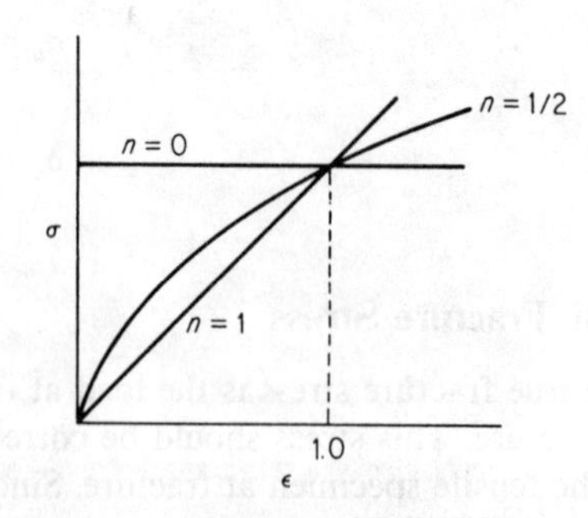

Figure 8-6 Various forms of power curve $\sigma = K\varepsilon^n$.

Table 8-3 Values for *n* and *K* for metals at room temperature

Metal	Condition	n	K, psi	Ref.
0.05% C steel	Annealed	0.26	77,000	†
SAE 4340 steel	Annealed	0.15	93,000	†
0.6% C steel	Quenched and tempered 1000°F	0.10	228,000	‡
0.6% C steel	Quenched and tempered 1300°F	0.19	178,000	‡
Copper	Annealed	0.54	46,400	†
70/30 brass	Annealed	0.49	130,000	‡

†J. R. Low and F. Garofalo, *Proc. Soc. Exp. Stress Anal.*, vol. 4, no. 2, pp. 16–25, 1947.
‡J. R. Low, "Properties of Metals in Materials Engineering," American Society for Metals, Metals Park, Ohio, 1949.

The flow curve of many metals in the region of uniform plastic deformation can be expressed by the simple power curve relation

$$\sigma = K\varepsilon^n \tag{8-21}$$

where n is the *strain-hardening exponent* and K is the *strength coefficient*. A log-log plot of true stress and true strain up to maximum load will result in a straightline if Eq. (8-21) is satisfied by the data (Fig. 8-5). The linear slope of this line is n and K is the true stress at $\varepsilon = 1.0$ (corresponds to $q = 0.63$). The strain-hardening exponent[1] may have values from $n = 0$ (perfectly plastic solid) to $n = 1$ (elastic solid) (see Fig. 8-6). For most metals n has values between 0.10 and 0.50 (see Table 8-3).

It is important to note that the *rate of strain hardening* $d\sigma/d\varepsilon$ is not identical with the strain-hardening exponent.

From the definition of n

$$n = \frac{d(\log\sigma)}{d(\log\varepsilon)} = \frac{d(\ln\sigma)}{d(\ln\varepsilon)} = \frac{\varepsilon}{\sigma}\frac{d\sigma}{d\varepsilon}$$

or

$$\frac{d\sigma}{d\varepsilon} = n\frac{\sigma}{\varepsilon} \tag{8-22}$$

There is nothing basic about Eq. (8-21) and deviations from this relationship frequently are observed, often at low strains (10^{-3}) or high strains ($\varepsilon \approx 1.0$). One common type of deviation is for a log-log plot of Eq. (8-21) to result in two straight lines with different slopes. Sometimes data which do not plot according to Eq. (8-21) will yield a straight line according to the relationship

$$\sigma = K(\varepsilon_0 + \varepsilon)^n \tag{8-23}$$

Datsko[2] has shown how ε_0 can be considered to be the amount of strain hardening that the material received prior to the tension test.

[1] Other techniques for measuring n are discussed by J. L. Duncan, *Sheet Met. Ind.*, July 1967, pp. 483–489; see also ASTM Standard E646-78.

[2] J. Datsko, "Material Properties and Manufacturing Processes," pp. 18–20, John Wiley & Sons, Inc., New York, 1966.

Another common variation in Eq. (8-21) is the Ludwik equation

$$\sigma = \sigma_0 + K\varepsilon^n \tag{8-24}$$

where σ_0 is the yield stress and K and n are the same constants as in Eq. (8-21). This equation may be more satisfying than Eq. (8-21) since the latter implies that at zero true strain the stress is zero. Morrison[1] has shown that σ_0 can be obtained from the intercept of the strain-hardening portion of the stress-strain curve and the elastic modulus line by

$$\sigma_0 = \left(\frac{K}{E^n}\right)^{1/1-n}$$

The true-stress–true-strain curve of metals such as austenitic stainless steel, which deviate markedly from Eq. (8-21) at low strains, can be expressed by[2]

$$\sigma = K\varepsilon^n + e^{K_1}e^{n_1\varepsilon}$$

where e^{K_1} is approximately equal to the proportional limit and n_1 is the slope of the deviation of stress from Eq. (8-21) plotted against ε. Still other expressions for the flow curve have been discussed in the literature.[3,4]

The true strain term in Eqs. (8-21) to (8-24) properly should be the *plastic strain* $\varepsilon_p = \varepsilon_{\text{total}} - \varepsilon_E = \varepsilon_{\text{total}} - \sigma/E$.

Example In the tension test of a metal fracture occurs at maximum load. The conditions at fracture were: $A_f = 100$ mm^2 and $L_f = 60$ mm. The initial values were: $A_0 = 150$ mm^2 and $L_0 = 40$ mm. Determine the true strain to fracture using changes in both length and area.

$$\varepsilon_f = \ln\left(\frac{L_f}{L_0}\right) = \ln\left(\frac{60}{40}\right) = 0.405$$

$$\varepsilon_f = \ln\left(\frac{A_0}{A_f}\right) = \ln\left(\frac{150}{100}\right) = 0.405$$

The same value of fracture strain was obtained from length and area measurements because constancy of volume could be applied because the specimen has not undergone necking.

If a more ductile metal is tested such that necking occurs and the final gage length is 83 mm and the final diameter is 8 mm, while $L_0 = 40$ mm and $D_0 = 12.8$ mm.

$$\varepsilon_f = \ln\left(\frac{L_f}{L_0}\right) = \ln\left(\frac{83}{40}\right) = 0.730$$

$$\varepsilon_f = \ln\left(\frac{D_0}{D_f}\right)^2 = 2\ln\left(\frac{12.8}{8}\right) = 0.940$$

[1] W. B. Morrison, *Trans. Am. Soc. Met.*, vol. 59, p. 824, 1966.
[2] D. C. Ludwigson, *Metall. Trans.*, vol. 2, pp. 2825–2828, 1971.
[3] H. J. Kleemola and M. A. Nieminen, *Metall. Trans.*, vol. 5, pp. 1863–1866, 1974.
[4] C. Adams and J. G. Beese, *Trans. ASME, Ser. H*, vol. 96, pp. 123–126, 1974.

In this case the fracture strain determined from length values is incorrect because necking produces a very nonuniform strain along the length of the specimen. The strain based on the diameter at the neck is the maximum strain experienced by the specimen.

8-3 INSTABILITY IN TENSION

Necking generally begins at maximum load during the tensile deformation of a ductile metal. An ideal plastic material in which no strain hardening occurs would become unstable in tension and begin to neck just as soon as yielding took place. However, a real metal undergoes strain hardening, which tends to increase the load-carrying capacity of the specimen as deformation increases. This effect is opposed by the gradual decrease in the cross-sectional area of the specimen as it elongates. Necking or localized deformation begins at maximum load, where the increase in stress due to decrease in the cross-sectional area of the specimen becomes greater than the increase in the load-carrying ability of the metal due to strain hardening. This condition of instability leading to localized deformation is defined by the condition $dP = 0$.

$$P = \sigma A$$

$$dP = \sigma\, dA + A\, d\sigma = 0$$

From the constancy-of-volume relationship,

$$\frac{dL}{L} = -\frac{dA}{A} = d\varepsilon$$

and from the instability condition

$$-\frac{dA}{A} = \frac{d\sigma}{\sigma}$$

so that at a point of tensile instability

$$\frac{d\sigma}{d\varepsilon} = \sigma \tag{8-25}$$

Therefore, the point of necking at maximum load can be obtained from the

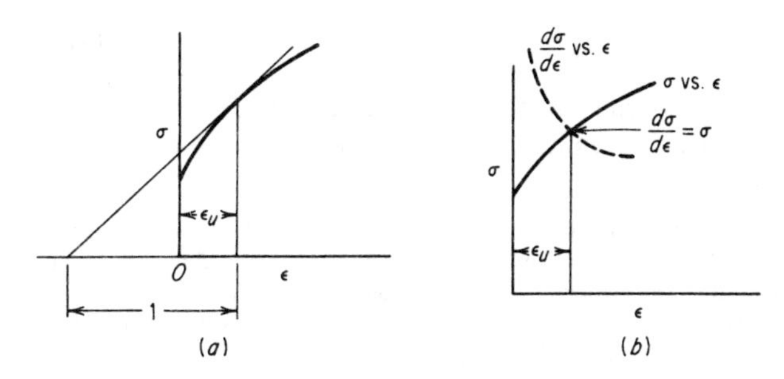

Figure 8-7 Graphical interpretation of necking criterion.

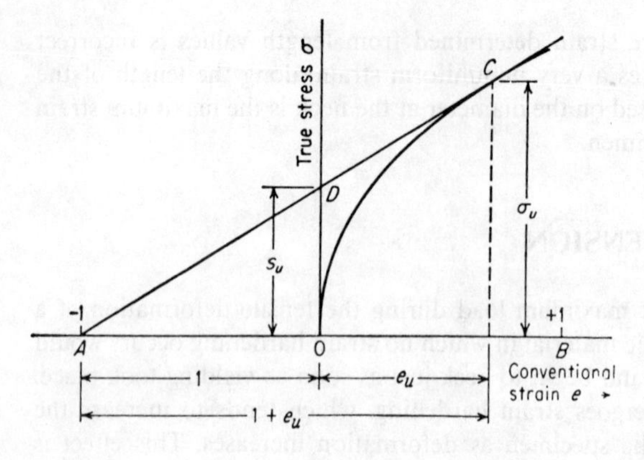

Figure 8-8 Considère's construction for the determination of the point of maximum load.

true-stress–true-strain curve[1] by finding the point on the curve having a subtangent of unity (Fig. 8-7*a*) or the point where the rate of strain hardening equals the stress (Fig. 8-7*b*).

The necking criterion can be expressed more explicitly if engineering strain is used. Starting with Eq. (8-25),

$$\frac{d\sigma}{d\varepsilon} = \frac{d\sigma}{de}\frac{de}{d\varepsilon} = \frac{d\sigma}{de}\frac{dL/L_0}{dL/L} = \frac{d\sigma}{de}\frac{L}{L_0} = \frac{d\sigma}{de}(1+e) = \sigma$$

$$\frac{d\sigma}{de} = \frac{\sigma}{1+e} \tag{8-26}$$

Equation (8-26) permits an interesting geometrical construction called Considère's construction[2] for the determination of the point of maximum load. In Fig. 8-8 the stress-strain curve is plotted in terms of true stress against conventional linear strain. Let point A represent a negative strain of 1.0. A line drawn from point A which is tangent to the stress-strain curve will establish the point of maximum load, for according to Eq. (8-26) the slope at this point is $\sigma/(1+e)$.

By substituting the necking criterion given in Eq. (8-25) into Eq. (8-22), we obtain a simple relationship for the strain at which necking occurs. This strain is the true uniform strain ε_u.

$$\varepsilon_u = n \tag{8-27}$$

Example If the true-stress–true-strain curve is given by $\sigma = 200{,}000\varepsilon^{0.33}$, where stress is in psi, what is the ultimate tensile strength of the material?

[1] An extension of Eq. (8-25) for the situation where the material transforms during tensile deformation has been given by J. R. C. Guimaraes and R. J. deAngelis, *Mater. Sci. Eng.*, vol. 13, pp. 109–111, 1974; for a more generalized discussion of stability in plastic deformation see A. S. Argon, "The Inhomogeneity of Plastic Deformation," American Society for Metals, Metals Park, Ohio, 1973.

[2] A Considère, *Ann. ponts et chaussées*, vol. 9, ser. 6, pp. 574–775, 1885.

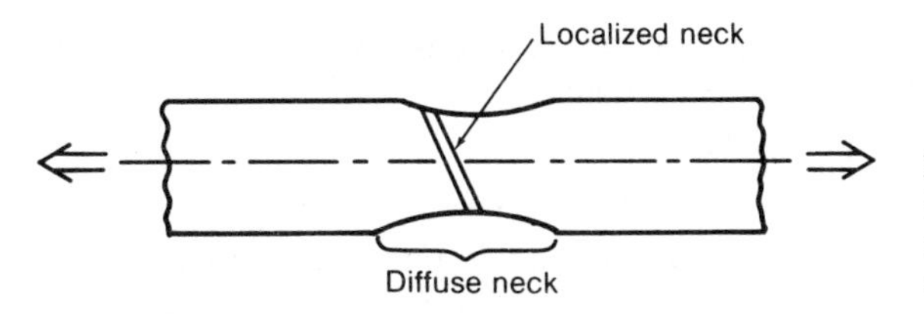

Figure 8-9 Illustration of diffuse necking and localized necking in a sheet tensile specimen.

The uniform elongation to maximum load is $\varepsilon_u = n = 0.33$.

The true stress at maximum load is $\sigma_u = 200{,}000(0.33)^{0.33} = 138{,}700$ psi.

From Eq. (8-12) $s_u = \sigma_u/e_u + 1$ and from Eq. (8-14) $\varepsilon_u = \ln(e_u + 1)$; $e_u + 1 = e^{\varepsilon_u} = \exp(\varepsilon_u) = \exp(0.33)$. Therefore, the ultimate tensile strength is

$$s_u = \frac{138{,}700}{e^{0.33}} = \frac{138{,}700}{1.391} = 99{,}700 \text{ psi}$$

Necking in a cylindrical tensile specimen is symmetrical around the tensile axis if the material is isotropic. However, a different type of necking behavior is found for a tensile specimen with rectangular cross section that is cut from a sheet. For a sheet tensile specimen where width is much greater than thickness there are two types of tensile flow instability. The first is *diffuse necking*, so called because its extent is much greater than the sheet thickness (Fig. 8-9). This form of unstable flow in a sheet tensile specimen is analogous to the neck formed in a cylindrical tensile specimen. Diffuse necking initiates according to the relationships discussed above. Diffuse necking may terminate in fracture but it often is followed by a second instability process called *localized necking*. In this mode the neck is a narrow band with a width about equal to the sheet thickness inclined at an angle to the specimen axis, across the width of the specimen (Fig. 8-9). In localized necking there is no change in width measured along the trough of the localized neck, so that localized necking corresponds to a state of plane-strain deformation.

With localized necking the decrease in specimen area with increasing strain (the geometrical softening) is restricted to the thickness direction. Thus, $dA = w\,dt$ where w is the constant length of the localized neck and t is the thickness of the neck.

$$\frac{dP}{d\varepsilon}\left(\frac{1}{A}\right) = -\frac{\sigma w\,dt}{d\varepsilon}\left(\frac{1}{A}\right) = -\frac{\sigma w\,dt}{wt\,d\varepsilon} = -\sigma\frac{dt/t}{d\varepsilon} \tag{8-28}$$

Let the direction of axial strain be ε_1, the width strain be ε_2, and the thickness strain be ε_3. From constancy of volume, $d\varepsilon_2 = d\varepsilon_3 = -d\varepsilon_1/2$, and $d\varepsilon_3 = dt/t$. Substituting into Eq. (8-28) gives

$$\frac{dP}{d\varepsilon}\left(\frac{1}{A}\right) = \frac{\sigma}{2} \tag{8-29}$$

The increase in load-carrying ability due to strain hardening is given by

$$\frac{dP}{d\varepsilon}\left(\frac{1}{A}\right) = \frac{A\,d\sigma}{d\varepsilon}\left(\frac{1}{A}\right) = \frac{d\sigma}{d\varepsilon} \tag{8-30}$$

As before, necking begins when the geometrical softening just balances the strain hardening, so equating Eq. (8-29) and (8-30) gives

$$\frac{d\sigma}{d\varepsilon} = \frac{\sigma}{2} \tag{8-31}$$

This criterion for localized necking expresses the fact that the specimen area decreases with straining less rapidly in this mode than in diffuse necking. Thus, more strain must be accumulated before the geometrical softening will cancel the strain hardening. For a power-law flow curve, $\varepsilon_u = 2n$ for localized necking.

8-4 STRESS DISTRIBUTION AT THE NECK

The formation of a neck in the tensile specimen introduces a complex triaxial state of stress in that region. The necked region is in effect a mild notch. As we discussed in Sec. 7-10 a notch under tension produces radial and transverse stresses which raise the value of longitudinal stress required to cause plastic flow. Therefore, the average true stress at the neck, which is determined by dividing the axial tensile load by the minimum cross-sectional area of the specimen at the neck, is higher than the stress which would be required to cause flow if simple tension prevailed. Figure 8-10 illustrates the geometry at the necked region and the stresses developed by this localized deformation. R is the radius of curvature of the neck, which can be measured either by projecting the contour of the necked region on a screen or by using a tapered, conical radius gage.

Bridgman[1] made a mathematical analysis which provides a correction to the average axial stress to compensate for the introduction of transverse stresses. This analysis was based on the following assumptions:

1. The contour of the neck is approximated by the arc of a circle.
2. The cross section of the necked region remains circular throughout the test.
3. The von Mises' criterion for yielding applies.
4. The strains are constant over the cross section of the neck.

According to Bridgman's analysis, the uniaxial flow stress corresponding to that which would exist in the tension test if necking had not introduced triaxial stresses is

$$\sigma = \frac{(\sigma_x)_{\text{avg}}}{(1 + 2R/a)[\ln(1 + a/2R)]} \tag{8-32}$$

where $(\sigma_x)_{\text{avg}}$ is the measured stress in the axial direction (load divided by minimum cross section). Figure 8-4 shows how the application of the Bridgman correction changes the true-stress–true-strain curve. A correction for the triaxial stresses in the neck of a flat tensile specimen has been considered by Aronofsky.[2]

[1] P. W. Bridgman, *Trans. Am. Soc. Met*, vol. 32, p. 553, 1944.
[2] J. Aronofsky, *J. Appl. Mech.*, vol. 18, pp. 75–84, 1951.

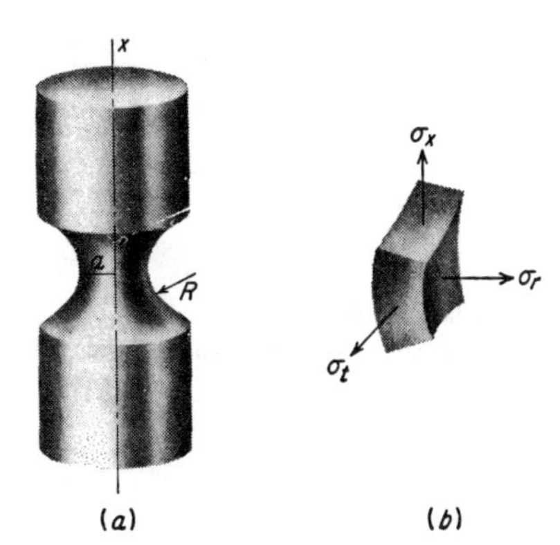

Figure 8-10 (*a*) Geometry of necked region; (*b*) stresses acting on element at point *O*.

The values of a/R needed for the analysis can be obtained[1] either by straining a specimen a given amount beyond necking and unloading to measure a and R directly or by measuring these parameters continuously past necking using photography or a tapered ring gage. Tegart[2] has shown that a good correlation is obtained between the Bridgman correction and the true strain in the neck minus the true strain at necking ε_u.

8-5 DUCTILITY MEASUREMENT IN TENSION TEST

Having discussed in Sec. 8-1 the standard measurements of ductility that are obtained from the tension test, i.e., percent elongation and reduction of area, we return again to this subject armed with an understanding of the phenomenon of necking. The measured elongation from a tension specimen depends on the gage length of the specimen or the dimensions of its cross section. This is because the total extension consists of two components, the uniform extension up to necking and the localized extension once necking begins. The extent of uniform extension will depend on the metallurgical condition of the material (through n) and the effect of specimen size and shape on the development of the neck. Figure 8-11 illustrates the variation of the local elongation, Eq. (8-7), along the gage length of a prominently necked tensile specimen. It readily can be seen that the shorter the gage length the greater the influence of localized deformation at the neck on the total elongation of the gage length.

The extension of a specimen at fracture can be expressed by

$$L_f - L_0 = \alpha + e_u L_0 \tag{8-33}$$

where α is the local necking extension and $e_u L_0$ is the uniform extension. The

[1] T. A. Trozera, *Trans. Am. Soc. Met.*, vol. 56, pp. 280–282, 1963.

[2] W. J. McG. Tegart, "Elements of Mechanical Metallurgy," p. 22, The Macmillan Company, New York, 1966.

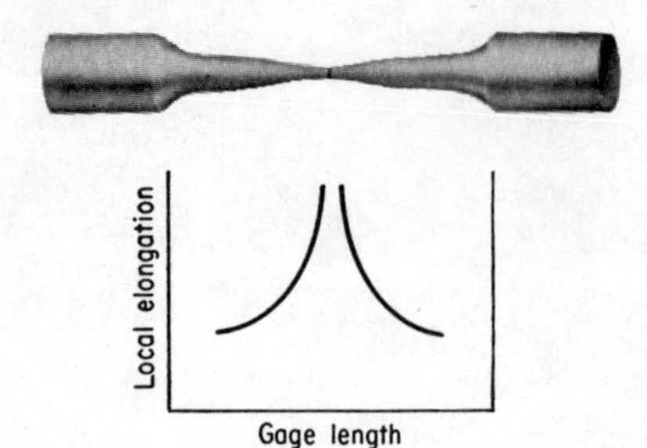

Figure 8-11 Schematic drawing of variation of local elongation with position along gage length of tensile specimen.

tensile elongation then is given by

$$e_f = \frac{L_f - L_0}{L_0} = \frac{\alpha}{L_0} + e_u \tag{8-34}$$

which clearly indicates that the total elongation is a function of the specimen gage length. The shorter the gage length the greater the percentage elongation.

Numerous attempts, dating back to about 1850, have been made to rationalize the strain distribution[1] in the tension test. Perhaps the most general conclusion that can be drawn is that geometrically similar specimens develop geometrically similar necked regions. According to Barba's law,[2] $\alpha = \beta\sqrt{A_0}$, and the elongation equation becomes

$$e_f = \beta\frac{\sqrt{A_0}}{L_0} + e_u \tag{8-35}$$

It is generally recognized that in order to compare elongation measurements of different-sized specimens, the specimens must be geometrically similar. Equation (8-35) shows that the critical geometrical factor for which similitude must be maintained is $L_0/\sqrt{A_0}$ for sheet specimens or L_0/D_0 for round bars. In the United States the standard tensile specimen has a 0.505-in diameter and a 2-in gage length. Subsize specimens have the following diameter and (gage length): 0.357(1.4), 0.252(1.0), and 0.160(0.634). Different values of $L_0/\sqrt{A_0}$ are specified by the standardizing agencies in different countries. Table 8-4 gives some appropriate values.

To a good approximation we can say that a given elongation will be produced in a material if $\sqrt{A_0}/L_0$ is maintained constant as predicted by Eq. (8-35). Thus, at a constant value of elongation $\sqrt{A_1}/L_1 = \sqrt{A_2}/L_2$, where A and L are the areas and gage lengths of two different specimens, 1 and 2 of the same metal. To predict the elongation using gage length L_2 on a specimen with area A_2 by means of measurements on a specimen with area A_1, it only is necessary to adjust the gage length of specimen 1 to conform with $L_1 = L_2\sqrt{A_1/A_2}$. For example,

[1] T. C. Hsu, G. S. Littlejohn and B. M. Marchbank, *Am. Soc. Test Mater. Proc.*, vol. 65, p. 874, 1965.

[2] M. J. Barba, *Mem. Soc. Ing. Civils*, pt. 1, p. 682, 1880.

Table 8-4 Dimensional relationships of tensile specimens used in different countries

Type specimen	United States (ASTM)	Great Britain		Germany
		Before 1962	Current	
Sheet ($L_0/\sqrt{A_0}$)	4.5	4.0	5.65	11.3
Round (L_0/D_0)	4.0	3.54	5.0	10.0

suppose that $\frac{1}{8}$-in-thick sheet is available and it is desired to predict the elongation with a 2-in gage length for the identical material but in 0.080-in thickness. Using $\frac{1}{2}$ in wide sheet specimens, we would predict that a test specimen with a gage length $L = 2\sqrt{0.125/0.080} = 2.5$-in made from the $\frac{1}{8}$-in-thick sheet. Experimental verification for this procedure has been shown by Kula and Fahey.[1]

Obviously, the occurrence of necking in the tension test makes any quantitative conversion between elongation and reduction of area impossible. While elongation and reduction of area usually vary in the same way, for example as a function of test temperature, tempering temperature, alloy content, etc., instances will be found where this is not the case. In the general sense, elongation and reduction of area measure different types of material behavior. Provided the gage length is not too short, percentage elongation is chiefly influenced by uniform elongation, and thus it is dependent on the strain-hardening capacity of the material. Reduction of area is more a measure of the deformation required to produce fracture and its chief contribution results from the necking process. Because of the complicated stress state in the neck, values of reduction of area are dependent on specimen geometry and deformation behavior, and they should not be taken as true material properties. However, reduction of area is the most structure-sensitive ductility parameter, and as such it is useful in detecting quality changes in the material.

8-6 EFFECT OF STRAIN RATE ON FLOW PROPERTIES

The rate at which strain is applied to a specimen can have an important influence on the flow stress. Strain rate is defined as $\dot{\varepsilon} = d\varepsilon/dt$, and is conventionally expressed in units of "per second." The spectrum of available strain rates is given in Table 8-5.

Figure 8-12 shows that increasing strain rate increases flow stress. Moreover, the strain-rate dependence of strength increases with increasing temperature. The yield stress and flow stress at lower plastic strains are more dependent on strain rate than the tensile strength. High rates of strain cause the yield point to appear

[1] E. G. Kula and N. N. Fahey, *Mater. Res. Stand.*, vol. 1, p. 631, 1961.

Table 8-5 Spectrum of strain rate

Range of strain rate	Condition or type test
10^{-8} to 10^{-5} s^{-1}	Creep tests at constant load or stress
10^{-5} to 10^{-1} s^{-1}	"Static" tension tests with hydraulic or screw-driven machines
10^{-1} to 10^{2} s^{-1}	Dynamic tension or compression tests
10^{2} to 10^{4} s^{-1}	High-speed testing using impact bars (must consider wave propagation effects)
10^{4} to 10^{8} s^{-1}	Hypervelocity impact using gas guns or explosively driven projectiles (shock-wave propagation)

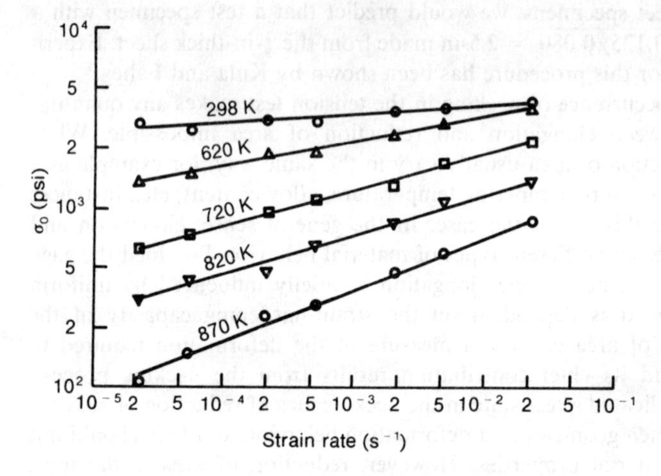

Figure 8-12 Flow stress at $\varepsilon = 0.002$ versus strain rate for 6063-O aluminum alloy. *(From R. Mignogna, R. D'Antonia, C. Maciag and K. Mukherjee, Met. Trans., vol. 1, p. 1771, 1970.)*

in tests on low-carbon steel that do not show a yield point under ordinary rates of loading.

Nadai[1] has presented a mathematical analysis of the conditions existing during the extension of cylindrical specimen with one end fixed and the other attached to the movable crosshead of the testing machine. The crosshead velocity is $v = dL/dt$. The strain rate expressed in terms of conventional linear strain is $\dot{e}$.

$$\dot{e} = \frac{de}{dt} = \frac{d(L - L_0)/L_0}{dt} = \frac{1}{L_0}\frac{dL}{dt} = \frac{v}{L_0} \tag{8-36}$$

Thus, the conventional strain rate is proportional to the crosshead velocity. In a

[1] A. Nadai, "Theory of Flow and Fracture of Solids," vol. I, pp. 74–75, McGraw-Hill Book Company, New York, 1950.

modern testing machine in which the crosshead velocity can be set accurately and controlled, it is a simple matter to carry out tension tests at constant conventional strain rate.

The true strain rate $\dot{\varepsilon}$ is given by

$$\dot{\varepsilon} = \frac{d\varepsilon}{dt} = \frac{d[\ln(L/L_0)]}{dt} = \frac{1}{L}\frac{dL}{dt} = \frac{v}{L} \tag{8-37}$$

The true strain rate is related to the conventional strain rate by the following equation:

$$\dot{\varepsilon} = \frac{v}{L} = \frac{L_0}{L}\frac{de}{dt} = \frac{1}{1+e}\frac{de}{dt} = \frac{\dot{e}}{1+e} \tag{8-38}$$

Equation (8-37) indicates that for a constant crosshead speed the true-strain rate will decrease as the specimen elongates. To maintain a constant true-strain rate using open-loop control the deformation velocity must increase in proportion to the increase in the length of the specimen[1] or must increase as[2]

$$v = \dot{\varepsilon} L_0 \exp(\dot{\varepsilon} t) \tag{8-39}$$

When plastic flow becomes localized or nonuniform along the gage length then open-loop control no longer is satisfactory. Now it is necessary to monitor the instantaneous cross section of the deforming region using closed-loop control. For deformation occurring at constant volume a constant true-strain rate is obtained if the specimen area changes as[3]

$$A = A_0 \exp(-\dot{\varepsilon} t) \tag{8-40}$$

As Fig. 8-12 indicates, a general relationship between flow stress and strain rate, at constant strain and temperature is

$$\sigma = C(\dot{\varepsilon})^m|_{\varepsilon, T} \tag{8-41}$$

where m is known as the *strain-rate sensitivity*. The exponent m can be obtained from the slope of a plot of $\log \sigma$ vs. $\log \dot{\varepsilon}$, like Fig. 8-12. However, a more sensitive way is a rate-change test in which m is determined by measuring the change in flow stress brought about by a change in $\dot{\varepsilon}$ at a constant ε and T (see Fig. 8-13).

$$m = \left(\frac{\partial \ln \sigma}{\partial \ln \dot{\varepsilon}}\right)_{\varepsilon, T} = \frac{\dot{\varepsilon}}{\sigma}\left(\frac{\partial \sigma}{\partial \dot{\varepsilon}}\right)_{\varepsilon, T} = \frac{\Delta \log \sigma}{\Delta \log \dot{\varepsilon}} = \frac{\log \sigma_2 - \log \sigma_1}{\log \dot{\varepsilon}_2 \cdot \log \dot{\varepsilon}_1} = \frac{\log(\sigma_2/\sigma_1)}{\log(\dot{\varepsilon}_2/\dot{\varepsilon}_1)} \tag{8-42}$$

Strain-rate sensitivity of metals is quite low (< 0.1) at room temperature but m increases with temperature, especially at temperatures above half of the absolute melting point. In hot-working conditions m values of 0.1 to 0.2 are common.

Equation (8-41) is not the best description of the strain-rate dependence of flow stress for steels. For these materials a semilogarithmic relationship between

[1] E. P. Lautenschlager and J. O. Brittain, *Rev. Sci. Instrum.*, vol. 39, pp. 1563–1565, 1968.
[2] C. S. Hartley and D. A. Jenkins, *J. Met.*, vol. 32, no. 7, pp. 23–28, 1980.
[3] C. G'Sell and J. J. Jonas, *J. Mater. Sci.*, vol. 14, pp. 583–591, 1979.

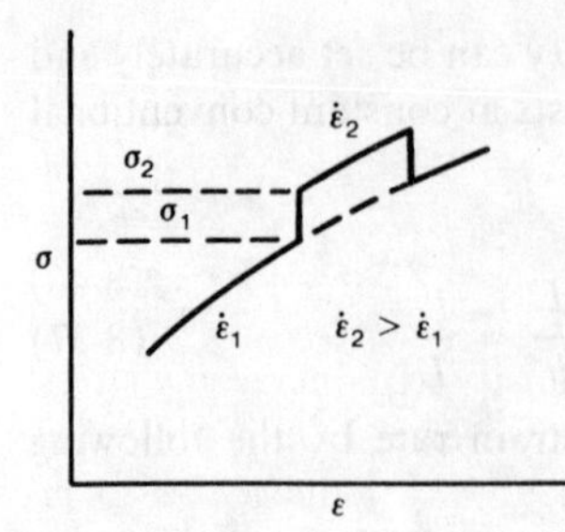

Figure 8-13 Strain-rate change test to determine strain-rate sensitivity.

flow stress and strain rate appears to hold.

$$\sigma = k_1 + k_2 \ln \frac{\dot{\varepsilon}}{\varepsilon_0} \tag{8-43}$$

where k_1, k_2, and ε_0 are constants.

Example The parameters for Eq. (8-41) for commercially pure aluminum are as follows at a true strain of 0.25.

	70°F	825°F
C:	10.2 ksi	2.1 ksi
m:	0.066	0.211

Determine the change in flow stress for a two order of magnitude change in strain rate at each of the temperatures.

At 70°F $\quad \sigma_a = C(\dot{\varepsilon})^m = 10.2(1)^{0.066} = 10.2$ ksi

$\sigma_b = 10.2(100)^{0.066} = 13.8$ ksi $\quad \sigma_b/\sigma_a = 1.35$

At 825°F $\quad \sigma_a = 2.1(1)^{0.211} = 2.10$ ksi

$\sigma_b = 2.1(100)^{0.211} = 5.55$ ksi $\quad \sigma_b/\sigma_a = 2.64$

Strain-rate sensitivity is a good indicator of changes in deformation behavior and measurements of m provide a key link[1] between dislocation concepts of plastic deformation and the more macroscopic measurements made in the tension test. In Sec. 4-6 we saw that the velocity of dislocation motion was very strongly dependent on stress according to

$$v = A\sigma^{m'} \tag{8-44}$$

Moreover, strain rate is related to velocity of mobile dislocations by

$$\dot{\varepsilon} = \rho b v \tag{8-45}$$

From Eqs. (8-42) and (8-45)

$$\frac{1}{m} = \frac{\partial \ln \dot{\varepsilon}}{\partial \ln \sigma} = \frac{\partial \ln v}{\partial \ln \sigma} + \frac{\partial \ln \rho}{\partial \ln \sigma}$$

[1] W. G. Johnston and D. F. Stein, *Acta Metall.*, vol. 11, pp. 317–318, 1963.

and from Eq. (8-44)

$$\frac{\partial \ln v}{\partial \ln \sigma} = m'$$

Therefore,

$$m' = \frac{1}{m} - \frac{\partial \ln \rho}{\partial \ln \sigma} \tag{8-46}$$

Thus, if there is no change in the mobile dislocation density with increasing stress, $m' = 1/m$. This hardly is a reasonable assumption. However, if $1/m$ is plotted as a function of strain, the curve extrapolated to zero gives a value close to m' determined from etch-pit measurements of dislocation velocity.

While strain-rate sensitivity may be quite low in metals at room temperature, in other materials it may be quite appreciable. The extreme case is a newtonian viscous solid, where the flow stress is described by

$$\sigma = \eta \dot{\varepsilon} \tag{8-47}$$

and by comparison with Eq. (8-41), we find that $m = 1$.

High strain-rate sensitivity is a characteristic of superplastic metals and alloys. *Superplasticity*[1] refers to extreme extensibility with elongations usually between 100 and 1,000 percent. Superplastic metals have a grain size or interphase spacing of the order of 1 μm. Testing at high temperature and low strain rates accentuates superplastic behavior. While the mechanism of superplastic deformation is not yet well established, it is clear that the large elongations result from the suppression of necking in these materials with high values of m. An extreme case is hot glass ($m = 1$) which can be drawn from the melt into glass fibers without the fibers necking down.

In a normal metal the geometric softening that constitutes the formation of a neck is opposed by strain hardening, and so long as $d\sigma/d\varepsilon > \sigma$, the tensile specimen will not neck down. With a superplastic material the rate of strain hardening is low (because of the high temperature or structural condition) but necking is prevented by the presence of strain-rate hardening,[2] and $d\sigma/d\varepsilon > \sigma$. Consider a superplastic rod with cross-sectional area A that is loaded with an axial force P.

$$\frac{P}{A} = \sigma = C(\dot{\varepsilon})^m$$

or

$$\dot{\varepsilon} = \left(\frac{P}{C}\right)^{1/m}\left(\frac{1}{A}\right)^{1/m} \tag{8-48}$$

From the definition of true strain rate

$$\dot{\varepsilon} = \frac{1}{L}\frac{dL}{dt} = -\frac{1}{A}\frac{dA}{dt} \tag{8-49}$$

[1] W. A. Backofen, *et al.*, in "Ductility," chap. 10, American Society for Metals, Metals Park, Ohio, 1968; J. W. Edington, K. N. Melton, and C. P. Cutler, "Superplasticity," *Prog. Mater. Sci.*, vol. 21, pp. 61–170, 1976.

[2] D. H. Avery and W. A. Backofen, *Trans. Am. Soc. Met.*, vol. 58, p. 551, 1968; E. W. Hart, *Acta. Metall.*, vol. 15, p. 351, 1967; T. Y. M. Al-Naib and J. L. Duncan, *Int. J. Mech. Sci.*, vol. 12, pp. 463–477, 1970.

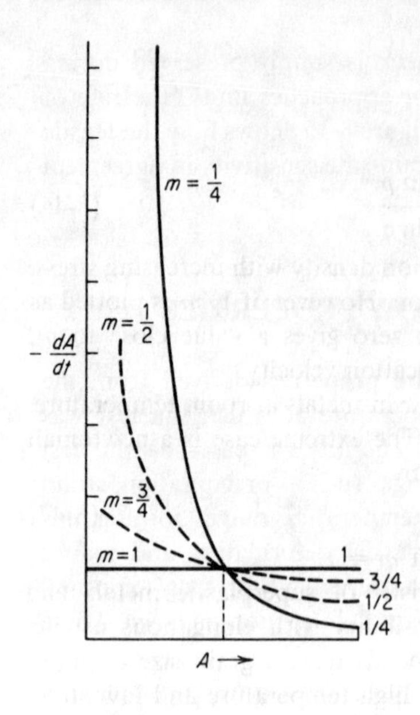

Figure 8-14 Graphical representation of Eq. (8-50). Dependence of rate of decrease of area on cross-sectional area for different values of m.

and combining Eqs. (8-48) and (8-49)

$$-\frac{dA}{dt} = A\dot{\varepsilon} = A^{1-1/m}\left(\frac{P}{C}\right)^{1/m}$$

and
$$-\frac{dA}{dt} = \left(\frac{P}{C}\right)^{1/m}\left(\frac{1}{A^{(1-m)/m}}\right) \tag{8-50}$$

Equation (8-50) states that so long as m is < 1 the smaller the cross-sectional area, the more rapidly the area is reduced. Figure 8-14 shows how the area decrease varies with m. When $m = 1$ the deformation is newtonian viscous and

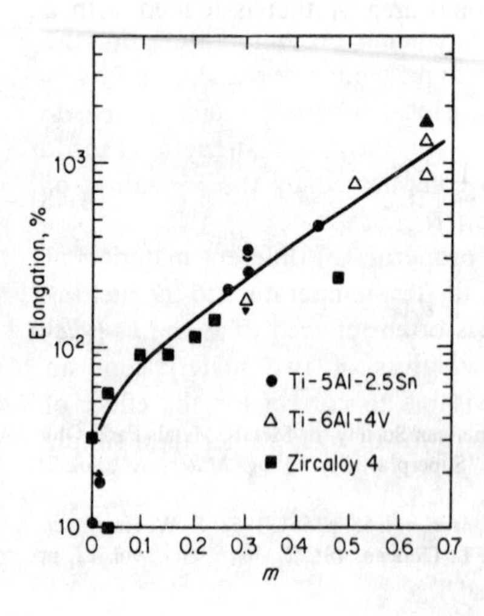

Figure 8-15 Dependence of tensile elongation on strain-rate sensitivity. *(From D. Lee and W. A. Backofen, Trans. Metall. Soc. AIME, vol. 239, p. 1034, 1967.)*

dA/dt is independent of A and any incipient neck is simply preserved during elongation and does not propagate inward. As m approaches unity, the rate of growth of incipient necks is drastically reduced. Figure 8-15 shows how the tensile elongation of superplastic alloys increases with strain-rate sensitivity in agreement with the above analysis.

8-7 EFFECT OF TEMPERATURE ON FLOW PROPERTIES

The stress-strain curve and the flow and fracture properties derived from the tension test are strongly dependent on the temperature at which the test was conducted.[1] In general, strength decreases and ductility increases as the test temperature is increased. However, structural changes such as precipitation, strain aging, or recrystallization may occur in certain temperature ranges to alter this general behavior. Thermally activated processes assist deformation and reduce strength at elevated temperatures. At high temperatures and/or long exposure, structural changes occur resulting in time-dependent deformation or *creep*. These topics are discussed in greater detail in Chap. 13.

The change with temperature of the engineering stress-strain curve in mild steel is shown schematically in Fig. 8-16. Figure 8-17 shows the variation of yield strength with temperature for body-centered cubic tantalum, tungsten, molybdenum, and iron and for face-centered cubic nickel. Note that for the bcc metals the yield stress increases rapidly with decreasing temperature, while for nickel (and other fcc metals) the yield stress is only slightly temperature-dependent. Based on the concept of fracture stress introduced in Sec. 7-11, and especially Fig. 7-17, it is easy to see why bcc metals exhibit brittle fracture at low temperatures. Figure 8-18 shows the variation of reduction of area with temperature for these same metals. Note that tungsten is brittle at 100°C, iron at −225°C, while nickel decreases little in ductility over the entire temperature interval.

In fcc metals flow stress is not strongly dependent on temperature but the strain-hardening exponent decreases with increasing temperature. This results in the stress-strain curve flattening out with increasing temperature and the tensile strength being more temperature-dependent than the yield strength. Tensile deformation at elevated temperature may be complicated by the formation of more than one neck[2] in the specimen.

The best way to compare the mechanical properties of different materials at various temperatures is in terms of the ratio of the test temperature to the melting point, expressed in degrees Kelvin. This ratio is often referred to as the *homologous temperature*. When comparing the flow stress of two materials at an equivalent homologous temperature, it is advisable to correct for the effect of temperature on elastic modulus by comparing ratios of σ/E rather than simple ratios of flow stress.

[1] For test techniques see ASTM E21-69, ASTM, *op. cit.*, pp. 265–270.

[2] P. J. Wray, *J. Appl. Phys.*, vol. 41, pp. 3347–3352, 1970.

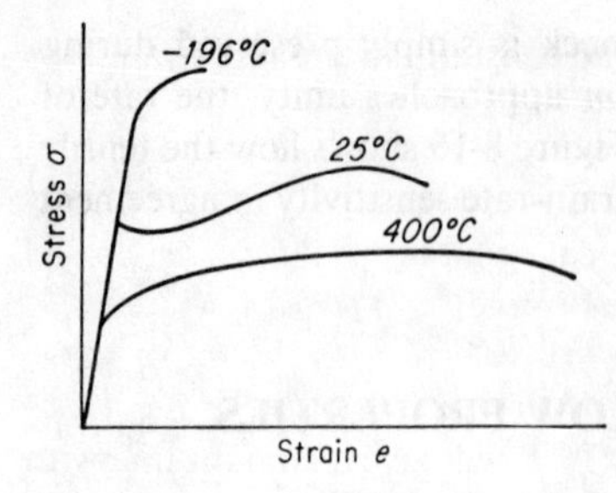

Figure 8-16 Changes in engineering stress-strain curves of mild steel with temperature.

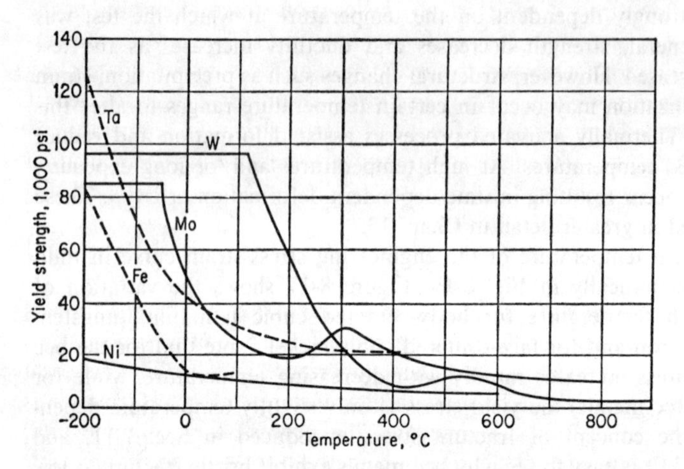

Figure 8-17 Effect of temperature on the yield strength of body-centered cubic Ta, W, Mo, Fe, and face-centered cubic Ni. *(From J. H. Bechtold, Acta Metall., vol. 3, p. 252, 1955.)*

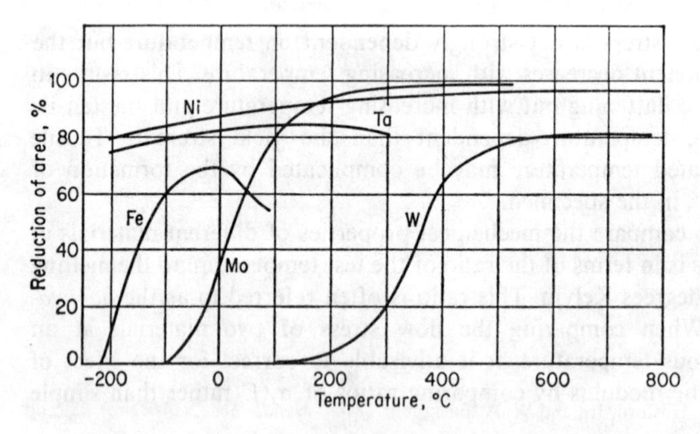

Figure 8-18 Effect of temperature on the reduction of area of Ta, W, Mo, Fe, and Ni. *(From J. H. Bechtold, Acta Metall., vol. 3, p. 253, 1955.)*

The temperature dependence of flow stress at constant strain and strain rate generally can be represented by

$$\sigma = C_2 e^{Q/RT}\Big|_{\varepsilon, \dot{\varepsilon}} \tag{8-51}$$

where Q = an activation energy for plastic flow, cal/g mole
R = universal gas constant, 1.987 cal/(deg)(mole)
T = testing temperature, °K

If this expression is obeyed, a plot of $\ln \sigma$ versus $1/T$ will give a straight line with a slope Q/R.

The value of flow stress depends on the dislocation structure existing at the time the flow stress is measured. Dislocation structure will change with temperature, strain rate, and strain. One way to separate these effects is to evaluate Q by a temperature change test (Fig. 8-22b). The stress-strain curve is conducted at constant strain rate, and at the desired value of plastic strain the temperature is changed from T_2 to T_1. After the temperature comes to equilibrium, the activation energy is given by

$$Q = R \ln\left(\frac{\sigma_1}{\sigma_2}\right)\frac{T_1 T_2}{T_2 - T_1} \tag{8-52}$$

Better than 90 percent of the energy expended in plastic deformation is converted to heat. In many metalworking operations, because of inhomogeneous flow, the deformation is localized and the temperature rises within this local region. Because the flow stress decreases with temperature, further deformation is concentrated preferentially in this zone and the process may continue until fracture occurs. At high rates of deformation there is not time for appreciable heat flow to occur and a near adiabatic condition results. This type of localized fracture is called *adiabatic shear fracture*. It is sometimes observed in the later stages of the tension test, particularly in low-temperature tests.[1] Frequently in low-temperature tests, the adiabatic heating results in a marked drop in the flow curve and serrated stress-strain curves like in Fig. 6-10. An analysis of the conditions to produce instability in the flow curve by adiabatic shear has been given by Backofen.[2] Starting with a uniaxially loaded specimen

$$P = \sigma A \qquad \sigma = f(\varepsilon, \dot{\varepsilon}, T)$$

$$dP = \sigma\, dA + A\, d\sigma \qquad \frac{d\sigma}{d\varepsilon} = \frac{\partial \sigma}{\partial \varepsilon} + \frac{\partial \sigma}{\partial T}\frac{dT}{d\varepsilon} + \frac{\partial \sigma}{\partial \dot{\varepsilon}}\frac{d\dot{\varepsilon}}{d\varepsilon}$$

and since $d\varepsilon = -dA/A$

$$\frac{dP}{d\varepsilon}\frac{1}{A} = -\sigma + \frac{d\sigma}{d\varepsilon}$$

[1] G. Y. Chin, W. F. Hosford, Jr., and W. A. Backofen, *Trans. Metall. Soc. AIME*, vol. 230, pp. 437–449, 1964.

[2] W. A. Backofen, Metallurgical Aspects of Ductile Fracture, in "Fracture of Engineering Materials," pp. 121–124, American Society for Metals, Metals Park, Ohio, 1964.

and

$$dP = A\,d\varepsilon\left(\frac{\partial\sigma}{\partial\varepsilon} + \frac{\partial\sigma}{\partial T}\frac{dT}{d\varepsilon} + \frac{\partial\sigma}{\partial\dot{\varepsilon}}\frac{d\dot{\varepsilon}}{d\varepsilon} - \sigma\right) \tag{8-53}$$

For adiabatic heating $dT/d\varepsilon = \sigma/c\rho$, where c is the specific heat and ρ is the density. For low-temperature deformation the strain-rate dependence of the flow curve can be neglected and an instability (load drop) in the flow curve will occur when[1]

$$\frac{\partial\sigma}{\partial\varepsilon} - \sigma \leq \frac{\partial\sigma}{\partial T}\frac{\sigma}{c\rho} \tag{8-54}$$

Equation (8-54) explains why load drops due to adiabatic heating are more pronounced at low temperatures: specific heat decreases at low temperatures (cryogenic) and there is also a strong temperature-dependence of flow stress in bcc metals

8-8 INFLUENCE OF TESTING MACHINE ON FLOW PROPERTIES

Two general types of machines are used in tension testing: (1) load controlled machines and (2) displacement controlled machines. In load controlled machines the operator adjusts the load precisely but must live with whatever displacement happens to be associated with the load. The older type hydraulic machines are of this type. In the displacement controlled machine the displacement is controlled and the load adjusts itself to that displacement. The popular screw-driven machines in which the crosshead moves to a predetermined constant velocity are of this type. The more recently developed servohydraulic testing machines provide both load or displacement control. These versatile machines are well adapted to computer control.[2] In a nonautomated testing system the servocontrol is limited to control of load, stroke, or strain. However, with modern computer control it is possible to conduct tests based on the control of calculated variables such as true strain or stress intensity factor.

All testing machines deflect under load. Therefore, we cannot directly convert the crosshead motion velocity into deformation of the specimen without appropriate corrections. A constant crosshead velocity testing machine applies a constant total strain rate that is the sum of (1) the elastic strain rate in the specimen, (2) the plastic strain rate in the specimen, and (3) the strain rate resulting from the elasticity of the testing machine. At any instant of time there is some distribution of strain rate between these components. If the crosshead velocity is v, then at a particular time t the total displacement is vt. The force P on the specimen causes an elastic machine displacement P/K. The

[1] W. A. Backofen, Metallurgical Aspects of Ductile Fracture, in "Fracture of Engineering Materials," pp. 121–124, American Society for Metals, Metals Park, Ohio, 1964.
[2] W. D. Cooper and R. B. Zweigoron, *J. Met.*, vol. 32, no. 7, pp. 17–21, 1980.

elastic displacement of the specimen (from Hooke's law) is $\sigma L/E$ and the plastic displacement of the specimen is $\varepsilon_p L$. Since the total displacement is the sum of its components

$$vt = \frac{P}{K} + \frac{\sigma L}{E} + \varepsilon_p L \tag{8-55}$$

Solving for ε_p, we see that the plastic strain taken from a load-time chart on a constant crosshead-velocity testing machine must be corrected for machine stiffness as well as specimen elasticity.

$$\varepsilon_p = \frac{vt}{L} - \frac{\sigma}{E} - \frac{P}{KL} \tag{8-56}$$

Testing machine-specimen interaction has a major influence on the strain rate.[1] If we eliminate time from Eq. (8-55) by treating the stress rate $\dot{\sigma}$ and the strain rate $\dot{\varepsilon}$, then after substituting $P = \sigma A$,

$$\frac{v}{L} = \frac{\dot{\sigma}}{E}\left(\frac{AE}{KL} + 1\right) + \dot{\varepsilon}_p \tag{8-57}$$

and since

$$\dot{\varepsilon} = \dot{\varepsilon}_E + \dot{\varepsilon}_p = \frac{\dot{\sigma}}{E} + \dot{\varepsilon}_p$$

$$\dot{\varepsilon} = \frac{(vK/AE) + \dot{\varepsilon}_p}{(KL/AE) + 1} \tag{8-58}$$

This shows that the specimen strain rate usually will differ from the preset crosshead velocity depending on the rate of plastic deformation and the relative stiffness of the testing machine and the specimen. Experimental values of machine stiffness K are in the range of 40,000 to 180,000 lb/in, while a 0.505-in-diameter steel specimen with 2.5-in reduced section has a spring constant of 2,400,000 lb/in. Thus, except for wire specimens or specially designed testing machines the specimen usually has a much greater spring constant than the testing machine.[2]

The characteristics of the testing machine can have a strong influence on the shape of the stress-strain curve and the fracture behavior.[3] A rigid testing machine with a high spring constant is known as a "hard machine." Screw-driven mechanical machines tend to be hard machines, while hydraulically driven testing machines are "soft machines." A hard testing machine will reproduce faithfully the upper and lower yield point, but in a soft machine these will be smeared out and only the extension at constant load will be recorded.

[1] M. A. Hamstad and P. P. Gillis, *Mater. Res. Stand.*, vol. 6, pp. 569–573, 1966.

[2] J. E. Hockett and P. P. Gillis, *Int. J. Mech. Sci.*, vol. 13, pp. 251–264, 1971.

[3] G. Y. Chin, W. F. Hosford, Jr., and W. A. Backofen, op. cit., This paper discusses fixtures for converting a machine to extra-hard or soft tension tests.

8-9 CONSTITUTIVE EQUATIONS

Constitutive equations describe the relations between stress and strain in terms of the variables of strain rate and temperature. The simple power-law relationship (Eq. 8-21) and its variants are elementary forms of a constitutive equation.

An early concept in developing constituent relationships was the idea[1] that the flow stress depended only on the instantaneous values of strain, strain rate, and temperature.

$$f(\sigma, \varepsilon, \dot{\varepsilon}, T) = 0 \tag{8-59}$$

This is directly analogous to the concept that a thermodynamic system, in equilibrium, can be expressed by the state variables pressure, volume, and temperature.

$$f(P, V, T) = 0 \tag{8-60}$$

Thus, early studies at developing constitutive relations were aimed at representing a *mechanical equation of state*, in the same way that the pressure of an ideal gas depends on an equation of state with the instantaneous values of volume and temperature. However, it was soon realized that plastic deformation is an irreversible process so that ε and $\dot{\varepsilon}$ are not state functions like P, V, and T. Instead, the flow stress depends fundamentally on the dislocation structure which depends chiefly on the "metallurgical history" of strain, strain rate, and temperature.

From Eq. (8-59), the general form of a constitutive equation is given by

$$d\sigma = \left(\frac{\partial\sigma}{\partial\varepsilon}\right)_{\dot{\varepsilon}, T} d\varepsilon + \left(\frac{\partial\sigma}{\partial\dot{\varepsilon}}\right)_{\varepsilon, T} d\dot{\varepsilon} + \left(\frac{\partial\sigma}{\partial T}\right)_{\varepsilon, \dot{\varepsilon}} dT \tag{8-61}$$

In evaluating Eq. (8-61), $(\partial\sigma/\partial\varepsilon)_{\dot{\varepsilon}, T}$ is obtained from a constant true strain-rate test (Sec. 8-6). The second term $(\partial\sigma/\partial\dot{\varepsilon})_{\varepsilon, T}$ can be obtained from strain-rate change tests or the stress-relaxation test described in Sec. 8-11. The last term $(\partial\sigma/\partial T)_{\varepsilon, \dot{\varepsilon}}$ is obtained from instantaneous temperature changes during the tension test.

A form of Eq. (8-61) which has been shown[2] to be valid to large strains is

$$\frac{\dot{\sigma}}{\sigma} = \dot{\varepsilon}\gamma + \left(\frac{\ddot{\varepsilon}}{\dot{\varepsilon}}\right)m + \dot{T}\frac{(\partial\sigma/\partial T)}{\sigma} \tag{8-62}$$

where $\dot{\sigma} = \partial\sigma/\partial t$
$\dot{\varepsilon} = \partial\varepsilon/\partial t$
$\ddot{\varepsilon} = \partial\dot{\varepsilon}/\partial t$
$\gamma = (\partial\sigma/\partial\varepsilon)/\sigma$
$m = (\dot{\varepsilon}/\sigma)(\partial\sigma/\partial\dot{\varepsilon})$
$\dot{T} = \partial T/\partial t$

Yet another form of constitutive relations developed for computer representation

[1] J. Holloman, *Trans. AIME*, vol. 171, p. 355, 1947.

[2] C. S. Hartley and R. Srinivasan, *Trans. ASME J. Eng. Mater. Tech.*, vol. 105, pp. 162–167, 1983.

of high-temperature deformation is called MATMOD.[1] Ghosh[2] has developed a set of constitutive relations that describe the constant true-strain-rate tension test, the strain-rate change test, the stress-relaxation test, and the creep test.

A much simpler but useful relation describes the combined temperature and strain-rate dependence of flow stress.

$$\sigma = f(Z) = f(\dot{\varepsilon} e^{\Delta H/RT})\Big|_{\varepsilon} \tag{8-63}$$

where ΔH is an activation energy, calorie per mole, that is related to the activation energy Q in Eq. (8-51) by $Q = m\,\Delta H$, where m is the strain-rate sensitivity. The quantity Z is called the *Zener-Hollomon parameter*.[3] It may also be referred to as a *temperature-modified strain rate*.

$$Z = \dot{\varepsilon} e^{\Delta H/RT} \tag{8-64}$$

A relationship which is particularly useful for correlating stress, temperature, and strain rate under hot-working conditions was proposed by Sellars and Tegart.[4]

$$\dot{\varepsilon} = A(\sinh \alpha\sigma)^{n'} e^{-Q/RT} \tag{8-65}$$

where A, α, and n' are experimentally determined constants and Q is an activation energy. At low stresses ($\alpha\sigma < 1.0$), Eq. (8-65) reduces to a power relation such as is used to describe creep behavior

$$\dot{\varepsilon} = A_1 \sigma^{n'} e^{-Q/RT} \tag{8-66}$$

and at high stresses ($\alpha\sigma > 1.2$) it reduces to an exponential relation

$$\dot{\varepsilon} = A_2 \exp(\beta\sigma) e^{-Q/RT} \tag{8-67}$$

The constants α and n' are related by $\beta = \alpha n'$, so that α and n' can be simply determined from tests at high and low stresses. If Eq. (8-65) is followed, the data will plot as a series of parallel straight lines (one for each temperature) when $\dot{\varepsilon}$ is plotted against $\sinh \alpha\sigma$ on log-log coordinates. The activation energy Q can be obtained from a plot of $\log \dot{\varepsilon}$, at constant $\sinh \alpha\sigma$, against $1/T$. Equation (8-66) has been shown to be valid[5] over 15 decades of $\dot{\varepsilon} \exp(-Q/RT)$.

8-10 FURTHER CONSIDERATION OF INSTABILITY

The discussion of necking in the tension test in Sec. 8-3 assumed that the flow stress depended only on strain. In this section we reexamine plastic instability in the tension test for a material that is strain-rate sensitive.[6]

[1] C. G. Schmidt and A. K. Miller, *Res. Mech*., vol. 3, pp. 108–193, 1981.

[2] A. K. Ghosh, *Acta Met*., vol. 28, pp. 1443–1465, 1980.

[3] C. Zener and J. H. Holloman, *J. Appl. Phys*., vol. 15, pp. 22–32, 1944.

[4] C. M. Sellars and W. J. McG. Tegart, *Mem. Sci. Rev. Metall*., vol. 63, p. 731, 1966; see also W. J. McG. Tegart, in "Ductility," chap. 5, American Society for Metals, Metals Park, Ohio, 1968.

[5] J. J. Jonas, *Trans. Q. ASM*, vol. 62, pp. 300–303, 1969.

[6] E. W. Hart, *Acta Met*., vol. 15, p. 351, 1967; E. W. Hart, *Trans. ASME, J. Eng. Metal. Tech*., vol. 98, p. 193, 1976; S. L. Semiatin and J. J. Jonas, "Formability and Workability of Metals: Plastic Instability and Flow Localization", chap. 5, American Society for Metals, Metals Park, Ohio, 1984.

Consider a tensile specimen loaded to a value P. At any point a distance L along the specimen, the cross-sectional area is A and $P = \sigma A$. Since P does not vary along the length of the specimen, and $\sigma = f(\varepsilon, \dot{\varepsilon})$

$$\frac{dP}{dL} = 0 = A\left\{\left(\frac{\partial \sigma}{\partial \varepsilon}\right)_{\dot{\varepsilon}} \frac{d\varepsilon}{dL} + \left(\frac{\partial \sigma}{\partial \dot{\varepsilon}}\right)_{\varepsilon} \frac{d\dot{\varepsilon}}{dL}\right\} + \sigma \frac{dA}{dL} \tag{8-68}$$

Because the volume of the specimen remains constant, the true strain can be written as

$$d\varepsilon = \frac{dL}{L} = -\frac{dA}{A} \tag{8-69}$$

and

$$\frac{d\varepsilon}{dL} = -\frac{1}{A}\frac{dA}{dL} \tag{8-70}$$

Also, from Eq. (8-69) we can express the strain rate $\dot{\varepsilon}$ by

$$d\dot{\varepsilon} = \frac{d\varepsilon}{dt} = -\frac{1}{A}\frac{dA}{dt} = -\frac{\dot{A}}{A} \tag{8-71}$$

so that

$$\frac{d\dot{\varepsilon}}{dL} = -\frac{1}{A}\frac{d\dot{A}}{dL} + \frac{\dot{A}}{A^2}\frac{dA}{dL} \tag{8-72}$$

The material quantities which are important to the necking process are:

$$\text{dimensionless strain-hardening coefficient:} \qquad \gamma = \frac{1}{\sigma}\frac{\partial \sigma}{\partial \varepsilon} \tag{8-73}$$

$$\text{strain-rate sensitivity:} \qquad m = \left(\frac{\partial \ln \sigma}{\partial \ln \dot{\varepsilon}}\right)_{\varepsilon} = \frac{\dot{\varepsilon}}{\sigma}\left(\frac{\partial \sigma}{\partial \dot{\varepsilon}}\right)_{\varepsilon} \tag{8-74}$$

When Eqs. (8-71) and (8-72) are substituted into Eq. (8-68) and the definitions for γ and m are added through Eqs. (8-73) and (8-74) the result is

$$\frac{dA}{dL}(\sigma - m\sigma - \gamma\sigma) = \frac{d\dot{A}}{dL}\frac{m\sigma A}{\dot{A}} \tag{8-75}$$

The final rearrangement gives

$$\frac{1/\dot{A}(d\dot{A}/dL)}{1/A(dA/dL)} = \frac{[d(\ln \dot{A})]/dL}{[d(\ln A)]/dL} = \frac{m + \gamma - 1}{m} \tag{8-76}$$

This equation describes the rate of change of area with length and gives the criterion for the onset of necking.

Any real tension specimen will have variations in cross-sectional area along its length. These can arise from an intentional taper, from machining errors, or from heterogeneities of structure which lead to weaker cross sections as a result of plastic deformation.[1] Deformation becomes unstable when the smallest cross

[1] I. H. Lin, J. P. Hirth, and E. W. Hart, *Acta Met.*, vol. 29, pp. 819–829, 1981.

section of the specimen shrinks faster than the rest. This occurs when $d\dot{A}/dA > 0$. Deformation will be uniform and stable so long as $d\dot{A}/dA < 0$. Since $\dot{A}/A$ is negative in tension, stable deformation in tension occurs when $d\dot{A}/dA \geq 0$. Therefore, from Eq. (8-76)

$$\frac{d\dot{A}}{dA} \geq 0 = m + \gamma - 1$$

and

$$\gamma + m \geq 1 \tag{8-77}$$

Equation (8-77) shows that both strain hardening and strain-rate hardening contribute to suppressing the onset of necking. For room temperature deformation, where $m \to 0$, the instability criterion reduces to $\gamma \geq 1$ or $(1/\sigma)(d\sigma/d\varepsilon) \geq 1$, which gives the familiar necking criterion $d\sigma/d\varepsilon \geq \sigma$. At the other extreme, for newtonian viscous flow without strain hardening, $\gamma = 0$ and $m \geq 1$.

8-11 STRESS-RELAXATION TESTING

In a stress-relaxation test the strain is held constant at a fixed temperature and the decrease in stress due to rearrangements in the material is monitored as a function of time. Stress-relaxation testing has been used to establish fundamental information on the deformation of metal by measuring such factors as internal stresses,[1] activation volume (Sec. 8-12), and the exponent in the dislocation velocity-stress relationship (Sec. 8-6). At the level of engineering design, stress relaxation is important in dealing with the relaxation of stress in bolted joints and press-fitted assemblies.[2]

Stress relaxation testing is best conducted on a hard testing machine, but even so, the influence of the testing machine stiffness must be considered.[3] Figure 8-19 shows a specimen loaded in tension. At time $t = 0$ the crosshead is stopped and the relaxation in stress is measured as a function of time. Up until $\varepsilon = \varepsilon_0$ the crosshead velocity is matched by the elastic and plastic deformation rate in the specimen and the elastic extension rate of the testing machine.

$$v = \dot{L}_E + \dot{L}_P + \dot{L}_M$$

When the crosshead stops, and stress relaxation begins, $v = 0$. If we express the stiffness of the testing machine by $K_M = dP/dL_M$, then on stress relaxation we can write

$$\dot{L}_E + \dot{L}_P + \frac{\dot{P}}{K_M} = 0 \tag{8-78}$$

[1] I. Gupta and J. C. M. Li, *Met. Trans.*, vol. 1, p. 2323, 1970; H. Conrad, *Mater. Sci. Engr.*, vol. 2, p. 265, 1970.

[2] Procedures for conducting stress-relaxation tests will be found in ASTM Standard E328. Data on stress-relaxation properties are given by M. J. Manjoine and H. R. Voorhees, "Compilation of Stress Relaxation Data for Engineers," ASTM, Philadelphia, 1982.

[3] F. Guiu and P. L. Pratt, *Phys. Stat. Sol.*, vol. 6, pp. 111–120, 1964.

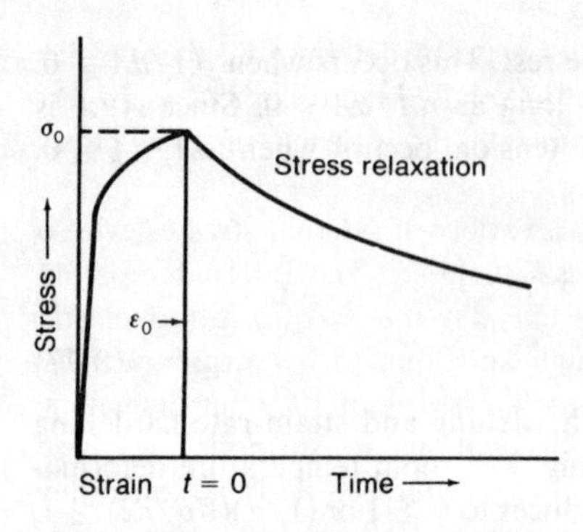

Figure 8-19 Relaxation curve determined at σ_0, ε_0.

If we divide Eq. (8-78) by the gage length L_0 at $\varepsilon = \varepsilon_0$, we can express (8-78) in terms of strain rate $\dot{\varepsilon}$ and stress rate $\dot{\sigma} = \dot{P}/A_0$.

$$\dot{\varepsilon}_E + \dot{\varepsilon}_P = -\frac{A_0}{K_M L_0}\dot{\sigma} \tag{8-79}$$

But the elastic strain rate is given by $\dot{\varepsilon}_E = \dot{\sigma}/E$, so that

$$\dot{\varepsilon}_P = -\left(\frac{A_0}{K_M L_0} + \frac{1}{E}\right)\dot{\sigma} = -M\dot{\sigma} \tag{8-80}$$

A given point on the stress-relaxation curve establishes $\dot{\sigma} = d\sigma/dt$. From Eq. (8-80) the plastic strain rate can be readily determined. With these two quantities the strain-rate sensitivity can be determined through Eq. (8-42).

Alternatively, since $m = d \ln \sigma / d \ln \dot{\varepsilon}$, and $\dot{\sigma} = k\dot{\varepsilon}$

$$d \ln \dot{\sigma} = d \ln \dot{\varepsilon}, \quad \text{so that}$$

$$m = d \ln \sigma / d \ln \dot{\sigma} \tag{8-81}$$

Therefore, if $\ln \sigma$ is plotted against $\ln \dot{\sigma}$ (both quantities are readily obtained from the stress-relaxation curve) then the slope of the curve at any point is the strain-rate sensitivity.

Note from Eq. (8-80) that different stress rates will be obtained for machines of different stiffness. For the same initial strain rate a soft machine (low K_M) will require a smaller value of stress rate.

8-12 THERMALLY ACTIVATED DEFORMATION

Plastic flow, especially at elevated temperature, can be considered a thermally activated process.[1] Experimental study of the activation energy of plastic deformation and its dependence on stress, temperature, deformation, strain rate,

[1] H. Conrad, *J. Metals.*, vol. 16, pp. 582–885, 1964; U. F. Kochs, A. S. Argon, and M. F. Ashby, Thermodynamics and Kinetics of Slip, "Progress in Materials Science," vol. 19, Pergamon Press, London, 1975; A. S. Krausz and H. Eyring, "Deformation Kinetics," Wiley-Interscience, New York, 1975.

impurity concentration, etc., is one of the chief ways of determining the validity of theoretical models for dislocation processes in solids. Macroscopic tension test measurements of the temperature and strain-rate dependence of flow stress provide a direct link with dislocation behavior.

The applied shear stress[1] τ is opposed by a variety of internal stresses, whose sum is given by τ_i. The effective shear stress is $\tau - \tau_i$. The internal resisting stresses can be grouped into two categories (1) *long-range obstacles* to plastic deformation, which represent barriers too high and long to be surmounted by thermal fluctuations, and (2) *short-range obstacles*, less than 10 atom diameters, for which thermal fluctuations can assist dislocations in surmounting these barriers. The first type of barriers are referred to as *athermal barriers* since they produce long-range internal stress fields which are not affected by temperature and strain rate except for the variation of flow stress due to the decrease in shear modulus[2] μ with increasing temperature. Accordingly, the contribution to the internal stress field from long-range obstacles is τ_μ. Short-range obstacles are called *thermal barriers*, and their contribution to the internal stress field is τ^*. τ^* is a strong function of temperature and strain rate. Major contributions to τ_μ arise from the stress fields of other dislocations and of large incoherent precipitates and massive second-phase particles. Basically, the level of τ_μ depends on the composition, heat treatment, and dislocation structure of the material. Contributions to τ^* come from many sources, such as the Peierls-Nabarro force, stress fields of coherent precipitates and solute atoms, cross slip, dislocation climb, and intersections of dislocations.

The situation described above is illustrated schematically by Fig. 8-20. Short-range stress fields are superimposed on a fluctuating long-range stress of wave-length λ. The magnitude of this curve will depend on direction in the lattice and will vary with plastic strain and strain rate. The total internal stress is given by

$$\tau_i = \tau_\mu + \tau^* \tag{8-82}$$

A positive value of τ_i opposes dislocation motion, while a negative τ_i assists motion of dislocations. The rate-controlling process will consist of overcoming the strongest short-range obstacle situated near the top of the opposing long-range stress field. At absolute zero, where thermal fluctuations do not exist, the applied stress τ would have to equal a value equivalent to the very top of the spike or curve to cause plastic flow at a value τ_0. However, at some temperature greater than 0°K, thermal fluctuations will assist the applied stress and plastic flow can occur at a stress τ which is less than τ_0. The energy that must be supplied by thermal fluctuations is shown shaded in Fig. 8-20. As the temperature is increased τ can decrease as more assist is received from thermal fluctuations, until, in the

[1] Shear stress and shear strain are usually obtained from tension test data by the approximation $\tau = \sigma/2$ and $\gamma = 1.4\varepsilon$.

[2] Elsewhere we have used G to represent shear modulus but the British usage of μ is so common to this subject that it has been retained.

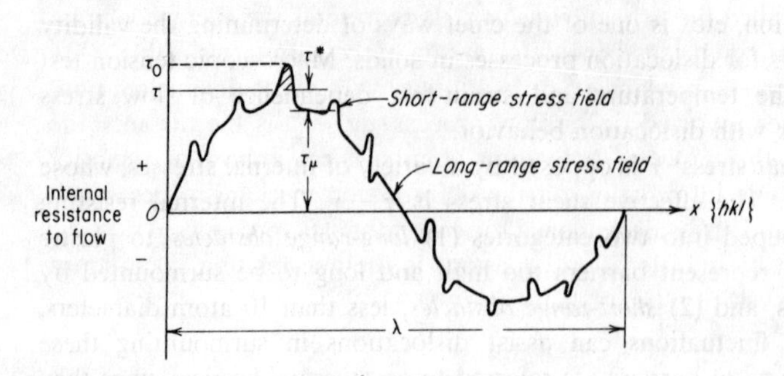

Figure 8-20 Schematic representation of long-range and short-range stress fields. *(After Conrad.)*

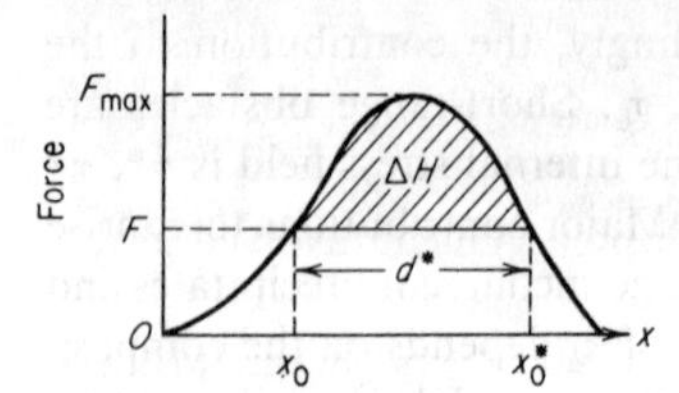

Figure 8-21 Thermal activation barrier.

limit, $\tau = \tau_\mu$, and all of the stress required to overcome the short-range stress field is supplied by thermal fluctuations. The establishment of the flat portion of the stress-temperature curve when $\tau = \tau_\mu$ was the original way of determining τ_μ, although a number of other techniques have been used since.[1]

Seeger[2] has considered the short-range barrier in detail for the common type of barrier where the activation energy is not a function of τ^*. This condition applies to the intersection of dislocations and the movement of jogs. Consider the barrier shown in Fig. 8-21. On the application of a stress $\tau > \tau_\mu$ a dislocation moves up the force barrier to a level $F = \tau^* \mathbf{b} l^*$, where $\mathbf{b}$ is the Burgers vector and l^* is the length of the dislocation segment involved in the thermal fluctuation. ΔH is the energy which must be supplied to overcome the barrier. The work done by the applied stress during thermal activation is

$$W = \tau^* \mathbf{b}(x_0^* - x_0) = \tau^* \mathbf{b} d^* \tag{8-83}$$

The term ΔH is the area under the force-distance curve from x_0 to x_0^* (designated ΔH^*) minus the work done by the applied stress during the thermal activation.

$$\Delta H = \int_{x_0}^{x_0^*} [F(x) - \tau^* \mathbf{b} l^*]\, dx = \Delta H^* - v^* \tau^* \tag{8-84}$$

The term ΔH^* represents the activation energy for zero applied stress.

[1] H. Conrad, *Mater Sci. Eng.*, vol. 6, pp. 265–273, 1970.
[2] A. Seeger, *Philos. Mag.*, vol. 1, p. 651, 1956.

The term v^* is called the *activation volume*. It represents the average volume of dislocation structure involved in the deformation process. In terms of Fig. 8-21 $v^* = l^*\mathbf{b}d^*$ for a process in which l^* does not vary with stress. In this relation, $l^*\mathbf{b}$ is the atomic area involved in the flow process and d^* is the distance the atoms move during this process. Activation volume is an important quantity which can give valuable information on the deformation mechanism because it has a definite value and stress dependence for each atomistic process.

One of the most basic dislocation equations is the one relating shear-strain rate with dislocation velocity, Eq. (4-12).

$$\dot{\gamma} = \rho \mathbf{b} \bar{v}$$

The strong temperature dependence of shear-strain rate can be expressed as

$$\dot{\gamma} = Ae^{-\Delta G/kT} = \rho \mathbf{b} s \nu^* e^{-\Delta G/kT} \tag{8-85}$$

where ΔG = the change in Gibbs free energy of the system
A = the overall frequency factor
ν^* = the frequency of vibration of the dislocation segment involved in the thermal activation process
s = the average distance a dislocation moves after every successful thermal fluctuation

Schoek[1] has shown that the activation volume v^* is related to Eq. (8-85) according to

$$v^* = kT\left[\frac{\partial \ln(\dot{\gamma}/A)}{\partial \tau}\right]_T = -\left(\frac{\partial \Delta G}{\partial \tau}\right)_T \tag{8-86}$$

The main object of the theory is to identify the mechanism which controls the thermally activated deformation process by comparing experimental values of ΔG, v^*, and A with values predicted from specific dislocation models. Most of the work in the literature neglects entropy changes and uses ΔH instead of ΔG. This results in a different formulation for the above equations.[2] The two approaches have been compared in detail[3] and found to be equally valid. The parameters are evaluated from macroscale measurements in tensile or creep (strain vs. time at constant stress) using incremental change tests to determine the relationships

$$\left(\frac{\partial \ln \dot{\gamma}}{\partial \tau}\right)_T \quad \left(\frac{\partial \tau}{\partial T}\right)_{\dot{\gamma}} \quad \text{and} \quad \left(\frac{\partial \ln \dot{\gamma}}{\partial T}\right)_\tau$$

at nearly constant dislocation structure. The procedure for determining these functions is shown in Fig. 8-22.

[1] G. Schoek, *Phys. Status Solidi*, vol. 8, pp. 499–507, 1965.

[2] Z. S. Basinski, *Acta Metall.*, vol. 5, p. 684, 1957; H. Conrad and H. Weidersich, *Acta Metall.*, vol. 8, p. 128, 1960.

[3] J. P. Hirth and W. D. Nix, *Phys. Status Solidi*, vol. 35, pp. 177–188, 1969.

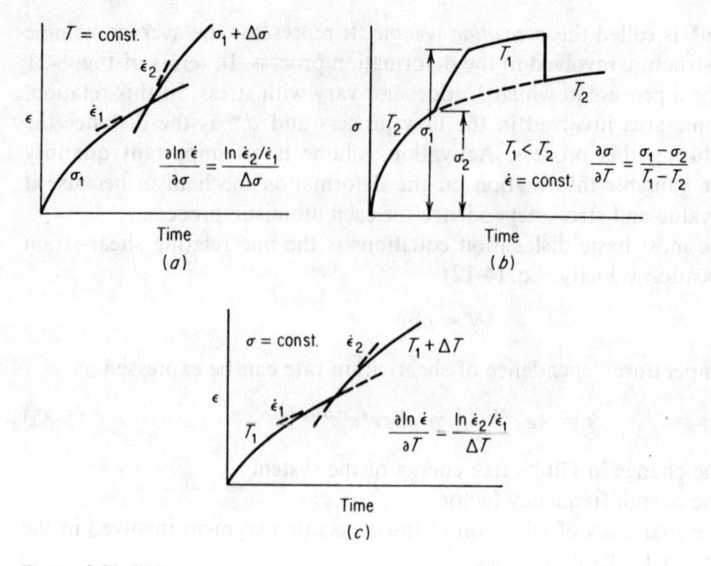

Figure 8-22 Differential tests to determine deformation parameters at constant structure.

8-13 NOTCH TENSILE TEST

Ductility measurements on standard smooth tensile specimens do not always reveal metallurgical or environmental changes that lead to reduced local ductility. The tendency for reduced ductility in the presence of a triaxial stress field and steep stress gradients (such as arise at a notch) is called *notch sensitivity*. A common way of evaluating notch sensitivity is a tension test using a notched specimen. The notch tensile test has been used extensively for investigating the properties of high-strength steels, for studying hydrogen embrittlement in steels and titanium, and for investigating the notch sensitivity of high-temperature alloys. More recently notched tension specimens have been used for fracture mechanics measurements (see Sec. 11-5). Notch sensitivity can also be investigated with the notched-impact test as described in Chap. 14.

The most common notch tensile specimen[1] uses a 60° notch with a root radius 0.001 in or less introduced into a round (circumferential notch) or flat (double-edge notch) tensile specimen. Usually the depth of the notch is such that the cross-sectional area at the root of the notch is one-half of the area in the unnotched section. The specimen is carefully aligned and loaded in tension until failure. The *notch strength* is defined as the maximum load divided by the original cross-sectional area at the notch. Because of the plastic constraint at the notch (see Sec. 7-10), this value will be higher than the tensile strength of an unnotched specimen if the material possesses some ductility. Therefore, the common way of

[1] Standard Method of Sharp-Notch Tension Testing of High-Strength Sheet Material, ASTM E338-68 op. cit., pp. 847–854.

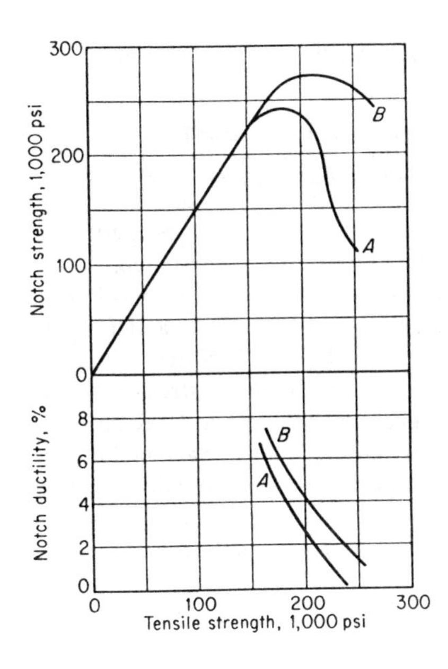

Figure 8-23 Notch tensile properties of two steels. Steel *A* has higher notch sensitivity than steel *B*.

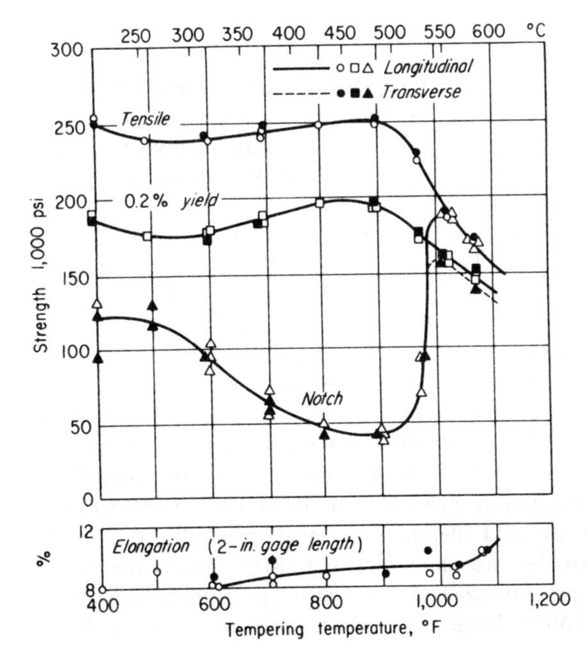

Figure 8-24 Notched and unnotched tensile properties of an alloy steel as a function of tempering temperature. *(From G. B. Espey, M. H. Jones, and W. F. Brown, Jr., Am. Soc. Test. Mater Proc., vol. 59, p. 837, 1959.)*

detecting notch brittleness (or high notch sensitivity) is by determining the *notch-strength ratio* NSR.

$$\text{NSR} = \frac{s_{\text{net}} \text{ (for notched specimen at maximum load)}}{s_u \text{ (tensile strength for unnotched specimen)}} \tag{8-87}$$

If the NSR is less than unity the material is notch brittle. The other property that is measured in the notch tensile test is the reduction of area at the notch.

As strength, hardness, or some metallurgical variable restricting plastic flow increases, the metal at the root of the notch is less able to flow, and fracture becomes more likely. A typical behavior is shown in Fig. 8-23. Notch brittleness may be considered to begin at the strength level where the notch strength begins to fall off, or more conventionally, at the strength level where the NSR becomes less than unity. The sensitivity of notch strength for detecting metallurgical embrittlement is illustrated in Fig. 8-24. Note that the conventional elongation measured on a smooth specimen was unable to detect the fall off in notch strength produced by tempering in the 600 to 900°F range. For a more detailed review of notch tensile testing the reader is referred to Lubahn.[1]

8-14 TENSILE PROPERTIES OF STEEL

Because of the commercial importance of ferrous materials, a great deal of work has been done in correlating their tensile properties with composition and microstructure. Moreover, from a scientific viewpoint the Fe-C system exhibits a great versatility in structure and properties. It has been clearly demonstrated that structure rather than composition *per se* is the chief variable which controls the properties of steel.

The tensile properties of annealed and normalized steels are controlled by the flow and fracture characteristics of the ferrite and by the amount, shape, and distribution of the cementite. The strength of the ferrite depends on the amount of alloying elements in solid solution and the ferrite grain size.[2] The carbon content has a very strong effect because it controls the amount of cementite present either as pearlite or as spheroidite. The strength increases and ductility decreases with increasing carbon content because of the increased amount of cementite in the microstructure. A normalized steel will have higher strength than an annealed steel because the more rapid rate of cooling used in the normalizing treatment causes the transformation to pearlite to occur at a lower temperature, and a finer pearlite spacing results. Differences in tensile properties due to the shape of the cementite particles are shown in Fig. 8-25, where the tensile properties of a spheroidized structure are compared with a pearlitic structure for a steel with the same carbon content. Empirical correlations have been worked out[3]

[1] J. D. Lubahn, *Trans. ASME*, vol. 79, pp. 111–115, 1957.
[2] C. E. Lacy and M. Gensamer, *Trans. Am. Soc. Met.*, vol. 32, p. 88, 1944.
[3] I. R. Kramer, P. D. Gorsuch, and D. L. Newhouse, *Trans. AIME*, vol. 172, pp. 244–272, 1947.

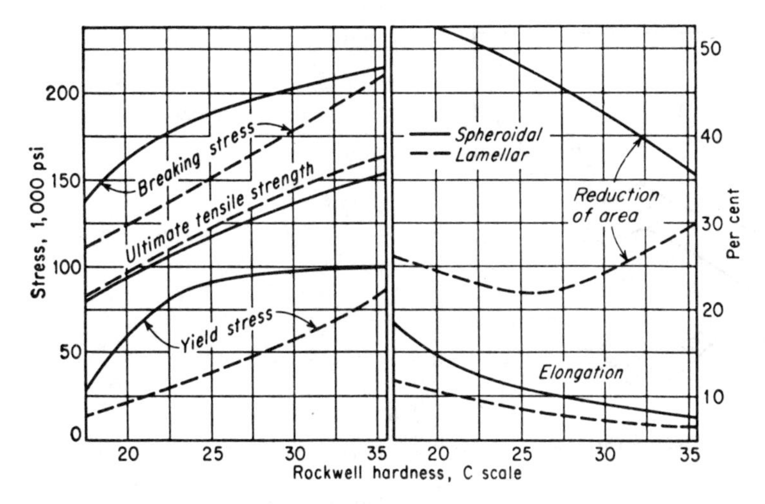

Figure 8-25 Tensile properties of pearlite and spheroidite in eutectoid steel. *(From E. C. Bain and H. W. Paxton, "Alloying Elements in Steel," 2d ed., p. 38, American Society of Metals, Metals Park, Ohio. Copyright American Society for Metals, 1961.)*

between composition and cooling rate for predicting the tensile properties of steels with pearlitic structures.

One of the best ways to increase the strength of annealed steel is by cold-working. Table 8-6 gives the tensile properties which result from the cold reduction of SAE 1016 steel bars by drawing through a die.

The pearlitic structure in steel can be controlled best by transforming the austenite to pearlite at a constant temperature instead of allowing it to form over a range of temperature on continuous cooling from above the critical tempera-

Table 8-6 Effect of cold-drawing on tensile properties of SAE 1016 steel†

Reduction of area by drawing, %	Yield strength, psi	Tensile strength, psi	Elongation, in 2 in, %	Reduction of area, %
0	40,000	66,000	34	70
10	72,000	75,000	20	65
20	82,000	84,000	17	63
40	86,000	95,000	16	60
60	88,000	102,000	14	54
80	96,000	115,000	7	26

† L. J. Ebert, "A Handbook on the Properties of Cold Worked Steels," PB 121662, Office of Technical Services, U.S. Department of Commerce, 1955.

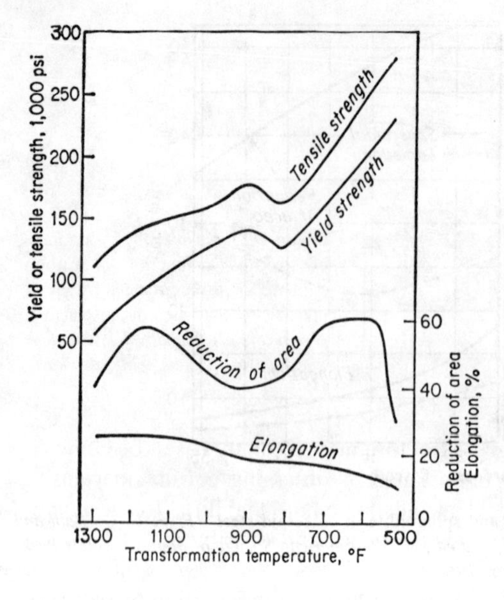

Figure 8-26 Relationship of tensile properties of Ni-Cr-Mo steel to isothermal-transformation temperature. *(From E. S. Davenport, Trans. AIME, vol. 209, p. 684, 1961.)*

ture. Although isothermal transformation is not in widespread commercial use, it is a good way of isolating the effect of certain microstructures on the properties of steel. Figure 8-26 shows the variation of tensile properties for a Ni-Cr-Mo eutectoid steel with isothermal-reaction temperature.[1] This is an extension of Gensamer's work,[2] which showed that the tensile strength varied linearly with the logarithm of the mean free ferrite path in isothermally transformed structures. In the region 1300 to 1000°F the transformation product is lamellar pearlite. The spacing between cementite platelets decreases with transformation temperature and correspondingly the strength increases. In the region 800 to 500°F the structure obtained on transformation is acicular bainite. The bainitic structure becomes finer with decreasing temperature, and the strength increases almost linearly to quite high values. Good ductility accompanies this high strength over part of the bainite temperature range. This is the temperature region used in the commercial heat-treating process known as austempering. The temperature region 1000 to 800°F is one in which mixed lamellar and acicular structures are obtained. There is a definite ductility minimum and a leveling off of strength for these structures. The sensitivity of the reduction of area to changes in microstructure is well illustrated by these results.

The best combination of strength and ductility is obtained in steel which has been quenched to a fully martensitic structure and then tempered. The best criterion for comparing the tensile properties of quenched and tempered steels is on the basis of an as-quenched structure of 100 percent martensite. However, the

[1] E. S. Davenport, *Trans. AIME*, vol. 209, pp. 677–688, 1957.

[2] M. Gensamer, E. B. Pearsall, W. S. Pellini, and J. R. Low, *Trans. Am. Soc. Met.*, vol. 30, pp. 983–1020, 1942.

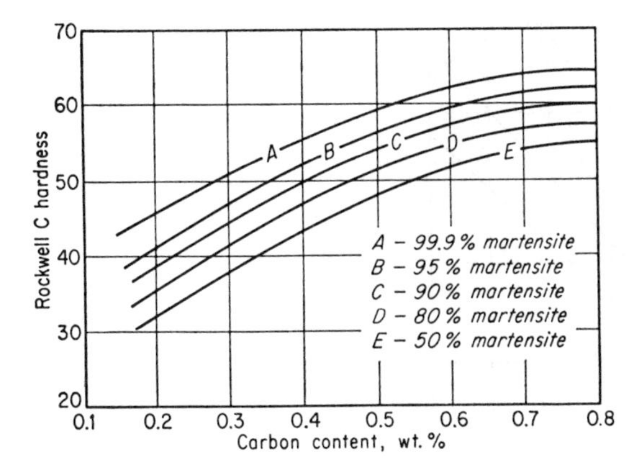

Figure 8-27 As-quenched hardness of steel as a function of carbon content for different percentages of martensite in the microstructure. *(From Metals Handbook, American Society for Metals, Metals Park, Ohio, p. 497, 1948.)*

attainment of a completely martensitic structure may, in many cases, be commercially impractical. Because of the importance of obtaining a fully martensitic structure, it is desirable that the steel have adequate hardenability. *Hardenability*, the property of a steel which determines the depth and distribution of hardness induced by quenching, should be differentiated from *hardness*, which is the property of a material which represents its resistance to indentation or deformation. (This subject is discussed in Chap. 9.) Hardness is associated with strength, while hardenability is connected with the transformation characteristics of a steel. Hardenability may be increased by altering the transformation kinetics by the addition of alloying elements, while the hardness of a steel with given transformation kinetics is controlled primarily by the carbon content. Figure 8-27 shows the hardness of martensite as a function of carbon content for different total amounts of martensite in the microstructure. These curves can be used to determine whether or not complete hardening was obtained after quenching. Hardness is used as a convenient measure of the strength of quenched and tempered steels. The validity of this procedure is based on the excellent correlation which exists between tensile strength and hardness for heat-treated, annealed, and normalized steels (Fig. 8-28).

The mechanical properties of a quenched and tempered steel may be altered by changing the tempering temperature. Figure 8-29 shows how hardness and the tensile properties vary with tempering temperature for an SAE 4340 steel. This is the typical behavior for heat-treated steel. Several methods for correlating and predicting the hardness change in different steels with tempering temperature have been proposed.[1-3] In using tempering diagrams like Fig. 8-29, it is important to know whether or not the data were obtained on specimens quenched to essentially 100 percent martensite throughout the entire cross section of the specimen. Because of the variability in hardenability from heat to heat, there is no assurance of reproducibility of data unless this condition is fulfilled.

[1] J. H. Hollomon and L. D. Jaffe, *Trans. Metall. Soc. AIME*, vol. 162, p. 223, 1945.
[2] R. A. Grange and R. W. Baughman, *Trans. Am. Soc. Met.*, vol. 48, pp. 165–197, 1956.
[3] L. D. Jaffe and E. Gordon, *Trans. Am. Soc. Met.*, vol. 49, pp. 359–371, 1957.

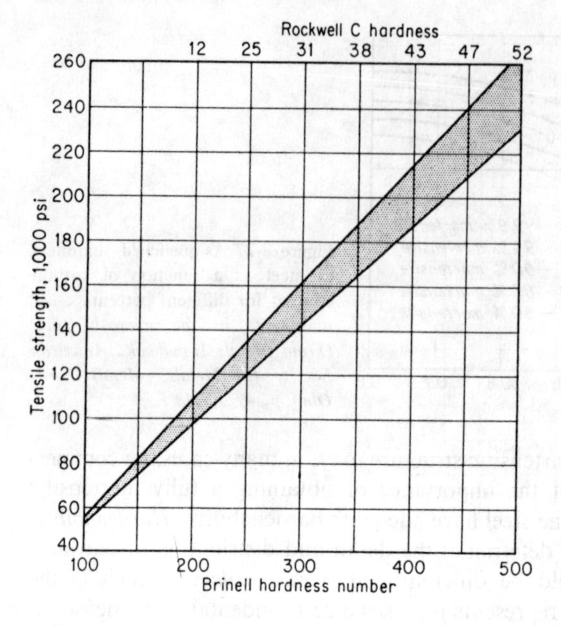

Figure 8-28 Relationship between tensile strength and hardness for quenched and tempered, annealed, and normalized steels. *(From SAE Handbook.)*

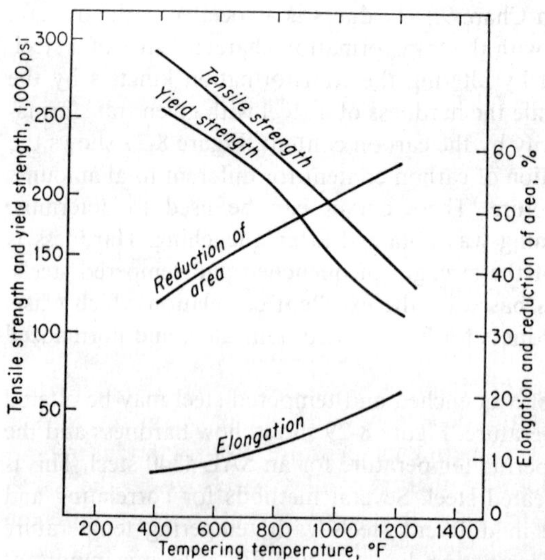

Figure 8-29 Tensile properties of quenched and tempered SAE-4340 steel as a function of tempering temperature (for fully hardened 1-in-diameter bars).

A great many low-alloy steels have been developed and are used in the quenched and tempered condition. A study of the tensile properties of these steels could lead to considerable confusion were it not for the fact that certain generalities can be made about their properties.[1,2] For low-alloy steels containing

[1] E. J. Janitsky and M. Baeyertz, in "Metals Handbook," pp. 515–518, American Society for Metals, Metals Park, Ohio, 1939.

[2] W. G. Patton, *Met. Prog.*, vol. 43, pp. 726–733, 1943.

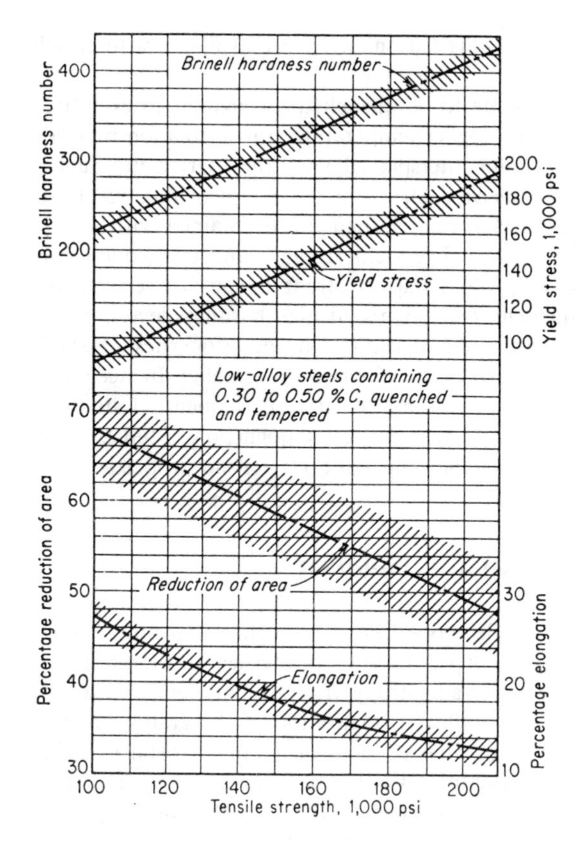

Figure 8-30 Relationships between tensile properties of quenched and tempered low-alloy steels. *(From W. G. Patton, Met. Progr., vol. 43, p. 726, 1943.)*

0.3 to 0.5 percent carbon which are quenched to essentially 100 percent martensite and then tempered back to any given tensile strength in the range 100,000 to 200,000 psi, the other common tensile properties will have a relatively fixed value depending only on the tensile strength. In other words, the mechanical properties of this important class of steels do not depend basically on alloy content, carbon content within the above limits, or tempering temperature. It is important to note that this generalization does not say that two alloy steels given the same tempering treatment will have the same tensile properties, because different tempering temperatures would quite likely be required to bring two different alloy steels to the same tensile strength. Figure 8-30 shows this relationship between the mechanical properties of steels with tempered martensitic structures. The expected scatter in values is indicated by the shading. Because of this similarity in properties, it is logical to ask why so many different alloy steels are used. Actually, as will be seen in Chap. 14, all low-alloy steels do not have the same fracture toughness or notch sensitivity, and they may differ considerably in these respects when heat-treated to tensile strengths in excess of 200,000 psi. Further, to minimize processing difficulties such as quench cracking and weld embrittlement, it is an advantage to use a steel with the lowest carbon content consistent with the required as-quenched hardness. For this reason, steels are available with closely spaced carbon contents.

Steel sections which are too large to be quenched throughout to essentially 100 percent martensite will contain higher-temperature transformation products such as ferrite, pearlite, and bainite interspersed with the martensite. Such a situation is known as a *slack-quenched* structure. Slack quenching results in tensile properties which are somewhat poorer than those obtained with a completely tempered martensitic structure. The yield strength and the reduction of area are generally most affected, while impact strength can be very greatly reduced. The effect of slack quenching is greatest at high hardness levels. As the tempering temperature is increased, the deviation in the properties of slack-quenched steel from those of tempered martensite becomes smaller. In steels with sufficient hardenability to form 100 percent martensite it is frequently found that not all the austenite is transformed to martensite on quenching. Studies[1] have shown that the greatest effect of retained austenite on tensile properties is in decreasing the yield strength.

8-15 ANISOTROPY OF TENSILE PROPERTIES

It is frequently found that the tensile properties of wrought-metal products are not the same in all directions. The dependence of properties on orientation is called *anisotropy*. Two general types of anisotropy are found in metals. *Crystallographic anisotropy* results from the preferred orientation of the grains which is produced by severe deformation. Since the strength of a single crystal is highly anisotropic, a severe plastic deformation which produces a strong preferred orientation will cause a polycrystalline specimen to approach the anisotropy of a single crystal. The yield strength, and to a lesser extent the tensile strength, are the properties most affected. The yield strength in the direction perpendicular to the main (longitudinal) direction of working may be greater or less than the yield strength in the longitudinal direction, depending on the type of preferred orientation which exists. This type of anisotropy is most frequently found in nonferrous metals, especially when they have been severely worked into sheet. Crystallographic anisotropy can be eliminated by recrystallization, although the formation of a recrystallization texture can cause the reappearance of a different type of anisotropy. A practical manifestation of crystallographic anisotropy is the formation of "ears," or nonuniform deformation in deep-drawn cups. Crystallographic anisotropy may also result in the elliptical deformation of a tensile specimen.

Mechanical fibering is due to the preferred alignment of structural discontinuities such as inclusions, voids, segregation, and second phases in the direction of working. This type of anisotropy is important in forgings and plates. The principal direction of working is defined as the *longitudinal direction*. This is the long axis of a bar or the rolling direction in a sheet or plate. Two transverse directions must be considered. The *short-transverse direction* is the minimum

[1] L. S. Castleman, B. L. Averbach, and M. Cohen, *Trans. Am. Soc. Met.*, vol. 44, pp. 240–263, 1952.

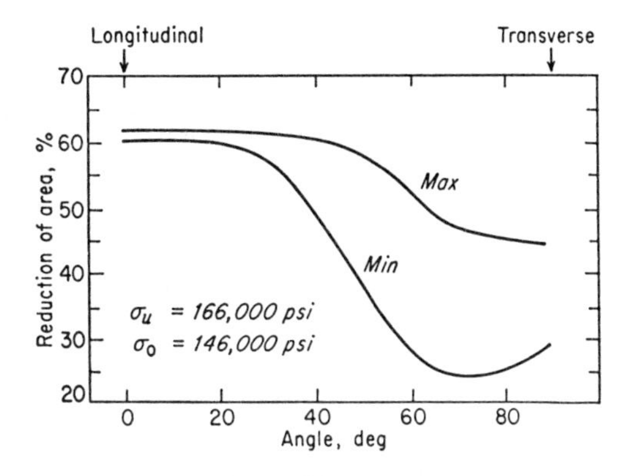

Figure 8-31 Relationship between reduction of area and angle between the longitudinal direction in forging and the specimen axis. *(From C. Wells and R. F. Mehl, Trans. Am. Soc. Met., vol. 41, p. 753, 1949.)*

dimension of the product, for example, the thickness of a plate. The *long-transverse direction* is perpendicular to both the longitudinal and short-transverse directions. In a round or square, both these transverse directions are equivalent, while in a sheet the properties in the short-transverse direction cannot be measured. In wrought-steel products mechanical fibering is the principal cause of directional properties. Measures of ductility like reduction of area are most affected. In general, reduction of area is lowest in the short-transverse direction, intermediate in the long-transverse direction, and highest in the longitudinal direction.

Transverse properties are particularly important in thick-walled tubes, like guns and pressure vessels, which are subjected to high internal pressures. In these applications the greatest principal stress acts in the circumferential direction, which corresponds to the transverse direction of a cylindrical forging. While there is no direct method for incorporating the reduction of area into the design of such a member, it is known that the transverse reduction of area (RAT) is a good index of steel quality for these types of applications. For this reason the RAT may be the limiting value in the design of a part. A great deal of work[1-3] on the transverse properties of gun tubes and large forgings has provided data in this field. Figure 8-31 shows the variation of reduction of area with the angle between the axis of the tensile specimen and the longitudinal direction in a forging of SAE 4340 steel. No similar variation with orientation is found for the yield strength or the tensile strength. This figure shows both the maximum and minimum values of reduction of area obtained for different specimen orientations. Because of the large scatter in measurements of RAT it is necessary to use statistical methods. The degree of anisotropy in reduction of area increases with strength level. In the region of tensile strength between 80,000 and 180,000 psi the RAT decreases by about 1.5 percent for each 5,000 psi increase in tensile strength. Figure 8-32 shows the way in which the longitudinal and transverse reduction of area varies with reduction by forging. The forging ratio is the ratio of the original to the final

[1] C. Wells and R. F. Mehl, *Trans. Am. Soc. Met.*, vol. 41, pp. 715–818, 1949.
[2] A. H. Grobe, C. Wells, and R. F. Mehl, *Trans. Am. Soc. Met.*, vol. 45, pp. 1080–1122, 1953.
[3] E. A. Loria, *Trans. Am. Soc. Met.*, vol. 42, pp. 486–498, 1950.

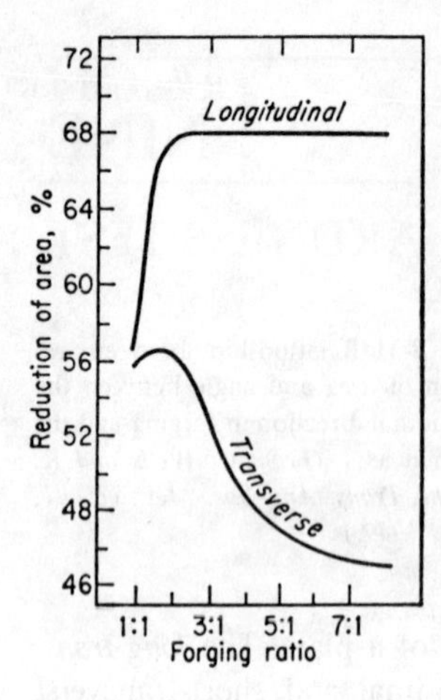

Figure 8-32 Effect of forging on longitudinal and transverse reduction of area. Tensile strength 118,000 psi. *(From C. Wells and R. F. Mehl, Trans. Am. Soc. Met., vol. 41, p. 755, 1949.)*

cross-sectional area of the forging. It is usually found that the optimum properties are obtained with a forging ratio of 2 to 3:1. Nonmetallic inclusions are considered to be a major source of low transverse ductility. This is based on the fact that vacuum-melted steels give higher RAT values than air-melted steel and on correlations which have been made[1] between inclusion content and RAT. Other factors such as microsegregation and dendritic structure may also be responsible for low transverse ductility in forgings.

An interesting graphic method for correlating the amount and direction of deformation with the resulting directionality of tensile properties has been presented by Hunsicker.[2] The procedure has been well documented for directional properties in aluminum alloys, but it has not yet been widely applied to other systems.

BIBLIOGRAPHY

Avery, D. H. and W. N. Findley, Quasistatic Mechanical Testing, in "Techniques of Metals Research," vol. 5, pt. 1, R. F. Bunshaw (ed.), Wiley-Interscience, New York, 1971.

Marin, J., "Mechanical Behavior of Engineering Materials," chap. 1, Prentice-Hall, Inc., Englewood Cliffs, N.J., 1962.

"Metals Handbook", 9th ed., vol. 8, Mechanical Testing, American Society for Metals, Metals Park, Ohio, 1985.

Meyers, M. A. and K. K. Chawla, "Mechanical Metallurgy," chap. 16, Prentice-Hall, Inc., Englewood Cliffs, N.J., 1984.

Polakowski, N. H. and E. J. Ripling, "Strength and Structure of Engineering Materials," chap. 10, Prentice-Hall, Inc., Englewood Cliffs, N.J., 1966.

[1] J. Welchner and W. G. Hildorf, *Trans. Am. Soc. Met.*, vol. 42, pp. 455–485, 1950; R. A. Cellitti and C. J. Carter, *Met. Prog.*, pp. 215–230, October 1969.

[2] H. Y. Hunsicker, *Trans. Metall. Soc. AIME*, vol. 245, pp. 29–42, 1969.

CHAPTER

NINE

THE HARDNESS TEST

9-1 INTRODUCTION

The hardness of a material is a poorly defined term which has many meanings depending upon the experience of the person involved. In general, hardness usually implies a resistance to deformation, and for metals the property is a measure of their resistance to permanent or plastic deformation. To a person concerned with the mechanics of materials testing, hardness is most likely to mean the resistance to indentation, and to the design engineer it often means an easily measured and specified quantity which indicates something about the strength and heat treatment of the metal. There are three general types of hardness measurements depending on the manner in which the test is conducted. These are (1) scratch hardness, (2) indentation hardness, and (3) rebound, or dynamic, hardness. Only indentation hardness is of major engineering interest for metals.

Scratch hardness is of primary interest to mineralogists. With this measure of hardness, various minerals and other materials are rated on their ability to scratch one another. Scratch hardness is measured according to the Mohs' scale. This consists of 10 standard minerals arranged in the order of their ability to be scratched. The softest mineral in this scale is talc (scratch hardness 1), while diamond has a hardness of 10. A fingernail has a value of about 2, annealed copper has a value of 3, and martensite a hardness of 7. The Mohs' scale is not well suited for metals since the intervals are not widely spaced in the high-hardness range. Most hard metals fall in the Mohs' hardness range of 4 to 8. A different type of scratch-hardness test[1] measures the depth or width of a scratch

[1] E. B. Bergsman, *ASTM Bull.* 176, pp. 37–43, September, 1951.

made by drawing a diamond stylus across the surface under a definite load. This is a useful tool for measuring the relative hardness of microconstituents, but it does not lend itself to high reproducibility or extreme accuracy.

In dynamic-hardness measurements the indenter is usually dropped onto the metal surface, and the hardness is expressed as the energy of impact. The Shore scleroscope, which is the commonest example of a dynamic-hardness tester, measures the hardness in terms of the height of rebound of the indenter.

9-2 BRINELL HARDNESS

The first widely accepted and standardized indentation-hardness test was proposed by J. A. Brinell in 1900. The Brinell hardness test consists in indenting the metal surface with a 10-mm-diameter steel ball at a load of 3,000 kg. For soft metals the load is reduced to 500 kg to avoid too deep an impression, and for very hard metals a tungsten carbide ball is used to minimize distortion of the indenter. The load is applied for a standard time, usually 30 s, and the diameter of the indentation is measured with a low-power microscope after removal of the load. The average of two readings of the diameter of the impression at right angles should be made. The surface on which the indentation is made should be relatively smooth and free from dirt or scale. The Brinell hardness number (BHN) is expressed as the load P divided by the *surface area* of the indentation. This is expressed by the formula[1]

$$\text{BHN} = \frac{P}{(\pi D/2)\left(D - \sqrt{D^2 - d^2}\right)} = \frac{P}{\pi D t} \tag{9-1}$$

where P = applied load, kg
D = diameter of ball, mm
d = diameter of indentation, mm
t = depth of the impression, mm

It will be noticed that the units of the BHN are kilograms per square millimeter. However, the BHN is not a satisfactory physical concept since Eq. (9-1) does not give the mean pressure over the surface of the indentation.

From Fig. 9-1 it can be seen that $d = D \sin \phi$. Substitution into Eq. (9-1) gives an alternate expression for Brinell hardness number.

$$\text{BHN} = \frac{P}{(\pi/2) D^2 (1 - \cos \phi)} \tag{9-2}$$

In order to obtain the same BHN with a nonstandard load or ball diameter it is necessary to produce geometrically similar indentations. Geometric similitude is achieved so long as the included angle 2ϕ remains constant. Equation (9-2) shows

[1] Tables giving BHN as a function of d for standard loads may be found in most of the references in the Bibliography at the end of this chapter. See *ASTM* Standard E10-66.

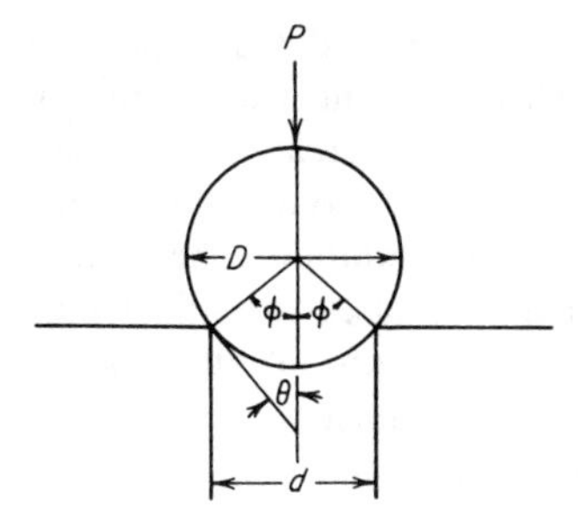

Figure 9-1 Basic parameters in Brinell test.

that for ϕ and BHN to remain constant the load and ball diameter must be varied in the ratio

$$\frac{P_1}{D_1^2} = \frac{P_2}{D_2^2} = \frac{P_3}{D_3^2} \tag{9-3}$$

Unless precautions are taken to maintain P/D^2 constant, which may be experimentally inconvenient, the BHN generally will vary with load. Over a range of loads the BHN reaches a maximum at some intermediate load. Therefore, it is not possible to cover with a single load the entire range of hardnesses encountered in commercial metals. The relatively large size of the Brinell impression may be an advantage in averaging out local heterogeneities. Moreover, the Brinell test is less influenced by surface scratches and roughness than other hardness tests. On the other hand, the large size of the Brinell impression may preclude the use of this test with small objects or in critically stressed parts where the indentation could be a potential site of failure.

9-3 MEYER HARDNESS

Meyer[1] suggested that a more rational definition of hardness than that proposed by Brinell would be one based on the *projected area* of the impression rather than the surface area. The mean pressure between the surface of the indenter and the indentation is equal to the load divided by the projected area of the indentation.

$$p_m = \frac{P}{\pi r^2}$$

Meyer proposed that this mean pressure should be taken as the measure of hardness. It is referred to as the *Meyer hardness*.

$$\text{Meyer hardness} = \frac{4P}{\pi d^2} \tag{9-4}$$

Like the Brinell hardness, Meyer hardness has units of kilograms per square

[1] E. Meyer, *Z. ver. Deut. Ing.*, vol. 52, pp. 645–654, 1908.

millimeter. The Meyer hardness is less sensitive to the applied load than the Brinell hardness. For a cold-worked material the Meyer hardness is essentially constant and independent of load, while the Brinell hardness decreases as the load increases. For an annealed metal the Meyer hardness increases continuously with the load because of strain hardening produced by the indentation. The Brinell hardness, however, first increases with load and then decreases for still higher loads. The Meyer hardness is a more fundamental measure of indentation hardness; yet it is rarely used for practical hardness measurements.

Meyer proposed an empirical relation between the load and the size of the indentation. This relationship is usually called *Meyer's law.*

$$P = kd^{n'} \tag{9-5}$$

where P = applied load, kg
d = diameter of indentation, mm
n' = a material constant related to strain hardening of metal
k = a material constant expressing resistance of metal to penetration

The parameter n' is the slope of the straight line obtained when log P is plotted against log d, and k is the value of P at $d = 1$. Fully annealed metals have a value of n' of about 2.5, while n' is approximately 2 for fully strain-hardened metals. This parameter is roughly related to the strain-hardening coefficient in the exponential equation for the true-stress–true-strain curve. The exponent in Meyer's law is approximately equal to the strain-hardening coefficient plus 2.

There is a lower limit of load below which Meyer's law is not valid. If the load is too small, the deformation around the indentation is not fully plastic and Eq. (9-5) is not obeyed. This load will depend on the hardness of the metal. For a 10-mm-diameter ball the load should exceed 50 kg for copper with a BHN of 100, and for steel with a BHN of 400 the load should exceed 1,500 kg. For balls of different diameter the critical loads will be proportional to the square of the diameter.

9-4 ANALYSIS OF INDENTATION BY AN INDENTER

The plastic zone beneath a hardness indentation is surrounded with elastic material which acts to hinder plastic flow in a manner similar to the die constraint forces in a closed-die forging. Therefore, the mean compressive stress required to cause plastic flow in the hardness test exceeds that in simple compression because of this constraint. The prediction of the load required to indent a solid is one of the classic problems in plasticity. Prandtl applied slip-line field theory to show that the constraint factor for plane-strain compression was 2.57 (see Sec. 3-12 and Fig. 3-13).

$$\frac{p_m}{\sigma} = 1 + \frac{\pi}{2} = 2.57$$

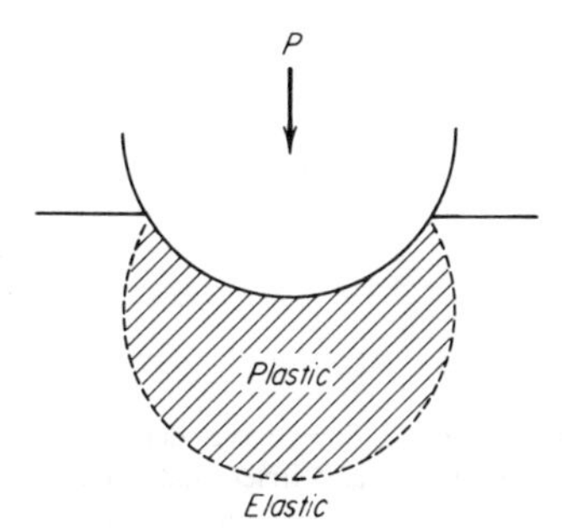

Figure 9-2 Plastic zone under a Brinell indenter. *(After Shaw and DeSalvo.)*

In this idealized model the material surrounding the deformed zone is rigid and upward flow of material compensates for the material displaced by the punch. However, the geometry of the Brinell test is axisymmetric as opposed to plane strain. Shaw and DeSalvo have shown[1] that the plastic region under this blunt indenter bears little resemblance to the slip-line field, but instead is very similar to an elastic-plastic boundary resembling the line of constant maximum shear stress beneath a sphere pressed against a flat surface (Fig. 9-2). This curve may be derived by applying Hertz's theory of contact stresses.[2] Using this plastic-elastic model the material displaced by the indenter is completely accounted for by the decrease in volume of the elastic material. This removes the need for an upward flow of material around the indenter, which agrees with observations that show only a small amount of upward flow. The plastic-elastic analysis results in a constraint factor of $C = 3.0$ for indentation with a spherical ball.

$$p_m = C\sigma_0 = 3\sigma_0 \tag{9-6}$$

9-5 RELATIONSHIP BETWEEN HARDNESS AND THE FLOW CURVE

Tabor[3] has suggested a method by which the plastic region of the true-stress–true-strain curve may be determined from indentation hardness measurements. The method is based on the fact that there is a similarity in the shape of the flow curve and the curve obtained when the Meyer hardness is measured on a number of specimens subjected to increasing amounts of plastic strain. The method is basically empirical, since the complex stress distribution at the hardness indentation precludes a straightforward relationship with the stress distribution in the tension or compression test. However, the method has been shown to

[1] M. C. Shaw and G. J. DeSalvo, *Trans. ASME, Ser. B*; *J. Eng. Ind.*, vol. 92, pp. 469–479, 1970; M. C. Shaw and G. J. DeSalvo, *Met. Eng. Q.*, vol. 12, pp. 1–7, May 1972.

[2] See for example S. Timoshenko and J. N. Goodier, "Theory of Elasticity," 2d ed., pp. 372–382, McGraw-Hill Book Company, New York, 1951.

[3] D. Tabor, "The Hardness of Metals," pp. 67–76, Oxford University Press, New York 1951; *J. Inst. Met.*, vol. 79, p. 1, 1951.

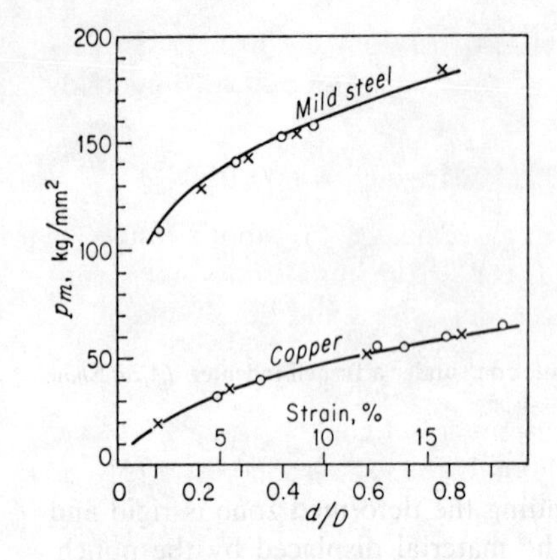

Figure 9-3 Comparison of flow curve determined from hardness measurements (circles and crosses) with flow curve determined from compression test (solid lines). *(From D. Tabor, "The Hardness of Metals," p. 74, Oxford University Press, New York, 1951.)*

give good agreement for several metals, and thus should be of interest as a means of obtaining flow data in situations where it is not possible to measure tensile properties. The true stress (flow stress) is obtained from Eq. (9-6), where σ_0 is to be considered the flow stress at a given value of true strain. From a study of the deformation at indentations, Tabor concluded that the true strain was proportional to the ratio d/D and could be expressed as

$$\varepsilon = 0.2\frac{d}{D} \tag{9-7}$$

Thus, if the Meyer hardness is measured under conditions such that d/D varies from the smallest value for full plasticity up to large values and Eqs. (9-6) and (9-7) are used, it is possible at least to approximate the tensile-flow curve. Figure 9-3 shows the agreement which has been obtained by Tabor between the flow curve and hardness versus d/D curve for mild steel and annealed copper. Tabor's results have been verified by Lenhart[1] for duralumin and OFHC copper. However, Tabor's analysis did not predict the flow curve for magnesium, which was attributed by Lenhart to the high anisotropy of deformation in this metal. This work should not detract from the usefulness of this correlation but, rather, should serve to emphasize that its limitations should be investigated for new applications.

The 0.2 percent offset yield strength can be determined[2] with good precision from Vickers hardness measurements (see Sec. 9-6) according to the relation

$$\sigma_0 = \frac{DPH}{3}(0.1)^{n'-2} \tag{9-8}$$

where σ_0 = the 0.2 percent offset yield strength, kg/mm²
DPH = the Vickers hardness number
$n' = n + 2$ = the exponent in Meyer's law

[1] R. E. Lenhart, *WADC Tech. Rept.*, 55–114, June, 1955.
[2] J. B. Cahoon, W. H. Broughton, and A. R. Kutzak, *Metall. Trans.*, vol. 2, pp. 1979–1983, 1971.

There is a very useful engineering correlation between the Brinell hardness and the ultimate tensile strength of heat-treated plain-carbon and medium-alloy steels (see Fig. 8-28).

$$\text{Ultimate tensile strength, in pounds per square inch} = 500(\text{BHN})$$

A brief consideration will show that this is in agreement with Tabor's results. If we make the simplifying assumption that this class of materials does not strain-harden, then the tensile strength is equal to the yield stress and Eq. (9-6) applies.

$$s_u = \tfrac{1}{3}p_m = 0.33p_m \text{ kg/mm}^2$$

The Brinell hardness will be only a few percent less than the value of Meyer hardness p_m. Upon converting to engineering units the expression becomes

$$s_u = 470(\text{BHN})$$

It should now be apparent why the same relationship does not hold for other metals. For example, for annealed copper the assumption that strain hardening can be neglected will be grossly in error. For a metal with greater capability for strain hardening the "constant" of proportionality will be greater than that used for heat-treated steel.

9-6 VICKERS HARDNESS

The Vickers hardness test uses a square-base diamond pyramid as the indenter. The included angle between opposite faces of the pyramid is 136°. This angle was chosen because it approximates the most desirable ratio of indentation diameter to ball diameter in the Brinell hardness test.[1] Because of the shape of the indenter, this is frequently called the *diamond-pyramid hardness test*. The diamond-pyramid hardness number (DPH), or Vickers hardness number (VHN, or VPH), is defined as the load divided by the surface area of the indentation. In practice, this area is calculated from microscopic measurements of the lengths of the diagonals of the impression. The DPH may be determined from the following equation

$$\text{DPH} = \frac{2P\sin(\theta/2)}{L^2} = \frac{1.854P}{L^2} \tag{9-9}$$

where P = applied load, kg
L = average length of diagonals, mm
θ = angle between opposite faces of diamond = 136°

The Vickers hardness test has received fairly wide acceptance for research work because it provides a continuous scale of hardness, for a given load, from very soft metals with a DPH of 5 to extremely hard materials with a DPH of

[1] In most Brinell tests d/D lies between 0.25 and 0.50. For the diamond-pyramid indenter a value of $d = 0.375D$ was used, which results in cone angle of 136°. As a result, DPH and BHN hardnesses are nearly identical so long as the Brinell impressions are of normal depth.

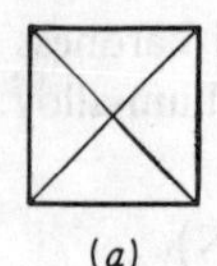

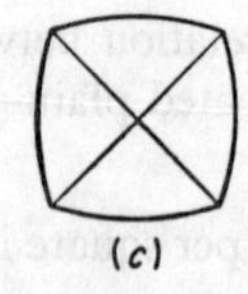

Figure 9-4 Types of diamond-pyramid indentations. (*a*) Perfect indentation; (*b*) pincushion indentation due to sinking in; (*c*) barreled indentation due to ridging.

1,500. With the Rockwell hardness test, described in Sec. 9-7, or the Brinell hardness test, it is usually necessary to change either the load or the indenter at some point in the hardness scale, so that measurements at one extreme of the scale cannot be strictly compared with those at the other end. Because the impressions made by the pyramid indenter are geometrically similar no matter what their size, the DPH should be independent of load. This is generally found to be the case, except at very light loads. The loads ordinarily used with this test range from 1 to 120 kg, depending on the hardness of the metal to be tested. In spite of these advantages, the Vickers hardness test has not been widely accepted for routine testing because it is slow, requires careful surface preparation of the specimen, and allows greater chance for personal error in the determination of the diagonal length. The Vickers hardness test is described in ASTM Standard E92-72.

A perfect indentation made with a perfect diamond-pyramid indenter would be a square. However, anomalies corresponding to those described earlier for Brinell impressions are frequently observed with a pyramid indenter (Fig. 9-4). The pincushion indentation in Fig. 9-4*b* is the result of sinking in of the metal around the flat faces of the pyramid. This condition is observed with annealed metals and results in an overestimate of the diagonal length. The barrel-shaped indentation in Fig. 9-4*c* is found in cold-worked metals. It results from ridging or piling up of the metal around the faces of the indenter. The diagonal measurement in this case produces a low value of the contact area so that the hardness numbers are erroneously high. Empirical corrections for this effect have been proposed.[1]

9-7 ROCKWELL HARDNESS TEST

The most widely used hardness test in the United States is the Rockwell hardness test. Its general acceptance is due to its speed, freedom from personal error, ability to distinguish small hardness differences in hardened steel, and the small size of the indentation, so that finished heat-treated parts can be tested without damage. This test utilizes the depth of indentation, under constant load, as a measure of hardness. A minor load of 10 kg is first applied to seat the specimen. This minimizes the amount of surface preparation needed and reduces the tendency for ridging or sinking in by the indenter. The major load is then applied, and the depth of indentation is automatically recorded on a dial gage in terms of

[1] T. B. Crowe and J. F. Hinsely, *J. Inst. Met.*, vol. 72, p. 14, 1946.

arbitrary hardness numbers. The dial contains 100 divisions, each division representing a penetration of 0.00008 in. The dial is reversed so that a high hardness, which corresponds to a small penetration, results in a high hardness number. This is in agreement with the other hardness numbers described previously, but unlike the Brinell and Vickers hardness designations, which have units of kilograms per square millimeter, the Rockwell hardness numbers are purely arbitrary.

One combination of load and indenter will not produce satisfactory results for materials with a wide range of hardness. A 120° diamond cone with a slightly rounded point, called a *Brale indenter*, and $\frac{1}{16}$- and $\frac{1}{8}$-in-diameter steel balls are generally used as indenters. Major loads of 60, 100, and 150 kg are used. Since the Rockwell hardness is dependent on the load and indenter, it is necessary to specify the combination which is used. This is done by prefixing the hardness number with a letter indicating the particular combination of load and indenter for the hardness scale employed. A Rockwell hardness number without the letter prefix is meaningless. Hardened steel is tested on the C scale with the diamond indenter and a 150-kg major load. The useful range for this scale is from about R_C 20 to R_C 70. Softer materials are usually tested on the B scale with a $\frac{1}{16}$-in-diameter steel ball and a 100-kg major load. The range of this scale is from R_B 0 to R_B 100. The A scale (diamond penetrator, 60-kg major load) provides the most extended Rockwell hardness scale, which is usable for materials from annealed brass to cemented carbides. Many other scales are available for special purposes.[1]

The Rockwell hardness test is a very useful and reproducible one provided that a number of simple precautions are observed. Most of the points listed below apply equally well to the other hardness tests:

1. The indenter and anvil should be clean and well seated.
2. The surface to be tested should be clean and dry, smooth, and free from oxide. A rough-ground surface is usually adequate for the Rockwell test.
3. The surface should be flat and perpendicular to the indenter.
4. Tests on cylindrical surfaces will give low readings, the error depending on the curvature, load, indenter, and hardness of the material. Theoretical[2] and empirical[3] corrections for this effect have been published.
5. The thickness of the specimen should be such that a mark or bulge is not produced on the reverse side of the piece. It is recommended that the thickness be at least 10 times the depth of the indentation. Tests should be made on only a single thickness of material.
6. The spacing between indentations should be three to five times the diameter of the indentation.
7. The speed of application of the load should be standardized. This is done by adjusting the dashpot on the Rockwell tester. Variations in hardness can be

[1] See ASTM Standard E18-74.

[2] W. E. Ingerson, *Am. Soc. Test. Mater. Proc.*, vol. 39, pp. 1281–1291, 1939.

[3] R. S. Sutton and R. H. Heyer, *ASTM Bull.* 193, pp. 40–41, October, 1953.

appreciable in very soft materials unless the rate of load application is carefully controlled. For such materials the operating handle of the Rockwell tester should be brought back as soon as the major load has been fully applied.

9-8 MICROHARDNESS TESTS

Many metallurgical problems require the determination of hardness over very small areas.[1] The measurement of the hardness gradient at a carburized surface, the determination of the hardness of individual constituents of a microstructure, or the checking of the hardness of a delicate watch gear might be typical problems. The use of a scratch-hardness test for these purposes was mentioned earlier, but an indentation-hardness test has been found to be more useful.[2] The development of the Knoop indenter by the National Bureau of Standards and the introduction of the Tukon tester for the controlled application of loads down to 25 g have made microhardness testing a routine laboratory procedure.

The Knoop indenter is a diamond ground to a pyramidal form that produces a diamond-shaped indentation with the long and short diagonals in the approximate ratio of 7 : 1 resulting in a state of plane strain in the deformed region. The Knoop hardness number (KHN) is the applied load divided by the unrecovered projected area of the indentation.

$$\mathrm{KHN} = \frac{P}{A_p} = \frac{P}{L^2 C} \tag{9-10}$$

where P = applied load, kg
A_p = unrecovered projected area of indentation, mm^2
L = length of long diagonal, mm
C = a constant for each indenter supplied by manufacturer.

The special shape of the Knoop indenter makes it possible to place indentations much closer together than with a square Vickers indentation, e.g., to measure a steep hardness gradient. Its other advantage is that for a given long diagonal length the depth and area of the Knoop indentation are only about 15 percent of what they would be for a Vickers indentation with the same diagonal length. This is particularly useful when measuring the hardness of a thin layer (such as an electroplated layer), or when testing brittle materials where the tendency for fracture is proportional to the volume of stressed material.

The low load used with microhardness tests requires that extreme care be taken in all stages of testing. The surface of the specimen must be carefully prepared. Metallographic polishing is usually required. Work hardening of the surface during polishing can influence the results. The long diagonal of the Knoop

[1] See ASTM Standard E334-69.

[2] For a review of microhardness testing see H. Bückle, *Metall. Rev.*, vol. 4, no. 3, pp. 49–100, 1959.

impression is essentially unaffected by elastic recovery for loads greater than about 300 g. However, for lighter loads the small amount of elastic recovery becomes appreciable. Further, with the very small indentations produced at light loads the error in locating the actual ends of the indentation become greater. Both these factors have the effect of giving a high hardness reading, so that it is usually observed that the Knoop hardness number increases as the load is decreased below about 300 g. Tarasov and Thibault[1] have shown that if corrections are made for elastic recovery and visual acuity the Knoop hardness number is constant with load down to 100 g.

9-9 HARDNESS-CONVERSION RELATIONSHIPS

From a practical standpoint it is important to be able to convert the results of one type of hardness test into those of a different test. Since a hardness test does not measure a well-defined property of a material and since all the tests in common use are not based on the same type of measurements, it is not surprising that no universal hardness-conversion relationships have been developed. It is important to realize that hardness conversions are empirical relationships. The most reliable hardness-conversion data exist for steel which is harder than 240 Brinell. The ASTM, ASM, and SAE (Society of Automotive Engineers) have agreed on a table[2] for conversion between Rockwell, Brinell, and diamond-pyramid hardness which is applicable to heat-treated carbon and alloy steel and to almost all alloy constructional steels and tool steels in the as-forged, annealed, normalized, and quenched and tempered conditions. However, different conversion tables are required for materials, with greatly different elastic moduli, such as tungsten carbide, or with greater strain-hardening capacity. Heyer[3] has shown that the indentation hardness of soft metals depends on the strain-hardening behavior of the material during the test, which in turn is dependent on the previous degree of strain hardening of the material before the test. As an extreme example of the care which is required in using conversion charts for soft metals, it is possible for Armco iron and cold-rolled aluminum each to have a Brinell hardness of 66; yet the former has a Rockwell B hardness of 31 compared with a hardness of R_B 7 for the cold-worked aluminum. On the other hand, metals, such as yellow brass and low-carbon sheet steel have a well-behaved Brinell-Rockwell conversion[4] relationship for all degrees of strain hardening. Special hardness-conversion tables for cold-worked aluminum, copper, and 18-8 stainless steel are given in the ASM Metals Handbook.

[1] L. P. Tarasov and N. W. Thibault, *Trans. Am. Soc. Met.*, vol. 38, pp. 331–353, 1947.

[2] This table may be found in ASTM Standard E-140-78, SAE Handbook, ASM Metals Handbook, and many other standard references.

[3] R. H. Heyer, *Am. Soc. Test. Mater. Proc.*, vol. 44, pp. 1027, 1944.

[4] The Wilson Mechanical Instrument Co. Chart 38 for metals softer than BHN 240 (see ASM Handbook, 1948 ed., p. 101) is based on tests on these metals.

9-10 HARDNESS AT ELEVATED TEMPERATURES

Interest in measuring the hardness of metals at elevated temperatures has been accelerated by the great effort which has gone into developing alloys with improved high-temperature strength. Hot hardness gives a good indication of the potential usefulness of an alloy for high-temperature strength applications. Some degree of success has been obtained in correlating hot hardness with high-temperature strength properties. This will be discussed in Chap. 13. Hot-hardness testers using a Vickers indenter made of sapphire and with provisions for testing in either vacuum or an inert atmosphere have been developed,[1] and a high-temperature microhardness test has been described.[2]

In an extensive review of hardness data at different temperatures Westbrook[3] showed that the temperature dependence of hardness could be expressed by

$$H = Ae^{-BT} \tag{9-11}$$

where H = hardness, kg/mm^2
T = test temperature, °K
A, B = constants

Plots of log H versus temperature for pure metals generally yield two straight lines of different slope. The change in slope occurs at a temperature which is about one-half the melting point of the metal being tested. Similar behavior is found in plots of the logarithm of the tensile strength against temperature. Figure 9-5 shows this behavior for copper. It is likely that this change in slope is due to a change in the deformation mechanism at higher temperature. The constant A derived from the low-temperature branch of the curve can be considered to be the intrinsic hardness of the metal, that is, H at 0°K. This value would be expected to be a measure of the inherent strength of the binding forces of the lattice. Westbrook correlated values of A for different metals with the heat content of the liquid metal at the melting point and with the melting point. This correlation was sensitive to crystal structure. The constant B, derived from the slope of the curve, is the temperature coefficient of hardness. This constant was related in a rather complex way to the rate of change of heat content with increasing temperature. With these correlations it is possible to calculate fairly well the hardness of a pure metal as a function of temperature up to about one-half its melting point.

Hardness measurements as a function of temperature will show an abrupt change at the temperature at which an allotropic transformation occurs. Hot-hardness tests on Co, Fe, Ti, U, and Zr have shown[4] that the body-centered cubic lattice is always the softer structure when it is involved in an allotropic transfor-

[1] F. Garofalo, P. R. Malenock and G. V. Smith, *Trans. Am. Soc. Met.*, vol. 45, pp. 377–396, 1953; M. Semchyshen and C. S. Torgerson, *Trans. Am. Soc. Met.*, vol. 50, pp. 830–837, 1958.

[2] J. H. Westbrook, *Am. Soc. Test. Mater. Proc.*, vol. 57, pp. 873–897, 1957; *ASTM Bull.* 246, pp. 53–58, 1960.

[3] J. H. Westbrook, *Trans. Am. Soc. Met.*, vol. 45, pp. 221–248, 1953.

[4] W. Chubb, *Trans. AIME*, vol. 203, pp. 189–192, 1955.

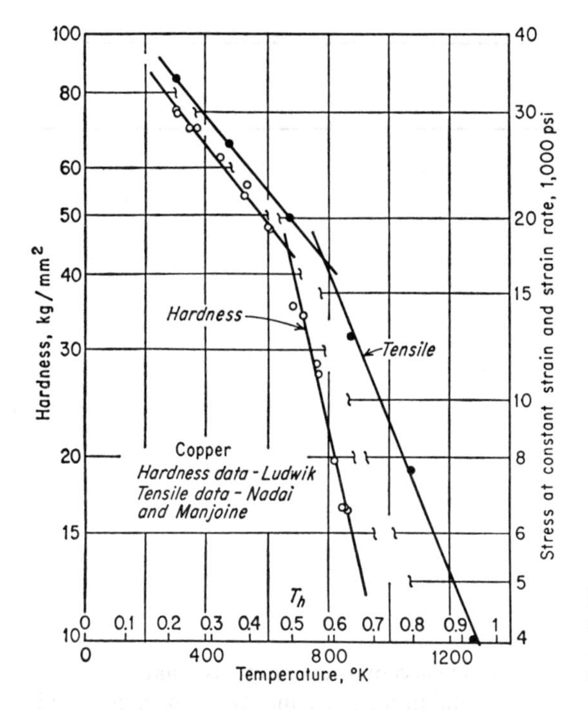

Figure 9-5 Temperature dependence of the hardness of copper. *(From J. H. Westbrook, Trans. Am. Soc. Met., vol. 45, p. 233, 1953.)*

mation. The face-centered cubic and hexagonal close-packed lattices have approximately the same strength, while highly complex crystal structures give even higher hardness. These results are in agreement with the fact that austenitic iron-based alloys have better high-temperature strength than ferritic alloys.

BIBLIOGRAPHY

Hardness Testing, "Metals Handbook," 9th ed., vol. 8, pp. 69–113, American Society for Metals, Metals Park, Ohio, 1985.

Mott, B. W.: "Micro-indentation Hardness Testing," Butterworth & Co. (Publishers), Ltd., London, 1956.

O'Neil, H.: "Hardness Measurement of Metals and Alloys," 2nd ed., Chapman and Hall, London, 1967.

Petty, E. R.: "Hardness Testing," in Techniques of Metals Research, vol. 5, pt. 2, R. F. Bunshaw (ed.), Wiley-Interscience, New York, 1971.

Tabor, D.: "The Hardness of Metals," Oxford University Press, New York, 1951.

Westbrook, J. H. and H. Conrad (eds.): "The Science of Hardness Testing and Its Research Applications," American Society for Metals, Metals Park, Ohio, 1973.

CHAPTER

TEN

THE TORSION TEST

10-1 INTRODUCTION

The torsion test has not met with the wide acceptance and the use that have been given the tension test. However, it is useful in many engineering applications and also in theoretical studies of plastic flow. Torsion tests are made on materials to determine such properties as the modulus of elasticity in shear, the torsional yield strength, and the modulus of rupture. Torsion tests also may be carried out on full-sized parts, such as shafts, axles, and twist drills, which are subjected to torsional loading in service. It is frequently used for testing brittle materials, such as tool steels, and has been used in the form of a high-temperature twist test to evaluate the forgeability of materials. The torsion test has not been standardized to the same extent as the tension test and is rarely required in materials specifications.

Torsion-testing equipment consists of a twisting head, with a chuck for gripping the specimen and for applying the twisting moment to the specimen, and a weighing head, which grips the other end of the specimen and measures the twisting moment, or torque. The deformation of the specimen is measured by a twist-measuring device called a troptometer. Determination is made of the angular displacement of a point near one end of the test section of the specimen with respect to a point on the same longitudinal element at the opposite end. A torsion specimen generally has a circular cross section, since this represents the simplest geometry for the calculation of the stress. Since in the elastic range the shear stress varies linearly from a value of zero at the center of the bar to a maximum value at the surface, it is frequently desirable to test a thin-walled tubular specimen. This results in a nearly uniform shear stress over the cross section of the specimen.

10-2 MECHANICAL PROPERTIES IN TORSION

Consider a cylindrical bar which is subjected to a torsional moment at one end (Fig. 10-1). The twisting moment is resisted by shear stresses set up in the cross section of the bar. The shear stress is zero at the center of the bar and increases linearly with the radius. Equating the twisting moment to the internal resisting moment,

$$M_T = \int_{r=0}^{r=a} \tau r \, dA = \frac{\tau}{r} \int_0^a r^2 \, dA \tag{10-1}$$

But $\int r^2 \, dA$ is the polar moment of inertia of the area with respect to the axis of the bar. Thus,

$$M_T = \frac{\tau J}{r}$$

or

$$\tau = \frac{M_T r}{J} \tag{10-2}$$

where τ = shear stress, psi
M_T = torsional moment, lb-in
r = radial distance measured from center of bar, in
J = polar moment of inertia, in^4

Since the shear stress is a maximum at the surface of the bar, for a solid cylindrical specimen where $J = \pi D^4/32$, the maximum shear stress is

$$\tau_{\max} = \frac{M_T D/2}{\pi D^4/32} = \frac{16 M_T}{\pi D^3} \tag{10-3}$$

For a tubular specimen the shear stress on the outer surface is

$$\tau = \frac{16 M_T D_1}{\pi \left(D_1^4 - D_2^4 \right)} \tag{10-4}$$

where D_1 = outside diameter of tube
D_2 = inside diameter of tube

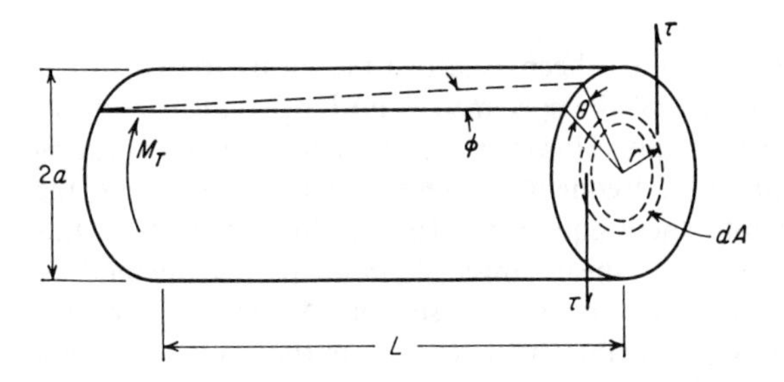

Figure 10-1 Torsion of a solid bar.

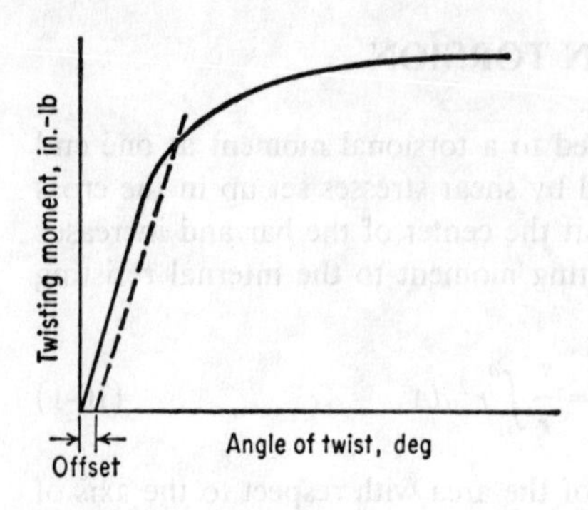

Figure 10-2 Torque-twist diagram.

The troptometer is used to determine the angle of twist θ, usually expressed in radians. If L is the test length of the specimen, from Fig. 10-1 it will be seen that the shear strain is given by

$$\gamma = \tan \phi = \frac{r\theta}{L} \tag{10-5}$$

During a torsion test measurements are made of the twisting moment M_T and the angle of twist θ. A torque-twist diagram is usually obtained, as shown in Fig. 10-2.

The elastic properties in torsion may be obtained by using the torque at the proportional limit or the torque at some offset angle of twist, frequently 0.001 rad/in of gage length, and calculating the shear stress corresponding to the twisting moment from the appropriate equations given above. A tubular specimen is usually required for a precision measurement of the torsional elastic limit or yield strength. Because of the stress gradient across the diameter of a solid bar, the surface fibers are restrained from yielding by the less highly stressed inner fibers. Thus, the first onset of yielding is generally not readily apparent with the instruments ordinarily used for measuring the angle of twist. The use of a thin-walled tubular specimen minimizes this effect because the stress gradient is practically eliminated. Care should be taken, however, that the wall thickness is not reduced too greatly, or the specimen will fail by buckling rather than torsion. Experience has shown that for determinations of the shearing yield strength and modulus of elasticity the ratio of the length of the reduced test section to the outside diameter should be about 10 and the diameter-thickness ratio should be about 8 to 10.

Once the torsional yield strength has been exceeded the shear-stress distribution from the center to the surface of the specimen is no longer linear and Eq. (10-3) or (10-4) does not strictly apply. However, an ultimate torsional shearing strength, or *modulus of rupture*, is frequently determined by substituting the maximum measured torque into these equations. The results obtained by this procedure overestimate the ultimate shear stress. A more precise method of calculating this value will be discussed in the next section. Although the procedure just described results in considerable error, for the purpose of comparing and selecting materials it is generally sufficiently accurate. For the determination of the modulus of rupture with tubular specimens, the ratio of gage length to diameter should be about 0.5 and the diameter-thickness ratio about 10 to 12.

Within the elastic range the shear stress can be considered proportional to the shear strain. The constant of proportionality G is the *modulus of elasticity in shear*, or the *modulus of rigidity*.

$$\tau = G\gamma \tag{10-6}$$

Substituting Eqs. (10-2) and (10-5) into Eq. (10-6) gives an expression for the shear modulus in terms of the geometry of the specimen, the torque, and the angle of twist.

$$G = \frac{M_T L}{J\theta} \tag{10-7}$$

10-3 TORSIONAL STRESSES FOR LARGE PLASTIC STRAINS

Beyond the torsional yield strength the shear stress over a cross section of the bar is no longer a linear function of the distance from the axis, and Eqs. (10-3) and (10-4) do not apply. Nadai[1] has presented a method for calculating the shear stress in the plastic range if the torque-twist curve is known. To simplify the analysis, we shall consider the angle of twist per unit length θ', where $\theta' = \theta/L$. Referring to Eq. (10-5), the shear strain will be

$$\gamma = r\theta' \tag{10-8}$$

Equation (10-1), for the resisting torque in a cross section of the bar, can be expressed as follows:

$$M_T = 2\pi \int_0^a \tau r^2 \, dr \tag{10-9}$$

Now the shear stress is related to the shear strain by the stress-strain curve in shear.

$$\tau = f(\gamma)$$

Introducing this equation into Eq. (10-9) and changing the variable from r to γ by means of Eq. (10-8) gives

$$M_T = 2\pi \int_0^{\gamma_a} f(\gamma) \frac{(\gamma^2)}{(\theta')^2} \frac{d\gamma}{\theta'}$$

$$M_T(\theta')^3 = 2\pi \int_0^{\gamma_a} f(\gamma)\gamma^2 \, d\gamma \tag{10-10}$$

[1] A. Nadai, "Theory of Flow and Fracture of Solids," 2d ed., vol. I, pp. 347–349, McGraw-Hill Book Company, New York, 1950.

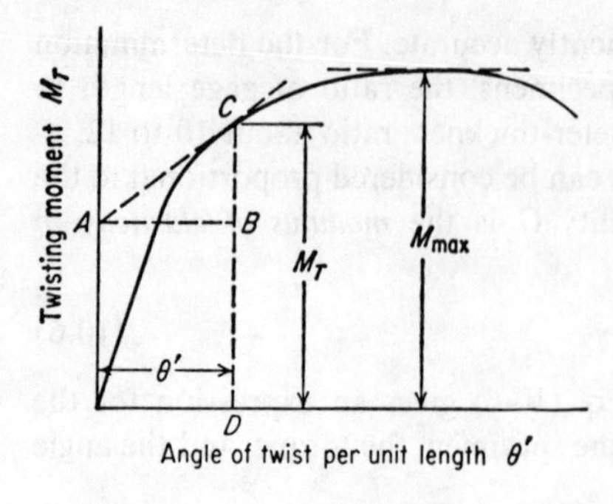

Figure 10-3 Method of calculating shear stress from torque-twist diagram.

where $\gamma_a = a\theta'$. Differentiating Eq. (10-10) with respect to θ'

$$\frac{d}{d\theta'}\left(M_T\theta'^3\right) = 2\pi a f(a\theta')a^2(\theta')^2 = 2\pi a^3(\theta')^2 f(a\theta')$$

But, the maximum shear stress in the bar at the outer fiber is $\tau_a = f(a\theta')$. Therefore,

$$\frac{d\left(M_T\theta'^3\right)}{d\theta'} = 2\pi a^3(\theta')^2\tau_a$$

$$3M_T(\theta')^2 + (\theta')^3\frac{dM_T}{d\theta'} = 2\pi a^3(\theta')^2\tau_a$$

Therefore,

$$\tau_a = \frac{1}{2\pi a^3}\left(\theta'\frac{dM_T}{d\theta'} + 3M_T\right) \tag{10-11}$$

If a torque-twist curve is available, the shear stress can be calculated with the above equation. Figure 10-3 illustrates how this is done. Examination of Eq. (10-11) shows that it can be written in terms of the geometry of Fig. 10-3 as follows:

$$\tau_a = \frac{1}{2\pi a^3}(BC + 3CD) \tag{10-12}$$

It will also be noticed from Fig. 10-3 that at the maximum value of torque $dM_T/d\theta' = 0$. Therefore, the ultimate torsional shear strength, or modulus of rupture, can be expressed by

$$\tau_u = \frac{3M_{\max}}{2\pi a^3} \tag{10-13}$$

Large plastic strains in torsion can result in considerable changes in the length of the specimen. If both ends of the specimen are fixed, these length changes lead to the superposition of longitudinal stresses of unknown magnitude on the torsional shearing stresses. While these longitudinal stresses usually are small compared to the shearing stresses and can be ignored, they may be important in influencing the torsional strain to fracture. Extraneous stresses due to length change can be greatly minimized, at some increase in experimental complexity, by using a torsion tester in which one end of the specimen is free to move.

10-4 TYPES OF TORSION FAILURES

Figure 10-4 illustrates the state of stress at a point on the surface of a bar subjected to torsion. The maximum shear stress occurs on two mutually perpendicular planes, perpendicular to the longitudinal axis *yy* and parallel with the longitudinal axis *xx*. The principal stresses σ_1 and σ_3 make an angle of 45° with the longitudinal axis and are equal in magnitude to the shear stresses. σ_1 is a tensile stress, and σ_3 is an equal compressive stress. The intermediate stress σ_2 is zero.

Torsion failures are different from tensile failures in that there is little localized reduction of area or elongation. A ductile metal fails by shear along one of the planes of maximum shear stress. Generally the plane of the fracture is normal to the longitudinal axis (see Fig. 10-5*a*). A brittle material fails in torsion along a plane perpendicular to the direction of the maximum tensile stress. Since this plane bisects the angle between the two planes of maximum shear stress and makes an angle of 45° with the longitudinal and transverse directions, it results in a helical fracture (Fig. 10-5*b*). Fractures are sometimes observed in which the test section of the specimen breaks into a large number of fairly small pieces. In these cases it can usually be determined that the fracture started on a plane of maximum shear stress parallel with the axis of the specimen. A study of torsion

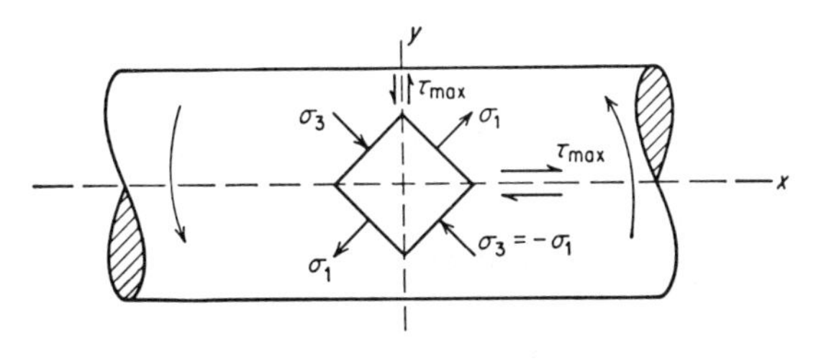

Figure 10-4 State of stress in torsion.

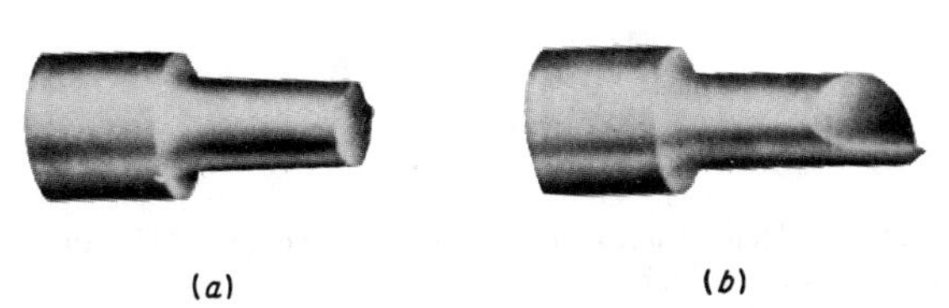

Figure 10-5 Typical torsion failures. (*a*) Shear (ductile) failure; (*b*) tensile (brittle) failure.

failures in a tool steel as a function of hardness[1] showed that fracture started on planes of maximum shear stress up to a Vickers hardness of 720 and that above this hardness tensile stresses were responsible for starting fracture.

10-5 TORSION TEST VS. TENSION TEST

A good case can be made for the position advanced by Sauveur[2] that the torsion test provides a more fundamental measure of the plasticity of a metal than the tension test. For one thing, the torsion test yields directly a shear-stress–shear-strain curve. This type of curve has more fundamental significance in characterizing plastic behavior than a stress-strain curve determined in tension. Large values of strain can be obtained in torsion without complications such as necking in tension or barreling due to frictional and effects in compression. Moreover, in torsion, tests can be made fairly easily at constant or high strain rates. On the other hand, unless a tubular specimen is used, there will be a steep stress gradient across the specimen. This will make it difficult to make accurate measurements of the yield strength.

The tension test and the torsion test are compared below in terms of the state of stress and strain developed in each test:

Tension test	*Torsion test*
$\sigma_1 = \sigma_{max};\ \sigma_2 = \sigma_3 = 0$	$\sigma_1 = -\sigma_3;\ \sigma_2 = 0$
$\tau_{max} = \frac{\sigma_1}{2} = \frac{\sigma_{max}}{2}$	$\tau_{max} = \frac{2\sigma_1}{2} = \sigma_{max}$
$\varepsilon_{max} = \varepsilon_1;\ \varepsilon_2 = \varepsilon_3 = -\frac{\varepsilon_1}{2}$	$\varepsilon_{max} = \varepsilon_1 = -\varepsilon_3;\ \varepsilon_2 = 0$
$\gamma_{max} = \frac{3\varepsilon_1}{2}$	$\gamma_{max} = \varepsilon_1 - \varepsilon_3 = 2\varepsilon_1$

$$\bar{\sigma} = \frac{\sqrt{2}}{2}\left[(\sigma_1 - \sigma_2)^2 + (\sigma_2 - \sigma_3)^2 + (\sigma_3 - \sigma_1)^2\right]^{1/2}$$

$$\bar{\varepsilon} = \frac{\sqrt{2}}{3}\left[(\varepsilon_1 - \varepsilon_2)^2 + (\varepsilon_2 - \varepsilon_3)^2 + (\varepsilon_3 - \varepsilon_1)^2\right]^{1/2}$$

$\bar{\sigma} = \sigma_1$	$\bar{\sigma} = \sqrt{3}\,\sigma_1$
$\bar{\varepsilon} = \sigma_1$	$\bar{\varepsilon} = \frac{2}{\sqrt{3}}\varepsilon_1 = \frac{\gamma}{\sqrt{3}}$

[1] R. D. Olleman, E. T. Wessel, and F. C. Hull, *Trans. Am. Soc. Met.*, vol. 46, pp. 87–99, 1954.
[2] A. Sauveur, *Am. Soc. Test. Mater. Proc.*, vol. 38, pt. 2, pp. 3–20, 1938.

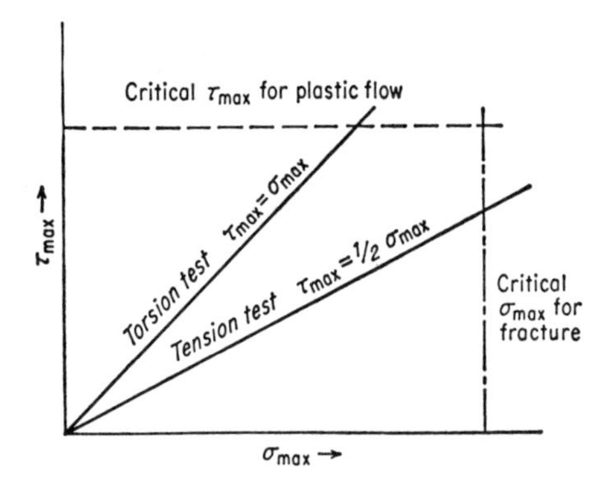

Figure 10-6 Effect of ratio τ_{max}/σ_{max} in determining ductility. *(After Gensamer.)*

This comparison shows that τ_{max} will be twice as great in torsion as in tension for a given value of σ_{max}. Since as a first approximation it can be considered that plastic deformation occurs on reaching a critical value of τ_{max} and brittle fracture occurs on reaching a critical value of σ_{max}, the opportunity for ductile behavior is greater in torsion than in tension. This is illustrated schematically in Fig. 10-6, which can be considered representative of the condition for a brittle material such as hardened tool steel. In the torsion test the critical shear stress for plastic flow is reached before the critical normal stress for fracture, while in tension the critical normal stress is reached before the shear stress reaches the shear stress for plastic flow. Even for a metal which is ductile in the tension test, where the critical normal stress is pushed far to the right in Fig. 10-6, the figure shows that the amount of plastic deformation is greater in torsion than in tension.

10-6 HOT TORSION TESTING

The torsion test is often used to obtain data on the flow properties[1] and fracture[2] of metals under hot-working conditions, that is, $T > 0.6T_m$ and $\bar{\varepsilon}$ up to $10^3\ s^{-1}$. Since it is easy to vary and control the speed of rotation, tests may be carried out over a wide range of strain rates. Moreover, by proper control of temperature and strain rate it is possible to simulate the metallurgical structures[3] produced in multiple-pass processes such as rolling. Because the torsion test specimen is not subject to necking, as in tension, or barreling, as in compression, it is possible to carry out the test to large plastic strains. Figure 10-7 shows typical torque-twist curves obtained with a hot torsion tester. The top curve (800°F) is for an

[1] W. J. McG. Tegart, The Role of Ductility in Hot Working, in "Ductility," chap. 5, American Society for Metals, Metals Park, Ohio, 1968.

[2] E. Shapiro and G. E. Dieter, *Metall. Trans.*, vol. 1, pp. 1711–1719, 1970; *Metall. Trans.*, vol. 2, pp. 1385–1391, 1971.

[3] M. M. Farag, C. M. Sellars, and W. J. McG. Tegart, in "Deformation Under Hot Working Conditions," pp. 60–67, Pub. No. 108, Iron and Steel Institute, London, 1968.

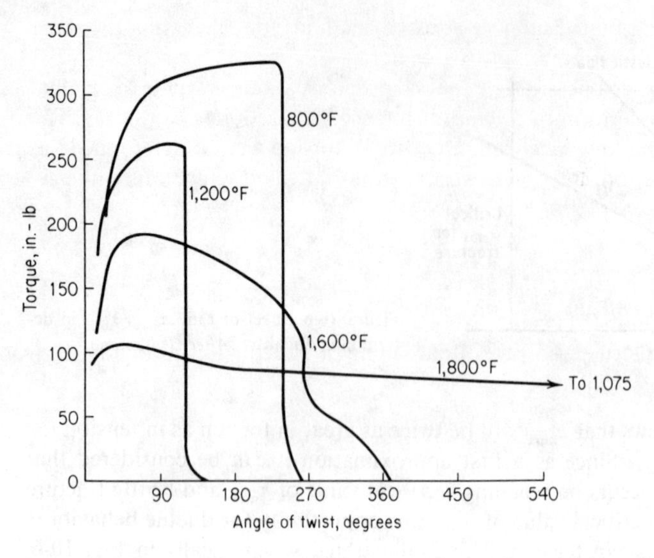

Figure 10-7 Torque-twist curves at different temperatures for a nickel-based alloy (Inconel 600) tested in torsion at a constant shear-strain rate of 2.5 s^{-1}. *(From E. Shapiro and G. E. Dieter, Metall. Trans., vol. 1, p. 1712, 1970.)*

essentially cold deformation condition in which the torque (flow stress) increases continuously up to fracture. The bottom curve (1800°F) is characteristic of hot-working deformation. The flow stress peaks at a relatively small plastic strain and then falls to an essentially constant value which is maintained until fracture occurs at a large plastic strain. The curve at 1600°F is characteristic of the deformation conditions known as warm-working.

The analysis discussed in Sec. 10-3 for converting torque-twist curves to shear-stress–shear-strain curves requires modification for use with hot torsion tests because it assumed that the stress is independent of strain rate. An analysis which considers the strong strain-rate dependence of flow stress at high temperature was proposed by Fields and Backofen.[1] It results in the relation

$$\frac{dM_T}{d\theta'} = \frac{M_T}{\theta'}(m + n)$$

where m is the strain-rate sensitivity and n is the strain-hardening exponent. At hot-working temperatures $m \gg n$. Substitution of the above expression into Eq. (10-11) results in

$$\tau_a = \frac{M_T}{2\pi a^3}(3 + m + n) \tag{10-14}$$

[1] D. S. Fields and W. A. Backofen, *Am. Soc. Test. Mater. Proc.*, vol. 57, pp. 1259–1272, 1957.

Further research[1] has resulted in more accurate methods for calculating the shear stress in torsion for a strain-rate sensitive material.

While necking is not possible in torsional deformation, a type of torsional plastic instability in the form of a band of intense localized flow can occur. The condition under which flow localization occurs in torsion can be determined[2] as follows. If the torque is a function of strain, strain rate, and temperature, then

$$M_T = M_T(\theta, \dot{\theta}, T)$$

and
$$dM_T = \left(\frac{\partial M_T}{\partial \theta}\right)_{\dot{\theta},T} d\theta + \left(\frac{\partial M_T}{\partial \dot{\theta}}\right)_{\theta,T} d\dot{\theta} + \left(\frac{\partial M_T}{\partial T}\right)_{\theta,\dot{\theta}} dT \qquad (10\text{-}15)$$

At a constant $\dot{\theta}$ the normalized rate of hardening or softening of torque is given by V, where

$$V \equiv \frac{1}{M_T}\left(\frac{dM_T}{d\theta}\right)_{\dot{\theta}} = \left[\left(\frac{\partial M_T}{\partial \theta}\right)_{\dot{\theta},T} d\theta + \left(\frac{\partial M_T}{\partial T}\right)_{\theta,\dot{\theta}} dT\right] \Big/ M_T\, d\theta \qquad (10\text{-}16)$$

so that Eq. (10-15) can be written as

$$dM_T = VM_T\, d\theta + \frac{M_T}{\dot{\theta}}\left(\frac{\partial \ln M_T}{\partial \ln \dot{\theta}}\right)_{\theta,T} d\dot{\theta} \qquad (10\text{-}17)$$

and $(\partial \ln M_T)/(\partial \ln \dot{\theta}) \approx (\partial \ln \tau_a)/(\partial \ln \dot{\gamma}_a) = m$, the strain-rate sensitivity obtained in uniaxial tension or compression. The instability occurs when $dM_T = 0$, so that

$$dM_T = 0 = VM_T\, d\theta + \frac{M_T}{\dot{\theta}} m\, d\dot{\theta}$$

and
$$\frac{1}{\dot{\theta}}\frac{d\dot{\theta}}{d\theta} = \frac{-V}{m} \qquad (10\text{-}18)$$

Flow localization in torsion is defined as the fractional change in local twisting rate per unit of twist, and this is expressed by the quantity $-V/m$. Materials tested under conditions which show high torque softening (large negative V) and low strain-rate sensitivity may be expected to undergo rapid flow localization.

BIBLIOGRAPHY

Marin, Joseph: "Mechanical Behavior of Engineering Materials," chap. 2, Prentice-Hall, Inc., Englewood Cliffs, N.J., 1962.

Tegart W. J. McG.: "Elements of Mechanical Metallurgy," chap. 3, The Macmillan Company, New York, 1966.

Torsion Testing, "Metals Handbook," 9th ed., vol. 8, pp. 137–184, American Society for Metals, Metals Park, Ohio, 1985.

[1] G. R. Canova, S. Shrivastava, J. J. Jones, and C. G'Sell, *ASTM STP* 753, pp. 189–210, 1982.
[2] S. L. Semiatin and G. D. Lahoti, *Met Trans.*, vol. 12A, pp. 1719–1728, 1981.

CHAPTER
ELEVEN
FRACTURE MECHANICS

11-1 INTRODUCTION

Chapter 7 provided a broad overview of the fracture of metals, particularly brittle fracture. It was shown that the theoretical cohesive stress is much greater than the observed fracture stress for metals. This led to the idea of defects or cracks which locally raise the stress to the level of the theoretical cohesive stress. It was shown that microcracks can be formed in metallurgical systems by a variety of mechanisms and that the critical step usually is the stress required to propagate the microcracks to a complete fracture. The first successful theoretical approach to this problem was the Griffith theory of brittle fracture (Sec. 7-4). Griffith's theory was modified by Orowan to allow for the degree of plasticity always present in the brittle fracture of metals. According to this approach the fracture stress is given by

$$\sigma_f \approx \left(\frac{E\gamma_p}{a} \right)^{1/2} \tag{11-1}$$

where E is Young's modulus and γ_p is the plastic work required to extend the crack wall for a crack of length $2a$. In this chapter we denote crack length by the symbol a, as is customary in the literature of fracture mechanics, rather than the symbol c.

Equation (11-1) was modified by Irwin[1] to replace the hard to measure γ_p with a term that was directly measurable.

$$\sigma_f = \left(\frac{E\mathscr{G}_c}{\pi a} \right)^{1/2} \tag{11-2}$$

[1] G. R. Irwin, Fracture, in "Encyclopedia of Physics," vol. VI, Springer, Berlin, 1958; G. R. Irwin, J. A. Kies, and H. L. Smith, *Am. Soc. Test. Mater. Proc.*, vol. 58, pp. 640–660, 1958.

where $\mathscr{G}_c$ corresponds to a critical value of the *crack-extension force*.

$$\mathscr{G} = \frac{\pi a \sigma^2}{E} \tag{11-3}$$

The crack-extension force has units of inch pounds per square inch (in-lb/in^2) or J/m^2. $\mathscr{G}$ also may be considered the *strain-energy release rate*, i.e., the rate of transfer of energy from the elastic stress field of the cracked structure to the inelastic process of crack extension. The critical value of $\mathscr{G}$ which makes the crack propagate to fracture, $\mathscr{G}_c$, is called the *fracture toughness* of the material.

This chapter shows how these concepts developed into the important tool of engineering analysis called *fracture mechanics*. Fracture mechanics makes it possible to determine whether a crack of given length in a material of known fracture toughness is dangerous because it will propagate to fracture at a given stress level. It also permits the selection of materials for resistance to fracture and a selection of the design which is most resistant to fracture.

The subject of fracture is discussed in other sections of this text. Fatigue crack propagation and its relationship to fracture mechanics is discussed in Chap. 12. Fracture occurring at elevated temperature is discussed in Chap. 13, while Chap. 14 considers a variety of engineering tests for evaluating the ductile-to-brittle transition in steel. Chapter 14 also considers environmental-assisted fracture processes such as stress corrosion cracking, hydrogen embrittlement, and liquid metal embrittlement.

11-2 STRAIN-ENERGY RELEASE RATE

In this section we will consider the significance of the strain-energy release rate in greater detail. Figure 11-1 shows how $\mathscr{G}$ can be measured. A single-edge notch specimen is loaded axially through pins. The sharpest possible notch is produced

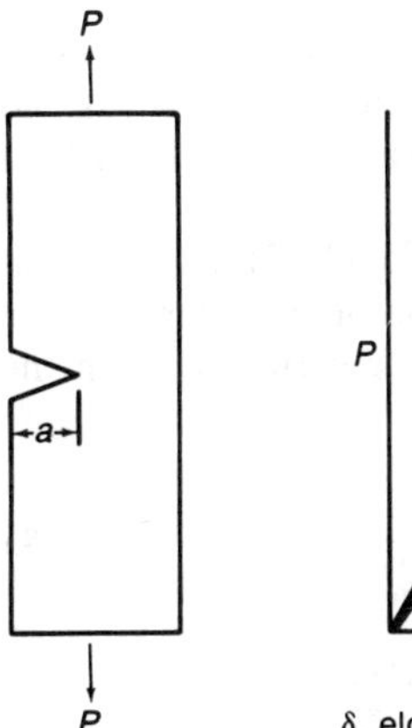

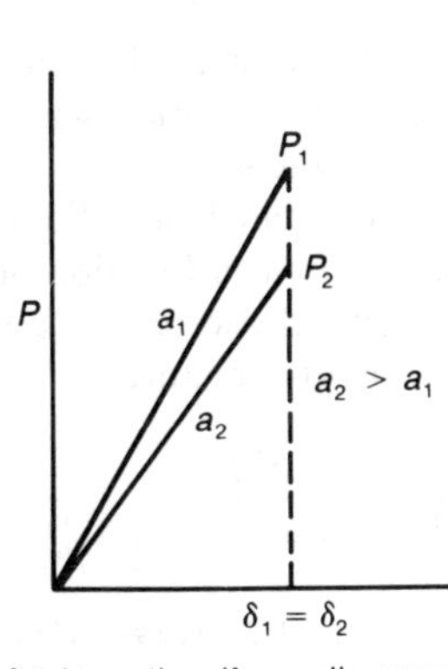

Figure 11-1 Determination of crack extension force $\mathscr{G}$.

by introducing a fatigue crack at the root of the machined notch. The displacement of this crack as a function of the axial force is measured with a strain-gage clip gage at the entrance to the notch. Load vs. displacement curves are determined for different length notches, where $P = M\delta$. M is the stiffness of a specimen with a crack of length a. The elastic strain energy is given by the area under the curve to a particular value of P and δ.

$$U_0 = \frac{1}{2}P\delta = \frac{P^2}{2M} \tag{11-4}$$

Consider the case shown in Fig. 11-1 where the specimen is rigidly gripped so that an increment of crack growth da results in a drop in load from P_1 to P_2.

$$\delta_1 = \delta_2 = \frac{P_1}{M_1} = \frac{P_2}{M_2}$$

Since P/M = constant

$$\left(\frac{\partial P}{\partial a}\right)\frac{1}{M} + P\frac{\partial(1/M)}{\partial a} = 0$$

$$\frac{\partial P}{\partial a} = -PM\frac{\partial(1/M)}{\partial a} \tag{11-5}$$

But, the crack extension force is defined as

$$\mathscr{G} = \left(\frac{\partial U_0}{\partial a}\right)_\delta = \frac{1}{2}\left[\frac{2P}{M}\frac{\partial P}{\partial a} + P^2\frac{\partial(1/M)}{\partial a}\right] \tag{11-6}$$

Upon substituting (11-5) into (11-6)

$$\mathscr{G} = -\frac{1}{2}P^2\frac{\partial(1/M)}{\partial a} \tag{11-7}$$

Note that the same equation would be derived[1] for a test condition of constant load except the sign of Eq. (11-7) would be reversed. For the fixed grip case no work is done on the system by the external forces $P\,d\delta$, while for the fixed load case external work equal to $P\,d\delta$ is fed into the system.

The strain-energy release rate can be evaluated from Eq. (11-7) by determining values of the specimen compliance $(1/M)$ as a function of crack length. The fracture toughness, or critical strain-energy release rate, is determined from the load, P_{max}, at which the crack runs unstably to fracture.

$$\mathscr{G}_c = \frac{P_{max}^2}{2}\frac{\partial(1/M)}{\partial a} \tag{11-8}$$

[1] G. R. Irwin and J. A. Kies, *Weld. J. Res. Suppl.*, vol. 33, p. 1935, 1954.

11-3 STRESS INTENSITY FACTOR

The stress distribution at the crack tip in a thin plate for an elastic solid in terms of the coordinates shown in Fig. 11-2 is given by Eqs. (11-9).

$$\begin{aligned}
\sigma_x &= \sigma\left(\frac{a}{2r}\right)^{1/2}\left[\cos\frac{\theta}{2}\left(1 - \sin\frac{\theta}{2}\sin\frac{3\theta}{2}\right)\right] \\
\sigma_y &= \sigma\left(\frac{a}{2r}\right)^{1/2}\left[\cos\frac{\theta}{2}\left(1 + \sin\frac{\theta}{2}\sin\frac{3\theta}{2}\right)\right] \\
\tau_{xy} &= \sigma\left(\frac{a}{2r}\right)^{1/2}\left[\sin\frac{\theta}{2}\cos\frac{\theta}{2}\cos\frac{3\theta}{2}\right]
\end{aligned} \tag{11-9}$$

where σ = gross nominal stress = P/wt. These equations are valid for $a > r > \rho$. For an orientation directly ahead of the crack ($\theta = 0$)

$$\sigma_x = \sigma_y = \sigma\left(\frac{a}{2r}\right)^{1/2} \quad \text{and} \quad \tau_{xy} = 0$$

Irwin pointed out that Eqs. (11-9) indicate that the *local stresses* near a crack depend on the product of the nominal stress σ and the square root of the half-flaw length. He called this relationship the *stress intensity factor* K, where for a sharp elastic crack in an infinitely wide plate, K is defined as

$$K = \sigma\sqrt{\pi a} \tag{11-10}$$

Note the K has the unusual dimensions of psi$\sqrt{\text{in}}$ or MN/m$^{3/2}$ or MPa$\sqrt{\text{m}}$. Using this definition for K, the equations for the stress field at the end of a crack can be written

$$\begin{aligned}
\sigma_x &= \frac{K}{\sqrt{2\pi r}}\left[\cos\frac{\theta}{2}\left(1 - \sin\frac{\theta}{2}\sin\frac{3\theta}{2}\right)\right] \\
\sigma_y &= \frac{K}{\sqrt{2\pi r}}\left[\cos\frac{\theta}{2}\left(1 + \sin\frac{\theta}{2}\sin\frac{3\theta}{2}\right)\right] \\
\tau_{xy} &= \frac{K}{\sqrt{2\pi r}}\left(\sin\frac{\theta}{2}\cos\frac{\theta}{2}\cos\frac{3\theta}{2}\right)
\end{aligned} \tag{11-11}$$

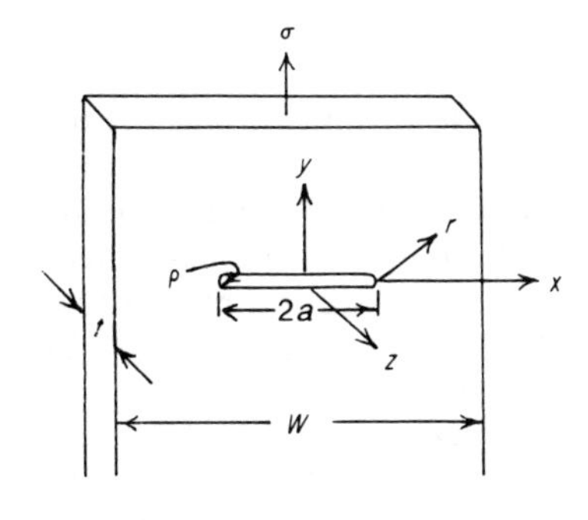

Figure 11-2 Model for equations for stresses at a point near a crack.

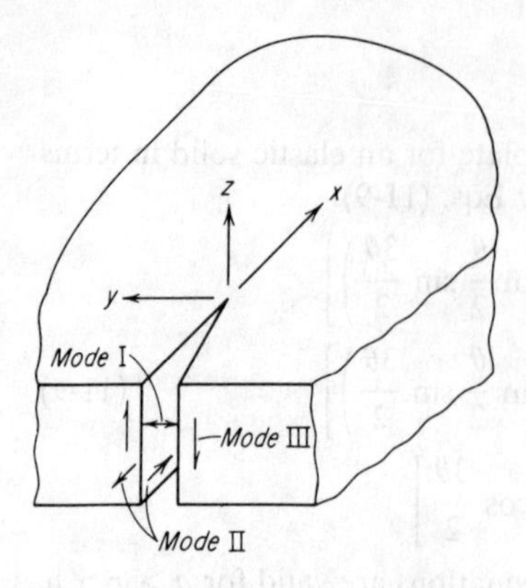

Figure 11-3 Crack-deformation modes.

It is apparent from Eq. (11-11) that the local stresses at the crack tip could rise to very high levels as r approaches zero. However, as discussed in Sec. 7-10, this does not happen because plastic deformation occurs at the crack tip.

The stress intensity factor K is a convenient way of describing the stress distribution around a flaw. If two flaws of different geometry have the same value of K, then the stress fields around each of the flaws are identical. Values of K for many geometrical cracks and many types of loading may be calculated[1] with the theory of elasticity. For the general case the stress intensity factor is given by

$$K = \alpha\sigma\sqrt{\pi a} \tag{11-12}$$

where α is a parameter that depends on the specimen and crack geometry. As an example, for a plate of width w loaded in tension with a centrally located crack of length $2a$

$$K = \sigma\sqrt{\pi a}\left(\frac{w}{\pi a}\tan\frac{\pi a}{w}\right)^{1/2} \tag{11-13}$$

In dealing with the stress intensity factor there are several modes of deformation that could be applied to the crack. These have been standardized as shown in Fig. 11-3. Mode I, the crack-opening mode, refers to a tensile stress applied in the y direction normal to the faces of the crack. This is the usual mode for fracture-toughness tests and a critical value of stress intensity determined for this mode would be designated K_{Ic}. Mode II, the forward shear mode, refers to a shear stress applied normal to the leading edge of the crack but in the plane of the crack. Mode III, the parallel shear mode, is for shearing stresses applied parallel to the leading edge of the crack. Mode I loading is the most important situation. There are two extreme cases for mode I loading. With thin plate-type specimens the stress state is plane stress, while with thick specimens there is a plane-strain condition. The plane-strain condition represents the more severe stress state and

[1] An excellent compendium of these relations is given by P. C. Paris and G. C. M. Sih, in J. E. Srawley and W. F. Brown (eds.), "Fracture Toughness Testing," p. 30, ASTM STP No. 381, Philadelphia, Pa. 1965.

the values of K_c are lower than for plane-stress specimens. Plane-strain values of critical stress intensity factor K_{Ic} are valid material properties, independent of specimen thickness, to describe the *fracture toughness* of strong materials like heat-treated steel, aluminum, and titanium alloys. Further details on fracture-toughness testing are given in Sec. 11-5.

While the crack-extension force $\mathcal{G}$ has a more direct physical significance to the fracture process, the stress intensity factor K is preferred in working with fracture mechanics because it is more amenable to analytical determination. By combining Eqs. (11-3) and (11-10), we see that the two parameters are simply related.

$$K^2 = \mathcal{G}E \quad \text{plane stress} \tag{11-14}$$

$$K^2 = \mathcal{G}E/(1 - \nu^2) \quad \text{plane strain} \tag{11-15}$$

11-4 FRACTURE TOUGHNESS AND DESIGN

A properly determined value of K_{Ic} represents the fracture toughness of the material independent of crack length, geometry, or loading system. It is a material property in the same sense that yield strength is a material property. Some values of K_{Ic} are given in Table 11-1.

The basic equation for fracture toughness illustrates the design tradeoff that is inherent in fracture mechanics design.

$$K_{Ic} = \sigma\sqrt{\pi a} \tag{11-16}$$

If the material is selected, K_{Ic} is fixed. Further, if we allow for the presence of a relatively large stable crack, then the design stress is fixed and must be less than K_{Ic}.

On the other hand, if the system is such that high strength and light weight are required, K_{Ic} is fixed because of limited materials with low density and high fracture toughness and the stress level must be kept high because of the need to maximize payload. Therefore, the allowable flaw size will be small, often below the level at which it can be easily detected with inspection techniques. These

Table 11-1 Typical values of K_{Ic}

Material	Yield strength, MPa	Fracture toughness K_{Ic}, $\text{MPa}\sqrt{\text{m}}$
4340 steel	1470	46
Maraging steel	1730	90
Ti-6Al-4V	900	57
2024-T3 Al alloy	385	26
7075-T6 Al alloy	500	24

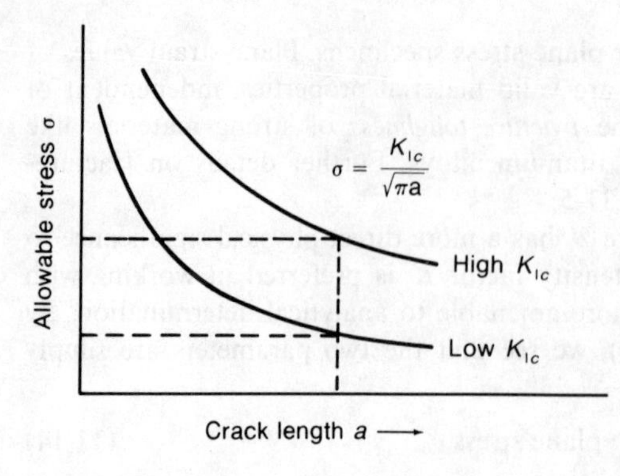

Figure 11-4 Relation between fracture toughness and allowable stress and crack size.

tradeoffs between fracture toughness, allowable stress, and crack size are illustrated in Fig. 11-4.

Example The stress intensity for a partial-through thickness flaw is given by $K = \sigma\sqrt{\pi a}\sqrt{\sec \pi a/2t}$ where a is the depth of penetration of the flaw through a wall thickness t. If the flaw is 5 mm deep in a wall 0.5 in thick, determine whether the wall will support a stress of 25,000 psi if it is made from 7075-T6 aluminum alloy.

From Table 11-1, $K_{Ic} = 24$ MPa$\sqrt{\text{m}}$. We will determine the critical stress level to make a 5-mm flaw propagate to failure in this material. $a = 5$ mm $= 5 \times 10^{-3}$ m; $2t = 1.0$ in $= 2.54$ cm $= 2.54 \times 10^{-2}$ m.

$$\sec \frac{\pi a}{2t} = \sec \frac{\pi(5 \times 10^{-3})}{2.54 \times 10^{-2}} = \sec 0.6184 = \frac{1}{\cos 0.6184} = \frac{1}{0.8148} = 1.227$$

$$\sigma = K_{Ic}/\sqrt{\pi a}\sqrt{\sec \pi a/2t} = \frac{24}{\sqrt{\pi \times 5 \times 10^{-3}}\sqrt{1.227}} = \frac{24}{\sqrt{0.019274}} = \frac{24}{0.139}$$

$$\sigma = 173 \text{ MPa}$$

But the applied stress is 25,000 psi = 172 MPa. Therefore, the flaw will propagate as a brittle fracture.

Example A thin-wall pressure vessel is made from Ti-6Al-4V with $K_{Ic} = 57$ MPa m$^{1/2}$ and $\sigma_0 = 900$ MPa. The internal pressure produces a circumferential hoop stress of 360 MPa. The crack is a semielliptical surface crack oriented with the major plane of the crack perpendicular to the uniform tensile hoop stress (see Fig. 11-5). For this type of loading and geometry the stress intensity factor is given by[1]

$$K_I^2 = \frac{1.21 a \pi \sigma^2}{Q}$$

where a = surface crack depth

σ = the applied nominal stress

$Q = \phi^2 - 0.212(\sigma/\sigma_0)^2$, where ϕ is an elliptic integral of the second kind

[1] G. R. Irwin, *J. Appl. Mech.*, vol. 24, pp. 109–114, 1957.

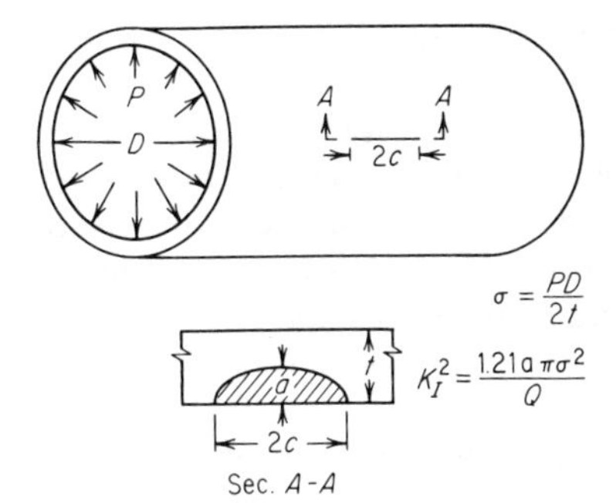

Figure 11-5 Flaw geometry and design of cylindrical pressure vessel.

Figure 11-6 shows the flaw-shape parameter Q plotted against the crack-shape ratio $a/2c$. These curves can be used both for surface cracks and internal cracks. Since internal cracks are less severe than surface cracks the previous equation becomes $K_I^2 = a\pi\sigma^2/Q$.

We want to find the size of the critical crack which will cause rupture of the pressure vessel with a wall thickness of 12 mm. $\sigma/\sigma_0 = 0.4$. If we assume $2c = 2a$, then $Q = 2.35$ and

$$a_c = \frac{K_I^2 Q}{1.21\pi\sigma^2} = \frac{(57)^2(2.35)}{1.21\pi(360)^2} = 0.01549 \text{ m} = 15.5 \text{ mm}$$

We note that the critical crack depth, 15.5 mm is greater than the thickness of the vessel wall, 12 mm. The crack will break through the vessel wall and the fluid would leak. This would be a "leak-before-break" condition.

However if the crack is very elongated, e.g., $a/2c = 0.05$, then $Q = 1.0$, and $a_c = 0.00659$ m $= 6.6$ mm. In this case the vessel would fracture when the crack had propagated about half-way through the wall thickness.

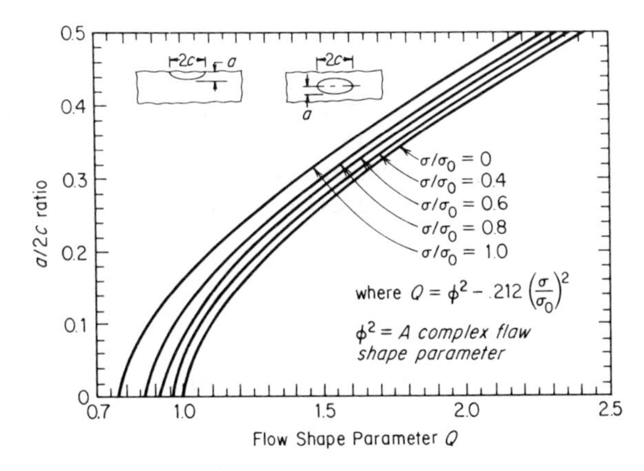

Figure 11-6 Flaw-shape parameter Q for surface and internal elliptical cracks.

11-5 K_{Ic} PLANE-STRAIN TOUGHNESS TESTING

The concepts of crack-extension force and stress intensity factor have been introduced and thoroughly explored. In this section we consider the testing procedures by which the fracture mechanics approach can be used to measure meaningful material properties. Since the methods of analysis are based on linear elastic fracture mechanics, these testing procedures are restricted to materials with limited ductility. Typical materials are high-strength steel, titanium, and aluminum alloys.

We have seen that the elastic stress field near a crack tip can be described by a single parameter called the *stress intensity factor* K. The magnitude of this stress intensity factor depends on the geometry of the solid containing the crack, the size and location of the crack, and the magnitude and distribution of the loads imposed on the solid. We saw that the criterion for brittle fracture in the presence of a crack-like defect was that unstable rapid failure would occur when the stresses at the crack tip exceeded a critical value. Since the crack-tip stresses can be described by the stress intensity factor K, a critical value of K can be used to define the conditions for brittle failure. As the usual test involves the opening mode of loading (mode I) the critical value of K is called K_{Ic}, *the plane-strain fracture toughness*. K_{Ic} can be considered a material property which describes the inherent resistance of the material to failure in the presence of a crack-like defect. For a given type of loading and geometry the relation is

$$K_{Ic} = \alpha\sigma\sqrt{\pi a_c} \tag{11-17}$$

where α is a parameter which depends on specimen and crack geometry and a_c is the critical crack length. If K_{Ic} is known, then it is possible to compute the maximum allowable stress for a given flaw size. While K_{Ic} is a basic material property, in the same sense as yield strength, it changes with important variables such as temperature and strain rate. For materials with a strong temperature and strain-rate dependence K_{Ic} usually decreases with decreased temperature and increased strain rate. For a given alloy, K_{Ic} is strongly dependent on such metallurgical variables as heat treatment, texture, melting practice, impurities, inclusions, etc.

There has been so much research activity and rapid development[1] in the field of fracture-toughness testing that, in a period of about 10 years, it has evolved from a research activity to a standardized procedure.[2] In the discussion of the influence of a notch on fracture (Sec. 7-10), we saw that a notch in a thick plate is far more damaging than in a thin plate because it leads to a *plane-strain* state of

[1] "Fracture Toughness Testing and its Applications," *ASTM STP* 381, 1965; "Plane Strain Crack Toughness Testing of High Strength Metallic Materials," *ASTM Spec. Tech. Publ.* 410, 1966; "Review of Developments in Plane Strain Fracture, Toughness Testing," *ASTM Spec. Tech. Publ.* 463, 1970.

[2] ASTM Standards, Designations E399-70T, A. S. Kobayashi (ed.), "Experimental Techniques in Fracture Mechanics," Society for Experimental Stress Analysis, Westport, Conn., 1973.

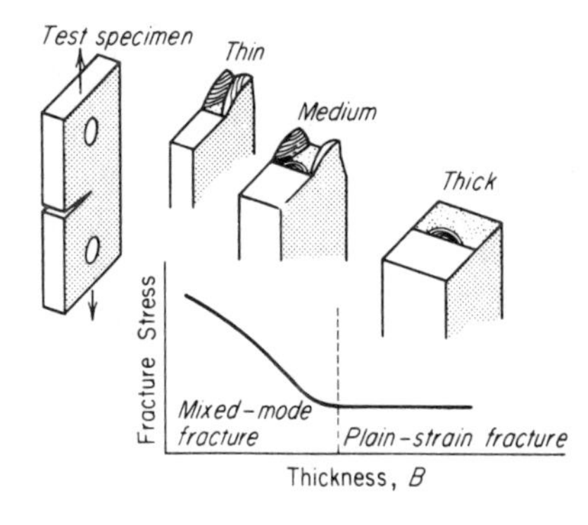

Figure 11-7 Effect of specimen thickness on stress and mode of fracture. *(From C. C. Osgood, Machine Design, August 5, 1971, p. 91.)*

stress with a high degree of triaxiality. The fracture toughness measured under plane-strain conditions is obtained under maximum constraint or material brittleness. The plane-strain fracture toughness is designated K_{Ic} and is a true material property. Figure 11-7 shows how the measured fracture stress varies with specimen thickness B. A mixed-mode, ductile brittle fracture with 45° shear lips is obtained for thin specimens. Once the specimen has the critical thickness for the toughness of the material, the fracture surface is flat and the fracture stress is constant with increasing specimen thickness. The minimum thickness to achieve plane-strain conditions and valid K_{Ic} measurements is

$$B = 2.5\left(\frac{K_{Ic}}{\sigma_0}\right)^2 \tag{11-18}$$

where σ_0 is the 0.2 percent offset yield strength.

A variety of specimens have been proposed for measuring K_{Ic} plane-strain fracture toughness. The three specimens shown in Fig. 11-8 represent the most common specimen designs. The compact tension specimen and the three-point loaded bend specimen have been standardized by ASTM.[1] Other useful specimen designs are the single-edge-cracked-plate specimen (Fig. 11-7), the center-cracked and double-edge-cracked plate.[2] After the notch is machined in the specimen, the sharpest possible crack is produced at the notch root by fatiguing the specimen in a low-cycle, high-strain mode (typically 1,000 cycles with a strain of 0.03 in/in). The initial crack length a_i includes both the depth of the notch and the length of the fatigue crack. Plane-strain toughness testing is unusual because there can be no advance assurance that a valid K_{Ic} will be determined in a particular test. Equation (11-18) should be used with an estimate of the expected K_{Ic} to determine the specimen thickness.

The test must be carried out in a testing machine which provides for a continuous autographic record of load P and relative displacement across the open end of the notch (proportional to crack displacement). The three types of

[1] Annual Book of ASTM Standards, vol. 03.01, Metal Test Methods, Designation E399-707.

[2] W. F. Brown and J. E. Srawley, *ASTM Spec. Tech. Publ.* 410, 1966; J. M. Kraft, "Techniques of Metals Research," vol. V, pt. 2, chap. 7, John Wiley & Sons, Inc., New York, 1971.

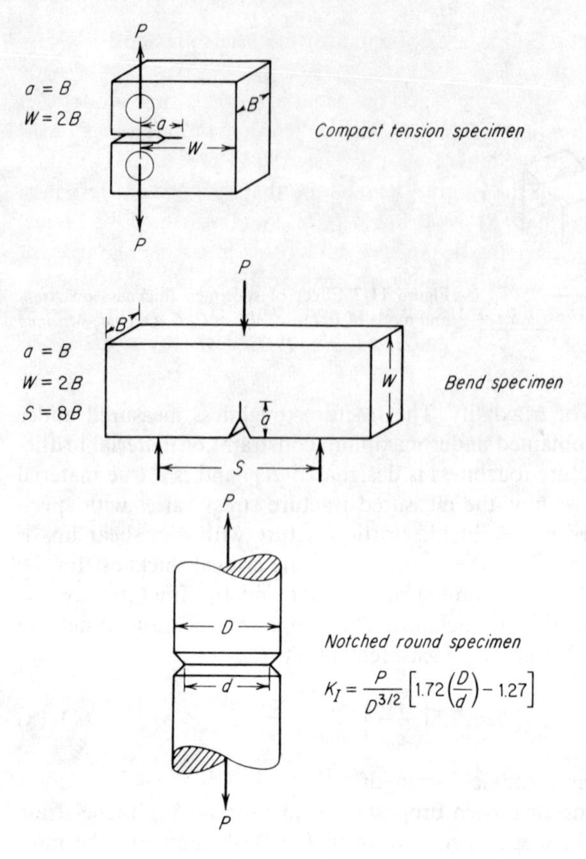

Figure 11-8 Common specimens for K_{Ic} testing.

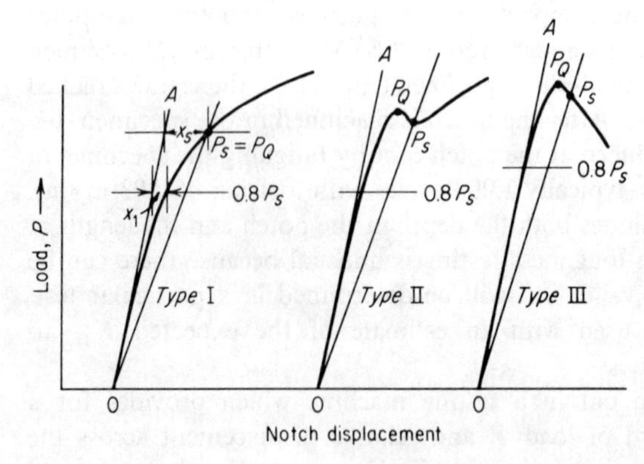

Figure 11-9 Load-displacement curves. (Note that slope OP_s is exaggerated for clarity.)

load–crack-displacement curves that are obtained for materials are shown in Fig. 11-9. The type I load-displacement curve represents the behavior for a wide variety of ductile metals in which the crack propagates by a tearing mode with increasing load. This curve contains no characteristic features to indicate the load corresponding to the onset of unstable fracture. The ASTM procedure is to first draw the secant line OP_s from the origin with a slope that is 5 percent less than the tangent OA. This determines P_s. Next draw a horizontal line at a load equal to 80 percent of P_s and measure the distance x_1 along this line from the tangent OA to the actual curve. If x_1 exceeds one-fourth of the corresponding distance x_s at P_s, the material is too ductile to obtain a valid K_{Ic} value. If the material is not too ductile, then the load P_s is designated P_Q and used in the calculations described below.

The type II load-displacement curve has a point where there is a sharp drop in load followed by a recovery of load. The load drop represents a "pop in" which arises from sudden unstable, rapid crack propagation before the crack slows-down to a tearing mode of propagation. The same criteria for excessive ductility is applied to type II curves, but in this case P_Q is the maximum recorded load.

The type III curve shows complete "pop in" instability where the initial crack movement propagates rapidly to complete failure. This type of curve is characteristic of a very brittle "elastic material."

The value of P_Q determined from the load-displacement curve is used to calculate a conditional value of fracture toughness denoted K_Q. The equations relating specimen geometry (see Fig. 11-8), crack length a, and critical load P_Q are Eqs. (11-19) and (11-20).

For the compact tension specimen:

$$K_Q = \frac{P_Q}{BW^{1/2}}\left[29.6\left(\frac{a}{W}\right)^{1/2} - 185.5\left(\frac{a}{W}\right)^{3/2} + 655.7\left(\frac{a}{W}\right)^{5/2} - 1017.0\left(\frac{a}{W}\right)^{7/2} + 638.9\left(\frac{a}{W}\right)^{9/2}\right] \tag{11-19}$$

For the bend specimen:

$$K_Q = \frac{P_Q S}{BW^{3/2}}\left[2.9\left(\frac{a}{W}\right)^{1/2} - 4.6\left(\frac{a}{W}\right)^{3/2} + 21.8\left(\frac{a}{W}\right)^{5/2} - 37.6\left(\frac{a}{W}\right)^{7/2} + 38.7\left(\frac{a}{W}\right)^{9/2}\right] \tag{11-20}$$

The crack length a used in the equations is measured after fracture. Next calculate the factor $2.5(K_Q/\sigma_0)^2$. If this quantity is less than both the thickness and crack length of the specimen, then K_Q is equal to K_{Ic} and the test is valid. Otherwise it is necessary to us a thicker specimen to determine K_{Ic}. The measured value of K_Q can be used to estimate the new specimen thickness through Eq. (11-18).

11-6 PLASTICITY CORRECTIONS

The expressions for the elastic stress field at a crack, Eq. (11-9), describe a stress singularity at the tip of the crack. In reality, metals will yield when $\sigma = \sigma_0$ and a plastic zone will exist at the crack tip. This is illustrated in Fig. 11-10. Out to a distance $r = r_p$ the elastic stress σ_y is greater than the yield stress σ_0. To a first approximation this distance r_p is the size of the plastic zone. From Eq. (11-11) for $\theta = 0$,

$$\sigma_y = \frac{K}{\sqrt{2\pi r_p}} = \sigma_0$$

and

$$r_p = \frac{K^2}{2\pi\sigma_0^2} = \frac{\sigma^2 a}{2\sigma_0^2}$$

However, it is evident that the plastic zone must be larger than r_p because it does not allow for yielding caused by the elastic stress distribution from $\sigma_y = \sigma_0$ out to $\sigma_y = \sigma_{max}$. The load-carrying capability that is represented by the shaded area in Fig. 11-10 must be compensated for by extending the dimension of the plastic zone. More detailed analysis shows that the diameter of the plastic zone is $2r_p$.

Irwin[1] proposed that the existence of a plastic zone makes the crack act as if it were longer than its physical size, i.e., as a result of crack-tip plasticity the displacements are larger and the stiffness is lower than for the strictly elastic situation. The usual correction is to assume that the effective crack length is the actual length plus the radius of the plastic zone.

$$a' = a_{\text{eff}} = a + r_p \tag{11-21}$$

where

$$r_p \approx \frac{1}{2\pi}\frac{K^2}{\sigma_0^2} \quad \text{(plane stress)}$$

$$r_p \approx \frac{1}{6\pi}\frac{K^2}{\sigma_0^2} \quad \text{(plane strain)}^2$$

The smaller value of r_p for plane strain is consistent with the fact that the triaxial stress field in plane strain suppresses the extent of plastic deformation.

If we apply Irwin's correction to the fracture toughness of a center-notched tensile specimen, Eq. (11-13), we will get

$$K = \sigma\sqrt{\pi a'}\left[\frac{w}{\pi a'}\tan\frac{\pi a'}{W} + \frac{1}{2W}\frac{K^2}{\sigma_0^2}\right] \tag{11-22}$$

Since a' is a function of K the value of the stress intensity factor must be determined by an iterative process.

[1] G. R. Irwin, Handbuch der Physik, op. cit.

[2] F. A. McClintock and G. R. Irwin, *ASTM STP* 381, p. 84, 1965.

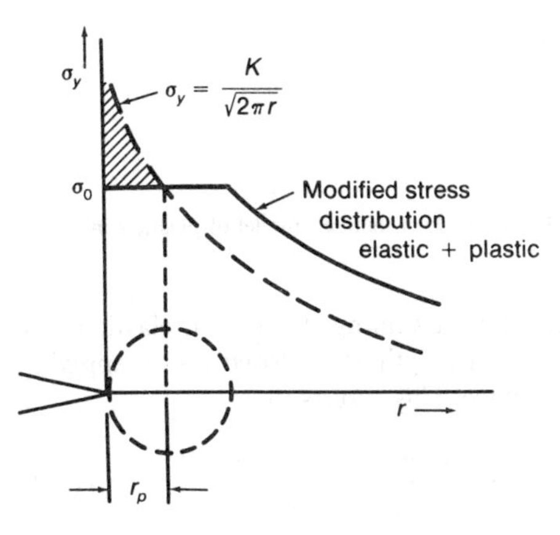

Figure 11-10 Estimation of plastic zone size.

Example A steel plate with a through thickness crack of length $2a = 20$ mm is subjected to a stress of 400 MPa normal to the crack. If the yield strength of the steel is 1500 MPa, what is the plastic zone size and the stress intensity factor for the crack. Assume that the plate is infinitely wide.

$$r_p = \frac{1}{2\pi}\frac{\sigma^2 a}{\sigma_0^2} = \frac{(400)^2(10 \times 10^{-3})}{2\pi(1500)^2} = 1.13 \times 10^{-4} \text{ m} = 0.113 \text{ mm}$$

For an infinite wide plate $K = \sigma\sqrt{\pi a}$ and $K_{\text{eff}} = \sigma\sqrt{\pi a'}$

$$K_{\text{eff}} = \sigma\sqrt{\pi}\sqrt{a + r_p} = \sigma\sqrt{\pi}\sqrt{a + \frac{K^2}{2\pi\sigma_0^2}} = \sigma\sqrt{\pi}\sqrt{a + \frac{\sigma^2\pi a}{2\pi\sigma_0^2}}$$

$$K_{\text{eff}} = \sigma\sqrt{\pi a}\sqrt{1 + \frac{\sigma^2}{2\sigma_0^2}} = 400\sqrt{\pi \times 10 \times 10^{-3}}\sqrt{1 + \frac{160{,}000}{2(2{,}250{,}000)}}$$

$$= 400(1.772 \times 10^{-1})(1 + 0.035)^{1/2} = 70.9(1.018) = 72.2 \text{ MPa m}^{1/2}$$

Note that the plastic zone has only a small effect for this high-strength steel, but if σ_0 is changed to 700 MPa then

$$K_{\text{eff}} = 70.9\left(1 + \frac{160{,}000}{980{,}000}\right)^{1/2}$$

$$= 70.9(1.1633)^{1/2} = 70.9(1.078) = 76.4 \text{ MPa m}^{1/2}$$

Another model of the plastic zone at a crack tip was proposed by Dugdale[1] for the case of plane stress. Dugdale considered that the plastic zone takes the form of narrow strips extending a distance from each crack tip

[1] D. S. Dugdale, *J. Mech. Phys. Sol.*, vol. 8, pp. 100–108, 1960; see also B. A. Bilby, A. H. Cottrell, and K. H. Swinden, *Proc. Roy. Soc.*, vol. A272, p. 304, 1963.

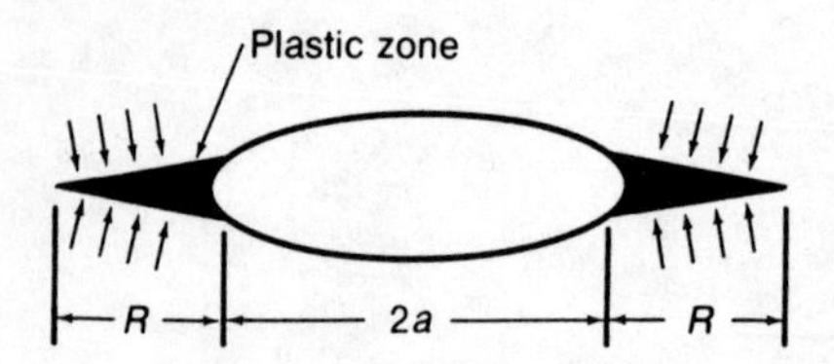

Figure 11-11 Dugdale's model of plastic zone.

(Fig. 11-11). The analysis is conducted by assuming there is an elastic crack of length $2(a + R)$ and that the region of length R is closed up by applying compressive stresses normal to the surface. This results in

$$\frac{a}{a + R} = \cos\left(\frac{\pi}{2}\frac{\sigma}{\sigma_0}\right) \tag{11-23}$$

and neglecting the higher-order terms in the cosine series

$$R = \frac{\pi^2\sigma^2 a}{8\sigma_0^2} = \frac{\pi K^2}{8\sigma_0^2} \tag{11-24}$$

11-7 CRACK-OPENING DISPLACEMENT

The linear elastic fracture mechanics (LEFM) approach works well for high-strength materials ($\sigma_0 > 200{,}000$ psi) but it is less universally applicable for low-strength structural materials. There is a limit to the extent to which K can be adjusted for crack-tip plasticity by the methods of the previous section. When r_p becomes an appreciable fraction of a other approaches become necessary.[1]

The *crack-tip displacement* concept[1] considers that the material ahead of the crack contains a series of miniature tensile specimens having a gage length l and a width w (Fig. 11-12). The length of the sample is determined by the root radius of the crack ρ, and the width is limited by those microstructural factors which control the ductility. In this model, crack growth occurs when the specimen adjacent to the crack is fractured. When the failure of the first specimen adjacent to the crack tip immediately causes the next specimen to fail, and so on, the overall fracture process is unstable and crack propagation occurs under decreasing stress. When each specimen does onto fail immediately in turn, we have a situation of slow crack growth where the applied stress must be increased for stable crack growth to continue.

If a thick plate is loaded in tension (mode I) so that plane-strain conditions prevail, the plastic deformation at the crack tip is confined to narrow bands with a thickness of the order of the diameter of the crack tip 2ρ. The displacement

[1] D. G. H. Latzko (ed.), "Post-yield Fracture Mechanics," Applied Sci. Publ., London, 1979.

[1] A. H. Cottrell, *Proc. R. Soc. London*, vol. 285, p. 10, 1965; see also A. S. Telleman and A. J. McEvily, Jr., op. cit., pp. 60–78.

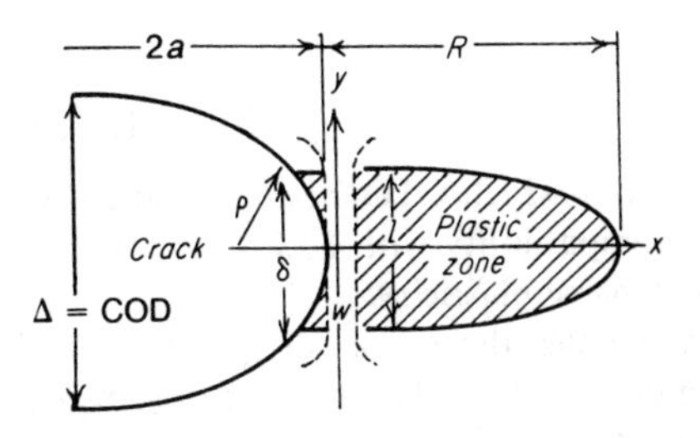

Figure 11-12 Model for crack-tip displacement concept and crack-opening displacement.

(deformation) of the miniature tensile specimen at the crack tip is

$$\delta = \varepsilon l = \varepsilon 2\rho \tag{11-25}$$

Unstable fracture occurs when the strain in the specimen adjacent to the crack tip reaches the tensile ductility of the specimen ε_f. Thus, the fracture criterion is

$$\delta_c = 2\rho\varepsilon_f \tag{11-26}$$

In *plane-stress* (thin-plate) tensile loading the strains at the crack tip are distributed over a distance of the order of the sheet thickness t. Thus,

$$\delta = \varepsilon l = \varepsilon t \tag{11-27}$$

and the critical crack-opening displacement for fracture is

$$\delta_c = \varepsilon_f t \tag{11-28}$$

Using the Dugdale crack model, Fig. 11-11, the crack-tip opening displacement (CTOD) for a crack of length $2a$ in an infinite thin plate subjected to uniform tension σ in a material where plastic deformation occurs at the crack tip is given by[1]

$$\delta = \frac{8\sigma_0 a}{\pi E} \ln\left(\sec\frac{\pi\sigma}{2\sigma_0}\right) \tag{11-29}$$

Expanding the secant term and keeping only the first term results in

$$\delta = \frac{\pi\sigma^2 a}{E\sigma_0} \tag{11-30}$$

However, remembering the definition of strain-energy release rate and dividing each by σ_0

$$\frac{\mathcal{G}}{\sigma_0} = \frac{\pi a\sigma^2}{E\sigma_0} \tag{11-31}$$

we readily see that

$$\mathcal{G} = \sigma_0\delta \tag{11-32}$$

[1] G. T. Hahn and A. R. Rosenfeld, *Acta Metall.*, vol. 13, p. 293, 1965.

Therefore, unstable fracture occurs when

$$\mathcal{G}_{\mathrm{I}c} = \lambda \sigma_0 \delta_c \tag{11-33}$$

The factor λ depends upon the exact location at which CTOD is determined. It has been suggested[1] that $\lambda \approx 2.1$ provides compatibility with LEFM conditions and that $\lambda = 1.0$ applies to conditions of extensive plasticity.

Widespread plasticity at the crack tip enables the crack surfaces to move apart at the crack tip without an increase in crack length. This relative movement of the two crack faces at a distance removed from the crack tip is called the crack-opening displacement (COD) (see Fig. 11-12). The COD, denoted by Δ, is measured with a clip gage. Measurement of Δ provides an indirect way to measure δ. If the origin of measurement is at the center of a crack of length $2a$, then

$$\mathrm{COD} = \Delta = \frac{4\sigma}{E}\left[(a + r_p)^2 - x^2\right]^{1/2} \tag{11-34}$$

where x is the distance from the center of the crack toward the crack tip. The crack-tip opening displacement (CTOD) is $\Delta = \delta$ and it is measured at $x = a$ and when $r_p \ll a$

$$\delta = \frac{4\sigma}{E}(2ar_p)^{1/2} \tag{11-35}$$

Expanding Eq. (11-34) and substituting Eq. (11-35) for the ar_p term, we get

$$\Delta = \frac{4\sigma}{E}\left(a^2 - x^2 + \frac{E^2}{16\sigma^2}\delta^2\right)^{1/2} \tag{11-36}$$

This equation can be used to determine δ_c from measured values of COD. Equation (11-36) also illustrates the basic problem with the COD approach. The strain fields and crack-opening displacements will vary with specimen geometry. Thus, it is difficult to define a single critical COD value for a material in the way that we can determine $K_{\mathrm{I}c}$.

11-8 *J* INTEGRAL

A more comprehensive approach to the fracture mechanics of lower-strength ductile materials is provided by the J integral.[2] Rice[3] showed that the line integral related to the energy in the vicinity of a crack can be used to solve two-dimensional crack problems in the presence of plastic deformation. Fracture occurs when the J integral reaches a critical value. J has units of MN/m or in-lb/in².

$$J = \int_\Gamma \left(W\,dy - T\frac{\partial u}{\partial x}\,ds \right) \tag{11-37}$$

[1] A. A. Wells, *Eng. Fract. Mech.*, vol. 1, p. 399, 1970.
[2] R. O. Ritchie, *Trans. ASME, J. Eng. Mater. Tech.*, vol. 105, pp. 1–7, 1983.
[3] J. R. Rice, *Trans. ASME Ser. E., J. Appl. Mech.*, vol. 90, pp. 379–386, 1968.

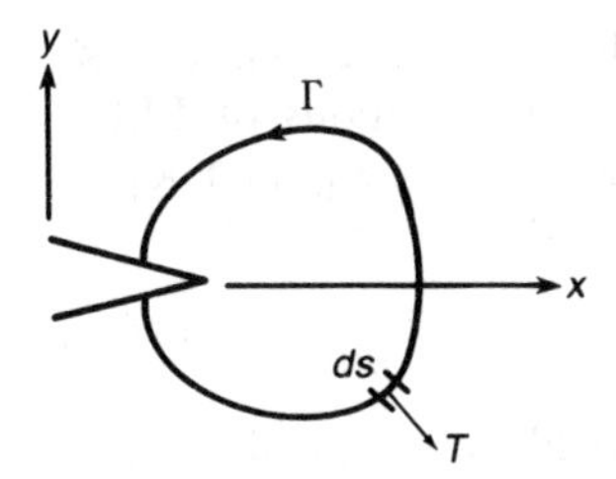

Figure 11-13 Sketch of Γ contour drawn around a crack tip to define the J integral.

where $W = \int \sigma_{ij}\, d\varepsilon_{ij}$ is the strain energy per unit volume due to loading
Γ is the path of the integral which encloses the crack (see Fig. 11-13).
T is the outward traction (stress) vector acting on the contour around the crack
u is the displacement vector
ds is an increment of the contour path
x, y are the rectangular coordinates
$T(\partial u/\partial x)\, ds$ is the rate of work input from the stress field into the area enclosed by Γ.

Rice has shown that the J integral is path-independent. Therefore J can be determined from a stress analysis where σ and ε are established by finite element analysis around the contour enclosing the crack. Since J is path-independent we may choose the most convenient path, usually the specimen boundary.

The J integral can be interpreted as the potential energy difference between two identically loaded specimens having slightly different crack lengths. This is illustrated in Fig. 11-14.

$$J = \frac{\partial U_0}{\partial a} = \mathscr{G} = \frac{K^2}{E'} \tag{11-38}$$

where $E' = E$ (plane stress)
$E' = E/1 - \nu^2$ (plane strain)

Equation (11-38) is at the heart of the J-integral approach. It says that the value of J (obtained under elastic-plastic conditions) is numerically equal to the strain-energy release rate (obtained under elastic conditions). This equivalence has

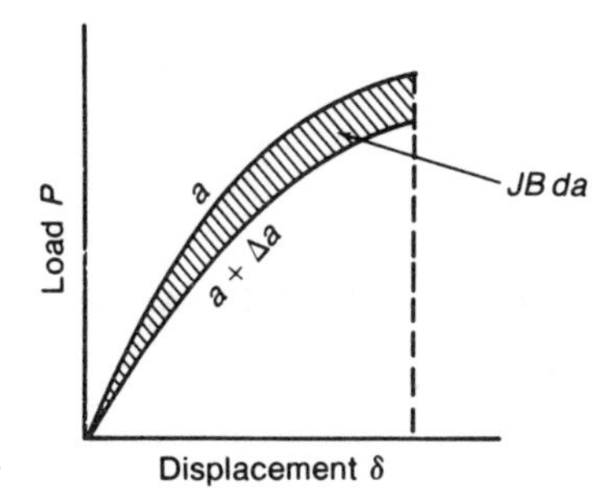

Figure 11-14 Physical interpretation of the J integral. Note that B is specimen thickness.

been demonstrated[1] by measuring J_{Ic} from small fully plastic specimens and $\mathcal{G}_{Ic}$ from large elastic specimens satisfying the plane-strain conditions of LEFM. Thus, J_{Ic} can be used as a fracture criterion in the same way as $\mathcal{G}_{Ic}$ and K_{Ic}.

The crack-tip opening displacement δ_t can be related[2] to J by

$$\delta_t = f(\varepsilon_0, n) J/\sigma_0 \tag{11-39}$$

where f is a proportionality factor dependent on the yield strain ε_0 and the strain hardening exponent n. f has been found to vary from 1 for $n = 1$ to 0.4 for $n = 0.3$ in plane stress and from 0.8 for $n = 1$ to 0.3 for $n = 0.3$ in plane strain.

The underlying assumption of the J-integral approach is that the material deformation can be described by the deformation theory of plasticity, where stresses and strains are functions only of the point of measurement and not of the path taken to get to that point. This is a good assumption for a stationary crack subjected to monotonically increasing load, where conditions will not depart far from proportional loading. However, for growing cracks where regions of elastic unloading and nonproportional plastic flow may be encountered, the behavior is not properly modeled by deformation theory. The J-integral approach is being used, apparently successfully, in these situations but there is no theoretical justification.

A standard test procedure for determining the fracture toughness of ductile materials with the J-integral method has been developed.[3] The three-point bend specimen or the compact tension specimen (see Fig. 11-8) generally are used. Using a series of identical test specimens (the multispecimen approach) or a single test specimen with an independent method of monitoring crack growth, values of J are determined at different amounts of crack extension Δa. The J integral is evaluated from the following:

For the three-point bend specimen[4]

$$J = \frac{2A}{Bb} \tag{11-40}$$

where A = area under the load vs. displacement curve
b = unbroken ligament ($W - a$ in Fig. 11-8)
B = specimen thickness.

For the compact tension specimen[5]

$$J = \frac{2A}{Bb}\left[\frac{(1+\alpha)}{(1+\alpha^2)}\right] \tag{11-41}$$

[1] J. A. Begley and J. D. Landes, *ASTM STP* 514, pp. 1–24, 1972.
[2] C. F. Shih, *J. Mech. Phys. Sol.*, vol. 29, pp. 305–330, 1981.
[3] ASTM Standards, pt. 31, Designation E813-81; J. D. Landes, H. Walker, and G. A. Clarke, Elastic-Plastic Fracture, *ASTM STP* 668, pp. 266–287, 1979.
[4] J. R. Rice, P. C. Paris, and J. G. Merkle, *ASTM STP* 532, p. 231, 1973.
[5] G. A. Clarke and J. D. Landis, *J. Test Eval.*, vol. 7, no. 5, pp. 264–269, 1979.

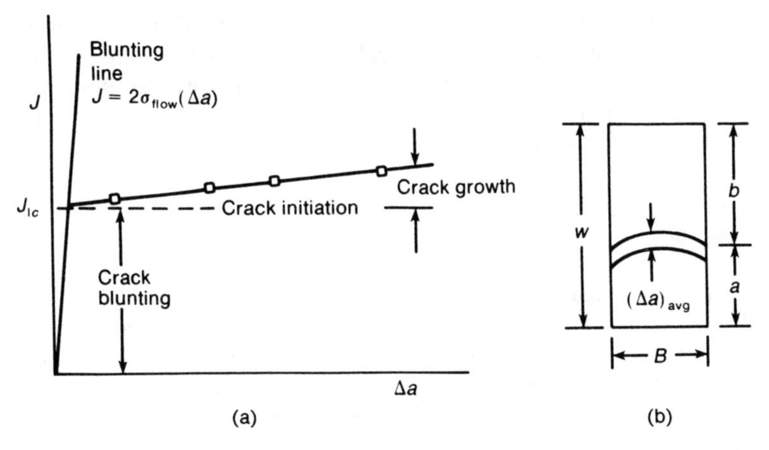

Figure 11-15 (*a*) J vs. Δa curve for establishing J_{Ic}; (*b*) sketch of a specimen fracture surface showing how Δa is determined.

where
$$\alpha = \left[\left(\frac{2a}{b}\right)^2 + 2\left(\frac{2a}{b}\right) + 2\right]^{1/2} - \left(\frac{2a}{b} + 1\right)$$

The data are plotted as a crack-resistance curve, J vs. Δa (Fig. 11-15*a*). Each specimen contains a crack-starter fatigue crack. The crack growth due to loading to a given displacement level (Δa) is "marked" by heat-tinting a steel specimen at 300°C or introducing a few cycles of fatigue crack growth in materials which do not discolor by heat-tinting. Then the specimen is broken and Δa is measured from the fracture surface (Fig. 11-15*b*). The beginning of stable crack extension is marked by the end of the flat fatigue precracked area. The end of crack extension is marked by the end of the heat-tint region or the beginning of the second flat fatigue area. The value of J_{Ic} is established by extrapolating the linear portion of the $J - \Delta a$ plot to its intersection with the *blunting line*. This is given by $J = 2\sigma_{flow}(\Delta a)$, where $\sigma_{flow} = (\sigma_0 + \sigma_u)/2$, and the line is defined based on the assumption that the crack advance is equal to one-half at the crack-tip opening displacement.

For J to be utilized as a geometry-independent parameter to describe crack extension the region ahead of the crack tip that is enclosed by the J integral must be large compared to the microstructural deformation and fracture events which are involved. This implies that certain specimen size requirements must be met for the J analysis to be relevant. Unlike the LEFM case, these size limitations can vary markedly for different specimen geometries. For the edge-cracked bend specimen

$$b \geq 25J_{Ic}/\sigma_0$$

but for the center-cracked tension specimen

$$b \geq 200J_{Ic}/\sigma_0$$

However, for the three-point bend specimen or compact tension specimen the J-integral approach requires a thickness much less than that required to determine a valid K_{Ic} in a ductile material. For example, an A533B nuclear pressure vessel steel required a 2-ft-thick specimen to obtain a valid K_{Ic} but a value of only $b = 0.5$ in was needed with the J_{Ic} test.

11-9 *R* CURVE

The R curve characterizes the resistance to fracture of a material during slow and stable crack propagation as the plastic zone grows as the crack extends from a sharp notch. It provides a record of the toughness of a material as the crack extends under increasing crack-extension forces. An R curve is a graphical representation of the resistance to crack propagation, R, versus the crack length a.

Irwin[1] suggested that failure (crack instability) will occur when the rate of change of strain-energy release rate $\partial \mathcal{G}/\partial a$ equals the rate of change in resistance to crack growth $\partial R/\partial a$. Thus, $\mathcal{G}_c$ is defined by the point in Fig. 11-16*a* where $\partial \mathcal{G}/\partial a = \partial R/\partial a$. By contrast, the R curve for a brittle material is shown in Fig. 11-16*b*. Note that the crack does not extend until the curve becomes tangent to the R curve, whereupon unstable fracture occurs ($\mathcal{G} = \mathcal{G}_c$). For the more general case of a material with some ductility, crack extension occurs when $\mathcal{G} > R$. Consider the $\mathcal{G}$ curve labelled $\mathcal{G}_1$ in Fig. 11-16*a*. At the load or stress corresponding to this value at $\mathcal{G}_1$, the crack will propagate stably from an original length a_0 to a_1 since $\mathcal{G} > R$. However, the crack will not extend beyond a_1 because $\mathcal{G} < R$. For additional crack extension to occur the value of $\mathcal{G}$ must increase, until $\mathcal{G} = R$, when unstable fracture occurs.

The ASTM has established a standard procedure[2] for determining R curves. A compact tension specimen is loaded incrementally, allowing time between steps for the crack to stabilize before measuring load P and crack length a. At each step calculate K_R (equivalent to R) from

$$K_R = P/B\sqrt{W} \times f(a/W) \tag{11-42}$$

where

$$f(a/W) = \left[(2 + a/W)/(1 - a/W)^{3/2}\right]\left[0.866 + 4.64(a/W) - 13.32(a/W)^2 + 14.72(a/W)^3 - 5.6(a/W)^4\right]$$

The crack length used in Eq. (11-42) is the *effective crack length* where the physical crack growth is corrected for the plastic zone, r_p.

$$a_{\text{eff}} = a_0 + \Delta a + r_p \tag{11-43}$$

For low strength, high-toughness materials the correction for the plastic zone given by Eq. (11-21) may not be sufficient and a compliance method, which uses

[1] G. R. Irwin, *ASTM Bull*, p. 29, Jan. 1960.
[2] Annual Book of ASTM Standards, Vol. 03.01, Metals Test Methods, Designation E561-81.

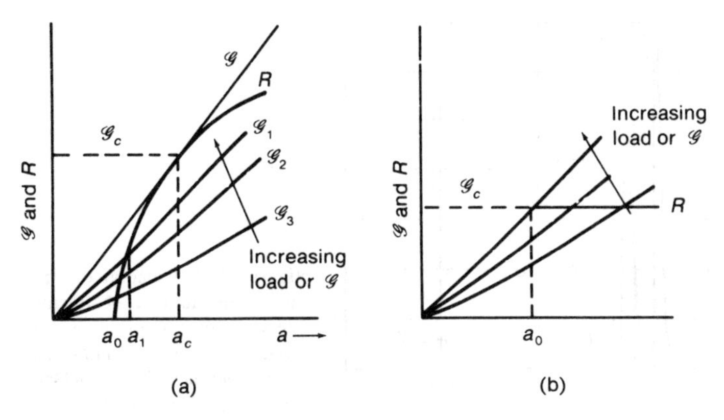

Figure 11-16 (*a*) *R* curve for a ductile material; (*b*) *R* curve for a brittle material.

the elastic spring characteristics of the specimen, should be employed.[1] The *J* integral may be used[2] to determine the *R* curve in high-toughness materials.

11-10 PROBABILISTIC ASPECTS OF FRACTURE MECHANICS

Mechanical-strength measurements of brittle materials, like glass and ceramics, or of metals under conditions where they behave in a brittle manner, show a high variability of results which requires statistical analysis. Therefore, the mechanical properties are not expressible by a single number, but instead we must think in terms of probability of failure at a given stress. An important consequence of the statistical behavior is that the strength of brittle solids shows a pronounced *size effect* in which the strength decreases with increasing size or volume of the specimen. One of the early observations with brittle materials was that the strength of a fine glass fiber or metal wire is usually higher than that of a large-diameter rod. Moreover, if a fiber is broken into two fragments and each piece is broken successively in turn, the fracture strength will increase as the length decreases. These manifestations of the size effects are the result of the fact that the strength of a brittle material is controlled by the stress at the root of its most dangerous crack (usually the longest crack). As the specimen size decreases there is less likelihood of finding a large crack, and the strength increases.

The statistical nature of fracture in brittle materials introduces some complications not present with ductile materials. Ordinarily we consider that fracture will initiate at the most highly stressed region in a structure, usually at a discontinuity or stress concentration. However, in a brittle material whose strength is determined by the distribution of flaws, the stress distribution may be so localized at a discontinuity that the probability of failure is low because the high stress acts over a small volume of material. Failure may occur elsewhere in the

[1] ASTM Standard E561-81.
[2] P. Albrecht, et al., *J. Test. Eval.*, pp. 245–251, Nov. 1982.

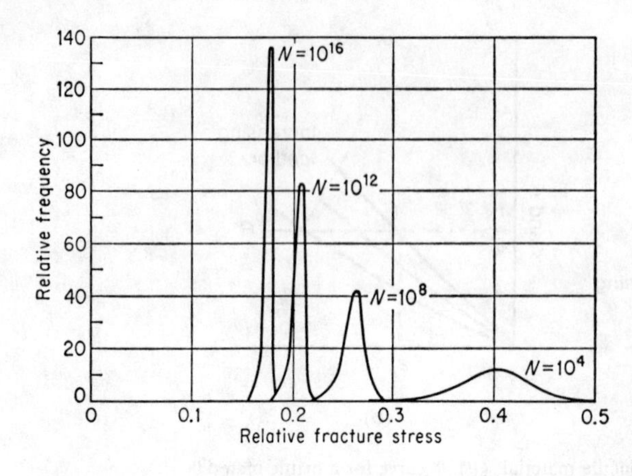

Figure 11-17 Calculated frequency distribution of fracture stress as a function of number of cracks *N*. *(From J. C. Fisher and J. H. Hollomon, Trans. AIME, vol. 171, p. 555, 1947.)*

structure where the peak stress is lower, but it acts over a larger volume of material.

A statistical theory of brittle fracture assumes that the specimen is divided into many volume elements, each containing a single crack. The usual simplifying assumption is that there is no interaction between the cracks in the different volume elements. The strength of the specimen is determined by the element with the longest crack, for this results in the lowest value of fracture stress. Therefore, the brittle-fracture strength is determined, not by an average value of the distribution of flaws, but by the single-most dangerous flaw. This concept of brittle fracture is called the *weakest-link concept* since it uses a model in which the flaws are arranged in series like the links of a chain. Figure 11-17 shows the calculated frequency distribution of fracture stress in a brittle material as a function of the number of cracks N. This shows that the scatter (spread) in fracture stress decreases with an increase in the number of cracks. Also, as the number of flaws increases the mean value of the fracture stress decreases, but as the number of flaws reaches large values there is less relative decrease in the mean value of fracture stress.

While the gaussian or normal frequency distribution is often taken as the accepted statistical distribution for failure strengths there is no theoretical or experimental justification for this situation. In fact, since the normal frequency distribution has long tails on each side of the mean it can be argued that this is unrealistic and does not represent the observed facts. Therefore, the most generally applicable frequency distribution in fracture problems is the *Weibull distribution*.[1]

$$P(x) = 1 - \exp\left[-\left(\frac{x - x_0}{\theta - x_0}\right)^m\right] \tag{11-44}$$

[1] W. Weibull, *J. Appl. Mech.*, vol. 18, pp. 293–297, 1951; *Mater. Res. Stds.*, pp. 405–411, May 1962.

where $P(x)$ = cumulative frequency distribution of random variable x
m = shape parameter, sometimes referred to as the Weibull modulus
θ = scale parameter, sometimes called the characteristic value
x_0 = minimum allowable value of x

By taking natural logarithms twice, Eq. (11-44) can be transformed into

$$\ln\left(\ln \frac{1}{1 - P(x)}\right) = m \ln (x - x_0) - m \ln \theta \tag{11-45}$$

which is of the form $y = bx + c$. Special Weibull probability paper is available to permit the direct plotting of cumulative failure probability vs. x. The Weibull modulus, m, is the slope of this line. The greater the slope the smaller the scatter in the random variable x.

When applying fracture mechanics to design it is common to ensure safety by assuming worst-case conditions. For example, the initial crack size might be assumed to be the largest crack size that can be expected to be undetected by nondestructive inspection and the fracture toughness might be assumed to be the lowest possible value to be expected in the material. While this certainly is a conservative approach, many satisfactory components are rejected for service by worst-case analysis. To overcome the obvious economic penalty from this situation and reduce the risk of failure *probabilistic fracture mechanics* analyses[1] have been developed. The component fracture probability is given by

$$P(F) = \exp(-PN_R)[1 - \exp(-PN_F)] \tag{11-46}$$

where PN_R is the probable number of rejection sites in a component
PN_F is the probable number of component failure sites in a component not rejected by inspection and causing failure

These quantities are determined by three probability distributions.

$$PN_R = \int_0^\infty pn(a) P(R/a, s)\, da \tag{11-47}$$

where $pn(a)\,da$ is the preinspection quality of the material, i.e., the probable number of defects of size $a \pm \frac{1}{2}\,da$
$P(R/a, s)$ is the probability of rejection of a component given that an actual defect of size a exists and that the inspection method can detect a defect size s or larger.

and

$$PN_F = \int_0^\infty pn(a)[1 - P(R/a, s)]\, P(F/a)\, da \tag{11-48}$$

where $P(F/a)$ is the probability of component failure given an existing defect of size a

[1] D. P. Johnson, *Nucl. Engr. Design*, vol. 43, p. 219, 1977; J. M. Bloom and J. C. Ekvall, "Probabilistic Fracture Mechanics and Fatigue Methods," *ASTM Std.*, p. 798, Amer. Soc. for Test and Materials, Philadelphia, 1983.

While the conceptual mathematical framework for probabilistic fracture mechanics is well developed the actual development of usable data is an arduous task.

11-11 TOUGHNESS OF MATERIALS

The objective of this section is to consider the general microstructural features which control the toughness of metals and alloys. More specific information on the role of metallurgical variables on toughness will be found in Sec. 14-4. For a given metallurgical structure toughness increases with increasing temperature and decreases with increasing strain rate or rate of loading.

It is not unusual to find that efforts to increase the strength of materials result in a decrease in toughness. Unfortunately the relationship shown in Fig. 11-18 appears to be the general situation. To escape from this inverse relationship between toughness and strength we need to move to a different parameter line which represents a different alloy system, usually with a somewhat different strengthening mechanism. This trend is consistent with a simple relationship, expressed in the more general fracture toughness parameter, J_{Ic}.

$$J_{Ic} = \sigma_0 \varepsilon_f l_0 \tag{11-49}$$

where σ_0 = yield stress
ε_f = true fracture strain
l_0 = the spacing of microvoids ahead of the crack which establishes the process zone of intense plastic deformation

The best microstructures for combining toughness with strength are strengthened with fine particles. Particles which are small in size and well bonded to the matrix so as to avoid microvoid formation by decohesion are best. Particles should be in the range of 100 μm or less. A small particle size minimizes the opportunity for multiple slip-band pileups, which would greatly increase the local pileup stress and create decohesion or particle fracture. Small particles should be as widely spaced as possible to avoid the overlap of stress fields. This imposes a volume fraction limit of about 10 percent. Particles should be rounded to avoid stress concentration associated with corners and should be as equiaxed as possible to avoid the multiple pileup effects previously discussed. Incoherent or semicoherent particles in which dislocations bypass the particles instead of cutting them are preferred for toughness, since this avoids planar slip localization and coherency strains. Coherent precipitates and solid-solution strengthening are desirable as secondary hardening effects when combined with fine dispersoids.

Inclusions should be avoided to preserve high toughness in strong materials since they readily debond. Large widely spaced inclusions usually are less damaging than fine closely spaced inclusions where inclusions are weakly bonded to the interface. The improvements in steel achieved by removing inclusions by

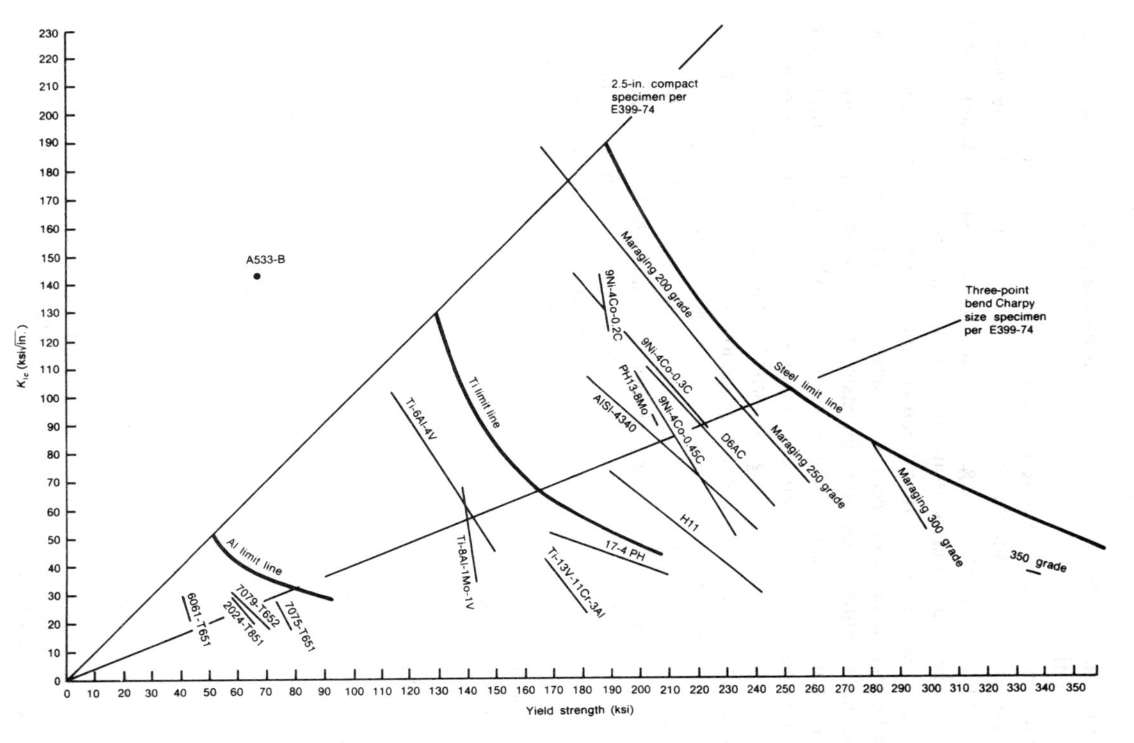

Figure 11-18 Range of toughness and yield strength values for a variety of alloys at room temperature *(From Rapid Inexpensive Tests for Determining Fracture Toughness, NMAB Report 328, 1976.)*

vacuum melting and in aluminum alloys by using high-purity melt stock and filtering are dramatic.[1]

Fine grain size helps to minimize dislocation pileup stresses, so it improves toughness. Grain boundaries should be free of large particles or particle networks. Frequently small solute additions, e.g., B, Mg, and Zr in iron and titanium are used to control adsorbed impurity atoms at grain boundaries.

Fractographic studies show that fractographic features occur in two size regimes. The large-scale features occur with dimensions of the order of 10 to 300 μm while the small features occur in a dimension typically $< 1\ \mu$m. These small features are related to the localized fracture processes which occur during crack extension. The large-scale features are associated with crack deflection or reorientation as it propagates. The total energy absorbed in the fracture process J_{Ic} is a product of the contribution due to crack deflection and the propagation energy required for local crack extension. For example, an irregular fracture which follows a tortuous path through a brittle constituent will result in low J_{Ic} because the unit propagation energy is very low. On the other hand, the zig-zag path of a crack following the interface of Widmanstatten alpha plates in an α-β titanium alloy multiplies a considerable crack deflection energy with an already high propagation energy. Layered microstructures, such as can be produced in hot-worked steels or fibrous composites have been used to increase toughness by increasing the fracture area that must be traversed.

BIBLIOGRAPHY

Broek, D.: "Elementary Engineering Fracture Mechanics," 3rd ed., Martinus Nijhoff Publishers, The Hague, 1982.

Burke, J. J. and V. Weiss (eds.): "Application of Fracture Mechanics to Design," Plenum Press, New York, 1979.

Campbell, J. E., J. H. Underwood, and W. W. Gerberich: "Application of Fracture Mechanics for the Selection of Metallic Structural Materials," American Society for Metals, Metals Park, Ohio, 1982.

Ewalds, H. L. and R. J. H. Wanhill: "Fracture Mechanics," Edward Arnold Ltd., Baltimore, Md., 1984.

Fracture Mechanics, "Metals Handbook," 9th ed., vol. 8, pp. 437–491, American Society for Metals, Metals Park, Ohio, 1985.

Hellan, K.: "Introduction to Fracture Mechanics," McGraw-Hil Book Company, New York, 1984.

Hertzberg, R. W.: "Deformation and Fracture Mechanics of Engineering Materials," 2d ed., John Wiley & Sons, Inc., New York, 1983.

Knott, J. F.: "Fundamentals of Fracture Mechanics," Butterworths, London, 1973.

Parker, A. P.: "The Mechanics of Fracture and Fatigue," E & F. N. Spon Ltd., London, 1981.

Philos. Trans. Roy. Soc. London, vol. 299A, pp. 1–239, 1981.

Rolfe, S. T. and J. M. Barsom: "Fracture and Fatigue Control in Structures," Prentice-Hall, Inc., Englewood Cliffs, N.J., 1977.

Tetleman, A. S. and A. J. McEvely: "Fracture of Structural Materials," John Wiley & Sons, Inc., New York, 1967.

[1] R. W. Hertzberg, "Deformation and Fracture Mechanics of Engineering Materials," 2d ed., pp. 363–382, John Wiley & Sons, New York, 1983.

CHAPTER

TWELVE

FATIGUE OF METALS

12-1 INTRODUCTION

It has been recognized since 1830 that a metal subjected to a repetitive or fluctuating stress will fail at a stress much lower than that required to cause fracture on a single application of load. Failures occurring under conditions of dynamic loading are called *fatigue failures*, presumably because it is generally observed that these failures occur only after a considerable period of service. Fatigue has become progressively more prevalent as technology has developed a greater amount of equipment, such as automobiles, aircraft, compressors, pumps, turbines, etc., subject to repeated loading and vibration, until today it is often stated that fatigue accounts for at least 90 percent of all service failures due to mechanical causes.[1]

A fatigue failure is particularly insidious because it occurs without any obvious warning. Fatigue results in a brittle-appearing fracture, with no gross deformation at the fracture. On a macroscopic scale the fracture surface is usually normal to the direction of the principal tensile stress. A fatigue failure can usually be recognized from the appearance of the fracture surface, which shows a smooth region, due to the rubbing action as the crack propagated through the section (top portion of Fig. 12-1), and a rough region, where the member has failed in a ductile manner when the cross section was no longer able to carry the load. Frequently the progress of the fracture is indicated by a series of rings, or "beach marks," progressing inward from the point of initiation of the failure. Figure 12-1 also illustrates another characteristic of fatigue, namely, that a failure usually occurs at a point of stress concentration such as a sharp corner or notch or at a metallurgical stress concentration like an inclusion.

[1] Numerous examples of fatigue failures are given in Failure Analysis and Prevention, "Metals Handbook," vol. 10, 8th ed., American Society for Metals, Metals Park, Ohio, 1975.

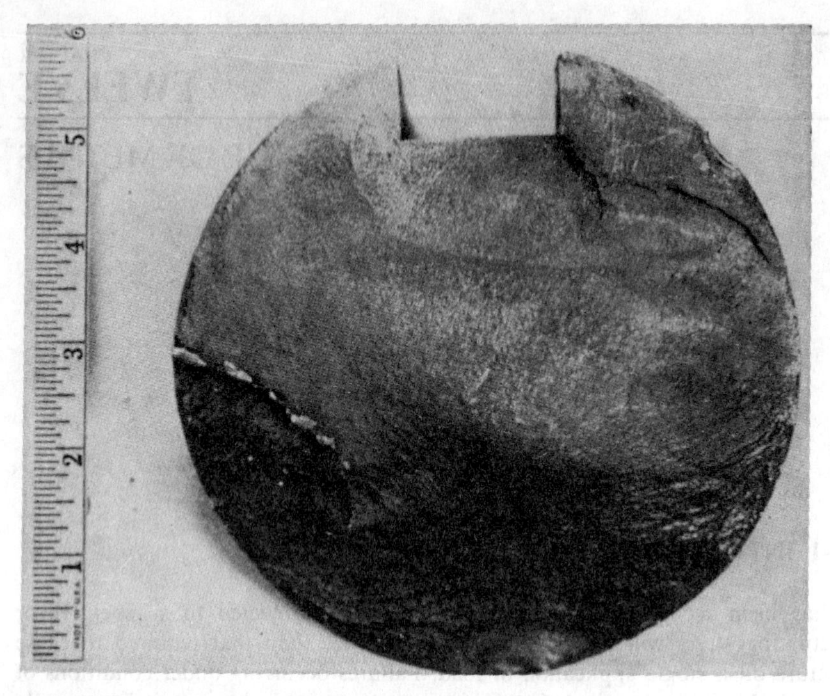

Figure 12-1 Fracture surface of fatigue failure which started at sharp corner of a keyway in a shaft (1×).

Three basic factors are necessary to cause fatigue failure. These are (1) a maximum tensile stress of sufficiently high value, (2) a large enough variation or fluctuation in the applied stress, and (3) a sufficiently large number of cycles of the applied stress. In addition, there are a host of other variables, such as stress concentration, corrosion, temperature, overload, metallurgical structure, residual stresses, and combined stresses, which tend to alter the conditions for fatigue. Since we have not yet gained a complete understanding of what causes fatigue in metals, it will be necessary to discuss each of these factors from an essentially empirical standpoint. Because of the mass of data of this type, it will be possible to describe only the highlights of the relationship between these factors and fatigue. For more complete details the reader is referred to the number of excellent publications listed at the end of this chapter.

12-2 STRESS CYCLES

At the outset it will be advantageous to define briefly the general types of fluctuating stresses which can cause fatigue. Figure 12-2 serves to illustrate typical fatigue stress cycles. Figure 12-2*a* illustrates a *completely reversed cycle of stress* of

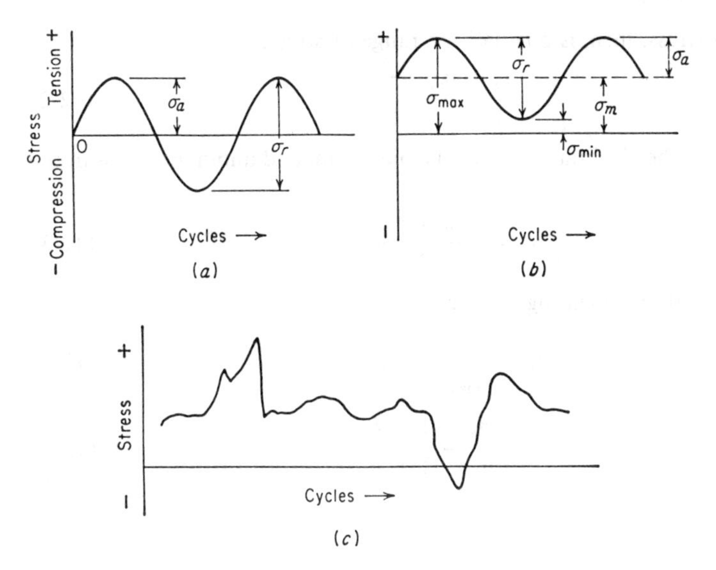

Figure 12-2 Typical fatigue stress cycles. (*a*) Reversed stress; (*b*) repeated stress; (*c*) irregular or random stress cycle.

sinusoidal form. This is an idealized situation which is produced by an R. R. Moore rotating-beam fatigue machine[1] and which is approached in service by a rotating shaft operating at constant speed without overloads. For this type of stress cycle the maximum and minimum stresses are equal. In keeping with the conventions established in Chap. 2, the minimum stress is the lowest algebraic stress in the cycle. Tensile stress is considered positive, and compressive stress is negative. Figure 12-2*b* illustrates a *repeated stress cycle* in which the maximum stress σ_{max} and minimum stress σ_{min} are not equal. In this illustration they are both tension, but a repeated stress cycle could just as well contain maximum and minimum stresses of opposite signs or both in compression. Figure 12-2*c* illustrates a complicated stress cycle which might be encountered in a part such as an aircraft wing which is subjected to periodic unpredictable overloads due to gusts.

A fluctuating stress cycle can be considered to be made up of two components, a *mean*, or steady, stress σ_m, and an *alternating*, or variable, stress σ_a. We must also consider the *range of stress* σ_r. As can be seen from Fig. 12-2*b*, the range of stress is the algebraic difference between the maximum and minimum stress in a cycle.

$$\sigma_r = \sigma_{max} - \sigma_{min} \tag{12-1}$$

[1] Common types of fatigue machines are described in the references listed at the end of this chapter and in the Manual on Fatigue Testing, *ASTM Spec. Tech. Publ.* 91, 1949.

The alternating stress, then, is one-half the range of stress.

$$\sigma_a = \frac{\sigma_r}{2} = \frac{\sigma_{max} - \sigma_{min}}{2} \tag{12-2}$$

The mean stress is the algebraic mean of the maximum and minimum stress in the cycle.

$$\sigma_m = \frac{\sigma_{max} + \sigma_{min}}{2} \tag{12-3}$$

Two ratios are used in presenting fatigue data:

Stress ratio $$R = \frac{\sigma_{min}}{\sigma_{max}} \tag{12-4}$$

Amplitude ratio $$A = \frac{\sigma_a}{\sigma_m} = \frac{1 - R}{1 + R} \tag{12-5}$$

12-3 THE *S-N* CURVE

The basic method of presenting engineering fatigue data is by means of the *S-N* curve, a plot of stress *S* against the number of cycles to failure *N*. A log scale is almost always used for *N*. The value of stress that is plotted can be σ_a, σ_{max}, or σ_{min}. The stress values are usually nominal stresses, i.e., there is no adjustment for stress concentration. The *S-N* relationship is determined[1] for a specified value of σ_m, *R*, or *A*. Most determinations of the fatigue properties of materials have been made in completed reversed bending, where the mean stress is zero. Figure 12-3 gives typical *S-N* curves from rotating-beam tests. Cases where the mean stress is not zero are of considerable engineering importance and will be considered in Sec. 12-5.

It will be noted that this *S-N* curve is concerned chiefly with fatigue failure at high numbers of cycles ($N > 10^5$ cycles). Under these conditions the stress, on a gross scale, is elastic, but as we shall see shortly the metal deforms plastically in a highly localized way. At higher stresses the fatigue life is progressively decreased, but the gross plastic deformation makes interpretation difficult in terms of stress. For the low-cycle fatigue region ($N < 10^4$ or 10^5 cycles) tests are conducted with controlled cycles of elastic plus plastic strain instead of controlled load or stress cycles. *Low-cycle fatigue* will be considered in Sec. 12-7.

As can be seen from Fig. 12-3, the number of cycles of stress which a metal can endure before failure increases with decreasing stress. Unless otherwise indicated, *N* is taken as the number of cycles of stress to cause complete fracture of the specimen. Fatigue tests at low stresses are usually carried out for 10^7 cycles and sometimes to 5×10^8 cycles for nonferrous metals. For a few important

[1] H. E. Boyer, ed., "Atlas of Fatigue Curves," American Society for Metals, Metals Park, Ohio, 1985.

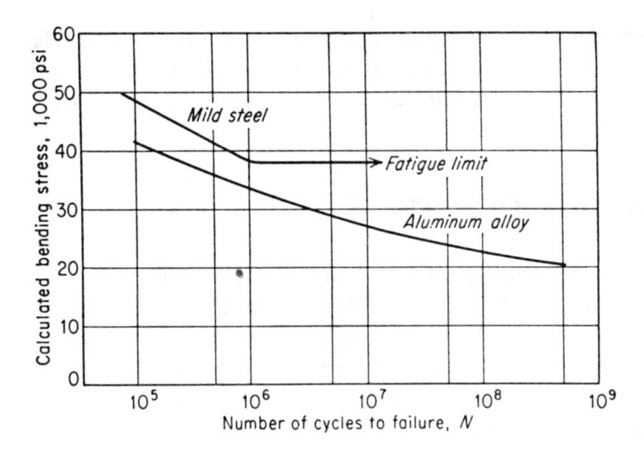

Figure 12-3 Typical fatigue curves for ferrous and nonferrous metals.

engineering materials such as steel and titanium, the *S-N* curve becomes horizontal at a certain limiting stress. Below this limiting stress, which is called the *fatigue limit*, or endurance limit, the material presumably can endure an infinite number of cycles without failure. Most nonferrous metals, like aluminum, magnesium, and copper alloys, have an *S-N* curve which slopes gradually downward with increasing number of cycles. These materials do not have a true fatigue limit because the *S-N* curve never becomes horizontal. In such cases it is common practice to characterize the fatigue properties of the material by giving the *fatigue strength* at an arbitrary number of cycles, for example, 10^8 cycles.

The *S-N* curve in the high-cycle region is sometimes described by the Basquin equation

$$N\sigma_a^p = C \tag{12-6}$$

where σ_a is the stress amplitude and p and C are empirical constants.

The usual procedure for determining an *S-N* curve is to test the first specimen at a high stress where failure is expected in a fairly short number of cycles, e.g., at about two-thirds the static tensile strength of the material. The test stress is decreased for each succeeding specimen until one or two specimens do not fail in the specified numbers of cycles, which is usually at least 10^7 cycles. The highest stress at which a runout (nonfailure) is obtained is taken as the fatigue limit. For materials without a fatigue limit the test is usually terminated for practical considerations at a low stress where the life is about 10^8 or 5×10^8 cycles. The *S-N* curve is usually determined with about 8 to 12 specimens. It will generally be found that there is a considerable amount of scatter in the results, although a smooth curve can usually be drawn through the points without too much difficulty. However, if several specimens are tested at the same stress, there is a great amount of scatter in the observed values of number of cycles to failure, frequently as much as one log cycle between the minimum and maximum value.

Further, it has been shown[1] that the fatigue limit of steel is subject to considerable variation and that a fatigue limit determined in the manner just described can be considerably in error. The statistical nature of fatigue will be discussed in the next section.

12-4 STATISTICAL NATURE OF FATIGUE

A considerable amount of interest has been shown in the statistical analysis of fatigue data and in reasons for the variability in fatigue-test results.[2] Since fatigue life and fatigue limit are statistical quantities, it must be realized that considerable deviation from an average curve determined with only a few specimens is to be expected. It is necessary to think in terms of the probability of a specimen attaining a certain life at a given stress or the probability of failure at a given stress in the vicinity of the fatigue limit. To do this requires the testing of considerably more specimens than in the past so that the statistical parameters[3] for estimating these probabilities can be determined. The basic method for expressing fatigue data should then be a three-dimensional surface representing the relationship between stress, number of cycles to failure, and probability of failure. Figure 12-4 shows how this can be presented in a two-dimensional plot.

A distribution of fatigue life at constant stress is illustrated schematically in this figure, and based on this, curves of constant probability of failure are drawn. Thus, at σ_1, 1 percent of the specimens would be expected to fail at N_1 cycles, 50 percent at N_2 cycles, etc. The figure indicates a decreasing scatter in fatigue life with increasing stress, which is usually found to be the case. The statistical distribution function which describes the distribution of fatigue life at constant stress is not accurately known, for this would require the testing of over 1,000 identical specimens under identical conditions at a constant stress. Muller-Stock[4] tested 200 steel specimens at a single stress and found that the frequency distribution of N followed the gaussian, or normal, distribution if the fatigue life was expressed as $\log N$. For engineering purposes it is sufficiently accurate to assume a logarithmic normal distribution of fatigue life at constant life in the region of the probability of failure of $P = 0.10$ to $P = 0.90$. However, it is frequently important to be able to predict the fatigue life corresponding to a probability of failure of 1 percent or less. At this extreme limit of the distribution the assumption of a log-normal distribution of life is no longer justified, although it is frequently used. Alternative approaches have been the use of the extreme-value distribution[5] or Weibull's distribution.[6]

[1] J. T. Ransom and R. F. Mehl, *Trans. AIME*, vol. 185, pp. 364–365, 1949.

[2] P. H. Armitage, *Metall. Rev.*, vol. 6, pp. 353–385, 1964; R. E. Little and E. H. Jebe, "Statistical Design of Fatigue Experiments," John Wiley & Sons, Inc., New York, 1975.

[3] The chief statistical parameters to be considered are the estimate of the mean (average) and standard deviation (measure of scatter) of the population.

[4] H. Muller-Stock, *Mitt. Kohle Eisenforsch. GmbH*, vol. 8, pp. 83–107, 1938.

[5] A. M. Freudenthal and E. J. Gumbel, *J. Am. Stat. Assoc.*, vol. 49, pp. 575–597, 1954.

[6] W. Weibull, *J. Appl. Mech.*, vol. 18, no. 3, pp. 293–297, 1951; W. Weibull, "Fatigue Testing and the Analysis of Results," Pergamon Press, New York, 1961.

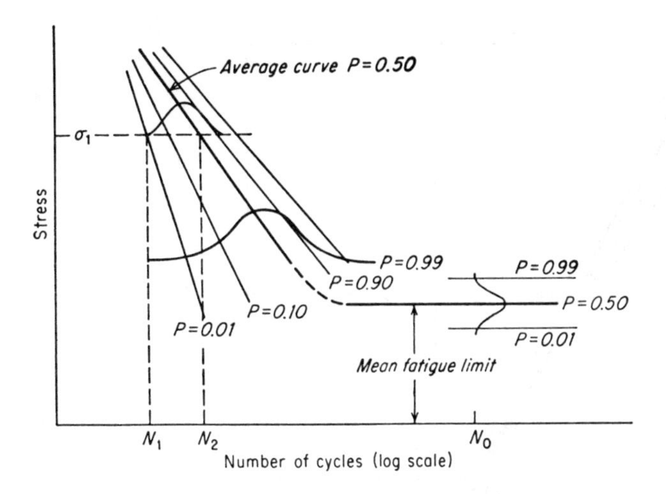

Figure 12-4 Representation of fatigue data on a probability basis.

For the statistical interpretation of the fatigue limit we are concerned with the distribution of stress at a constant fatigue life. The fatigue limit of steel was once considered to be a sharp threshold value, below which all specimens would presumably have infinite lives. However, it is now recognized that the fatigue limit is really a statistical quantity which requires special techniques for an accurate determination. For example, in a heat-treated alloy forging steel the stress range which would include the fatigue limits of 95 percent of the specimens could easily be from 40,000 to 52,000 psi. An example of the errors which can be introduced by ordinary testing with a few specimens is illustrated in Fig. 12-5. This figure summarizes[1] 10 *S-N* curves determined in the conventional manner for the *same* bar of alloy steel, each curve being based on 10 specimens. The specimens were as identical as it was possible to make them, and there was no excessive scatter or uncertainty as to how to draw the *S-N* curves. Yet, as can be seen from the figure, there is considerable difference in the measured values of the fatigue limit for the steel due to the fact that the curves were based on insufficient data.

In determining the fatigue limit of a material, it should be recognized that each specimen has its own fatigue limit, a stress above which it will fail but below which it will not fail, and that this critical stress varies from specimen to specimen for very obscure reasons. It is known that inclusions in steel have an important effect on the fatigue limit and its variability, but even vacuum-melted steel shows appreciable scatter in fatigue limit. The statistical problem of accurately determining the fatigue limit is complicated by the fact that we cannot measure the individual value of the fatigue limit for any given specimen. We can only test a specimen at a particular stress, and if the specimen fails, then the stress was somewhere above the fatigue limit of the specimen. Since the specimen cannot be retested, even if it did not fail at the test stress, we have to estimate the statistics

[1] J. T. Ransom, discussion in *ASTM Spec. Tech. Publ.* 121, pp. 59–63, 1952.

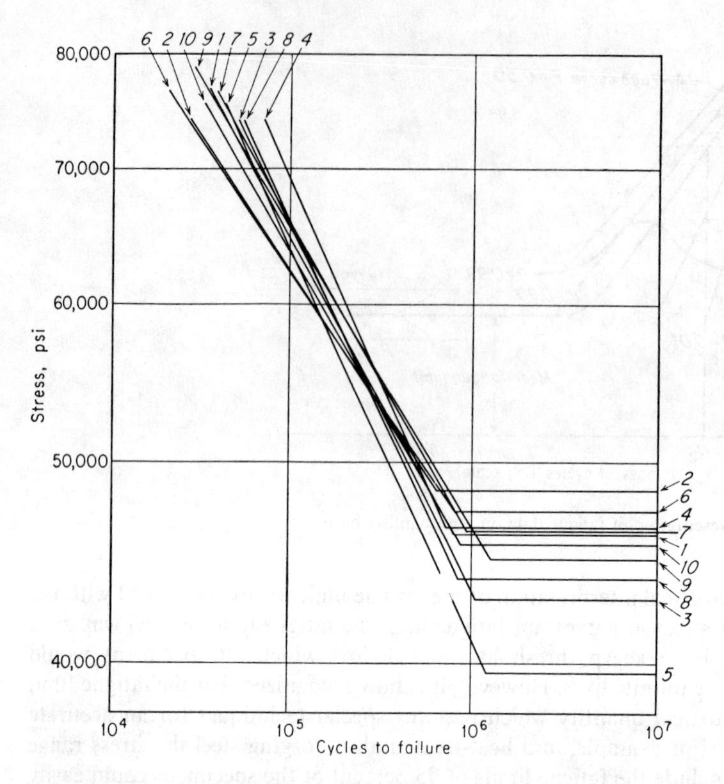

Figure 12-5 Summary of *S-N* curves, each based on 10 specimens, drawn from the same bar of steel. *(From J. T. Ransom, ASTM Spec. Tech. Publ. 121, p. 61, 1952.)*

of the fatigue limit by testing groups of specimens at several stresses to see how many fail at each stress. Thus, near the fatigue limit fatigue is a "go–no-go" proposition, and all that we can do is to estimate the behavior of a universe of specimens by means of a suitable sample. The two statistical methods which are used for making a statistical estimate of the fatigue limit are called *probit analysis* and the *staircase method*. The procedures for applying these methods of analysis to the determination of the fatigue limit have been well established.[1]

12-5 EFFECT OF MEAN STRESS ON FATIGUE

Much of the fatigue data in the literature have been determined for conditions of completely reversed cycles of stress, $\sigma_m = 0$. However, conditions are frequently met in engineering practice where the stress situation consists of an alternating

[1] "A Guide for Fatigue Testing and the Statistical Analysis of Fatigue Data," *ASTM Spec. Tech. Publ.*, 91-A, 2d ed., 1963; J. A. Collins, "Failure of Materials in Mechanical Design," chaps. 9 and 10, John Wiley & Sons, New York, 1981.

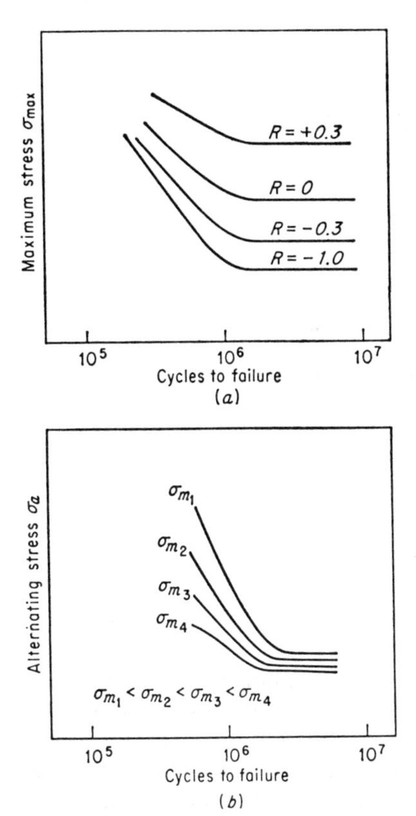

Figure 12-6 Two methods of plotting fatigue data when the mean stress is not zero.

stress and a superimposed mean, or steady, stress. The possibility of this stress situation has already been considered in Sec. 12-2, where various relationships between σ_m and σ_a have been given.

There are several possible methods of determining an *S-N* diagram for a situation where the mean stress is not equal to zero. Figure 12-6 shows the two most common methods of presenting the data. In Fig. 12-6*a* the maximum stress is plotted against log *N* for constant values of the stress ratio $R = \sigma_{min}/\sigma_{max}$. This is achieved by applying a series of stress cycles with decreasing maximum stress and adjusting the minimum stress in each case so that it is a constant fraction of the maximum stress. The case of completely reversed stress is given at $R = -1.0$. Note that as *R* becomes more positive, which is equivalent to increasing the mean stress, the measured fatigue limit becomes greater. Figure 12-6*b* shows the same data potted in terms of the alternating stress vs. cycles to failure at constant values of mean stress. Note that as the mean stress becomes more positive the allowable alternating stress decreases. Other ways of plotting these data are maximum stress vs. cycles to failure at constant mean stress and maximum stress vs. cycles to failure at constant minimum stress.

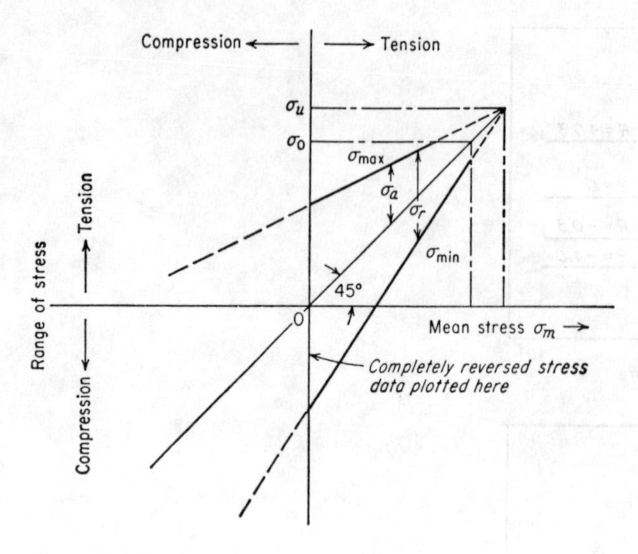

Figure 12-7 Goodman diagram.

For each value of mean stress there is a different value of the limiting range of stress, $\sigma_{max} - \sigma_{min}$, which can be withstood without failure. Early contributions to this problem were made by Goodman,[1] so that curves which show the dependence of limiting range of stress on mean stress are frequently called *Goodman diagrams*. Figure 12-7 shows one common type of Goodman diagram which can be constructed from fatigue data of the type illustrated in Fig. 12-6. Basically, this diagram shows the variation of the limiting range of stress, $\sigma_{max} - \sigma_{min}$ with mean stress. This relationship is established for a fixed number of cycles or at the fatigue limit for steel. Note that as the mean stress becomes more tensile the allowable range of stress is reduced, until at the tensile strength σ_u the stress range is zero. However, for practical purposes testing is usually stopped when the yield stress σ_0 is exceeded. The test data usually lie somewhat above and below the σ_{max} and σ_{min} lines, respectively, so that these lines shown on Fig. 12-7 may actually be curves. A conservative approximation of the Goodman diagram may be obtained, in lieu of actual test data, by drawing straight lines from the fatigue limit for completely reversed stress (which is usually available from the literature) to the tensile strength. A diagram similar to Fig. 12-7 may be constructed for the fatigue strength at any given number of cycles. Very few test data exist for conditions where the mean stress is compressive. Data[2] for SAE 4340 steel tested in axial fatigue indicate that the allowable stress range increases with increasing compressive mean stress up to the yield

[1] John Goodman, "Mechanics Applied to Engineering," Longmans, Green & Co., Ltd., London, 1899.

[2] J. T. Ransom, discussion in *Am. Soc. Test. Mater. Proc.*, vol. 54, pp. 847–848, 1954.

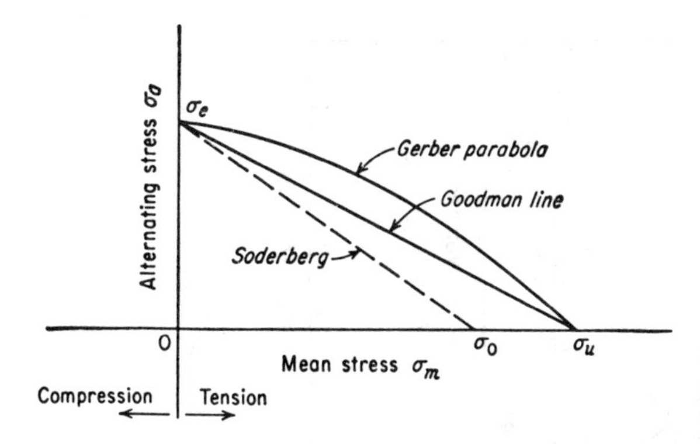

Figure 12-8 Alternative method of plotting the Goodman diagram.

stress in compression. This is in agreement with the fact that compressive residual stress increases the fatigue limit.

An alternative method of presenting mean-stress data is shown in Fig. 12-8. This is sometimes known as the Haig-Soderberg diagram.[1] The alternating stress is plotted against the mean stress. A straight-line relationship follows the suggestion of Goodman, while the parabolic curve was proposed by Gerber. Test data for ductile metals generally fall closer to the parabolic curve. However, because of the scatter in the results and the fact that tests on notched specimens fall closer to the Goodman line, the linear relationship is usually preferred in engineering design. These relationships may be expressed by the following equation,

$$\sigma_a = \sigma_e\left[1 - \left(\frac{\sigma_m}{\sigma_u}\right)^x\right] \tag{12-7}$$

where $x = 1$ for the Goodman line, $x = 2$ for the Gerber parabola, and σ_e is the fatigue limit for completely reversed loading. If the design is based on the yield strength, as indicated by the dashed Soderberg line in Fig. 12-8, then σ_0 should be substituted for σ_u in Eq. (12-7).

Figure 12-8 is obtained for alternating axial or bending stresses with static tension or compression or alternating torsion with static tension. However, for alternating torsion with static torsion or alternating bending with static torsion there is no effect of the static stress on the allowable range of alternating stress provided that the static yield strength is not exceeded.[2]

For design purposes master diagrams like Fig. 12-9 are used to represent the influence of mean stress on fatigue. For example, a stress state of $\sigma_{max} = 60$ ksi, $\sigma_{min} = 0$ gives a fatigue life in the notched condition of less than 10^6 cycles. (In the unnotched condition the stress is below the fatigue limit.) If $\sigma_{min} = -60$ ksi ($R = -1.0$) the fatigue life is less than 10^4 cycles, but if $\sigma_{min} = +20$ ksi ($R = +0.33$) the stress range is reduced and the fatigue life moves to greater than 10^6 cycles.

[1] C. R. Soderberg, *Trans. ASME*, vol. 52, APM-52-2, pp. 13–28, 1930.

[2] G. Sines, Failure of Materials under Combined Repeated Stresses with Superimposed Static Stresses, *NACA Tech. Note* 3495, 1955.

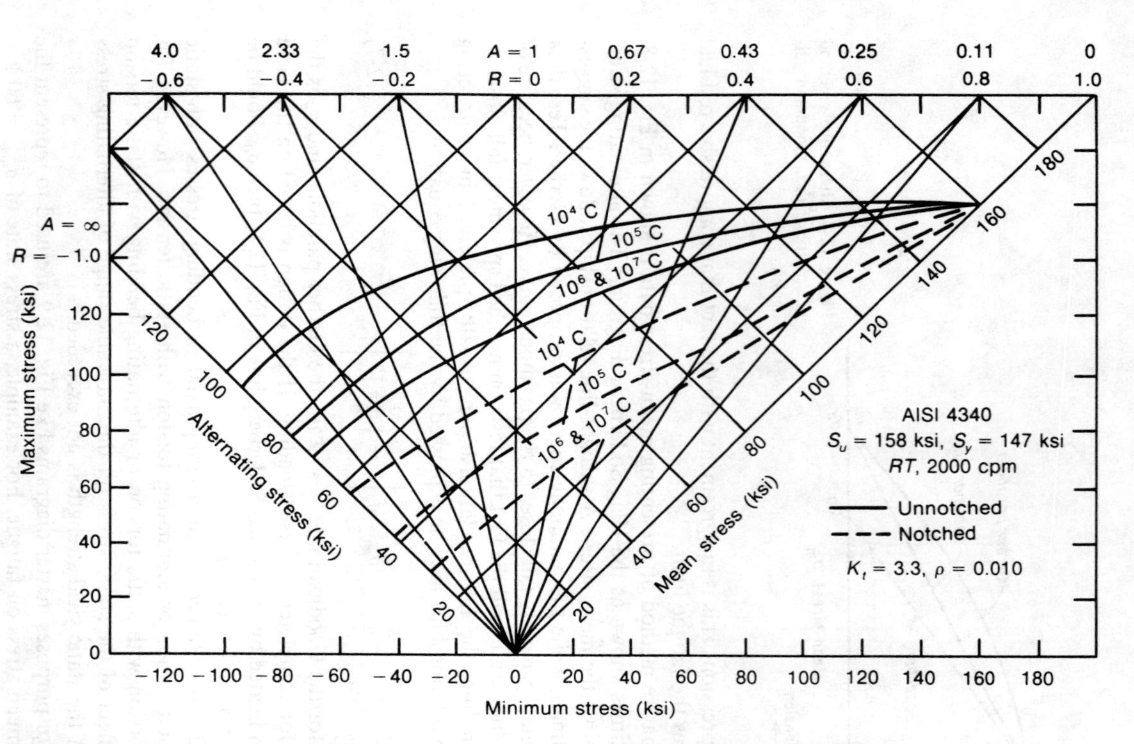

Figure 12-9 Master diagram for establishing influence of mean stress in fatigue. *(From MIL-HBDK-5, U.S. Dept. of Defense.)*

Example A 4340 steel bar is subjected to a fluctuating axial load that varies from a maximum of 75,000-lb tension to a minimum of 25,000-lb compression. The mechanical properties of the steel are:

$$\sigma_u = 158{,}000 \text{ psi} \qquad \sigma_0 = 147{,}000 \text{ psi} \qquad \sigma_e = 75{,}000 \text{ psi}$$

Determine the bar diameter to give infinite fatigue life based on a safety factor of 2.5.

Since the bar has a constant cylindrical cross section A, the variation in stress will be proportional to the load

$$\sigma_{\max} = \frac{75}{A} \text{ ksi} \qquad \sigma_{\min} = -\frac{25}{A} \text{ ksi}$$

$$\sigma_m = \frac{\sigma_{\max} + \sigma_{\min}}{2} = \frac{75/A + (-25/A)}{2} = \frac{25}{A}$$

$$\sigma_a = \frac{\sigma_{\max} - \sigma_{\min}}{2} = \frac{75/A - (-25/A)}{2} = \frac{50}{A}$$

using the conservative Goodman line and Eq. (12-7)

$$\sigma_a = \sigma_e\left(1 - \frac{\sigma_m}{\sigma_u}\right) \qquad \sigma_e = \frac{75}{2.5} = 30 \text{ ksi}$$

$$\frac{50/A}{30} = 1 - \frac{25/A}{158} \qquad A = \frac{50}{30} + \frac{25}{158} = 1.667 + 0.158 = 1.825 \text{ in}^2$$

$$D = \sqrt{\frac{4A}{\pi}} = 1.52 \text{ in}$$

12-6 CYCLIC STRESS-STRAIN CURVE

Cyclic strain controlled fatigue, as opposed to our previous discussion of cyclic stress controlled fatigue, occurs when the strain amplitude is held constant during cycling. Strain controlled cyclic loading is found in thermal cycling, where a component expands and contracts in response to fluctuations in the operating temperature, or in reversed bending between fixed displacements. In a more general view, the localized plastic strains at a notch subjected to *either* cyclic stress or strain conditions result in strain controlled conditions near the root of the notch due to the constraint effect of the larger surrounding mass of essentially elastically deformed material.

Figure 12-10 illustrates a stress-strain loop under controlled constant strain cycling. During initial loading the stress-strain curve is *O-A-B*. On unloading yielding begins in compression at a lower stress *C* due to the Bauschinger effect. In reloading in tension a hysteresis loop develops. The dimensions of the hysteresis loop are described by its width $\Delta\varepsilon$, the total strain range, and its height $\Delta\sigma$, the stress range. The total strain range $\Delta\varepsilon$ consists of the elastic strain

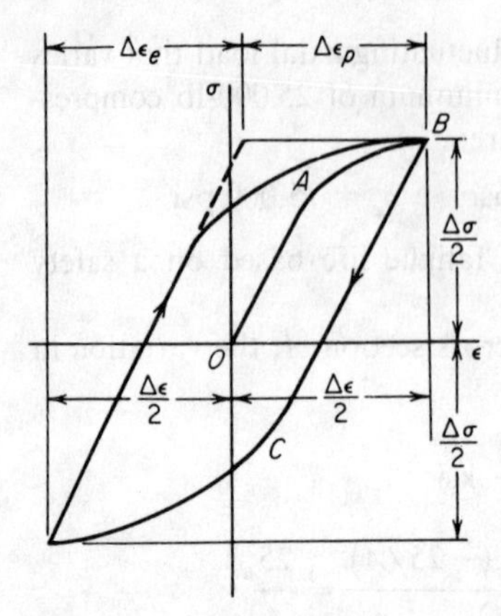

Figure 12-10 Stress-strain loop for constant strain cycling.

component $\Delta\varepsilon_e = \Delta\sigma/E$ plus the plastic strain component $\Delta\varepsilon_p$. The width of the hysteresis loop will depend on the level of cyclic strain.

Since plastic deformation is not completely reversible, modifications to the structure occur during cyclic straining and these can result in changes in the stress-strain response. Depending on the initial state a metal may undergo cyclic hardening, cyclic softening, or remain cyclically stable. It is not uncommon for all

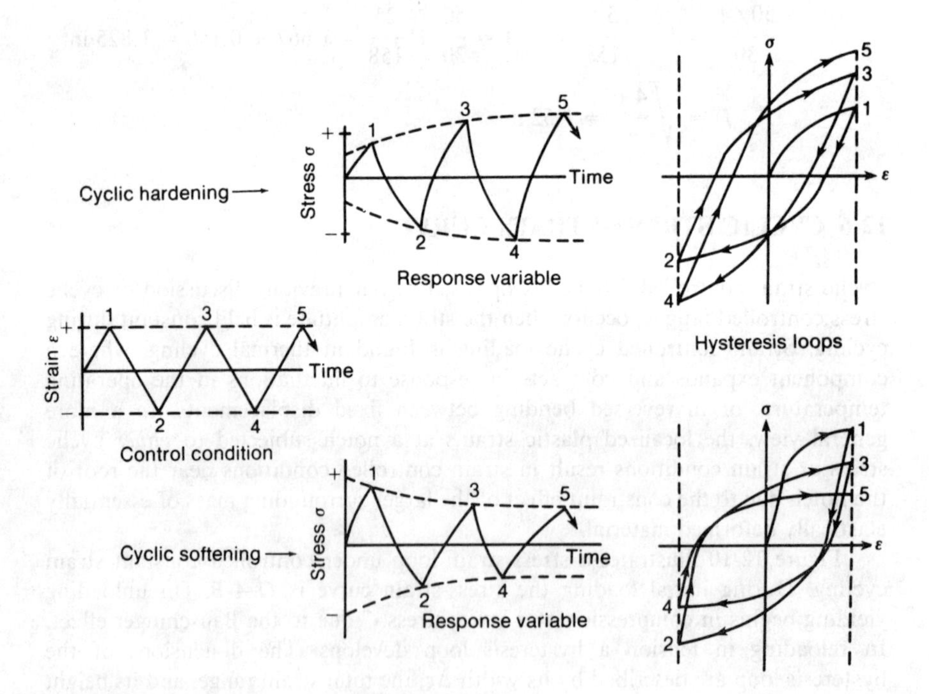

Figure 12-11 Response of metals to cyclic strain cycles.

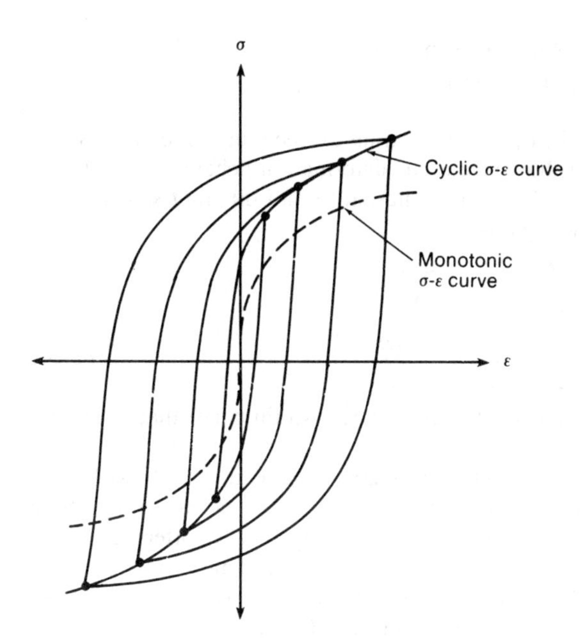

Figure 12-12 Comparison of monotonic and cyclic stress-strain curves for a material that cyclic hardens. Points on cyclic stress-strain curve represent tips of stable hysteresis loops.

three behaviors to occur in a given material depending on the initial state of the material and the test conditions. Figure 12-11 illustrates cyclic hardening and softening for strain controlled cycling. Note how the hysteresis loops change with successive cycles. Note that with cyclic stressing under stress control the behavior would be reversed. Cyclic hardening would lead to a decreasing peak strain with increasing cycles and cyclic softening would lead to a continually increasing strain range and early fracture.

Generally the hysteresis loop stabilizes after about 100 cycles and the material arrives at an equilibrium condition for the imposed strain amplitude. The cyclically stabilized stress-strain curve may be quite different from the stress-strain curve obtained on monotonic static loading. The cyclic stress-strain curve is usually determined by connecting the tips of stable hysteresis loops from constant-strain-amplitude fatigue tests of specimens cycled at different strain amplitudes (Fig. 12-12). Under conditions where saturation of the hysteresis loop is not obtained, the maximum stress amplitude for hardening or the minimum stress amplitude for softening is used. Sometimes the stress is taken at 50 percent of the life to failure. Several shortcut procedures have been developed.[1]

The cycle stress-strain curve may be described by a power curve in direct analogy with Eq. (8-21)

$$\Delta\sigma = K'(\Delta\varepsilon_p)^{n'} \tag{12-8}$$

[1] R. W. Landgraf, J. D. Morrow, and T. Endo, *J. Mater.*, vol. 4, no. 1, p. 176, 1969.

where n' is the cyclic strain-hardening exponent
K' is the cyclic strength coefficient

Equation (12-8) implies that the cyclic stress-strain can be represented by a single straight line on a log-log plot, but since cyclic deformation behavior may change appreciably with strain amplitude it is not unusual to find different slopes in the long life and short life fatigue regions.

Since
$$\frac{\Delta\varepsilon}{2} = \frac{\Delta\varepsilon_e}{2} + \frac{\Delta\varepsilon_p}{2}$$
$$\frac{\Delta\varepsilon}{2} = \frac{\Delta\sigma}{2E} + \frac{1}{2}\left(\frac{\Delta\sigma}{K'}\right)^{1/n'} \tag{12-9}$$

This is the form of the equation for the cyclic stress-strain curve that is usually given.

For metals n' varies between 0.10 and 0.20. In general, metals with high monotonic strain-hardening exponents ($n > 0.15$) undergo cyclic hardening; those with a low strain-hardening exponent ($n < 0.15$) undergo cyclic softening. Cyclic hardening is to be expected[1] when the ratio of the monotonic ultimate tensile strength to the 0.2 percent offset yield strength is greater than 1.4. When this ratio is less than 1.2 cyclic softening is to be expected. Between 1.2 and 1.4 large changes in hardness are not to be expected. While this rule is useful in predicting where hardening or softening is to be expected, the magnitude of the cyclically induced change can only be found from tests.

12-7 LOW-CYCLE FATIGUE

Although historically fatigue studies have been concerned with conditions of service in which failure occurred at more than 10^4 cycles of stress, there is growing recognition of engineering failures which occur at relatively high stress and low numbers of cycles to failure.[2] This type of fatigue failure must be considered in the design of nuclear pressure vessels, steam turbines, and most other types of power machinery. Low-cycle fatigue conditions frequently are created where the repeated stresses are of thermal origin.[3] Since thermal stresses arise from the thermal expansion of the material, it is easy to see that in this case fatigue results from cyclic strain rather than from cyclic stress.[4]

The usual way of presenting low-cycle fatigue test results is to plot the plastic strain range $\Delta\varepsilon_p$ against N. Figure 12-13 shows that a straight line is obtained

[1] R. W. Smith, M. H. Hirschberg, and S. S. Manson, NASA TN D-1574, April 1963.

[2] L. F. Coffin, Fatigue in Machines and Structures—Power Generation, "Fatigue and Microstructure," pp. 1–27, American Society for Metals, Metals Park, Ohio, 1979.

[3] S. S. Manson, "Thermal Stress and Low-Cycle Fatigue," McGraw-Hill Book Company, New York, 1960.

[4] "Manual on Low-Cycle Fatigue Testing," *ASTM Spec. Tech. Publ.* 465, 1969; Constant Amplitude Low-Cycle Fatigue Testing, ASTM Standard E606-80.

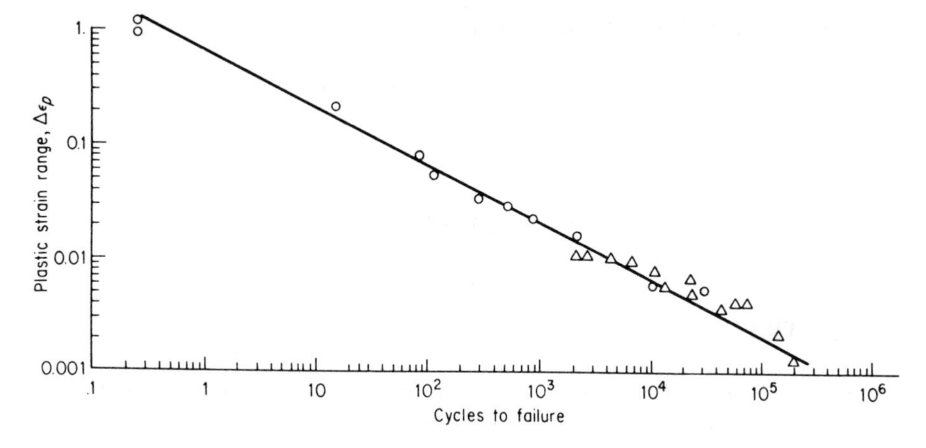

Figure 12-13 Low-cycle fatigue curve ($\Delta\varepsilon_p$ vs. N) for Type 347 stainless steel. *(From L. F. Coffin, Jr., Met. Eng. Q., vol. 3, p. 22, 1963.)*

when plotted on log-log coordinates. This type of behavior is known as the Coffin-Manson relation, which is best described by

$$\frac{\Delta\varepsilon_p}{2} = \varepsilon_f'(2N)^c \tag{12-10}$$

where $\Delta\varepsilon_p/2 =$ plastic strain amplitude

$\varepsilon_f' =$ fatigue ductility coefficient defined by the strain intercept at $2N = 1$. ε_f' is approximately equal to the true fracture strain ε_f for many metals.

$2N =$ number of strain reversals to failure (one cycle is two reversals)

$c =$ fatigue ductility exponent, which varies between -0.5 and -0.7 for many metals.

A smaller value of c results in larger values of fatigue life. Morrow[1] has shown that $c = -(1/1 + 5n')$, so that materials with larger values of n' have longer fatigue lives.

Example For the cyclic stress-strain curve shown in Fig. 12-10, $\sigma_B = 75$ MPa and $\varepsilon_B = 0.000645$. If $\varepsilon_f = 0.30$ and $E = 22 \times 10^4$ MPa determine (a) $\Delta\varepsilon_e$ and $\Delta\varepsilon_p$; (b) the number of cycles to failure.

(a) $$\Delta\varepsilon_e = \frac{\Delta\sigma}{E} = \frac{2(75)}{22 \times 10^4} = 6.818 \times 10^{-4}$$

$$\Delta\varepsilon_p = \Delta\varepsilon - \Delta\varepsilon_e = (2 \times 0.000645) - 0.0006818 = 6.082 \times 10^{-4}$$

[1] J. D. Morrow, in "Internal Friction, Damping and Cyclic Plasticity," ASTM STP No. 378, p. 72, ASTM, Philadelphia, 1965.

(b) From the Coffin-Manson relation

$$\frac{\Delta\varepsilon_p}{2} = \varepsilon_f'(2N)^c \quad \text{if } c = -0.6 \quad \text{and} \quad \varepsilon_f \approx \varepsilon_f'$$

$$\frac{6.082 \times 10^{-4}}{2} = 0.30(2N)^{-0.6}$$

$$N = 49{,}000 \text{ cycles}$$

12-8 STRAIN-LIFE EQUATION

Equation (12-10) describes the relationship between plastic strain and fatigue life in the low-cycle (high strain) fatigue regime. For the high-cycle (low strain) regime, where the nominal strains are elastic, Basquin's equation [Eq. (12-6)] can be reformulated to give

$$\sigma_a = \frac{\Delta\varepsilon_e}{2}E = \sigma_f'(2N)^b \tag{12-11}$$

where σ_a = alternating stress amplitude
$\Delta\varepsilon_e/2$ = elastic strain amplitude
E = Young's modulus
σ_f' = fatigue strength coefficient defined by the stress intercept at $2N = 1$. σ_f' is approximately equal to the monotonic true fracture stress, σ_f.
$2N$ = number of load reversals to failure (N = number of cycles to failure)
b = fatigue strength exponent, which varies between -0.05 and -0.12 for most metals

A smaller value of b results in longer fatigue lives. Morrow[1] has shown that $b = -n'/(1 + 5n')$, so that lower values of n' result in increased fatigue life for fatigue controlled by Eq. (12-11).

An equation valid for the entire range of fatigue lives can be obtained[2] by superposition of Eqs. (12-10) and (12-11).

$$\frac{\Delta\varepsilon}{2} = \frac{\Delta\varepsilon_e}{2} + \frac{\Delta\varepsilon_p}{2}$$

$$\frac{\Delta\varepsilon}{2} = \frac{\sigma_f'}{E}(2N)^b + \varepsilon_f'(2N)^c \tag{12-12}$$

As shown in Fig. 12-14 the fatigue strain-life curve tends toward the plastic curve at large total strain amplitudes and tends toward the elastic curve at small total strain amplitudes. Ductile materials give the best performance for high cyclic

[1] J. D. Morrow, op. cit.

[2] S. S. Manson and M. H. Hirschberg, "Fatigue: An Interdisciplinary Approach," p. 133, Syracuse University Press, Syracuse, N.Y., 1964.

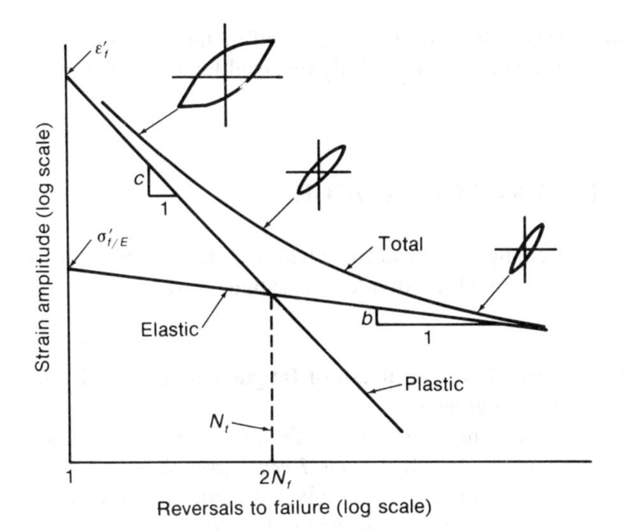

Figure 12-14 Fatigue strain-life curve obtained by superposition of elastic and plastic strain-life equations (schematic).

strain conditions while strong materials give the best results for low strain situations. The value of fatigue life at which this transition occurs is

$$2N_t = \left(\frac{\varepsilon'_f E}{\sigma'_f}\right)^{1/(b-c)} \tag{12-13}$$

It can further be shown[1] that $n' = b/c$ and $K' = \sigma'_f/(\varepsilon'_f)^{n'}$, where n' and K' are as defined in Eq. (12-9).

When the mean stress is not zero, this correction is introduced into Eq. (12-11) as

$$\sigma_a = (\sigma'_f - \sigma_m)(2N)^b \tag{12-14}$$

However, mean stresses will usually be relaxed after a short number of cycles in strain controlled fatigue tests.[2]

Manson[3] has proposed a simplified version of Eq. (12-12) which is known as the method of universal slopes

$$\Delta\varepsilon = 3.5\frac{S_u}{E}N^{-0.12} + \varepsilon_f^{0.6}N^{-0.6} \tag{12-15}$$

where S_u = ultimate tensile strength
ε_f = true strain at fracture in tension
E = elastic modulus in the tension test

[1] ASTM Standard E606-80, Annual Book of ASTM Standards.
[2] F. Lorenzo and C. Laird, *Mater. Sci. Eng.*, vol. 62, pp. 206–210, 1984.
[3] S. S. Manson, *Exp. Mech.*, vol. 5, no. 7, p. 193, 1965.

This equation is based on average values for a large class of metals and is used as a first approximation for the strain-life curve for fully reversed fatigue cycles in an unnotched specimen.

12-9 STRUCTURAL FEATURES OF FATIGUE

Studies of the basic structural changes[1] that occur when a metal is subjected to cyclic stress have found it convenient to divide the fatigue process into the following stages:

1. *Crack initiation*—includes the early development of fatigue damage which can be removed by a suitable thermal anneal.
2. *Slip-band crack growth*—involves the deepening of the initial crack on planes of high shear stress. This frequently is called *stage I crack growth*.
3. *Crack growth on planes of high tensile stress*—involves growth of well-defined crack in direction normal to maximum tensile stress. Usually called *stage II crack growth*.
4. *Ultimate ductile failure*—occurs when the crack reaches sufficient length so that the remaining cross section cannot support the applied load.

The relative proportion of the total cycles to failure that are involved with each stage depends on the test conditions and the material. However, it is well established that a fatigue crack can be formed before 10 percent of the total life of the specimen has elapsed. There is, of course, considerable ambiguity in deciding when a deepened slip band should be called a crack. In general, larger proportions of the total cycles to failure are involved with the propagation of stage II cracks in low-cycle fatigue than in long-life fatigue, while stage I crack growth comprises the largest segment for low-stress, high-cycle fatigue. If the tensile stress is high, as in the fatigue of sharply notched specimens, stage I crack growth may not be observed at all.

An overpowering structural consideration in fatigue is the fact that fatigue cracks usually are initiated at a free surface. In those rare instances where fatigue cracks initiate in the interior there is always an interface involved, such as the interface of a carburized surface layer and the base metal. Fatigue has certain things in common with plastic flow and fracture under static or unidirectional deformation. The work of Gough[2] has shown that a metal deforms under cyclic strain by slip on the same atomic planes and in the same crystallographic directions as in unidirectional strain. Whereas with unidirectional deformation slip is usually widespread throughout all the grains, in fatigue some grains will show slip lines while other grains will give no evidence of slip. Slip lines are generally formed during the first few thousand cycles of stress. Successive cycles

[1] W. J. Plumbridge and D. A. Ryder, *Metall. Rev.*, vol. 14, no. 136, 1969.
[2] H. J. Gough, *Am. Soc. Test. Mater. Proc.*, vol. 33, pt. 2, pp. 3–114, 1933.

produce additional slip bands, but the number of slip bands is not directly proportional to the number of cycles of stress. In many metals the increase in visible slip soon reaches a saturation value, which is observed as distorted regions of heavy slip. Cracks are usually found to occur in the regions of heavy deformation parallel to what was originally a slip band. Slip bands have been observed at stresses below the fatigue limit of ferrous materials. Therefore, the occurrence of slip during fatigue does not in itself mean that a crack will form.

A study of crack formation in fatigue can be facilitated by interrupting the fatigue test to remove the deformed surface by electropolishing. There will generally be several slip bands which are more persistent than the rest and which will remain visible when the other slip lines have been polished away. Such slip bands have been observed after only 5 percent of the total life of the specimen.[1] These persistent slip bands are embryonic fatigue cracks, since they open into wide cracks on the application of small tensile strains. Once formed, fatigue cracks tend to propagate initially along slip planes, although they later take a direction normal to the maximum applied tensile stress. Fatigue-crack propagation is ordinarily transgranular.

An important structural feature which appears to be unique to fatigue deformation is the formation on the surface of ridges and grooves called *slip-band extrusions* and *slip-band intrusions*.[2] Extremely careful metallography on tapered sections through the surface of the specimen has shown that fatigue cracks initiate at intrusions and extrusions.[3]

W. A. Wood,[4] who made many basic contributions to the understanding of the mechanism of fatigue, suggested a mechanism for producing slip-band extrusions and intrusions. He interpreted microscopic observations of slip produced by fatigue as indicating that the slip bands are the result of a systematic buildup of fine slip movements, corresponding to movements of the order of 10^{-7} cm rather than steps of 10^{-5} to 10^{-4} cm, which are observed for static slip bands. Such a mechanism is believed to allow for the accommodation of the large total strain (summation of the microstrain in each cycle) without causing appreciable strain hardening. Figure 12-15 illustrates Wood's concept of how continued deformation by fine slip might lead to a fatigue crack. The figures illustrate schematically the fine structure of a slip band at magnifications obtainable with the electron microscope. Slip produced by static deformation would produce a contour at the metal surface similar to that shown in Fig. 12-15*a*. In contrast, the back-and-forth fine slip movements of fatigue could build up notches (Fig. 12-15*b*) or ridges (Fig. 12-15*c*) at the surface. The notch would be a stress raiser with a notch root of atomic dimensions. Such a situation might well be the start of a fatigue crack. This mechanism for the initiation of a fatigue crack is in agreement with the facts

[1] G. C. Smith, *Proc. R. Soc. London*, vol. 242A, pp. 189–196, 1957.

[2] P. J. E. Forsyth and C. A. Stubbington, *J. Inst. Met.*, vol. 83, p. 395, 1955–1956.

[3] W. A. Wood, Some Basic Studies of Fatigue in Metals, in "Fracture," John Wiley & Sons, Inc., New York, 1959.

[4] W. A. Wood, *Bull. Inst. Met.*, vol. 3, pp. 5–6, September, 1955.

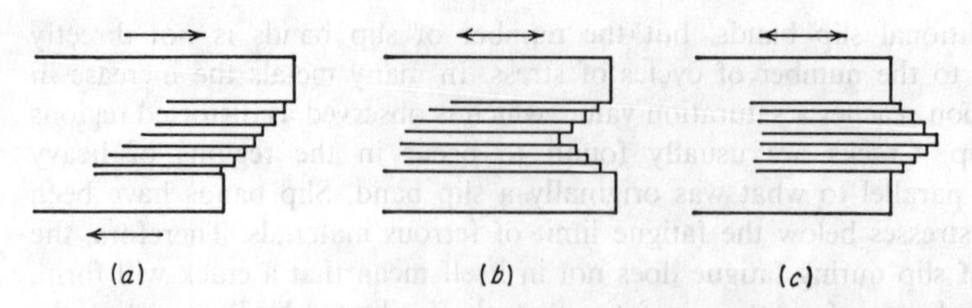

Figure 12-15 W. A. Wood's concept of microdeformation leading to formation of fatigue crack. (*a*) Static deformation; (*b*) fatigue deformation leading to surface notch (intrusion); (*c*) fatigue deformation leading to slip-band extrusion.

that fatigue cracks start at surfaces and that cracks have been found to initiate at slip-band intrusions and extrusions.

Extensive structural studies[1] of dislocation arrangements in persistent slip bands have brought much basic understanding to the fatigue fracture process.

The stage I crack propagates initially along the persistent slip bands. In a polycrystalline metal the crack may extend for only a few grain diameters before the crack propagation changes to stage II. The rate of crack propagation in stage I is generally very low, on the order of angstroms per cycle, compared with crack propagation rates of microns per cycle for stage II. The fracture surface of stage I fractures is practically featureless.

By marked contrast the fracture surface of stage II crack propagation frequency shows a pattern of ripples or fatigue fracture striations (Fig. 12-16). Each striation represents the successive position of an advancing crack front that is normal to the greatest tensile stress. Each striation was produced by a single cycle of stress. The presence of these striations unambiguously defines that failure was produced by fatigue, but their absence does not preclude the possibility of fatigue fracture. Failure to observe striations on a fatigue surface may be due to a very small spacing that cannot be resolved with the observational method used, insufficient ductility at the crack tip to produce a ripple by plastic deformation that is large enough to be observed or obliteration of the striations by some sort of damage to the surface. Since stage II cracking does not occur for the entire fatigue life, it does not follow that counting striations will give the complete history of cycles to failure.

Stage II crack propagation occurs by a plastic blunting process[2] that is illustrated in Fig. 12-17. At the start of the loading cycle the crack tip is sharp (Fig. 12-17*a*). As the tensile load is applied the small double notch at the crack tip concentrates the slip along planes at 45° to the plane of the crack (Fig. 12-17*b*). As the crack widens to its maximum extension (Fig. 12-17*c*) it grows longer by plastic shearing and at the same time its tip becomes blunter. When the load is changed to compression the slip direction in the end zones is reversed (Fig.

[1] C. Laird, Mechanisms and Theories of Fatigue, in "Fatigue and Microstructure," pp. 149–203, American Society for Metals, Metals Park, Ohio, 1979.

[2] C. Laird, in "Fatigue Crack Propagation," ASTM *Spec. Tech. Publ.* 415, pp. 131–168, 1967.

Figure 12-16 Fatigue striations in beta-annealed Ti-6Al-4V alloy (2000 ×). *(Courtesy of R. A. Bayles, Naval Research Laboratory.)*

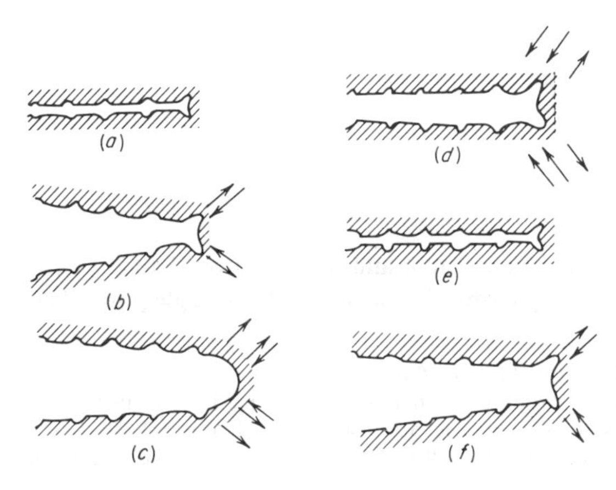

Figure 12-17 Plastic blunting process for growth of stage II fatigue crack. *(From C. Laird, ASTM Spec. Tech. Publ. 415, p. 136, 1967.)*

12-17*d*). The crack faces are crushed together and the new crack surface created in tension is forced into the plane of the crack (Fig. 12-17*e*) where it partly folds by buckling to form a resharpened crack tip. The resharpened crack is then ready to advance and be blunted in the next stress cycle.

There are a number of indications that cyclic deformation results in a higher concentration of vacancies than cold-working by unidirectional deformation. The difference in the release of stored energy between fatigued and cold-worked copper is in line with what would be expected from a large concentration of point defects. The fact that initially cold-worked copper becomes softer as a result of fatigue[1] can be explained by the generation of point defects which allow the metal partly to recover by permitting dislocations to climb out of the slip plane. Age-hardening aluminum alloys in the precipitation-hardened condition can be overaged by fatigue deformation at room temperature. This suggests that vacancies produced by fatigue are available to accomplish the diffusion required for the overaging process.[2] Moreover, the fatigue strength increases markedly on going from 20 to −190°C, where vacancy movement is negligible. However, the fact that fatigue fracture can be produced at 4°K indicates that a temperature-activated process such as the diffusion of vacancies is not essential for fatigue failure.[3]

12-10 FATIGUE CRACK PROPAGATION

Considerable research has gone into determining the laws of fatigue crack propagation for stage II growth.[4] Reliable crack propagation relations permit the implementation of a fail-safe design philosophy which recognizes the inevitability of cracks in engineering structures and aims at determining the safe load and crack length which will preclude failure in a conservatively estimated service life. The crack propagation rate da/dN is found to follow an equation

$$\frac{da}{dN} = C\sigma_a^m a^n \tag{12-16}$$

where C = a constant
σ_a = the alternating stress
a = the crack length

In different investigations m ranges from 2 to 4 and n varies from 1 to 2. Crack propagation can also be expressed in terms of total strain[5] by a single power-law

[1] N. H. Polakowski and A. Palchoudhuri, *Am. Soc. Test Mater. Proc.*, vol. 54, p. 701, 1954.
[2] T. Broom, J. H. Molineux, and V. N. Whittaker, *J. Inst. Met.*, vol. 84, pp. 357–363, 1955–1956.
[3] R. D. McCammon and H. M. Rosenberg, *Proc. R. Soc. London*, vol. 242A, p. 203, 1957.
[4] "Fatigue Crack Propagation," ASTM *Spec. Tech. Publ.* 415, 1967; D. Walton and E. G. Ellison, *Int. Metall. Rev.*, vol. 17, pp. 100–116, 1972.
[5] T. W. Crooker and E. A. Lange, *op. cit.* p. 94.

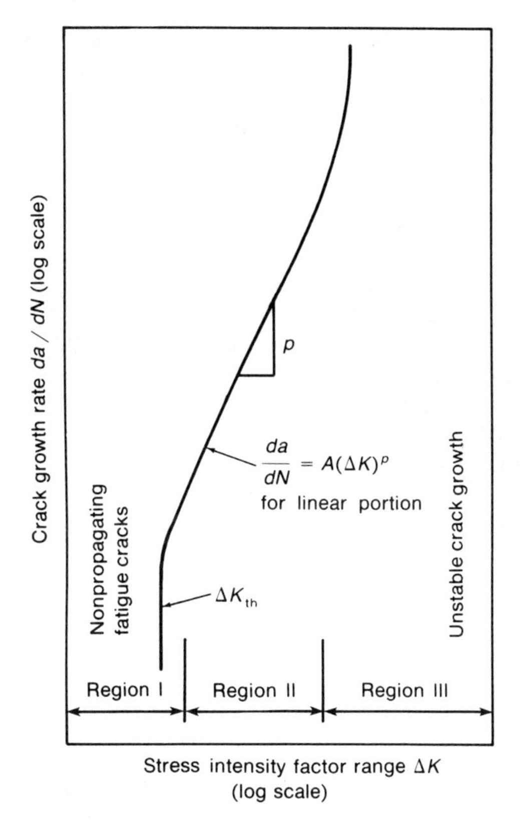

Figure 12-18 Schematic representation of fatigue crack growth behavior in a non-agressive environment.

expression which extends from elastic to plastic strain region.

$$\frac{da}{dN} = C_1 \varepsilon^{m_1} \tag{12-17}$$

The most important advance in placing fatigue crack propagation into a useful engineering context was the realization that crack length versus cycles at a series of different stress levels could be expressed by a general plot of da/dN versus ΔK. da/dN is the slope of the crack growth curve at a given value of a and ΔK is the range of the stress intensity factor, defined as

$$\Delta K = K_{\text{max}} - K_{\text{min}}$$

$$\Delta K = \sigma_{\text{max}}\sqrt{\pi a} - \sigma_{\text{min}}\sqrt{\pi a} = \sigma_r\sqrt{\pi a} \tag{12-18}$$

Since the stress intensity factor is undefined in compression, K_{min} is taken as zero if σ_{min} is compression.

The relationship between fatigue crack growth rate and ΔK is shown in Fig. 12-18. This curve has a sigmoidal shape that can be divided into three regions. Region I is bounded by a threshold value ΔK_{th}, below which there is no observable fatigue crack growth.[1] At stresses below ΔK_{th} cracks behave as nonpropating cracks. ΔK_{th} occurs at crack propagation rates of the order of 2.5×10^{-10} m/cycle (10^{-8} in/cycle) or less.

Region II represents an essentially linear relationship between log da/dN and log ΔK

$$\frac{da}{dN} = A(\Delta K)^p \tag{12-19}$$

For this empirical relationship p is the slope of the curve and A is the value found by extending the straight line to $\Delta K = 1$ MPa$\sqrt{\text{m}}$ (ksi$\sqrt{\text{in}}$). The value of p is approximately 3 for steels and in the range 3 to 4 for aluminum alloys. Equation (12-19) is often referred to as Paris' law.[2]

Region III is a region of accelerated crack growth. Here K_{max} approaches K_c, the fracture toughness of the material.

Increasing the mean stress in the fatigue cycle ($R = \sigma_{min}/\sigma_{max} = K_{min}/K_{max}$) has a tendency to increase the crack growth rates in all portions of the sigmoidal curve. Generally the effect of increasing R is less in Region II than in Regions I and III. The influence of R on the Paris relationship is given by[3]

$$\frac{da}{dN} = \frac{A(\Delta K)^p}{(1 - R)K_c - \Delta K} \tag{12-20}$$

where K_c = the fracture toughness applicable to the material and thickness
R = stress ratio = $\sigma_{min}/\sigma_{max} = K_{min}/K_{max}$

Fatigue life testing is usually carried out under conditions of fully reversed stress or strain ($R = -1$). However, fatigue crack growth data is usually determined for conditions of pulsating tension ($R = 0$). Compression loading cycles are not used because during compression loading the crack is closed and the stress intensity factor is zero. While compression loading generally is considered to be of little influence in crack propagation, under variable amplitude loading compression cycles can be important.

Equation (12-19) provides an important link between fracture mechanics and fatigue.[4] The elastic stress intensity factor is applicable to fatigue crack growth even in low-strength, high ductility materials because the K values needed to cause fatigue crack growth are very low and the plastic zone sizes at the tip are small enough to permit an LEFM approach. When K is known for the component under relevant loading conditions the fatigue crack growth life of the component can be obtained by integrating Eq. (12-19) between the limits of initial

[1] R. O. Ritchie, Near Threshold Fatigue Crack Propagation in Steels, *Int. Met. Rev.*, vol. 24, pp. 205–230, 1979.

[2] P. C. Paris and F. Erdogan, *Trans. ASME J. Basic Eng.*, vol. 85, pp. 528–534, 1963.

[3] R. G. Forman, V. E. Kearney, and R. M. Engle, *Trans. ASME J. Basic Eng.*, vol. 89, p. 459, 1967.

[4] R. P. Wei, *Trans. ASME J. Eng. Mater. Tech.*, vol. 100, pp. 113–120, 1978.

crack size and final crack size

$$\Delta K = \alpha \, \Delta\sigma\sqrt{\pi a} = \alpha\sigma_r\sqrt{\pi a} \tag{12-21}$$

$$\frac{da}{dN} = A(\Delta K)^p = A\left(\alpha\sigma_r\sqrt{\pi a}\right)^p$$

$$= A(\alpha)^p(\sigma_r)^p(\pi a)^{p/2} \tag{12-22}$$

$$a_f = \frac{1}{\pi}\left(\frac{K_c}{\sigma_{\max}\alpha}\right)^2 \tag{12-23}$$

$$N_f = \int_0^{N_f} dN = \int_{a_i}^{a_f} \frac{da}{A(\alpha)^p(\sigma_r)^p(\pi a)^{p/2}}$$

$$= \frac{1}{A\alpha^p(\sigma_r)^p\pi^{p/2}}\int_{a_i}^{a_f}\frac{da}{a^{p/2}} \tag{12-24}$$

If $p \neq 2$,

$$\int_{a_i}^{a_f}\frac{da}{a^{p/2}} = \left.\frac{a^{-(p/2)+1}}{-(p/2)+1}\right|_{a_i}^{a_f} = \frac{a_f^{-(p/2)+1} - a_i^{-(p/2)+1}}{-(p/2)+1}$$

$$N_f = \frac{a_f^{-(p/2)+1} - a_i^{-(p/2)+1}}{(-(p/2)+1)A\sigma_r^p\pi^{p/2}\alpha^p} \tag{12-25}$$

Equation (12-25) is the appropriate integration of the Paris equation when $p \neq 2$ and α is independent of crack length, which unfortunately is not the usual case. For the more general case $\alpha = f(a)$ and Eq. (12-24) must be written

$$N_f = \frac{1}{A\sigma_r^p\pi^{p/2}}\int_{a_i}^{a_f}\alpha(a)^{-p}a^{-p/2}\,da \tag{12-26}$$

This is usually solved by an iterative process in which ΔK and ΔN are determined for successive increments of crack growth.

Example A mild steel plate is subjected to constant amplitude uniaxial fatigue loads to produce stresses varying from $\sigma_{\max} = 180$ MPa to $\sigma_{\min} = -40$ MPa. The static properties of the steel are $\sigma_0 = 500$ MPa, $S_u = 600$ MPa, $E = 207$ GPa, and $K_c = 100 \text{ MPa}\sqrt{\text{m}}$. If the plate contains an initial through thickness edge crack of 0.5 mm, how many fatigue cycles will be required to break the plate?

We can assume an infinite wide plate, for which $\alpha = 1.12$. For ferritic-pearlitic steels, a general correlation gives

$$\frac{da}{dN}(\text{m/cycle}) = 6.9 \times 10^{-12}(\Delta K)^3 \; (\text{MPa}\sqrt{\text{m}})^3$$

Thus $A = 6.9 \times 10^{-12}$ $(\text{MPa}\sqrt{\text{m}})$, $p = 3.0$, and $\sigma_r = 180 - 0$ (since compressive stresses are ignored), and we shall neglect the small influence of mean stress on the crack growth

$$a_i = 0.0005 \text{ m} \qquad a_f = \frac{1}{\pi}\left(\frac{K_c}{\sigma_{\max}\alpha}\right)^2 = \frac{1}{\pi}\left(\frac{100}{180 \times 1.12}\right)^2$$

$$a_f = 0.078 \text{ m} = 78 \text{ mm}$$

$$N_f = \frac{(0.078)^{-(3/2)+1} - (0.0005)^{-(3/2)+1}}{(-(3/2)+1)(6.9 \times 10^{-12})(180)^3(\pi)^{3/2}(1.12)^3}$$

$$= \frac{(0.078)^{-0.5} - (0.0005)^{-0.5}}{-157.4 \times 10^{-6}}$$

$$N_f = 261{,}000 \text{ cycles}$$

12-11 EFFECT OF STRESS CONCENTRATION ON FATIGUE

Fatigue strength is seriously reduced by the introduction of a stress raiser such as a notch or hole. Since actual machine elements invariably contain stress raisers like fillets, keyways, screw threads, press fits, and holes, it is not surprising to find that fatigue cracks in structural parts usually start at such geometrical irregularities. One of the best ways of minimizing fatigue failure is by the reduction of avoidable stress raisers through careful design and the prevention of accidental stress raisers by careful machining and fabrication. While this section is concerned with stress concentration resulting from geometrical discontinuities, stress concentration can also arise from surface roughness and metallurgical stress raisers such as porosity, inclusions, local overheating in grinding, and decarburization.

The effect of stress raisers on fatigue is generally studied by testing specimens containing a notch, usually a V notch or a circular notch. It has been shown in Chap. 7 that the presence of a notch in a specimen under uniaxial load introduces three effects: (1) there is an increase or concentration of stress at the root of the notch; (2) a stress gradient is set up from the root of the notch in toward the center of the specimen; (3) a triaxial state of stress is produced.

The ratio of the maximum stress to the nominal stress is the *theoretical stress-concentration factor* K_t. As was discussed in Sec. 2-15, values of K_t can be computed from the theory of elasticity for simple geometries and can be determined from photoelastic measurements for more complicated situations. Most of the available data on stress-concentration factors have been collected by Peterson.[1]

The effect of notches on fatigue strength is determined by comparing the *S-N* curves of notched and unnotched specimens. The data for notched specimen are

[1] R. E. Peterson, "Stress-Concentration Design Factors," John Wiley & Sons, Inc., New York, 1974.

usually plotted in terms of nominal stress based on the net section of the specimen. The effectiveness of the notch in decreasing the fatigue limit is expressed by the *fatigue-strength reduction factor*, or *fatigue-notch factor*, K_f. This factor is simply the ratio of the fatigue limit of unnotched specimens to the fatigue limit of notched specimens. For materials which do not exhibit a fatigue limit the fatigue-notch factor is based on the fatigue strength at a specified number of cycles. Values of K_f have been found to vary with (1) severity of the notch, (2) the type of notch, (3) the material, (4) the type of loading, and (5) the stress level. The values of K_f published in the literature are subject to considerable scatter and should be carefully examined for their limitations and restrictions. However, two general trends are usually observed for test conditions of completely reversed loading. First, K_f is usually less than K_t, and second, the ratio of K_f/K_t decreases as K_t increases. Thus, very sharp notches (high K_t) have less effect on fatigue strength than would be expected from their high value of K_t.

The notch sensitivity of a material in fatigue is expressed by a notch-sensitivity factor q

$$q = \frac{K_f - 1}{K_t - 1} \tag{12-27}$$

Equation (12-27) was chosen so that a material which experiences no reduction in fatigue due to a notch ($K_f = 1$) has a factor of $q = 0$, while a material in which the notch has its full theoretical effect ($K_f = K_t$) has a factor of $q = 1$. However, q is not a true material constant since it varies with the severity and type of notch (Fig. 12-19), the size of specimen, and the type of loading. As Fig. 12-19 indicates notch sensitivity increases with tensile strength. Thus, it is possible in certain circumstances to decrease fatigue performance by increasing the hardness or tensile strength of a material.

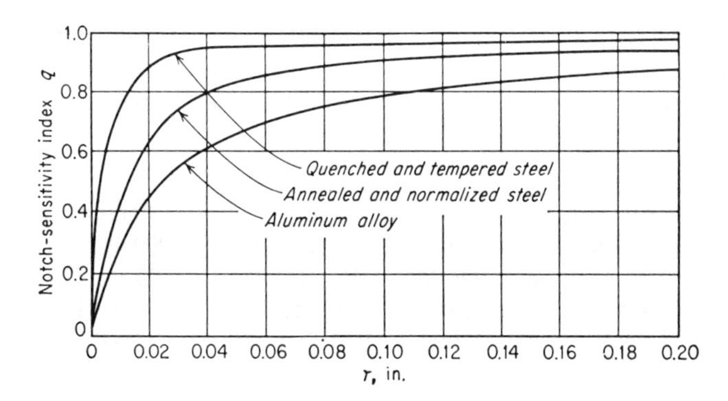

Figure 12-19 Variation of notch-sensitivity index with notch radius for materials of different tensile strength. *(From R. E. Peterson, in G. Sines and J. L. Waisman (eds.), "Metal Fatigue," p. 301, McGraw-Hill Book Company, New York, 1959. By permission of the publishers.)*

Table 12-1 Some values of Neuber's constant ρ'

Material	strength level, MPa (ksi)	ρ', mm (0.001 in)
Steel	$S_u = 552$ (80)	0.15 (6)
	$S_u = 896$ (130)	0.07 (3)
	$S_u = 1310$ (190)	0.01 (0.4)
Aluminum	$S_u = 150$ (22)	2 (80)
alloy	$S_u = 300$ (43)	0.6 (25)
	$S_u = 600$ (87)	0.4 (15)

Another approach to notch sensitivity in fatigue was advanced by Neuber.[1] He expressed the fatigue stress concentration factor as

$$K_f = 1 + \frac{K_t - 1}{1 + \sqrt{\rho'/r}} \tag{12-28}$$

where r = contour radius at the root of the notch
ρ' = a material constant related to the strength of the metal. See Table 12-1 for some values of ρ'.

The above relationships express the fact that, for large notches with large radii, K_f is almost equal to K_t, but for small notches we find $K_f \ll K_t$ for soft ductile metals and K_f higher for stronger metals. We say these are more notch sensitive.

In the absence of data, an estimate of the notched fatigue strength can be made by constructing a straight line on a semilogarithmic S-N plot from the ultimate tensile strength at a life of one cycle to the unnotched fatigue strength divided by K_f at 10^6 cycles.

For design purposes, in ductile materials K_f should be applied only to the alternating component of stress and not to the steady component of stress when $\sigma_m \neq 0$. For fatigue of brittle materials K_f should also be applied to the steady component of stress.

Example A shaft with a transverse oil hole is subjected to a fluctuating bending moment of ± 200 in-lb and an axial steady load of 5000 lb. The diameter of the shaft is 0.5 in and the diameter of the hole is 0.05 in. This situation is shown in Fig. 2-22*e*, where $a/d = 0.10$ and $K_t = 2.2$. If the shaft is made from steel with $S_u = 190$ ksi, then from Table 12-1 $\rho' = 0.0004$ and

[1] H. Neuber, "Theory of Notch Stresses," J. W. Edwards, Publisher, Incorporated, Ann Arbor, Mich. 1946.

$\sqrt{\rho'} = 0.02$. Since $r = 0.025$

$$K_f = 1 + \frac{K_t - 1}{1 + \sqrt{\rho'/r}} = 1 + \frac{2.2 - 1}{1 + \sqrt{0.0004/0.025}} = 1 + \frac{1.1}{1 + 0.126} = 1.98$$

$$q = \frac{K_f - 1}{K_t - 1} = \frac{0.98}{1.2} = 0.82$$

The mean or steady stress is

$$\sigma_m = \frac{P_m}{A} = \frac{5000}{(\pi/4)(0.5)^2} = 25{,}471 \text{ psi}$$

The fluctuating bending stress is

$$\sigma_a = K_f\left[\frac{M(D/2)}{I}\right] = 1.98\left[\frac{200(0.25)}{\pi\dfrac{(0.50)^4}{64}}\right] = 1.98(33{,}112) = 65{,}561 \text{ psi}$$

The effective maximum stress is given by

$$\sigma_{\max} = \sigma_a + \sigma_m = 65{,}561 + 25{,}471 = 91{,}032 \text{ psi}$$

and $$\sigma_{\min} = \sigma_a - \sigma_m = 65{,}561 - 25{,}471 = 40{,}090 \text{ psi}$$

A good estimate of the unnotched fatigue limit is $S_u/2 = 95{,}000$ psi. The S-N curve can be estimated as follows

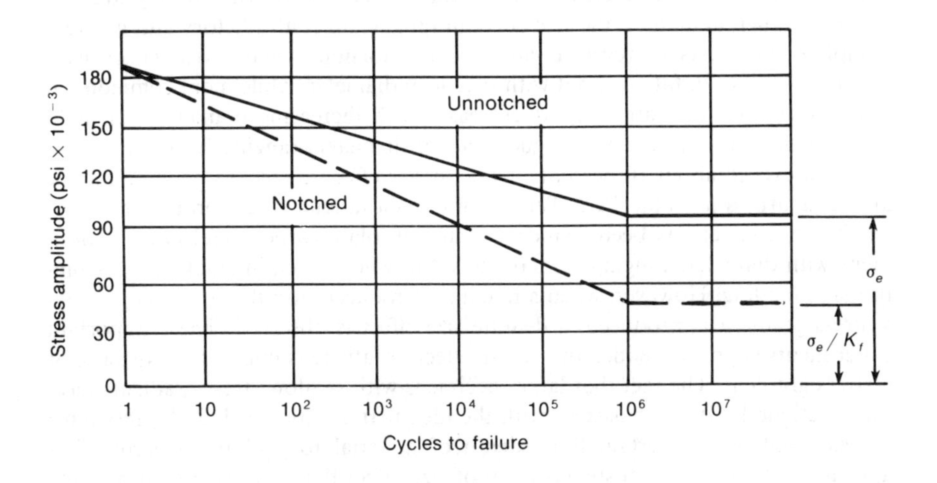

Using the equation for the Goodman line, Eq. (12-7)

$$\sigma_a = \frac{\sigma_e}{K_f}\left(1 - \frac{\sigma_m}{\sigma_u}\right) = \frac{95{,}000}{1.98}\left(1 - \frac{25{,}400}{190{,}000}\right) = 41{,}500 \text{ psi}$$

This is the value of σ_a which will give infinite life for a shaft with a hole in it subjected to a mean stress of 25,471 psi. Since the actual fluctuating stress is greater than this, the shaft will fail at about 10^5 cycles, as estimated from the above diagram.

12-12 SIZE EFFECT

An important practical problem is the prediction of the fatigue performance of large machine members from the results of laboratory tests on small specimens. Experience has shown that in most cases a *size effect* exists; i.e., the fatigue strength of large members is lower than that of small specimens. A precise study of this effect is difficult for several reasons. It is extremely difficult, if not altogether impossible, to prepare geometrically similar specimens of increasing diameter which have the same metallurgical structure and residual stress distribution throughout the cross section. The problems in fatigue testing large-sized specimens are considerable, and there are few fatigue machines which can accommodate specimens having a wide range of cross sections.

Changing the size of a fatigue specimen usually results in a variation in two factors. First, increasing the diameter increases the volume or surface area of the specimen. The change in amount of surface is of significance, since fatigue failures usually start at the surface. Second, for plain or notched specimens loaded in bending or torsion, an increase in diameter usually decreases the stress gradient across the diameter and increases the volume of material which is highly stressed.

Experimental data on the size effect in fatigue are contradictory and not very complete. For tests in reversed bending and torsion, some investigators have found no change in fatigue limit with specimen diameter, while more commonly it is observed that the fatigue limit decreases with increasing diameter. For mild steel the decrease in bending fatigue limit for diameters ranging from 0.1 to 2 in does not exceed about 10 percent.[1] Data on size effect in steel in bending fatigue are summarized in Table 12-2. The factor C_S is a fatigue reduction factor.

No size effect has been found[2] for smooth plain-carbon-steel fatigue specimens with diameters ranging from 0.2 to 1.4 in when tested in axial tension-compression loading. However, when a notch is introduced into the specimen, so that a stress gradient is produced, a definite size effect is observed. These important experiments support the idea that a size effect in fatigue is due to the existence of a stress gradient. The fact that large specimens with shallow stress gradients have lower fatigue limits is consistent with the idea that a critical value of stress must be exceeded over a certain finite depth of material for failure to occur. This appears to be a more realistic criterion of size effect than simply the ratio of the change in surface area to the change in specimen diameter. The importance of

[1] O. J. Horger, Fatigue Characteristics of Large Sections, in "Fatigue," American Society for Metals, Metals Park, Ohio, 1953.

[2] C. E. Phillips and R. B. Heywood, *Proc. Inst. Mech. En.* (*London*), vol. 165, pp. 113–124, 1951.

Table 12-2 Fatigue reduction factor due to size effect

Diameter, in	C_S
$D \leq 0.4$	1.0
$0.4 \leq D \leq 2.0$	0.9
$2.0 \leq D \leq 9.0$	$1 - \dfrac{D - 0.03}{15}$

stress gradients in size effect helps explain why correlation between laboratory results and service failure is often rather poor. Actual failures in large parts are usually directly attributable to stress concentrations, either intentional or accidental, and it is usually impossible to duplicate the same stress concentration and stress gradient in a small-sized laboratory specimen.

Analysis of considerable data for steels has shown[1] a size-effect relationship between fatigue limit and the critically stressed volume of the material.

$$\sigma_{f1} = \sigma_{f0}\left(\frac{V}{V_0}\right)^{-0.034} \tag{12-29}$$

where σ_{f1} is the fatigue limit for a critical volume V and σ_{f0} is the known fatigue limit for a specimen with volume V_0. Critically stressed volume is defined as the volume near the surface of the specimen that is stressed to at least 95 percent of σ_{max}.

A fruitful approach to size effect has been made by using Weibull statistics.[2]

12-13 SURFACE EFFECTS AND FATIGUE

Practically all fatigue failures start at the surface. For many common types of loading, like bending and torsion, the maximum stress occurs at the surface so that it is logical that failure should start there. However, in axial loading the fatigue failure nearly always begins at the surface. There is ample evidence that fatigue properties are very sensitive to surface condition. The factors which affect the surface of a fatigue specimen can be divided roughly into three categories, (1) surface roughness or stress raisers at the surface, (2) changes in the fatigue strength of the surface metal, and (3) changes in the residual stress condition of the surface. In addition, the surface is subjected to oxidation and corrosion.

[1] R. Kuguel, *Am. Soc. Test. Mater. Proc.*, vol. 61, pp. 732–748, 1961.
[2] G. G. Trantina, *J. Test. Eval.*, vol. 9, no. 1, pp. 44–49, 1981.

Table 12-3 Fatigue life of SAE 3130 steel specimens tested under completely reversed stress at 95,000 psi†

Type of finish	Surface roughness, μin	Median fatigue life, cycles
Lathe-formed	105	24,000
Partly hand-polished	6	91,000
Hand-polished	5	137,000
Ground	7	217,000
Ground and polished	2	234,000
Superfinished	7	212,000

† P. G. Fluck, *Am. Soc. Test. Mater. Proc.*, vol. 51, pp. 584–592, 1951.

Surface Roughness

Since the early days of fatigue investigations, it has been recognized that different surface finishes produced by different machining procedures can appreciably affect fatigue performance. Smoothly polished specimens, in which the fine scratches (stress raisers) are oriented parallel with the direction of the principal tensile stress, give the highest values in fatigue tests. Such carefully polished specimens are usually used in laboratory fatigue tests and are known as "par bars." Table 12-3 indicates how the fatigue life of cantilever-beam specimens varies with the type of surface preparation. Extensive data on this subject have been published by Siebel and Gaier.[1]

Figure 12-20 shows the influence of various surface finishes on steel in reducing the fatigue limit of carefully polished "par bars." Note that the surface finish is characterized by the process used to form the surface. The extreme sensitivity of high-strength steel to surface conditions is well illustrated.

Changes in Surface Properties

Since fatigue failure is so dependent on the condition of the surface, anything that changes the fatigue strength of the surface material will greatly alter the fatigue properties. Decarburization of the surface of heat-treated steel is particularly detrimental to fatigue performance. Similarly, the fatigue strength of aluminum-alloy sheet is reduced when a soft aluminum coating is applied to the stronger age-hardenable aluminum-alloy sheet. Marked improvements in fatigue properties can result from the formation of harder and stronger surfaces on steel parts by carburizing and nitriding.[2] However, since favorable compressive residual stresses are produced in the surface by these processes, it cannot be considered that the higher fatigue properties are due exclusively to the formation of higher-strength material on the surface. The effectiveness of carburizing and nitriding in improv-

[1] E. Siebel and M. Gaier, *VDIZ.*, vol. 98, pp. 1715–1723, 1956; abstracted in *Engineer's Digest*, vol. 18, pp. 109–112, 1957.

[2] "Fatigue Durability of Carburized Steel," American Society for Metals, Metals Park, Ohio, 1957; T. B. Cameron, D. E. Diesburg, and C. Kim, *J. Metals*, pp. 37–41, July 1983.

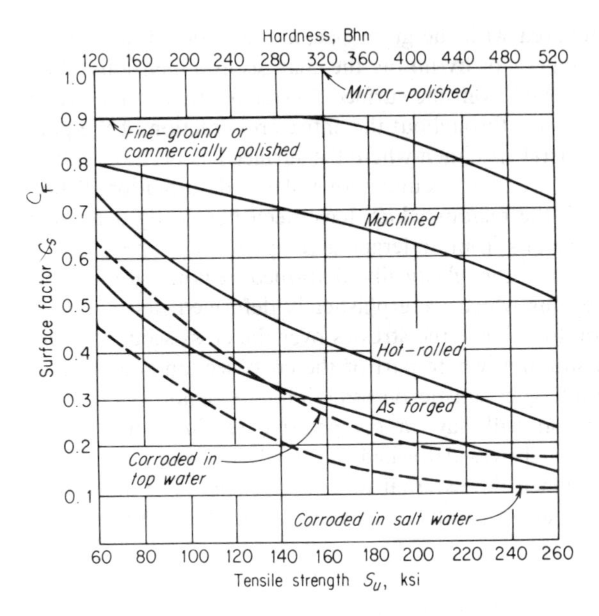

Figure 12-20 Reduction factor for fatigue limit of steel due to various surface treatments. *(From R. C. Juvinall, "Stress, Strain, and Strength," p. 234, McGraw-Hill Book Company, New York, 1967. By permission of the publishers.)*

ing fatigue performance is greater for cases where a high stress gradient exists, as in bending or torsion, than in an axial fatigue test. The greatest percentage increase in fatigue performance is found when notched fatigue specimens are nitrided. The amount of strengthening depends on the diameter of the part and the depth of surface hardening. Improvements in fatigue properties similar to those caused by carburizing and nitriding may also be produced by flame hardening and induction hardening. It is a general characteristic of fatigue in surface-hardened parts that the failure initiates at the interface between the hard case and the softer case, rather than at the surface.

Electroplating of the surface generally decreases the fatigue limit of steel. Chromium plating is particularly difficult to accomplish without impairment of fatigue properties, while a softer cadmium plating is believed to have little effect on fatigue strength. The particular plating conditions used to produce an electroplated surface can have an appreciable effect on the fatigue properties, since large changes in the residual stress, adhesion, porosity, and hardness of the plate can be produced.[1]

Surface Residual Stress

The formation of a favorable compressive residual-stress pattern at the surface is probably the most effective method of increasing fatigue performance. It can be considered that residual stresses are locked-in stresses which are present in a part which is not subjected to an external force. Only macrostresses, which act over

[1] A detailed review of the effect of electroplating on fatigue strength is given by R. A. R. Hammond and C. Williams, *Metall. Rev.*, vol. 5, pp. 165–223, 1960.

regions which are large compared with the grain size, are considered here. They can be measured by x-ray methods or by noting the changes in dimensions when a thin layer of material is removed from the surface. Residual stresses arise when plastic deformation is not uniform throughout the entire cross section of the part being deformed. Consider a metal specimen where the surface has been deformed in tension by bending so that part of it has undergone plastic deformation. When the external force is removed, the regions which have been plastically deformed prevent the adjacent elastic regions from undergoing complete elastic recovery to the unstrained condition. Thus, the elastically deformed regions are left in residual tension, and the regions which were plastically deformed must be in a state of residual compression to balance the stresses over the cross section of the specimen. In general, for a situation where part of the cross section is deformed plastically while the rest undergoes elastic deformation, the region which was plastically deformed in tension will have a compressive residual stress after unloading, while the region which was deformed plastically in compression will have a tensile residual stress when the external force is removed. The maximum value of residual stress which can be produced is equal to the elastic limit of the metal.

For many purposes residual stresses can be considered identical to the stresses produced by an external force. Thus, the addition of a compressive residual stress, which exists at a point on the surface, to an externally applied tensile stress on that surface decreases the likelihood of fatigue failure at that point. Figure 12-21 illustrates this effect. Figure 12-21*a* shows the elastic stress distribution in a beam with no residual stress. A typical residual-stress distribution, such as would be produced by shot peening, is shown in Fig. 12-21*b*. Note that the high compressive residual stresses at the surface must be balanced by tensile residual stresses over the interior of the cross section. In Fig. 12-21*c* the stress distribution due to the algebraic summation of the external bending stresses and the residual stresses is shown. Note that the maximum tensile stress at the surface is reduced by an amount equal to the surface compressive residual stress. The peak tensile stress is displaced to a point in the interior of the specimen. The magnitude of this stress depends on the gradient of applied stress and the residual-stress distribution. Thus, subsurface initiation of failure is possible under these conditions. It should also be apparent that the improvements in fatigue performance which result from the introduction of surface compressive residual stress will be greater when the loading is one in which a stress gradient exists than when no stress gradient is present. However, some improvement in the fatigue performance of axial-loaded fatigue specimens results from surface compressive residual stresses, presumably because the surface is such a potential source of weakness.

The chief commercial methods of introducing favorable compressive residual stresses in the surface are by surface rolling with contoured rollers and by shot peening.[1] Although some changes in the strength of the metal due to strain

[1] J. O. Almen and P. H. Black, "Residual Stresses and Fatigue in Metals," McGraw-Hill Book Company, New York, 1963.

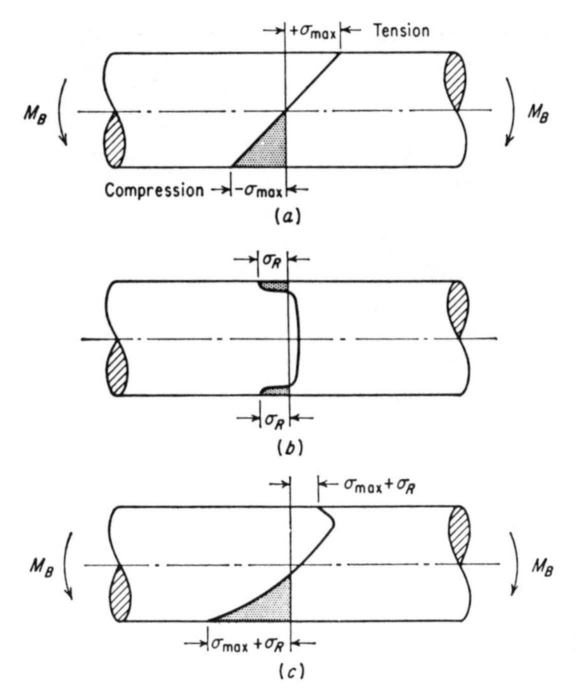

Figure 12-21 Superposition of applied and residual stresses.

hardening occur during these processes, the improvement in fatigue performance is due chiefly to the formation of surface compressive residual stress. Surface rolling is particularly adapted to large parts. It is frequently used in critical regions such as the fillets of crankshafts and the bearing surface of railroad axles. Shot peening consists in projecting fine steel or cast-iron shot against the surface at high velocity. It is particularly adapted to mass-produced parts of fairly small size. The severity of the stress produced by shot peening is frequently controlled by measuring the residual deformation of shot-peened beams called Almen strips. The principal variables in this process are the shot velocity and the size, shape, and hardness of the shot. Care must be taken to ensure uniform coverage over the area to be treated. Frequently an additional improvement in fatigue properties can be obtained by carefully polishing the shot-peened surface to reduce the surface roughness. Other methods of introducing surface compressive residual stresses are by means of thermal stresses produced by quenching steel from the tempering temperature and from stresses arising from the volume changes accompanying the metallurgical changes resulting from carburizing, nitriding, and induction hardening.

It is important to recognize that improvements in fatigue properties do not automatically result from the use of shot peening or surface rolling. It is possible to damage the surface by excessive peening or rolling. Experience and testing are required to establish the proper conditions which produce the optimum residual-stress distribution. Further, certain metallurgical processes yield surface tensile residual stresses. Thus, surface tensile stresses are produced by quenching deep-hardening steel, and this stress pattern may persist at low tempering temperatures. Grinding of hardened steel requires particular care to prevent a large decrease in fatigue properties. It has been shown[1] that either tensile or compressive surface residual stresses can be produced, depending on the grinding conditions. Further, the polishing methods ordinarily used for preparing fatigue specimens can result in appreciable surface residual stress.

Residual-stress distributions may be modified by plastic deformation or by thermal activation. When gross plastic deformation occurs, the stress drops to the yield stress. Thus, periods of overloading in the fatigue cycle or testing at high stresses in the low-cycle region may alter the residual-stress distribution by plastic deformation. This is referred to as *fading of residual stress*. Residual stresses have their greatest influence near the fatigue limit, where little fading occurs. Conversely, the fatigue life at high applied stresses depends little on residual stresses.

12-14 FATIGUE UNDER COMBINED STRESSES

Although data on combined stress fatigue failure are fewer than for static yielding certain generalizations can be made. Fatigue tests with different combinations of bending and torsion show that for ductile metals a distortion-energy (von Mises) criterion provides the best representation of the data. For brittle materials a maximum principal stress theory provides the best criterion of fatigue failure.

Sines[2] has proposed the following expression for fatigue failure under combined stress for conditions of low strain, long life fatigue ($N > 10^5$ cycles).

$$\left[(\sigma_{a1} - \sigma_{a2})^2 + (\sigma_{a2} - \sigma_{a3})^2 + (\sigma_{a3} - \sigma_{a1})^2\right]^{1/2} + C_2(\sigma_{m1} + \sigma_{m2} + \sigma_{m3}) \geq \frac{\sqrt{2}\,\sigma_e}{K_f} \tag{12-30}$$

where σ_{a1} = alternating component of principal stress in "1" direction, etc.
σ_{m1} = static component of principal stress in "1" direction, etc.
σ_e = fatigue strength (or fatigue limit) for completely reversed stress
C_2 = material constant giving influence of σ_m on σ_a, that is, the negative slope of Goodman line on Fig. 12-8. As first approximation $C_2 = 0.5$.

The effect of residual stresses is included by adding their value to the appropriate σ_m term. Since compressive residual stresses will substract from the σ_m term, they will allow a greater alternating stress for the same fatigue life.

While Eq. (12-30) is written to include a triaxial state of stress it has been tested only against data for biaxial stresses. Since fatigue cracks invariably start at

[1] L. P. Tarasov, W. S. Hyler, and H. R. Letner, *Am. Soc. Test. Mater. Proc.*, vol. 57, pp. 601–622, 1957.

[2] G. Sines and G. Ohgi, *Trans. ASME J. Eng. Mater. Tech.*, vol. 103, pp. 82–90, 1981.

a free surface where the stress is biaxial this is not a serious limitation. The failure criterion in Eq. (12-30) when plotted for the plane-stress condition (like in Fig. 3-5) will plot as an ellipse. Any combination of alternating stresses within the ellipse represents a safe condition. The size of the ellipse will depend upon the fatigue properties and the static (mean) stresses. The more positive (tensile) the sum of the static stresses, the smaller the ellipse.

Sines proposes that combined stress fatigue under high strain, low-cycle conditions can be correlated with the octahedral shear strain

$$\varepsilon_q = \frac{\gamma_{\text{oct}}}{2} = \frac{1}{3}\left[(\varepsilon_1 - \varepsilon_2)^2 + (\varepsilon_2 - \varepsilon_3)^2 + (\varepsilon_3 - \varepsilon_1)^2\right]^{1/2} \tag{12-31}$$

The combined strain state is consolidated by the above equation and ε_q is used to enter the Coffin-Manson plot. Note that mean stress is not considered because the plastic deformation in the high strain cycling will cause it to be relaxed. Also note that, while biaxial stress is commonly found in engineering applications, biaxial strain is not common. The strain in most engineering applications is triaxial.

12-15 CUMULATIVE FATIGUE DAMAGE AND SEQUENCE EFFECTS

The conventional fatigue test subjects a specimen to a fixed amplitude until the specimen fails. Tests may be made at a number of different values of stress to determine the *S-N* curve, but each time the maximum stress is held constant until the test is completed. However, there are many practical applications where the cyclic stress does not remain constant, but instead there are periods when the stress either is above or below some average design level. Further, there are applications involving complex loading conditions where it is difficult to arrive at an average stress level and the loading cannot be assumed to vary sinusoidally.

Overstressing is the process of testing a virgin specimen for some number of cycles less than failure at a stress above the fatigue limit, and subsequently running the specimen to failure at another test stress. The ratio of the cycles of overstress at the prestress to the virgin fatigue life at this same stress is called the *cycle ratio*. The damage produced by a cycle ratio of overstress may be evaluated by the reduction in fatigue life at the test stress. Alternatively, the damage due to overstressing may be measured by subjecting a number of specimens to a certain cycle ratio at the prestress and then determining the fatigue limit of the damaged specimens. Bennett[1] has shown that increasing the cycle ratio at the prestress produces a greater decrease in the fatigue limit of damaged specimens, while similar experiments[2] which employed a statistical determination of the fatigue limit showed a much greater reduction in the fatigue limit due to the overstressing. In line with this point, it is important to note that, because of the statistical

[1] J. A. Bennett, *Am. Soc. Test. Mater Proc.*, vol. 46, pp. 693–714, 1946.

[2] G. E. Dieter, G. T. Horne, and R. F. Mehl, *NACA Tech. Note* 3211, 1954.

nature of fatigue, it is very difficult to reach reliable conclusions by overstressing tests, unless statistical methods are used.

If a specimen is tested below the fatigue limit, so that it remains unbroken after a long number of cycles, and if then it is tested at a higher stress, the specimen is said to have been *understressed*. Understressing frequently results in either an increase in the fatigue limit or an increase in the number of cycles of stress to fracture over what would be expected for virgin specimens. It has frequently been considered that the improvements in fatigue properties due to understressing are due to localized strain hardening at sites of possible crack initiation. A different interpretation of the understressing effect has resulted from experiments on the statistical determination of the fatigue limit.[1] Specimens which did not fail during the determination of the fatigue limit showed greater than normal fatigue lives when retested at a higher stress. By means of statistical analysis it was possible to show that the observed lives at the higher stress were to be expected owing to the elimination of the weaker specimens during the prior testing below the fatigue limit. Thus, it was concluded that understressing was at least partly due to a statistical selectivity effect.

If a specimen is tested without failure for a large number of cycles below the fatigue limit and the stress is increased in small increments after allowing a large number of cycles to occur at each stress level, it is found that the resulting fatigue limit may be as much as 50 percent greater than the initial fatigue limit. This procedure is known as *coaxing*. An extensive investigation of coaxing[2] showed a direct correlation between a strong coaxing effect and the ability for the material to undergo strain aging. Thus, mild steel and ingot iron show a strong coaxing effect, while brass, aluminum alloys, and heat-treated low-alloy steels show little improvement in properties from coaxing.

Considerable data indicates that the percentage of life consumed by operation at one overstress level depends on the magnitude of subsequent stress levels. However, the linear cumulative damage rule,[3] also called *Miner's rule*, assumes that the total life of a part can be estimated by adding up the percentage of life consumed by each overstress cycle. If $n_1, n_2, \ldots, n_k$ represent the number of cycles of operation at specific overstress levels and $N_1, N_2, \ldots, N_k$ represent the life (in cycles) at these same overstress levels, then

$$\frac{n_1}{N_1} + \frac{n_2}{N_2} + \cdots + \frac{n_k}{N_k} = 1 \quad \text{or} \quad \sum_{j=1}^{j=k} \frac{n_j}{N_j} = 1 \tag{12-32}$$

While many deviations from Miner's rule have been observed, and numerous modifications to this relationship have been proposed,[4] none has been proven better or gained wide acceptance.

[1] E. Epremian and R. F. Mehl, *ASTM Spec. Tech. Publ.* 137, 1952.
[2] G. M. Sinclair, *Am. Soc. Test. Mater. Proc.*, vol. 52, pp. 743–758, 1952.
[3] M. A. Miner, *J. Appl. Mech.*, vol. 12, pp. A159–A164, 1945.
[4] J. P. Collins, "Failure of Materials in Mechanical Design," pp. 240–275, John Wiley & Sons, New York, 1981.

The fatigue strength of smooth specimens is reduced more than would be predicted by Miner's linear damage rule if a few cycles of high stress are applied before testing at lower stresses. This effect is particularly pronounced with notched specimens where the fatigue life might be changed by a factor of 10 or 100 if an overload is applied at the beginning of a sequence of cycles rather than at the end of the sequence.[1] This effect is due to residual stresses produced at the notch by overload stresses in the plastic region. Even a small tensile load can produce a plastic zone at the tip of the crack, which upon unloading forms compressive residual stresses that retard the growth of a crack. However, compressive overloads do not open a crack, so large loads must be present before yielding occurs at the crack tip. When this happens the residual stresses produced are tensile in nature, and these accelerate the growth of cracks. Therefore, infrequent tensile overloads produce crack arrest, while compressive overloads large enough to produce yielding can accelerate crack growth.

It is apparent that there is a need for methods of analyzing loading histories that vary irregularly with time. A number of techniques for deducing the appropriate number of cycles and stress range have been proposed and found wanting.[2] However, the cycle counting methods known as the *range pair method* and the *rainflow method* have found wide acceptance.[3]

12-16 EFFECT OF METALLURGICAL VARIABLES ON FATIGUE

The fatigue properties of metals are quite structure-sensitive. However, at the present time there are only a limited number of ways in which the fatigue properties can be improved by metallurgical means. By far the greatest improvements in fatigue performance result from design changes which reduce stress concentration and from the intelligent use of beneficial compressive residual stress, rather than from a change in material. Nevertheless, there are certain metallurgical factors which must be considered to ensure the best fatigue performance from a particular metal or alloy. Fatigue tests designed to measure the effect of some metallurgical variable, such as special heat treatments, on fatigue performance are usually made with smooth, polished specimens under completely reversed stress conditions. It is usually assumed that any changes in fatigue properties due to metallurgical factors will also occur to about the same extent under more complex fatigue conditions, as with notched specimens under combined stresses. That this is not always the case is shown by the notch-sensitivity results discussed previously.

Fatigue properties are frequently correlated with tensile properties. In general, the fatigue limit of cast and wrought steels is approximately 50 percent of

[1] R. I. Stephens, D. K. Chen, and B. W. Hom, "Fatigue Crack Growth Under Spectrum Loads," *ASTM STP* 595, p. 27, 1976.

[2] N. E. Dowling, *J. Mater.*, vol. 7, no. 1, pp. 71–87, 1972.

[3] N. E. Dowling, *Trans. ASME J. Eng. Mater. Tech.*, vol. 105, pp. 206–214, 1983.

the ultimate tensile strength. The ratio of the fatigue limit (or the fatigue strength at 10^8 cycles) to the tensile strength is called the *fatigue ratio*. Several nonferrous metals such as nickel, copper, and magnesium have a fatigue ratio of about 0.35. While the use of correlations of this type is convenient, it should be clearly understood that these constant factors between fatigue limit and tensile strength are only approximations and hold only for the restricted condition of smooth, polished specimens which have been tested under zero mean stress at room temperature. For notched fatigue specimens the fatigue ratio for steel will be around 0.20 to 0.30. However, as yield strength is increased by the various strengthening mechanisms, the fatigue limit usually does not increase proportionately. Most high-strength materials are fatigue-limited.

Several parallels can be drawn between the effect of certain metallurgical variables on fatigue properties and the effect of these same variables on tensile properties. The effect of solid-solution alloying additions on the fatigue properties of iron[1] and aluminum[2] parallels nearly exactly their effect on the tensile properties. Gensamer[3] showed that the fatigue limit of a eutectoid steel increased with decreasing isothermal-reaction temperature in the same fashion as did the yield strength and the tensile strength. However, the greater structure sensitivity of fatigue properties, compared with tensile properties, is shown in tests comparing the fatigue limit of a plain carbon eutectoid steel heat-treated to coarse pearlite and to spheroidite of the same tensile strength.[4] Even though the steel in the two structural conditions had the same tensile strength, the pearlitic structure resulted in a significantly lower fatigue limit due to the higher notch effects of the carbide lamellae in pearlite.

There is good evidence[5] that high fatigue resistance can be achieved by homogenizing slip deformation so that local concentrations of plastic deformation are avoided. This is in agreement with the observation that fatigue strength is directly proportional to the difficulty of dislocation cross slip. Materials with high stacking-fault energy permit dislocations to cross slip easily around obstacles, which promotes slip-band formation and large plastic zones at the tips of cracks. Both of these phenomena promote the initiation and propagation of fatigue cracks. In materials with low stacking-fault energy, cross slip is difficult and dislocations are constrained to move in a more planar fashion. This limits local concentrations of plastic deformation and suppresses fatigue damage. Feltner and Laird[6] have referred to these two extremes of deformation as "wavy" and "planar" slip.

While the concept has been useful in understanding fatigue mechanisms, the ability to control fatigue strength by altering stacking-fault energy has practical

[1] E. Epremian, and E. F. Nippes, *Trans. Am. Soc. Met.*, vol. 40, pp. 870–896, 1948.

[2] J. W. Riches, O. D. Sherby, and J. E. Dorn, *Trans. Am. Soc. Met.*, vol. 44, pp. 852–895, 1952.

[3] M. Gensamer, E. B. Pearsall, W. S. Pellini, and J. R. Low, Jr., *Trans. Am. Soc. Met.*, vol. 30, pp. 983–1020, 1942.

[4] G. E. Dieter, R. F. Mehl, and G. T. Horne, *Trans. Am. Soc. Met.*, vol. 47, pp. 423–439, 1955.

[5] J. C. Grosskreutz, *Metall. Trans.*, vol. 3, pp. 1255–1262, 1972.

[6] C. E. Feltner and C. Laird, *Acta Metall.*, vol. 15, p. 1621, 1967.

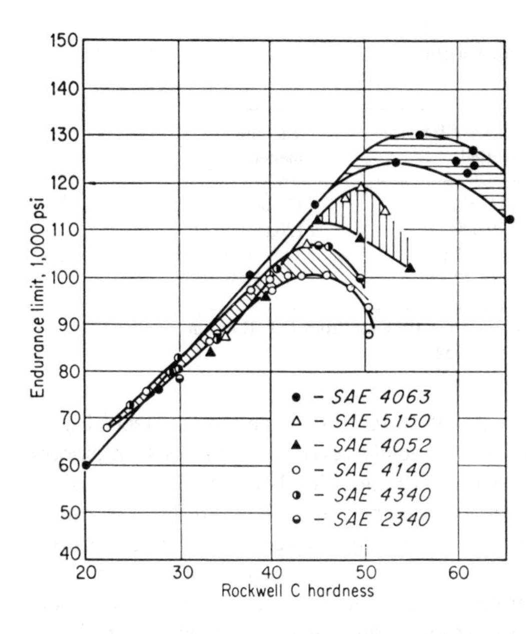

Figure 12-22 Fatigue limit of alloy steels as a function of Rockwell hardness. *(From M. F. Garwood, H. H. Zurburg, and M. A. Erickson, in "Interpretation of Tests and Correlation with Service," p. 12, American Society for Metals, Metals Park, Ohio, 1951.)*

limitations. A more promising approach to increasing fatigue strength appears to be the control of microstructure through thermomechanical processing to promote homogeneous slip with many small regions of plastic deformation as opposed to a smaller number of regions of extensive slip.

The dependence of fatigue life on grain size varies also depending on the deformation mode.[1] Grain size has its greatest effect on fatigue life in the low-stress, high-cycle regime in which stage I cracking predominates. In high stacking-fault-energy materials (such as aluminum and copper) cell structures develop readily and these control the stage I crack propagation. Thus, the dislocation cell structure masks the influence of grain size, and fatigue life at constant stress is insensitive to grain size. However, in a low stacking-fault-energy material (such as alpha brass) the absence of cell structure because of planar slip causes the grain boundaries to control the rate of cracking. In this case, fatigue life is proportional to (grain diameter)$^{-1/2}$.

In general, quenched and tempered microstructures result in the optimum fatigue properties in heat-treated low-alloy steels. However, at a hardness level above about R_C 40 a bainitic structure produced by austempering results in better fatigue properties than a quenched and tempered structure with the same hardness.[2] Electron micrographs indicate that the poor performance of the quenched and tempered structure is the result of the stress-concentration effects of the thin carbide films that are formed during the tempering of martensite. For quenched and tempered steels the fatigue limit increases with decreasing temper-

[1] A. W. Thompson and W. A. Backofen, *Acta Metall.*, vol. 19, pp. 597–606, 1971.

[2] F. Borik and R. D. Chapman, *Trans. Am. Soc. Met.*, vol. 53, 1961.

Table 12-4 Influence of inclusions on fatigue limit of SAE 4340 steel†

	Electric-furnace-melted	Vacuum-melted
Longitudinal fatigue limit, psi	116,000	139,000
Transverse fatigue limit, psi	79,000	120,000
Ratio transverse/longitudinal	0.68	0.86
Hardness, R_C	27	29

† Determined in repeated-bending fatigue test ($R = 0$). Data from J. T. Ransom, *Trans. Am. Soc. Met.*, vol. 46, pp. 1254–1269, 1954.

ing temperature up to a hardness of R_C 45 to R_C 55, dependon the steel.[1] Figure 12-22 shows the results obtained for completely reversed stress tests on smooth specimens. The fatigue properties at high hardness levels are extremely sensitive to surface preparation, residual stresses, and inclusions. The presence of only a trace of decarburization on the surface may drastically reduce the fatigue properties. Only a small amount of nonmartensitic transformation products can cause an appreciable reduction in the fatigue limit.[2] The influence of small amounts of retained austenite on the fatigue properties of quenched and tempered steels has not been well established.

The results, which are summarized in Fig. 12-22, indicate that below a tensile strength of about 200,000 psi the fatigue limits of quenched and tempered low-alloy steels of different chemical composition are about equivalent when the steels are tempered to the same tensile strength. This generalization holds for fatigue properties determined in the longitudinal direction of wrought products. However, tests have shown[3] that the fatigue limit in the transverse direction of steel forgings may be only 60 to 70 percent of the longitudinal fatigue limit. It has been established[4] that practically all the fatigue failures in transverse specimens start at nonmetallic inclusions. Nearly complete elimination of inclusions by vacuum melting produces a considerable increase in the transverse fatigue limit (Table 12-4). The low transverse fatigue limit in steels containing inclusions is generally attributed to stress concentration at the inclusions, which can be quite high when an elongated inclusion stringer is oriented transverse to the principal tensile stress. However, the fact that nearly complete elimination of inclusions by vacuum melting still results in appreciable anisotropy of the fatigue limit indicates that other factors may be important. Further investigations[5] of this subject have

[1] M. F. Garwood, H. H. Zurburg, and M. A. Erickson, "Interpretation of Tests and Correlation with Service," American Society for Metals, Metals Park, Ohio, 1951.

[2] F. Borik, R. D. Chapman, and W. E. Jominy, *Trans. Am. Soc. Met*, vol. 50, pp. 242–257, 1958.

[3] J. T. Ransom and R. F. Mehl, *Am. Soc. Test Mater Proc.*, vol. 52, pp. 779–790, 1952.

[4] J. T. Ransom, *Trans. Am. Soc. Met.*, vol. 46, pp. 1254–1269, 1954.

[5] G. E. Dieter, D. L. Macleary, and J. T. Ransom, Factors Affecting Ductility and Fatigue in Forgings, "Metallurgical Society Conferences," vol. 3, pp. 101–142, Interscience Publishers, Inc., New York, 1959.

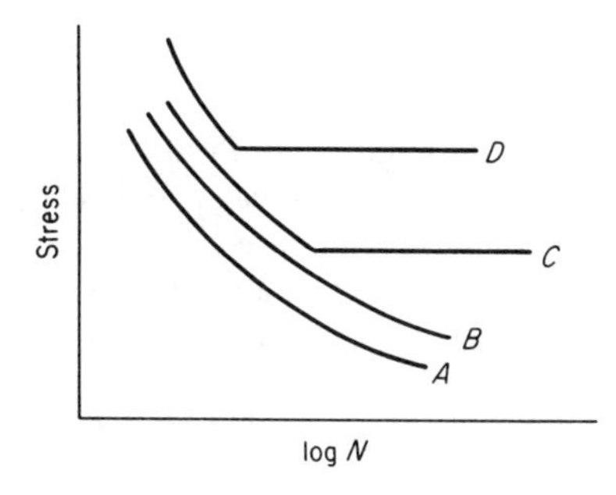

Figure 12-23 Steps in the development of a material with a fatigue limit (schematic): *A* (pure metal), *B* (effect of solid solution elements on *A*), *C* (fatigue limit due to strain aging from interstitials), *D* (increased fatigue limit from enhanced strain aging.) *(After Tegart.)*

shown that appreciable changes in the transverse fatigue limit which cannot be correlated with changes in the type, number, or size of inclusions are produced by different deoxidation practices. Transverse fatigue properties appear to be one of the most structure-sensitive engineering properties.

The existence of a fatigue limit in certain materials, especially iron and titanium alloys, has been shown to depend on the presence of interstitial elements.[1] The mechanism by which this occurs is shown schematically in Fig. 12-23. The *S-N* curve for a pure metal (*A*) is a monotonic function with *N* increasing as stress decreases. The introduction of a solute element raises the yield strength and since it is more difficult to initiate a slip band, the *S-N* curve is shifted upward and to the right to curve *B*. If the alloy has suitable interstitial content so it undergoes strain aging, there is an additional strengthening mechanism. Since strain aging will not be a strong function of applied stress, there will be some limiting stress at which a balance occurs between fatigue damage and localized strengthening due to strain aging. This results in the fatigue limit of curve *C*. With enhanced strain aging, brought about by higher interstitial content or elevated temperature the fatigue limit is raised and the break in the curve occurs at a lower number of cycles, curve *D*. In quenched and tempered steels, which do not normally exhibit strain aging in the tension test, the existence of a pronounced fatigue limit presumably is the result of localized strain aging at the tip of the crack.

12-17 DESIGN FOR FATIGUE

A considerable technical literature has been developed on methods and procedures for designing against fatigue failure. The most appropriate references,

[1] J. C. Levy and G. M. Sinclair, *Am. Soc. Test. Mater Proc.*, vol. 55, p. 866, 1955; H. A. Lipsitt and G. T. Horne, *Am. Soc. Test. Mater. Proc.*, vol. 57, pp. 587–600, 1957; J. C. Levy and S. L. Kanitkar, *J. Iron Steel Inst. London*, vol. 197, pp. 296–300, 1961; H. A. Lipsitt and D. Y. Wang, *Trans. Metall. Soc. AIME*, vol. 221, p. 918, 1961.

which consider extensive design examples are:

Ruiz, C., and F. Koenigsberger: "Design for Strength and Production," Gordon and Breach Science Publishers, Inc., New York, 1970. Pages 106–120 give a concise discussion of the general fatigue design procedure.
Juvinall, R. C.: "Engineering Considerations of Stress, Strain, and Strength," McGraw-Hill Book Company, New York, 1967. Chapters 11 to 16 cover in considerable detail the machine design aspects of fatigue design.
Graham, J. A. (ed.): "Fatigue Design Handbook," Society of Automotive Engineers, New York, 1968.
Heywood, R. B.: "Designing Against Fatigue of Metals," Reinhold Publishing Corporation, New York, 1962. Heavily oriented toward stress-concentration calculations.
Osgood, C. C.: "Fatigue Design," 2d ed., Pergamon Press, New York, 1982. An encyclopedic collection of data and design examples.
Sors, L.: "Fatigue Design of Machine Components," Pergamon Press, New York, 1971.

Fatigue data on various materials are given in the references listed above. In addition, the following are useful sources of fatigue data.

MIL-HDBK-5D, "Metallic Materials and Elements for Flight Vehicle Structures," Department of Defense, 1983.
H. E. Boyer, ed., "Atlas of Fatigue Curves," American Society for Metals, Metals Park, Ohio, 1985.
Properties and Selection of Metals, "Metals Handbook," vol. 1, 9th ed., American Society for Metals, Metals Park, Ohio, 1978.
Aerospace Structural Metals Handbook, Battelle Columbus Laboratories, 1983.

There are several distinct philosophies concerning design for fatigue that must be understood to put this vast subject into proper perspective.

1. Infinite-life design. This design criterion is based on keeping the stresses at some fraction of the fatigue limit of steel. This is the oldest fatigue design philosophy. It has largely been supplanted by the other criteria discussed below. However, for situations in which the part is subjected to very large cycles of uniform stress it is a valid design criterion.
2. Safe-life design. Safe-life design is based on the assumption that the part is initially flaw-free and has a finite life in which to develop a critical crack. In this approach to design one must consider that fatigue life at a constant stress is subject to large amounts of statistical scatter. For example, the Air Force historically designed aircraft to a safe life that was one-fourth of the life demonstrated in full-scale fatigue tests of production aircraft. The factor of 4 was used to account for environmental effects, material property variations, and variations in as-manufactured quality. Bearings are another good example

of parts that are designed to a safe-life criterion. For example, the bearing may be rated by specifying the load at which 90 percent of all bearings are expected to withstand a given lifetime. Safe-life design also is common in pressure vessel and jet engine design.

3. Fail-safe design. In fail-safe design the view is that fatigue cracks will not lead to failure before they can be detected and repaired. This design philosophy developed in the aircraft industry, where the weight penalty of using large safety factors could not be tolerated but neither could the danger to life from very small safety factors. Fail-safe designs employ multiple-load paths and crack stoppers built into the structure along with rigid regulations and criteria for inspection and detection of cracks.
4. Damage-tolerant design. The latest design philosophy is an extension of the fail-safe design philosophy. In damage-tolerant design the assumption is that fatigue cracks will exist in an engineering structure. The techniques of fracture mechanics are used to determine whether the cracks will grow large enough to cause failure before they are sure to be detected during a periodic inspection. The emphasis in this design approach is on using materials with high fracture toughness and slow crack growth. The success of the design approach depends upon having a reliable nondestructive evaluation (NDE) program and in being able to identify the critical damage areas in the design.

Although much progress has been made in designing for fatigue, especially through the merger of fracture mechanics and fatigue, the interaction of many variables that is typical of real fatigue situations makes it inadvisable to depend on a design based solely on analysis. Simulated service testing[1] should be part of all critical fatigue applications. The failure areas not recognized in design will be detected by these tests. Simulating the actual service loads requires great skill and experience. Often it is necessary to accelerate the test, but doing so may produce misleading results. For example, when time is compressed in that way, the full influence of corrosion or fretting is not measured, or the overload stress may appreciably alter the residual stresses. It is common practice to eliminate many small load cycles from the load spectrum, but they may have an important influence on fatigue crack propagation.

12-18 MACHINE DESIGN APPROACH—INFINITE-LIFE DESIGN

The design approach that is used with conventional machine components is based on the fatigue limit of smooth laboratory specimens and reduces this value because of the effects of size, surface finish, and statistical scatter.[2] Consider the following example.

[1] R. M. Wetzel (ed.), "Fatigue Under Complex Loading," Society of Automotive Engineers, Warrendale, Pa., 1977.

[2] R. C. Juvinall, "Engineering Consideration of Stress, Strain and Strength," McGraw-Hill Book Company, New York, 1967.

Example A steel shaft is heat-treated to BHN200. The shaft has a major diameter of 1.5 in and a small diameter of 1.0 in. There is a 0.10-in radius at the shoulder between the diameters. The shaft is subjected to completely reversed cycles of stress of pure bending. The fatigue limit for the steel measured on polished specimens of 0.2-in diameter is 42,000 psi. The shaft is manufactured by machining from bar stock. What is the best estimate of the fatigue limit of the shaft?

We first need to determine the stress concentration due to the shoulder in the shaft. From curves like those in Sec. 2-15 we find that $K_t = 1.68$. From Fig. 12-19, for $r = 0.10$ we find $q = 0.90$. Therefore, $K_f = q(K_t - 1) + 1 = 0.9(0.68) + 1 = 1.6$.

The fatigue limit of the small polished unnotched specimen is $\sigma_e' = 42{,}000$ psi. We need to reduce this value because of size effect, surface finish, and statistical scatter.

$$\sigma_e = \sigma_e' C_S C_F C_Z$$

where C_S = factor for size effect. From Table 12-2, $C_S = 0.9$.
C_F = factor for surface finish. From Fig. 12-20, $C_F = 0.75$.
C_Z = factor for statistical scatter.

As discussed in Sec. 12-4, fatigue tests show considerable scatter in results. Fatigue limit determinations are normally distributed. If the test value of σ_e' is taken as the mean value of the fatigue limit (which is a big assumption), then this value is reduced by a statistical factor according to the reliability that is desired.[1] For a 99 percent reliability $C_Z = 0.81$. Therefore, the unnotched fatigue limit corrected for these factors is

$$\sigma_e = \sigma_e' C_S C_F C_Z = 42{,}000(0.9)(0.75)(0.81) = 22{,}900 \text{ psi}$$

Since $K_f = 1.6$ the fatigue limit of the notched shaft is estimated to be

$$\sigma_{e_n} = \frac{22{,}900}{1.6} = 14{,}300 \text{ psi}$$

We note that the working stress is 34 percent of the laboratory value of the fatigue limit.

12-19 LOCAL STRAIN APPROACH

An important use of the strain-life curve, Sec. 12-8, is to predict the life to crack initiation at notches in machine parts where the nominal stresses are elastic but the local stresses and strains at the root of a notch are inelastic. When there is plastic deformation both a strain concentration K_ε and a stress concentration K_σ must be considered. These are related by Neuber's rule,[2] which states that the geometric mean of the stress and strain concentration factors equals K_t.

$$K_t = (K_\sigma K_\varepsilon)^{1/2} \tag{12-33}$$

[1] G. Castleberry, *Machine Design*, pp. 108–110, Feb. 23, 1978.
[2] H. Neuber, *Trans. ASME Ser E., J. Appl. Mech.*, vol. 28, pp. 544–550, 1961.

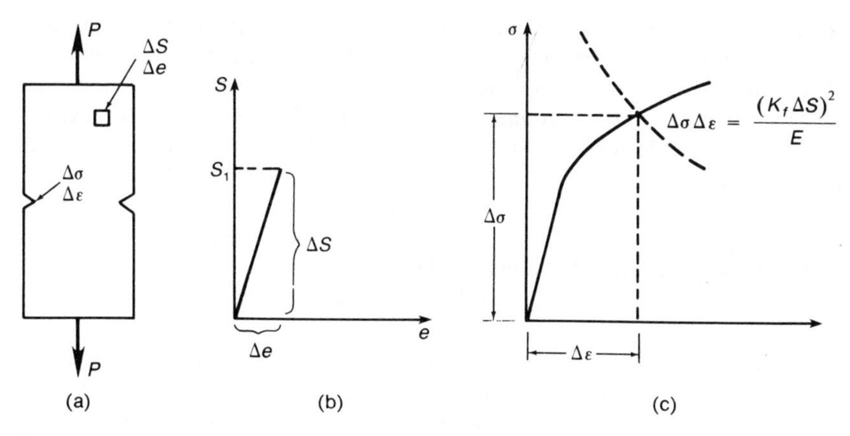

Figure 12-24 Notch stress analysis based on Neuber's analysis.

The situation is depicted in Fig. 12-24*a*, where Δs and Δe are the elastic stress and strain increments at a location remote from the notch, and $\Delta\sigma$ and $\Delta\varepsilon$ are the local stress and strain at the root of the notch.

$$K_\sigma = \frac{\Delta\sigma}{\Delta s} \quad \text{and} \quad K_\varepsilon = \frac{\Delta\varepsilon}{\Delta e}$$

and from Neuber's rule

$$K_t \approx K_f = \left(\frac{\Delta\sigma}{\Delta s}\frac{\Delta\varepsilon}{\Delta e}\right)^{1/2} = \left(\frac{\Delta\sigma\,\Delta\varepsilon\,E}{\Delta s\,\Delta e\,E}\right)^{1/2}$$

and $$K_f(\Delta s\,\Delta e\,E)^{1/2} = (\Delta\sigma\,\Delta\varepsilon\,E)^{1/2} \tag{12-34}$$

For nominally elastic loading, $\Delta s = \Delta e\,E$ and

$$K_f\,\Delta s = (\Delta\sigma\,\Delta\varepsilon\,E)^{1/2} \tag{12-35}$$

This relation shows[1] that a function of nominal stress remotely measured from the notch can be used to predict failure at the notch. Rearranging Eq. (12-35) gives

$$\Delta\sigma\,\Delta\varepsilon = \frac{(K_f\,\Delta s)^2}{E} = \text{const} \tag{12-36}$$

which is the equation of a rectangular hyperbola (Fig. 12-24*c*).The cyclic stress-strain curve is plotted in Fig. 12-24*c* and its intersection with Eq. (12-36) gives the local stress and strain at the notch tip. This strain can then be used with the strain-life equation [Eq. (12-12)] or Fig. 12-14 to estimate the fatigue life.

Actual application of the local strain technique is often much more complex than this simple example.[2] However, it shows great promise for fatigue analysis of

[1] T. H. Topper, R. M. Wetzel, and J. Morrow, *J. Mater.*, vol. 4, pp. 200–209, 1969.

[2] N. E. Dowling, *Trans. ASME J. Eng. Mater. Tech.*, vol. 105, pp. 206–214, 1983.

structural components. The technique[1] utilizes the cyclic stress-strain curve and low-cycle fatigue curve. The complex strain-time history that the part experiences is determined experimentally, and this strain signal is separated into individual cycles by a cycle-counting routine such as the rainflow method. This method separates the strain history into stress-strain hysteresis loops that are comparable with those found in constant-amplitude strain-controlled cycling. Then the notch analysis described above is used in a computer-aided procedure to determine the stresses and strains at the notch for each cycle and the fatigue life is assessed with a cumulative damage rule. Although the design procedures still are evolving, they show great promise of placing fatigue life prediction on a firm basis.

Example A splined shaft drives a large pump. Its loading history consists of both high- and low-cycle stresses. Each start-up includes an overload peak stress due to motor overshoot. During steady operation the shaft vibrates at 2 Hz due to gear misalignment which produces a 10 percent variation around the mean torque. The system operates for 8 hr/day for 250 days per year with a single start-up and shutdown each day. Using the local strain method predict the fatigue life of the shaft.

The cyclic fatigue properties of the alloy steel from which the shaft is made are:

Cyclic strength coefficient $K' = 189$ ksi.
Cyclic strain-hardening exponent $n' = 0.12$.
Fatigue ductility coefficient $\varepsilon_f' = 1.06$.
Fatigue strength coefficient $\sigma_f' = 190$ ksi.
Fatigue strength exponent $b = -0.08$.
Fatigue ductility exponent $c = -0.66$.
Elastic modulus $E = 30 \times 10^6$ psi.
The start-up overload stress is 200 ksi.

$\Delta s = 200$ ksi (elastic stress at the spline) includes K_t; $\Delta s = \sqrt{\sigma \varepsilon E}$ from Eq. (12-35), and $\sigma\varepsilon = \Delta s^2/E = (200)^2/30{,}000 = 1.333$ ksi. From Eq. (12-9)

$$\frac{\Delta\varepsilon}{2} = \frac{\Delta\sigma}{2E} + \left(\frac{\Delta\sigma}{2K'}\right)^{1/n'}$$

$$\frac{\Delta\sigma\,\Delta\varepsilon}{2} = \frac{\Delta\sigma^2}{2E} + \Delta\sigma\left(\frac{\Delta\sigma}{2K'}\right)^{1/n'}$$

$$\frac{1.333}{2} = \frac{\Delta\sigma^2}{2(30{,}000)} + \Delta\sigma\left(\frac{\Delta\sigma}{2 \times 189}\right)^{1/0.12}$$

[1] R. W. Landgraf and N. R. LaPointe, Society of Automotive Engineers, Paper 740280, 1974; A. R. Michetti, *Exp. Mech.*, pp. 69–76, Feb. 1977.

and by trial and error we find $\Delta\sigma = 168$ ksi.

$$\Delta\sigma\,\Delta\varepsilon = 1.333 \quad \text{so} \quad \Delta\varepsilon = \frac{1.333}{168} = 0.00794 \text{ in/in}$$

and
$$\frac{\Delta\varepsilon}{2} = 0.00397 \text{ in/in}$$

From Eq. (12-12)

$$\frac{\Delta\varepsilon}{2} = \frac{\sigma_f'}{E}(2N)^b + \varepsilon_f'(2N)^c$$

$$0.00397 = \frac{190}{30{,}000}(2N)^{-0.08} + 1.06(2N)^{-0.66}$$

$$2N = 29{,}600 \qquad N = 14{,}800 \text{ cycles}$$

and the fatigue damage per cycle is $1/N = 6.75 \times 10^{-5}$. The start-up stress is 20 percent greater than the mean stress in the high-cycle component, or $\sigma_m = 167$ ksi. The alternating stress around this mean is $\sigma_a = 17$ ksi.

From Eq. (12-11) we obtain the elastic strain amplitude high-cycle low strain condition $\Delta\varepsilon_e/2 = \sigma_a/E = 17/30{,}000 = 0.000567$. Substituting into Eq. (12-12), with a correction for mean stress

$$0.000567 = \frac{190 - 167}{30{,}000}(2N)^{-0.08} + 1.06(2N)^{-0.66}$$

$$2N = 270{,}000 \qquad N = 135{,}000 \text{ cycles}$$

and
$$1/N = 7.407 \times 10^{-6} \text{ damage per year}$$

However, the cycles produced in 1 day are $2 \times 60 \times 60 \times 8 = 5.76 \times 10^4$, so that the shaft will fail in $13.5 \times 10^4/5.76 \times 10^4 = 2.3$ days. The chief contribution to this unacceptably short life is the very high mean stress imposed in this situation.

12-20 CORROSION FATIGUE

The simultaneous action of cyclic stress and chemical attack is known as corrosion fatigue.[1] Corrosive attack without superimposed stress often produces pitting of metal surfaces. The pits act as notches and produce a reduction in fatigue strength. However, when corrosive attack occurs simultaneously with fatigue loading, a very pronounced reduction in fatigue properties results which is greater than that produced by prior corrosion of the surface. When corrosion and fatigue occur simultaneously, the chemical attack greatly accelerates the rate at

[1] A. J. McEvily and R. W. Staehle, eds., "Corrosion Fatigue," Nat. Assoc. Corrosion Eng., Houston, 1972; D. J. Duquette, in "Fatigue and Microstructure," pp. 335–363, American Society for Metals, Metals Park, Ohio, 1979.

which fatigue cracks propagate. Materials which show a definite fatigue limit when tested in air at room temperature show no indication of a fatigue limit when the test is carried out in a corrosive environment. While ordinary fatigue tests in air are not affected by the speed of testing, over a range from about 1,000 to 12,000 cycles/min, when tests are made in a corrosive environment there is a definite dependence on testing speed. Since corrosive attack is a time-dependent phenomenon, the higher the testing speed, the smaller the damage due to corrosion. Corrosion-fatigue tests may be carried out in two ways. In the usual method the specimen is continuously subjected to the combined influences of corrosion and cyclic stress until failure occurs. In the two-stage test the corrosion fatigue test is interrupted after a certain period and the damage which was produced is evaluated by determining the remaining life in air. Tests of the last type have helped to establish the mechanism of corrosion fatigue.[1] The action of the cyclic stress causes localized disruption of the surface oxide film so that corrosion pits can be produced. Many more small pits occur in corrosion fatigue than in corrosive attack in the absence of stress. The cyclic stress will also tend to remove or dislodge any corrosion products which might otherwise stifle the corrosion. The bottoms of the pits are more anodic than the rest of the metal so that corrosion proceeds inward, aided by the disruption of the oxide film by cyclic strain. Cracking will occur when the pit becomes sharp enough to produce a high stress concentration.

There is evidence to indicate that even fatigue tests in air at room temperature are influenced by corrosion fatigue. Fatigue tests on copper showed that the fatigue strength was higher in a partial vacuum than in air.[2] Separate tests in oxygen and water vapor showed little decrease over the fatigue strength in vacuum. It was concluded that water vapor acts as a catalyst to reduce the fatigue strength in air, indicating that the relative humidity may be a variable to consider in fatigue testing. Subsequent work with copper[3] showed that the fatigue life was much longer in oxygen-free nitrogen than in air. Metallographic observation showed that the development of persistent slip bands was slowed down when tests were made in nitrogen.

Aggressive environments can have a major effect on fatigue crack propagation. As Fig. 12-25 illustrates, corrosion processes have strong influences on the fatigue life of a structure. The existence of a regime in which slow crack growth occurs according to Paris' law is eliminated, and small cracks grow quickly into large cracks. A reduction of 50 percent in fatigue life due to environmental effects is to be expected.

A number of methods are available for minimizing corrosion-fatigue damage. In general, the choice of a material for this type of service should be based on its corrosion-resistant properties rather than the conventional fatigue properties. Thus, stainless steel, bronze, or beryllium copper would probably give better

[1] U. R. Evans and M. T. Simnad, *Proc. R. Soc. London*, vol. 188A, p. 372, 1947.
[2] J. J. Gough and D. G. Sopwith, *J. Inst. Met.*, vol. 72, pp. 415–421, 1946.
[3] N. Thompson, N. Wadsworth, and N. Louat, *Phil. Mag.*, vol. 1, pp. 113–126, 1956.

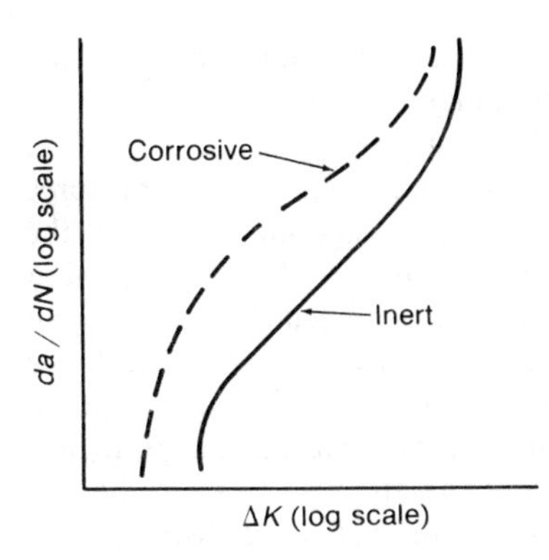

Figure 12-25 Schematic representation of role of a corrosive environment on fatigue crack propagation.

service than heat-treated steel. Protection of the metal from contact with the corrosive environment by metallic or nonmetallic coatings is successful provided that the coating does not become ruptured from the cyclic strain. Zinc and cadmium coatings on steel and aluminum coatings on Alclad aluminum alloys are successful for many corrosion-fatigue applications, even though these coatings may cause a reduction in fatigue strength when tests are conducted in air. The formation of surface compressive residual stresses tends to keep surface notches from opening up and giving ready access to the corrosive medium. Nitriding is particularly effective in combating corrosion fatigue, and shot peening has been used with success under certain conditions. In closed systems it is possible to reduce the corrosive attack by the addition of a corrosion inhibitor. Finally, the elimination of stress concentrators by careful design is very important when corrosion fatigue must be considered.

Fretting

Fretting is the surface damage which results when two surfaces in contact experience slight periodic relative motion. The phenomenon is more related to wear than to corrosion fatigue. However, it differs from wear by the facts that the relative velocity of the two surfaces is much lower than is usually encountered in wear and that since the two surfaces are never brought out of contact there is no chance for the corrosion products to be removed. Fretting is frequently found on the surface of a shaft with a press-fitted hub or bearing. Surface pitting and deterioration occur, usually accompanied by an oxide debris (reddish for steel and black for aluminum). Fatigue cracks often start in the damaged area, although they may be obscured from observation by the surface debris. Fretting is caused by a combination of mechanical and chemical effects. Metal is removed from the surface either by a grinding action or by the alternate welding and tearing away of the high spots. The removed particles become oxidized and form an abrasive powder which continues the destructive process. Oxidation of the metal surface occurs and the oxide film is destroyed by the relative motion of the surfaces. Although oxidation is not essential to fretting, as is demonstrated by relative motion between two nonoxidizing gold surfaces, when conditions are such that oxidation can occur fretting damage is many times more severe.

There are no completely satisfactory methods of preventing fretting. If all relative motion is prevented, then fretting will not occur. Increasing the force normal to the surfaces may accomplish this, but the damage increases with the normal force up to the point where relative motion is stopped. If relative motion cannot be completely eliminated, then reduction of the coefficient of friction between the mating parts may be beneficial. Solid lubricants such as MoS are most successful, since the chief problem is maintaining a lubricating film for a long period of time. Increasing the wear resistance of the surfaces so as to reduce surface welding is another approach. Exclusion of the atmosphere from the two surfaces will reduce fretting, but this is frequently difficult to do with a high degree of effectiveness. Several excellent reviews of this subject have been published.[1,2,3]

12-21 EFFECT OF TEMPERATURE ON FATIGUE

Low-Temperature Fatigue

Fatigue tests on metals at temperatures below room temperature show that the fatigue strength increases with decreasing temperature. Although steels become more notch-sensitive in fatigue at low temperatures, there is no evidence to indicate any sudden change in fatigue properties at temperatures below the ductile-to-brittle transition temperature. The fact that fatigue strength exhibits a proportionately greater increase than tensile strength with decreasing temperature has been interpreted as an indication that fatigue failure at room temperature is associated with vacancy formation and condensation.

High-Temperature Fatigue

In general, the fatigue strength of metals decreases with increasing temperature above room temperature. An exception is mild steel, which shows a maximum in fatigue strength at 400 to 600°F. The existence of a maximum in the tensile strength in this temperature range due to strain aging has been discussed previously. As the temperature is increased well above room temperature, creep will become important and at high temperatures (roughly at temperatures greater than half the melting point) it will be the principal cause of failure. The transition from fatigue failure to creep failure with increasing temperature will result in a change in the type of failure from the usual transcrystalline fatigue failure to the intercrystalline creep failure. Local grain-boundary oxidation can contribute

[1] R. B. Waterhouse, *Proc. Inst. Mech. Eng. London*, vol. 169, pp. 1157–1172, 1955.
[2] P. L. Teed, *Metall. Rev.*, vol. 5, pp. 267–295, 1960.
[3] R. B. Waterhouse (ed.), "Fretting Fatigue," Elsevier Applied Science Publishers, Essex, Great Britain, 1981.

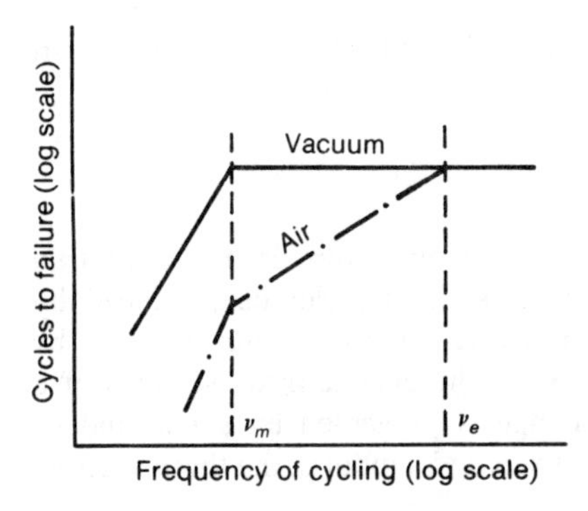

Figure 12-26 Effect of frequency on fatigue life at constant elevated temperature. *(After Coffin.)*

significantly to crack initiation.[1] At any given temperature the amount of creep will increase with increasing mean stress.

Ferrous materials, which ordinarily exhibit a sharp fatigue limit in room-temperature tests, will no longer have a fatigue limit when tested at temperatures above approximately 800°F. Also, fatigue tests at high temperature will depend on the frequency of stress application. It is customary to report the total time to failure as well as the number of cycles to failure.

In general, the higher the creep strength of a material, the higher its high-temperature fatigue strength. However, the metallurgical treatment which produces the best high-temperature fatigue properties does not necessarily result in the best creep of stress-rupture properties. Fine grain size results in better fatigue properties at lower temperatures. As the test temperature is increased the difference in fatigue properties between coarse- and fine-grain material decreases until at high temperatures, where creep predominates, coarse-grain material has higher strength. Procedures which are successful in reducing fatigue failures at room temperature may not be effective in high-temperature fatigue. For example, compressive residual stresses may be annealed out before the operating temperature is reached.

Coffin[2] has extended the analysis of low-cycle fatigue represented by Eq. (12-10) to frequency-dependent high-temperature fatigue. The frequency dependence of fatigue life is illustrated in Fig. 12-26. At high frequencies, $\nu > \nu_e$, the damage process is independent of frequency and failure occurs by the usual transgranular fatigue mechanism. In the intermediate range ($\nu_m < \nu < \nu_e$) an air environment is capable of interacting with the fatigue cracks. In this regime the fracture mode changes from transgranular to intergranular with decreasing frequency of cycling. At lower frequencies, $\nu < \nu_m$, microstructural instabilities

[1] C. H. Wells, in "Fatigue and Microstructure," pp. 307–333, American Society for Metals, Metals Park, Ohio, 1979.

[2] L. F. Coffin, Jr., *Metall. Trans.*, vol. 2, pp. 3105–3113, 1971; H. D. Solomon and L. F. Coffin, Jr., "Fatigue at Elevated Temperatures," *ASTM STP* 520, pp. 112–122, ASTM, Philadelphia, Pa., 1973.

and intergranular fracture may result. This complex subject of high-temperature fatigue has been reviewed recently.[1]

Thermal Fatigue

The stresses which produce fatigue failure at high temperature do not necessarily need to come from mechanical sources. Fatigue failure can be produced by fluctuating thermal stresses under conditions where no stresses are produced by mechanical causes. Thermal stresses result when the change in dimensions of a member as the result of a temperature change is prevented by some kind of constraint. For the simple case of a bar with fixed end supports, the thermal stress developed by a temperature change ΔT is

$$\sigma = \alpha E \Delta T \tag{12-37}$$

where α = linear thermal coefficient of expansion
E = elastic modulus

If failure occurs by one application of thermal stress, the condition is called *thermal shock*. However, if failure occurs after repeated applications of thermal stress, of a lower magnitude, it is called *thermal fatigue*.[2] Conditions for thermal-fatigue failure are frequently present in high-temperature equipment. Austenitic stainless steel is particularly sensitive to this phenomenon because of its low thermal conductivity and high thermal expansion. Extensive studies of thermal fatigue in this material have been reported.[3] The tendency for thermal-fatigue failure appears related to the parameter $\sigma_f k/E\alpha$, where σ_f is the fatigue strength at the mean temperature and k is the thermal conductivity. A high value of this parameter indicates good resistance to thermal fatigue. An excellent review of the entire subject of high-temperature fatigue has been prepared by Allen and Forrest[4] and Ellison.[5]

BIBLIOGRAPHY

American Society for Testing and Materials: "Metal Fatigue Damage-Mechanism, Detection, Avoidance and Repair," *ASTM Spec. Tech. Publ.* 495, 1971.

Battelle Memorial Institute: "Prevention of Fatigue of Metals," John Wiley & Sons, Inc., New York, 1941. (Contains very complete bibliography up to 1941.)

"Fatigue and Microstructure," American Society for Metals, Metals Park, Ohio, 1979.

[1] J. J. Burke and V. Weiss (eds.), "Fatigue: Environment and Temperature Effects," Plenum Press, New York, 1983; Skelton, R. P. (ed.), "Fatigue at High Temperatures," Elsevier Applied Science Publishers, Essex, Great Britain, 1983.

[2] Failure of metals like uranium which have highly anisotropic thermal-expansion coefficients under repeated heating and cooling is also called thermal fatigue.

[3] L. F. Coffin, Jr. *Trans. ASME*, vol. 76, pp. 931–950, 1954.

[4] N. P. Allen and P. G. Forrest, *Proc. Int. Conf. Fatigue of Metals*, London, 1956, pp. 327–340.

[5] E. G. Ellison, *J. Mech. Eng. Sci.*, vol. 11, pp. 318–339, 1969.

Fatigue Testing, "Metals Handbook," 9th ed., vol. 8, pp. 361–435, American Society for Metals, Metals Park, Ohio, 1985.

Frost, N. E., K. J. Marsh, and L. P. Pook: "Metal Fatigue," Oxford University Press, London, 1974.

Fuchs, H. O. and R. I. Stephens: "Metal Fatigue in Engineering," John Wiley & Sons, New York, 1980.

Parker, A. P.: "The Mechanics of Fracture and Fatigue," E. & F. N. Spon Ltd., London, 1981.

Rolfe, S. T. and J. M. Barsom: "Fracture and Fatigue Control, in Structures," Prentice-Hall, Inc., Englewood Cliffs, N.J., 1977.

Sandor, B. I.: "Fundamentals of Cyclic Stress and Strain," The University of Wisconsin Press, Madison, 1972.

Sines, G. and J. L. Waisman (eds.): "Metal Fatigue," McGraw-Hill Book Company, New York, 1959.

CHAPTER
THIRTEEN
CREEP AND STRESS RUPTURE

13-1 THE HIGH-TEMPERATURE MATERIALS PROBLEM

In several previous chapters it has been mentioned that the strength of metals decreases with increasing temperature. Since the mobility of atoms increases rapidly with temperature, it can be appreciated that diffusion-controlled processes can have a very significant effect on high-temperature mechanical properties. High temperature will also result in greater mobility of dislocations by the mechanism of climb. The equilibrium concentration of vacancies likewise increases with temperature. New deformation mechanisms may come into play at elevated temperatures. In some metals the slip system changes, or additional slip systems are introduced with increasing temperature. Deformation at grain boundaries becomes an added possibility in the high-temperature deformation of metals. Another important factor to consider is the effect of prolonged exposure at elevated temperature on the metallurgical stability of metals and alloys. For example, cold-worked metals will recrystallize and undergo grain coarsening, while age-hardening alloys may overage and lose strength as the second-phase particles coarsen. Another important consideration is the interaction of the metal with its environment at high temperature. Catastrophic oxidation and intergranular penetration of oxide must be avoided.

Thus, it should be apparent that the successful use of metals at high temperatures involves a number of problems. Greatly accelerated alloy-development programs have produced a number of materials with improved high-temperature properties, but the ever-increasing demands of modern technology require materials with even better high-temperature strength and oxidation resistance.

For a long time the principal high-temperature applications were associated with steam power plants, oil refineries, and chemical plants. The operating temperature in equipment such as boilers, steam turbines, and cracking units seldom exceeded 1000°F. With the introduction of the gas-turbine engine, requirements developed for materials to operate in critically stressed parts, like turbine buckets, at temperatures around 1500°F. The design of more powerful engines has pushed this limit to around 1800°F. Rocket engines and ballistic-missile nose cones present much greater problems, which can be met only by the most ingenious use of the available high-temperature materials and the development of still better ones. There is no question that the available materials of construction limit rapid advancement in high-temperature technology.

An important characteristic of high-temperature strength is that it must always be considered with respect to some time scale. The tensile properties of most engineering metals at room temperature are independent of time, for practical purposes. It makes little difference in the results if the loading rate of a tension test is such that it requires 2 h or 2 min to complete the test. However, at elevated temperature the strength becomes very dependent on both strain rate and time of exposure. A number of metals under these conditions behave in many respects like viscoelastic materials. A metal subjected to a constant tensile load at an elevated temperature will *creep* and undergo a time-dependent increase in length.

A strong time dependence of strength becomes important in different materials at different temperatures. What is high temperature for one material may not be so high for another. To compensate for this, temperature often is expressed as a homologous temperature, i.e., the ratio of the test temperature to the melting temperature on an absolute temperature scale. Generally, creep becomes of engineering significance at a homologous temperature greater than 0.5.

The tests which are used to measure elevated-temperature strength must be selected on the basis of the time scale of the service which the material must withstand. Thus, an elevated-temperature tension test can provide useful information about the high-temperature performance of a short-lived item, such as a rocket engine or missle nose cone, but it will give only the most meager information about the high-temperature performance of a steam pipeline which is required to withstand 100,000 h of elevated-temperature service. Therefore, special tests are required to evaluate the performance of materials in different kinds of high-temperature service. The *creep test* measures the dimensional changes which occur from elevated-temperature exposure, while the *stress-rupture test* measures the effect of temperature on the long-time load-bearing characteristics. Other tests may be used to measure special properties such as thermal-shock resistance and stress relaxation. These high-temperature tests will be discussed in this chapter from two points of view. The engineering significance of the information obtained from the tests will be discussed, and information which is leading to a better understanding of the mechanism of high-temperature deformation will be considered.

13-2 TIME-DEPENDENT MECHANICAL BEHAVIOR

Before proceeding with a discussion of high-temperature mechanical behavior we shall digress to consider time-dependent mechanical behavior in a more general context. Creep is one important manifestation of *anelastic behavior*. In metals anelastic effects usually are very small at room temperature, but they can be large in the same temperature region for polymeric materials. A second material behavior discussed in this section is *internal friction*, which arises from a variety of anelastic effects in crystalline solids.

In our discussion of elastic behavior in Chap. 2 it was implicitly assumed that elastic strain was a single-valued function of stress alone. This is a valid assumption for the engineering analysis of metals by the theory of elasticity. However, under certain circumstances there is a time dependence to elastic strain which is called *anelasticity*. In Fig. 13-1, an elastic strain e_1 is applied to an anelastic material. With increasing time the strain gradually increases to a value e_2, the completely relaxed strain. The amount of anelastic strain is $e_2 - e_1$. If the load is suddenly removed at $t = t_1$, the material undergoes an immediate elastic contraction equal in magnitude to e_1 and with the passage of time the strain decays to zero. This behavior is known as an *elastic aftereffect*.

If a rod of material is loaded to an elastic stress so rapidly that there is not time for any thermal effects to equilibrate with the surroundings, the loading is done at constant entropy and under adiabatic conditions. For uniaxial loading the change in temperature of the material with strain is given by

$$\left.\frac{\partial T}{\partial \varepsilon}\right|_s = -\frac{V_m \alpha E T}{c_v} \tag{13-1}$$

where V_m = the molar volume of the material
α = the coefficient of linear thermal expansion
E = the isothermal Young's modulus
T = the absolute temperature
c_v = the specific heat at constant volume

Since α is positive for most materials and the other terms in Eq. (13-1) are also

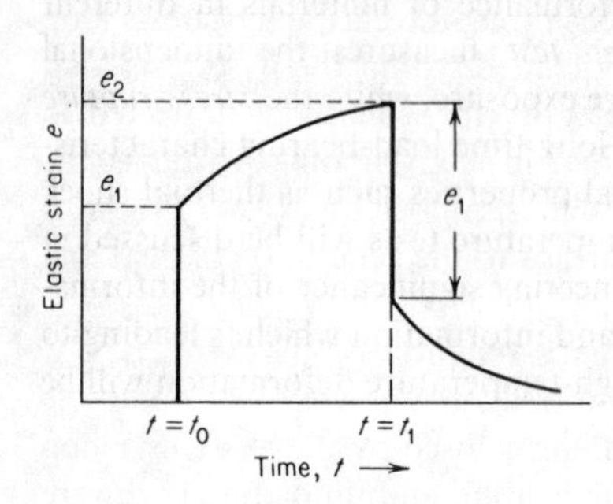

Figure 13-1 Anelastic behavior and the elastic aftereffect.

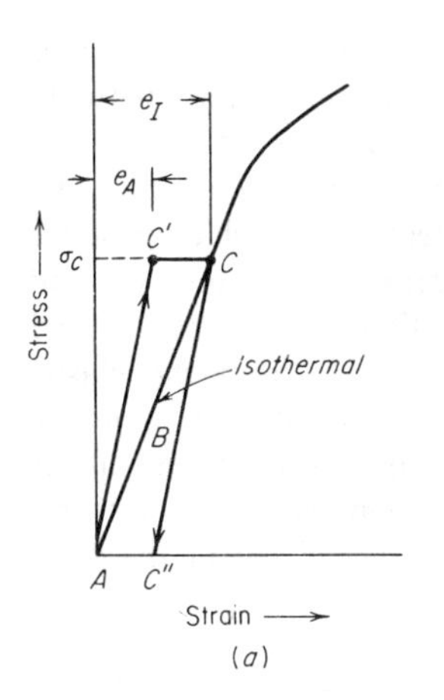

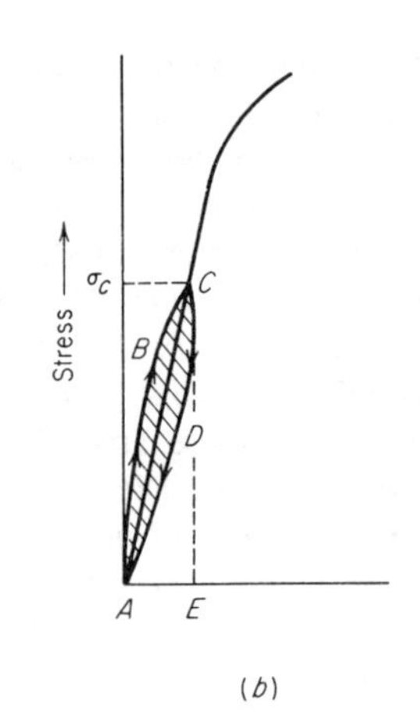

Figure 13-2 (*a*) Idealized adiabatic and isothermal stress-strain curves; (*b*) elastic hysteresis loop.

positive, it follows that an adiabatic elastic tension lowers the temperature of the material and an adiabatic compression increases the temperature. However, these temperature changes associated with the *thermoelastic effect* are usually small.

If a specimen is stretched elastically at a slow rate so that it is in thermal equilibrium with its surroundings, it will follow a path *A-B-C* along the stress-strain curve (Fig. 13-2*a*) and the path *C-B-A* on unloading. However, if the specimen is loaded rapidly to σ_c so that there is not time for thermal equilibration with the surroundings, the temperature of the specimen will decrease a small amount and the specimen follows path *A-C′*. The strain that accumulates in the specimen will be only e_A instead of the strain e_I that would result from an isothermal path. Also, the elastic modulus under dynamic (rapid) loading condition is higher than the modulus under static (slow loading) conditions. With the passing of time the specimen will warm up and elongate due to thermal expansion along the path *C′-C*. If now the load is removed suddenly, the specimen will follow the adiabatic path *C-C″* and the specimen temperature will rise a small amount. With the passage of time the specimen transfers heat to the surroundings and the strain decreases by thermal contraction along the path *C″-A*. The result is a closed hysteresis loop *A-C′-C-C″-A*. However, in an actual material where the specimen is loaded and unloaded in a continuous cycle, the elastic hysteresis loop would resemble Fig. 8-2*b*. Although the area included within the hysteresis loop may be very small, it can be an important quantity for a material subjected to rapid vibration since the total energy dissipated in a given time period is the product of the hysteresis area per cycle times the number of cycles. From an engineering point of view this elastic hysteresis leads to the generation of heat and the damping of vibrations.

The area under the hysteresis loop will be a function of the frequency at which the stress is applied and removed. If the frequency is very low, the cycle will be almost completely isothermal and the area enclosed by the hysteresis loop will be very small. At very high frequencies the loading and unloading paths are almost completely adiabatic and once again the area enclosed by the hysteresis loop is very small. However, there will be some frequency intermediate between these extremes for which the area of the hysteresis loop is a maximum.

The thermoelastic effect is but one of many mechanisms by which vibrational energy is dissipated internally by the material. Other mechanisms which produce anelastic effects in crystalline solids are stress-induced ordering of interstitial and substitutional solute atoms, grain-boundary sliding, motion of dislocations, and intercrystalline and transcrystalline thermal currents arising from the elastic anisotropy of crystals. In polymers these arise from bond rotation, moisture absorbtion, and a variety of other effects. Thes various energy-dissipating effects can be grouped under the generic heading of *internal friction*.[1] Precision measurements of the energy dissipation under very small strains[2] are one of the most sensitive tools for detecting changes in solid-state structure such as precipitation, diffusion, and impurity concentration.

Internal friction is usually expressed by the *logarithmic decrement*, δ. Different expressions for δ are developed depending on the type of experiment used for studying internal friction. Internal friction frequently is measured by a system which is set into motion with a certain amplitude A_0 and then allowed to decay freely. The amplitude at any time A_t is given by

$$A_t = A_0 e^{-\beta t} \tag{13-2}$$

where β is the attenuation coefficient. The logarithmic decrement is the logarithm of the ratio of successive amplitudes.

$$\delta = \ln \frac{A_n}{A_{n+1}} \tag{13-3}$$

If the internal friction is independent of amplitude,[3] a plot of $\ln A$ versus the number of cycles of vibration will be linear and the slope of the curve is the decrement. With anelastic behavior the stress and strain are not *in phase* (Fig. 13-3*a*). The phase angle by which the strain lags behind the stress in time is α. The logarithmic decrement is related to the lag angle by

$$\delta = \pi\alpha \tag{13-4}$$

For a condition of forced vibration in which the specimen is driven at a constant amplitude, a measure of internal friction is the fractional decrease in

[1] C. Zener, "Elasticity and Anelasticity of Metals," University of Chicago Press, Chicago, 1948; K. M. Entwistle, *Metall. Rev.*, vol. 7, p. 175, 1962.

[2] C. Wert, "Modern Research Techniques in Physical Metallurgy," pp. 225–250, American Society for Metals, Metals Park, Ohio, 1953.

[3] True anelastic behavior is independent of amplitude, although some internal friction mechanisms are not.

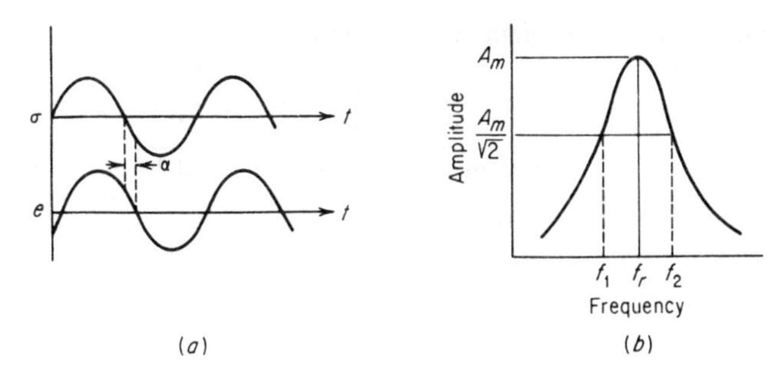

Figure 13-3 (*a*) Phase lag between stress and strain; (*b*) resonance curve.

vibrational energy per cycle.

$$\delta = \frac{\Delta W}{2W} \tag{13-5}$$

where ΔW is the energy lost per cycle (area *ABCD* in Fig. 13-2*b*) and W is the total vibrational energy per cycle (area *ACE* in Fig. 13-2*b*). If a fixed amount of energy per cycle is supplied to a specimen, a maximum amplitude of vibration occurs at the resonant frequency (Fig. 13-3*b*). The logarithmic decrement for a resonance curve is given approximately by

$$\delta = \frac{\pi(\text{bandwidth})}{f_r} \approx \frac{\pi(f_2 - f_1)}{f_r} \tag{13-6}$$

A measure of internal friction that often is used is the Q, where $Q = \pi/\delta$. Since in electrical circuit theory the reciprocal of this value is called the Q of the circuit, the symbol Q^{-1} has been adopted as a measure of internal friction

$$Q^{-1} = \frac{f_2 - f_1}{f_r} \tag{13-7}$$

Internal friction studies generally are conducted at low stress levels (10 to 100 psi) and small strains. Internal energy dissipation at larger stresses and strains generally is called *damping*.[1] High damping capacity is important in minimizing noises in machinery or in suppressing vibrations in high-speed machinery. A high damping capacity is of practical engineering importance in limiting the amplitude of vibration at resonance conditions and thereby reducing the likelihood of fatigue failure. While damping capacity is not very dependent on frequency of vibration, Table 13-1 indicates that it depends on stress level (or strain amplitude) and it varies significantly from material to material. For example, the exceptionally high damping capacity of cast iron arises from the presence of graphite flakes which do not readily transmit elastic waves.

[1] Internal Friction, Damping, and Cyclic Plasticity, *ASTM Spec. Tech. Pub.* 378, 1965.

Table 13-1 Damping capacity of some engineering materials†

Material	Specific damping capacity at various stress levels $\Delta W/W$		
	4,500 psi	6,700 psi	11,200 psi
Carbon steel (0.1% C)	2.28	2.78	4.16
Ni-Cr steel—quenched and tempered	0.38	0.49	0.70
12% Cr stainless steel	8.0	8.0	8.0
18-8 stainless steel	0.76	1.16	3.8
Cast iron	28.0	40.0	
Yellow brass	0.50	0.86	

† S. L. Hoyt, "Metal Data," rev. ed., Reinhold Publishing Corporation, New York, 1952.

13-3 THE CREEP CURVE

The progressive deformation of a material at constant stress is called *creep*. To determine the engineering creep curve of a metal, a constant load is applied to a tensile specimen maintained at a constant temperature, and the strain (extension) of the specimen is determined as a function of time. Although the measurement of creep resistance is quite simple in principle, in practice it requires considerable laboratory equipment.[1,2] The elapsed time of such tests may extend to several months, while some tests have been run for more than 10 years. The general procedures for creep testing are covered in ASTM Specification E139-70.

Curve *A* in Fig. 13-4 illustrates the idealized shape of a creep curve. The slope of this curve ($d\varepsilon/dt$ or $\dot{\varepsilon}$) is referred to as the *creep rate*. Following an initial rapid elongation of the specimen, ε_0, the creep rate decreases with time, then reaches essentially a steady state in which the creep rate changes little with time, and finally the creep rate increases rapidly with time until fracture occurs. Thus, it is natural to discuss the creep curve in terms of its three stages. It should be noted, however, that the degree to which these three stages are readily distinguishable depends strongly on the applied stress and temperature.

In making an engineering creep test, it is usual practice to maintain the load constant throughout the test. Thus, as the specimen elongates and decreases in cross-sectional area, the axial stress increases. The initial stress which was applied to the specimen is usually the reported value of stress. Methods of compensating for the change in dimensions of the specimen so as to carry out the creep test under constant-stress conditions have been developed.[3,4] When constant-stress tests are made it is found that the onset of stage III is greatly delayed. The dashed

[1] Creep and Creep Rupture Tests, ASM Metals Handbook, Supplement, August, 1955, *Met. Prog.*, vol. 68, no. 2A, pp. 175–184, Aug. 15, 1955.

[2] W. V. Green, Creep and Stress Rupture Testing, in R. F. Bunshaw (ed.), "Techniques of Metals Research," vol. V, pt. 1, chap. 6, Interscience Publishers, Inc., New York, 1971.

[3] E. N. da C. Andrade and B. Chalmers, *Proc. R. Soc. London*, vol. 138A, p. 348, 1932.

[4] R. L. Fullman, R. P. Carreker, and J. C. Fisher, *Trans. AIME*, vol. 197, pp. 657–659, 1953.

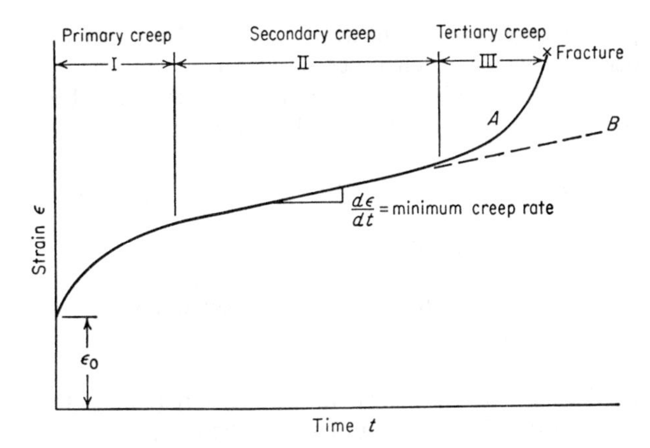

Figure 13-4 Typical creep curve showing the three steps of creep. Curve *A*, constant-load test; curve *B*, constant-stress test.

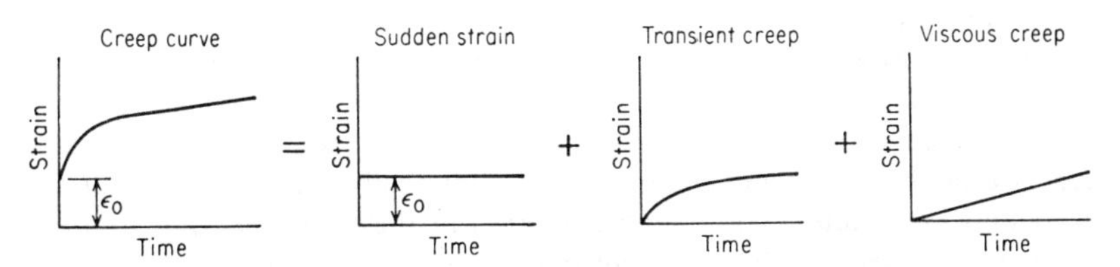

Figure 13-5 Andrade's analysis of the competing processes which determine the creep curve.

line (curve *B*) shows the shape of a constant-stress creep curve. In engineering situations it is usually the load not the stress that is maintained constant, so a constant-load creep test is more important. However, fundamental studies of the mechanism of creep should be carried out under constant-stress conditions.

Andrade's pioneering work[1] on creep has had considerable influence on the thinking on this subject. He considered that the constant-stress creep curve represents the superposition of two separate creep processes which occur after the sudden strain which results from applying the load. The first component of the creep curve is a *transient creep* with a creep rate decreasing with time. Added to this is a constant-rate *viscous creep* component. The superposition of these creep processes is shown in Fig. 13-5. Andrade found that the creep curve could be represented by the following empirical equation:

$$\varepsilon = \varepsilon_0\left(1 + \beta t^{1/3}\right)e^{(\kappa t)} \tag{13-8}$$

where ε is the strain in time t and β and κ are constants.[2] The transient creep is

[1] E. N. da C. Andrade, *Proc. R. Soc. London*, vol. 90A, pp. 329–342, 1914; "Creep and Recovery," pp. 176–198, American Society for Metals, Metals Park, Ohio, 1957.

[2] For a least squares method for evaluating β and κ see J. B. Conway, *Trans. Metall. Soc. AIME*, vol. 223, pp. 2018–2019, 1965.

represented by β and Eq. (13-8) reverts to this form when $\kappa = 0$. The constant κ describes an extension per unit length which proceeds at a constant rate. An equation which gives better fit than Andrade's equation, although it has been tested on a limited number of materials, was proposed by Garofalo.[1]

$$\varepsilon = \varepsilon_0 + \varepsilon_t(1 - e^{-rt}) + \dot{\varepsilon}_s t \tag{13-9}$$

where ε_0 = the instantaneous strain on loading
ε_t = the limit for transient creep
r = the ratio of transient creep rate to the transient creep strain
$\dot{\varepsilon}_s$ = the steady-state creep rate

The various stages of the creep curve shown in Fig. 13-4 require further explanation. It is generally considered in this country that the creep curve has three stages. In British terminology the instantaneous strain designated by ε_0 in Fig. 13-4 is often called the first stage of creep, so that with this nomenclature the creep curve is considered to have four stages. The strain represented by ε_0 occurs practically instantaneously on the application of the load. Even though the applied stress is below the yield stress, not all the instantaneous strain is elastic. Most of this strain is instantly recoverable upon the release of the load (elastic), while part is recoverable with time (anelastic) and the rest is nonrecoverable (plastic). Although the instantaneous strain is not really creep, it is important because it may constitute a considerable fraction of the allowable total strain in machine parts. Sometimes the instantaneous strain is subtracted from the total strain in the creep specimen to give the strain due only to creep. This type of creep curve starts at the origin of coordinates.

The first stage of creep, known as *primary creep*, represents a region of decreasing creep rate. Primary creep is a period of predominantly transient creep in which the creep resistance of the material increases by virtue of its own deformation. For low temperatures and stresses, as in the creep of lead at room temperature, primary creep is the predominant creep process. The second stage of creep, known also as *secondary creep*, is a period of nearly constant creep rate which results from a balance between the competing processes of strain hardening and recovery. For this reason, secondary creep is usually referred to as *steady-state creep*. The average value of the creep rate during secondary creep is called the *minimum creep rate*. Third-stage or *tertiary creep* mainly occurs in constant-load creep tests at high stresses at high temperatures. Tertiary creep occurs when there is an effective reduction in cross-sectional area either because of necking or internal void formation. Third-stage creep is often associated with metallurgical changes such as coarsening of precipitate particles, recrystallization, or diffusional changes in the phases that are present.

Figure 13-6 shows the effect of applied stress on the creep curve at constant temperature. It is apparent that a creep curve with three well-defined stages will

[1] F. Garofalo, "Properties of Crystalline Solids," *ASTM Spec. Tech. Publ.*, 283, p. 82, 1960.

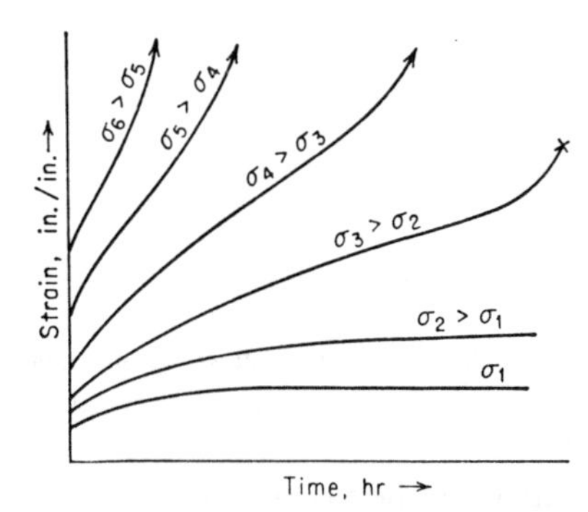

Figure 13-6 Schematic representation of the effect of stress on creep curves at consant temperature.

be found for only certain combinations of stress and temperature. A similar family of curves is obtained for creep at constant stress for different temperatures. The higher the temperature, the greater the creep rate.

The minimum creep rate is the most important design parameter derived from the creep curve. Two standards of this parameter are commonly used in this country: (1) the stress to produce a creep rate of 0.0001 percent per hour or 1 percent/10,000 h or (2) the stress for a creep rate of 0.00001 percent per hour or 1 percent/100,000 h (about $11\frac{1}{2}$ years). The first criterion is more typical of the requirements for jet-engine alloys, while the last criterion is used for steam turbines and similar equipment. A log-log plot of stress vs. minimum creep rate frequently results in a straight line. This type of plot is very useful for design purposes, and its use will be discussed more fully in a later part of this chapter.

13-4 THE STRESS-RUPTURE TEST

The stress-rupture test is basically similar to the creep test except that the test is always carried out to the failure of the material. Higher loads are used with the stress-rupture test than in a creep test, and therefore the creep rates are higher. Ordinarily the creep test is carried out at relatively low stresses so as to avoid tertiary creep. Emphasis in the creep test is on precision determination of strain, particularly as to the determination of the minimum creep rate. Creep tests are frequently conducted for periods of 2,000 h and often to 10,000 h. In the creep test the total strain is often less than 0.5 percent, while in the stress-rupture test the total strain may be around 50 percent. Thus, simpler strain-measuring devices, such as dial gages, can be used. Stress-rupture equipment is simpler to build, maintain, and operate than creep-testing equipment, and therefore it lends itself more readily to multiple testing units. The higher stresses and creep rates of the stress-rupture test cause structural changes to occur in metals at shorter times than would be observed ordinarily in the creep test, and therefore stress-rupture tests can usually be terminated in 1,000 h. These factors have contributed to the increased use of the stress-rupture test. It is particularly well suited to determining the relative high-temperature strength of new alloys for jet-engine applications.

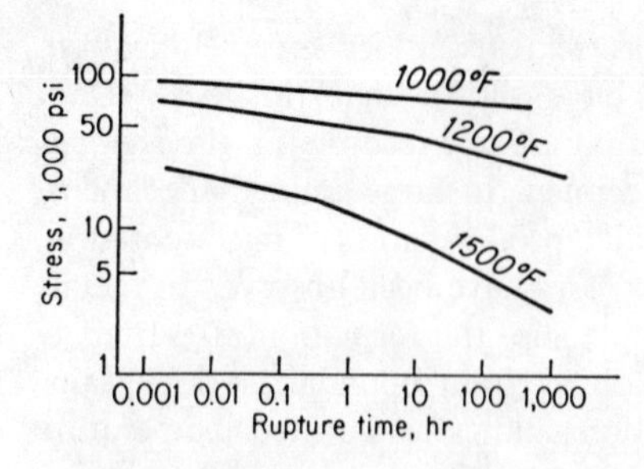

Figure 13-7 Method of plotting stress-rupture data (schematic).

Further, for applications where creep deformation can be tolerated but fracture must be prevented, it has direct application in design.

The basic information obtained from the stress-rupture test is the time to cause failure at a given nominal stress for a constant temperature. The elongation and reduction of area at fracture are also determined. If the test is of suitable duration, it is customary to make elongation measurements as a function of time and from this to determine the minimum creep rate. The stress is plotted against the rupture time on a log-log scale (Fig. 13-7). A straight line will usually be obtained for each test temperature. Changes in the slope of the stress-rupture line are due to structural changes occurring in the material, e.g., changes from transgranular to intergranular fracture, oxidation, recrystallization and grain growth, or other structural changes such as spheroidization, graphitization, or sigma-phase formation. It is important to know about the existence of such instabilities, since serious errors in extrapolation of the data to longer times can result if they are not detected.

13-5 STRUCTURAL CHANGES DURING CREEP

If we plot the slope of a creep curve (Fig. 13-4) vs. strain, we obtain a curve of creep rate vs. total strain (Fig. 13-8). This curve dramatically illustrates the large change in creep rate which occurs during the creep test. Since the stress and temperature are constant, this variation in creep rate is the result of changes in the internal structure of the material with creep strain and time.

The principal deformation processes at elevated temperature are slip, subgrain formation, and grain-boundary sliding. High-temperature deformation is

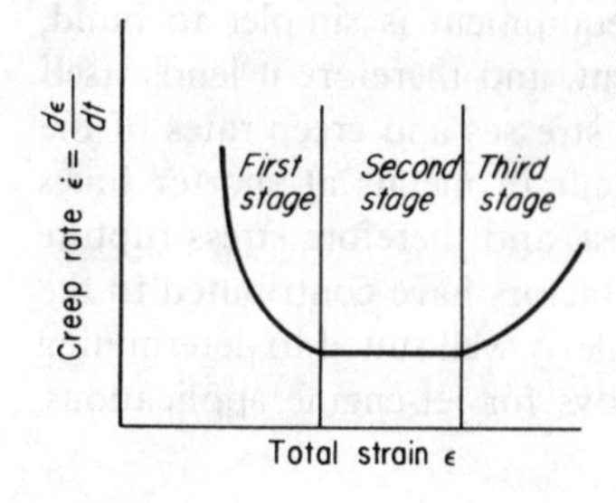

Figure 13-8 Strain rate in creep test as function of total strain.

characterized by extreme inhomogeneity. Measurements of local creep elongation[1] at various locations in a creep specimen have shown that the local strain undergoes many periodic changes with time that are not recorded in the changes in strain of the total gage length of the specimen. In large-grained specimens, local regions may undergo lattice rotations which produce areas of misorientation.

A number of secondary deformation processes have been observed in metals at elevated temperature. These include multiple slip, the formation of extremely coarse slip bands, kink bands, fold formation at grain boundaries, and grain-boundary migration. Many of the deformation studies at elevated temperature have been made with large-grain-size sheet specimens of aluminum. (Aluminum is favored for this type of study because its thin oxide skin eliminates problems from oxidation). Studies have also been made of creep deformation in iron, magnesium, and lead. It is important to remember that all the studies of high-temperature deformation have been made under conditions which give a creep rate of several percent in 100 or 1,000 h, while for many engineering applications a creep rate of less than 1 percent in 100,000 h is required. Because the deformation processes which occur at elevated temperature depend on the rate of strain as well as on the temperature, it is not always possible to extrapolate the results obtained for high strain-rate conditions to conditions of greater practical interest. Much of the work on deformation processes during creep has been reviewed by Sully[2] and Grant and Chaudhuri[3] and others.[4]

Deformation by Slip

New slip systems may become operative when metals are deformed at elevated temperature. Slip in aluminum[5] occurs on the $\{111\}$, $\{100\}$, or $\{211\}$ planes above 500°F. Zinc and magnesium undergo nonbasal slip at high temperature. The ease with which fresh slip systems operate in hcp metals with increasing temperature is shown by some data on zirconium.[6] The critical resolved shear stress for slip on the $(10\bar{1}0)[\bar{1}2\bar{1}0]$ system is 1.0 kg/mm^2 at 77°K, 0.2 kg/mm^2 at 575°K, and 0.02 kg/mm^2 at 1075°K. The slip bands produced at high temperature are coarser and more widely spaced than for room-temperature deformation. At high temperature and low strain rates it may be difficult to detect any slip lines at all. However, McLean[7] was able to detect very fine slip bands within the creeping grains of aluminum using phase contrast microscopy. Slip under high-temperature creep conditions occurs on many slip planes for small slip distances. Weertman[8] has

[1] H. C. Chang and N. J. Grant, *Trans. AIME*, vol. 197, p. 1175, 1953.

[2] A. H. Sully, "Progress in Metal Physics," vol. 6, pp. 135–180, Pergamon Press, Ltd., London, 1956.

[3] N. J. Grant and A. R. Chaudhuri, Creep and Fracture, in "Creep and Recovery," pp. 284–343, American Society for Metals, Metals Park, Ohio, 1957.

[4] L. Bendersky, A. Rosen, and A. K. Mukherjee, Creep and Dislocation Substructure, *Int. Met. Rev.*, vol. 30, pp. 1–15, 1985.

[5] I. S. Servi, J. T. Norton, and N. J. Grant, *Trans. Metall. Soc. AIME*, vol. 194, p. 965, 1952.

[6] E. J. Rapperport and C. S. Hartley, *Trans. Metall. Soc. AIME*, vol. 218, p. 869, 1960.

[7] D. McLean, *J. Inst. Met.*, vol. 81, p. 133, 1952–1953.

[8] J. Weertman, *Trans. Metall. Soc. AIME*, vol. 218, p. 207, 1960.

suggested that this arises from the operation of many dislocation sources which under low-temperature conditions would stop operating because the dislocations from adjacent loops would repel each other. Since at high temperature the loops can climb and annihilate each other, there can be a steady stream of new dislocations from many sources.

Subgrain Formation

Creep deformation is quite inhomogeneous and many opportunities for lattice bending occur, especially near grain boundaries. Bending results in the formation of excess dislocations of one sign, and since dislocation climb can occur readily at high temperature, the dislocations arrange themselves into a low-angle grain boundary as was shown in Fig. 6-6. The formation of a subgrain structure, or cell structure during primary creep has been studied by x-rays, metallography, and thin-film electron microscopy. The dislocation density of the subgrain network increases during primary creep to a level that remains essentially constant during steady-state creep. The size of the subgrain depends on the stress and the temperature. Large subgrains are produced by high temperature and a low stress or creep rate.

The formation of a subgrain structure occurs most readily in metals of high stacking-fault energy. There is less experimental information for metals of low stacking-fault energy which tend to recrystallize rather than form a cell structure during primary creep, but the cell's walls are less regular and well-defined. Thus, they serve as less effective barriers to moving dislocations.

Grain-Boundary Sliding

At elevated temperature the grains in polycrystalline metals are able to move relative to each other. *Grain-boundary sliding* is a shear process which occurs in the direction of the grain boundary. It is promoted by increasing the temperature and/or decreasing the strain rate. Although most investigations indicate that sliding occurs along the grain boundary as a bulk movement of the two grains,[1] other observations[2] indicate that flow occurs in a softened area a finite distance into one grain away from the grain boundary. Grain-boundary sliding occurs discontinuously with time, and the amount of shear displacement is not uniform along the grain boundary.

Grain-boundary sliding may be demonstrated by scribing a line on a polished and etched surface and observing the shear offset where the line crosses the grain boundary. More sophisticated techniques have been developed[3] for measuring grain-boundary shear strain, but the measurements are not at all routine and there is considerable doubt whether surface measurements represent the bulk behavior of the specimen. Studied indicate that the strain due to grain-boundary sliding represents from only a few percent to as high as 50 percent of the total

[1] H. C. Chang, and N. J. Grant, *Trans. AIME*, vol. 206, p. 169, 1956.
[2] R. N. Rhines, W. E. Bond, and M. A. Kissel, *Trans. Am. Soc. Met.*, vol. 48, p. 919, 1956.
[3] R. N. Stevens, *Metall. Rev.*, vol. 11, pp. 129–142, 1966.

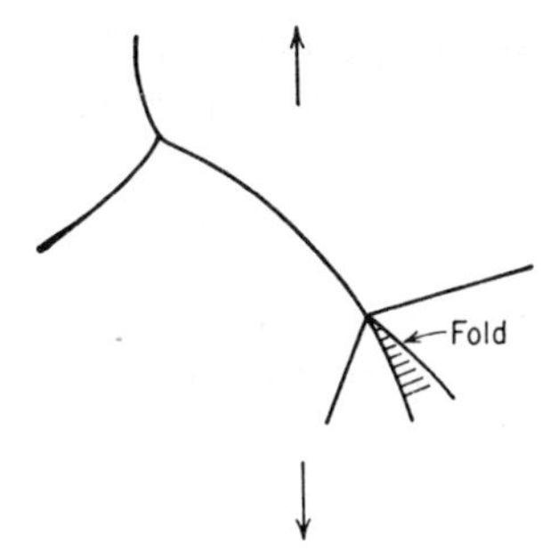

Figure 13-9 Fold formation at a triple point (schematic).

strain, depending on the metal and test conditions. However, a linear relation between the total grain-boundary sliding distance and the total elongation has been found for most systems studied. Since the total strain is the sum of the strain due to slip within the grains and grain-boundary sliding, the close relationship between sliding distance and total strain means that there is an equally close relationship between crystallographic slip and grain-boundary sliding.

The main import of grain-boundary sliding is on the initiation of grain-boundary fracture (Sec. 13-10). For grain-boundary deformation to occur without producing cracks at the grain boundaries, there must be a deformation mechanism available to achieve continuity of strain along the grain boundary. One way of accommodating grain-boundary strain at high temperature is by the formation of folds at the end of a grain boundary[1] (Fig. 13-9). Another recovery process is *grain-boundary migration*, in which the grain boundary moves normal to itself under the influence of shear stress and relieves the strain concentration. The wavy grain boundaries which frequently are observed during high-temperature creep are a result of inhomogeneous grain-boundary deformation and grain-boundary migration.

13-6 MECHANISMS OF CREEP DEFORMATION

A great deal of research has gone into developing mechanisms for creep deformation.[2] The chief creep deformation mechanisms can be grouped as follows:

Dislocation glide—involves dislocations moving along slip planes and overcoming barriers by thermal activation. This mechanism occurs at high stress, $\sigma/G > 10^{-2}$.

Dislocation creep—involves the movement of dislocations which overcome barriers by thermally assisted mechanisms involving the diffusion of vacancies or interstitials. Occurs for $10^{-4} < \sigma/G < 10^{-2}$.

[1] H. C. Chang and N. J. Grant, *Trans. Metall. Soc. AIME*, vol. 194, p. 619, 1952.

[2] For summaries see J. Gittus, "Creep, Viscoelasticity and Creep Fracture in Solids," Wiley-Halsted Press, New York, 1975; R. Lagneborg, *Int. Metall. Rev.*, vol. 17, no. 165, p. 130, 1972.

Diffusion creep—involves the flow of vacancies and interstitials through a crystal under the influence of applied stress. Occurs for $\sigma/G < 10^{-4}$. This category includes Nabarro-Herring and Coble creep.
Grain boundary sliding—involves the sliding of grains past each other.

Frequently, more than one creep mechanism will operate at the same time. If several mechanisms are operating in parallel, i.e., they are operating independently of each other, then the steady-state creep rate is given by

$$\dot{\varepsilon} = \sum_i \dot{\varepsilon}_i \tag{13-10}$$

where $\dot{\varepsilon}_i$ is the creep rate for the ith mechanism. For mechanisms operating in parallel the fastest mechanism will dominate the creep behavior. If there are i mechanisms operating in series, i.e., the mechanisms operate in sequence,

$$\frac{1}{\dot{\varepsilon}} = \sum_i (1/\dot{\varepsilon}_i) \tag{13-11}$$

and the slowest mechanism will control the creep deformation.

Dislocation Glide

Creep resulting from a dislocation glide mechanism occurs at stress levels which are high relative to those normally considered in creep deformation. The creep rate is established by the ease with which dislocations are impeded by obstacles such as precipitates, solute atoms, and other dislocations. The concepts of thermally activated deformation discussed in Sec. 8-12 are used in the theoretical treatment of the dislocation glide mechanism.

Dislocation Creep

Dislocation creep occurs by dislocation glide aided by vacancy diffusion. The framework on which the various theories are hung is the idea of Orowan and Bailey that the steady-state creep rate represents a balance between the competing factors of rate of strain hardening $h = \partial\sigma/\partial\varepsilon$ and the rate of thermal recovery by rearrangement and annihilation of dislocations, $r = -\partial\sigma/\partial t$. A steady-state creep condition occurs when the rate of recovery is fast enough and the rate of strain hardening is slow enough that a balance is reached between these competing factors.

$$\dot{\varepsilon}_s = \frac{r}{h} = -\frac{\partial\sigma/\partial t}{\partial\sigma/\partial\varepsilon} \tag{13-12}$$

Physical models of dislocation creep must predict the h and r terms. The mechanism proposed by Gittus[1] gives good agreement with experiment. It is based on a model of stress and diffusion-aided movement of dislocations in a

[1] J. H. Gittus, *Trans. ASME J. Eng. Mater. Tech.*, vol. 98, p. 52, 1976.

three-dimensional network (substructure).

$$\dot{\varepsilon}_s = \frac{16\pi^3 c_j D_v G b}{kT}\left(\frac{\sigma}{G}\right)^3 \tag{13-13}$$

where c_j = concentration of jogs
D_v = bulk or lattice self-diffusion coefficient
G = shear modulus
b = Burgers vector of the dislocation
σ = applied stress
k = Boltzmann's constant
T = temperature, absolute scale

The earliest models of dislocation creep were advanced by Weertman[1] based on a mechanism in which dislocation climb plays a major role. At elevated temperature, if a gliding dislocation is held up by an obstacle, a small amount of climb may permit it to surmount the obstacle, allowing it to glide to the next set of obstacles where the process is repeated. The glide step produces almost all of the strain but the climb step controls the velocity. Since dislocation climb requires diffusion of vacancies or interstitials, the rate controlling step is atomic diffusion. This model again predicts an equation for creep rate in which stress is raised to the third power. However, creep experiments with a range of metals show that the exponent on stress varies from 3 to 8, with a value of 5 most common. Thus, for intermediate to high stress levels at temperatures above $0.5T_m$ the steady-state creep rate is described by a *power-law relation*

$$\dot{\varepsilon}_s = \frac{A D_v G b}{kT}\left(\frac{\sigma}{G}\right)^n \tag{13-14}$$

where A and n are material constants. Since the diffusion coefficient D_v can be described by

$$D_v = D_0 \exp\left(-Q/kT\right)$$

we can rearrange Eq. (13-14) into

$$\dot{\varepsilon}_s = B\sigma^n \exp\left(-Q/kT\right) \tag{13-15}$$

A modification[2] to Eq. (13-14) allows it to be used for both high-temperature creep, where lattice diffusion predominates, and for low-temperature creep, where diffusion along dislocation cores is the predominant mechanism.

Figure 13-10 shows the steady-state creep rate, normalized with diffusion coefficient, plotted against the stress, (normalized with the shear modulus). Power-law creep occupies the central region of stress. At low stresses, below $\sigma/G = 5 \times 10^{-6}$, a linear dependence on stress ($n = 1$) is observed. This is known as Harper-Dorn creep.[3] It is believed to be due to climb-controlled creep under conditions where the dislocation density does not change with the stress. At

[1] J. Weertman, *J. Mech. Phys. Sol.*, vol. 4, p. 230, 1956; *Trans. AIME*, vol. 218, p. 207, 1960; *Trans. AIME*, vol. 227, p. 1475, 1963.

[2] S. L. Robinson and O. D. Sherby, *Acta Met.*, vol. 17, p. 109, 1969.

[3] J. G. Harper and J. E. Dorn, *Acta Met.*, vol. 5, p. 654, 1957.

high stresses, $\sigma/G > 10^{-3}$, the power law breaks down and the measured creep rates are greater than predicted by Eq. (13-14). This region is described by the Sellars-Tegart equation (Sec. 8-9) or by the equation proposed by Wu and Sherby.[1]

$$\dot{\varepsilon}_s = \frac{AD}{\alpha^n b^2}\left(\sinh \alpha \frac{\sigma}{E}\right)^n \tag{13-16}$$

where $\alpha = (\sigma/E)^{-1}$ at the point where the power law breaks down.

Diffusion Creep

Diffusion creep becomes the controlling mechanism at high temperatures and relatively low stresses, $\sigma/G < 10^{-4}$. Nabarro[2] and Herring[3] proposed that the creep process was controlled by stress-directed atomic diffusion. Stress changes the chemical potential of the atoms on the surfaces of the grains in a polycrystal in such a way that there is a flow of vacancies from grain boundaries experiencing tensile stresses to those which have compressive stresses. Simultaneously, there is a corresponding flow of atoms in the opposite direction, and this leads to elongation of the grain. The Nabarro-Herring creep equation is

$$\dot{\varepsilon}_s \approx \frac{14\sigma b^3 D_v}{kTd^2} \tag{13-17}$$

where d is the grain diameter and D_v is the lattice diffusion coefficient. We note that increasing the grain size reduces the creep rate.

At lower temperatures grain-boundary diffusion predominates.[4] Coble-type creep is described by

$$\dot{\varepsilon}_s \approx \frac{50\sigma b^4 D_{gb}}{kTd^3} \tag{13-18}$$

where d is the grain diameter and D_{gb} is the grain-boundary diffusion coefficient. Note that Nabarro-Herring creep scales as D_v/d^2, while Coble creep scales as D_{gb}/d^3.

Grain-Boundary Sliding

While grain-boundary sliding does not contribute significantly to steady-state creep, it is important in initiating intergranular fracture (see Sec. 13-10). However,

[1] M. Y. Wu and O. D. Sherby, *Acta Met.*, vol. 32, p. 1561, 1984.

[2] F. R. N. Nabarro, "Report of a Conference on Strength of Solids," Physical Society, London, p. 75, 1948.

[3] C. Herring, *J. Appl. Phys.*, vol. 21, p. 437, 1950.

[4] R. L. Coble, *J. Appl. Phys.*, vol. 34, p. 1679, 1963.

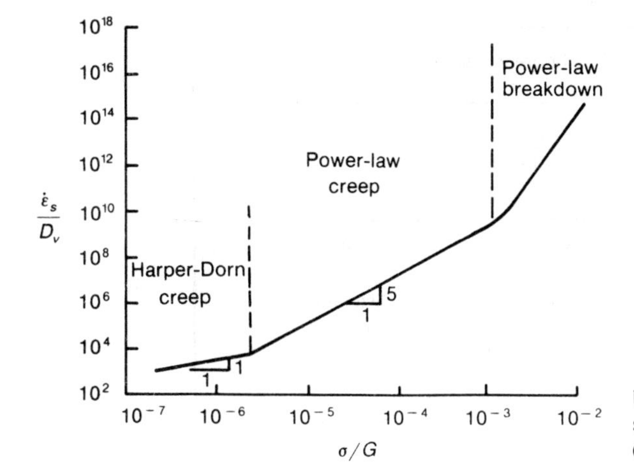

Figure 13-10 Influence of stress on steady-state creep rate (schematic).

it has been shown[1] that grain-boundary sliding must be present to maintain grain contiguity during diffusional flow mechanisms.

13-7 DEFORMATION MECHANISM MAPS

A practical way of illustrating and utilizing the constitutive equations for the various creep deformation mechanisms is with deformation mechanism maps. Ashby and co-workers[2] have developed these maps in stress-temperature space. The various regions of the map (Fig. 13-11) indicate the dominant deformation mechanism for that stress-temperature combination. The boundaries of these regions are obtained by equating the appropriate equations in Sec. 13-6 and solving for stress as a function of temperature. The boundaries represent combinations of stress and temperature where the respective strain rates for the two deformation mechanisms are equal. We see for example, for the metal shown in Fig. 13-11 that at a homologous temperature of 0.8 and a low stress the deformation occurs by diffusional flow (Nabarro-Herring creep). Keeping the temperature constant and increasing the stress we enter a region of power-law creep (dislocation creep) and at still higher stress the metal deforms by thermally activated dislocation glide. The upper bound on the diagram is the stress to produce slip in a perfect (dislocation free) lattice.

Contours of isostrain rate can be calculated from the constitutive equations and plotted on the deformation mechanism map (Fig. 13-12). Thus, in addition to

[1] R. Raj and M. F. Ashby, *Met. Trans.*, vol. 2, p. 1113, 1971.

[2] M. F. Ashby, *Acta Met.*, vol. 20, pp. 887–897, 1972; H. J. Frost and M. F. Ashby, "Deformation-Mechanism Maps," Pergamon Press, New York, 1982.

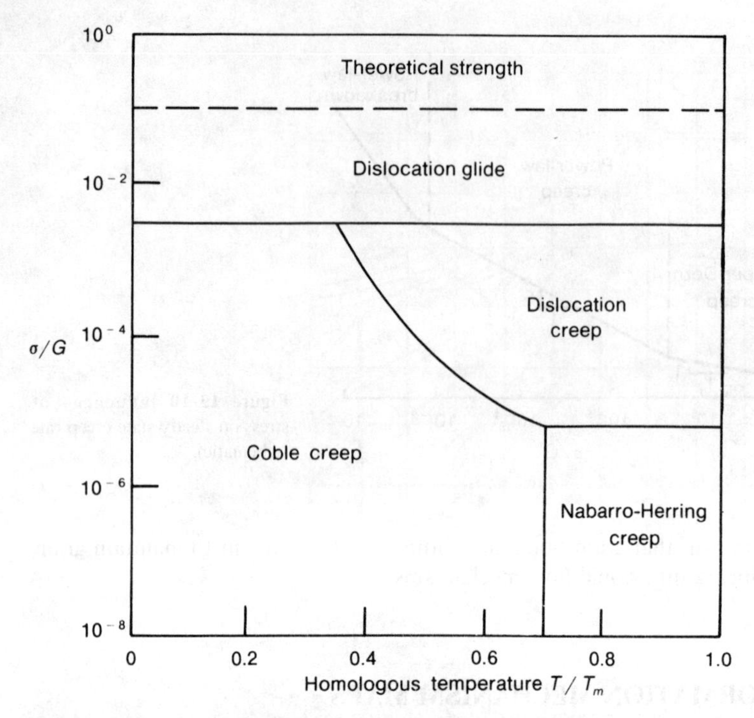

Figure 13-11 Simplified deformation mechanism map. (*After Ashby.*)

identifying the dominant deformation mechanism the map allows selection of any two of the three variables, σ, $\dot{\varepsilon}$, or T, and establishment of the third value. A deformation mechanism map is not only a useful pedagogical tool but it can be helpful in decisions involving alloy design and selection.

13-8 ACTIVATION ENERGY FOR STEADY-STATE CREEP

Steady-state creep predominates at temperatures above about $0.5T_m$. The simplest assumption is that creep is a singly activated process which can be expressed by an Arrhenius-type rate equation

$$\dot{\varepsilon}_s = Ae^{-Q/RT} \tag{13-19}$$

where Q = the activation energy for the rate-controlling process

A = the preexponential complex constant containing the frequency of vibration of the flow unit, the entropy change, and a factor that depends on the structure of the material

T = the absolute temperature

R = the universal gas constant. Note that $R = kN$ where k is Boltzmann's constant and N is Avogadro's number.

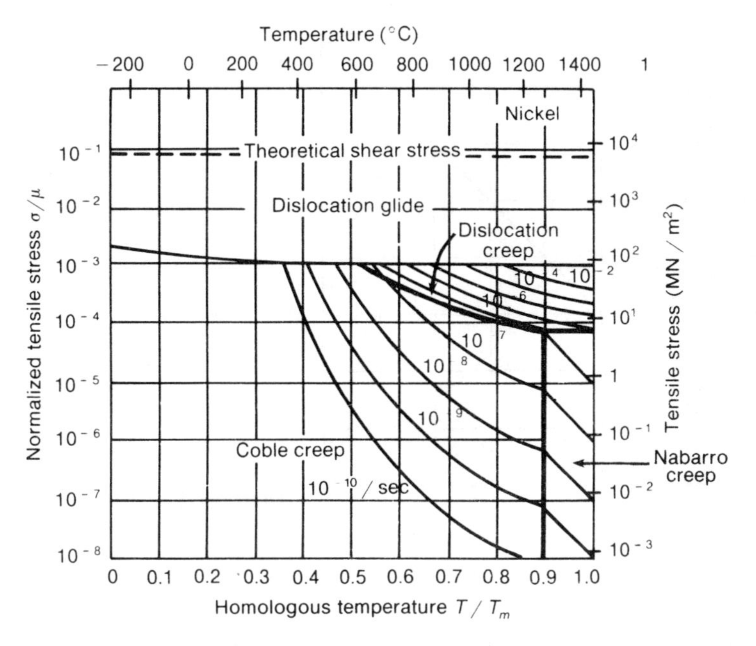

Figure 13-12 Deformation mechanism map for pure nickel with grain size of 32 μm. *(M. F. Ashby, Acta Met., vol. 20, p. 3, 1972.)*

If we plot log $\dot{\varepsilon}_s$ vs. $1/T$ a series of parallel straight lines, one for each stress level, will be obtained if Eq. (13-19) is operative. The slope of these lines is $Q/2.3R$.

A temperature differential creep test is often used to measure the activation energy for creep. If the temperature interval is small so that creep mechanism would not be expected to change, we can write

$$A = \dot{\varepsilon}_1 e^{Q/RT_1} = \dot{\varepsilon}_2 e^{Q/RT_2}$$

and

$$Q = \frac{R \ln(\dot{\varepsilon}_1/\dot{\varepsilon}_2)}{(1/T_2 - 1/T_1)} \tag{13-20}$$

An extensive correlation[1] of creep and diffusion data for pure metals shows that the activation energy for high-temperature creep is equal to the activation energy for self-diffusion (Fig. 13-13). Since the activation energy for self-diffusion is the sum of the energies for the formation and movement of vacancies, this gives a strong support to the view that dislocation climb is the rate-controlling step in high-temperature creep. The formation of a dislocation subgrain structure is another factor in support of this view. We would expect that metals in which vacancies move rapidly would have low creep resistance. Sherby[2] has pointed out

[1] O. D. Sherby, R. L. Orr, and J. E. Dorn, *Trans. AIME*, vol. 200, pp. 71–80, 1954.
[2] O. D. Sherby, *Acta Metall.*, vol. 10, pp. 135–147, 1962.

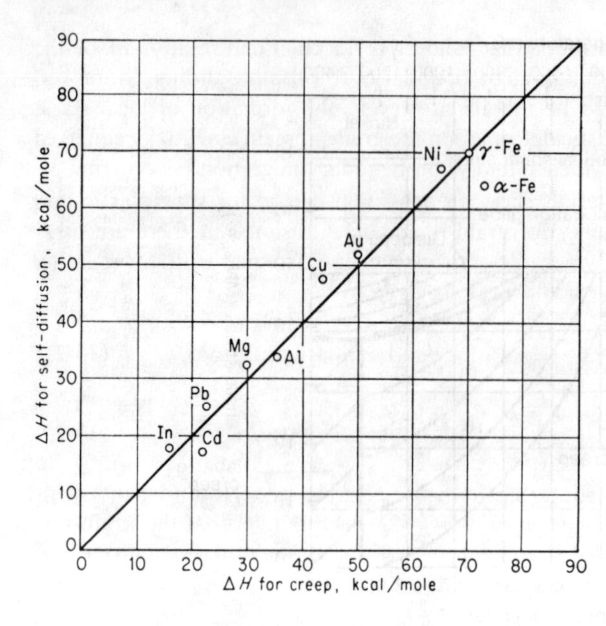

Figure 13-13 Correlation between activation energies for high-temperature creep and self-diffusion.

that the creep rates for alpha and gamma iron at the transition temperature differ by $\dot{\varepsilon}_\alpha/\dot{\varepsilon}_\gamma = 200$, which correlates with the much higher self-diffusion in alpha iron in the ratio $D_\alpha/D_\gamma = 350$. As another example, the self-diffusion coefficient in alpha iron increases with carbon content and so does the steady-state creep rate.

13-9 SUPERPLASTICITY

Superplasticity is the ability of a material to withstand very large deformations in tension without necking. Elongations in excess of 1,000 percent are observed. In Sec. 8-6 we showed that superplasticity was related to the existence of a high strain-rate sensitivity. In this section we deal more broadly with superplastic behavior and relate it to the appropriate high-temperature deformation mechanisms.

Superplastic behavior occurs at $T > 0.5T_m$. Not only does the material show large extensibility without fracture but at low strain rates the flow stress is very low. Thus, complex shapes may be readily formed under superplastic conditions.

The requirements for a material to exhibit superplasticity[1] are a fine grain size (less than 10 μm) and the presence of a second phase which inhibits grain growth at the elevated temperature. Most superplastic alloys are eutectic or eutectoid compositions. The strength of the second phase should be similar to that of the matrix phase to avoid extensive internal cavity formation. If the second phase is harder than the matrix it should be fine and well distributed in

[1] J. W. Edington, K. N. Melton, and C. P. Cutler, *Prog. Mater. Sci.*, vol. 21, p. 61, 1976.

the matrix phase. Since grain-boundary sliding is the chief deformation mode the grain boundaries should be high-angle boundaries to promote sliding. Moreover, the grain boundaries should be mobile to prevent the formation of local stress concentrations. In superplastic deformation the grains remain essentially equiaxed after large deformations, evidence that grain-boundary migration is occurring.

Most superplastic materials show an activation energy for superplastic flow equal to the activation energy for grain-boundary diffusion, but there are other superplastic materials for which the activation energy is equal to that for lattice self-diffusion. The superplastic flow rate is given by[1]

$$\dot{\varepsilon} = 10^8 \left(\frac{\sigma}{E}\right)^2 \frac{bD_{gb}}{\bar{L}^3} \quad \text{for grain-boundary diffusion} \tag{13-21}$$

and

$$\dot{\varepsilon} = 2 \times 10^9 \left(\frac{\sigma}{E}\right)^2 \frac{D_v}{\bar{L}^2} \quad \text{for lattice self-diffusion} \tag{13-22}$$

where $\bar{L}$ is the mean linear intercept measure of grain size. Note that the stress dependence of the power law is given by $n = 2$, which implies that the strain-rate sensitivity is $m = 0.5$. The predominant mechanism for superplastic deformation is grain-boundary sliding accommodated by slip.

13-10 FRACTURE AT ELEVATED TEMPERATURE

It has been known since the early work of Rosenhain and Ewen[2] that metals undergo a transition from transgranular fracture to intergranular fracture as the temperature is increased. When transgranular fracture occurs, the slip planes are weaker than the grain boundaries, while for intergranular fracture the grain boundary is the weaker component. Jeffries[3] introduced the concept of the *equicohesive temperature* (ECT), which was defined as that temperature at which the grains and grain boundaries have equal strength. Like the recrystallization temperature, the ECT is not a fixed one. In addition to the effect of stress and temperature on the ECT, the strain rate has an important influence. Decreasing the strain rate lowers the ECT and therefore increases the tendency for intergranular fracture. The effect of strain rate on the strength-temperature relationship is believed to be much larger for the grain-boundary strength than for the strength of the grains. Since the amount of grain-boundary area decreases with increasing grain size, a material with a large grain size will have higher strength above the ECT than a fine-grain material. Below the ECT the reverse is true.

The spectrum of fracture mechanisms can be displayed on a fracture mechanism map[4] (Fig. 13-14). Ordinary ductile fracture, initiated at inclusions and

[1] O. D. Sherby and J. Wadsworth, "Deformation, Processing and Structure," chap. 8, American Society for Metals, Metals Park, Ohio, 1984.

[2] W. Rosenhain and D. Ewen, *J. Inst. Met.*, vol. 10, p. 119, 1913.

[3] Z. Jeffries, *Trans. AIME*, vol. 60, pp. 474–576, 1919.

[4] M. F. Ashby, C. Gandhi, and D. M. R. Taplin, *Acta Met.*, vol. 27, pp. 699–729, 1979.

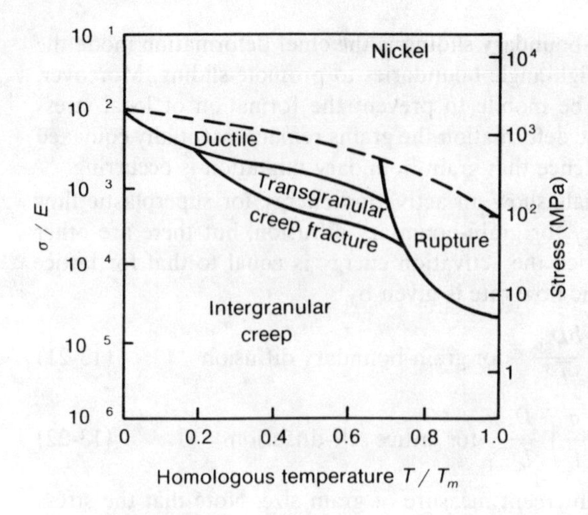

Figure 13-14 Fracture mechanism map for nickel. (*After Ashby.*)

second-phase particles, can extend to rather high temperatures. At the lower flow stresses existing under power-law creep the same type of ductile fracture mechanism exists. However, since the metal is creeping, the stresses within it tend to be lower than for ductile fracture and the nucleation of voids is postponed to larger strains. Also, because of the strain-rate dependence of creep, the flow can be stabilized and void coalescence is postponed. Fracture in this limited range of creep is transgranular. At lower stresses a transition from transgranular to intergranular fracture occurs. Within this region grain-boundary sliding becomes prevalent and wedge cracks and voids grow on the grain boundaries lying roughly normal to the tensile axis. At the highest temperature, if no other fracture mechanism intervenes, the material becomes mechanically unstable and necks down to a point. This requires that the nucleation of voids be suppressed or that coalescence of voids not occur. This type of ductile rupture behavior is observed when dynamic recrystallization prevents void linkup and coalescence.

It is well documented that cavities form continuously throughout the creep process. However, in the third stage of creep the nucleation and growth of grain-boundary voids and cracks occurs at an accelerated rate. This process frequently is called *creep cavitation*. Analysis of the problem[1] indicates that extremely high stresses normal to a grain boundary, e.g., on the order of $E/100$, are required to nucleate a cavity. Thus, very large stress concentrations are required. These large stress concentrations can be produced at particles in sliding grain boundaries, at the intersection of slip bands and grain-boundary particles, and at grain-boundary triple points. However, under creep conditions the generation of high stress concentrations is moderated by both power-law creep and diffusional creep processes. This explains why large number of cavities are not formed at the beginning of the creep test, but rather they accumulate with time.

[1] W. D. Nix, *Scripta Met*, vol. 17, pp. 1–4, 1983.

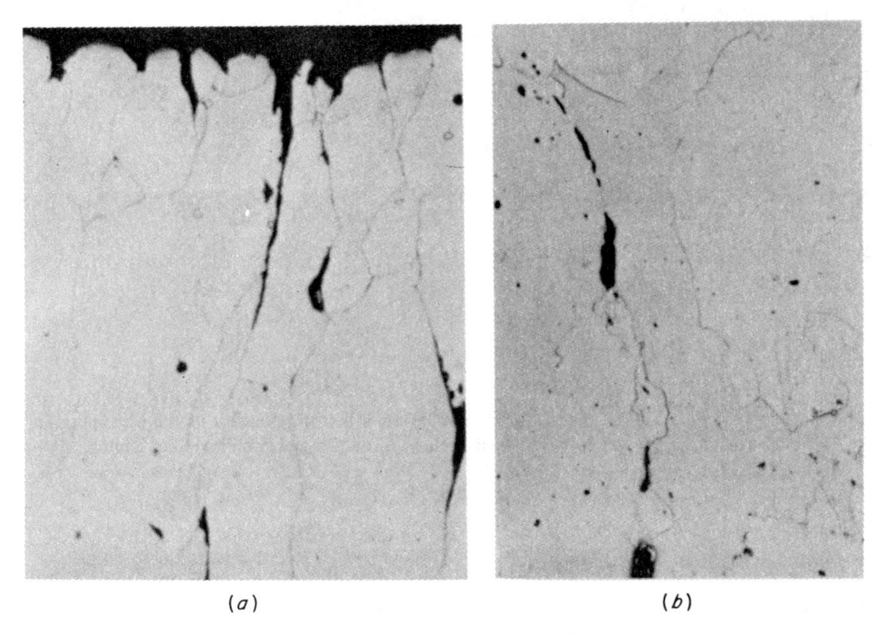

Figure 13-15 (*a*) *w*-type cracks; (*b*) *r*-type crack.

The nucleated cavity must be large enough to grow in the face of capillarity forces which tend to make it shrink.[1] For a particle of diameter a in a boundary which slides a distance δ the volume of the cavity formed is about $a^2\delta$. When surface diffusion acts on the cavity its radius is $r \simeq (a^2\delta)^{1/3}$. If the capillarity stress is taken to be $2\gamma_s/r$, cavity growth will occur only if the grain-boundary sliding displacement exceeds a critical value

$$\delta \geq \frac{1}{a^2}\left(\frac{2\gamma_s}{\sigma}\right)^3 \tag{13-23}$$

where σ is the applied stress. If the sliding distance does not exceed δ capillarity forces would gradually cause the cavity to sinter closed by diffusion. Equation (13-23) shows that while high normal stresses are necessary for nucleation of cavities they are not sufficient. Stable cavities require that large sliding displacements occur.

Two types of intergranular cracks occur in creep (Fig. 13-15). Wedge-shaped cracks (*w*-type cracking) initiate mostly at grain boundaries which are aligned for maximum shear. They initiate by grain-boundary[2] sliding in the manner il-

[1] B. F. Dyson, *Scripta Met.*, vol. 17, p. 31, 1983.
[2] E. Smith and J. T. Barnby, *Met. Sci.*, vol. 1, p. 56, 1967.

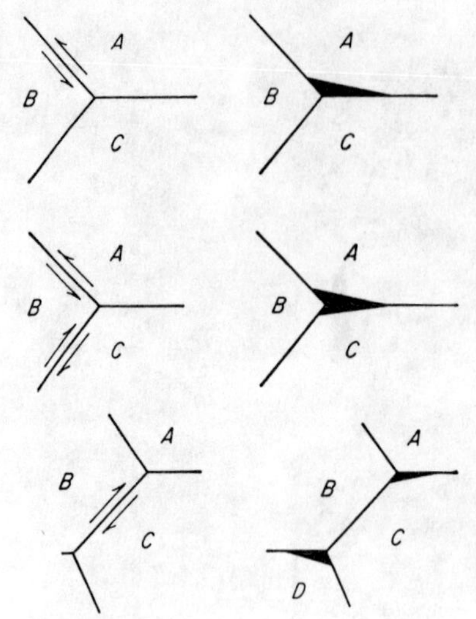

Figure 13-16 Schematic drawings of the way intergranular cracks form owing to grain-boundary sliding. *(From H. C. Chang and N. J. Grant, Trans. Metall. Soc. AIME, vol. 206, p. 545, 1956.)*

lustrated in Fig. 13-16. Round or elliptical cavities (r-type cracks) form in the grain boundaries that are aligned normal to the tensile stress. Extensive studies of their growth mechanism[1] show that they grow by diffusion when they are small, but as they become larger power-law creep becomes the dominant growth mechanism. As the strain rate is decreased or the temperature is raised the r-type cavitation is favored over the w-type. There is theoretical evidence[2] for the existence of a bound on strain rate above which wedge cracking is not possible.

If the elongation or reduction of area is measured with increasing temperature from room temperature, there is usually an intermediate temperature region of low ductility.[3] For aluminum the ductility minimum falls near room temperature but for many alloys, especially nickel-based, it falls within the hot-working region. The region of minimum ductility typically occurs just below the recrystallization temperature in a region where grain-boundary sliding can occur to produce intergranular cracking. When the temperature is raised beyond the ductility minimum into a region where recrystallization can occur readily, these new grains isolate the grain-boundary cracks and the ductility is increased. Especially large losses in ductility can occur in certain materials when they are subjected both to elevated temperature and nuclear irradiation.[4]

[1] A. C. F. Cocks and M. F. Ashby, *Prog. Mater. Sci.*, vol. 27, pp. 189–244, 1982.
[2] C. Gandhi and R. Raj, *Met. Trans.*, vol. 12A, pp. 515–520, 1981.
[3] F. N. Rhines and P. J. Wray, *Trans. Am. Soc. Met.*, vol. 54, p. 117, 1961.
[4] J. O. Stiegler and J. R. Weir, Jr., in "Ductility," chap. 11, American Society for Metals, Metals Park, Ohio, 1968.

13-11 HIGH-TEMPERATURE ALLOYS

High-temperature alloys are a class of complex materials developed for a specific application. Some appreciation of the metallurgical principles behind the development of these alloys is important to an understanding of how metallurgical variables influence creep behavior. The development of high-temperature alloys has, in the main, been the result of painstaking empirical investigation, and it is really only in retrospect that principles underlying these developments have become evident.

The nominal composition of a number of high-temperature alloys is given in Table 13-2. Only a few of the many available alloys[1] could be included in this table. A number of general guidelines have developed for increasing high-temperature strength. In general, the creep resistance is higher the greater the melting point of the metal since the rate of self-diffusion is lower in metals with a high T_m. Since dislocation cross slip is an important step in dislocations climbing to avoid obstacles, metals with a low stacking-fault energy have higher creep resistance because extended partial dislocations have difficulty in cross slipping. Solid-solution alloy additions of high valence are the most effective because they produce a strong decrease in the stacking-fault energy. Solid-solution additions can increase the strength by a variety of mechanisms, such as:

1. Segregation to stacking faults (Suzuki interaction)
2. Elastic interactions of solute atoms with moving dislocations to increase the Peierls-Nabarro stress or the friction stress
3. Interaction with vacancies and dislocation jogs
4. Segregation to grain boundaries so as to influence grain-boundary sliding and migration

Fine dispersed precipitates are necessary for high creep resistance. Many nickel-based superalloys contain small amounts of Al and/or Ti which combine with the matrix to form fine precipitates of the intermetallic compounds Ni_3Al, Ni_3Ti, or $Ni_3(Al, Ti)$. In creep-resistant steels the precipitates are carbides (such as VC, TiC, NbC, Mo_2C, or $Cr_{23}C_6$) which can be precipitated by heat treatment prior to creep service or may be precipitated preferentially at dislocations during creep deformation. *Thermal stability* of the precipitate or strengthening phases is important. Since the greatest strengthening is achieved with the finest particles which are the most unstable, there is usually a critical heat treatment and thermomechanical processing for optimum high-temperature strength. Coarsening of the precipitate is minimized by using a dispersed phase which is nearly

[1] W. F. Simmons, Rupture Strengths of Selected High-Iron, Nickel-Base, Cobalt-Base and Refractory Metal Alloys, *DMIC memo* 234, Battelle Memorial Institute, Columbus, Ohio, May 1, 1968; Metals Handbook, 9th ed., vol. 3, pp. 207–268, American Society for Metals, Metals Park, Ohio, 1980.

Table 13-2 Compositions of some high-temperature alloys

Alloy	C	Cr	Ni	Mo	Co	W	Cb	Ti	Al	Fe	Other
					Ferritic steels						
1.25 Cr-Mo	0.10	1.25	—	0.50						Bal.	
5 Cr-Mo	0.20	5.00	—	0.50						Bal.	
Greek Ascoloy	0.12	13.0	2.0			3.0				Bal.	
					Austenitic steels						
316	0.08	17.0	12.0	2.50						Bal.	
16-25-6	0.10	16.0	25.0	6.00						Bal.	
A-286	0.05	15.0	26.0	1.25				1.95	0.2	Bal.	
					Nickel-based alloys						
Astroloy	0.06	15.0	56.5	5.25	15.0			3.5	4.4		
Inconel	0.04	15.5	76.0							7.0	
Inconel 718	0.04	19.0	Bal.	3.0			5.0	0.80	0.60	18.0	
René 41	0.10	19.0	Bal.	10.0	11.0			3.2	1.6	2.0	
Mar-M-200	0.15	9.0	Bal.	—	10.0	12.5	1.0	2.0	5.0		
TRW 1900	0.11	10.3	Bal.	—	10.0	9.0	1.5	1.0	6.3		
Udimet 700	0.15	15.0	Bal.	5.2	18.5			3.5	4.25	1.0	
In-100	0.15	10.0	Bal.	3.0	15.0			4.7	5.5		1.0 V
TD Nickel	—	—	Bal.								2.0 ThO_2
					Cobalt-based alloys						
Vitallium (HS-21)	0.25	27.0	3.0	5.0	Bal.					1.0	
S-816	0.40	20.0	20.0	4.0	Bal.	4.0		4.0		3.0	

insoluble in the matrix so that the re-solution of fine particles and the growth of coarser particles is slow. This situation is achieved best in dispersion-strengthened alloys in which inert oxide particles (such as Al_2O_3, ThO_2, or SiO_2) are mixed with metal matrix powders and the mixture is consolidated by powder metallurgy techniques. TD-nickel and sintered aluminum powder (SAP) are examples of this type of alloy.

While emphasis has been given to strengthening processes in high-temperature alloys, other considerations are important. An alloy with high-temperature strength may be difficult to fabricate by hot-working, cold-shaping, or welding. Some of the strongest alloys are so highly alloyed that they can only be produced by precision casting. In addition to strength, high-temperature materials must be resistant to the thermal environment in which they operate. Some of the strongest alloys, such as molybdenum, have inadequate oxidation resistance and must be coated with an oxidation-resistant material if used at elevated temperature.

Unfortunately, this introduces problems with reliability, quality control, and fabrication. In other cases the alloy composition must be adjusted to produce a balance between high strength and oxidation resistance.

13-12 PRESENTATION OF ENGINEERING CREEP DATA

Although much progress has been made on the theory of creep, our understanding has not advanced to the point where creep and stress-rupture behavior can be predicted reliably from theoretical models. For design purposes it is essential that decisions be based on reliable experimental data. In reporting high-temperature strength data it is common practice to speak of *creep strength* or *rupture strength*. Creep strength is defined as the stress at a given temperature which produces a steady-state creep rate[1] of a fixed amount, usually taken as 0.00001 or 0.001 percent per hour. Alternatively, the creep strength may be defined as the stress to cause a creep strain of 1 percent at the given temperature. Rupture strength refers to the stress at a given temperature to produce a life to rupture of a certain amount, usually 1,000, 10,000 or 100,000 h.

It frequently is important to be able to extrapolate creep or stress-rupture data into regions where data are not available. Therefore, the common methods of plotting creep data are based on plots which yield reasonable straight lines. Figure 13-17 shows the common method of presenting the influence of stress on the steady-state or minimum creep rate. Note that a log-log plot is used, so that extrapolation of one log-cycle represents a tenfold change. A change in slope of the line will sometimes occur. It has been shown[2] that the value of the minimum creep rate depends on the length of time the creep test has been carried out. For precise determination of $\dot{\varepsilon}_s$, creep curves which have entered the third stage of creep should be used. When $\dot{\varepsilon}_s$ is based on creep curves of shorter duration, the value of minimum creep rate obtained is larger than the true value. Thus, the error is on the conservative side. It has also been shown for long-time creep tests ($t > 10{,}000$ h) that the creep strength based on 1 percent creep strain is essentially equal to the creep strength based on true minimum creep rate. This may provide an economy of time in determining design values of $\dot{\varepsilon}_s$.

Another method of presenting creep data is a plot of stress vs. the time to produce different amounts of total strain (instantaneous strain plus creep strain) (Fig. 13-18). The uppermost curve in Fig. 13-18 is the stress-rupture curve. The percentages beside each data point are the percentage reduction of area at failure.

For short-time high-temperature applications (such as in missiles and high-speed aircraft), data are needed at higher temperatures and stresses and at shorter times than are usually determined for creep tests. From a set of creep curves at constant temperature and various stresses (Fig. 13-6) it is possible to construct

[1] In engineering usage $\dot{\varepsilon}_s$ is often called the *minimum creep rate*. The reason for the terminology is evident from Fig. 13-8.

[2] R. F. Gill and R. M. Goldhoff, *Met. Eng. Q.*, vol. 10, pp. 30–39, August 1970.

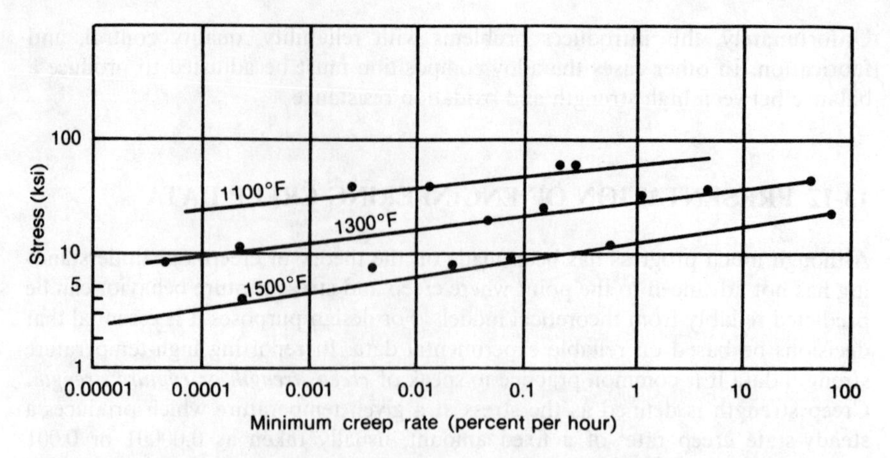

Figure 13-17 Stress vs. minimum creep rate for type 316 (18-8Mo) stainless steel. Typical data, not for design purposes.

stress-strain curves by drawing lines at fixed times, for example, $t = 0, 1, 10, 100$ h. The resulting fictitious stress-strain curves are called *isochronous stress-strain curves*.

It often is cheaper and more convenient to conduct stress-rupture tests than creep tests, and it would be very useful if creep strength could be estimated from rupture strength with sufficient accuracy for design purposes. Monkman and

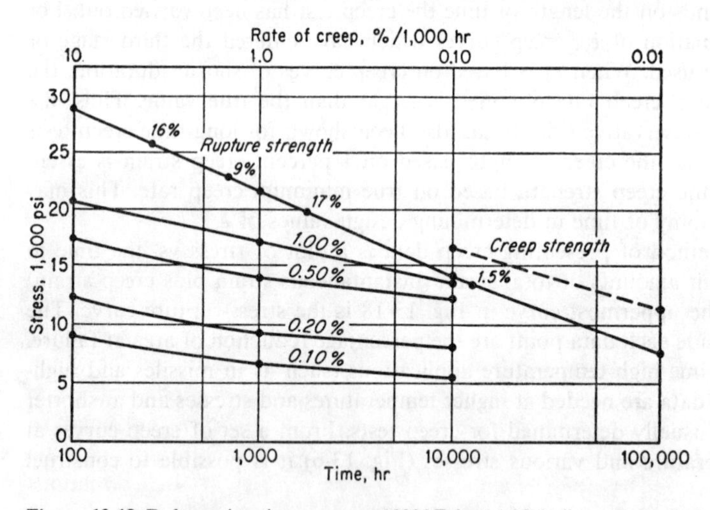

Figure 13-18 Deformation-time curves at 1300°F for 16-25-6 alloy. *(From C. L. Clark, in "Utilization of Heat Resistant Alloys," p. 40, American Society for Metals, Metals Park, Ohio, 1954.)*

Grant[1] showed empirically that a log-log plot of minimum creep rate vs. time to rupture results in a straight line

$$\log t_r + C \log \dot{\varepsilon}_s = K \qquad (13\text{-}24)$$

where C and K are constants for a given alloy. Equation (13-24) shows that rupture life, t_r, is inversely proportional to minimum creep rate, $\dot{\varepsilon}_s$. Thus, we can estimate one from the other if the constants are known for the material. The fact that fracture time correlates well with minimum creep rate supports the view that cavity formation and growth is basic to the creep process.

Example Determine the working stress at 1100°F and 1500°F for type 316 stainless steel if the design criterion is a creep strength based on 1 percent extension in 1000 hr. Use a factor of safety of 3.

$$1\% \text{ creep in } 1000 \text{ hr} = \frac{1\%}{1000 \text{ hr}} = 1 \times 10^{-3}\% \text{ per hr}$$

From Fig. 13-17

Temperature	Creep strength, psi	Working stress, psi
1100°F	30,000	10,000
1500°F	4,000	1,333

Example For the data presented in Fig. 13-17, determine the activation energy for creep at a stress of 10,000 psi

at $T_2 = 1300 + 460 = 1760°\text{R} = 978°\text{K}$; $\quad \dot{\varepsilon}_2 = 0.0001\%$ per hr

at $T_1 = 1500 + 460 = 1960°\text{R} = 1089°\text{K}$; $\quad \dot{\varepsilon}_1 = 0.4\%$ per hr

From Eq. (13-20)

$$Q = \frac{R \ln(\dot{\varepsilon}_1/\dot{\varepsilon}_2)}{(1/T_2 - 1/T_1)} = \frac{(1.987 \text{ cal/mol °K}) \ln(0.4/0.0001)}{1/978 - 1/1089} = 158 \text{ cal/mol}$$

13-13 PREDICTION OF LONG-TIME PROPERTIES

Frequently high-temperature strength data are needed for conditions for which there is no experimental information. This is particularly true of long-time creep and stress-rupture data, where it is quite possible to find that the creep strength to give 1 percent deformation in 100,000 h (11.4 years) is required, although the alloy has been in existence for only 2 years. Obviously, in such situations

[1] F. C. Monkman and N. J. Grant, *Am. Soc. Test. Mater. Proc.*, vol. 56, pp. 593–620, 1956.

extrapolation[1] of the data to long times is required. Reliable extrapolation of creep and stress-rupture curves to longer times can be made only when it is certain that no structural changes occur in the region of extrapolation which would produce a change in the slope of the curve. Since structural changes generally occur at shorter times for higher temperatures, one way of checking on this point is to examine the log-stress–log-rupture life plot at a temperature several hundred degrees above the required temperature. For example, if in 1,000 h no change in slope occurs in the curve at 200°F above the required temperature, extrapolation of the lower temperature curve as a straight line to 10,000 h is probably safe and extrapolation even to 100,000 h may be possible. A logical method of extrapolating stress-rupture curves which takes into consideration the changes in slope due to structural changes has been proposed by Grant and Bucklin.[2]

As an aid in extrapolation of stress-rupture data several time-temperature parameters have been proposed for trading off temperature for time.[3] The basic idea of these parameters is that they permit the prediction of long-time rupture behavior from the results of shorter time tests at higher temperatures at the same stress. When properly developed these time-temperature parameters can be used to represent creep-rupture data in a compact form, allowing for analytical representation and interpolation of data. They also can be used to provide a simple means of comparing the behavior of materials and rating them on a relative basis. Finally, these parameters are useful to extrapolate experimental data to ranges ordinarily difficult to evaluate directly because of test limitations.

The Sherby-Dorn temperature-compensated time parameter provides a rational basis for the development of these parameters.

$$\theta = t \exp(-Q/RT) \tag{13-25}$$

Taking the logarithm of both sides

$$\log \theta = \log t - M\frac{Q}{RT}$$

$$\log t = \log \theta + \frac{M}{R}\frac{Q}{T} \tag{13-26}$$

where $M = \log e = 0.434$. If θ and Q/R are functions of stress only, then Eq. (13-26) is linear in $\log t$ and $1/T$. In Eq. (13-26) t can be the time to rupture or it can be the time to reach a given creep strain. Larson and Miller[4] showed that experimental stress-rupture data plotted in accordance with Eq. (13-26). See Fig. 13-19. Since the line for each constant stress converges to a common point on the $\log t$ axis this plot indicates that Q varies with stress but θ does not. The point of

[1] A. Mendelson and S. S. Manson, *Trans. ASME, Ser. D: J. Basic Eng.*, vol. 82, pp. 839–847, 1960.

[2] N. J. Grant and A. G. Bucklin, *Trans. Am. Soc. Met.*, vol. 42, pp. 720–761, 1950.

[3] The development of time-temperature parameters from reaction-rate theory is considered by M. Grounes, *Trans. ASME, Ser. D: J. Basic Eng.*, vol. 91, pp. 59–62, March 1969.

[4] F. R. Larson and J. Miller, *Trans. ASME*, vol. 74, pp. 765–771, 1952.

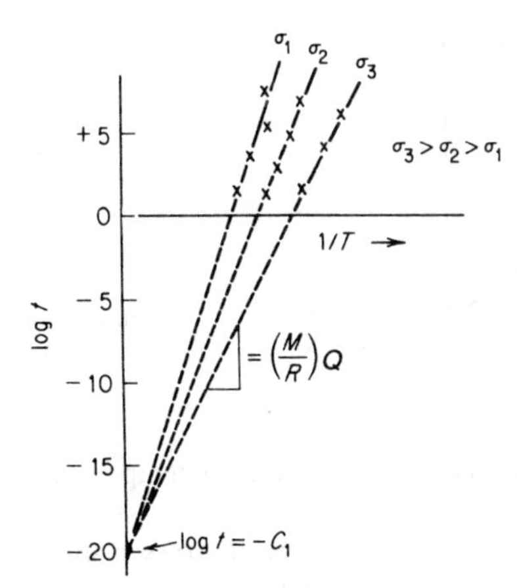

Figure 13-19 Stress-rupture data at various stresses plotted according to Eq. (13-26).

Table 13.3 Time compression of operating conditions based on Larson-Miller parameter $C_1 = 20$

Operating conditions	Larson-Miller test conditions
10,000 h at 1000°F	13 h at 1200°F
1,000 h at 1200°F	12 h at 1350°F
1,000 h at 1350°	17 h at 1500°F

convergence is $\log \theta = -C_1$. For most alloys C_1 varies from about 15 to 25 depending on material. Figure 13-19 also indicates that the slope $b = (M/R)Q$ is a function of stress. From Eq. (13-26) the Larson-Miller parameter[1] can be formulated as

$$T(\log t + C_1) = b = f(\sigma) = P_1 \tag{13-27}$$

or

$$T(\log t + C_1) = P_1$$

where T = the test temperature °R = °F + 460
t = the time to rupture h
C_1 = the Larson-Miller constant, often assumed to be 20

Table 13-3 illustrates how Eq. (13-27) provides a time compression at the expense of increased temperature. When "trading-up" in temperature through the

[1] The Larson-Miller parameter expressed in terms of minimum creep rate instead of time to rupture is $T(C_1 - \log \dot{\varepsilon}_s) = P_1$.

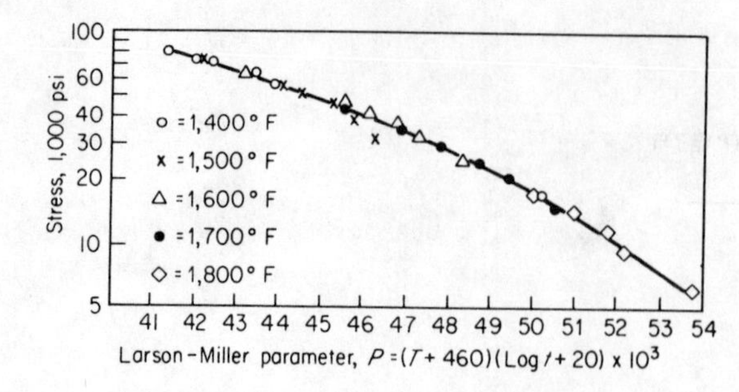

Figure 13-20 Master curve for Larson-Miller parameter for Astroloy.

Larson-Miller parameter, it is important to be sure that the higher-temperature does not cause a change in metallurgical structure.

If P_1 is evaluated for pairs of t and T obtained over a range of stress, a single master curve is obtained for the material (Fig. 13-20). A plot such as this is a good way of summarizing for comparison the stress-rupture characteristics of several materials.

Example Determine the stress required for failure of Astroloy in 100,000 hr at temperatures of 1200 and 1600°F.

Using Eq. (13-27) we determine values of the Larson-Miller parameter.

$$P_{1200} = (1200 + 460)(\log 10^5 + 20) = 1660(5 + 20) = 41.5 \times 10^3$$

$$P_{1600} = (1600 + 460)(\log 10^5 + 20) = 2060(25) = 51.5 \times 10^3$$

From the master plot for Astroloy (Fig. 13-20) we find the stresses corresponding to the values of the parameters.

At 1200°F, $P = 41.5 \times 10^3$ and $\sigma = 78{,}000$ psi

At 1600°F, $P = 51.5 \times 10^3$ and $\sigma = 11{,}000$ psi

Not all high-temperature materials agree well with the Larson-Miller parameter. Therefore, a variety of other time-temperature parameters have been developed.[1] In all, over 30 parameters have been suggested. Figure 13-21 summarizes three of the most popular. Because there is no overwhelming preference for any one of these parameters, there is a growing feeling that the problem should be treated experimentally. This has been termed the minimum commitment method[2] (MCM). The approach avoids attempting to fit the data to one of the time-temperature parameters and instead aims at determining by regression methods[3] the

[1] S. S. Manson and C. R. Ensign, *Trans. ASME, J. Eng. Mater. Tech.*, vol. 101, pp. 317–325, 1979; I. LeMay, *Trans. ASME, J. Eng. Mater. Tech.*, vol. 101, pp. 326–330, 1979.

[2] S. S. Manson and C. R. Ensign, NASA TMX-52999, 1971.

[3] J. B. Conway, *Stress-Rupture Parameters: Origin, Calculation and Use*, Gordon and Breach, Science Publishers, Inc., New York, 1969.

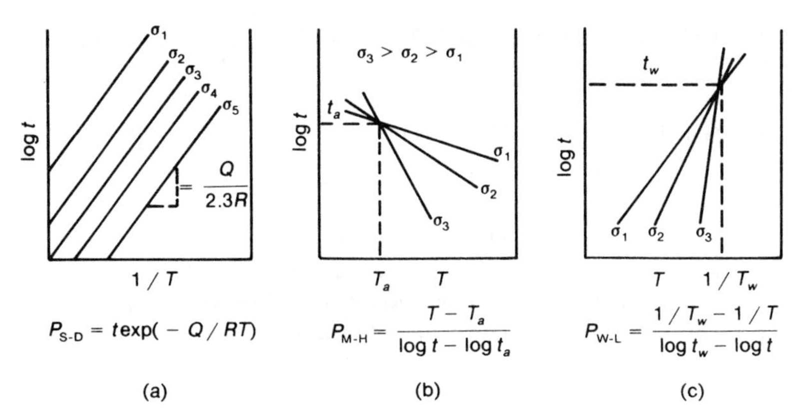

Figure 13-21 Three additional time-temperature parameters. (*a*) Orr-Sherby-Dorn [R. L. Orr, O. D. Sherby, and J. E. Dorn, *Trans. ASM*, vol. 46, p. 113, 1954]; (*b*) Manson-Haferd [S. S. Manson and A. M. Haferd, NACA TN 2890, 1953]; (*c*) White-LeMay [W. E. White and I. LeMay, *Trans. ASME, J. Eng. Matls. and Tech.*, vol. 100, p. 319, 1978].

best fitting parameter from the data. The MCM parameter has the form.

$$\log t + AP(T)\log t + P(T) = Z(\log \sigma) \qquad (13\text{-}28)$$

where A is a constant relating to structural stability.

13-14 CREEP UNDER COMBINED STRESSES

Considerable attention has been given to the problem of design for combined stress conditions during steady-state creep.[1] In the absence of metallurgical changes, the basic simplifying assumptions of plasticity theory hold reasonably well for these conditions. The assumption of incompressible material leads to the familiar relationship $\dot{\varepsilon}_1 + \dot{\varepsilon}_2 + \dot{\varepsilon}_3 = 0$. The assumption that principal shear-strain rates are proportional to principal shear stresses gives

$$\frac{\dot{\varepsilon}_1 - \dot{\varepsilon}_2}{\sigma_1 - \sigma_2} = \frac{\dot{\varepsilon}_2 - \dot{\varepsilon}_3}{\sigma_2 - \sigma_3} = \frac{\dot{\varepsilon}_3 - \dot{\varepsilon}_1}{\sigma_3 - \sigma_1} = C \qquad (13\text{-}29)$$

Combining these equations results in

$$\dot{\varepsilon}_1 = \frac{2}{3C}\left[\sigma_1 - \tfrac{1}{2}(\sigma_2 + \sigma_3)\right] \qquad (13\text{-}30)$$

Similar expressions are obtained for $\dot{\varepsilon}_2$ and $\dot{\varepsilon}_3$.

For engineering purposes the stress dependence of the creep rate can be expressed by Eq. (13-15). For combined stress conditions $\dot{\varepsilon}$ and σ must be

[1] The original analysis of this problem was given by C. R. Soderberg, *Trans. ASME*, vol. 58, p. 733, 1936. Subsequent analysis has been made by I. Finnie, *Trans. ASME, Ser. D; J. Basic Eng.*, vol. 82, pp. 462–464, 1960. For a critical review see A. E. Johnson, *Met. Rev.*, vol. 5, pp. 447–506, 1960.

replaced by the effective strain rate $\dot{\bar{\varepsilon}}$ and the effective stress $\bar{\sigma}$. Thus we can write

$$\dot{\bar{\varepsilon}} = B\bar{\sigma}^{n'} \tag{13-31}$$

Combining Eqs. (13-30) and (13-31) results in

$$\dot{\bar{\varepsilon}}_1 = B\bar{\sigma}^{n'-1}\left[\sigma_1 - \tfrac{1}{2}(\sigma_2 + \sigma_3)\right] \tag{13-32}$$

The effective stress and the effective strain rate are useful parameters for correlating steady-state creep data. When plotted on log-log coordinates, they give a straight-line relationship. Correlation has been obtained between uniaxial creep tests, creep of thick-walled tubes under internal pressure, and tubes stressed in biaxial tension.[1]

13-15 CREEP-FATIGUE INTERACTION

Most deformation at elevated temperature cannot be considered to arise from pure creep or pure fatigue but from an interaction between the two damage processes.[2] The main approaches that have been suggested for design under conditions of creep-fatigue interaction are: (1) linear damage accumulation rules; (2) modification to the low-temperature fatigue relationships, and (3) strain-range partitioning.

Damage Accumulation Rule

The damage accumulation approach considers the damage due to creep and fatigue separately. These independent damages are summed and compared against the limit of damage which the material can withstand. This approach is used in the ASME nuclear pressure vessel codes. It is essentially an extension of Miner's rule that is used for predicting cumulative fatigue damage (Sec. 12-15).

$$\sum_{j=1}^{j=k}\left(\frac{n}{N_d}\right)_j + \sum_{l=1}^{l=m}\left(\frac{t}{t_d}\right)_l \le D \tag{13-33}$$

where n = number of fatigue cycles applied at loading condition j
N_d = number of design allowable cycles at loading condition j
t = time under the applied creep load condition l
t_d = allowable time at load condition l
D = total allowable creep-fatigue damage. Often $D = 1$, but sometimes $D < 1.0$

[1] E. A. Davis, *Trans. ASME, Ser. D: J. Basic Eng.*, vol. 82, pp. 453–461, 1960; A. I. Smith and A. M. Nicholson, (eds.): "Advances in Creep Design," John Wiley & Sons, Inc., New York, 1972; R. K. Penny, "Design for Creep," McGraw-Hill Book Company, New York, 1971.

[2] L. F. Coffin, *Proc. I. Mech. E.*, vol. 188, p. 109, 1974; S. S. Manson, in "Fatigue at Elevated Temperature," *ASTM STP* 520, p. 774, Amer. Soc. for Testing and Materials, Philadelphia, 1973.

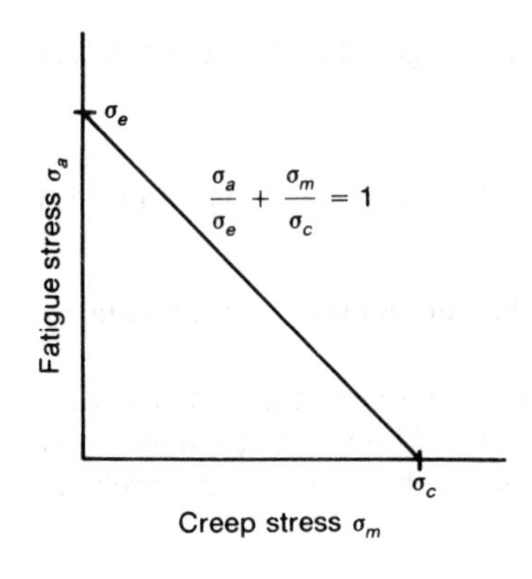

Figure 13-22 Fatigue-creep interaction diagram for constant temperature.

A related approach is to assume that fatigue is caused by the alternating stress component, σ_a, and that creep is caused by the mean stress component, σ_m. Then, the concept of the Goodman line (Sec. 12-5) is modified in the manner shown in Fig. 13-22. In this diagram σ_e represents the limiting fatigue strength at some design designated number of cycles and σ_c represents the creep strength based either on a design limited creep strain or rupture life. This damage relationship tends to give conservative values.[1]

Modification of Fatigue Relationships

Because of the importance of environmental effects in high-temperature fatigue it is important to introduce time dependency. Coffin[2] did this through a frequency term, ν,

$$\nu^k t_f = \nu^k (N/\nu) = N\nu^{k-1} \tag{13-34}$$

where t_f is the time to failure and k is a constant that depends on temperature. $N\nu^{k-1}$ is called the *frequency modified fatigue life*. The Coffin-Manson relationship is given by

$$\frac{\Delta\varepsilon_p}{2} = \varepsilon_f'(2N)^c$$

Substituting Eq. (13-34) into the above, results in

$$\Delta\varepsilon_p (N\nu^{k-1})^{-c} = C_1 \tag{13-35}$$

The high-temperature equivalent of the Basquin equation is given by[3]

$$\Delta\varepsilon_e = \frac{\Delta\sigma}{E} = \frac{A'}{E} N^{-b} \nu k' \tag{13-36}$$

[1] A. J. Kennedy, "Processes of Creep and Fatigue in Metals," John Wiley & Sons, New York, 1963.

[2] L. F. Coffin, op. cit.

[3] I. LeMay, "Principles of Mechanical Metallurgy," p. 378, Elsevier North-Holland, Inc., New York, 1981.

and when Eqs. (13-35) and (13-36) are combined to give the high-temperature equivalent of the strain-life equation

$$\Delta\varepsilon = \Delta\varepsilon_e + \Delta\varepsilon_p = \left(\frac{A'}{E}\right) C_1^{n'} N^{-(1/c)n'} \nu^{-(1/c)n'(k-1)+k'} + C_1\left(N\nu^{k-1}\right)^{-c} \quad (13\text{-}37)$$

where A', C_1, c, k, k', and n' are constants to be determined from the data by regression analysis.

While this approach is useful for many high-temperature fatigue situations, it is difficult to apply to complex creep-fatigue cycles. Moreover, the number of constants which must be determined presents a formidable experimental problem

Strain-range Partitioning

The strain-range partitioning method[1] is based on the idea that any cycle of completely reversed inelastic strain can be broken down into four distinct strain-range components. They are: completely reversed plasticity, $\Delta\varepsilon_{pp}$; tensile plasticity reversed by compressive creep, $\Delta\varepsilon_{pc}$; tensile creep reversed by compressive plasticity, $\Delta\varepsilon_{cp}$; and completely reversed creep, $\Delta\varepsilon_{cc}$. In this notation the first letter of the subscript designates the type of deformation in the tensile part of the cycle. The second letter designates the type of strain in the compression part of the cycle. Two types of deformation are considered, creep (c) or time-dependent deformation and plastic flow (p) (time-independent deformation). Examples of the four types of partitioned strain-range cycles are shown in Fig. 13-23.

Consider a completely reversed strain hysteresis cycle (Fig. 13-24). The plastic flow and creep segments of the cycle are identified. In this figure the total tensile inelastic strain $\Delta\varepsilon_i = AD$ is made up of a plastic strain AC plus a creep strain CD. The compressive inelastic strain is DA, which equals the sum of a plastic strain DB and a creep strain BA. The completely reversed portion of the plastic strain range, $\Delta\varepsilon_{pp}$, is the smaller of the two plastic flow components, i.e., $\Delta\varepsilon_{pp} = DB$. Likewise the completely reversed portion of the creep strain range $\Delta\varepsilon_{cc}$, is the smaller of the two creep components, i.e., $\Delta\varepsilon_{cc} = CD$. We now need to find $\Delta\varepsilon_{pc}$ or $\Delta\varepsilon_{cp}$. In any hysteresis cycle one but not both of these components will exist, depending on whether plasticity or creep is the larger component. For Fig. 13-24 the difference between the two plastic flow components must equal the difference between the creep components, i.e., $AC - DB = BA - CD$. This difference is either $\Delta\varepsilon_{pc}$ or $\Delta\varepsilon_{cp}$. In Fig. 13-24 it is $\Delta\varepsilon_{pc}$ since the tensile plastic strain component is greater than the compressive plastic strain component. In addition, the sum of the partioned strain ranges equals the total inelastic strain range, i.e., $\Delta\varepsilon_i = \Delta\varepsilon_{pp} + \Delta\varepsilon_{cc} + \Delta\varepsilon_{pc}$.

[1] S. S. Manson, G. R. Halford, and M. H. Hirschberg, in "Design for Elevated Temperature Environment," p. 12, ASME, New York, 1971; S. S. Manson, G. R. Halford, and M. H. Hirschberg, NASA TMX-67838, May 1971.

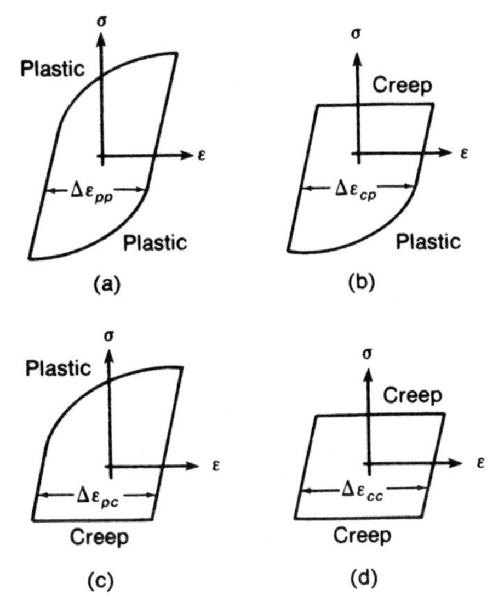

Figure 13-23 The four types of partitioned strain-range cycles.

A Coffin-Manson-type relation exists for each of the strain-range components (Fig. 13-25). At a given value of $\Delta\varepsilon_i$ the components can be written as fractions.

$$F_{pp} = \frac{\Delta\varepsilon_{pp}}{\Delta\varepsilon_i} \qquad F_{pc} = \frac{\Delta\varepsilon_{pc}}{\Delta\varepsilon_i} \qquad F_{cc} = \frac{\Delta\varepsilon_{cc}}{\Delta\varepsilon_i} \qquad F_{cp} = \frac{\Delta\varepsilon_{cp}}{\Delta\varepsilon_i}$$

Then the predicted life for the combined creep and fatigue cycle, N_{pred}, is given

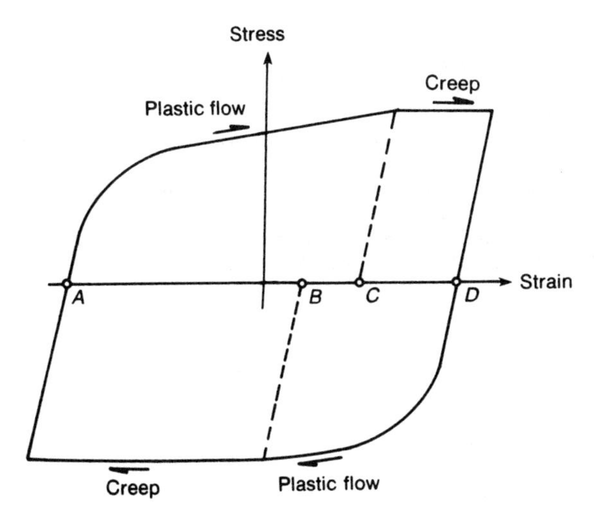

Figure 13-24 A typical creep-fatigue strain hysteresis cycle.

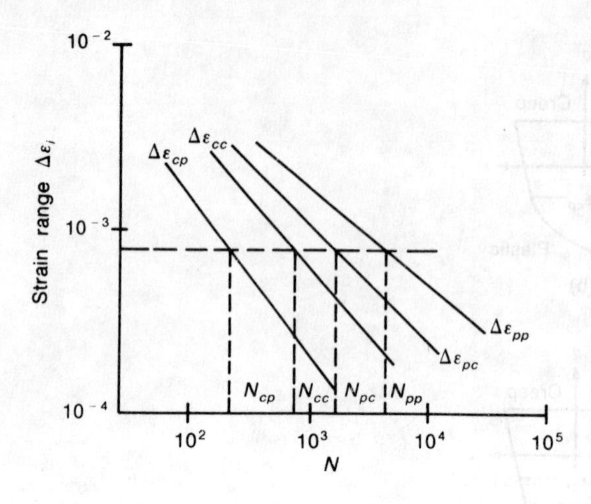

Figure 13-25 Coffin-Manson plots for strain-range components.

by the *interaction damage rule*

$$\frac{1}{N_{\text{pred}}} = \frac{F_{pp}}{N_{pp}} + \frac{F_{pc}}{N_{pc}} + \frac{F_{cc}}{N_{cc}} + \frac{F_{cp}}{N_{cp}} \tag{13-38}$$

BIBLIOGRAPHY

Clauss, F. J.: "Engineer's Guide to High Temperature Materials," Addison-Wesley, Reading, Mass., 1969.

Conway, S. B.: "Numerical Methods for Creep and Rupture," Gordon and Breach, Science Publishers, Inc., New York, 1967.

Creep, Stress-Rupture and Stress-Relaxation Testing, "Metals Handbook," 9th ed., vol. 8, pp. 299–360, American Society for Metals, Metals Park, Ohio, 1985.

Evans, H. E.: "Mechanisms of Creep Fracture," Elsevier Applied Science Publishers, New York, 1984.

Evans, R. W. and B. Wilshire: "Creep of Metals and Alloys," The Institute of Metals, London, 1985.

Garofalo, F.: "Fundamentals of Creep and Creep-Rupture in Metals," The Macmillan Company, New York, 1965.

Gittus, J.: "Creep, Viscoelasticity and Creep Fracture in Solids," John Wiley-Halsted Press, New York, 1975.

R. Raj (ed.): "Flow and Fracture at Elevated Temperatures," American Society for Metals, Metals Park, Ohio, 1985.

Wilshire, B. and D. R. J. Owen (eds.): "Recent Advances in Creep and Fracture of Engineering Materials and Structures," Pineridge Press, Swansea, U.K., 1982.

CHAPTER

FOURTEEN

BRITTLE FRACTURE AND IMPACT TESTING

14-1 THE BRITTLE-FRACTURE PROBLEM

During World War II a great deal of attention was directed to the brittle failure of welded Liberty ships and T-2 tankers.[1] Some of these ships broke completely in two, while, in other instances, the fracture did not completely disable the ship. Most of the failure occurred during the winter months. Failures occurred both when the ships were in heavy seas and when they were anchored at dock. These calamities focused attention on the fact that normally ductile mild steel can become brittle under certain conditions. A broad research program was undertaken to find the causes of these failures and to prescribe the remedies for their future prevention. In addition to research designed to find answers to a pressing problem, other research was aimed at gaining a better understanding of the mechanism of brittle fracture and fracture in general. Many of the results of this basic work are described in Chap. 7, which should be reviewed before proceeding further with this chapter. While the brittle failure of ships concentrated great attention to brittle failure in mild steel, it is important to understand that this is not the only application where brittle fracture is a problem. Brittle failures in tanks, pressure vessels, pipelines, and bridges have been documented[2] as far back as the year 1886.

Three basic factors contribute to a brittle-cleavage type of fracture. They are (1) a triaxial state of stress, (2) a low temperature, and (3) a high strain rate or

[1] M. L. Williams, Analysis of Brittle Behavior in Ship Plates, Symposium on Effect of Temperature on the Brittle Behavior of Metals with Particular Reference to Low Temperatures, *ASTM Spec. Tech. Publ.*, 158, pp. 11–44, 1954.

[2] M. E. Shank, A Critical Survey of Brittle Failure in Carbon Plate Steel Structures Other than Ships, *ASTM Spec. Tech. Publ.*, 158, pp. 45–110, 1954.

rapid rate of loading. All three of these factors do not have to be present at the same time to produce brittle fracture. A triaxial state of stress, such as exists at a notch, and low temperature are responsible for most service failures of the brittle type. However, since these effects are accentuated at a high rate of loading, many types of impact tests have been used to determine the susceptibility of materials to brittle behavior. Steels which have identical properties when tested in tension or torsion at slow strain rates can show pronounced differences in their tendency for brittle fracture when tested in a notched-impact test.

Since the ship failures occurred primarily in structures of welded construction, it was considered for a time that this method of fabrication was not suitable for service where brittle fracture might be encountered. A great deal of research has since demonstrated that welding, per se, is not inferior in this respect to other types of construction. However, strict quality control is needed to prevent weld defects which can act as stress raisers or notches. New electrodes have been developed that make it possible to make a weld with better properties than the mild-steel plate. The design of a welded structure is more critical than the design of an equivalent riveted structure, and much effort has gone into the development of safe designs for welded structures. It is important to eliminate all stress raisers and to avoid making the structure too rigid. To this end, riveted sections, known as crack arresters, were incorporated in some of the wartime ships so that, if a brittle failure did occur, it would not propagate completely through the structure.

14-2 NOTCHED-BAR IMPACT TESTS

Various types of notched-bar impact tests are used to determine the tendency of a material to behave in a brittle manner. This type of test will detect differences between materials which are not observable in a tension test. The results obtained from notched-bar tests are not readily expressed in terms of design requirements, since it is not possible to measure the components of the triaxial stress condition at the notch. Furthermore, there is no general agreement on the interpretation or significance of results obtained with this type of test.

A large number of notched-bar test specimens of different design have been used by investigators of the brittle fracture of metals. Two classes of specimens have been standardized[1] for notched-impact testing. Charpy bar specimens are used most commonly in the United States, while the Izod specimen is favored in Great Britain. The Charpy specimen has a square cross section (10 × 10 mm) and contains a 45° V notch, 2 mm deep with a 0.25-mm root radius. The specimen is supported as a beam in a horizontal position and loaded behind the notch by the impact of a heavy swinging pendulum (the impact velocity is approximately 16 ft/s). The specimen is forced to bend and fracture at a high strain rate on the order of 10^3 s^{-1}. The Izod specimen, which is used rarely today, has either a

[1] ASTM Standards pt. 31, Designation E23-82.

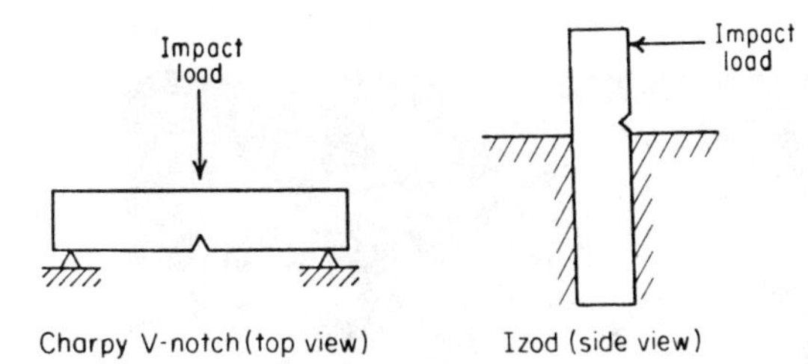

Figure 14-1 Sketch showing method of loading in Charpy and Izod impact tests.

circular or square cross section and contains a V notch near the clamped end. The difference in loading between the Charpy and Izod tests is shown in Fig. 14-1.

Plastic constraint at the notch produces a triaxial state of stress similar to that shown in Fig. 7-15. The maximum plastic stress concentration is given by

$$K_\sigma = \left(1 + \frac{\pi}{2} - \frac{\omega}{2}\right) \tag{14-1}$$

where ω is the included flank angle of the notch. The relative values of the three principal stresses depend strongly on the dimensions of the bar and the details of the notch. The standard specimen is thick enough to ensure a high degree of plane-strain loading and triaxiality across almost all of the notched cross section. Thus, the standard Charpy V-notch specimen provides a severe test for brittle fracture. Therefore, nonstandard specimens[1] should be used with great care.

The principal measurement from the impact test is the energy absorbed in fracturing the specimen. After breaking the test bar, the pendulum rebounds to a height which decreases as the energy absorbed in fracture increases. The energy absorbed in fracture, usually expressed in foot-pounds or joules, is read directly from a calibrated dial on the impact tester. The energy required for fracture of a Charpy specimen is often designed C_V, for example, C_V 25 ft-lb. In Europe impact test results are frequently expressed in energy absorbed per unit cross-sectional area of the specimen. It is important to realize that fracture energy measured by the Charpy test is only a relative energy and cannot be used directly in design equations.

Another common measurement obtained from the Charpy test results from the examination of the fracture surface to determine whether the fracture is fibrous (shear fracture), granular (cleavage fracture), or a mixture of both. These different modes of failure are readily distinguishable even without magnification. The flat facets of cleavage fracture provide a high reflectivity and bright appearance, while the dimpled surface of a ductile fibrous fracture provides a light-absorptive surface and dull appearance. Usually an estimate is made of the percentage of the fracture surface that is cleavage (or fibrous) fracture. Figure 14-2 shows how the fracture appearance changes from 100 percent flat cleavage (left) to 100 percent fibrous fracture (right) as the test temperature is increased. Note that the fibrous fracture appears first around the outer surface of the

[1] R. C. McNicol, *Weld. J.*, vol. 44, pp. 1s–9s, 1965.

Figure 14-2 Fracture surfaces of Charpy specimens tested at different temperatures. *Left*, 40° F; *center*, 100° F; *right*, 212° F. Note gradual decrease in the granular region and increase in lateral contraction at the notch with increasing temperature.

specimen (shear lip) where the triaxial constraint is the least. A third measurement that is sometimes made in the Charpy test is the ductility, as indicated by the percent contraction of the specimen at the notch.

The notched-bar impact test is most meaningful when conducted over a range of temperature so that the temperature at which the ductile-to-brittle transition takes place can be determined. Figure 14-3 illustrates the type of curves which are obtained. Note that the energy absorbed decreases with decreasing temperature but that for most cases the decrease does not occur sharply at a certain temperature. This makes it difficult to determine accurately the transition temperature. In selecting a material from the standpoint of notch toughness or tendency for brittle failure the important factor is the transition temperature. Figure 14-3 illustrates how reliance on impact resistance at only one temperature can be misleading. Steel *A* shows higher notch toughness at room temperature; yet its transition temperature is higher than that of steel *B*. The material with the lowest transition temperature is to be preferred.

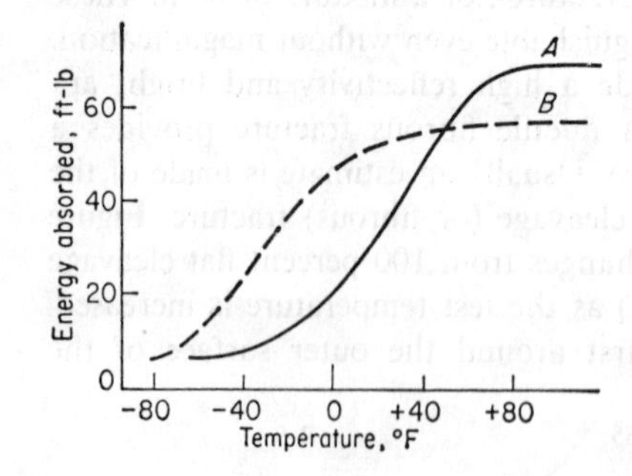

Figure 14-3 Transition-temperature curves for two steels, showing fallacy of depending on room-temperature results.

Notched-bar impact tests are subject to considerable scatter,[1] particularly in the region of the transition temperature. Most of this scatter is due to local variations in the properties of the steel, while some is due to difficulties in preparing perfectly reproducible notches. Both notch shape and depth are critical variables, as is the proper placement of the specimen in the impact machine.

The principal advantage of the Charpy V-notch impact test is that it is a relatively simple test that utilizes a relatively cheap, small test specimen. Tests can readily be carried out over a range of subambient temperatures. Moreover, the design of the test specimen is well suited for measuring differences in notch toughness in low-strength materials such as structural steels. The test is used for comparing the influence of alloy studies and heat treatment on notch toughness. It frequently is used for quality control and material acceptance purposes. The chief difficulty is that the results of the Charpy test are difficult to use in design. Since there is no measurement in terms of stress level, it is difficult to correlate C_V data with service performance. Moreover, there is no correlation of Charpy data with flaw size. In addition, the large scatter inherent in the test may make it difficult to determine well-defined transition-temperature curves.

14-3 INSTRUMENTED CHARPY TEST

The ordinary Charpy test measures the total energy absorbed in fracturing the specimen. Additional information can be obtained if the impact tester is instrumented to provide a load-time history of the specimen during the test.[2] Figure 14-4 shows an idealized load-time curve for an instrumented Charpy test. With this kind of record it is possible to determine the energy required for initiating fracture and the energy required for propagating fracture. It also yields information on the load for general yielding, the maximum load, and the fracture load.

If the velocity of the impacting pendulum can be assumed constant throughout the test then

$$E' = v_0 \int_0^t P\,dt \tag{14-2}$$

where v_0 = initial pendulum velocity
P = instantaneous load
t = time

However, the assumption of a constant pendulum velocity v is not valid, for v decreases in proportion to the load on the specimen. It is usually assumed that[3]

$$E_t = E'(1 - \alpha) \tag{14-3}$$

where E_t = the total fracture energy
$\alpha = E'/4E_0$, where E_0 is the initial energy of the pendulum

[1] N. H. Fahey, Impact Testing of Metals, *ASTM Spec. Tech. Publ.*, 466, pp. 76–92, 1970; T. A. Bishop, A. J. Markworth, and A. R. Rosenfield, *Met. Trans.*, vol. 14A, pp. 687–693, 1983.

[2] "Instrumented Impact Testing," *ASTM Spec. Tech. Pub.* 563, ASTM, Philadelphia, 1973.

[3] B. Augland, *Brit. Weld. J.*, vol. 9, p. 434, 1962.

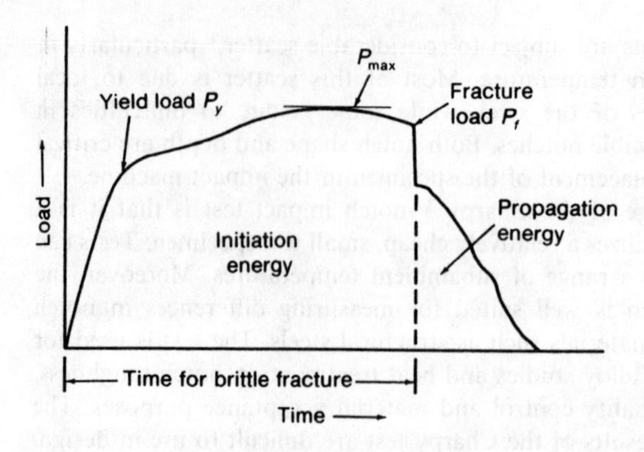

Figure 14-4 Load-time history for an instrumented Charpy test.

Good correlation is obtained between values obtained from Eq. (14-3) and those read directly from the dial of an impact tester.

Because the root of the notch in a Charpy specimen is not as sharp as is used in fracture mechanics tests, there has been a trend toward using standard Charpy specimens which are precracked by the introduction of a fatigue crack at the tip of the V notch. These precracked specimens have been used in the instrumented Charpy test to measure dynamic fracture toughness values (K_{Id}).

14-4 SIGNIFICANCE OF TRANSITION-TEMPERATURE CURVE

The chief engineering use of the Charpy test is in selecting materials which are resistant to brittle fracture by means of transition-temperature curves. The design philosophy is to select a material which has sufficient notch toughness when subjected to severe service conditions so that the load-carrying ability of the structural member can be calculated by standard strength of materials methods without considering the fracture properties of the material or stress concentration effects of cracks or flaws.

The transition-temperature behavior of a wide spectrum of materials falls into the three categories shown in Fig. 14-5. Medium- and low-strength fcc metals and most hcp metals have such high notch toughness that brittle fracture is not a problem unless there is some special reactive chemical environment. High-strength materials ($\sigma_0 > E/150$) have such low notch toughness that brittle fracture can occur at nominal stresses in the elastic range at all temperatures and strain rates when flaws are present. High-strength steel, aluminum and titanium alloys fall into this category. At low temperature fracture occurs by brittle cleavage, while at higher temperatures fracture occurs by low-energy rupture. It is under these conditions that fracture mechanics analysis is useful and appropriate. The notch toughness of low- and medium-strength bcc metals, as well as Be, Zn, and ceramic materials is strongly dependent on temperature. At low temperature the fracture occurs by cleavage while at high temperature the fracture occurs by ductile

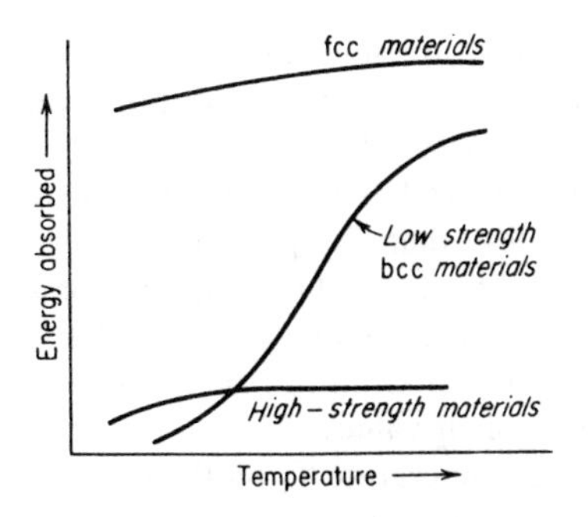

Figure 14-5 Effect of temperature on notch toughness (schematic).

rupture. Thus, there is a transition from notch brittle to notch tough behavior with increasing temperature. In metals this transition occurs at 0.1 to 0.2 of the absolute melting temperature T_m, while in ceramics the transition occurs at about 0.5 to $0.7T_m$.

The design philosophy using transition-temperature curves centers about the determination of a temperature above which brittle fracture will not occur at elastic stress levels. Obviously, the lower this transition temperature, the greater the fracture toughness of the material. The shape of the typical C_V versus temperature curve (Fig. 14-3) shows that there is no single criterion that defines the transition temperature. The various definitions of transition temperature obtained from an energy vs. temperature curve or a fracture appearance vs. temperature curve are illustrated in Fig. 14-6. The most conservative criterion for transition temperature is to select T_1, corresponding to the upper shelf in fracture energy and the temperature above which the fracture is 100 percent fibrous (0 percent cleavage). This transition temperature criterion is called the *fracture transition plastic* (FTP). The FTP is the temperature at which the fracture changes from totally ductile to substantially brittle. The probability of brittle fracture is negligible above the FTP.

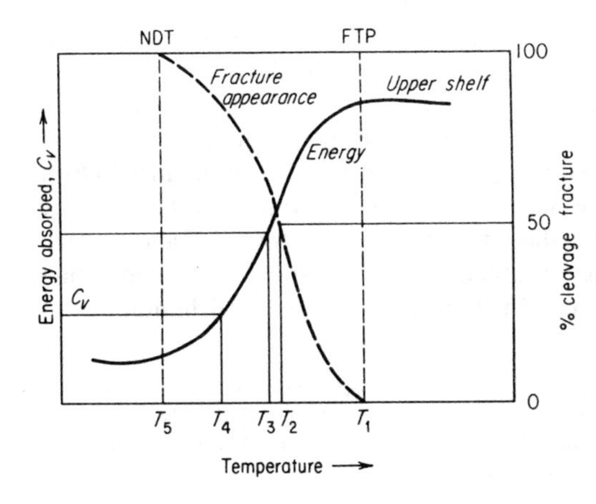

Figure 14-6 Various criteria of transition temperature obtained from Charpy test.

The use of the FTP is very conservative and in many applications it may be impractical. An arbitrary, but less conservative criterion is to base the transition temperature on 50 percent cleavage-50 percent shear, T_2. This is called a *fracture-appearance transition temperature* (FATT). Correlations between Charpy impact tests and service failures indicate[1] that less than 70 percent cleavage fracture in the Charpy bar indicates a high probability that failure will not occur at or above the temperature if the stress does not exceed about one-half of the yield stress. Roughly similar results are obtained by defining the transition temperature as the average of the upper and lower shelf values, T_3.

A common criterion is to define the transition temperature T_4 on the basis of an arbitrary low value of energy absorbed C_V. This is often called the *ductility transition temperature*. As a result of extensive tests on World War II steel ship plates, it was established that brittle fracture would not initiate if C_V was equal to 15 ft-lb at the test temperature. A 15 ft-lb C_V transition temperature has become an accepted criterion for low-strength ship steels. However, it is important to realize that for other materials C_V 15 bears no special significance. For higher strength steel the corresponding value would exceed C_V 15, but in general this value is not known.

A well-defined criterion is to base the transition temperature on the temperature at which the fracture becomes 100 percent cleavage, T_5. This point is known as *nil ductility temperature* (NDT). The NDT is the temperature at which fracture initiates with essentially no prior plastic deformation. Below the NDT the probability of ductile fracture is negligible.

14-5 METALLURGICAL FACTORS AFFECTING TRANSITION TEMPERATURE

Changes in transition temperature of over 100°F can be produced by changes in the chemical composition or microstructure of mild steel. The largest changes in transition temperature result from changes in the amount of carbon and manganese.[2] The 15 ft-lb transition temperature for V-notch Charpy specimens (ductility transition) is raised about 25°F for each increase of 0.1 percent carbon. This transition temperature is lowered about 10°F for each increase of 0.1 percent manganese. Increasing the carbon content also has a pronounced effect on the maximum energy and the shape of the energy transition-temperature curves (Fig. 14-7). The Mn : C ratio should be at least 3 . 1 for satisfactory notch toughness. A maximum decrease of about 100°F in transition temperature appears possible by going to higher Mn : C ratios. The practical limitations to

[1] T. J. Hodgson and G. M. Boyd, *Trans. Inst. Nav. Archit. London*, vol. 100, p. 141, 1958.

[2] J. A. Rinebolt and W. J. Harris, Jr., *Trans. Am. Soc. Met.*, vol. 43, pp. 1175–1214, 1951; for a broad review see A. S. Tetelman and A. J. McEvily, Jr., "Fracture of Structural Materials," chap. 10, John Wiley & Sons, Inc., New York, 1967.

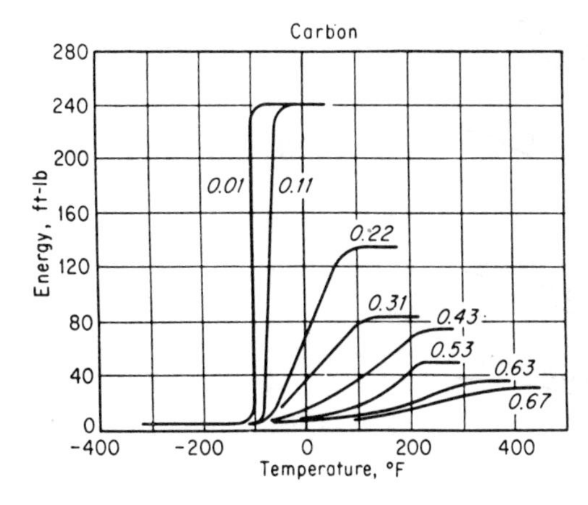

Figure 14-7 Effect of carbon content on the energy-transition-temperature curves for steel. *(From J. A. Rinebolt and W. J. Harris, Jr., Trans. Am. Soc. Met., vol. 43, p. 1197, 1951.)*

extending this beyond 7 : 1 are that manganese contents above about 1.4 percent lead to trouble with retained austenite, while about 0.2 percent carbon is needed to maintain the required tensile properties.

Phosphorus also has a strong effect in raising the transition temperature. The 15 ft-lb V-notch Charpy transition temperature is raised about 13°F for each 0.01 percent phosphorus. The role of nitrogen is difficult to assess because of its interaction with other elements. It is, however, generally considered to be detrimental to notch toughness. Nickel is generally accepted to be beneficial to notch toughness in amounts up to 2 percent and seems to be particularly effective in lowering the ductility transition temperature. Silicon, in amounts over 0.25 percent, appears to raise the transition temperature. Molybdenum raises the transition almost as rapidly as carbon, while chromium has little effect.

Notch toughness is particularly influenced by oxygen. For high-purity iron[1] it was found that oxygen contents above 0.003 percent produced intergranular fracture and corresponding low energy absorption. When the oxygen content was raised from 0.001 percent to the high value of 0.057 percent, the transition temperature was raised from 5 to 650°F. In view of these results, it is not surprising that deoxidation practice has an important effect on the transition temperature. Rimmed steel, with its high iron oxide content, generally shows a transition temperature above room temperature. Semikilled steels, which are deoxidized with silicon, have a lower transition temperature, while for steels which are fully killed with silicon plus aluminum the 15 ft-lb transition temperature will be around −75°F. Aluminum also has the beneficial effect of combining with nitrogen to form insoluble aluminum nitride.

Grain size has a strong effect on transition temperature. An increase of one ASTM number in the ferrite grain size (actually a decrease in grain diameter) can

[1] W. P. Rees, B. E. Hopkins, and H. R. Tipler, *J. Iron Steel Inst. London*, vol. 172, pp. 403–409, 1952.

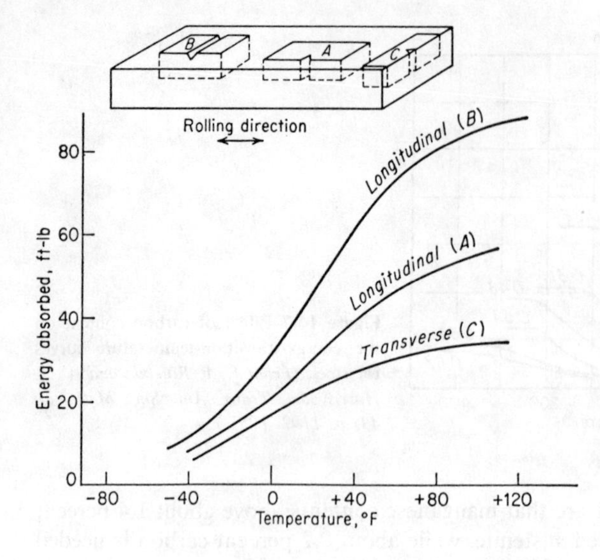

Figure 14-8 Effect of specimen orientation of Charpy transition-temperature curves.

result in a decrease in transition temperature of 30°F for mild steel. Decreasing the grain diameter from ASTM grain size 5 to ASTM grain size 10 can change the 10 ft-lb Charpy V-notch transition temperature[1] from about 70 to −60°F. A similar effect of decreasing transition temperature with decreasing austenite grain size is observed with higher alloyed heat-treated steels. Many of the variables concerned with processing mild steel affect the ferrite grain size and therefore affect the transition temperature. Since normalizing after hot rolling results in a grain refinement, if not carried out at too high a temperature, this treatment results in reduced transition temperature. The cooling rate from the normalizing treatment and the deoxidation practice are variables which also must be considered. Air cooling and aluminum deoxidation result in a lower transition temperature. Using the lowest possible finishing temperature for hot rolling of plate is also beneficial. Spray cooling from the rolling temperature before coiling[2] can lower the transition temperature 100°F.

An important development which has resulted in high-strength, low-alloy ferritic steels ($\sigma_0 = 60{,}000$ psi) with good impact properties (FATT below −20°F) is the addition of small amounts of Cb or V. This causes CbC precipitates to form during transformation. The dispersion strengthening raises the yield strength, while at the same time retarding grain growth and improving the impact resistance.

For a given chemical composition and deoxidation practice, the transition temperature will be appreciably higher in thick hot-rolled plates than in thin plates. This is due to the difficulty of obtaining uniformly fine pearlite and grain

[1] W. S. Owen, D. H. Whitmore, M. Cohen, and B. L. Averbach, *Weld. J.*, vol. 36, pp. 503s–511s, 1957.

[2] J. H. Bucher and J. D. Grozier, *Met. Eng. Q.*, vol. 5, no. 1, 1965.

size in a thick section. Generally speaking, allowance for this effect must be made in plates greater than $\frac{1}{2}$in in thickness.

The notched-impact properties of rolled or forged products vary with orientation in the plate or bar. Figure 14-8 shows the typical form of the energy-temperature curves for specimens cut in the longitudinal and transverse direction of a rolled plate. Specimens A and B are oriented in the longitudinal direction of the plate. In specimen A the notch is perpendicular to the plate, while in B the notch lies parallel to the plate surface. The orientation of the notch in specimen A is generally preferred. In specimen C this notch orientation is used, but the specimen is oriented transverse to the rolling direction. Transverse specimens are used in cases where the stress distribution is such that the crack would propagate parallel to the rolling direction. Reference to Fig. 14-8 shows that quite large differences can be expected for different specimen orientations at high energy levels, but the differences become much less at energy levels below 20 ft-lb. Since ductility transition temperatures are evaluated in this region of energy, it seems that specimen and notch orientation are not a very important variable for this criterion. If, however, materials are compared on the basis of room-temperature impact properties, orientation can greatly affect the results.

Low-carbon steels can exhibit two types of aging phenomena which produce an increase in transition temperature. *Quench aging* is caused by carbide precipitation in a low-carbon steel which has been quenched from around 1300°F (600°C). *Strain aging* occurs in low-carbon steel which has been cold-worked. Cold-working by itself will increase the transition temperature, but strain aging results in a greater increase, usually around 40 to 60°F. Quench aging results in less loss of impact properties than strain aging. The phenomenon of *blue brittleness*, in which a decrease in impact resistance occurs on heating in the range 450 to 700°F (230 to 370°C) is due to an accelerated form of strain aging.

It has been demonstrated many times that a tempered martensitic structure produces the best combination of strength and impact resistance of any microstructure that can be produced in steel. In Chap. 8 it was shown that the tensile properties of tempered martensites of the same hardness and carbon content are alike, irrespective of the amount of other alloy additions. This generalization holds approximately for the room-temperature impact resistance of heat-treated steels, but it is not valid for the variation of impact resistance with temperature. Figure 14-9 shows the temperature dependence of impact resistance for a number of different alloy steels, all having about 0.4 percent carbon and all with a tempered martensite structure produced by quenching and tempering to a hardness of R_C35. Note that a maximum variation of about 200°F in the transition temperature at the 20 ft-lb level is possible. Even greater spread in transition temperature would be obtained if the tempering temperature were adjusted to give a higher hardness.[1] Slack quenching so that the microstructure consists of a mixture of tempered martensite, bainite, and pearlite results in even greater

[1] H. J. French, *Trans. AIME*, vol. 206, pp. 770–782, 1956.

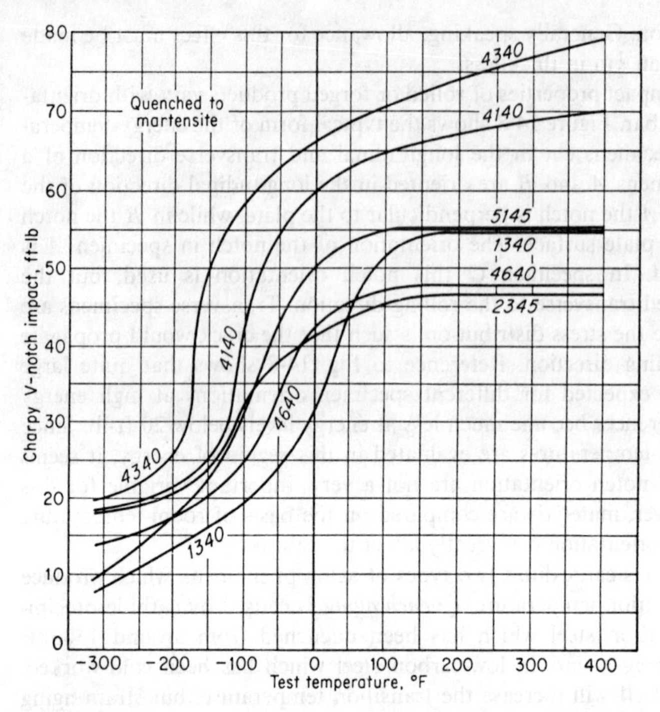

Figure 14-9 Temperature dependence of impact resistance for different alloy steels of same carbon content, quenched and tempered to R_C 35. *(From H. J. French, Trans. Metall. Soc. AIME, vol. 206, p. 770, 1956.)*

differences between alloy steels and in a general increase in the transition temperature.

The energy absorbed in the impact test of an alloy steel at a given test temperature generally increases with increasing tempering temperature. However, there is a minimum in the curve in the general region of 400 to 600°F. This has been called 500°F *embrittlement*, but because the temperature at which it occurs depends on both the composition of the steel and the tempering time, a more appropriate name is *tempered-martensite embrittlement*. Embrittlement of steel in this tempering region is one of the chief deterrents to using steels at strength levels much above 200,000 psi. Studies of this embrittlement phenomenon have shown that it is due to the precipitation of platelets of cementite from ε-carbide during the second stage of tempering. These platelets have no effect on the reduction of area of a tensile specimen, but they severely reduce the room temperature impact resistance. They can be formed at temperatures as low as 212°F and as high as 800°F, depending on the time allowed for the reaction. Silicon additions of around 2.25 percent are effective in increasing the temperature at which the platelets precipitate, and this permits tempering in the region 500 to 600°F, without severe embrittlement.

14-6 DROP-WEIGHT TEST AND OTHER LARGE-SCALE TESTS

Probably the chief deficiency of the Charpy impact test is that the small specimen is not always a realistic model of the actual situation. Not only does the small specimen lead to considerable scatter, but a specimen with a thickness of 0.394 in cannot provide the same constraint as would be found in a structure with a much greater thickness. The situation that can result is shown in Fig. 14-10. At a particular service temperature the standard Charpy specimen shows a high shelf energy, while actually the same material in a thick-section structure has low toughness at the same temperature.

The most logical approach to this problem is the development of tests that are capable of handling specimens at least several inches thick. The development of such tests and their rational method of analysis has been chiefly the work of Pellini[1] and his coworkers at the Naval Research Laboratory. The basic need for large specimens resulted from the inability to produce fracture in small laboratory specimens at stresses below the gross yield stress, whereas brittle fractures in ship structures occur at service temperatures at elastic stress levels.

The first development[2] was the explosion-crack-starter test which featured a short, brittle weld bead deposited on the surface of a $14 \times 14 \times 1$ in steel plate. The plate was placed over a circular die and dynamically loaded with an explosive charge. The brittle weld bead introduces a small natural crack in the test plate similar to a weld-defect crack. Tests are carried out over a range of temperature and the appearance of the fracture determines the various transition temperatures (Fig. 14-11). Below the NDT the fracture is a flat (elastic) fracture running completely to the edges of the test plate. Above the nil ductility temperature a plastic bulge forms in the center of the plate, but the fracture is still a flat elastic fracture out to the plate edge. At still higher temperature the fracture does not propagate outside of the bulged region. The temperature at which elastic fracture no longer propagates to the edge of the plate is called the *fracture transition elastic* (FTE). The FTE marks the highest temperature of fracture propagation by purely elastic stresses. At yet higher temperature the extensive plasticity results in a helmet-type bulge. The temperature above which this fully ductile tearing occurs is the *fracture transition plastic* (FTP).

The drop weight test (DWT) was developed[3] specifically for the determination of the NDT on full thickness plates. A short bead of brittle weld metal is deposited on the surface of a plate, typically $3\frac{1}{2} \times 14 \times \frac{5}{8}$ to 1 in thick. A small notch is introduced in the weld bead and the specimen supported as a simple beam in a constant temperature bath (Fig. 14-12). The brittle weld bead is fractured at near yield-stress levels as a result of dynamic loading from a falling weight. The anvil stop restricts the deflection of the test specimen. Since the specimen is a wide beam loaded in three-point bending, this restriction limits the

[1] W. S. Pellini, *Weld. J.*, vol. 50, pp. 915–1095, 147s–162s, 1971.

[2] P. P. Puzak, M. E. Shuster, and W. S. Pellini, *Weld. J.*, vol. 33, p. 481s, 1954.

[3] P. P. Puzak, and W. S. Pellini, *NRL Rept.* 5831, Aug. 21, 1962; ASTM Standards, pt. 31, 1969, pp. 582–601, Designation E208-69.

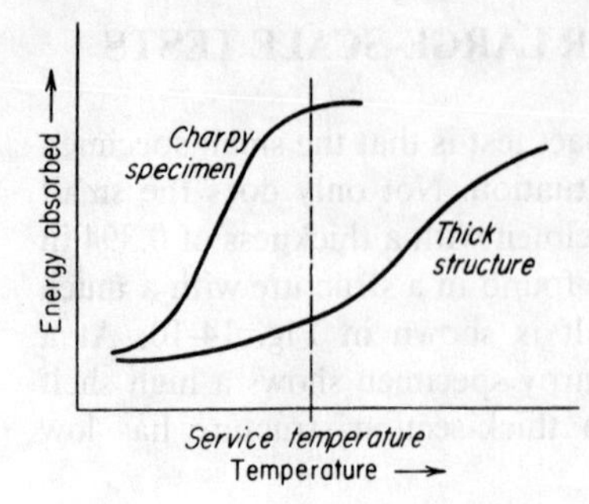

Figure 14-10 Effect of section thickness on transition-temperature curves.

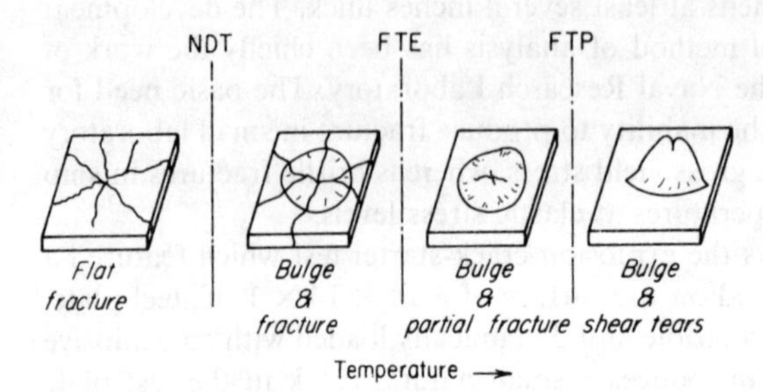

Figure 14-11 Fracture appearance vs. temperature for explosion-crack-starter test.

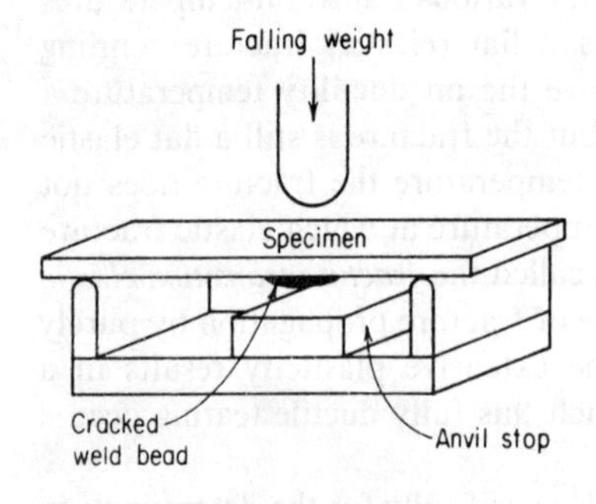

Figure 14-12 Drop-weight test (DWT).

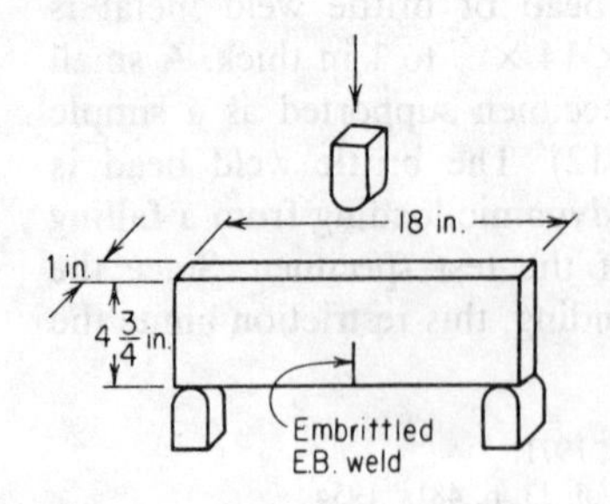

Figure 14-13 Dynamic-tear test (DT).

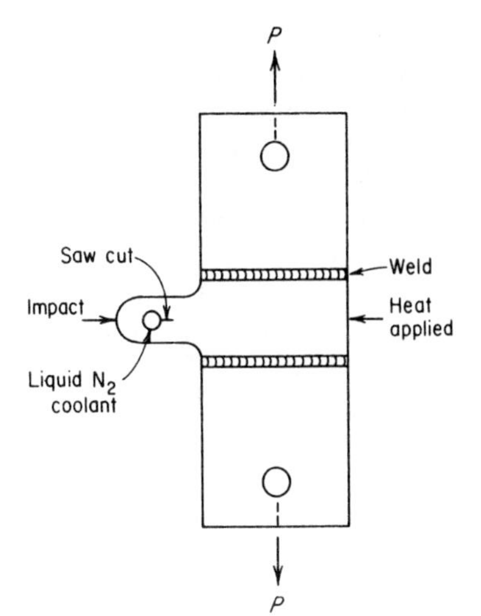

Figure 14-14 Robertson crack-arrest test.

stress on the tension face of the plate to a value that does not exceed the yield stress. If the starter-crack propagates across the width of the plate on the tension surface to the edges, the test temperature is below the NDT. Complete separation on the compression side of the specimen is not required. The NDT is the highest temperature at which a nil ductility break is produced. The test is quite reproducible and the NDT can be determined to the nearest 10°F.

The *dynamic tear test*[1] (DT) is in effect a giant Charpy test (Fig. 14-13). While specimens are usually $\frac{5}{8}$ and 1 in thick, DT tests have been made on specimens up to 12 in thick. The notch is an electron-beam weld which is embrittled metallurgically by alloying (Ti is added to produce a brittle Fe-Ti alloy). The narrow weld is fractured easily providing a reproducible sharp crack. As in the Charpy test, specimens are fractured over a range of temperature in a pendulum-type machine[2] and the energy absorbed in fracture is measured. However, while the maximum energy capacity of a standard Charpy impact tester is 240 ft-lb, the pendulum tester used with the DT test has a capacity of 10,000 ft-lb. The use of this test in fracture analysis will be discussed in Sec. 14-7.

Another important type of test is the *crack-arrest test*[3] which provides a relationship between the stress level and the ability of the material to arrest a rapidly propagating crack. Figure 14-14 illustrates the Robertson crack-arrest test. A uniform elastic tensile stress is applied to a plate specimen 6 in wide. A rapidly moving brittle fracture is initiated by impact loading at a starter crack on the cold

[1] Early versions of this test were known as the drop-weight tear test (DWTT).

[2] P. P. Puzak, and E. A. Lange, *NRL Rept.* 6851, Feb. 13, 1969; E. A. Lange and F. J. Loss, "Impact Testing of Metals," pp. 241–258, *ASTM Spec. Tech. Publ.* 466, 1970.

[3] T. S. Robertson, *Engineering*, vol. 172, pp. 445–448; *J. Iron Steel Inst.* London, vol. 175, p. 361, 1953; F. J. Feely, M. S. Northrup, S. R. Kleepe, and M. Gensamer, *Weld. J.*, vol. 34, pp. 596s–607s, 1955.

side of the specimen. The crack propagates up a temperature gradient toward the hot side. The point across the specimen width at which the temperature is high enough to give enough ductility to blunt the crack is called the *crack-arrest temperature* (CAT). In an alternative form of the test, the temperature across the specimen is constant and tests are carried out with successive specimens at increasing temperature until the CAT is reached. Crack-arrest tests on mild steel below the NDT show that the CAT is independent of temperature but the stress level for crack arrest is very low. If the stress is greater than 5,000 to 8,000 psi, brittle fracture will occur. Obviously, this stress level is too low for practical engineering design, so that steels cannot be used below the NDT. While the crack-arrest tests are among the most quantitative of brittle-fracture tests, they are not used extensively because they require large testing machines and large specimens.

14-7 FRACTURE ANALYSIS DIAGRAM

The previous sections have introduced a number of terms dealing with brittle fracture, such as NDT, FTE, FTP, etc. The tests for determining these transition temperatures have been described. Before seeing how they are used in engineering design through the fracture analysis diagram, we redefine these transition points through reference to basic properties of the tension test. The subambient temperature dependence of yield strength σ_0 and ultimate tensile strength σ_u in a bcc metal are shown in Fig. 14-15. For an unnotched specimen without flaws the material is ductile until a very low temperature, point A, where $\sigma_0 = \sigma_u$. Point A represents the NDT temperature for a flaw-free material. The curve BCD represents the fracture strength of a specimen containing a small flaw ($a < 0.1$). The temperature corresponding to point C is the highest temperature at which the fracture strength $\sigma_f \approx \sigma_0$. Point C represents the NDT for a specimen with a small crack or flaw. The presence of a small flaw raises the NDT of a steel by about 200°F. Increasing the flaw size decreases the fracture stress curve, as in curve EF,

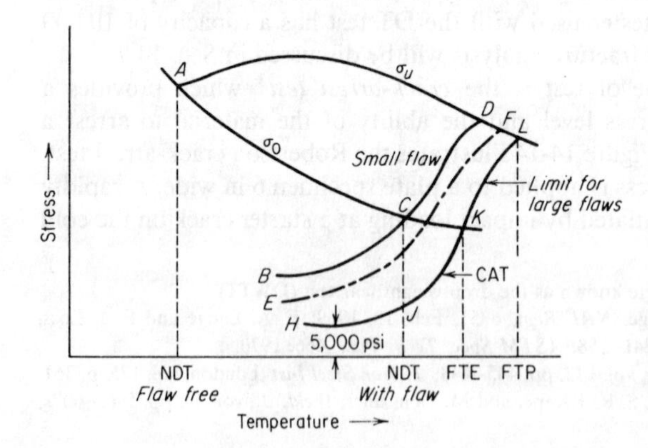

Figure 14-15 Temperature dependence of yield strength (σ_0), tensile strength (σ_u), and fracture strength for a steel containing flaws of different sizes.

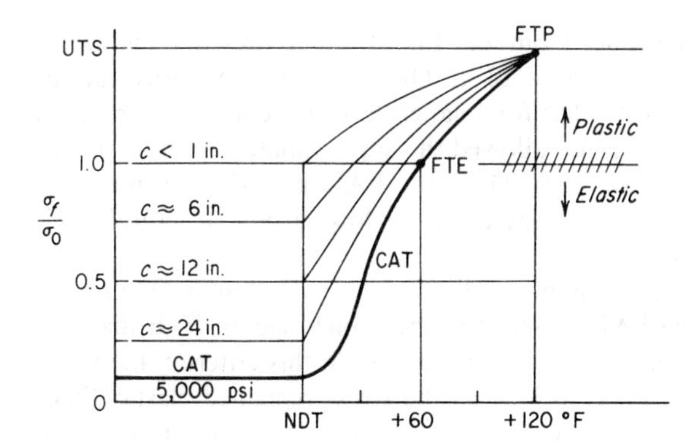

Figure 14-16 Fracture-analysis diagram showing influence of various initial flaw sizes.

until with increasing flaw size a limiting curve of fracture stress *HJKL* is reached. Below the NDT the limiting safe stress is 5,000 to 8,000 psi. Above the NDT the stress required for the unstable propagation of a long flaw (*JKL*) rises sharply with increasing temperature. This is the crack-arrest temperature curve (CAT). The CAT defines the highest temperature at which unstable crack propagation can occur at any stress level. Fracture will not occur for any point to the right of the CAT curve.

The temperature above which *elastic* stresses cannot propagate a crack is the *fracture transition elastic* (FTE). This is defined by the temperature when the CAT curve crosses the yield-strength curve (point *K*). The *fracture transition plastic* (FTP) is the temperature where the CAT curve crosses the tensile-strength curve (point *L*). Above this temperature the material behaves as if it were flaw-free, for any crack, no matter how large, cannot propagate as an unstable fracture.

Data obtained from the DWT and other large-scale fracture tests have been assembled by Pellini and coworkers[1] into a useful design procedure called the *fracture analysis diagram* (FAD). The NDT as determined by the DWT provides a key data point to start construction of the fracture analysis diagram (Fig. 14-16). For mild steel below the NDT the CAT curve is flat. A stress level in excess of 5,000 to 8,000 psi causes brittle fracture regardless of the size of the initial flaw. Extensive correlation between the NDT and Robertson CAT tests for a variety of structural steels have shown that the CAT curve bears a fixed relationship to the NDT temperature. Thus, the NDT + 30°F provides a conservative estimate of the CAT curve at stress of $\sigma_0/2$. NDT + 60°F provides an estimate of the CAT at $\sigma = \sigma_0$, that is, the FTE and NDT + 120°F provides an estimate of the FTP. Therefore, for structural steels, once the NDT has been determined, the entire scope of the CAT curve can be established well enough for engineering design.[2]

[1] W. S. Pellini, and P. P. Puzak, *NRL Rept*. 5920, Mar. 15, 1963 and *Trans. ASME, Ser. A: J. Eng. Power*, vol. 86, pp. 429–443, 1964.

[2] Tests on very thick sections (6 to 12 in.) show that the transition region is expanded from NDT + 120°F to NDT + 180°F to NDT + 220°F.

The curve that has been traced out on Fig. 14-16 represents the worse possible case for large flaws in excess of 24 in. One can envision a spectrum of curves translated upward and to the left for smaller, less severe flaws. Correlation with service failures and other tests has allowed the approximate determination of curves for a variety of initial flaw sizes. Thus, the FAD provides a generalized relationship of flaw size, stress, and temperature for low-carbon structural steels of the type used in ship construction.

The fracture analysis diagram can be used several ways in design. One simple approach would be to use the FAD to select a steel which had an FTE that was lower than the lowest expected service temperature. With this criterion the worst expected flaw would not propagate so long as the stress remained elastic. Since the assumption of elastic behavior is basic in structural design, this design philosophy would be tantamount to being able to ignore the presence of flaws and brittle fracture. However, this procedure may prove to be too expensive and overconservative. A slightly less conservative design against brittle fracture, but still a practical approach, would be to design on the basis of an allowable stress level not exceeding $\sigma_0/2$. From Fig. 14-16 we see that any crack will not propagate under this stress so long as the temperature is not below NDT + 30°F. If for example, the service temperature is not expected to be below 10°F, we would select a steel whose NDT is 10° − 30° = −20°F.

The dynamic tear test (DT) can be used to construct the FAD as shown in Fig. 14-17. Below the NDT the fracture is brittle and has a flat, featureless surface devoid of any shear lips. At temperatures above the NDT there is a sharp rise in energy for fracture and the fracture surfaces begin to develop shear lips. The shear lips become progressively more prominent as the temperature is increased to the FTE. Above the FTE the fracture is ductile, void coalescence-type fracture. The fracture surface is a fibrous slant fracture. The upper shelf of energy represents the FTP. The lower half of the DT energy curve traces the temperature course of the CAT curve from NDT to FTE. The DT test is a highly versatile test because it is equally useful with low-strength ductile materials which show a high upper energy shelf and with high-strength low-toughness materials which have a low value of upper shelf energy. The large size of the DT specimen provides a

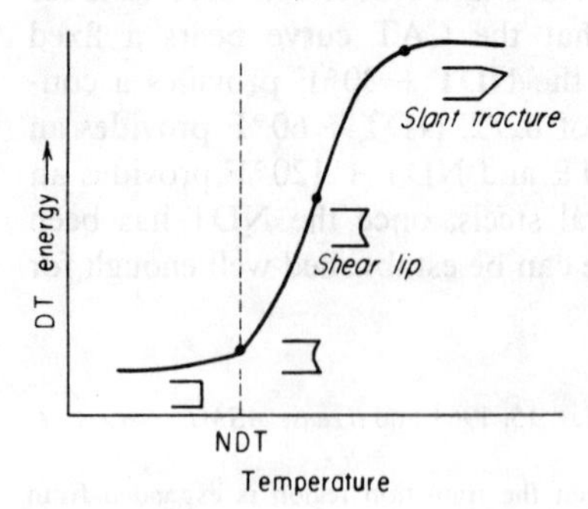

Figure 14-17 Test results from DT test.

high degree of triaxial constraint and results in a minimum of scatter. Extensive correlations are being developed between DT results and fracture toughness and C_V test data.

14-8 TEMPER EMBRITTLEMENT

Temper embrittlement[1] refers to the decrease in notch toughness which often occurs in alloy steels when heated in, or cooled slowly through, the temperature range 800 to 1100°F. This can become a particularly important problem with heavy sections that cannot be cooled through this temperature region rapidly enough to suppress embrittlement. Temper embrittlement also can be produced by isothermal treatments in this temperature region.

The presence of temper embrittlement is determined by measuring the transition temperature with the notched-bar impact test. The hardness and tensile properties are not sensitive to the embrittlement, except for very extreme cases, but the transition temperature can be raised around 200°F by ordinary embrittling heat treatments. The fracture in a temper-embrittled steel is intergranular and propagates along prior austenitic grain boundaries. The embrittlement occurs only in the presence of specific impurities and increases in severity with the concentration of these impurities. The chief embrittling elements are antimony, phosphorus, tin, and arsenic (in decreasing importance). Plain carbon steels with less than 0.5 percent manganese are not susceptible to temper embrittlement. Susceptibility is enhanced greatly by Cr and Mn and less strongly by Ni. Small amounts of Mo and W inhibit embrittlement, but larger amounts enhance it.

The above facts strongly support a mechanism in which the embrittling elements segregate to austenite grain boundaries during austenitization. This segregation has been verified for Sb and P by Auger electron spectroscopy. The segregation is retained on quenching, and on heating to the tempering temperature the segregates weaken the cohesion at the ferrite-carbide interfaces. Fracture initiates along the ferrite-cementite interface[2] but propagates along prior austenite grain boundaries because they have a high embrittling element content and they contain a large density of carbide platelets.

Steels that have suffered temper embrittlement can be restored to their original toughness by heating to above 1100°F (600°C) and cooling rapidly to below about 570°F (300°C). However, the best procedure is to reduce the susceptibility to temper embrittlement by reducing the embrittling impurities through control of raw materials and melting practice.

[1] The extensive literature in this field has been reviewed by C. J. McMahon, Jr., in "Temper Embrittlement in Steel," *ASTM Spec. Tech. Publ.* 407, 1967; I. Olefjord, *Int. Met Rev.*, vol. 23, no. 4, pp. 149–163, 1978.

[2] C. L. Briant and S. K. Banerji, *Int. Met. Rev.*, vol. 23, no. 4, pp. 164–199, 1978.

14-9 ENVIRONMENT SENSITIVE FRACTURE

There is a group of metallurgical phenomena which involve the interaction between a static stress (often a residual stress) and the environment to reduce the toughness. These phenomena are hydrogen embrittlement, stress corrosion cracking, and liquid metal embrittlement. Yet another specialized but important behavior is neutron irradiation embrittlement. Because the first three phenomena involve the interaction of a chemical species with a crack they are beginning to be studied in a generic way under the subject of *environment assisted cracking*.[1]

Hydrogen Embrittlement

Severe embrittlement can be produced in many metals by very small amounts of hydrogen. Body-centered cubic and hexagonal close-packed metals are most susceptible to hydrogen embrittlement. As little as 0.0001 weight percent of hydrogen can cause cracking in steel. Face-centered cubic metals are not generally susceptible to hydrogen embrittlement.[2] Hydrogen may be introduced during melting and entrapped during solidification, or it may be picked up during heat treatment, electroplating, acid pickling, or welding. Hydrogen also can be introduced by the cathodic reaction during corrosion.

The chief characteristics of hydrogen embrittlement are its strain-rate sensitivity, its temperature dependence, and its susceptibility to delayed fracture. Unlike most embrittling phenomena, hydrogen embrittlement is enhanced by slow strain rates. At low temperatures and high temperatures hydrogen embrittlement is negligible, but it is most severe in some intermediate temperature region. For steel the region of greatest susceptibility to hydrogen embrittlement is in the vicinity of room temperature. Slow bend tests and notched and unnotched tension tests will detect hydrogen embrittlement by a drastic decrease in ductility, but notched-impact tests are of no use for detecting the phenomenon.

A common method of studying hydrogen embrittlement is to charge notched tensile specimens with known amounts of hydrogen, load them to different stresses in a dead-weight machine, and observe the time to failure. A typical delayed-fracture curve is shown in Fig. 14-18. Note that the notched tensile strength of a charged specimen may be much lower than the strength of a hydrogen-free specimen. There is a region in which the time to fracture depends only slightly on the applied stress. There is also a minimum critical value below which delayed fracture will not occur. The similarity of the delayed fracture curve to the fatigue *S-N* curve has led to the use of the term "static fatigue" for the delayed-fracture phenomenon. The minimum critical stress, or "static fatigue limit," increases with a decrease in hydrogen content or a decrease in the severity

[1] C. L. Briant and S. K. Banerji (eds.), "Embrittlement of Engineering Alloys," Academic Press, New York, 1983; R. Gibala and R. F. Hehnemann (eds.), "Hydrogen Embrittlement and Stress Corrosion Cracking," American Society for Metals, Metals Park, Ohio, 1984.

[2] The familiar example of the embrittlement of copper by hydrogen at elevated temperature is due to the reaction of hydrogen with oxygen to form internal pockets of steam.

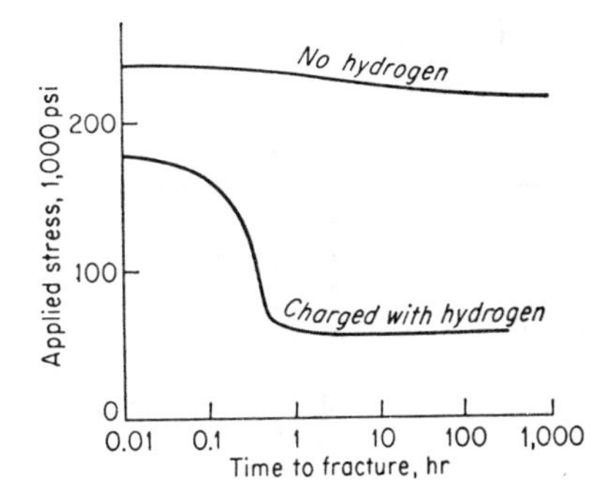

Figure 14-18 Delayed-fracture curve.

of the notch. The hydrogen content of steel may be reduced by "baking," or heating at around 300 to 500°F.

Hydrogen is present in the metal as monatomic hydrogen due to dissociation of molecular hydrogen by chemisorption on a metal surface. Because it is a small interstitial atom it can diffuse very rapidly at temperatures above room temperature. Moreover, hydrogen transport rates in association with dislocation motion can be several orders of magnitude greater than for diffusion alone. However, there is not an unequivocal mechanism for hydrogen embrittlement.[1] The planar pressure mechanism suggests that embrittlement is caused by the high pressures developed as hydrogen diffuses into voids and microcracks. However, this cannot explain the embrittlement from low-pressure hydrogen gas. Troiano[2] has suggested that hydrogen diffuses under the influence of a stress gradient to regions of high tensile triaxiality where the hydrogen interacts with the metal lattice to reduce its cohesive strength. Still other proposals suggest that hydrogen reduces the fracture strength by acting to decrease the surface energy at the tip of a crack subjected to tensile stress. From this viewpoint, hydrogen embrittlement is a special case of stress-corrosion cracking (discussion follows).

The fracture process in hydrogen embrittlement can involve cleavage, intergranular, or ductile fracture depending upon the stress level.[3] Thus, there is no single fracture mode that is characteristic of hydrogen embrittlement.

If a part is not under stress when it contains hydrogen then the hydrogen can be driven off by thermal treatment without damage to the part. Annealing in vacuum is particularly effective. However, if the part is subjected to tensile stress (either applied or residual) while containing hydrogen then local cracking will have occurred and the part will have been permanently damaged. Hydrogen embrittlement is most prevalent in high-strength materials.

[1] I. M. Bernstein and A. W. Thompson (eds.), "Hydrogen in Metals," Amer Soc. for Metals, Metals Park, Ohio, 1974; I. M. Bernstein and A. W. Thompson (ed.), "Hydrogen Effects in Metals," AIME, Warrendale, Pa., 1980.

[2] A. R. Troiano, *Trans. Am. Soc. Met.*, vol. 52, pp. 54–80, 1960.

[3] C. D. Beachem, *Met. Trans.*, vol. 3, p. 437, 1972.

Table 14-1 Combinations of chemical species and alloy that cause stress-corrosion cracking

Chemical species	Alloy	Temperature
Chlorides in aqueous solution	Austenitic stainless steels	Above room temp.
	High-strength steels	Room temp.
	High-strength aluminum alloys	Room temp.
Fluoride ions	Sensitized austenitic stainless steel	Room temp.
Fused chloride salt	Titanium and Zr alloys	Above m.p. of salt
O_2 dissolved in water	Sensitized stainless steels	300°C
Hydroxides (NaOH, KOH)	Carbon steels	100°C
Nitrates in aqueous solutions	Carbon steels	100°C
H_2S gas	High-strength low alloy steels	Room temp.
N_2O_4 liquid	Titanium alloys	50°C
Nitrogen oxides plus moisture	Copper alloys	Room temp.
Aqueous NH_3 and ammonium salts in aqueous solution	Copper alloys	Room temp.

Stress-Corrosion Cracking

Stress-corrosion cracking (SCC) is the failure of an alloy from the combined effects of a corrosive environment and a static tensile stress. The stress may result from applied forces or "locked-in" residual stress. Only specific combinations of alloys and chemical environment lead to stress-corrosion cracking. Usually only a few chemical species in the environment are effective in causing SCC of a particular alloy and these species need not be present in large quantities or in high concentrations. Moreover, the chemical environment which causes SCC does not produce general chemical corrosion of the alloy. With some alloy/chemical species combinations temperatures substantially above room temperature are required to produce SCC. Common situations of SCC are aluminum alloys and sea water, copper alloys and ammonia, mild steel and caustic, and austenitic stainless steel and salt water. Over 80 combinations of alloys and corrosive environments have been reported to cause SCC.[1] Some of these combinations are given in Table 14-1.

SCC begins with the rupturing of the protective oxide film on the metal surface by either mechanical means or by the action of chemical species, such as a chloride ion. Localized rupture of the surface film leads in time to the formation of corrosion pits: cracking initiates at the root of the pit. The porous corrosion product cap which forms over the pit restricts the interchange between the local environment within the pit and the bulk environment outside it. Thus, the pH

[1] H. L. Logan, "The Stress Corrosion of Metals," John Wiley & Sons, New York, 1966.

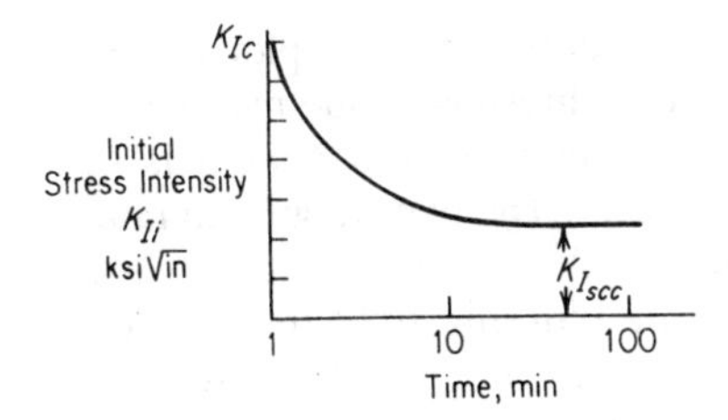

Figure 14-19 Fracture mechanics plot for stress-corrosion test data.

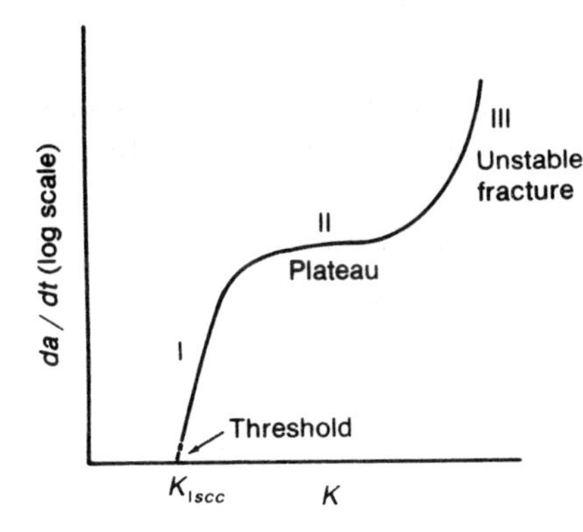

Figure 14-20 Three stages of environmentally assisted cracking under constant load in an aggressive environment (schematic).

within the pit is quite different from the bulk environment and is more conducive to the initiation of cracking. The crack grows slowly under the SCC conditions until it reaches a critical size and brittle fracture occurs.

The dependence of the time to failure on stress is very similar to the curve for a hydrogen-charged steel (Fig. 14-18). The threshold stress below which cracking does not occur invariably is less than the yield stress of the alloy. It is useful to employ a fracture mechanics[1] approach in studying SCC. A major advantage of using precracked specimens is that the incubation period for the initiation of a crack is eliminated and that the characterization of a material's response to SCC can be placed on a more quantitative basis. The basic stress-time curve can be expressed in terms of the initial level of K_I (Fig. 14-19). The threshold value of stress intensity is designated $K_{I_{SCC}}$. Tests should be conducted for at least 1000 hr to establish $K_{I_{SCC}}$, and in some alloy systems like high-strength aluminum alloys there does not appear to be a threshold stress. When the stress is $K_{I_{SCC}} < K < K_{IC}$ subcritical crack growth occurs with time. Thus, the value of K increases continuously until it reaches K_{IC}. Regardless of the initial value of K, failure occurs when $K = K_{IC}$.

Measurement of the rate of crack growth in SCC as a function of the stress intensity factor at the crack tip shows three regions of crack development (Fig.

[1] B. F. Brown, *Metal. Rev.*, vol. 13, p. 171–183, 1968; B. F. Brown (ed.), "Stress-Corrosion Cracking in High Strength Steels and in Titanium and Aluminum Alloys," U.S. Govt. Printing Office, Washington, 1972; D. O. Sprowls, Tests for Stess-Corrosion Cracking, "Metals Handbook," 9th ed., vol. 8, pp. 495–536, American Society for Metals, Metals Park, Ohio, 1985.

14-20). In region I the crack growth is a strong function of K. This plot serves to aid in establishing $K_{I_{SCC}}$ but the magnitude of this threshold value may be very low. In region II crack growth is essentially independent of K, but it is still strongly effected by temperature and the environment. The plateau growth rate is the maximum that it is possible to sustain in the alloy environment system by environmental crack growth alone. These rates are typically 10^{-9} to 10^{-6} m/s, which is too fast to provide a reasonable design life. Region III represents a situation here da/dt varies strongly with K. As K approaches K_{IC} the crack growth rate becomes unstable.

Example In an alloy steel used in water at 110°C the SCC growth rate for the plateau (region II) is 10^{-8} m/s. If the allowable crack growth is 10 mm, what is the estimated life?

$$10 \text{ mm} = \frac{0.01 \text{ m}}{10^{-8} \text{ m/s}} = \frac{10^6 \text{ s}}{3.6 \times 10^3 \text{ s/hr}} = \frac{2.8 \times 10^2 \text{ hr}}{24 \text{ hr/day}} = 10 \text{ days}$$

For this steel $K_{I_{SCC}} = 10$ MN m$^{-3/2}$. For a long thin surface crack

$$K_{I_{SCC}}^2 = \frac{1.21 a \pi \sigma^2}{1.0}$$

and
$$a\sigma^2 = \frac{100}{3.801} = 26.31 \text{ MN}^2 \text{ m}^{-3} = 26.31 \text{ Pa}^2 \text{ m}$$

The crack length corresponding to various stress levels is given below.

Stress, MPa (ksi)	Crack length, mm
500 (72.5)	0.10
300 (43.5)	0.29
100 (14.5)	2.63

Since the crack length that is detectable by nondestructive evaluation methods is no smaller than 1 mm we see that NDE is not good enough to avoid the presence of stress-corrosion cracks at useful design stresses.

Stress-corrosion cracks result in a nonductile type of failure. Cracking may be intergranular, transgranular, or mixed mode depending on the alloy, its microstructure, and the environment. On a microscopic scale cracks frequently undergo extensive branching. Because of the specific nature of stress corrosion it is difficult to identify a general mechanism. Surely successive formation and rupture of a passive layer at the crack tip is an important mechanism. It is widely believed that electrochemical dissolution plays a major role. The crack tip can act as a local anode so that metal ions go into solution at the anode and electrons flow from the anode to a cathodic region near the surface. In some alloys preferential dissolution occurs in the grain-boundary region. There is also the possibility of the adsorption of damaging ions which weakens the atomic bonding at the crack tip. Hydrogen is generated at local cathodes as a result of corrosion reactions. It is then able to enter the metal, diffuse to the crack tip and cause crack propagation. There is growing evidence of the link between SCC and hydrogen embrittlement. Finally, it must be recognized that the corrosion products that are formed have a large volume and can generate high stresses due to a wedging action in a narrow crack.

Liquid-Metal Embrittlement

Liquid-metal embrittlement results when a solid metal surface is wetted by a lower melting liquid metal. For example, highly ductile 70-30 brass will fail in an intergranular brittle fracture when it is wetted by mercury.

Like SCC, liquid-metal embrittlement is unique to certain metal-liquid combinations.[1] Aluminum is embrittled by liquid gallium, mercury, sodium, tin, and zinc. Carbon steels are embrittled by liquid cadmium, indium, lithium, mercury, and zinc. Molten bismuth embrittles copper alloys but has no effect on aluminum or carbon steels.

Liquid-metal embrittlement occurs by direct interaction of the liquid-metal atoms and the highly strained atoms at the crack tip.[2] Therefore, wetting is the first requirement to cause liquid-metal embrittlement. Adsorption of atoms from the liquid metal greatly reduces the surface energy, γ_s in Eq. (7-8), and through this the fracture stress is reduced.

Neutron Embrittlement

While not an environmentally assisted embrittlement in the same sense as hydrogen embrittlement, stress-corrosion cracking and liquid-metal embrittlement, this type of embrittlement occurs as a result of a high neutron flux in a nuclear reactor. It is chiefly concerned with structural steel, where neutron irradiation can increase the ductile-to-brittle transition temperature as much as 200°C. In addition, neutron irradiation reduces the upper shelf energy value.[3] The degree of embrittlement increases with neutron fluence (neutron flux × time) and decreases with temperature of exposure. Steels with low initial transition temperature, and fine grain size offer greater resistance to neutron embrittlement. Steels with a tempered martensite structure are more resistant than ferritic steels. Vacuum degassing during steelmaking and control of copper and phosphorus levels help to reduce susceptibility to neutron embrittlement. While the mecha-

[1] W. Rostoker, J. M. McCaughey, and H. Markus, "Embrittlement by Liquid Metals," Reinhold, New York, 1960.

[2] A. R. C. Westwood, C. M. Preece, and M. H. Kamdar, in "Fracture," vol. III, chap. 10, H. Liebowitz (ed.), Academic Press, New York, 1971.

[3] S. H. Bush, *J. Test. Eval.*, vol. 2, p. 435, 1974.

nism of neutron embrittlement is not well understood, it is thought to be related to the interaction of dislocations with aggregates of defects like solute atom-vacancy clusters produced by neutron bombardment. These defects can be annealed out at high temperature, thus reducing most of the embrittlement. However, since neutron embrittlement is chiefly a problem in large steel nuclear reactor vessels, annealing may not be a practical solution.

Another form of irradiation damage occurs in stainless steel and zirconium used in nuclear fuel rods. Hydrogen and helium are produced by the nuclear reactions and these segregate to vacancy clusters and internal voids. The result is undesirable swelling of the cladding for the fuel elements.

14-10 FLOW AND FRACTURE UNDER VERY RAPID RATES OF LOADING

The flow and fracture of metals generally is found to be very sensitive to the rate of deformation. The spectrum of available strain rates is shown in Fig. 14-21. Also shown is a characteristic time, corresponding to the time required to produce 1 percent strain at the corresponding strain rate.[1] Creep and quasi-static testing is done with dead-weight, hydraulic, or screw testing machines. Inertia effects may be neglected and tests are carried out under essentially isothermal conditions. In the intermediate strain-rate range fast pneumatic or mechanical testing machines are used, and the mechanical resonance of the specimen and the machine must be considered. In the range 10^2 to 10^4 s^{-1} impact bars are usually used. Elastic and elastic-plastic wave propagation must be considered in detail. At the highest strain rates testing usually involves explosively driven plate impact where measurements are made of the propagation and reflection of shock waves. From a strain rate of about 10^0 s^{-1} on up, the deformation becomes essentially adiabatic because the internal heat generated during plastic deformation does not have time to dissipate. The influence of the resulting temperature rise on the flow stress must be considered when comparing properties as a function of strain rate.

Material behavior under high strain-rate conditions is of practical importance in such problems as the response of structures to blast and impulsive loads, contact stresses under high-speed bearings, high-speed machining, explosive forming,[2] and ballistics.

A rapidly applied load is not instantaneously transmitted to all parts of the solid body. At a brief instant after the load has been applied the remote portions of the body remain undisturbed. The deformation and stress produced by the load move through the body in the form of a wave.[3] Since, except for the highest

[1] U. S. Lindholm, High Strain Rate Tests, chap. 4 in R. F. Bunshah (ed.), "Techniques of Metals Research," vol. V, pt. I, Interscience Publishers, Inc., New York, 1971.

[2] J. S. Rinehart and J. Pearson, "Explosive Working of Metals," The Macmillan Company, New York, 1963.

[3] H. Kolsky, "Stress Waves in Solids," Oxford University Press, New York, 1953.

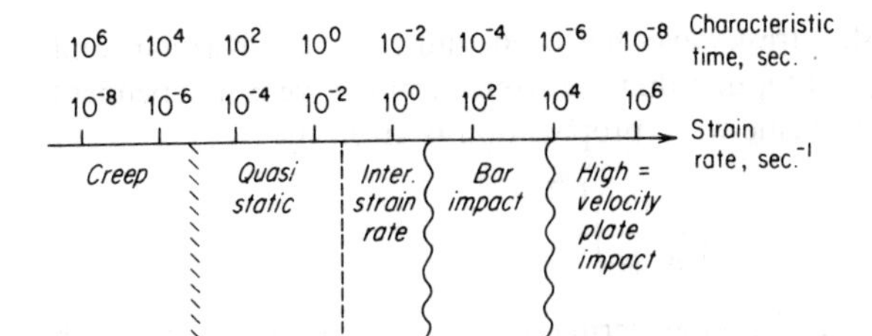

Figure 14-21 Spectrum of strain rates for mechanical testing. *(After Lindholm.)*

strain rates, most dynamic testing is done with circular cylindrical specimens of small lateral dimensions, it is permissible to consider only the case of one-dimensional wave propagation. An element of the bar is subjected to an axial stress σ_x giving rise to a particle velocity of v_x in the axial direction. The governing first-order partial differential equations are

Equation of motion:
$$\frac{\partial \sigma_x}{\partial x} = \rho \frac{\partial v_x}{\partial t} \tag{14-4}$$

Equation of continuity:
$$\frac{\partial v_x}{\partial x} = \frac{\partial \varepsilon_x}{\partial t} \tag{14-5}$$

where $v_x = \dfrac{\partial u_x}{\partial t}$

$\varepsilon_x = \dfrac{\partial u_x}{\partial x}$

u_x = the axial particle displacement
ρ = the mass density

For elastic waves the constitutive equation is

$$\sigma_x = E\varepsilon_x \tag{14-6}$$

If Eqs. (14-4) to (14-6) are expressed in terms of u_x, they can be combined to form the well-known wave equation

$$\frac{\partial^2 u_x}{\partial x^2} = \frac{1}{c_0^2}\frac{\partial^2 u_x}{\partial t^2} \tag{14-7}$$

where $c_0 = (E/\rho)^{1/2}$ is the *elastic wave velocity*. The simplest solution of Eq. (14-7) yields

$$\frac{\partial u_x}{\partial t} = c_0\left(\frac{\partial u_x}{\partial x}\right)$$

or

$$v_x = c_0 \varepsilon_x \tag{14-8}$$

Substituting Eq. (14-6) into Eq. (14-8) gives

$$\sigma_x = \rho c_0 v_x \tag{14-9}$$

The propagation of plastic stress waves has been studied by von Karman[1] and Wood.[2] For the simplifying assumption that the stress-strain curve is independent of strain rate, the velocity of plastic wave propagation is given by

$$c_p = \left(\frac{1}{\rho}\frac{ds}{de}\right)^{1/2} \tag{14-10}$$

where ds/de is the slope of the engineering stress-strain curve. Thus each increment of plastic strain propagates at a different velocity since the slope of the curve (ds/de) depends on the magnitude of plastic strain. Therefore, the shape of the plastic wave changes as it propagates along the bar. The particle velocity is related to the plastic wave velocity by

$$v_x = \int_0^e c_p\,de = \frac{1}{\rho}\int_0^s \frac{ds}{c_p} = \int_0^e \left(\frac{ds/de}{\rho}\right)^{1/2} de \tag{14-11}$$

The first of these equations is used to determine the strain or the stress in a dynamically loaded material. The last version of Eq. (14-11) leads to the important conclusion that at the tensile strength of the material $ds/de = 0$ so that $v_x = 0$. Thus, there is a *critical impact velocity* v_c. If a bar is struck with an impact velocity greater than v_c, it ruptures at the impacted end at the instant of impact with a weak plastic strain propagating along the bar. The value of critical impact velocity for most metals lies in the range 100 to 500 ft/s.

We have already seen that if ds/de decreases with strain (the normal case) the wave front becomes diffuse. However if ds/de increases with strain, large strains travel faster than small strains and the wave front sharpens into a plastic *shock wave*. This can occur in materials which shows pronounced yield point[3] or in metal cutting with a localized shear plane.

The highest strain rates are produced by impulsive loading with explosive charges. Particle velocities are very high and stresses can easily exceed 100 kbar (1.47×10^6 psi). Much work has been done under these conditions in determining thermodynamic equations of state for metals. Because of high shock pressures, the shear strength of the metal is neglected and the problem is treated by compressible hydrodynamic theory.

The impact of explosively driven plates has been developed into a fairly well-understood test. It is useful for studying the hardening and transformations in metals[4] subjected to shock pressures in excess of 500 kbar. Shock-wave propagation through a thin plate is essentially one dimensional in the direction of the shock front. For small plastic strains this produces a state of near triaxial

[1] T. von Karman and P. E. Duwez, *J. Appl. Phys.*, vol. 21, pp. 987–994, 1950.

[2] D. S. Wood, *J. Appl. Mech.*, vol. 19, p. 521, 1952.

[3] A. H. Cottrell, *Proc. Conf. Properties of Materials at High Rates of Strain*, Institute of Mechanical Engineers, London, 1957.

[4] G. E. Dieter Hardening Effect of Shock Waves, in "Strengthening Mechanisms in Solids," American Society for Metals, Metals Park, Ohio, 1962; M. A. Meyers and L. E. Murr (eds.), "Shock Waves and High-Strain Rate Phenomena in Metals," Plenum Press, New York, 1981.

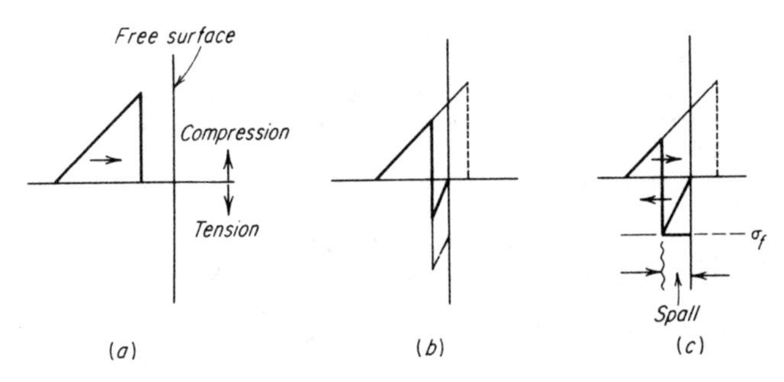

Figure 14-22 Reflection of a shock wave from a free surface and development of a spall (fracture).

loading.[1] Thus shock loading provides an environment for subjecting materials to high stresses without corresponding large plastic deformations.

There are marked differences between fracture under impulsive loads and under static loads. With impulsive loads there is not time for the stress to be distributed throughout the entire body, so that fracture can occur in one part of the solid independently of what happens in another part. The velocity of propagation of stress waves in solids is in the range 3,000 to 20,000 ft/s, while the velocity of crack propagation is about 6,000 ft/s. Therefore, with impulsive loads it often is found that cracks have formed but did not have time to propagate before the stress state changed.

Reflections of stress waves occur at the free surfaces and fixed ends, at changes in cross section, and at discontinuities within the solid. Figure 14-22*a* shows a compressive shock wave approaching the free surface of a plate. The wave moves from left to right. At the free surface the wave is reflected back as a tensile wave from right to left. The heavy lines indicate the net stress after summing the incident (compressive) and reflected (tensile) waves. The dashed line corresponds to the portion of the pulse that has already been reflected. There is a rapid buildup of tensile stress as the wave is reflected back from the free surface. When the tensile stress reaches a high enough value, fracture occurs and the plate is said to have *spalled*. If the stress distribution within the pulse is known from velocity measurements, then the fracture stress can be determined[2] by a measurement of the position of the fracture relative to the free surface.

BIBLIOGRAPHY

"A Review of Engineering Approaches to Design Against Fracture," American Society of Mechanical Engineers, New York, 1964.

[1] J. L. O'Brien and R. S. Davis, "Fracture in Engineering Materials," chap. 12, American Society for Metals, Metals Park, Ohio, 1964.

[2] J. S. Rinehart, *J. Appl. Phys.*, vol. 22, p. 555, 1951.

High Strain Rate Testing "Metals Handbook," 9th ed., vol. 8, pp. 187–297, American Society for Metals, Metals Park, Ohio, 1985.
Knott, J. F.: "Fundamentals of Fracture Mechanics," John Wiley & Sons, Inc., New York, 1973.
"Rapid Inexpensive Tests for Determining Fracture Toughness," National Materials Advisory Board, Nat. Academy of Sciences, Washington, D.C., 1976.
Rolfe, S. T. and J. M. Barsom, "Fracture and Fatigue Control in Structures," Prentice-Hall, Inc., Englewood Cliffs, N.J., 1977.
Späth, W.: "Impact Testing of Materials," Thames-Hudson, London, 1962.
Tetelman, A. S. and A. J. McEvily, Jr.: "Fracture of Structural Materials," John Wiley & Sons, Inc., New York, 1967.
Tipper, C. F.: "The Brittle Fracture Story," Cambridge University Press, London, 1962.

PART FOUR

PLASTIC FORMING OF METALS

CHAPTER
FIFTEEN
FUNDAMENTALS OF METALWORKING

15-1 CLASSIFICATION OF FORMING PROCESSES

The importance of metals in modern technology is due, in large part, to the ease with which they may be formed into useful shapes such as tubes, rods, and sheets. Useful shapes may be generated in two basic ways:

1. By *plastic deformation processes* in which the volume and mass of metal are conserved and the metal is displaced from one location to another.
2. By *metal removal* or *machining processes* in which material is removed in order to give it the required shape.

Major emphasis in Part Four is given to plastic deformation processes. However, the fundamentals of metal removal are considered in the last chapter of this book.

Equal in importance to the creation of useful shapes by plastic forming processes is the control of mechanical properties by the metalworking process. For example, blowholes and porosity in a cast ingot may be eliminated by hot-forging or hot-rolling, to the improvement of ductility and fracture toughness. In many products the mechanical properties depend on the control of strain hardening during processing, while in other instances precise control of deformation, temperature, and strain rate during processing is required to develop the optimum structure and properties. The *thermomechanical processing* in ausforming (Sec. 6-12) is a good example.

Hundreds of processes have been developed for specific metalworking applications. However, these processes may be classified into only a few categories on the basis of the type of forces applied to the workpiece as it is formed into shape.

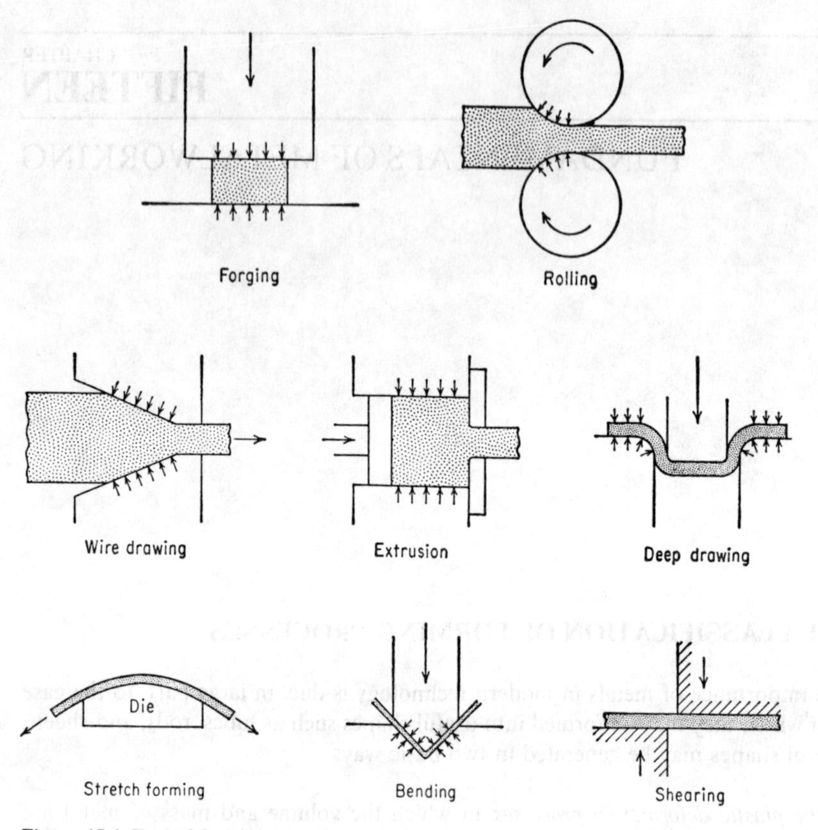

Figure 15-1 Typical forming operations.

These categories are:

1. Direct-compression-type processes
2. Indirect-compression processes
3. Tension type processes
4. Bending processes
5. Shearing processes

In direct-compression processes the force is applied to the surface of the workpiece, and the metal flows at right angles to the direction of the compression. The chief examples of this type of process are forging and rolling (Fig. 15-1). Indirect-compression processes include wiredrawing and tube drawing, extrusion, and the deep drawing of a cup. The primary applied forces are frequently tensile, but the indirect compressive forces developed by the reaction of the workpiece

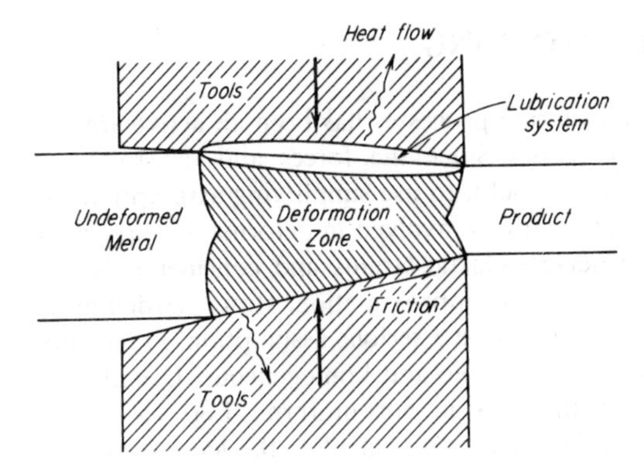

Figure 15-2 Deformation processing system.

with the die reach high values. Therefore, the metal flows under the action of a combined stress state which includes high compressive forces in at least one of the principal directions. The best example of a tension-type forming process is stretch forming, where a metal sheet is wrapped to the contour of a die under the application of tensile forces. Bending involves the application of bending moments to the sheet, while shearing involves the application of shearing forces of sufficient magnitude to rupture the metal in the plane of shear. Figure 15-1 illustrates these processes in a very simplified way.

Plastic working processes which are designed to reduce an ingot or billet to a standard mill product of simple shape, such as sheet, plate, and bar, are called *primary mechanical working processes*. Forming methods which produce a part to a final finished shape are called *secondary mechanical working processes*. Most sheet-metal forming operations, wire drawing, and tube drawing are secondary processes. The terminology in this area is not very precise. Frequently the first category is referred to as *processing operations*, and the second is called *fabrication*.

The deformation processing situation is best viewed as a total system (Fig. 15-2). The deformation zone is concerned with the distribution of stress, strain, and particle velocities, and with the overall pressure required to perform the operation. Obviously, the applied forces must develop yielding in the material but the stresses must not locally create fracture. Such metallurgical phenomena as strain hardening, recrystallization and fracture are important, but often under specialized conditions of high strain rates and/or high temperature. The flow stress of the material will be a strong function of strain, strain rate, and temperature, often at values of these variables that are difficult to simulate in the laboratory. The workpiece will be in contact with nondeforming (elastic) tools or dies. The friction along this interface and the heat transfer from the workpiece to the die are important considerations, as are such practical concerns as tool wear and surface finish of the product.

15-2 MECHANICS OF METALWORKING

The analysis of the stresses in metalworking processes has been an active area of applied plasticity for the past 50 years.[1] Since the forces and deformations generally are quite complex, it is usually useful to use simplifying assumptions to obtain a tractable solution. Because the strains involved in deformation processes are large, it is usually possible to neglect elastic strains and consider only the plastic strains (plastic-rigid solid). As a first approximation, strain hardening is often neglected. In hot-working this may not be a serious assumption. Practically all analyses consider the material to be isotropic and homogeneous. Usually the deformation of the metal between the dies is not uniform, Fig. 15-11. To accurately predict the nonuniform deformation and calculate the local stresses is the chief analytical problem. In recent years great progress has been made by applying finite element methods to this large scale plasticity problem.

The principal use of analytical studies of metalworking processes is for determining the forces required to produce a given deformation for a certain geometry prescribed by the process. Such calculations are useful for selecting or designing the equipment to do a particular job. In this area existing theory is generally adequate for the task. A mechanics analysis of a process may also be used to develop information on the frictional conditions in the process. An important problem area is predicting the limiting deformation at which fracture will occur. In general, theory concerning formability limits and fracture criteria are not well developed.

Before proceeding further it would be desirable to review Chap. 3, Elements of the Theory of Plasticity, especially Secs. 3-1 to 3-4. The most basic relationship for plastic deformation is the *constant-volume relationship*. Because of the large deformations, it is important to express strain in terms of true strain or natural strain. From Eq. (3-5) we recall that the constant-volume relationship is expressed by

$$\varepsilon_1 + \varepsilon_2 + \varepsilon_3 = 0 \tag{15-1}$$

Compressive stresses and strains often predominate in metalworking processes. If a block of initial height h_0 is compressed to h_1, the axial compressive strain (using the usual sign convention) will be
True strain:

$$\varepsilon = \int_{h_0}^{h_1} \frac{dh}{h} = \ln \frac{h_1}{h_0} = -\ln \frac{h_0}{h_1} \qquad h_0 > h_1 \tag{15-2}$$

Conventional strain:

$$e = \frac{h_1 - h_0}{h_0} = \frac{h_1}{h_0} - 1 \tag{15-3}$$

The calculated strains will be negative, indicating compressive strains. However,

[1] W. Johnson, *Appl. Mech. Rev.*, vol. 24, pp. 977–989, 1971

in metalworking problems it is common practice to reverse the convention so that compressive stresses and strains are defined as *positive*. When using this convention we shall use the subscript c. Thus, the strains described above would be calculated from

$$\varepsilon_c = \ln \frac{h_0}{h_1} \qquad e_c = \frac{h_0 - h_1}{h_0} = 1 - \frac{h_1}{h_0}$$

Metalworking deformations frequently are expressed as reduction in cross-sectional area. The fractional reduction is given by

$$r = \frac{A_0 - A_1}{A_0} \tag{15-4}$$

From the constancy-of-volume relation,

$$A_1 L_1 = A_0 L_0$$

$$r = 1 - \frac{A_1}{A_0} \quad \text{or} \quad \frac{A_1}{A_0} = 1 - r$$

$$\varepsilon = \ln \frac{L_1}{L_0} = \ln \frac{A_0}{A_1} = \ln \frac{1}{1 - r} \tag{15-5}$$

Example Determine the engineering strain, true strain, and reduction for (a) a bar which is doubled in length; (b) a bar which is halved in length.
(a) For a bar which is doubled in length, $L_2 = 2L_1$

$$e = \frac{L_2 - L_1}{L_1} = \frac{2L_1 - L_1}{L_1} = 1.0$$

$$\varepsilon = \ln \frac{L_2}{L_1} = \ln \frac{2L_1}{L_1} = 0.693$$

$$r = \frac{A_1 - A_2}{A_1} = 1 - \frac{A_2}{A_1} = 1 - \frac{L_1}{L_2} = 1 - \frac{L_1}{2L_1} = 0.5$$

(b) For a bar which is decreased in length so $L_2 = L_1/2$

$$e = \frac{L_1/2 - L_1}{L_1} = -0.5 \quad \text{or} \quad e_c = \frac{L_1 - L_1/2}{L_1} = 0.5$$

$$\varepsilon = \ln \frac{L_1/2}{L_1} = \ln \frac{1}{2} = -\ln 2 = -0.693$$

$$r = 1 - \frac{L_1}{L_1/2} = -1.0$$

The requirement of a theory describing the mechanics of a metalworking process is the ability to make an accurate prediction of the stresses, strains, and velocities at every point in the deformed region of the workpiece. The various approaches differ in complexity and in the degree to which they meet this requirement. In general, such a theory consists of three sets of equations:

1. The static equilibrium of force equations
2. The Levy-Mises equations (see Sec. 3-11) expressing a relationship between stress and strain rate
3. The yield criterion (see Sec. 3-4)

In the general case there are nine independent equations containing nine unknowns; six stress components and three velocity (strain-rate) components. While an analytical solution is possible if a sufficient number of boundary conditions are specified, the mathematical difficulties in a general solution are formidable. Thus, most analyses of actual metalworking processes are limited to two-dimensional or symmetric three-dimensional problems.

The methods of analysis, in increasing order of complexity and ability to predict fine detail, are:

1. *The slab method*—assumes homogeneous deformation
2. *Uniform-deformation energy method*—calculates average forming stress from the work of plastic deformation
3. *Slip-line field theory*—permits point-by-point calculation of stress for plane-strain conditions only
4. *Upper- and lower-bound solutions*—based on the theory of limit analysis, uses reasonable stress and velocity fields to calculate the bounds within which the actual forming load must lie
5. *Finite element methods*—a technique called the matrix method allows large increments of deformation for rigid plastic materials, with considerable reduction in computational time

The general framework of each of these approaches will be illustrated below using the problem of the drawing of a strip through wedge dies (Fig. 15-3). While strip drawing is not a common production process it is probably the most often studied problem in the mechanics of metalworking theory. At this stage in the presentation we will ignore the vital practical consideration of friction between the die and workpiece interface. Methods of considering friction will be discussed in Sec. 15-7.

The *slab method* assumes that the metal deforms uniformly in the deformation zone. A square grid placed in the deformation zone would be distorted uniformly into rectangular elements. This was the earliest approach developed in the 1920s by von Kármán, Hencky, Siebel, and later by Sachs.[1] The approach essentially is that used in strength of materials.

[1] O. Hoffman and G. Sachs, "Introduction to the Theory of Plasticity for Engineers," McGraw-Hill Book Company, New York, 1953.

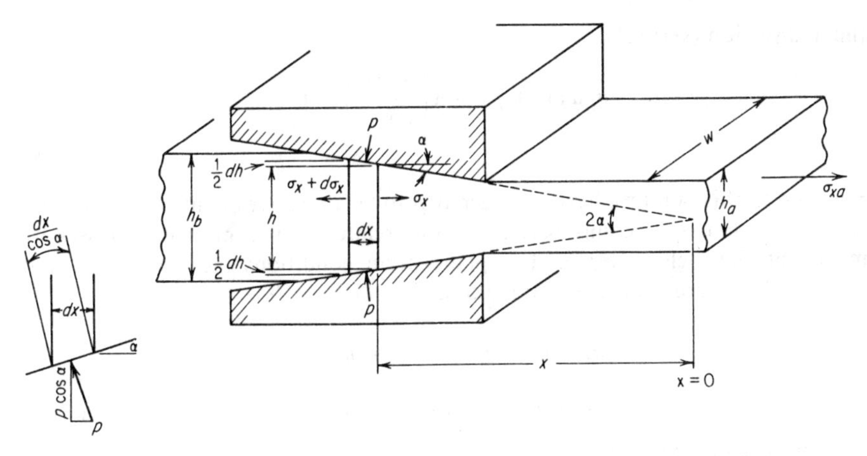

Figure 15-3 Stresses acting on an element during strip drawing of a wide sheet.

In Fig. 15-3, a wide strip is being drawn through a frictionless die with a total included angle of 2α. Because the original thickness of the strip is assumed to be much less than the width, the conditions approximate plane-strain deformation where there is no strain in the width direction. An element under the die of length dx is subjected to a longitudinal stress σ_x and a vertical stress σ_y due to the die pressure p. A characteristic of the slab method is that these stresses are assumed to be principal stresses. The equilibrium of forces in the x direction is made up of two components

1. Due to the change in longitudinal stress with x increasing positively to the left

$$(\sigma_x + d\sigma_x)(h + dh)w - \sigma_x hw$$

2. Due to the die pressure at the two interfaces

$$2p\left(w\frac{dx}{\cos\alpha}\right)\sin\alpha$$

Taking the equilibrium of forces in the x direction and neglecting the product $d\sigma_x dh$, we have

$$\sigma_x\,dh + h\,d\sigma_x + 2p\tan\alpha\,dx = 0 \tag{15-6}$$

From the geometry, h may be expressed in terms of x as

$$h = 2x\tan\alpha \quad \text{and} \quad dh = 2\,dx\tan\alpha$$

so that Eq. (15-6) simplifies to

$$\sigma_x\,dh + h\,d\sigma_x + p\,dh = 0 \tag{15-7}$$

To find a relationship between σ_x and p, we turn to the equilibrium of forces in the y direction (vertical).

$$\sigma_y \, dx \, w + p \cos\alpha \left(\frac{w \, dx}{\cos\alpha} \right) = 0$$

$$\sigma_y = -p \tag{15-8}$$

Since up to this point we have considered p to be a positive quantity, we see from Eq. (15-8) that σ_y is compressive. We now relate the two principal stresses by means of the yield criterion. For plane-strain conditions the von Mises' and Tresca criteria are equivalent, (Eq. 3-55).

$$\sigma_1 - \sigma_3 = 2k = \frac{2\sigma_0}{\sqrt{3}} = \sigma_0'$$

$$\sigma_x + p = \sigma_0' \quad \text{or} \quad p = \sigma_0' - \sigma_x \tag{15-9}$$

Substituting into Eq. (15-7), we get

$$\sigma_0' \, dh + h \, d\sigma_x = 0$$

$$\frac{d\sigma_x}{\sigma_0'} = -\frac{dh}{h}$$

On integrating

$$\frac{\sigma_x}{\sigma_0'} = -\ln h + \text{constant}$$

The boundary condition is that at the entry to the die, $x = b$; $h = h_b$, the axial stress equals zero. Therefore, constant $= \ln h_b$.

$$\sigma_x = \sigma_0' \int_{h_b}^{h} -\frac{dh}{h} = \sigma_0' \ln \frac{h_b}{h} = \frac{2}{\sqrt{3}} \sigma_0 \ln \frac{h_b}{h} \tag{15-10}$$

The axial stress at the die exit needed to cause deformation under this ideal frictionless condition is

$$\sigma_{xa} = \frac{2}{\sqrt{3}} \sigma_0 \ln \frac{h_b}{h_a} = \frac{2}{\sqrt{3}} \sigma_0 \ln \frac{1}{1-r} \tag{15-11}$$

The above result can be obtained much more simply by means of the *uniform-deformation energy method*. Consider the simplest case of a cylinder loaded in tension. The increment of work for an increment in length δL is

$$\delta W = \delta P \, \delta L = \bar{\sigma} A \, \delta L$$

where $\bar{\sigma}$ is the mean yield stress. The deformation energy per unit volume is

$$\frac{\delta W}{V} = \frac{\bar{\sigma} A}{A} \frac{\delta L}{L} = \bar{\sigma} \frac{\delta L}{L}$$

The plastic work of deformation per unit volume is

$$U_p = \bar{\sigma} \int d\varepsilon = \bar{\sigma} \int \frac{\delta L}{L} = \bar{\sigma} \ln \frac{L_1}{L_0} \tag{15-12}$$

If we apply this to drawing a cylindrical wire from area A_b down to area A_a,

$$W = P_a L_a = U_p V = A_a L_a \bar{\sigma} \ln \frac{L_b}{L_a}$$

$$P_a = A_a \bar{\sigma} \ln \frac{L_b}{L_a} = A_a \bar{\sigma} \ln \frac{A_a}{A_b}$$

The draw stress

$$\sigma_{xa} = \frac{P_a}{A_a} = \bar{\sigma} \ln \frac{A_a}{A_b}$$

or

$$\sigma_{xa} = \bar{\sigma}_0 \ln \frac{1}{1-r} \tag{15-13}$$

Equation (15-13) not only neglects friction, but it neglects the influence of transverse stresses (plastic constraint). Notice that it differs from Eq. (15-11) by the magnitude of the average yield-stress term.

Example The strain hardening of an annealed metal is expressed by $\bar{\sigma} = 200{,}000\bar{\varepsilon}^{0.5}$, where stress is in psi. A 25-mm-diam bar is drawn down to 20 mm and 15 mm in two steps using tapered cylindrical dies. Determine the plastic work per unit volume for each reduction.

$$dU = \bar{\sigma}\, d\bar{\varepsilon}$$

Due to symmetry in the rod drawing process $\varepsilon_2 = \varepsilon_3$ and from $\varepsilon_1 + \varepsilon_2 + \varepsilon_3 = 0$, $\varepsilon_2 = \varepsilon_3 = -\varepsilon_1/2$

$$\bar{\varepsilon} = \frac{\sqrt{2}}{3}\left[\left(\varepsilon_1 + \frac{\varepsilon_1}{2}\right)^2 + \left(-\frac{\varepsilon_1}{2} + \frac{\varepsilon_1}{2}\right)^2 + \left(-\frac{\varepsilon_1}{2} - \varepsilon_1\right)^2\right]^{1/2}$$

$\bar{\varepsilon} = \varepsilon_1$, where ε_1 is the strain in the longitudinal direction.

$$\bar{\varepsilon} = \varepsilon_1 = \ln \frac{L_1}{L_0} = \ln \frac{A_0}{A_1} = \ln \frac{(\pi/4) D_0^2}{(\pi/4) D_1^2} = \ln\left(\frac{D_0}{D_1}\right)^2 = 2 \ln D_0/D_1$$

$$\bar{\varepsilon} = 2 \ln \frac{25}{20} = 0.446 \quad \text{for the first reduction}$$

$$U_1 = \int dU = \int_0^{\bar{\varepsilon}_1} \bar{\sigma}\, d\bar{\varepsilon} = \int_0^{0.446} 200{,}000(\bar{\varepsilon})^{0.5}\, d\bar{\varepsilon}$$

$$U_1 = \frac{200{,}000(\bar{\varepsilon})^{0.5+1}}{0.5+1}\bigg|_0^{0.446} = \frac{200{,}000}{1.5}(0.446)^{1.5} = 39{,}714 \frac{\text{in-lb}}{\text{in}^3}$$

For the second reduction $\bar{\varepsilon} = 2 \ln 20/15 = 0.575$

$$U_2 = \int_{0.446}^{1.021} \bar{\sigma}\, d\bar{\varepsilon} = \frac{200{,}000}{1.5} \bar{\varepsilon}^{1.5}\bigg|_{0.446}^{1.021} = 97{,}842 \frac{\text{in-lb}}{\text{in}^3}$$

The total work per unit volume in an actual deformation process is appreciably in excess of that given by Eq. (15-12). In addition to the ideal work of deformation U_p, we have the work to overcome the friction at the metal-tool interface U_f and the redundant work U_r. The redundant work is the work involved in internal shearing processes due to nonuniform deformation that does not contribute to the change in shape of the body. The amount of U_r will depend on the geometry of the process and the friction. Thus, the total work is given by

$$U_T = U_p + U_f + U_r \tag{15-14}$$

The efficiency of a process can be expressed as

$$\eta = \frac{U_p}{U_T} \tag{15-15}$$

Typical values of efficiency are 30 to 60 percent for extrusion and 75 to 95 percent for rolling.

The first approach to the analysis of metalworking processes that did not assume homogeneous uniform deformation was *slip-line field theory*. The mathematics of slip-line fields was introduced in Sec. 3-12 where the technique was used to analyze the quite simple problem of the pressure under the indentation of a narrow frictionless punch. The presentation given there was incomplete since it did not discuss the Geiringer velocity equations[1] and the construction of a *hodograph* describing the admissible velocity fields[2] corresponding to the slip-line field. These topics are aptly discussed in a monograph[3] which presents complete bibliographic information on nearly all available slip-line fields.

The reason for using slip-line fields is that they account for inhomogeneous deformation in the calculation of the overall forming loads. Figure 15-4 shows the slip-line field proposed by Hill and Tupper[4] for frictionless strip drawing through wedge dies of included angle 2α. This solution is valid for reductions up to $r = 2\sin\alpha/(1 + 2\sin\alpha)$. It will be recalled that the shear stress k is everywhere constant within the slip-line field but that the hydrostatic pressure normal to the slip line changes from point to point as the line curves. The Hencky equations, Eq. (3-56), are used to find the pressure p at any point along ADF. Integrating over the pressure distribution gives the average draw stress for the strip

$$\sigma_{xa} = \frac{2}{\sqrt{3}}\sigma_0\left[\frac{2(1+\alpha)\sin\alpha}{1 + 2\sin\alpha}\right] \tag{15-16}$$

For many metalworking processes there are no slip-line fields to allow prediction of stresses. Moreover, slip-line fields are valid only for plane-strain conditions. *Upper- and lower-bound techniques*[1] have been developed that have general applicability. An upper-bound solution provides an overestimation of the required deformation force while a lower-bound solution provides an underesti-

[1] J. Halling, *Engineer*, vol. 207, p. 250, 1959.

[2] A. P. Green, *J. Mech. Phys. Solids*, vol. 2, p. 73, 1954.

[3] W. Johnson, R. Sowerby, and R. D. Venter, "Plane-Strain Slip-Line Fields for Metal Deformation Processes: A Source Book and Bibliography," Pergamon Press, New York, 1982.

[4] R. Hill and S. J. Tupper, *J. Iron Steel Inst. London*, vol. 159, p. 353, 1948.

[1] W. Johnson and H. Kudo, "The Mechanics of Metal Extrusion," Manchester University Press, 1962; B. Avitzur, "Metal Forming: Processes and Analysis," McGraw-Hill Book Company, New York, 1968.

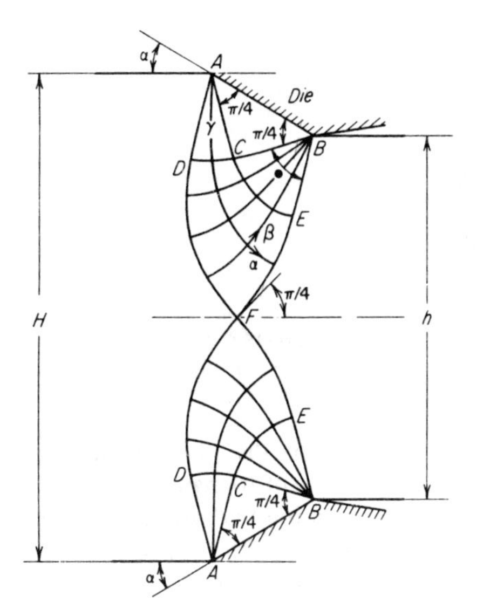

Figure 15-4 Slip-line field for strip drawing through a frictionless wedge-shaped die at a small reduction. (*After Hill and Tupper.*)

mate of the force. The degree of agreement between the upper- and lower-bound predictions is an indication of how close the prediction is to the exact value. A lower-bound analysis requires that a statically admissible stress field be found throughout the entire material without making an attempt to ensure that the velocity conditions are satisfied at every point in the material. The term *statically admissible stress field* means that the assumed stress field satisfies the equilibrium equations and the stress boundary conditions and that it does not violate the yield criterion.

From a practical viewpoint the upper-bound technique is more important than the lower bound since calculations based on upper bound will always result in an overestimate of the load that the press or machine will be called upon to deliver. The upper-bound technique[1] has been used widely by civil engineers for the plastic design of steel structures (theory of limit design). In the upper-bound analysis a *kinematically admissible velocity field* is constructed, i.e., the hodograph must be satisfied, and the loads are calculated to cause the velocity field to operate. No attempt is made to satisfy the stress equilibrium conditions at any point in the field. In a useful form of the upper-bound analysis[2] the deformation is assumed to take place by rigid-body movements of triangular elements in which all particles in a given element move with the same velocity. Figure 15-5 shows an admissible velocity field and hodograph for the strip-drawing problem.[3] Material

[1] D. C. Drucker, H. J. Greenberg, and W. Prager, *J. Appl. Mech.*, vol. 18, p. 371, 1951; J. Baker, M. R. Horne, and J. Heyman, "The Steel Skeleton," vols. I and II, Cambridge University Press, London, 1956.

[2] H. Kudo, *Int. J. Mech. Sci.*, vol. 1, pp. 57–83, 229–252, 336–368, 1960.

[3] S. Kobayashi and E. G. Thomsen, Methods of Solution of Metal-forming Problems, in W. A. Backofen, et al. (eds.), "Fundamentals of Deformation Processing," Syracuse University Press, 1964.

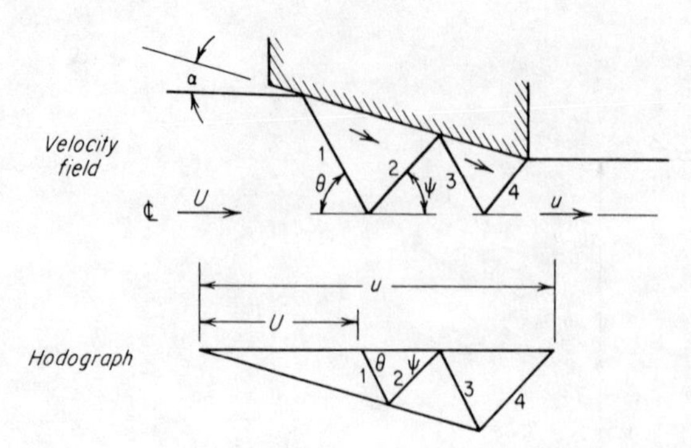

Figure 15-5 An upper-bound solution or strip drawing.

enters the die with an initial velocity U and leaves the die with an increased velocity u. In constructing the velocity field, the number of triangular elements n and the angle of inclination of the first velocity discontinuity θ are the independent variables. These have been selected so as to minimize the average drawing stress. The calculation of stress is based on Drucker's upper-bound theorem. This theorem states that the calculated load is above the critical upper-bound load if a velocity field can be found for which the rate of work due to the external load exceeds the rate of internal energy dissipation. Mathematically stated the theorem is

$$Pu = \sigma_{\text{avg}} uA \leq \int_V \bar{\sigma}\dot{\bar{\varepsilon}}\, dV + \int_S \tau_s v^*\, dS \tag{15-17}$$

where P = the upper-bound load at a forming velocity u
$\dot{\bar{\varepsilon}}$ = the effective strain rate in an element of volume dV
v^* = the rate of relative slip (velocity discontinuity) along the surface S
τ_s = the shear stress due to friction along S

The average draw stress for the velocity field in Fig. 15-5 as developed by the upper-bound theorem is

$$\sigma_{xa} = \frac{2\sigma_0}{\sqrt{3}} \frac{n \sin \alpha}{2} \left[\frac{1}{\sin \theta \sin(\theta - \alpha)} + \frac{1}{\sin(\psi + a) \sin \psi} \right] \tag{15-18}$$

The geometry of the net is given, for each reduction r, by the following equation:

$$(1 - r)^{1/n} = \frac{\sin(\theta - \alpha) \sin \psi}{\sin \theta \sin(\psi + \alpha)} \tag{15-19}$$

Figure 15-6 compares the average draw stress for the upper-bound solution with that obtained from uniform-deformation energy, Eq. (15-13), and with the draw stress obtained from slip-line field theory.

We return now to a simple derivation of the upper-bound equation[1] and a discussion of the velocity diagram called the *hodograph*. Consider a rigid element

[1] W. Johnson and P. B. Mellor, "Engineering Plasticity," Van Nostrand Reinhold Company, New York, 1973.

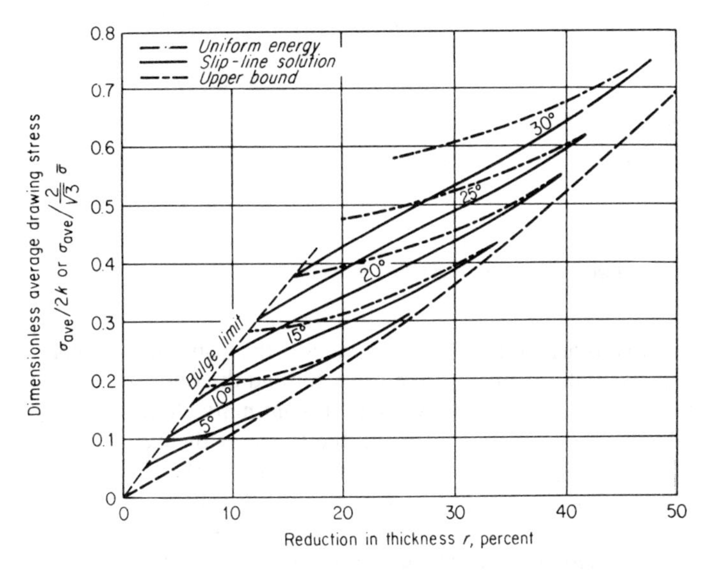

Figure 15-6 Comparison of the average drawing stresses calculated by various methods. *(From S. Kobayashi and E. G. Thomsen, in "Fundamentals of deformaton Processing," p. 61, Syracuse University Press, 1964.)*

of material $ABCD$ of unit height and thickness moving to the left with unit velocity (Fig. 15-7). All material to the right of the shear plane X-X is undeformed, while all material which crosses the shear plane is deformed. When the element crosses the interface X-X it is forced to change direction and velocity, and $ABCD$ is distorted to $A'B'C'D'$. However, AD and $A'D'$ remain parallel to the interface X-X.

The deformation of the element involves velocity changes. These velocities are represented on a diagram called a *hodograph* (Fig. 15-8). The original unit velocity v_1 is resolved into components perpendicular to X-X, v_p and parallel to X-X, v_a. A particle of material on crossing X-X is constrained to flow with a velocity v_2 at an angle α to the direction of v_1. Likewise, the velocity v_2 is resolved into components perpendicular and parallel to the interface X-X. However, $v_{p1} = v_{p2}$, since the volume of material entering and leaving the interface

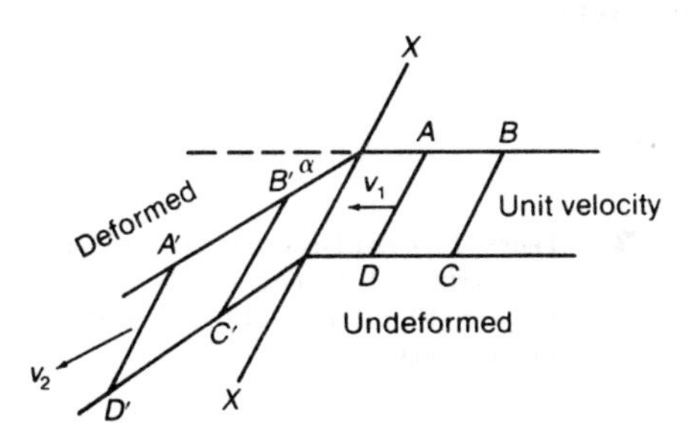

Figure 15-7 Deformation of an element at shear plane.

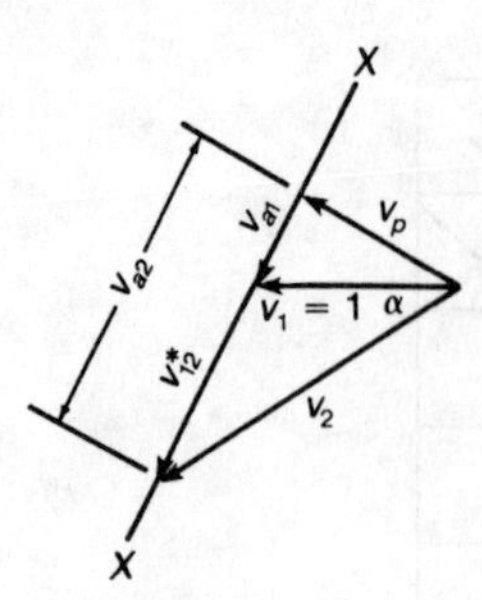

Figure 15-8 Hodograph for Fig. 15-7.

must be the same. Any other situation would violate the concept of incompressibility. This establishes the location of the v_2 vector and the value of v_{a2}. The vector difference between v_1 and v_2 is the velocity discontinuity along X-X, v^*_{12}.

We now consider the work done in changing the shape of the element from $ABCD$ to $A'B'C'D'$ (Fig. 15-9). The shear stress on the sides of the element is τ. The work done in distorting the element is $W = P\delta = [\tau \overline{BC}(1)]CC'$, where the factor of unity comes from assuming the element has unit thickness. This can be recast as $dW/dt = Pv = (\tau \overline{BC})CC'/t$, where t is the time for $\overline{DC}$ to cross X-X. Since $v_1 = 1$, $t = \overline{DC}$ and $dW/dt = (\tau \overline{BC})CC'/\overline{DC}$. Comparing the triangle $C'CD$ with the hodograph (Fig. 15-9b) we see that from similar triangles $1/v^* = CD/CC'$. Therefore,

$$\frac{dW}{dt} = Pv = \tau \overline{BC} v^* \tag{15-20}$$

For plane-strain deformation, $\tau = k$, and if X-X is a straight line

$$\frac{dW}{dt} = kSv^* \tag{15-21}$$

where S is the length of the line of tangential velocity discontinuity. For an upper-bound velocity field consisting of a number of straight lines of tangential velocity discontinuity

$$\frac{dW}{dt} = \sum_{1}^{i} kS_i v_i^* \tag{15-22}$$

Most of the velocity fields for upper-bound solutions consist of a number of polygons which are viewed as rigid blocks. In other words, all of the material inside of the polygon moves with the same velocity.

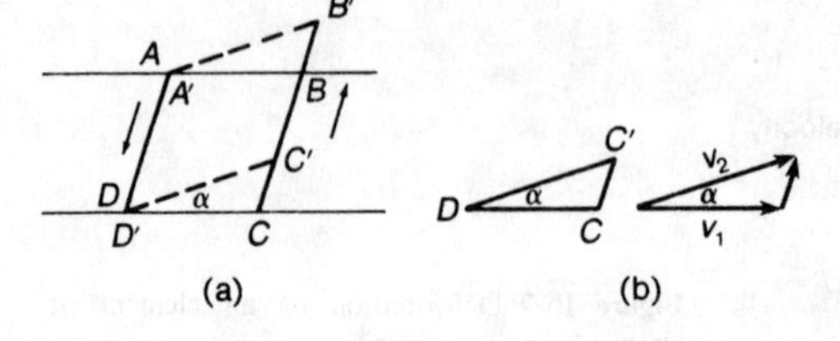

Figure 15-9 (a) Distortion of an element at interface X-X; (b) similar triangles between the distortion of the element and the velocities.

Example The upper-bound velocity field for the plane-strain indentation of a semiinfinite slab with a flat indenter is given below. Determine the hodograph and the estimate of the pressure to cause this indentation.

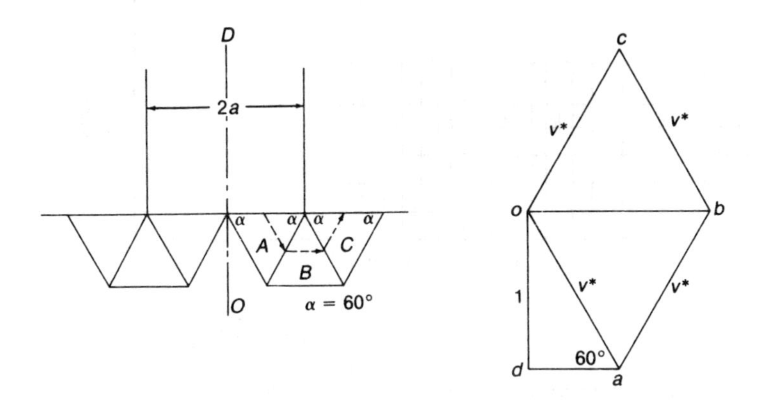

There is no friction between the indenter and the slab. Because of symmetry we need consider only the net to the right of the center line DO. All particles in the region A under the indenter are constrained to move vertically downward with the velocity of the indenter, $v_{OD} = 1$. But the particles also slide parallel to the surface OA and parallel to the interface DA. This gives the first triangle oda on the hodograph, from which the velocity discontinuity $v^* = 1/\sin 60° = 2/\sqrt{3}$. In a similar way the hodograph is completed to give the remaining values of v^*.

$$\frac{dW}{dt} = \sum kSv^*$$

$$pa(1) = k\left[(OA \cdot v_{oa}) + (AB \cdot v_{ab}) + (OB \cdot v_{bc}) + (BC \cdot v_{bc}) + (CO \cdot v_{co})\right]$$

$$pa = 5k\left(a \cdot \frac{2}{\sqrt{3}}\right) = \frac{10ka}{\sqrt{3}}$$

$$\frac{p}{2k} = \frac{5}{\sqrt{3}} = 2.89$$

The *finite element method* (Sec. 2-16), which has had such a major impact on the analysis of complex problems concerned with elastic stress analysis, is beginning to be applied to problems in metalworking plasticity. The FEM is a very powerful technique for determining the distribution of stresses and strains in plane stress, plane strain, and axisymmetric conditions for both steady-state and non-steady-state deformation problems. Early applications of FEM to plasticity

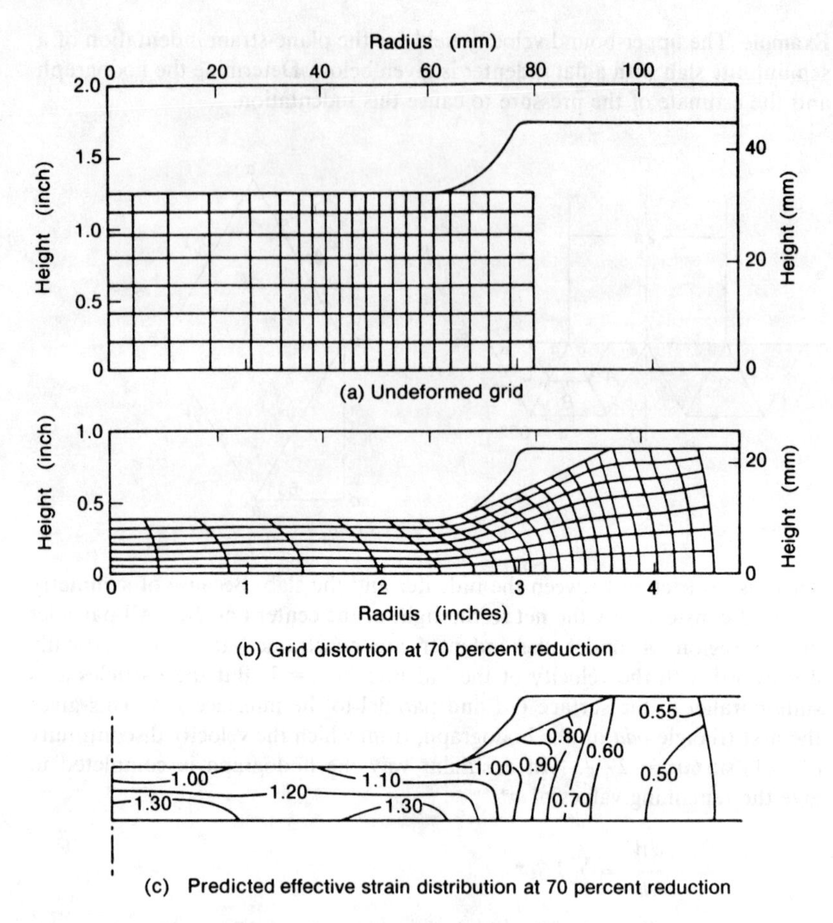

Figure 15-10 Distortion of FEM grid in forging of a compressor disk. Because of symmetry only one-quarter of the cross section need be considered. *(From T. Altan, S. I. Oh, and H. L. Gegel, "Metal Forming," p. 336, American Society for Metals, Metals Park, Ohio, 1983.)*

problems concentrated on elastic-plastic solutions. Because these problems require the use of very small increments of strain, with elastic calculations made at each increment, a very large amount of computer capacity is required.

A practical adaptation of the FEM to metalworking analysis was achieved by Kobayashi[1] in a technique called the *matrix method*. This method neglects elastic strains compared with the larger plastic strains and assumes rigid plastic behavior. Therefore, relatively large strain increments can be used and the computer

[1] C. H. Lee and S. Kobayashi, *Trans. ASME Ser. B., J. Eng. Ind.*, vol. 93, pp. 445–454, 1971; S. Kobayashi and S. N. Shah, "Advances in Deformation Processing," J. J. Burke and V. Weiss (eds.), pp. 51–98, Plenum Press, New York, 1978.

requirements are reduced considerably. The matrix method has been incorporated into a computer code called ALPID[1] (analysis of large plastic incremental deformation). It assumes a rigid viscoplastic material in which the flow stress is a function of strain, strain rate, and temperature. ALPID predicts stress, strain, strain rate, particle velocity, and temperature at any location within a deforming material. Figure 15-10 shows the distortion calculated by ALPID after a 70 percent reduction. Effective strain has been calculated at each point and displayed as contours of $\bar{\varepsilon}$ at the bottom of the figure. Because of the large deformations in many metalworking processes the initial FEM mesh may become too distorted to give reliable information. When this happens it is necessary to remesh by generating a new mesh within the boundary of the deformed material and interpolate the values of ε, $\dot{\varepsilon}$, v, and T to the new grid system.

FEM analysis can be used to simulate deformation very effectively when combined with real time computer graphics output. The influence of die geometry, preform design, friction, and material properties on such important factors as die fill and defect formation can be investigated quickly and reliably. While ALPID has been applied to two-dimensional metal deformation processes, it will surely be expanded to simulate three-dimensional deformation operations.

15-3 FLOW-STRESS DETERMINATION

The various expressions that will be developed in succeeding chapters to describe the forming stress, or pressure, in a particular metalworking process invariably consist of three terms:

$$p = \bar{\sigma}_0 g(f) h(c)$$

where $\bar{\sigma}_0$ = the flow resistance of the material for the appropriate stress state, i.e., uniaxial, plane strain, etc. It is a function of strain, temperature, and strain rate.

$g(f)$ = an expression for the friction at the tool-workpiece interface.

$h(c)$ = a function of the geometry of the tooling and the geometry of the deformation. This term may or may not include a contribution from redundant deformation.

It is obvious from the above relationship that if we are to make accurate predictions of forming loads and stresses, we need accurate values of flow resistance (flow stress). The experimental problems in measuring the flow curve under metalworking conditions are more severe than in the usual stress-strain test determined for structural or mechanical design applications. Since metalworking processes involve large plastic strains, it is desirable to measure the flow curve out

[1] S. I. Oh, *Int. J. Mech. Sci.*, vol. 17, pp. 479–493, 1982; S. I. Oh, G. D. Lahoti, and T. Altan, Proc. NAMRC-X, 1982; T. Altan, H. L. Gegel, and S. I. Oh, *Metal Forming*, chap. 20, American Society for Metals, Metals Park, Ohio, 1983.

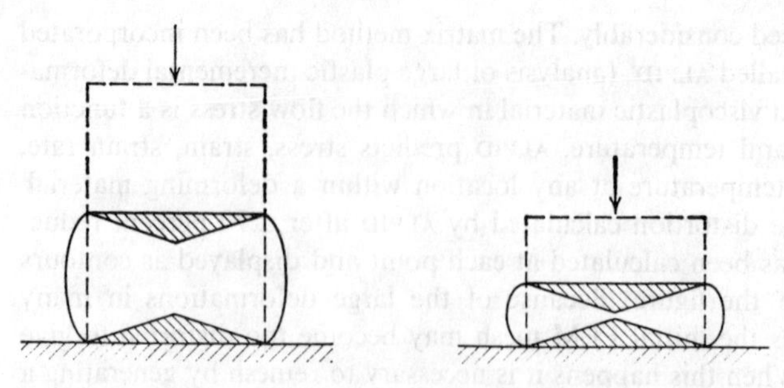

Figure 15-11 Undeformed regions (shaded) due to friction at ends of a compression specimen.

to a true strain of 2.0 to 4.0. In addition, many of these processes involve high strain rates ($\dot{\varepsilon} \approx 100\ s^{-1}$), which may not be obtained easily with ordinary test facilities. Further, many metalworking processes are carried out at elevated temperatures where the flow stress is strongly strain-rate sensitive but nearly independent of strain. Thus, tests for determining flow stress must be carried out under controlled conditions of temperature and constant true-strain rate.[1]

The true-stress–true-strain curve determined from the *tension test* is of limited usefulness because necking limits uniform deformation to true strains less than 0.5 (see Sec. 8-3). This is particularly severe in hot-working, where the low rate of strain hardening allows necking to occur at $\varepsilon \approx 0.1$. The formation of a necked region in the tension specimen introduces a complex stress state and locally raises the strain rate.

The *compression* of a short cylinder between anvils is a much better test for measuring the flow stress in metalworking applications. There is no problem with necking and the test can be carried out to strains in excess of 2.0 if the material is ductile. However, the friction between the specimen and anvils can lead to difficulties unless it is controlled. In the *homogeneous upset test* a cylinder of diameter D_0 and initial height h_0 would be compressed in height to h and spread out in diameter to D according to the law of constancy of volume:

$$D_0^2 h_0 = D^2 h$$

During deformation, as the metal spreads over the compression anvils to increase its diameter, frictional forces will oppose the outward flow of metal. This frictional resistance occurs in that part of the specimen in contact with the anvils, while the metal at specimen midheight can flow outward undisturbed. This leads to a *barreled* specimen profile, and internally a region of undeformed metal is created near the anvil surfaces (Fig. 15-11). As these cone-shaped zones approach and overlap, they cause an increase in force for a given increment of deformation and the load-deformation curve bends upward (Fig. 15-12). For a fixed diameter,

[1] H. J. McQueen and J. J. Jonas, Hot Workability Testing Techniques, in A. L. Hoffmanner (ed.), "Metal Forming: Interrelation Between Theory and Practice," Plenum Publishing Corporation, New York, 1971.

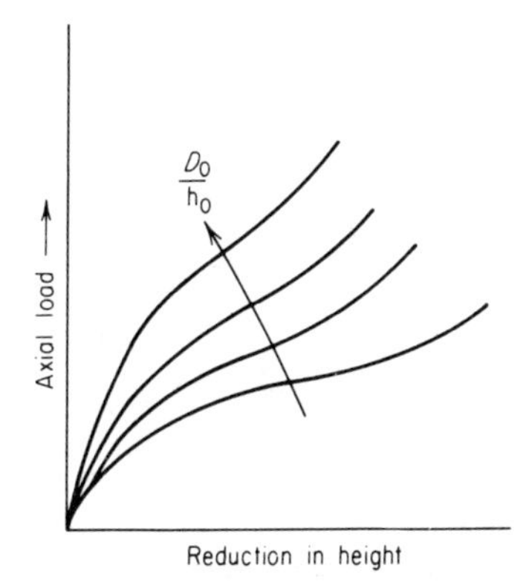

Figure 15-12 Load-deformation curves for compression tests with different values of D_0/h_0.

a shorter specimen will require a greater axial force to produce the same percentage reduction in height because of the relatively larger undeformed region (Fig. 15-11). Thus, one way to minimize the barreling and nonuniform deformation is to use a low value of D_0/h_0. However, there is a practical limit of $D_0/h_0 \approx 0.5$, for below this value the specimen buckles instead of barreling. The true flow stress in compression without friction can be obtained[1] by plotting load versus D_0/h_0 for several values of reduction and extrapolating each curve to $D_0/h_0 = 0$.

The friction at the specimen-platen interface can be minimized[2] by using smooth, hardened platens, grooving the ends of the specimen to retain lubricant, and carrying out the test in increments so that the lubricant can be replaced at intervals.[3] Teflon sheet for cold deformation and glass for hot deformation are especially effective lubricants. With these techniques it is possible to reach a strain of about $\varepsilon = 1.0$ with only slight barreling. When friction is not present the uniaxial compressive force required to produce yielding is

$$P = \sigma_0 A$$

The true compressive stress p produced by this force P is

$$p = \frac{4P}{\pi D^2}$$

and using the constancy-of-volume relationship

$$p = \frac{4Ph}{\pi D_0^2 h_0} \tag{15-23}$$

[1] M. Cooke and E. C. Larke, *J. Inst. Met.*, vol. 71, pp. 371–390, 1945.

[2] G. T. van Rooyen and W. A. Backofen, *Int. J. Mech. Sci.*, vol. 1, pp. 1–27, 1960; G. W. Pearsall and W. A. Backofen, *Trans. ASME, Ser. D: J. Basic Eng.*, vol. 85, p. 68, 1963.

[3] Standard Methods of Compression Testing of Metallic Materials at Room Temperature, ASTM Standards, pt. 31, Designation E9-70; for elevated temperature compression tests see ASTM E209; T. C. Hsu, *Mater. Res. Stand.*, vol. 9, pp. 20–25, 47–53, 1969.

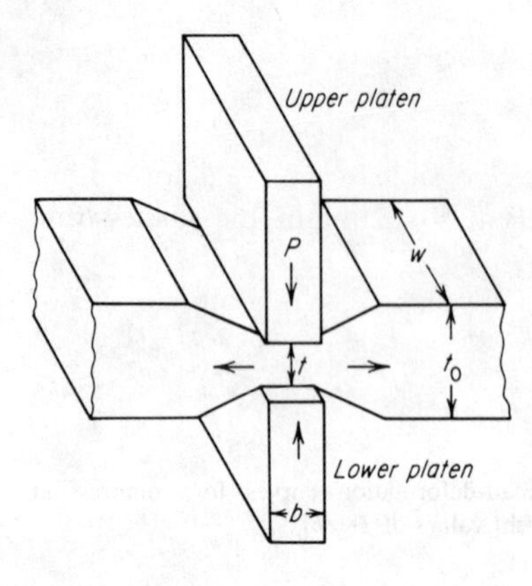

Figure 15-13 Plane-strain compression test.

where D_0 and h_0 are the initial diameter and height, and h is the height of the cylindrical sample at any instant during compression. The true *compressive* strain is given by

$$\varepsilon_c = \ln \frac{h_0}{h} \tag{15-24}$$

In a compression test at constant crosshead speed, the true-strain rate continuously increases with deformation [see Eq. (9-33)]. At strain rates up to $10\ \mathrm{s}^{-1}$ a servocontrolled testing machine can be modified to maintain a constant true-strain rate. For metalworking strain rates of 1 to $10^3\ \mathrm{s}^{-1}$ the only equipment capable of providing a constant true-strain rate is the cam plastometer.[1]

A compression test, such as that shown in Fig. 15-11, would be very difficult to conduct on a thin sheet, since it might be impossible to machine the specimens. A much more suitable test for sheet metal is the *plane-strain compression test*.[2] In this test a narrow band across the width of a strip is compressed by narrow platens which are wider than the strip (Fig. 15-13). The constraints of the undeformed shoulders of material on each side of the platens prevent extension of the sheet in the width dimension. There is deformation in the direction of platen motion and in the direction normal to the length of the platen, as occurs in the rolling process. In addition to its suitability with thin sheet, other advantages of this test are that it simulates the stress state in rolling, eliminates problems with barreling, and because the area under the platen is constant, the total deformation force does not rise as rapidly as in the compression of a cylinder. On the other hand, unless good lubrication is maintained, a dead metal zone will form in the specimen next to the face of the platens. Therefore, the test usually is carried out incrementally, measuring the sheet thickness after each increment and relubricating for the next higher load. Slip-line field theory has been used to analyze the

[1] J. F. Alder and V. A. Phillips, *J. Inst. Met.*, vol. 83, pp. 80–86, 1954–1955; J. E. Hockett, *Am. Soc. Test. Mater. Proc.*, vol. 59, pp. 1309–1319, 1959.

[2] A. B. Watts and H. Ford, *Proc. Inst. Mech. Eng.*, vol. 169, pp. 1141–1156, 1955.

plane-strain compression test. It has been found that the axial compressive stress simply is the applied load divided by the contact area of the platens so long as the ratio of indentation thickness to platen width t/b is in the range $\frac{1}{4}$ to $\frac{1}{2}$. By changing platens to maintain this t/b range, it is possible to achieve deformations of about 90 percent. An additional requirement to maintain the plane-strain condition is that w/b must be greater than 5.

True stress and true strain determined in the plane strain compression test may be expressed by

$$p = \frac{P}{wb} \qquad \varepsilon_{pc} = \ln \frac{t_0}{t}$$

The mean pressure on the platens is 15.5 percent higher than it would be in the corresponding uniaxial compression test (see Prob. 15-8). The true-stress–true-strain curve in uniaxial compression (σ_0 versus ε_c) may be obtained from the corresponding plane-strain compression curve (p versus ε_{pc}) by the following relations:

$$\sigma_0 = \frac{\sqrt{3}}{2} p = \frac{p}{1.155} \tag{15-25}$$

$$\varepsilon_c = \frac{2}{\sqrt{3}} \varepsilon_{pc} = 1.155 \varepsilon_{pc} \tag{15-26}$$

The *hot torsion test* (see Sec. 10-6) is capable of producing very large strains of the order of 20. Since the dimensions of the specimen do not change, the strain rate remains constant for a constant rpm.[1] Strain rates varying from 10^{-5} to 10^3 s^{-1} are readily achieved. The chief difficulty with this test is the fact that stress and strain vary with radial distance in the specimen. Usually the maximum values at the surface are reported, but this may lead to problems in interpretation if the surface strain-hardens more than the core. The problem of nonuniform stress and strain is largely eliminated by using a tubular specimen,[2] but care must be exercised to use a short specimen or buckling will occur. Also, because of the excessive material reorientation that may occur at large strains the torsion test is not an accurate simulation of metalworking processes. This affects the use of the test for workability studies, but not for flow stress determination. The uniaxial stress and strain are obtained from the torsional shear stress τ and shear strain γ by the following relations based on the von Mises' criterion.

$$\sigma_0 = \sqrt{3}\,\tau \qquad \varepsilon = \frac{\gamma}{\sqrt{3}}$$

The variation of flow stress with strain can be neglected for true hot-working (see Fig. 10-8) or for a highly cold-worked metal. In cases where strain hardening is present, the best approach to selecting a flow stress for use in forming load

[1] J. A. Bailey and S. L. Haas, *J. Mater.*, vol. 7, pp. 8–13, 1972.
[2] F. Hodierne, *J. Inst. Met.*, vol. 91, pp. 267–273, 1963.

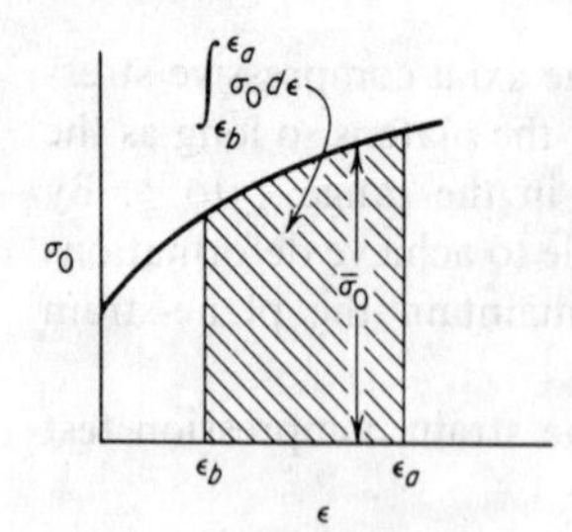

Figure 15-14 Definition of mean-flow stress.

calculations is to use the mean flow stress as given by

$$\bar{\sigma}_0 = \frac{1}{\varepsilon_a - \varepsilon_b} \int_{\varepsilon_b}^{\varepsilon_a} \sigma_0 \, d\varepsilon \tag{15-27}$$

Figure 15-14 shows the interpretation of this equation. This is considered a better choice than the flow stress based on $\varepsilon = (\varepsilon_a + \varepsilon_b)/2$. When analytical expressions for the flow curve are required, it is usually possible to fit the data to either Eq. (8-21) or Eq. (8-24). Extensive data for flow stress under metalworking conditions have been published.[1] However for large strains ($\varepsilon > 1$) it has been proposed[2] that the flow stress is best given by

$$\sigma_0 = A + B\varepsilon \tag{15-28}$$

where $A = K(1 - n)$
$B = Kn$

15-4 TEMPERATURE IN METALWORKING

Forming processes are commonly classified into *hot-working* and *cold-working* operations. Hot-working is defined as deformation under conditions of temperature and strain rate such that recovery processes take place simultaneously with the deformation. On the other hand, cold-working is deformation carried out under conditions where recovery processes are not effective. In hot-working the strain hardening and distorted grain structure produced by deformation are very rapidly eliminated by the formation of new strain-free grains as the result of recrystallization. Very large deformations are possible in hot-working because the recovery processes keep pace with the deformation. Hot-working occurs at an essentially constant flow stress, and because the flow stress decreases with increasing temperature, the energy required for deformation is generally much less for hot-working than for cold-working. Since strain hardening is not relieved in cold-working, the flow stress increases with deformation. Therefore, the total

[1] T. Altan, S. I. Oh, and H. L. Gegel, "Metal Forming," pp. 56–72, American Society for Metals, Metals Park, Ohio, 1983.

[2] M. C. Shaw, *Int. J. Mach. Tool Des. Res.*, vol. 22, no. 3, pp. 215–226, 1982.

deformation that is possible without causing fracture is less for cold-working than for hot-working, unless the effects of cold-work are relieved by annealing.

It is important to realize that the distinction between cold-working and hot-working does not depend on any arbitrary temperature of deformation. For most commercial alloys a hot-working operation must be carried out at a relatively high temperature in order that a rapid rate of recrystallization be obtained. However, lead and tin recrystallize rapidly at room temperature after large deformations, so that the working of these metals at room temperature constitutes hot-working. Similarly, working tungsten at 2000°F, in the hot-working range for steel, constitutes cold-working because this high-melting metal has a recrystallization temperature above this working temperature.

The temperature of the workpiece in metalworking depends on: (1) the initial temperature of the tools and the material, (2) heat generation due to plastic deformation, (3) heat generated by friction at the die/material interface, and (4) heat transfer between the deforming material and the dies and surrounding environment. For a frictionless deformation process the maximum increase in temperature is

$$T_d = \frac{U_p}{\rho c J} = \frac{\bar{\sigma}\bar{\varepsilon}\beta}{\rho c J} \tag{15-29}$$

where U_p = the work of plastic deformation per unit volume
ρ = the density of workpiece
c = the specific heat of the workpiece
J = the mechanical equivalent of heat, 778 ft-lb/Btu or 4185 J/kcal
β = fraction of deformation work converted into heat. Typically $\beta = 0.95$. The remainder is stored in the material as energy associated with the defect structure.

The temperature increase due to friction is given by[1]

$$T_f = \frac{\mu p v A \, \Delta t}{\rho c V J} \tag{15-30}$$

where μ = friction coefficient at material/tool interface (see Sec. 15-7)
p = stress normal to interface
v = velocity at the material/tool interface
A = surface area at the material/tool interface
Δt = time interval of consideration
V = volume subjected to the temperature rise

Usually the temperature is highest at the material/tool interface where friction generates the heat and it falls off toward the inside of the workpiece and into the die. For simplicity we can neglect these temperature gradients and consider the deforming material to be a thin plate between a workpiece initially at

[1] T. Altan, S. I. Oh, and H. L. Gegel, op. cit. p. 90.

T_0 and the die at a temperature T_1. Then the average instantaneous temperature of the deforming material at the interface is given by[1]

$$T = T_1 + (T_0 - T_1)\exp\left(\frac{-ht}{\rho c \delta}\right) \tag{15-31}$$

where h = heat transfer coefficient between the material and the dies
δ = material thickness between the dies

This equation describes the variation of the average material temperature during cooling of the material, which is assumed to be a thin plate cooled between two die surfaces. It does not include the temperature increase due to deformation and friction. Thus, the final average material temperature at a time t is

$$T_m = T_d + T_f + T \tag{15-32}$$

Example Compare the temperature rise when a cylinder of aluminum and titanium is quickly deformed to $\bar{\varepsilon} = 1.0$ at room temperature.

	$\bar{\sigma}$, MPa	$\bar{\varepsilon}$	ρ, g/cm³	c, cal/g °C
Al	200	1.0	2.69	0.215
Ti	400	1.0	4.50	0.124

For aluminum

$$\Delta T_d = \frac{\bar{\sigma}\bar{\varepsilon}\beta}{\rho c J} = \frac{(200)(1.0)(0.95)}{(2.69)(0.215)4.186} = 78°\text{C}$$

Check of units

$$\frac{\text{MN/m}^2}{\text{g/cm}^3 \times \text{cal/g °C} \times \text{J/cal}} = 10^6 \frac{\text{N}}{\text{m}^2} \times \text{cm}^3 \times \frac{\text{m}^3}{10^6\ \text{cm}^3} \times \frac{°\text{C}}{\text{J}}$$

$$= \frac{(\text{N} - \text{m})°\text{C}}{\text{J}} = \frac{(\text{J})°\text{C}}{\text{J}}$$

For titanium

$$\Delta T_d = \frac{(400)(1.0)(0.95)}{(4.50)(0.124)4.186} = 162°\text{C}$$

Hot-Working

Hot-working refers to deformation carried out under conditions of temperature and strain rate such that recovery processes occur substantially during the deformation process so that large strains can be achieved with essentially no

[1] T. Altan, S. I. Oh, and H. L. Gegel, op. cit., p. 91.

strain hardening. Hot-working processes such as rolling, extrusion, or forging typically are used in the first step of converting a cast ingot into a wrought product. The strain in hot-working is large ($\varepsilon \approx 2$ to 4) compared with tension or creep tests. Hot-working usually is carried out at temperatures above $0.6T_m$ and at high strain rates in the range 0.5 to 500 s^{-1}. The hot torsion test (Sec. 10-6) is a convenient laboratory method for studying metallurgical changes in hot-working.

Not only does hot-working result in a decrease in the energy required to deform the metal and an increased ability to flow without cracking, but the rapid diffusion at hot-working temperatures aids in decreasing the chemical inhomogeneities of the cast-ingot structure. Blowholes and porosity are eliminated by the welding together of these cavities, and the coarse columnar grains of the casting are broken down and refined into smaller equiaxed recrystallized grains. These changes in structure from hot-working result in an increase in ductility and toughness over the cast state.

However, there are certain disadvantages to hot-working. Because high temperatures are usually involved, surface reactions between the metal and the furnace atmosphere become a problem. Ordinarily hot-working is done in air, oxidation results, and a considerable amount of metal may be lost. Reactive metals like titanium are severely embrittled by oxygen, and therefore they must be hot-worked in an inert atmosphere or protected from the air by a suitable barrier. Surface decarburization of hot-worked steel can be a serious problem, and frequently extensive surface finishing is required to remove the decarburized layer. Rolled-in oxide makes it difficult to produce good surface finishes on hot-rolled products, and because allowance must be made for expansion and contraction, the dimensional tolerances for hot-worked mill products are greater than for cold-worked products. Further, the structure and properties of hot-worked metals are generally not so uniform over the cross section as in metals which have been cold-worked and annealed. Since the deformation is always greater in the surface layers, the metal will have a finer recrystallized grain size in this region. Because the interior will be at higher temperatures for longer times during cooling than will be the external surfaces, grain growth can occur in the interior of large pieces, which cool slowly from the working temperature.

The lower temperature limit for the hot-working of a metal is the lowest temperature at which the rate of recrystallization is rapid enough to eliminate strain hardening in the time when the metal is at temperature. For a given metal or alloy the lower hot-working temperature will depend on such factors as the amount of deformation and the time that the metal is at temperature. Since the greater the amount of deformation the lower the recrystallization temperature, the lower temperature limit for hot-working is decreased for large deformations. Metal which is rapidly deformed and cooled rapidly from temperature will require a higher hot-working temperature for the same degree of deformation than will metal slowly deformed and slowly cooled.

The upper limit for hot-working is determined by the temperature at which either melting or excessive oxidation occurs. Generally the maximum working temperature is limited to 100°F below the melting point. This is to allow for the

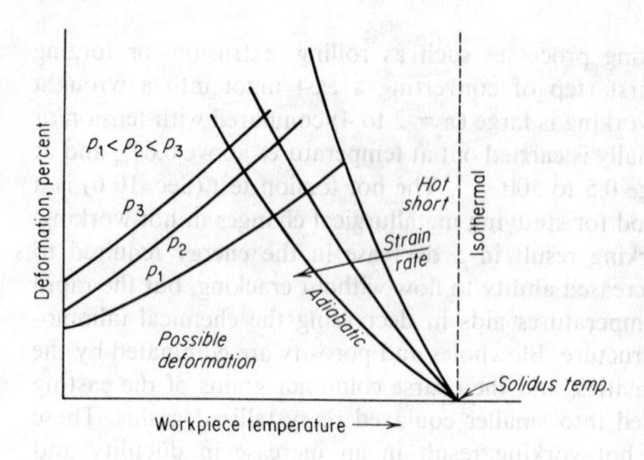

Figure 15-15 Schematic effect of temperature, pressure, and strain rate on the allowable working range. *(After S. Hirst and D. H. Ursell, Metal Treatment and Drop Forging, vol. 25, p. 409, 1958.)*

possibility of segregated regions of lower-melting-point material. Only a very small amount of a grain-boundary film of a lower-melting constituent is needed to make a material crumble into pieces when it is deformed. Such a condition is known as *hot shortness*, or *burning*.

Most hot-working operations are carried out in a number of multiple passes, or steps. Generally the working temperature for the intermediate passes is kept well above the minimum working temperature in order to take advantage of the economies offered by the lower flow stress. It is likely that some grain growth will occur subsequent to the recrystallization at these temperatures. Since a fine-grain-sized product is usually desired, common practice is to lower the working temperature for the last pass to the point where grain growth during cooling from the working temperature will be negligible. This *finishing temperature* is usually just above the minimum recrystallization temperature. In order to ensure a fine recrystallized grain size, the amount of deformation in the last pass should be relatively large.

The temperature range over which an alloy can be worked can be visualized with the aid of Fig. 15-15. For a given working pressure and temperature there will be a maximum amount of deformation that can be imparted to the workpiece. We are assuming here that this limitation is based on the flow resistance of the material and not on its ductility. As the preheat temperature of the billet increases the flow stress falls off and the amount of possible deformation increases. This gives the left-hand series of curves that increase to the right. The obvious limit on the temperature scale is the solidus temperature of the workpiece. Hot shortness will also be a limitation at somewhat lower temperatures. As the strain rate of deformation increases more heat will be retained in the workpiece and the workpiece temperature will have to be decreased to keep its final temperature from reaching the hot-shortness temperature.

The temperature and stress dependence of the steady-state hot-working rate is given by

$$\dot{\varepsilon}_s = A(\sinh \alpha\sigma)^{n'} e^{-Q/RT} \tag{15-33}$$

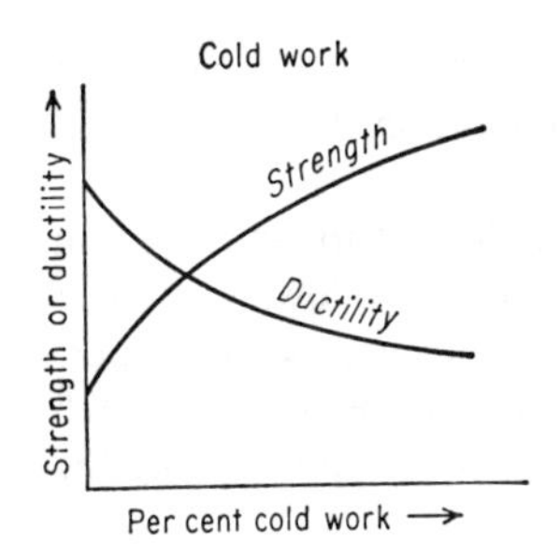

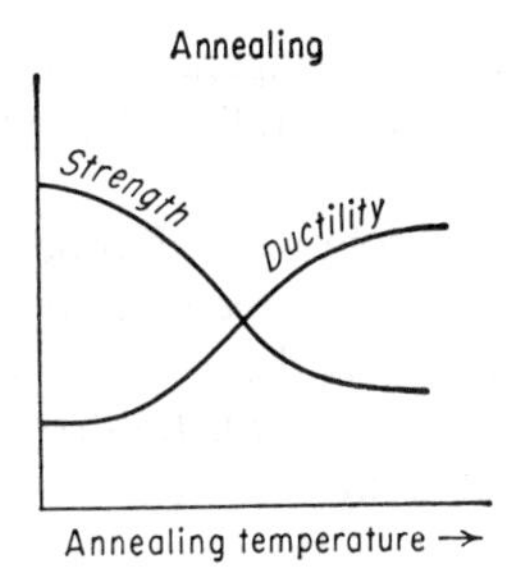

Figure 15-16 Typical variation of strength and ductility in the cold-work-anneal cycle.

where A, α, and n' are experimentally determined constants. There are two mechanisms[1] responsible for softening in hot-working, depending on the metal. In aluminum and alpha iron *dynamic recovery* is the softening mechanism. This occurs by the formation of a well-developed subgrain structure by cross slip and climb, as occurs in creep deformation. The activation energy for hot-working is equal to that for creep and for self-diffusion. In metals with a lower stacking-fault energy (copper, nickel, and austenitic stainless steel), the activation energy for softening in hot-work is higher than for creep. For these metals the softening process is *dynamic recrystallization*.

Cold-Working

As was shown in Sec. 6-13, cold-working of a metal results in an increase in strength or hardness and a decrease in ductility. When cold-working is excessive, the metal will fracture before reaching the desired size and shape. Therefore, in order to avoid such difficulties, cold-working operations are usually carried out in several steps, with intermediate annealing operations introduced to soften the cold-worked metal and restore the ductility. This sequence of repeated cold-working and annealing is frequently called the *cold-work–anneal cycle*. Figure 15-16 schematically illustrates the property changes involved in this cycle.

Although the need for annealing operations increases the cost of forming by cold-working, particularly for reactive metals which must be annealed in vacuum or inert atmospheres, it provides a degree of versatility which is not possible in hot-working operations. By suitably adjusting the cold-work–anneal cycle the part can be produced with any desired degree of strain hardening. If the finished product must be stronger than the fully annealed material, then the final operation must be a cold-working step with the proper degree of deformation to produce the desired strength. This would probably be followed by a stress relief to remove residual stresses. Such a procedure to develop a certain combination of strength and ductility in the final product is more successful than trying to achieve the same combinations of properties by partially softening a fully cold-worked material, because the recrystallization process proceeds relatively

[1] J. J. Jonas, C. M. Sellers, and W. J. McG. Tegart, *Metall. Rev.*, vol. 14, pp. 1–24, 1969.

rapidly and is quite sensitive to small temperature fluctuations in the furnace. If it is desired to have the final part in the fully softened condition, then an anneal follows the last cold-working step.

It is customary to produce cold-worked products like strip and wire in different *tempers*, depending on the degree of cold reduction following the last anneal. The cold-worked condition is described as the annealed (soft) temper: quarter-hard, half-hard, three-quarter-hard, full-hard, and spring temper. Each temper condition indicates a different percentage of cold reduction following the annealing treatment.

Warm Working

Warm working is the plastic deformation of a metal at temperatures below the temperature range for recrystallization and above room temperature. It attempts to combine the advantages of both hot and cold working into one operation. Warm working has been applied most extensively to the forging of steel, where it offers the potential of fewer forging steps, reduced forging loads, and energy savings (due to the elimination of in-process anneals) compared with cold forging. When compared with hot working it offers the advantage of improved dimensional control, higher quality surfaces, and lower energy costs. Successful implementation of warm working depends critically on using the proper lubricant and selecting a material and die design that are optimized for the warm working conditions.

15-5 STRAIN-RATE EFFECTS

Strain-rate, or deformation velocity, has three principal effects in metalworking: (1) the flow stress of the metal increases with strain rate, (2) the temperature of the workpiece is increased because of adiabatic heating, and (3) there is improved lubrication at the tool-metal interface, so long as the lubricant film can be maintained.

The usual definition of true-strain rate for a cylinder upset in compression is

$$\dot{\varepsilon} = \frac{d\varepsilon}{dt} = \frac{1}{h}\frac{dh}{dt} = \frac{v}{h} \tag{15-34}$$

where h = the instantaneous height
v = the deformation velocity

Since in many processes the workpiece is deformed between converging dies where h varies with axial distance (as in Fig. 15-3), it is convenient to define a mean strain rate by

$$\bar{\dot{\varepsilon}}_x = \frac{1}{L}\int_0^L \dot{\varepsilon}\, dx \tag{15-35}$$

Table 15-1 Typical values of velocity encountered in different testing and forming operations

Operation	Velocity, ft/s
Tension test	2×10^{-6} to 2×10^{-2}
Hydraulic extrusion press	0.01 to 10
Mechanical press	0.5 to 5
Charpy impact test	10 to 20
Forging hammer	10 to 30
Explosive forming	100 to 400

where L is the length of contact between the tool and the workpiece (see Fig. 15-27).

It is more usual to evaluate the mean strain rate in terms of the time for an element to travel through the die t_f.

$$\dot{\bar{\varepsilon}}_t = \frac{1}{t_f}\int_0^{t_f} \dot{\varepsilon}\, dt \tag{15-36}$$

However, Chandra and Jonas[1] have shown that the appropriate measure of strain rate for strain-rate sensitive materials at large reductions (as in hot extrusion) is the root-mean-power strain rate

$$\dot{\varepsilon}_{\text{rmp}} = \left[\frac{1}{\ln R}\int_0^{\ln R} (\dot{\varepsilon})^m \, d\varepsilon\right]^{1/m} \tag{15-37}$$

where $R = A_0/A$ is the deformation ratio
m = strain-rate sensitivity [see Eq. (8-42)]

Table 15-1 list typical values of deformation velocity for different testing and forming operations. It is important to note that the forming velocity of most commercial equipment is appreciably faster than the crosshead velocity used in the standard tension test. However, they are not high enough that stress-wave effects become important. When these forming velocities are combined with a situation in which the deformation zone is small, it is possible to produce very high local strain rates. The drawing of a fine wire at a speed of 120 ft/s can result in a strain rate in excess of 10^5 s^{-1}. A mean strain rate of around 2×10^3 s^{-1} is achieved in rolling thin tinplate. These high strain rates are attained by concentrating deformation into a narrow zone rather than by producing very high particle velocities.

However, there is a group of newer metalworking processes which utilize velocities as high as 700 ft/s to carry out forging, extrusion, sheet forming, etc.

[1] T. Chandra and J. J. Jonas, *Metall. Trans.*, vol. 2, pp. 877–881, 1971.

These processes are known as *high-energy-rate forming* (HERF) since the energy of deformation is delivered at a much higher rate than in conventional practice. These high-velocity forming processes[1] utilize the energy from exploding gas or a conventional explosive to produce high particle velocities. For many materials the elongation to fracture increases with strain rate beyond the usual metalworking range until a critical strain rate is reached where the ductility falls off sharply. Explosive forming has developed as a useful process for producing large parts in limited numbers where the force from the explosive energy is substituted for a large hydraulic press.[2] Explosive forming usually is conducted with water filling the "stand-off distance" between the workpiece and the explosive charge. As such there are no unusual hardening effects or structural changes in the metal. However, when the explosive charge is placed in direct contact with the metal, very high transient stresses are produced in the metal and unusual hardening and structural changes may be produced depending on the metal.[3] A remarkable aspect of explosive hardening is that quite high hardness can be produced with essentially no grain distortion or gross distortion of the workpiece.

At the other extreme of the strain-rate spectrum is *superplasticity forming*. In Sec. 8-6 we saw that materials with a high strain-rate sensitivity ($0.3 < m < 1.0$) exhibit pronounced resistance to necking. Generally this occurs in material with a very fine grain size, of the order of 1 μm, and at deformation temperatures above $0.4T_m$. Also, for any superplastic material there is a limiting strain rate above which it is no longer superplastic. Generally, superplastic forming must be done at strain rates below 0.01 s^{-1}. The high resistance to plastic instability of a superplastic metal means that forming processes generally restricted to hot polymers are possible. Biaxial stretching of sheet under small pressure differentials (vacuum forming) and blow modeling are two examples. However, the chief advantage of superplasticity probably is the low flow stress, on the order of 1,000 to 5,000 psi, that exists at superplastic conditions. This has been utilized in the forging of difficult-to-work superalloys (the gatorizing process) or in embossing fine details in other applications.

15-6 METALLURGICAL STRUCTURE

The marked strengthening that results from cold-drawing an iron wire is shown in Fig. 15-17. Here the stress-strain curves determined after the wire had been drawn to a specific true strain are superimposed to show the overall strain-hardening

[1] G. E. Dieter, Strain Rate Effects in Deformation Processing, in W. A. Backofen (ed.), "Fundamentals of Deformation Processing," chap. 7, Syracuse University Press, 1964; R. N. Orava and R. H. Whittman, High-Energy Rate Deformation Processing, in "Advances in Deformation Processing," J. B. Burke and V. Weiss (eds.), pp. 485–533, Plenum Press, New York, 1978.

[2] T. Z. Blazynski (ed.), "Explosive Welding, Forming and Compaction," Applied Science Publishers, London, 1983.

[3] G. E. Dieter, Hardening Effect Produced With Shock Waves, in "Strengthening Mechanisms in Solids," American Society for Metals, Metals Park, Ohio, 1962.

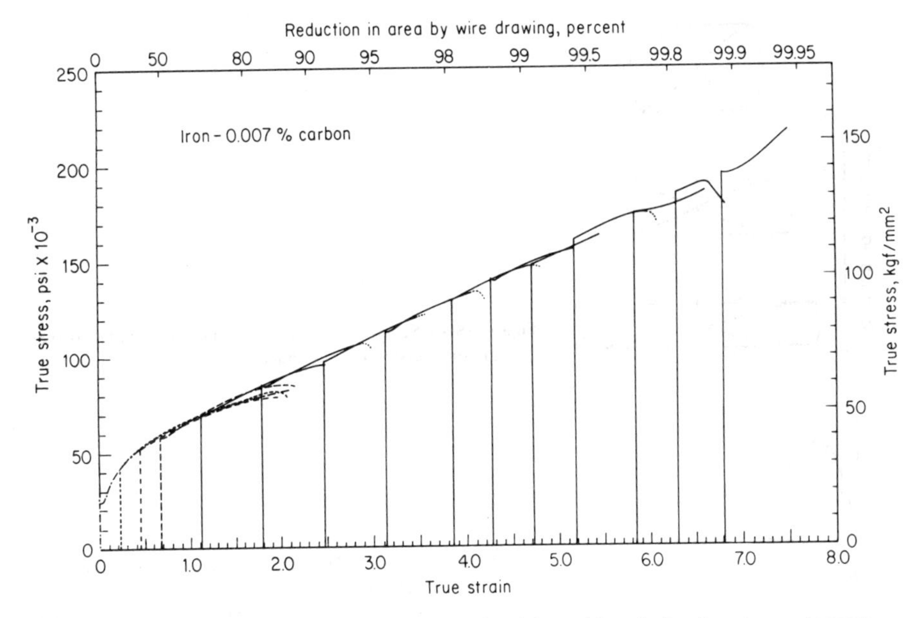

Figure 15-17 True-stress–true-strain curve for iron wire deformed by wiredrawing at room temperature. *(From G. Lankford and M. Cohen, Trans. Am. Soc. Met., vol. 62, p. 623, 1969. Copyright American Society for Metals, 1969.)*

effect. It is quite remarkable that the rate of strain hardening does not diminish significantly even at $\varepsilon > 6.0$.

The dislocation structure of strain-hardened metal was shown (in Sec. 6-13) to consist of a cellular substructure with the cell walls composed of tight-packed tangles of dislocations (Fig. 6-28). For the large plastic strains characteristic of wiredrawing and rolling, the cold-worked structure consists of highly elongated grains containing a relatively equiaxed dislocation cell structure.[1] The smaller strain-hardening ability of fcc based metals compared with iron-based alloys is explained by their development of an almost stable cell size, while with iron the cell size continues to decrease with plastic strain. The relationship between flow stress and cell size d for cold-drawn iron wire has been found to be[2]

$$\sigma_0 = a + md^{-1} \tag{15-38}$$

On a more macroscopic scale, the structure of severely cold-worked metal is characterized by the development of a strong crystallographic texture[3] (see Sec.

[1] J. D. Embury, A. S. Keh, and R. M. Fisher, *Trans. Metall. Soc. AIME*, vol. 236, pp. 1252–1260, 1966.

[2] G. Lankford and M. Cohen, *Trans. Am. Soc. Met.*, vol. 62, pp. 623–638, 1969.

[3] J. Gil Sevillano, P. von Houtte and A. Aernoudt, Large Strain Workhardening and Textures, *Prog. Mater. Sci.*, vol. 25, pp. 69–412, 1980.

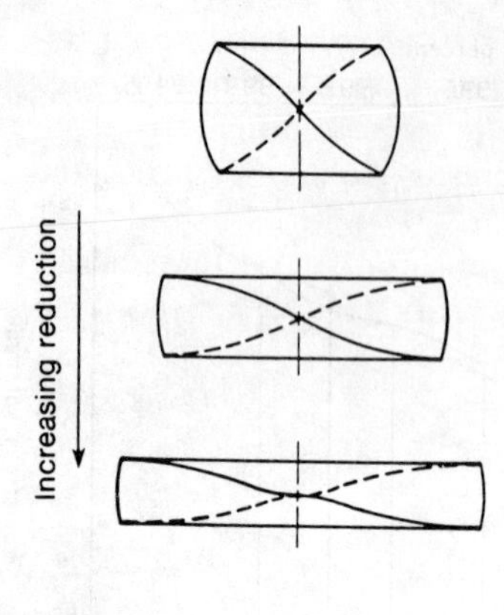

Figure 15-18 Schematic representation of shear band formation in compression of a cylinder.

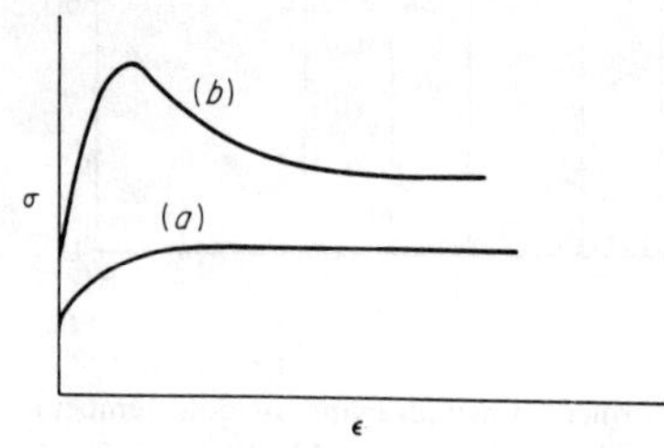

Figure 15-19 (*a*) Hot-working stress-strain curve for a metal which shows dynamic recovery; (*b*) metal which undergoes dynamic recrystallization after initial period of dynamic recovery.

6-17). The presence of preferred orientation causes anisotropy of mechanical properties. This is of particular importance in determining the deep-drawing properties of rolled sheet (see Sec. 20-6). Closely allied with the development of texture is the formation of *deformation bands* and shear bands.[1] Deformation bands are regions of distortion where a portion of a grain has rotated towards another orientation to accommodate the applied strain. When these regions extend across many grains they are called shear bands. Exhaustive shear along these bands is a common type of failure in worked products.

Figure 15-18 illustrates the development of shear bands in the compression of a cylinder. This type of deformation is associated with the occurrence of plastic instability in compression.[2] This usually accompanies hot deformation where some of the strengthening mechanisms become unstable and the rate of flow softening exceeds the rate of area increase in the compression specimen. A flow curve like that shown in curve (*b*) in Fig. 15-19 is the result. The analysis for plastic instability in the tension test that was presented in Sec. 8-10 can be applied in the case of compressive deformation provided the following sign convention is followed: σ, ε, and $\dot{\varepsilon}$ are negative; m, A, and $\dot{A}$ are positive. This leads to stability equations identical to those given by Hart for tension except that

[1] K. Brown, J. Inst. Met., vol. 100, pp. 341–345, 1972.

[2] J. J. Jonas, R. A. Holt, and C. E. Coleman, *Acta Met.*, vol. 24, p. 911, 1976; S. L. Semiatin and J. J. Jonas, "Formability and Workability of Metals: Plastic Instability and Flow Localization," American Society for Metals, Metals Park, Ohio, 1984.

the inequalities associated with stable and unstable flow are reversed. In compression, deformation is stable as long as $(\gamma + m) \leq 1$ and is unstable when $(\gamma + m) \geq 1$. While in tension a material with a high strain-rate sensitivity is more resistant to flow localization; in compression, the greater the rate sensitivity, the sooner flow localization is initiated.

The tendency for a material to produce localized shear-band deformation is given by the fractional change of strain rate with strain. This can be related[1] to basic material parameters by

$$\alpha = \frac{\gamma - 1}{m} \tag{15-39}$$

where

$$\gamma = \frac{1}{\sigma} \frac{d\sigma}{d\varepsilon}\bigg|_{\dot{\varepsilon}, T}$$

$$m = \frac{d \ln \sigma}{d \ln \dot{\varepsilon}}\bigg|_{\varepsilon, T}$$

Extensive studies on titanium alloys[2] have shown that, when α exceeds about 5, shear band formation is prevalent.

Considerable research in recent years has led to increased understanding of the mechanisms of hot-working.[3] At the outset it is important to distinguish between those processes which occur during deformation (called dynamic processes) and those which take place between intervals of deformation or after deformation is completed (called static processes). The two dynamic processes involved in hot working are *dynamic recovery* and *dynamic recrystallization*.

Dynamic recovery is the basic mechanism that leads to the annihilation of pairs of dislocations during straining. It results in a flow curve about one order of magnitude lower than in cold working. The flow curve is essentially exponential [curve (*a*) in Fig. 15-19]. The low dislocation densities associated with the deformation are due to the ease of cross slip, climb, and dislocation unpinning at nodes in this temperature region. These deformation mechanisms result in a microstructure consisting of elongated grains, inside of which is a well developed fine-subgrain structure, typically of the order of 1 to 10 μm. Although the grains are elongated and flattened the subboundaries within them are being continuously broken up and reformed during the deformation. Thus, they maintain their equiaxed shape. Dynamic recovery occurs in metals of high stacking-fault energy such as aluminum, alpha iron, and most bcc metals.

[1] S. L. Semiatin and G. D. Lahoti, *Met. Trans.*, vol. 12A, pp. 1705, 1981.

[2] S. L. Semiatin and G. D. Lahoti, *Met. Trans.*, vol. 13A, pp. 275–288, 1982.

[3] J. J. Jonas, C. M. Sellars, and W. J. McG. Tegart, *Metall. Rev.*, vol. 14, pp. 1–24, 1969: H. J. McQueen and J. J. Jonas, Recovery and Recrystallization during High Temperature Deformation," Treatise on Materials Science and Technology, R. J. Arsenault (ed.), Academic Press, New York, 1975; W. Roberts, Dynamic Changes During Hot Working, "Deformation, Processing and Structure," G. Krauss (ed.), American Society for Metals, Metals Park, Ohio, 1984; T. Sakai and J. J. Jonas, *Acta Met.*, vol. 32, pp. 189–209, 1984.

In dynamic recrystallization dislocation annihilation only occurs when the dislocation density reaches such high levels that recrystallization occurs. As a result, the rate of strain hardening is high until recrystallization begins. This causes the peak in curve (*b*) in Fig. 15-19. As discussed above, the drop in flow stress (flow softening) promotes flow localization and unstable flow in the regions of the metal which are the first to recrystallize. If the dynamic recrystallization is accompanied by localized deformation heating it can lead to catastrophic strain localization. Because the dislocation densities and flow stresses are higher in dynamic recrystallization there is more susceptability to internal crack and cavity formation during deformation. However, once dynamic recrystallization begins, the process of grain-boundary migration tends to isolate the cavities which form on the previous static grain boundaries, thus preventing them from joining up to cause a catastrophic failure. Dynamic recrystallization is the predominant softening mechanism in the hot working of all fcc metals except aluminum.

Static recovery that occurs between intervals of deformation leads to a decrease in the density of dislocations, but the decrease in the flow stress is relatively small. However, static recovery is responsible for the formation of recrystallization nuclei, which is a precursor to softening by recrystallization: Static recrystallization at the highest working temperatures can take place in $\frac{1}{10}$ to $\frac{1}{100}$ s. As the temperature of deformation is reduced the recrystallization time gradually increases to hundreds of seconds. Therefore, it is often possible to cool rapidly enough from the working temperature to avoid recrystallization and retain the strengthening benefits of a deformed structure. Although temperature has the greatest effect on the kinetics of the recrystallization process, other important parameters are the degree of prestrain, the strain rate, and the alloy composition (see Sec. 6-15). The two softening mechanisms which produce the low flow stresses characteristic of hot working are static recrystallization or dynamic recovery, depending on the stacking-fault energy.

In metals which soften only by dynamic recovery, it is often possible to cool them sufficiently rapidly so that static recrystallization is prevented and the dynamic recovery structure is retained. This results in improved strength for non-heat-treatable Al-Mg alloys. For metals which soften by dynamic recrystallization it usually is not possible to retain the as-worked structure because static recrystallization occurs very rapidly when the deformation has been completed.

Hot-working greatly accelerates diffusional processes. Two examples of practical importance are the elimination of compositional inhomogeneities, such as a cored structure, by hot-working and the coarsening of a second-phase structure, such as the spheroidization of a pearlitic steel. For example,[1] a degree of spheroidization at 700°C which would take 660 h of annealing can be achieved in 210 s by deformation at 0.016 s^{-1}. Diffusion is greatly enhanced by the formation during deformation of a dynamic recovery substructure in which the subboundaries provide paths for pipe diffusion which is considerably more rapid than volume diffusion.

[1] E. A. Chojnowski and W. J. McG. Tegart, *Met. Sci. J.*, vol. 2, pp. 14–18, 1968; S. Chattopadhyay and C. M. Sellars, *Acta Met.*, vol. 30, pp. 157–170, 1982.

The plastic working characteristics of two-phase alloys depend on the microscopic distribution of the two phases. The presence of a high volume fraction of hard uniformly dispersed particles, such as are found in many superalloys, greatly increases the flow stress and makes working difficult. If the second-phase particles are hard and more massive, they will tend to fracture on deformation with the softer matrix extruding into the voids created by this fracturing. If the second-phase particles are ductile, failure usually occurs by the matrix pulling apart between the particles. Alloys which contain a hard second phase located primarily in the grain boundaries are difficult to work because of the tendency for fracture to occur along the grain boundaries. Cast ingots usually contain voids or porosity that arise because of the shrinkage associated with solidification or because of the liberation of gas. If these cavities have clean surfaces, they may be closed shut by pressure welding from mechanical working. Even if the cavity contains a gas, it may be closed if the gas is readily soluble in the metal at the working temperature.

As the result of a mechanical working operation second-phase particles will tend to assume a shape and distribution which roughly correspond to the deformation of the body as a whole. Second-phase particles or inclusions which are originally spheroidal will be distorted in the principal working direction into an ellipsoidal shape if they are softer and more ductile than the matrix. If the inclusions or particles are brittle, they will be broken into fragments which will be oriented parallel to the working direction, while if the particles are harder and stronger than the matrix, they will be essentially undeformed.[1] The orientation of second-phase particles during hot- or cold-working and the preferred fragmentation of the grains during cold-working are responsible for the fibrous structure which is typical of wrought products. This fiber structure can be observed after macroetching. Microscopic examination of wrought products frequently shows the results of this *mechanical fibering* (Fig. 15-20). An important consequence of mechanical fibering is that the mechanical properties may be different for different orientations of the test specimen with respect to the fiber (working) direction. In general, the tensile ductility, fatigue properties, and impact properties will be lower in the transverse direction (normal to the fiber) than in the longitudinal direction.

The forming characteristics of an alloy can be affected if it undergoes a strain-induced precipitation or strain-induced phase transformation. If a *precipitation reaction* occurs in a metal while it is being formed, it will produce an increase in the flow stress but, more important, there will be an appreciable decrease in ductility, which can result in cracking. When brittleness is caused by precipitation, it usually results when the working is carried out at a temperature just below the solvus line[2] or from cold-working after the alloy had been heated to the same temperature region. Since precipitation is a diffusion-controlled process, difficulty from this factor is more likely when forming is carried out at a

[1] H. Unkel, *J. Inst. Met*, vol. 61, pp. 171–196, 1937; F. B. Pickering, *J. Iron Steel Inst. London*, vol. 189, pp. 148–159, 1958.

[2] The boundary on the phase diagram of the limit of solid solubility of a solid solution.

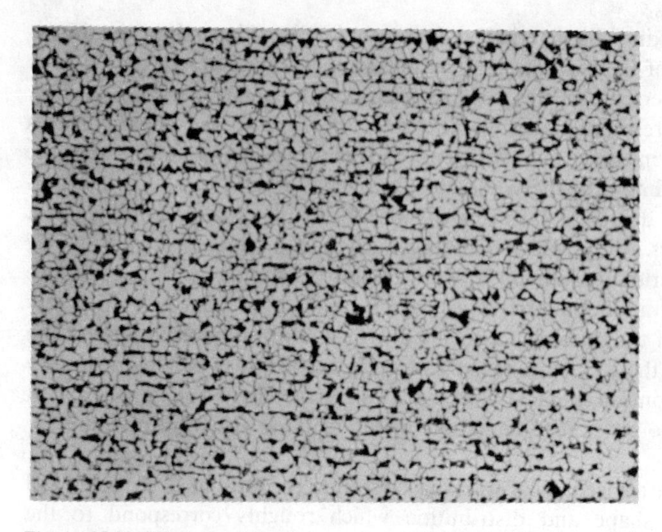

Figure 15-20 Fibered and banded structure in longitudinal direction of a hot-rolled mild-steel plate (100×).

slow speed at an elevated temperature. To facilitate forming, age-hardenable aluminum alloys are frequently refrigerated just before forming in order to suppress the precipitation reaction.

A most outstanding practical example of a *strain-induced phase transformation* occurs in certain austenitic stainless steels where the Cr : Ni ratio results in an unstable austenite phase. When these alloys are cold-worked, the austenite transforms to ferrite along the slip lines and produces an abnormal increase in the flow stress for the amount of deformation received. While this phase transformation is often used to increase the yield strength by cold rolling, it can also result in cracking during forming if the transformation occurs in an extreme amount in regions of highly localized strain.

The importance of grain size in determining strength and fracture toughness focuses special attention on the role of processing in creating a fine grain size.[1] Although every stage in the processing chain from ingot casting to cooling from the final working or heat treatment step may exert some influence, by far the most important role is played by the temperature at which hot-working is completed or the final annealing temperature for cold-worked material. In hot-working processes fine grain size is favored by a low finishing temperature and a rapid cooling rate from the working temperature. In processing steels which undergo a structural transformation on cooling from the finishing temperature, the ferrite grain size depends on achieving a fine austenite grain size. A fine austenite grain size is promoted by taking large reductions in the final pass and working at the lowest possible temperature in the austenite phase field.

[1] J. J. Burke and V. Weiss (eds.), "Ultra-fine Grain Metals," Sagamore Army Materials Research Conference Proceedings, vol. 16, Syracuse University Press, Syracuse, N.Y., 1970.

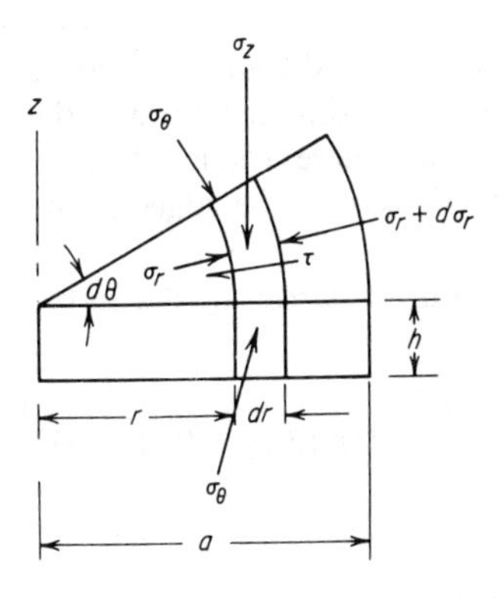

Figure 15-21 Stresses acting in compression of sector of flat circular disk.

The prior discussion has emphasized that the interaction of the deformation processing conditions of strain, strain rate, and temperature with metallurgical considerations such as phase transformation, precipitation, recrystallization, or grain growth determine the final properties of the product. The optimization of mechanical properties through appropriate *thermomechanical treatments*[1] is one of the areas of greatest promise in metalworking.

15-7 FRICTION AND LUBRICATION

The friction forces developed between the workpiece and the forming tools are an important consideration in metalworking. In the analysis of the forces in metalworking (Sec. 15-2), we neglected the friction forces in the interest of clarity. This certainly is not a realistic assumption, for in many real metalworking processes friction is the predominant factor.

As an example of an analysis with friction included, consider the slab analysis of the homogeneous compression of a flat circular disk (Fig. 15-21). We shall invoke the simplifying assumptions that there is no barreling of the edges of the disk and that the thickness is small enough that the axial compressive stress σ_z is constant through the thickness. The frictional conditions on the top and bottom faces of the disk are described by a constant coefficient of Coulomb friction

$$\mu = \frac{\tau}{p} \tag{15-40}$$

where τ = the shearing stress at the interface
p = the stress normal to the interface

[1] R. J. McElroy and Z. C. Szkopiak, *Int. Metall. Rev.*, vol. 17, pp. 175–202, 1972; E. B. Kula and M. Azrin, in "Advances in Deformation Processing," J. J. Burke and V. Weiss (eds.), Plenum Press, New York, pp. 245–300; J. C. Williams and E. A. Starke Jr., Role of Thermomechanical Processing in Tailoring the Properties of Aluminum and Titanium Alloys, in "Deformation Processing and Structure," G. Krauss (ed.), pp. 279–354, American Society for Metals, Metals Park, Ohio, 1984.

Lateral flow of the metal outward as it is being compressed leads to shearing stresses at the die contact surfaces. This surface shear is directed toward the center of the disk, opposing the outward radial flow. These frictional shear stresses lead to lateral pressure in the material, which is zero at the edges of the disk and builds up to a peak value at the center.

Referring to Fig. 15-21 and taking the equilibrium of *forces* in the radial direction:

$$\sigma_r hr\, d\theta - (\sigma_r + d\sigma_r)h(r + dr)\, d\theta + 2\sigma_\theta h\, dr \sin\frac{d\theta}{2} - 2\tau r\, d\theta\, dr = 0 \quad (15\text{-}41)$$

Using the approximation $\sin(d\theta/2) \approx d\theta/2$, this reduces to

$$\sigma_r h\, dr + d\sigma_r rh - \sigma_\theta h\, dr + 2\tau r\, dr = 0$$

From the axial symmetery of the disk, $d\varepsilon_\theta = d\varepsilon_r$, and from Eq. (3-47), $\sigma_\theta = \sigma_r$. Making this substitution, we arrive at

$$\frac{d\sigma_r}{dr} + \frac{2\tau}{h} = 0$$

and from the chosen friction law, $\tau = \mu p = \mu\sigma_z$

$$\frac{d\sigma_r}{dr} + \frac{2\mu\sigma_z}{h} = 0 \quad (15\text{-}42)$$

By making the assumption that σ_z, σ_θ, and σ_r are principal stresses, we can utilize the von Mises' yield criterion to develop a relation between σ_r and σ_z.

$$2\sigma_0^2 = (\sigma_1 - \sigma_2)^2 + (\sigma_2 - \sigma_3)^2 + (\sigma_3 - \sigma_1)^2$$

$$2\sigma_0^2 = (\sigma_r - \sigma_r)^2 + (\sigma_r - \sigma_z)^2 + (\sigma_z - \sigma_r)^2$$

$$\sigma_0^2 = (\sigma_r - \sigma_z)^2$$

and

$$\sigma_0 = \sigma_r - \sigma_z$$

If we define p as a *positive* compressive stress normal to the interface, then $p = -\sigma_z$, and $\sigma_0 = \sigma_r + p$, so that $d\sigma_r = -dp$. Making these substitutions into Eq. (15-42) gives

$$\frac{dp}{p} = -\frac{2\mu\, dr}{h} \quad (15\text{-}43)$$

Integration yields

$$\ln p = -\frac{2\mu r}{h} + C$$

At the outersurface of the disk, $r = a$, $\sigma_r = 0$, and $p = \sigma_0$, so that

$$C = \ln\sigma_0 + 2\frac{\mu a}{h}$$

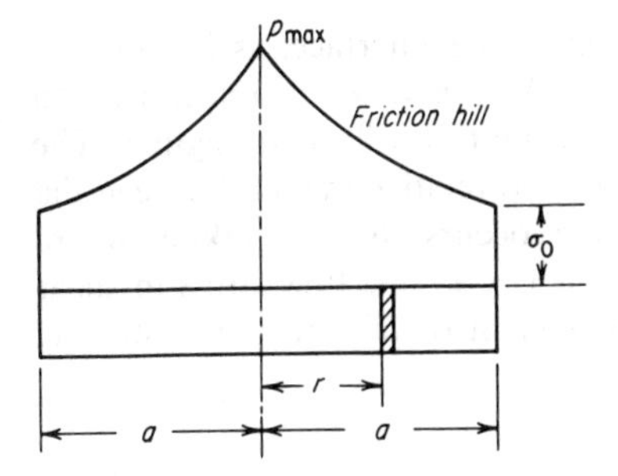

Figure 15-22 Friction hill for homogeneous compression of a disk with Coulomb friction.

Therefore,

$$\ln \frac{p}{\sigma_0} = \frac{2\mu}{h}(a - r) \tag{15-44}$$

or

$$p = \sigma_0 e^{\frac{2\mu}{h}(a-r)} \tag{15-45}$$

Figure 15-22 shows the axial compressive stress from Eq. (15-45) plotted over the diameter of the disk. The pressure distribution is symmeterical about the centerline and rises to a sharp peak at the center of the disk. This characteristic rise in deformation pressure with distance is often called a *friction hill*. The average deformation pressure (mean height of the friction hill) is given by

$$\bar{p} = \frac{\int_0^a 2\pi pr\, dr}{\pi a^2} = \frac{\sigma_0}{2}\left(\frac{h}{\mu a}\right)^2\left[e^{2\mu a/h} - \frac{2\mu a}{h} - 1\right] \tag{15-46}$$

Figure 15-23 plots the ratio of $\bar{p}$ to the uniaxial compressive flow stress as a function of the length-to-thickness ratio a/h of the disk and the coefficient of friction. For a given a/h increasing friction easily leads to more than a doubling of the deformation pressure. The role of friction in raising deformation forces should be clear from this figure. Note also that the role of friction becomes particularly important at large values of a/h.

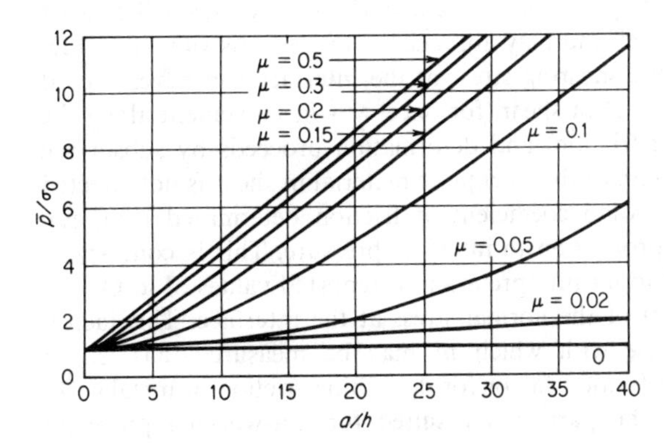

Figure 15-23 Average deformation pressure in compression of a disk as function of μ and a/h.

The above analysis was for sliding friction at the interface, as in our first encounter with friction in elementary physics. At the other extreme, we can envision a situation where the interface has a constant film shear strength τ_i. The most usual case is *sticking friction*, where there is no relative motion between the workpiece and the tools. This condition often occurs in hot-working where lubrication may be difficult. For sticking friction $\tau_i = \tau_0$, the flow stress in shear (k). With a von Mises' yield criterion, the coefficient of friction under sticking conditions is

$$\mu = \frac{\tau_i}{p} = \frac{k}{\sigma_0} = \frac{\sigma_0/\sqrt{3}}{\sigma_0} = \frac{1}{\sqrt{3}} = 0.577 \tag{15-47}$$

Using the sticking condition $\tau = k = \sigma_0/\sqrt{3}$ in the analysis for the compression of a disk results in

$$p = \sigma_0\left[1 + \frac{2}{\sqrt{3}}\frac{(a-r)}{h}\right] \tag{15-48}$$

Thus, the pressure distribution still peaks at $r = 0$, but now the sides of the hill are straight lines. Often a condition exists where sliding occurs at the outer periphery of the disk and sticking occurs at the interior of the disk.

The analysis for the condition of *sticking friction* suggests an alternative to the concept of Coulomb sliding friction as a way of looking at friction in metalworking processes. This approach considers that the workpiece in contact with the tools can be represented as a material of constant shear strength τ_i. However, instead of letting $\tau_i = k$, as for sticking friction, we consider that the interface shear strength may be some constant fraction m of the yield strength in shear where m is called the *interface friction factor*.

$$m = \frac{\tau_i}{k} = \frac{\text{interface shear strength}}{\text{yield stress in shear}} \tag{15-49}$$

Values of m vary from 0 (perfect sliding) to 1 (sticking).

There are basic differences between these two methods of treating friction. The interface pressure p developed in most metalworking processes will at least equal the uniaxial yield stress σ_0 and may appreciably exceed this value (see Fig. 15-23). On the other hand, the shearing stress at the interface can never exceed the yield strength of the material in shear, for once $\tau_i = k$, movement along the interface is arrested (sticking friction) and deformation proceeds by subsurface shearing. Since the yield strength of the workpiece material in shear is not affected by normal pressure, the Coulomb coefficient of friction determined from Eq. (15-40) decreases in inverse proportion to interface pressure. This is contrary to physical reality and can lead to misinterpretation of reported values of μ. On the other hand, m is independent of the normal stress at the interface. This factor, coupled with the relative ease with which m may be measured, has led to increased use of the interface friction factor for describing friction in metalworking processes. It appears to be particularly suited for hot-working processes

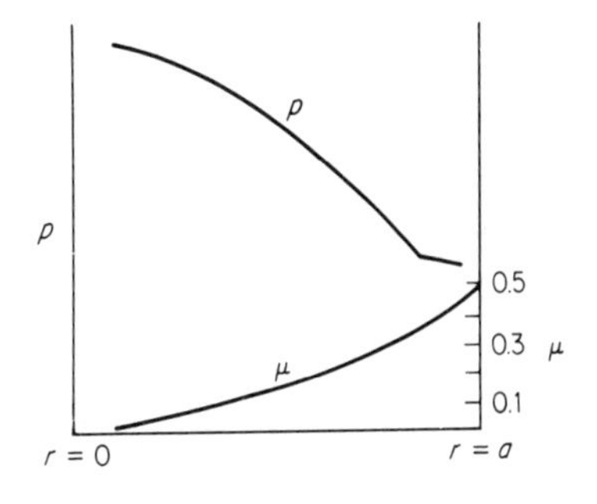

Figure 15-24 Variation of normal pressure p and μ with radial distance in compression of an aluminum disk without lubrication. *(After W. A. Backofen, "Deformation Processing," Addison-Wesley Publishing Company, Inc., Reading, Mass., 1972.)*

involving large deformations, such as forging and extrusion. In processes such as cold-rolling and wiredrawing the use of coefficient of friction appears to be well established.

Most analyses of metalworking processes have been carried out using μ. However, the use of m introduces mathematical simplification. Thus, if the compression of a disk is analyzed using m instead of μ, we replace Eq. (15-40) by $\tau = mk = m\sigma_0/\sqrt{3}$. Equation (15-42) now becomes, after simplification,

$$h\,dp = -\frac{2m\sigma_0}{\sqrt{3}}\,dr$$

and

$$p = -\frac{2m\sigma_0}{\sqrt{3}}\frac{r}{h} + C$$

which, when the boundary conditions are applied, reduces to

$$p = \sigma_0\left(1 + \frac{2m}{\sqrt{3}}\frac{a - r}{h}\right) \tag{15-50}$$

The similarity with Eq. (15-48) for sticking friction ($m = 1$) is to be expected.

It is usually assumed that the coefficient of friction has a constant value over the interface between the workpiece and the tools. Although there is little experimental data, it seems clear that μ varies with normal stress. Figure 15-24 shows the variation of p and μ for the compression of an aluminum disk. These data were obtained by painstaking experiments in which small pressure pins were buried in the tool under the workpiece.[1] Other factors which influence the local value of μ are the strain rate and the temperature at the interface.

The understanding of friction between two surfaces is at the heart of the science of *tribology*, an interdisciplinary field that involves surface chemistry, materials science, and mechanics. Most basic studies of friction[2] have been

[1] G. T. van Rooyen and W. A. Backofen, *Int. J. Mech. Sci.*, vol. 1, pp. 1–27, 1960; G. W. Pearsall and W. A. Backofen, *Trans. ASME, Ser. D: J. Basic Eng.*, vol. 85B, pp. 68–75, 1963.

[2] F. P. Bowden and D. Tabor, "The Frictional Lubrication of Solids," Oxford University Press, London, pt. I, 1950; pt. II, 1964; J. A. Schey, "Tribology in Metalworking: Friction, Lubrication and Wear," American Society for Metals, Metals Park, Ohio, 1983; E. Nachtman and S. Kalpakjian, "Lubricants and Lubrication in Metalworking Operations," Marcel Dekker Inc., New York, 1985.

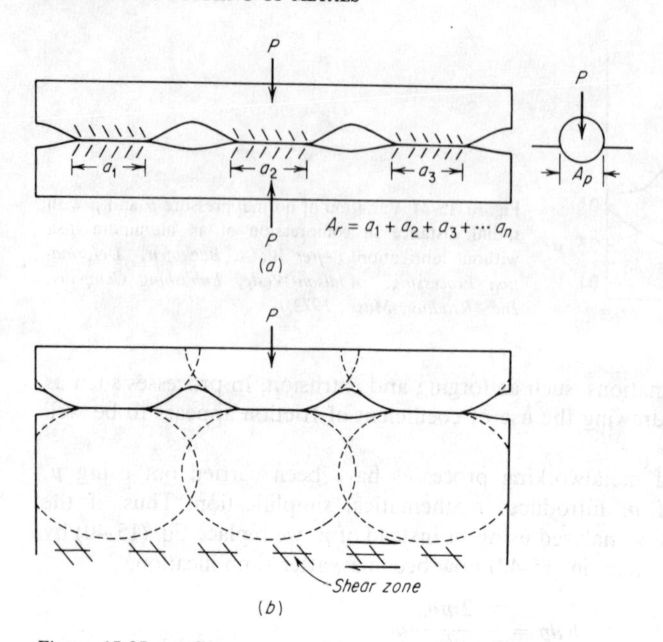

Figure 15-25 (*a*) Contact at asperities; (*b*) overlap of deformation zones to produce subsurface shear zone.

concerned with light loads where there is no plastic deformation of the workpiece. The running of a shaft in a bearing would be a typical example. A basic premise of the theory of friction as that apparently flat, smooth surfaces are not smooth when viewed on a microscopic scale. Real surfaces contain hills and valleys which are large when compared on an atomic scale. Therefore, when two surfaces are brought into contact they meet where the high spots, *asperities*, come into contact (Fig. 15-25). The real area of contact A_r is the sum of the area of contact at all of the asperities. Also, the surfaces are rarely clean. They contain layers of gas molecules or water molecules and often a thin oxide layer. Of course, frequently, special molecules are added to these surfaces. We call these lubricants.

As the load normal to the interface increases, the metal at the asperities deforms plastically, increasing the area of contact. Cold welding may also occur at these points. The situation at an asperity is very similar to indentation in a hardness test, where the hardness is the normal load P divided by the projected area of the indentation. $H = P/A_p \approx P/A_r$. Since hardness will be independent of the sum of the area of contact of the asperities and $A_r = P/H$, the real area of contact varies directly with the normal load. We shall let F be the shearing force parallel to the interface that tears apart the welded junctions. Thus, the coefficient of friction is given by

$$\mu = \frac{F}{P} = \frac{\tau_i A_r}{H A_r} = \frac{\tau_i}{H} \tag{15-51}$$

where τ_i is the shear stress to shear through the asperities. For light normal loads τ_i and H are independent of P. The coefficient of friction is thus independent of normal load. This is known as *Amonton's law* and is the criteria for Coulomb sliding friction to exist. As the normal force increases, the real area of contact increases. At high normal loads the asperities will undergo extensive plastic deformation. The "plastic zones" below each asperity will overlap (Fig. 15-25*b*). Because of this plastic deformation and the accompanying plastic constraint, the bond at the workpiece-tool interface is stronger than the shear strength of the material in a subsurface zone τ_0. For subsurface shear the coefficient of friction is

$$\mu = \frac{\tau_0}{H} = \tau_0 \frac{A}{P} \tag{15-52}$$

Since τ_0 and A are approximately constant, this shows that μ decreases as P increases, provided that bulk plastic deformation is taking place.

There are basically two approaches for measuring either μ or m under metalworking conditions. Tests developed for evaluating lubricants under light load (slider) conditions are not relevant to metalworking. The most common approach is to measure the average deformation pressure and the flow stress so that μ can be calculated from the analytical relation between these factors. For example, in the compression of a disk, if $\bar{p}$ is measured and σ_0 is known for the $\dot{\varepsilon}$ and T of the test conditions, then μ can be calculated with Eq. (15-46). Since information on the flow stress at the appropriate condition may not be available, it is useful to be able to determine μ without knowing σ_0 or k. For example, in the plane-strain compression test,[1] if the identical reduction is made with two values of b/t (see Fig. 15-13) and the average pressures are measured, then μ can be found by solving simultaneously the two sets of Eq. (15-53)

$$\frac{\bar{p}}{2k} = \frac{e^{\mu b/t} - 1}{\mu b/t} \tag{15-53}$$

A particularly useful technique for measuring m (or μ) is the *ring-compression test*.[2] The test uses a thin ring with the dimensions OD : ID : thickness in the ratio 6 : 3 : 1. The friction factor is determined by measuring the percentage of change in the inside diameter of the ring. For low values of m the ID increases, while for higher values of m the ID of the ring decreases. Calibration curves between ΔID, percent reduction, and m can be calculated from the analytical analysis of Avitzur.[3] Figure 15-26 shows a typical calibration curve.[4] It is important to note that it is not necessary to make a direct measurement of the deformation force or to know the flow stress in order to measure m with the ring-compression test. Thus, m may conveniently be measured at the high temperature and strain rate typical of hot-working processes.

[1] H. Takahashi and J. M. Alexander, *J. Inst. Met.*, vol. 90, pp. 72–79, 1961–1962.

[2] A. T. Male and M. G. Cockcroft, *J. Inst. Met.*, vol. 93, pp. 38–46, 1964–1965.

[3] B. Avitzur, "Metal Forming," McGraw-Hill Book Company, New York, 1968.

[4] C. H. Lee and T. Altan, *Trans. ASME, Ser. B: J. Eng. Ind.*, vol. 94, p. 775, 1972.

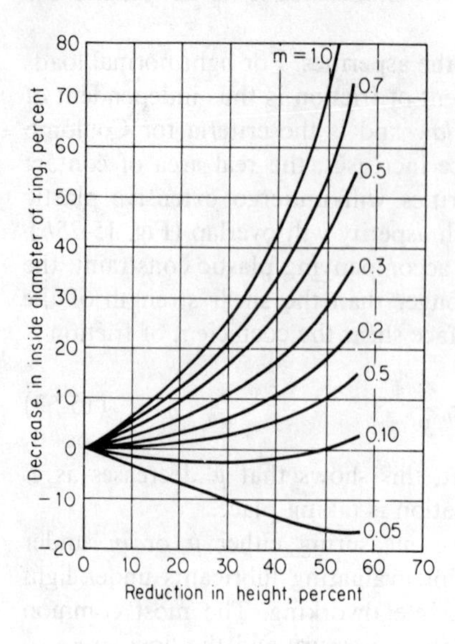

Figure 15-26 Calibration curve for upset ring test with outside diameter, inside diameter, and thickness in a ratio of 6 : 3 : 1. *(From C. H. Lee and T. Altan, Trans. ASME Ser. B: J. Eng. Ind.*, vol. 94, *p. 775, 1972.)*

However, if deformation pressure is measured during compression of the ring, it is possible to calculate[1] the flow stress as well as to determine m. This is particularly attractive for hot forming processes where die chilling of the workpiece makes it difficult to simulate the test conditions to measure the flow stress.

Example In a ring compression test a specimen 10 mm high with outside diameter 60 mm and inside diameter 30 mm is reduced in height by 50 percent. Determine the friction factor if (a) the OD after deformation is 70 mm and (b) the OD after deformation is 81.4 mm. Determine the peak pressure at the center of a 60-mm-diameter cylinder compressed 50 percent for each case.

(a)
$$\text{Volume} = \frac{\pi}{4}(60^2 - 30^2)10 = \frac{\pi}{4}(70^2 - D_i^2)5$$

$$D_i = 22.3 \text{ mm}$$

Percent change in inside diameter = $(30 - 22.3)/30 \times 100 = 25.6$ percent decrease. From Fig. 15-26, $m = 0.27$.

(b)
$$\text{Volume} = \frac{\pi}{4}(60^2 - 30^2)10 = \frac{\pi}{4}(81.4^2 - D_i^2)5$$

$$D_i = 35.1 \text{ mm}$$

[1] A. T. Male, V. dePierre, and G. Sand, *ASLE Trans.*, vol. 16, pp. 177–184, 1973; J. R. Douglas and T. Altan, *ASME*, Paper No. 73-WA/Prod-13, 1973.

Percent change in inside diameter = (30 − 35.1)/30 × 100 = 17 percent increase. From Fig. 15-26, $m = 0.05$. From Eq. (15-50)

$$\frac{p}{\sigma_0} = 1 + \frac{2m}{\sqrt{3}}\frac{a}{h} = 1 + \frac{2m}{\sqrt{3}}\left(\frac{30}{10}\right)$$

$$m = 0.27 \qquad p/\sigma_0 = 1.935$$

$$m = 0.05 \qquad p/\sigma_0 = 1.173$$

Boundary lubricants are thin organic films physically absorbed or chemisorbed on the metal surface. They are typically polar substances, fatty acids, alcohols, and fatty oil derivatives, which attach themselves to metal surfaces in an oriented fashion and react with the metal surface to form a metal soap. These films have a low shear strength which is pressure- and temperature-dependent. Boundary films form rapidly and as the film thickness decreases metal-to-metal contact occurs. Depending on the boundary film strength and thickness the coefficient of friction can vary from about 0.1 to 0.4.

Full-fluid film lubrication occurs when the surfaces are fully separated by a fluid film, as in a hydrodynamic bearing. These conditions are somewhat rare in metalworking processes, but they do occur when the geometery of the process creates a converging gap and the sliding speeds are high. High speed wiredrawing and rolling are the best examples. A thick fluid film is about 10 times greater than the surface roughness. For this condition there is essentially no wear and the coefficient of friction is between 0.001 and 0.02. As the normal force increases, or as the speed and viscosity of the fluid decrease, the film thickness is reduced to about 3 to 5 times the surface roughness. This produces some metal-to-metal contact,with an increase in coefficient of friction and wear rate.

In most liquid lubricated metalworking operations some asperity contact is unavoidable. A significant portion of the load is carried by the metal-to-metal contact and the rest is carried by pockets of liquid in the valleys of the asperities. This *mixed film lubrication* is common in metalworking. Boundary lubricants with extreme pressure (EP) additives should be added to the fluid to form a boundary film at the points of metal contact.

Conversion coatings (oxides, phosphates, or chromates) are often applied to the workpiece to serve as a base for the retention of the lubricant. Some of these coatings also have lubricating properties.

The surface finish of a wrought product depends on the frictional conditions. In cold-working the surface asperities are flattened by a smooth polished tool to produce a burnished or shiny surface. A rough tool will not produce a finish better than the finish of the tool. With thick-film lubrication and smooth tools a dull or matte finish is produced. When the workpiece is formed out of contact with the tools, as in certain sheet-forming operations, then the surface finish is controlled by the properties of the workpiece. Defects such as orange peel and stretcher strains may result.

Although the emphasis in this section has been on the role of friction in increasing the deformation forces, from a practical point of view many production machines have excess load capacity. Thus, while high forming loads per se may not be of practical significance, the high stresses arising from high friction may result in various limits to the attainable deformation. Many of these will be described in subsequent chapters dealing with specific processes. It is obvious that tearing and cracking are accentuated by high friction. In many practical situations the chief effect of friction is in influencing the surface finish of the product or in affecting the wear of the dies and tooling.

When one solid surface in a sliding pair is much harder than the other (as in the typical tool-workpiece combination) the asperities of the harder surface will penetrate the softer surface and displace a volume of metal proportional to the total length of sliding and the cross-sectional area of the asperity. This is known as ploughing. Frictional resistance due to ploughing is in addition to that arising from the shearing of asperity contacts. The ploughing force[1] is related to the flow properties of the workpiece and the size and shape of the asperities. Thus, smooth dies are important in reducing the friction contribution from ploughing.

The most serious result of inadequate lubrication is the transfer of workpiece material to the tools. This is called pickup. Metal transfer occurs in two ways. If the lubricant film is depleted at the interface of a rough tool surface, the workpiece is forced into crevices in the tool surface. Subsequent tangential motion shears off the projecting soft metal and results in tool pickup and a poor surface finish on the workpiece. If the lubricant breaks down under high pressure, there is local cold welding between tool and workpiece. When a piece is torn away from the workpiece surface, it exposes fresh clean metal which is even more susceptible to cold welding. Thus, when pickup begins it often becomes progressively worse and leads to galling and seizure.

Since the tool surfaces are subjected to many sliding cycles, there is continual wear of these surfaces.[2] The chief wear mechanism is abrasion by hard oxide particles on the workpiece. Another source of wear is surface fatigue resulting from the buildup and release of the interface pressure. Surface cracking of tools may also result from thermal stresses arising from heating and cooling of the tools.

In selecting a lubricant, the workpiece, the die, and the lubricant should be considered a total system.[3] The functions of a metalworking lubricant are many:

1. Reduces deformation load
2. Increases limit of deformation before fracture
3. Controls surface finish
4. Minimizes metal pickup on tools
5. Minimizes tool wear
6. Thermally insulates the workpiece and the tools
7. Cools the workpiece and/or the tools

[1] J. Goddard and H. Wilman, *Wear*, vol. 5, pp. 114–135, 1962.

[2] A. D. Sarkar, "Wear of Metals," Pergamon Press, New York, 1976.

[3] J. A. Schey (ed.), "Metal Deformation Processes: Friction and Lubrication," Marcel Dekker, Inc., New York, 1970.

There are many, often contradictory, attributes required of a good metalworking lubricant. It must be capable of functioning over a wide range of pressure, temperature, and sliding velocities. Since one of the characteristics of most plastic working processes is the generation of a large amount of new surface area, the lubricant must have favorable spreading and wetting characteristics. It must be compatible with both the die and workpiece material with regard to wetting and chemical attack. It should have good thermal stability and resistance to bacteriological attack and minor contaminants. A good lubricant produces a harmless residue that does not cause staining on subsequent heat treatment or welding, and which is easily removed. Finally, a lubricant should be nontoxic, free of fire hazard, and inexpensive.

Metalworking lubricants perform their functions in a number of ways. Certain lubricants function as a film of low shear strength. Some oxide films on a steel surface behave this way, as do soft metal coatings like lead or tin. Layer-lattice compounds such as graphite and molybdenum sulfide can be deposited as continuous low shear stress layers.

Extreme pressure lubricants are organic compounds of phosphorous and sulfur. They react with a steel surface to form a thin surface film which prevents adhesion. This reaction occurs chiefly at the contact points where high temperature and pressure are developed. As the film wears away it reforms if sufficient time is available. EP films are very thin and not continuous.

15-8 DEFORMATION-ZONE GEOMETERY

Most deformation processes require the material to flow through a converging channel, which we call dies. Figure 15-27 shows some typical situations. The basic feature of each channel is the ratio of the mean thickness to the length of the deformation zone.[1] This ratio is designated Δ, and for the simple case of parallel (nonconverging) dies is $\Delta = h/L$. Values of Δ for more complex situations are given in Fig. 15-27. Δ_s is based on a plane-strain reduction $r_s = 1 - (h_1/h_0)$ for strip. Δ_w is based on an axisymmeteric reduction $r_w = 1 - (d_1/d_0)^2$ for wire or rod. To decrease Δ, we can reduce the semidie angle α, thereby increasing L, or we can increase the reduction r which decreases h.

The importance of deformation-zone geometery Δ on the yield pressure for frictionless plane-strain indentation[2] is shown in Fig. 15-28. Examples of various deformation-zone geometries and typical processes which tend to have that value of Δ are given in this useful illustration. The increase in $p/2k$ with Δ is chiefly the result of increasing redundant work.[3]

[1] W. A. Backofen, "Deformation Processing," Addison-Wesley Publishing Company, Inc., Reading, Mass., 1972.

[2] R. Hill, "The Mathematical Theory of Plasticity," p. 257, Oxford University Press, London, 1950.

[3] W. A. Backofen, op. cit., p. 137.

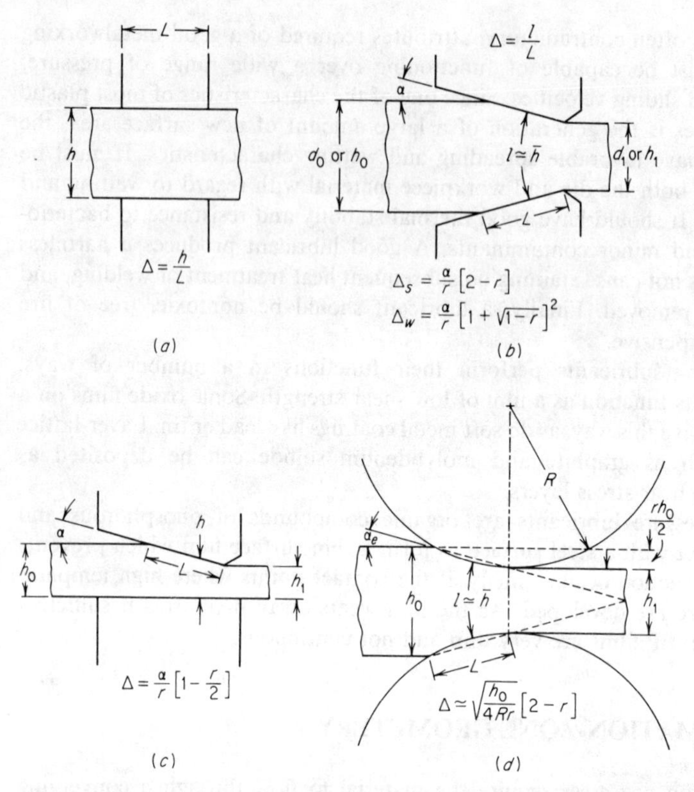

Figure 15-27 Converging channels for different deformation processes showing calculation of Δ *(a)* parallel indenters (plane-strain compression); *(b)* drawing or extrusion of strip or wire; *(c)* tube drawing over a mandrel; *(d)* rolling. *(From W. A. Backofen, "Deformation Processing," Addison-Wesley Publishing Company, Inc., Reading, Mass., 1972.)*

Figure 15-28 shows that the die pressure increases the greater the ratio h/L. However, this neglects friction. The smaller the h/L ratio the greater the effect of friction at the tool-workpiece interfaces. This factor has been considered by Shaw[1] in a reexamination of deformation-zone analysis.

15-9 HYDROSTATIC PRESSURE

An important factor in achieving a successful forming operation without fracture is the level of hydrostatic pressure achieved in the process. The presence of a high hydrostatic pressure reduces the tensile stresses below the critical value for cracking, while at the same time the flow stress is unaffected. As an extra bonus, deformation carried out under high hydrostatic pressure creates less damage to the material during deformation. The effect of hydrostatic pressure in increasing ductility in the tension test has already been discussed in Sec. 7-13.

[1] M. C. Shaw, *Int. J. Mach. Tool Des. Res.*, vol. 22, no. 3, pp. 215–226, 1982.

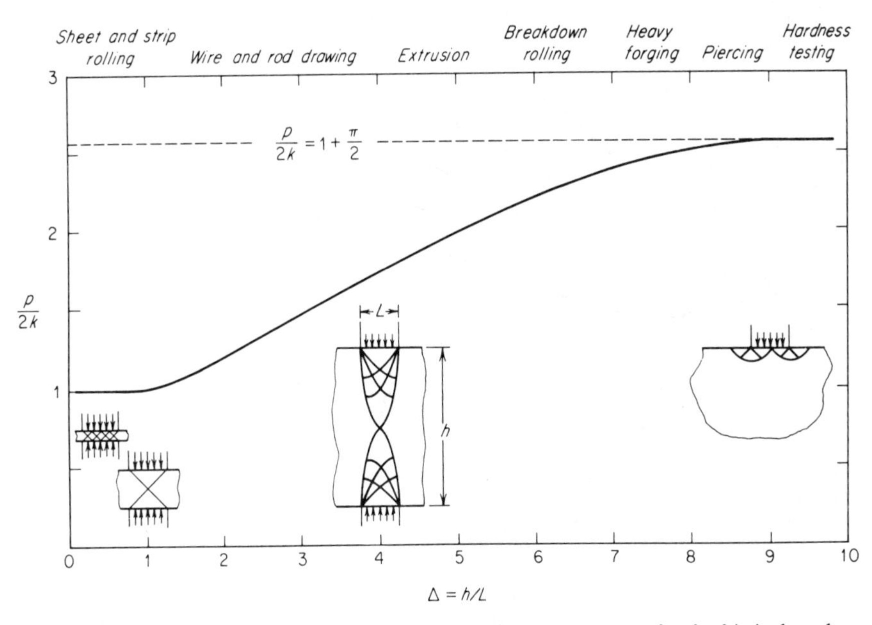

Figure 15-28 Dependence of yield pressure on deformation-zone geometry for the frictionless plane-strain indentation of a rigid-ideal plastic material. *(From W. A. Backofen, "Deformation Processing," p. 135, Addison-Wesley Publishing Company, Inc., Reading, Mass., 1972.)*

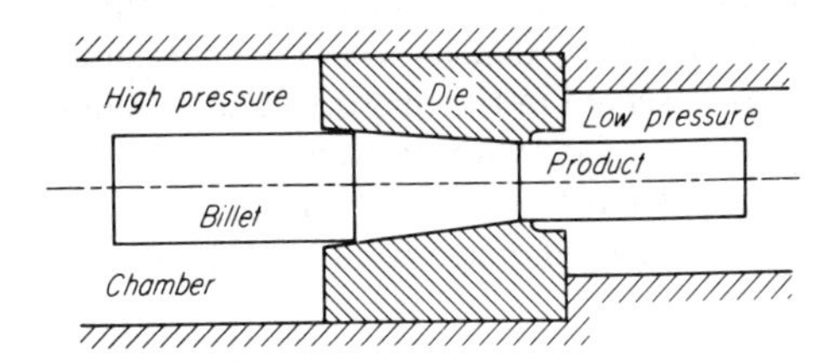

Figure 15-29 Typical arrangement for hydrostatic extrusion.

The hydrostatic pressure may arise from the hydrostatic component of the stress state, chiefly by the interaction of the workpiece and the tooling as in extrusion, or it may be introduced externally to the process, as in hydrostatic extrusion. Figure 15-29 shows the elements of the hydrostatic extrusion process in which the billet is forced through the die by a high hydrostatic fluid pressure. The chief advantages of this process are the elimination of the large drag friction force between the billet and the container wall and the achievement of hydrodynamic lubrication in the die.

The influence of hydrostatic stress state on the structural damage produced in cold-drawing of strip was studied by Coffin and Rogers.[1] Using the slip-line field

[1] L. F. Coffin, Jr., and H. C. Rogers, *Trans. Am. Soc. Met.*, vol. 60, pp. 672–686, 1967.

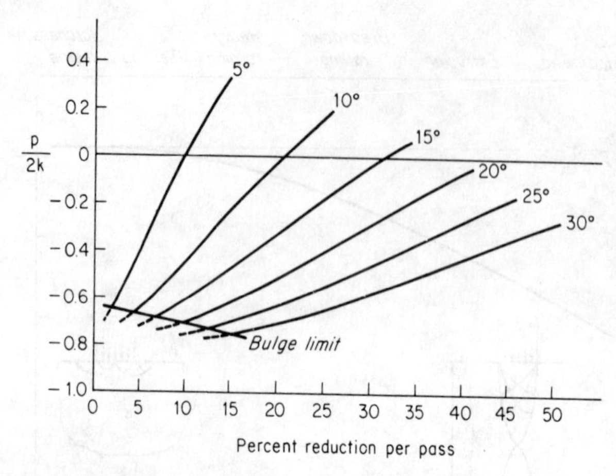

Figure 15-30 Ratio of hydrostatic pressure p to tensile yield stress $2k$ at centerline of cold-drawn strip. Each curve for dies with different semiangle die. Positive value of $p/2k$ denotes compressive hydrostatic stress. *(From L. F. Coffin and H. C. Rogers, Trans. Am. Soc. Met., vol. 60, p. 674, 1967. Copyright American Society for Metals, 1967.)*

for frictionless[1] strip drawing (Fig. 15-4), it is possible to map out the hydrostatic pressure at each point in the deformation zone. Figure 15-30 shows the hydrostatic pressure developed at the centerline (midplane) of the sheet as a function of semidie angle and reduction per pass. The damage produced by this deformation was measured by precision density measurement. Extreme damage also results in reduced ductility in the tension test. Decreases in density could be measured before any void formation or cracking was observed in the microstructure. Figure 15-31 shows that large decreases in density correlate with drawing conditions in which the hydrostatic stress at the centerline is most tensile. The strong influence of external hydrostatic pressure in greatly reducing damage from cold-drawing is shown in Fig. 15-32.

Hydrostatic pressure may be used to advantage in working brittle materials such as bismuth or cast iron. For brittle materials it is necessary to extrude into a pressurized chamber in order to avoid cracking.[2] Figure 15-33 shows that the hydrostatic pressure required to avoid cracking varies with extrusion ratio. This reflects the way the hydrostatic component of stress produced in the deformation zone varies with reduction, since the pressure plotted is the sum of this stress plus the fluid hydrostatic pressure. While the curve given is specific to a particular situation, it generally is found that cracking problems are less severe at very low or very large reductions and that there is an intermediate range of reduction for which cracking is more prevalent. The beneficial effects of hydrostatic pressure

[1] The influence of friction was subsequently covered in H. C. Rogers and L. F. Coffin, Jr., *Int. J. Mech. Sci.*, vol. 13, pp. 141–155, 1971.

[2] H. L. D. Pugh and D. Gunn, *Symp. Physics and Chemistry of High Pressure*, Society of Chemical Industry, London, 1963, pp. 157–162.

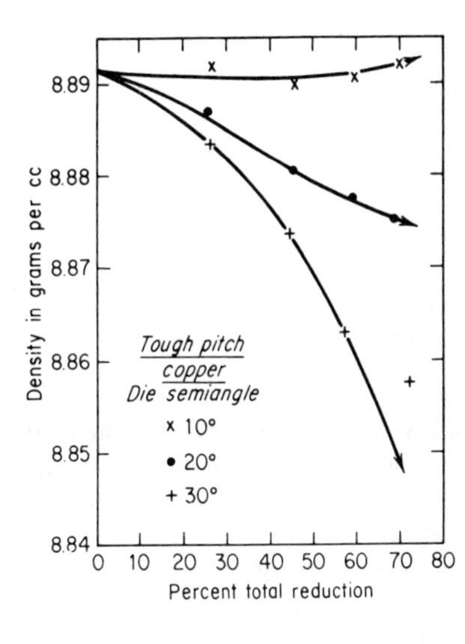

Figure 15-31 Structural damage produced in strip drawing of tough pitch copper, as measured by density. Increasing die semiangle corresponds to greater hydrostatic tension. (Compare Fig. 15-26.) (*L. F. Coffin and H. C. Rogers, Trans. Am. Soc. Met., vol. 60, p. 678, 1967. Copyright American Society for Metals, 1967.*)

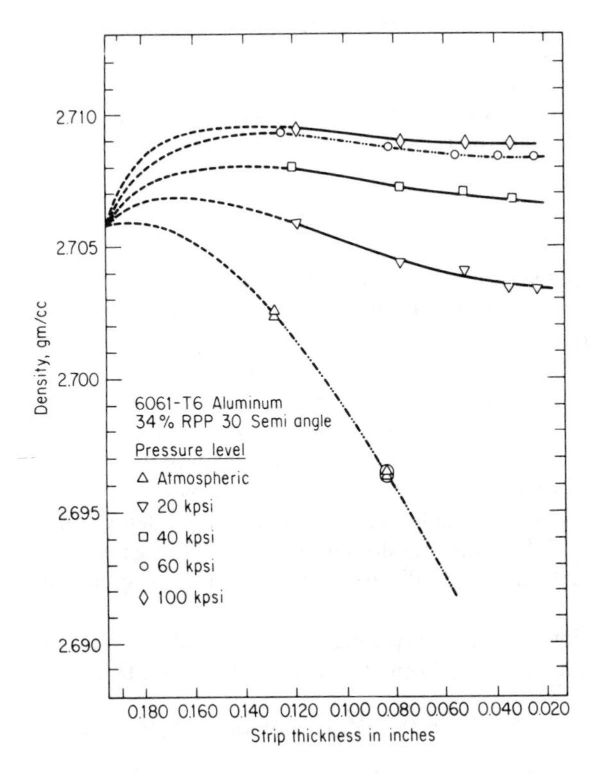

Figure 15-32 Influence of external hydrostatic pressure on the density change in 6061 aluminum alloy deformed in strip drawing under highly tensile conditions. (*From L. F. Coffin and H. C. Rogers, Trans. Am. Soc. Met., vol. 60, p. 681, 1967. Copyright American Society for Metals, 1967.*)

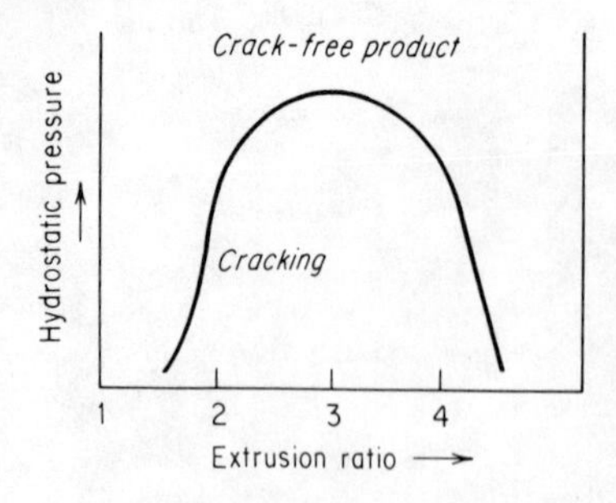

Figure 15-33 Effect of reduction on critical pressure to inhibit cracking (schematic).

may be achieved without the difficulty of employing high-pressure equipment if the workpiece is encased with a metal of low flow stress compared with the workpiece. For example, brittle beryllium billets are "canned" in mild steel to produce hydrostatic pressure support during forging.

While the presence of a high hydrostatic compressive stress enhances the limit of workability in processes which are limited by cracking, it will not improve workability in processes which are limited by the formation of a local necked region. Examples of metalworking processes where local plastic instability limits the extent of deformation are wiredrawing and deep drawing of sheet. The reason for this is that hydrostatic pressure does not affect the tensile instability of metals,[1] and therefore, the initiation of local necking in a metal under tension is independent of the hydrostatic state of stress.

15-10 WORKABILITY

Workability is concerned with the extent to which a material can be deformed in a specific metalworking process without the formation of cracks. In some processes the limit of workability is determined by the formation of local necking (plastic instability) rather than by the occurrence of fracture. Workability is a complex technological concept that depends not only on the fracture resistance (ductility) of the material but also on the specific details of the deformation process. Therefore, workability should be thought of as a function made up of two factors: f_1 depends on the properties of the material and is evaluated by some sort of small-scale laboratory test[2]; f_2 is a function of the parameters of the deformation process such as die geometry, lubrication conditions, and workpiece geometery. These two functions are brought together with the workability-limit diagram (see Fig. 15-36).

From the above discussion, it is not surprising that a generally acceptable laboratory test for workability has not been developed.[3] The upsetting of a

[1] H. L. D. Pugh, *ASTM Spec. Tech. Publ.* 374, p. 68, 1965; J. M. Alexander, *J. Inst. Met.*, vol. 93, p. 366, 1965.

[2] G. E. Dieter (ed.), "Workability Testing Techniques," American Society for Metals, Metals Park, Ohio, 1984.

[3] G. E. Dieter, Bulk Workability Testing, "Metals Handbook," 9th ed., vol. 8, American Society for Metals, Metals Park, Ohio, 1985, pp. 571–597.

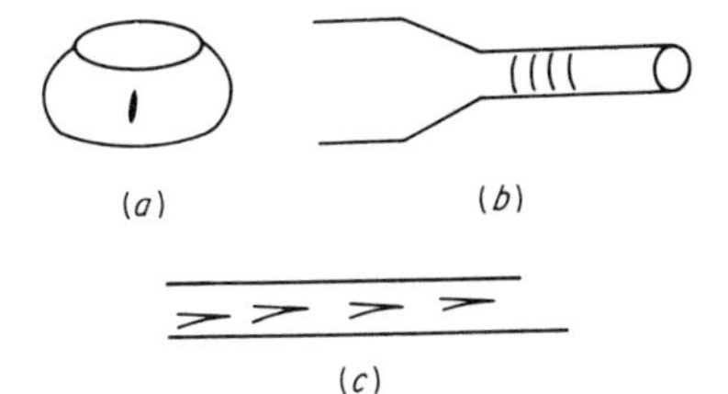

Figure 15-34 Examples of cracks in metalworking processes. (*a*) Free surface crack in upsetting; (*b*) surface cracks from heavy die friction in extrusion; (*c*) center burst or chevron cracks in a drawn rod.

cylinder under controlled strain rate conditions comes the closest to such a standard test. However, there is still a requirement for many very specialized and often empirical tests for evaluating the workability of materials.

The cracks that occur in metalworking processes can be grouped into three broad categories (see Fig. 15-34):

1. Cracks at a free surface, such as at the bulge in upsetting a cylinder or in edge cracking in rolling
2. Cracks that develop in a surface where interface friction is high, such as in extrusion
3. Internal cracks, such as chevron cracks in drawn bars

Internal cracks develop as a result of "secondary tensile stresses" which typically occur with large values of $\Delta = h/L$. Temperature gradients often occur in the workpiece, and this produces a situation very conducive to fracture. A particularly common situation is where the workpiece is chilled by the die, and since the flow stress is usually strongly temperature-dependent, a chilled region produces a local nondeforming zone. The presence of regions of "hard and easy flow" leads to the development of shear bands, and the localization of flow into these bands results in very high shear strains and often shear fracture (see p. 534–535).

Cracks that form in metalworking occur by ductile fracture (see Sec. 7-9). Ductile fracture is characterized by the formation and growth of holes around second-phase particles and the localization of shear strain in sharply defined bands. Void formation in ductile fracture in the tension test has been associated with the plastic instability of necking, and it appears that tensile cracking due to bulging in upsetting is similarly related to plastic instability.[1]

To predict workability, a fracture criterion for ductile fracture must be established. The most generally applicable ductile-fracture criterion is that proposed by Cockcroft and Latham.[2] They reasoned that a criterion of ductile fracture will be based on some combination of stress *and* strain and that only the highest local tensile stress σ^* will be important. They proposed that ductile fracture will occur when

$$\int_0^{\bar{\varepsilon}_f} \sigma^* \, d\bar{\varepsilon} = \text{constant} \tag{15-54}$$

[1] H. A. Kuhn and P. W. Lee, *Metall. Trans.*, vol. 2, pp. 3197–3202, 1971.

[2] M. G. Cockcroft and D. J. Latham, *J. Inst. Met.*, vol. 96, pp. 33–39, 1968.

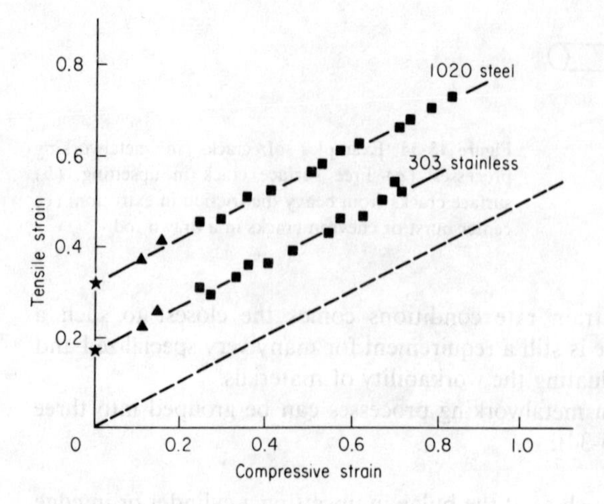

Figure 15-35 Fracture criterion expressed in terms of tensile and compressive strains at fracture for upset and bend tests at room temperature. *(From P. W. Lee and H. A. Kuhn, Metall. Trans., vol. 4, pp. 969–974, 1973.)*

for a given temperature and strain rate. This criterion was shown to provide good correlation between fracture strain in tension and torsion tests and helped predict safe ranges of reduction in cold extrusion. However, more extensive studies[1] of free surface fracture in room-temperature deformation have shown consistent fracture criteria in the form of a unique correlation for each material between the tensile and compressive strains at fracture (Fig. 15-35). The ratio of these strains is varied by changing the D/h ratio and the frictional conditions for upset cylinders and by using other tests such as bending and edge cracking in rolling. The dashed curve shows the strain ratio which would be obtained for ideal (nonbarreling) compression of the cylinders. Fracture occurs when the tensile and compressive strains at fracture place the strain state above the line with a slope of $\frac{1}{2}$. Other studies[2] have shown that this fracture criterion is sensitive to the anisotropy in rolled steel plate. The local surface strains at fracture in upset tests correlate well with the true zero-gage-length fracture strain in the tension test provided that proper consideration is given to fiber orientation. The constant slope of $\frac{1}{2}$ and the use of a strain correlation of this type have been shown[3] to be consistent with the Cockcroft and Latham fracture criterion.

The workability of a material in a given process is portrayed by the workability limit diagram (Fig. 15-36). The material factor (f_1) in workability is represented by a strain limit diagram for the material, such as Fig. 15-35. The process factor (f_2) is represented by strain paths at potential fracture sites in the material. If the strain path in the deformation of the material exceeds the fracture limit line then fracture will occur. The strain paths are determined experimentally by placing grids on the surface of a model material such as plasticene, lead, or pure

[1] P. W. Lee and H. A. Kuhn, *Metall. Trans.*, vol. 4, pp. 969–974, 1973.

[2] T. Erturk, W. L. Otto, and H. A. Kuhn, *Metall. Trans.*, vol. 5, pp. 1883–1886, 1974.

[3] H. A. Kuhn, P. W. Lee, and T. Erturk, *Trans. ASME, Ser. H: J. Eng. Mater. Technol.* vol. 95, pp. 213–218, 1973; S. I. Oh, C. C. Chen, and S. Kobayashi, *Trans. ASME J. Eng. Ind.*, vol. 101, pp. 36–44, 1979.

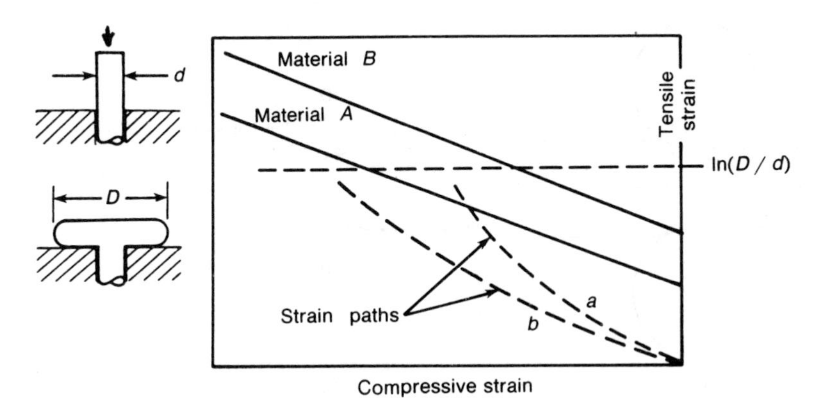

Figure 15-36 Workability limit diagram for cold upsetting of a bolt head. *(After H. A. Kuhn.)*

aluminum. Alternatively, the strain paths might be obtained from finite element models of the deformation process. In the example shown in Fig. 15-36 a bolt head is formed by cold upsetting. To form a head of diameter D from a rod of diameter d requires the material to withstand a circumferential surface strain of $\ln D/d$. Thus, the strain path must reach this limiting value of strain without crossing the fracture limit line. Strain path a does not meet this criterion, but by improving the lubrication to shift the strain path to curve b a bolt head can be made without fracture. Alternatively, a successful bolt head could be made with strain path a if the material were changed from A to the more workable material with curve B. This workability concept has been applied to complex forged parts.[1]

A much different approach to workability was taken by Avitzur[2] to predict internal cracking in extrusion and wiredrawing. Velocity flow fields are constructed for sound deformation without a defect and for deformation which produces a defect-like center burst or chevron cracks. These flow fields are functions of process conditions like die angle, friction, and total reduction. The prevailing flow field is that for which the total energy is minimum. The analysis predicts increased occurrence of center burst with increasing die angle and decreasing reduction, in agreement with earlier experimental results.

Extensive work has been done in developing fracture diagrams for sheet forming operations. This is described in detail in Chap. 20.

15-11 RESIDUAL STRESSES

Residual stresses are a system of stresses which can exist in a body when it is free from external forces. Since residual stresses are generated by nonuniform plastic deformation, it is important to consider the residual stresses developed in each

[1] C. L. Downey and H. A. Kuhn, *Trans. ASME J. Eng. Mater. Tech.*, vol. 97, pp. 121–125, 1975.
[2] B. Avitzur, *Trans. AMSE J. Eng. Ind.*, vol. 90, pp. 79–99, 1968.

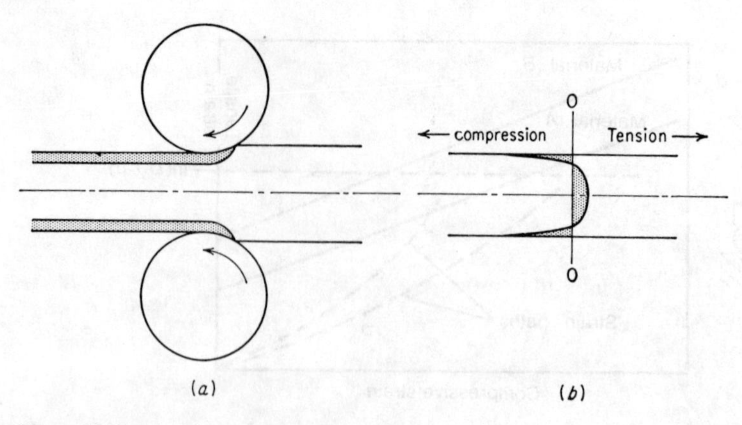

Figure 15-37 (*a*) Inhomogeneous deformation in rolling of sheet; (*b*) resulting distribution of longitudinal residual stress over thickness of sheet (schematic).

metalworking process. The general method by which residual stresses are produced in a metalworking process is shown in Fig. 15-37. In this example the rolling conditions are such that plastic flow occurs only near the surfaces of the sheet. The surface grains in the sheet are deformed and tend to elongate, while the grains in the center of the sheet are unaffected. Since the sheet must remain a continuous whole, the surface and center regions of the sheet must undergo a strain accommodation. The center fibers tend to restrain the surface fibers from elongating, while the surface fibers seek to stretch the central fibers of the sheet. The result is a residual-stress pattern in the sheet which consists of a high compressive stress at the surface and a tensile residual stress at the center of the sheet (Fig. 15-37*b*). In general, the sign of the residual stress which is produced by inhomogeneous deformation will be opposite to the sign of the plastic strain which produced the residual stress. For the case of the rolled sheet, the surface fibers which were elongated in the longitudinal direction by rolling are left in a state of compressive residual stress when the external rolling load is removed.

The residual-stress system existing in a body must be in static equilibrium. Thus, the total force acting on any plane through the body and the total moment of forces on any plane must be zero. For the longitudinal residual-stress pattern in Fig. 15-37*b*, the area under the curve subjected to compressive residual stresses must balance the area subjected to tensile residual stresses. However, it should be kept in mind that the actual situation is made complex by the presence of a three-dimensional state of residual stress.

Residual stresses are only elastic stresses. The maximum value which a residual stress can reach is the yield stress of the material. For purposes of analysis, residual stresses can be considered the same as ordinary applied stresses. Thus, a compressive residual stress effectively subtracts from the applied tensile stress, and a tensile residual stress adds to an applied tensile stress.

Metals containing residual stresses can be *stress relieved* by heating to a temperature where the yield strength of the material is the same or less than the value of the residual stress. Thus, the material can deform and release the stress. Creep deformation is important in thermal stress relief. It is important to realize that nonuniform thermal expansion or contraction due to nonuniform heating or cooling can produce residual stresses in the same way as nonuniform plastic deformation. Therefore, slow cooling from the stress-relief temperature may be an important consideration.

The differential strains that produce residual stresses also can be reduced substantially at room temperature by plastic deformation. For example, products such as sheet, plate, and extrusions are often stretched several percent beyond the yield stress to relieve differential strains by plastic deformation. Residual stresses in cold-drawn rod and tube are relieved by roller straightening.

Residual stresses are difficult to calculate with precision by analytical methods,[1] and therefore, they are usually determined by a variety of experimental techniques.[2] Most methods of measuring residual stresses are destructive because they involve removing part of the stressed material to cause a redistribution of stress in the remaining body. This may be done by machining layers from a cylinder or drilling a small hole and measuring the redistribution of strain with suitably placed strain gages. The chief nondestructive method of measuring residual stresses is x-ray analysis.[3] In this method x-rays are used to determine the interatomic spacing of a particular set of lattice planes in the strained material. The residual stresses can be calculated by comparing the spacing measured on the material with residual stress with the spacing in a stress-free sample.

15-12 EXPERIMENTAL TECHNIQUES FOR METALWORKING PROCESSES

Experimental studies of metalworking processes require techniques for measuring the forming loads and the deformations. Measurement of the forces presents no fundamental problems, although considerable ingenuity may be required to instrument many pieces of production equipment. Force usually is measured with strain-gage load cells. The upsetting of standardized copper cylinders[4] permits presses to be calibrated so that loads can be measured by strain gages attached to

[1] Conference on Calculation of Internal Stresses, *Mat. Sci. and Tech*. vol. 1, pp. 754–857, 1985.

[2] A. A. Denton, *Metall. Rev*., vol. 11, pp. 1–23, 1966; W. M. Baldwin, *Am. Soc. Test. Mater. Proc*., vol. 49, pp. 539–583, 1949; "Residual Stress Measurements," American Society for Metals, Metals Park, Ohio, 1952.

[3] C. S. Barrett and T. B. Massalski, "Structure of Metals," 3rd ed., chap. 17, McGraw Hill Book Company, New York, 1966; E. Macherauch, *Exp. Mech*., vol. 6, pp. 140–153, 1966; C. O. Ruud, *J. Metals*, pp. 35–40, July 1981.

[4] T. Altan and D. E. Nichols, *Trans. ASME, Ser. B: J. Eng. Ind*., vol. 94, pp. 769–774, 1972.

the press frame. Other instruments[1] frequently needed are transducers to measure ram displacement, velocity, and acceleration. High-speed photography may be useful for measuring free surface displacement and many other kinematic variables.

Most studies[2] of the deformation in forming operations have employed grid networks. A number of identical specimens are deformed by different amounts, and the progress of the deformation is determined from the distortion of the network. Either rectangular or polar-coordinate grids are used, depending on the application. The grid network may be applied either by scribing or by photographic methods. However, only a limited amount of information can be obtained from grids placed on the surface of the workpiece. The flow in the interior of the workpiece can be studied by cutting the billet in half, affixing a grid network to the two faces, fastening them back together, and then machining to a symmetrical shape. Under certain circumstances the interior deformation can be determined by embedding a lead grid in the casting. The distortion of the lead grid is obtained by radiographing the deformed billet. Sometimes plugs are inserted in the workpiece, and their distortion is measured by sectioning and examining under the microscope.

Metallographic techniques are useful for determining regions of heavy deformation. The direction of flow can be determined from the grain distortion and the preferred alignment of second-phase particles and inclusions. An initially banded structure provides a unidirectional internal grid system. Etches are available for most metals which selective attack the plastically deformed regions. Since recrystallization will begin first in the most heavily deformed grains, the examination of the microstructure of a deformed metal after annealing will indicate the presence of nonuniform deformation.

Experimentation in metalworking processes may be expensive and time-consuming when dealing with large production equipment. When many trial-and-error experiments are required, as may be necessary in optimizing tooling geometery, it may be advantageous to conduct a preliminary series of experiments in which the real workpiece material is simulated with a model material. Model materials (such as plasticine, wax, lead, or clay) have low flow stresses compared with metals so that low-cost plastic or wooden tooling can be used in the experiments. Thus, it is possible to study a wide range of tooling geometries or to construct a model of a complex piece of production equipment. Strict similarity between the real and model experiments is very difficult to achieve,[3] but useful results can be obtained if the friction conditions are similar and if appropriate

[1] H. N. Norton, "Handbook of Transducers for Electronic Measuring Systems," Prentice-Hall, Inc., Englewood Cliffs, N. J., 1969. J. W. Dally, W. F. Riley and K. G. McConnell, "Instrumentation for Engineering Measurements," John Wiley & Sons, New York, 1984.

[2] H. P. Tardiff, *Steel Process. Conver.*, vol. 43, pp. 626–632, 643–644, 650, 1957; G. L. Baraya, J. Parker and J. W. Flowett, *J. Mech. Sci.*, vol. 5, pp. 365–367, 1963; R. L. Bell and T. G. Langdon, *J. Sci. Instrum.*, vol. 42, p. 896, 1965.

[3] T. Altan, H. J. Henning, and A. M. Sabroff, *Trans. ASME, Ser. B: J. Eng. Ind.*, vol. x, pp. 444–452, 1970.

temperatures and strain rates are used for the model material. Plasticine[1] is a popular model material because its room-temperature flow curve approximates that of metals under hot-working conditions. Plasticine is often used to check theoretical analysis of plastic forming because it deforms similar to an ideal plastic material. Grid networks can easily be developed by working with a billet built up from layers of black and white plasticine. It is a simple matter to slice through the deformed material to observe the deformation pattern in any desired direction.

15-13 COMPUTER-AIDED MANUFACTURING

The potential of the computer for greatly optimizing manufacturing processes and reducing costs has been clearly demonstrated and is being implemented in many parts of the metalworking industry.[2] Computer-aided manufacturing is the use of computers to assist in all phases of manufacturing a part, starting with the design step. Usually we speak of computer-aided design/computer-aided manufacturing, CAD/CAM. Figure 15-38 shows in a simplified way the steps in a computer-aided system for selecting the most economical part and selecting the most cost effective manufacturing process. Group technology (GT) is used to classify part shapes to assist in the decisions on process selection and detailed process planning. A computerized process model which takes cognizance of the critical role of material behavior is a key part of the system.

In metal deformation processes the use of finite element codes, such as ALPID, has great promise. When combined with interactive graphics it permits a direct visualization of the metal deformation as well as determination of the velocities, strain rates, strains, and stresses throughout the deforming metal. For example, an interactive forging design program called DIEFRG starts with coordinate data for the various cross sections of the final forging and information on the flow stress of the material and the friction. The program calculates the stress distribution over the cross section, the forging load (for press selection), the part volume, and the forging flash dimensions. Next the preform cross section is defined and displayed. The program can also display the dies at various stroke positions. Another important output of many such design programs is the generation of numerical control tapes for machining the dies or for machining the electrical discharge machining electrode for producing the dies.

For maximum efficiency and cost it is important that manufacturing be treated as a total system. This involves not only the tie-in between design and

[1] A. P. Green, *Philos. Mag.*, vol. 42, pp. 365–373, 1951; K. Chijiiwa, Y. Hatamura, and N. Hasegawa, *Trans. Iron Steel Inst. Japan* (English edition), vol. 21, pp. 178–186, 1981.

[2] J. Harrington Jr., "Understanding the Manufacturing Process: Key to Successful CAD/CAM Implementation", Marcel Dekker, Inc., New York, 1984; U. Rembold, C. Blume and R. Dillman, "Computer-Integrated Manufacturing Technology and Systems", Marcel Dekker, Inc., New York, 1985.

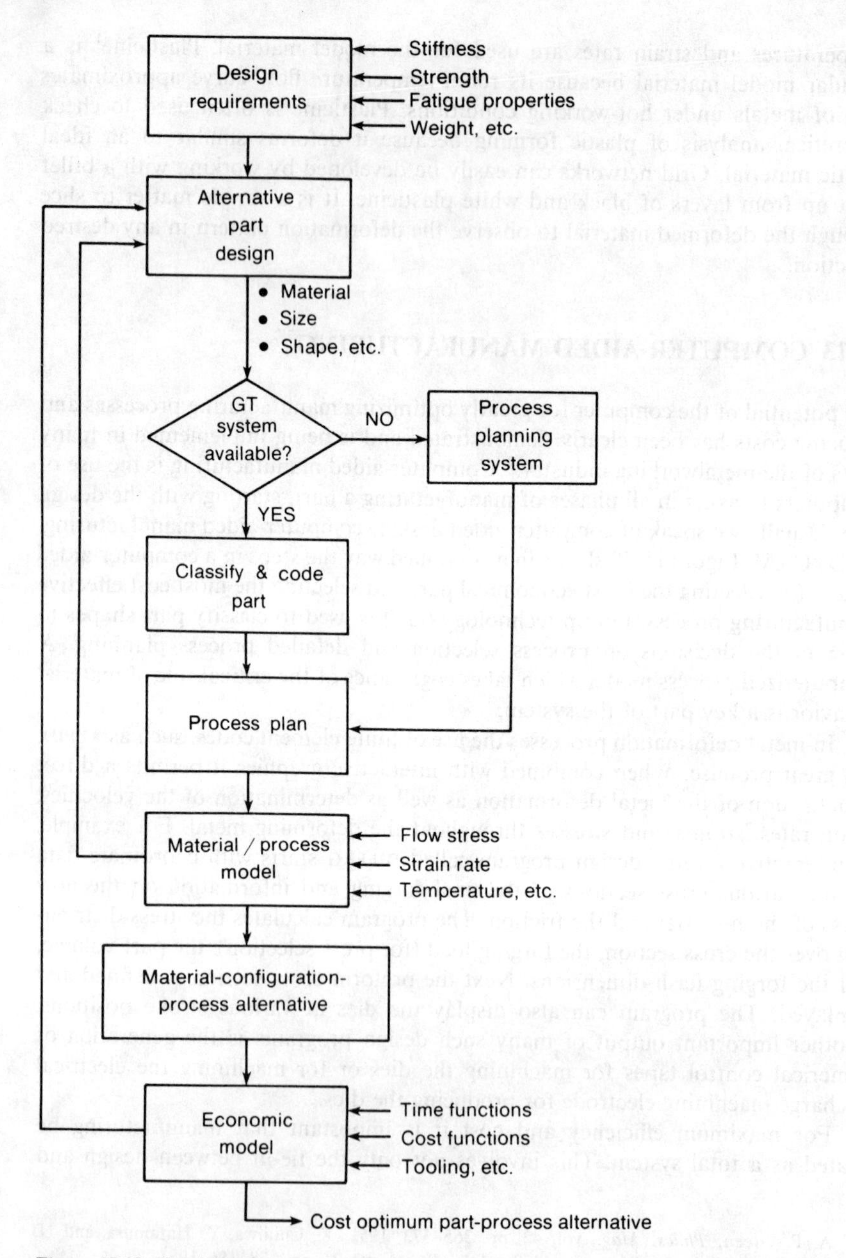

Figure 15-38 Simplified flow sheet for a computer-aided process to select the least cost manufacturing process.

manufacturing with computers but also the expansion of the computer to include inspection, quality control, and inventory control. This wider concept is called *computer-integrated manufacturing*, CIM.

BIBLIOGRAPHY

Altan, T. S. Oh, and H. L. Gegel: "Metal Forming," American Society for Metals, Metals Park, Ohio, 1983.

Alting, L.: "Manufacturing Engineering Processes," Marcel Dekker, Inc., New York, 1982.

Avitzur, B.: "Handbook of Metal Forming Processes," John Wiley & Sons, New York, 1983.

Backofen, W. A.: "Deformation Processing," Addison-Wesley Publishing Co., Reading, Mass., 1972.

Blazynski, T. Z.: "Metal Forming: Tool Profiles and Flow," John Wiley & Sons, New York, 1976.

Gopinathan, V.: "Plasticity Theory and Its Application in Metal Forming," John Wiley-Halsted Press, New York, 1982.

Harris, J. N.: Mechanical Working of Metals," Pergamon Press, Inc., New York, 1983.

Hosford, W. F. and Caddell, R. M.: "Metal Forming: Mechanics and Metallurgy," Prentice-Hall, Inc., Englewood Cliffs, N.J., 1983.

Johnson, W. and Mellor, P. B.: "Engineering Plasticity," Van Nostrand Reinhold Company, New York, 1973.

Kalpakjian, S.: "Manufacturing Processes for Engineering Materials," Addison-Wesley Publishing Co., Reading, Mass., 1984.

Lange, K. (ed.): "Handbook of Metal Forming", McGraw-Hill Book Company, New York 1985

Parkins, R. N.: "Mechanical Treatment of Metals," American Elsevier Publishing Company, New York, 1968.

Rowe, G. W.: "Principles of Industrial Metal Working Process," Edward Arnold, London, 1977.

Schey, J. A.: "Introduction to Manufacturing Processes," McGraw-Hill Book Company, New York, 1977.

Slater, R. A.: "Engineering Plasticity: Theory and Application to Metal Forming Processes," John Wiley & Sons, New York, 1974.

CHAPTER

SIXTEEN

FORGING

16-1 CLASSIFICATION OF FORGING PROCESSES

Forging is the working of metal into a useful shape by hammering or pressing. It is the oldest of the metalworking arts, having its origin with the primitive blacksmith of Biblical times. The development of machinery to replace the arm of the smith occurred early during the Industrial Revolution. Today there is a wide variety of forging machinery which is capable of making parts ranging in size from a bolt to a turbine rotor or an entire airplane wing.

Most forging operations are carried out hot, although certain metals may be cold-forged. Two major classes of equipment are used for forging operations. The forging hammer, or drop hammer, delivers rapid impact blows to the surface of the metal, while the forging press subjects the metal to a slow-speed compressive force.

The two broad categories of forging processes are *open-die forging* and *closed-die forging*. Open-die forging is carried out between flat dies or dies of very simple shape. The process is used mostly for large objects or when the number of parts produced is small. Often open-die forging is used to preform the workpiece for closed-die forging. In closed-die forging the workpiece is deformed between two die halves which carry the impressions of the desired final shape. The workpiece is deformed under high pressure in a closed cavity, and thus precision forgings with close dimensional tolerances can be produced.

The simplest open-die forging operation is the *upsetting* of a cylindrical billet between two flat dies. The compression test (Fig. 15-7) is a small-scale prototype of this process. As the metal flows laterally between the advancing die surfaces, there is less deformation at the die interfaces because of the friction forces than at

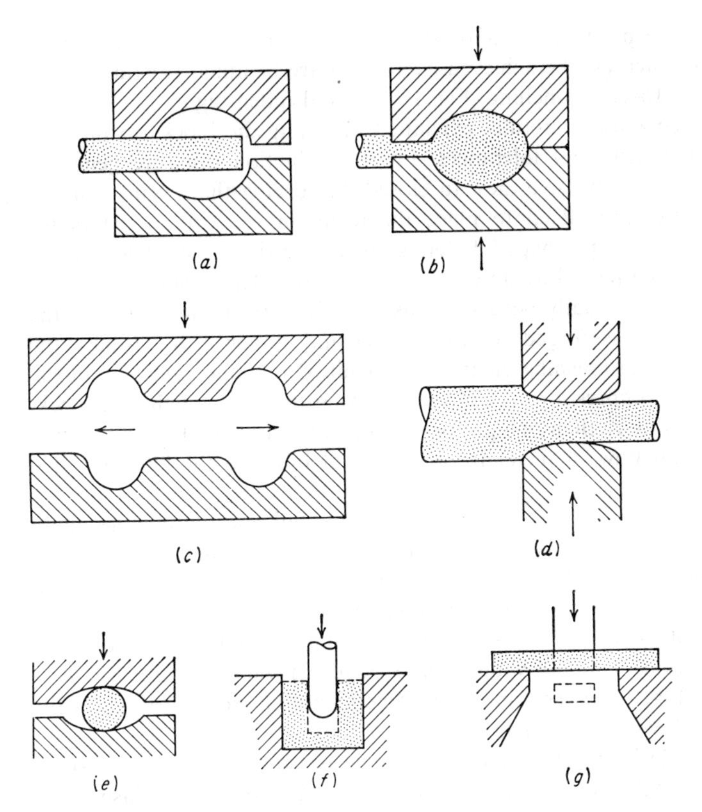

Figure 16-1 Forging operations. (*a*, *b*) Edging; (*c*) fullering; (*d*) drawing; (*e*) swaging; (*f*) piercing; (*g*) punching.

the midheight plane. Thus, the sides of the upset cylinder becomes barreled.[1] As a general rule, metal will flow most easily toward the nearest free surface because this represents the lowest frictional path.

The effect of friction in restraining metal flow is used to produce shapes with simple dies. *Edging* dies are used to shape the ends of the bars and to gather metal. As is shown in Fig. 16-1*a* and *b*, the metal is confined by the die from flowing in the horizontal direction but it is free to flow laterally to fill the die. *Fullering* is used to reduce the cross-sectional area of a portion of the stock. The metal flow is outward and away from the center of the fullering die (Fig. 16-1*c*). An example of the use of this type of operation would be in the forging of a connecting rod for an internal-combustion engine. The reduction in cross section of the work with concurrent increase in length is called *drawing down*, or drawing out (Fig. 16-1*d*). If the drawing-down operation is carried out with concave dies

[1] K. M. Kulkarni and S. Kalpakjian, *Trans. ASME, Ser. B: J. Eng. Ind.*, vol. 91, pp. 743–754, 1969.

(Fig. 16-1*e*) so as to produce a bar of smaller diameter, it is called *swagging*. Other operations which can be achieved by forging are bending, twisting, extrusion, piercing (Fig. 16-1*f*), punching (Fig. 16-1*g*), and indenting.

Closed-die forging uses carefully machined matching die blocks to produce forgings to close dimensional tolerances. Large production runs are generally required to justify the expensive dies. In closed-die forging the forging billet is usually first fullered and edged to place the metal in the correct places for subsequent forging. The preshaped billet is then placed in the cavity of the *blocking die* and rough-forged to close to the final shape. The greatest change in the shape of the metal usually occurs in this step. It is then transferred to the *finishing die*, where it is forged to final shape and dimensions. Usually the blocking cavity and the finishing cavity are machined into the same die block. Fullering and edging impressions are often placed on the edges of the die block. For complex shapes more than one preforming or blocking operation is required to achieve a gradual flow of metal from the initial billet to the complex final shape.

It is important to use enough metal in the forging billet so that the die cavity is completely filled. Because it is difficult to put just the right amount of metal in the correct places during fullering and edging, it is customary to use a slight excess of metal. When the dies come together for the finishing step, the excess metal squirts out of the cavity as a thin ribbon of metal called *flash*. In order to prevent the formation of a very wide flash, a ridge, known as a *flash gutter*, is usually provided (Fig. 16-2). The final step in making a closed-die forging is the removal of the flash with a *trimming die*.

Because of the flash, the term *closed-die forging* is a bit of a misnomer, and a better description for the process would be *impression-die forging*. The flash serves two purposes. As described above, it acts as a "safety valve" for excess metal in the closed-die cavity. Of more importance, the flash regulates the escape of metal, and thus the thin flash greatly increases the flow resistance of the system so that the pressure builds up to high values to ensure that metal fills all recesses of the die cavity. Figure 16-3 shows a typical curve of forging load vs. die advance (press stroke) for a closed-die forging process. The trick in designing the flash is to adjust its dimensions so that the extrusion of metal through the narrow flash opening is more difficult than the filling of the most intricate detail in the die. But, this must not be done to excess so as to create very high forging loads with attendent problems of die wear and breakage. The ideal is to design for the

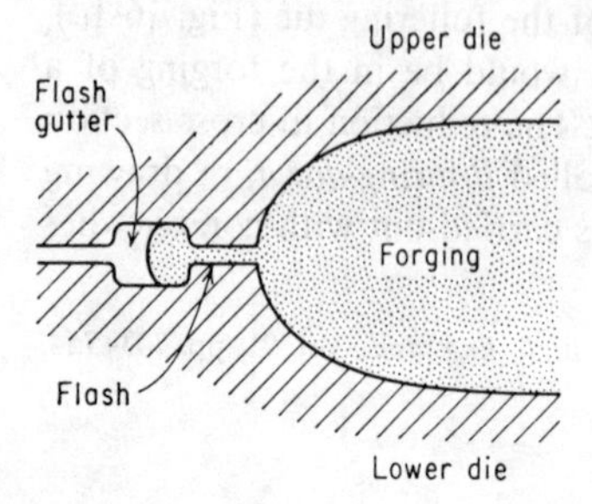

Figure 16-2 Sectional view through closed-die forging.

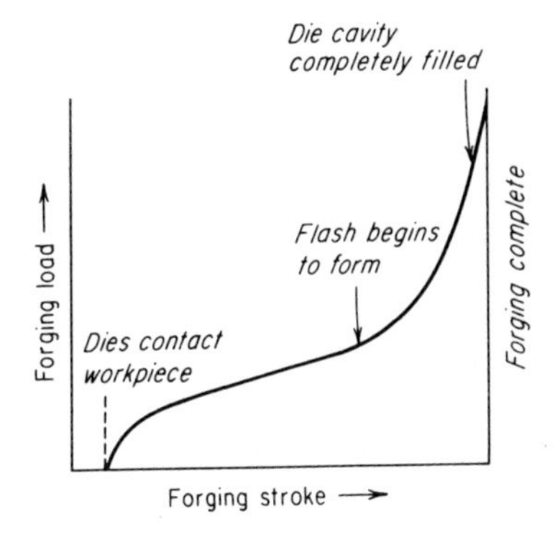

Figure 16-3 Typical curve of forging load vs. stroke for closed-die forging.

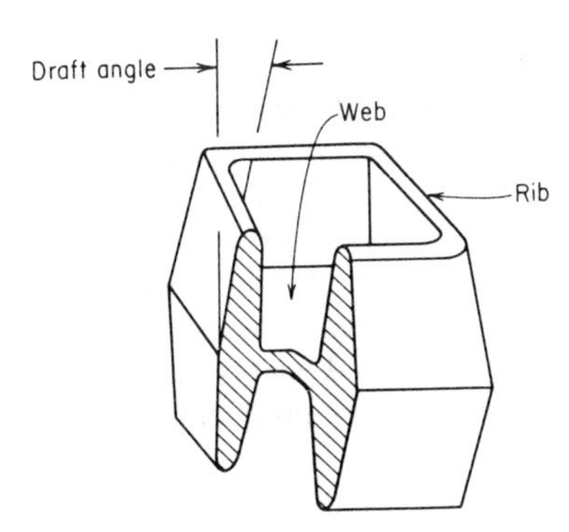

Figure 16-4 Some typical forging nomenclature.

minimum flash needed to do the job. Forging pressure increases with decreasing flash thickness and increasing flash land width.

The metal flow is greatly influenced by the part geometry. Spherical or blocklike shapes are easiest to forge in impression dies. Shapes with thin and long sections or projections (ribs and webs) are more difficult because they have higher surface area per unit volume, and therefore friction and temperature effects are enhanced. It is particularly difficult to produce parts with sharp fillets, wide thin webs, and high ribs (Fig. 16-4). Moreover, forging dies must be tapered to facilitate removal of the finished piece. This *draft allowance* is approximately 5° for steel forgings.

16-2 FORGING EQUIPMENT

Forging equipment may be classified with respect to the principle of operation.[1] In *forging hammers* the force is supplied by a falling weight or *ram*. These are

[1] T. Altan, "Forging Equipment, Materials, and Practice", chap. 1, Metals and Ceramics Information Center, MCIC-HB-03, Battelle Memorial Laboratories, Columbus, Ohio, 1973; Tool and Manufacturing Engineers Handbook, 4th ed., vol. 2, pp. 15-18 to 15-40, Soc. of Manufacturing Engineers, Dearborn, Mich., 1984.

Table 16-1 Typical values of velocity for different forging equipment†

Forging machine	Velocity range	
	ft/s	m/s
Gravity drop hammer	12–16	3.6–4.8
Power drop hammer	10–30	3.0–9.0
HERF machines	20–80	6.0–24.0
Mechanical press	0.2–5	0.06–1.5
Hydraulic press	0.2–1.0	0.06–0.30

† T. Altan, "Forging Equipment, Materials, and Practices," Metals and Ceramics Information Center, MCIC-HB-03, 1973.

energy-restricted machines since the deformation results from dissipating the kinetic energy of the ram. *Mechanical forging presses* are stroke-restricted machines since the length of the press stroke and the available load at various positions of the stroke represent their capability. *Hydraulic presses* are load-restricted machines since their capability for carrying out a forming operation is limited chiefly by the maximum load capacity. Each of these classes of forging equipment needs to be examined with respect to its load and energy characteristics, its time-dependent characteristics, and its capability for producing parts to dimension with high accuracy.

To successfully complete a forging operation, the available machine load must exceed the required load at any point in the process and the available machine energy must exceed the energy required by the process for the entire stroke. The most important characteristic of any machine is the number of strokes per minute, for this determines the production rate. The velocity under pressure v_p is the velocity of the machine slide under load. This variable determines the strain rate (which influences the flow stress) and the contact time under pressure t_p. Typical values of v_p are given in Table 16-1. The contact time is the time that the workpiece remains in the die under load. Since heat transfer between the hotter forging and the cooler dies is so effective when the interface is under high pressure, die wear increases with t_p. Dimensional accuracy of the parts produced by a forging machine is directly related to the stiffness of the equipment. In general terms, stiffness is increased by using larger components in the construction of the machine, and this is directly related to increased equipment cost. A particularly critical problem is the tilting of the ram under load.

The most commonly used piece of forging equipment is the forging hammer. The two basic types of hammers are the *board hammer* and the *power hammer* (Fig. 16-5). In the board hammer the upper die and ram are raised by friction rolls gripping the board. When the board is released, the ram falls under the influence of gravity to produce the blow energy. The board is immediately raised again for another blow. Forging under a hammer usually is done with repeated blows. Hammers can strike between 60 and 150 blows per minute depending on

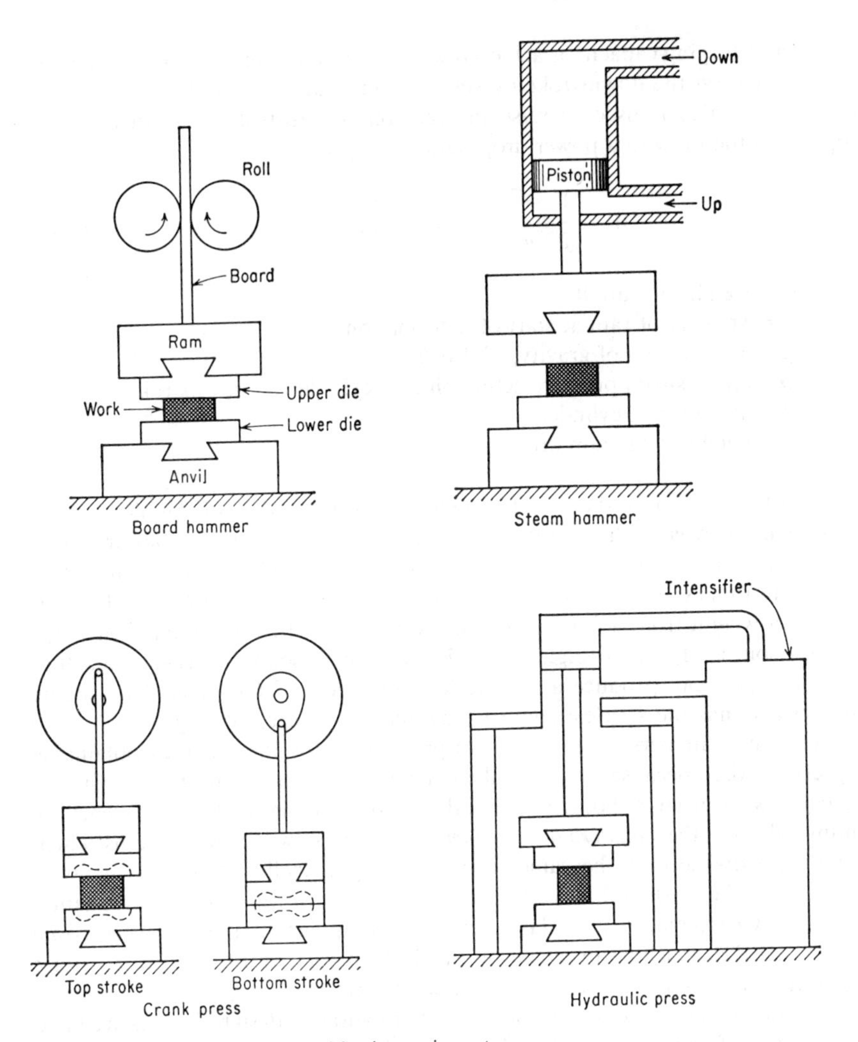

Figure 16-5 Schematic drawings of forging equipment.

size and capacity. The energy supplied by the blow is equal to the potential energy due to the weight of the ram and the height of the fall. Forging hammers are rated[1] by the weight of the ram. However, since the hammer is an energy-restricted machine, in which the deformation proceeds until the total kinetic energy is dissipated by plastic deformation of the workpiece or elastic deformation of the dies and machine, it is more correct to rate these machines in terms of energy delivered.

[1] "Metals Handbook," 8th ed., vol. 5, pp. 12–13, American Society for Metals, Metals Park, Ohio, 1970; H. W. Haller, *Trans. ASME, Jnl. Eng. for Ind.*, vol. 105, pp. 270–275, 1983.

Greater forging capacity is achieved with the power hammer in which the ram is accelerated on the downstroke by steam or air pressure in addition to gravity. Steam or air also is used to raise the ram on the upstroke. The total energy supplied to the blow in a power drop hammer is given by

$$W = \frac{1}{2}\frac{mv^2}{g} + pAH = (m + pA)H \tag{16-1}$$

where m = weight of ram, lb
v = velocity of ram at start of deformation
g = acceleration of gravity, 32.2 ft/s^2
p = air or steam pressure acting on ram cylinder on downstroke
A = area of ram cylinder
H = height of the ram drop

An important feature of the power hammer is that the energy of the blow can be controlled, whereas in the board hammer the mass and height of fall are fixed. Power hammers are preferred over board hammers for closed-die forging. This equipment ranges in size from 1,000 to 50,000 lb and can produce forgings ranging in weight from a few pounds to several tons. The forging hammer is the cheapest source of a high forging load. For example, (see Prob. 16-2) a 3,000 lb power hammer can produce a forging load in excess of 600 tons. The forging hammer also has the shortest contact time under pressure, ranging from 1 to 10 ms. However, hammers generally do not provide the forging accuracy obtainable in presses. Also, because of their inherent impact character, problems must be overcome with ground shock, noise, and vibration. Some of these problems are minimized with the *counterblow hammer* which uses two opposed rams which strike the workpiece at the same time so that practically all of the energy is absorbed by the work and very little energy is lost as vibration in the foundation and the environment. A new class of forging equipment is the *high-energy-rate forging* (HERF) machine in which unusually high velocities (30 to 80 ft/s) are substituted for mass to achieve the required energy level.[1]

Forging presses are of either mechanical or hydraulic design. Presses are rated on the basis of the force developed at the end of the stroke. Next to hammers, mechanical presses are the most widely used equipment for closed-die forging in the United States. Most mechanical presses utilize an eccentric crank (Fig. 16-5) to translate rotary motion into reciprocating linear motion of the press slide. The ram stroke is shorter than in a hammer or hydraulic press, so that mechanical presses are best suited for low-profile forgings. The maximum load is attained when the ram is about $\frac{1}{8}$ in of the bottom dead center position. Mechanical presses with load ratings from 300 to 12,000 tons are available. The blow of a press is more like a squeeze than like the impact of a hammer. Because of this,

[1] T. Altan, op. cit., chap. 2.

dies can be less massive and die life is longer than with a hammer. However, the initial cost of a press is much higher than that of a hammer, so that large production runs are needed. The production rate is comparable to that of a hammer, but since each blow is of equal force, a press may be less suitable for carrying out preliminary shaping and finishing operations in the same piece of equipment.

The total energy supplied during the stoke of a press is given by

$$W = \frac{1}{2}I\left(\omega_0^2 - \omega_f^2\right) = \frac{1}{2}I\left(\frac{\pi}{30}\right)^2\left(n_0^2 - n_f^2\right) \qquad (16\text{-}2)$$

where I = moment of inertia of the flywheel
ω = angular velocity, rad/s
n_0 = initial speed of flywheel, rpm
n_f = speed of flywheel after deformation, rpm

Hydraulic presses are load-restricted machines in which hydraulic pressure moves a piston in a cylinder. A chief feature is that the full press load is available at any point during the full stroke of the ram. This feature makes the hydraulic press ideally suited for extrusion-type forging operations. The ram velocity can be controlled and even varied during the stroke. The hydraulic press is a relatively slow-speed machine. This results in longer contact time, which may lead to problems with heat loss from the workpiece and die deterioration. On the other hand, the slow squeezing action of a hydraulic press results in close-tolerance forgings. Hydraulic presses are available in ratings from 500 to 18,000 tons, although several presses with ratings of 50,000 tons have been built. The initial cost of a hydraulic press is higher than that of a mechanical press of equal capacity. Factors for converting between the capacity of presses and hammers are available.[1]

Screw presses are widely used in Europe for both hot and cold closed-die forging. In a screw press the ram is connected by a rotary joint to a spindle, which is in effect a large screw. The rotary motion of a flywheel is transformed into linear motion by the multiple thread on the spindle and its nut.

Forging machines, also known as *upsetters* or *headers*, are horizontal mechanical presses which are used for the high-production forging of symmetrical shapes from bar stock, such as bolts, rivets, and gear blanks.[2] *Forging rolls*[3] are used for initial forming prior to closed-die forging and for producing tapered or long slender sections like leaf springs and long bolts. Forging rolls have matching grooves over part of their circumference to produce the shape change.

[1] "Metals Handbook," 8th ed., vol. 5, pp. 14–18, American Society for Metals, Metals Park, Ohio, 1970.

[2] *Ibid.*, pp. 69–95.

[3] *Ibid.*, pp. 95–99.

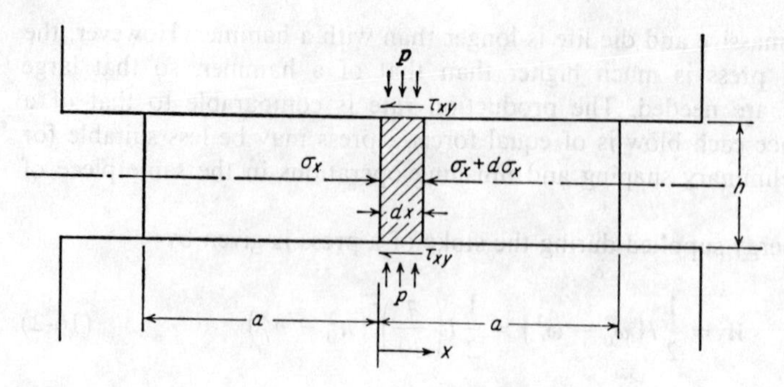

Figure 16-6 Stresses acting on a plate forged in plane strain.

16-3 FORGING IN PLANE STRAIN

Figure 16-6 shows the stresses in the forging of a plate of constant thickness under conditions of plane strain. The analogous problem of forging a disk was considered in Sec. 15-7. Lateral flow perpendicular to the ram travel leads to frictional shear stresses at the die contact surfaces. This surface shear is directed toward the centerline, opposing the metal flow. The presence of friction causes an imbalance of force on the element in the x direction which must be accommodated by a change in lateral pressure σ_x from one side of the element to another. It is assumed that the plate has unit width normal to the plane of the paper and that this width remains constant.

Taking the equilibrium of forces in the x direction,

$$\sigma_x h - (\sigma_x + d\sigma_x)h - 2\tau_{xy}\,dx = 0$$

$$\frac{d\sigma_x}{dx} = -\frac{2\tau_{xy}}{h} \tag{16-3}$$

The von Mises' yield criterion for a condition of plane strain is given by Eq. (3-54).

$$\sigma_1 - \sigma_3 = \frac{2}{\sqrt{3}}\sigma_0 = \sigma_0' \tag{16-4}$$

If we define p and σ_x as positive compressive principal stresses, then $p = \sigma_z$ and

$$\sigma_1 - \sigma_3 = \sigma_0' = p - \sigma_x \tag{16-5}$$

Since σ_0' does not change with x, $dp/dx = d\sigma_x/dx$ and on substituting into Eq. (16-3) the differential equation of equilibrium becomes

$$\frac{dp}{dx} = \frac{-2\tau_{xy}}{h} \tag{16-6}$$

If the shearing stress is related to the normal pressure by *Coulomb's law of sliding friction*, $\tau_{xy} = \mu p$, then Eq. (16-6) becomes

$$\frac{dp}{p} = -\frac{2\mu}{h}\,dx \tag{16-7}$$

Integrating both sides gives

$$\ln p = -\frac{2\mu x}{h} + \ln C$$

The constant of integration C is evaluated by the boundary condition that a free surface $x = a$, the lateral stress $\sigma_x = 0$ and $p = \sigma_0'$. Therefore,

$$\ln C = \ln \sigma_0' + 2\frac{\mu a}{h} \quad \text{and}$$

$$p = \sigma_0' \exp\left[\frac{2\mu}{h}(a - x)\right] \tag{16-8}$$

Since μ usually is a small number, we can use the expansion $e^{(y)} = 1 + y + y^2/2! + y^3/3! + \cdots$ to simplify the above equation.

$$p = \sigma_0'\left[1 + \frac{2\mu(a - x)}{h}\right] \tag{16-9}$$

The mean forging pressure is

$$\bar{p} = \int_0^a \frac{p\,dx}{a} = \sigma_0' \frac{e^{(2\mu a/h)} - 1}{2\mu a/h} \tag{16-10}$$

The total forging load P can now be established, since $P = \bar{p}(2a)w$, where w is the width in the direction normal to the plane of the paper.

A convenient way to write Eq. (16-8) is

$$p = \sigma_0' e^{\mu L/h(1 - 2x/L)} \tag{16-11}$$

where $L = 2a$. This equation shows that as the ratio of length to thickness L/h increases, the resistance to compressive deformation increases rapidly. This fact is used to advantage in closed-die forging where the deformation resistance of the flash must be very high so that the pressure in the die will be high enough to ensure complete filling of the die cavity. Figure 16-7 shows the variation of p and σ_x over $L = 2a$. Both stresses build up to a maximum at the center of the plate, illustrating again the *friction hill.* In this simple case the centerline of the plate defines a *neutral surface*. For the vertical downward motion of the ram the metal flow is in the lateral, or horizontal, direction. The metal is stationary at the neutral surface, but the flow is outward away from the neutral surface. In forgings of more complex geometry the neutral surface may be more difficult to establish.

Another way other than Coulomb's law to describe the interface shear stress is with the *friction factor m*, where $\tau_{xy} = \tau_i = mk$. Substituting $\tau_{xy} = mk$ into Eq. (16-6) gives

$$dp = -\frac{2mk}{h}\,dx = -\frac{2\sigma_0}{\sqrt{3}}m\frac{dx}{h} = -\sigma_0' m\frac{dx}{h} \tag{16-12}$$

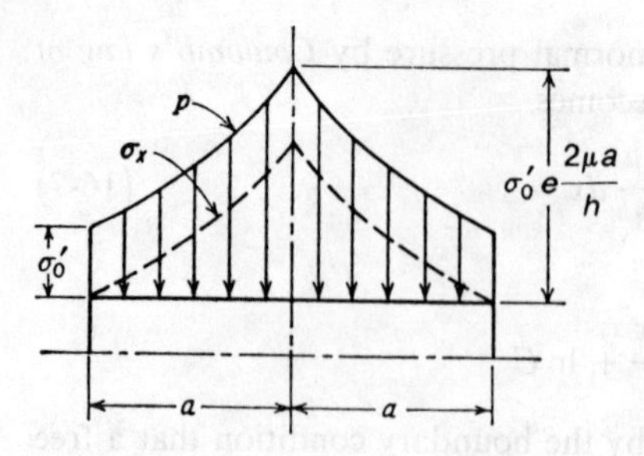

Figure 16-7 Distribution of normal stress and longitudinal stress for compression between plates.

Integration is straightforward

$$p = -\sigma_0' m \frac{x}{h} + C$$

and since $p = \sigma_0'$ at $x = a$, $C = \sigma_0' + \sigma_0' ma/h$ and

$$p = \sigma_0' \frac{m}{h}(a - x) + \sigma_0' \tag{16-13}$$

This pressure distribution is linear with distance from the centerline of the slab.

In Sec. 15-7 we saw there is a limit below which sliding friction can exist at the die-workpiece interface. When that limit is reached interfacial shear of the workpiece occurs at a value at flow stress $\tau_i = k$, where k is the yield stress in shear $= \sigma_0/\sqrt{3}$. In general $\tau_i = mk$, so that $m = 1.0$ for a condition of sticking friction. For the special case of *sticking friction*, $m = 1$, and Eq. (16-13) becomes

$$p = \sigma_0'\left(\frac{a - x}{h} + 1\right) \tag{16-14}$$

The mean forging pressure is

$$\bar{p} = \sigma_0'\left(\frac{a}{2h} + 1\right) \tag{16-15}$$

Frequently the frictional conditions are intermediate between full sticking and slipping so that there may be sliding friction at the edges of the plate (near $x = a$) where the pressure is lower, but at some distance closer in to the neutral surface the pressure increases to a point where $\tau_{xy} = k = \sigma_0/\sqrt{3} = \sigma_0'/2$. The distance, measured from the centerline where this transition occurs is x_1. At this point $\tau_{xy} = \mu p = k = \sigma_0'/2$. Substituting for p with Eq. (16-8) gives a solution for the location of the boundary between slipping and sticking friction.

$$x_1 = a - \frac{h}{2\mu} \ln \frac{1}{2\mu} \tag{16-16}$$

Example A block of lead 1 in × 1 in × 6 in is pressed between flat dies to a size $\frac{1}{4}$ in × 4 in × 6 in. If the uniaxial flow stress is $\sigma_0 = 1{,}000$ psi and $\mu = 0.25$ determine the presure distribution over the 4-in dimension and the total forging load.

Since the 6-in dimension does not change, the deformation is plane strain. To establish the pressure distribution use Eq. (16-8)

$$p = \frac{2\sigma_0}{\sqrt{3}} \exp\left[\frac{2\mu}{h}(a - x)\right]$$

At the centerline of the slab

$$p_{max} = \frac{2(1000)}{\sqrt{3}} \exp\left[\frac{2(0.25)}{0.25}(2 - 0)\right] = 63{,}000 \text{ psi}$$

The pressure distribution out from the centerline is:

x	0	0.25	0.5	0.75	1.0	1.25	1.5	1.75	2.0
p (ksi)	63.0	38.2	23.2	14.0	8.5	5.17	3.1	1.9	1.05
$\tau_i = \mu p$ (ksi)	15.75	9.55	5.80	3.50	2.13	1.29	0.78	0.48	0.26

We note that $k = \sigma_0/\sqrt{3} = 1{,}000/\sqrt{3} = 0.577$ ksi. Since $\mu p > k$ for values of x less than about 1.75 we see that sticking friction prevails over most of the 4-in length of the slab.

For sticking friction $p = 2\sigma_0/\sqrt{3}((a - x)/h + 1)$, Eq. (16-14) p_{max} at $x = 0$ is

$$p_{max} = \frac{2(1{,}000)}{\sqrt{3}}\left(\frac{2 - 0}{0.25} + 1\right) = 10.4 \text{ ksi}$$

Note that the peak pressure for the friction hill is much reduced for the case of sticking friction. The pressure distribution is linear from p_{max} out to $x_1 = a - h/2\mu \ln(1/2\mu)$, Eq. (16-16).

$$x_1 = 2 - \frac{\frac{1}{4}}{2 \cdot \frac{1}{4}} \ln \frac{1}{2 \cdot \frac{1}{4}} = 2 - \tfrac{1}{2}\ln 2 = 2 - 0.5(0.693) = 1.654 \text{ in}$$

For $1.654 < x < 2.000$, $\tau_i = \mu p$ and sliding friction occurs.

For simplicity (it is a slight overestimate) we calculate the forging load on the assumption that the pressure distribution is based on 100 percent sticking friction. Then, from Eq. (16-15)

$$\bar{p} = \frac{2\sigma_0}{\sqrt{3}}\left(\frac{a}{2h} + 1\right) = \frac{2(1{,}000)}{\sqrt{3}}\left(\frac{2}{2 \cdot \frac{1}{4}} + 1\right) = \frac{10{,}000}{\sqrt{3}} = 5.77 \text{ ksi}$$

The forging load is $P = 2(5{,}770)(2)(6) = 138{,}000$ lb $\approx$ 69 tons.

16-4 OPEN-DIE FORGING

Open-die forging typically deals with large, relatively simple shapes that are formed between simple dies in a large hydraulic press or power hammer.

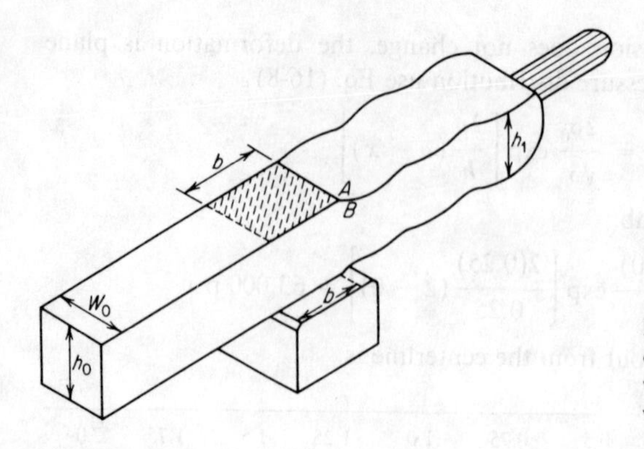

Figure 16-8 Cogging operation in open-die forging. Shaded area shows where contact would occur between workpiece and upper die.

Examples of parts made in open-die forging are ship propeller shafts, rings, gun tubes, and pressure vessels. Since the workpiece is usually larger than the tool, at any point in time deformation is confined to a small portion of the workpiece. The chief mode of deformation is compression, accompanied by considerable spreading in the lateral directions.

Probably the simplest open-die forging operation is *cogging* a billet between flat tools to reduce the cross-sectional area, usually without changing the final shape of the cross section. Figure 16-8 helps to define the nomenclature in dealing with spread in cogging. Tomlinson and Stringer[1] defined a coefficient of spread S

$$S = \frac{\text{width elongation}}{\text{thickness contraction}} = \frac{\ln (w_1/w_0)}{\ln (h_0/h_1)} \tag{16-17}$$

Because of barreling of the bar, it is difficult to measure the width natural strain, but the increase in length can be measured accurately. Using the constancy-of-volume relationship, we can write

$$\frac{h_1 w_1 l_1}{h_0 w_0 l_0} = 1 \tag{16-18}$$

or

$$\ln (h_1/h_0) + \ln (w_1/w_0) + \ln (l_1/l_0) = 0$$

Substituting into Eq. (16-17) gives the coefficient of elongation.

$$1 - S = \frac{\text{length elongation}}{\text{thickness contraction}} = \frac{\ln (l_1/l_0)}{\ln (h_0/h_1)} \tag{16-19}$$

If $S = 1$, then all of the deformation could manifest itself as spread, while if $S = 0$, all of the deformation would go into elongation. It was found that S

[1] A. Tomlinson and J. D. Stringer, *J. Iron Steel Inst. London*, vol. 193, pp. 157–162, 1969.

depended chiefly on the *bite ratio* b/w_0 according to

$$S = \frac{b/w_0}{1 + b/w_0} \tag{16-20}$$

Equation (16-17) often is expressed in terms of the "spread law"

$$\beta = \left(\frac{1}{\gamma}\right)^S \tag{16-21}$$

where β = spread ratio = w_1/w_0
γ = squeeze ratio = h_1/h_0

There are certain limiting ranges of these variables which must be considered. Since only that part of the surface under the bite is being deformed at any one time, there is danger of causing surface *laps* at the step separating the forged from the unforged portion of the workpiece. For a given geometry of tooling there will be a critical deformation which will produce laps. Wistreich and Shutt[1] recommend that the squeeze ratio h_0/h_1 should not exceed 1.3. Since open-die forging is done frequently on large sections, it is important to ensure that the billet is deformed through to the center. It is recommended that the bite ratio b/h should not be less than $\frac{1}{3}$ to minimize inhomogeneous deformation. Using these criteria Wistreich and Shutt developed optimization techniques for selecting the forging schedule from the thousands of possible combinations which would require the least number of steps.

The load required to forge a flat section in open dies may be estimated[2] by

$$P = \bar{\sigma} A C \tag{16-22}$$

where C is a constraint factor to allow for inhomogeneous deformation. It will be recalled that deformation resistance increases with $\Delta = h/L$ (Fig. 15-24). Hill[3] constructed slipline fields for forging with various conditions of Δ, and the results can be summarized by the relation $C = 0.8 + 0.2h/b = 0.8 + 0.2\Delta$.

16-5 CLOSED-DIE FORGING

The description of the closed-die forging process in Sec. 16-2 emphasized the important role of the flash in controlling die fill and in creating high forging loads. Usually the deformation in closed-die forging is very complex and the design of the intermediate steps to make a final precision part requires considerable experience and skill. Overall success of the forging operation requires an understanding of the flow stress of the material, the frictional conditions, and the flow of the material in order to develop the optimum geometry for the dies. A special problem in closed-die forging is preventing rapid cooling of the workpiece by the colder dies. Toward this end *isothermal forging* in heated superalloy dies is

[1] J. G. Wistreich and A. Shutt, *J. Iron Steel Inst. London*, vol. 193, pp. 163–176, 1959.
[2] *Ibid.*
[3] R. Hill, "The Mathematical Theory of Plasticity," Clarendon Press, Oxford, 1950.

being practiced with difficult to forge aerospace materials. The elimination of die quenching results in a lower flow stress and forging loads, and permits complete die fill and closer dimensional tolerances.

An important step toward rationalizing the design of closed-die forgings is the classification of the shapes[1] commonly produced by this process. The shape classification shown in Fig. 16-9 has been widely adopted. The degree of difficulty increases as the geometry moves down and toward the right in this illustration. Roughly 70 percent of forgings fall in the third shape class with one dimension significantly longer than the other two. Although this classification system is useful in cost estimating and designing preforming steps, it is not quantitative. More quantitative methods of defining the shape difficulty factor have been developed.[2]

The design of a part for production by closed-die forging involves the prediction of:

1. Workpiece volume and weight
2. Number of preforming steps and their configuration
3. Flash dimensions in preforming and finishing dies
4. The load and energy requirements for each forging operation

Steps 1 and 3 have been discussed in detail by Altan and Henning[1] and the procedures reduced to a working computer program that is applicable to axisymmetric shapes.

Preform design is the most difficult and critical step in forging design. Proper preform design assures defect-free flow, complete die fill, and minimum flash loss. Success here depends on a thorough understanding of the metal flow during forging. Although metal flow consists only of two basic types, extrusion (flow parallel to the direction of die motion) and upsetting (flow perpendicular to the direction of die motion), in most forgings both types of flow occur simultaneously. An important step in understanding metal flow is to identify the neutral surfaces. Metal flows away from the neutral surface in a direction perpendicular to the die motion.

In designing a preform it is usual practice to take key cross sections through the forging and design the preform on the basis of the metal flow. Some general considerations[3] are:

1. The area at each cross section along the length must equal the area in the finished cross section plus the flash.

[1] K. Spies, "The Preforms in Closed-Die Forging and Their Preparation by Reducer Rolling," (in German), doctoral dissertation, Technical University, Hanover, Germany, 1957.

[2] T. Altan and H. J. Henning, *Metallurgia and Metal Forming*, vol. 39, pp. 83–88, March 1972; W. A. Knight and C. Poli, *Machine Design*, Jan 24, 1985, pp. 94–99.

[3] K. Lange, "Closed-Die Forging of Steel" (in German,) Springer-Verlag OHG, Berlin, 1958; *Met. Treat. Drop Forg.*, May 1965, pp. 184–195; June 1965, pp. 210–230; July 1965, pp. 264–270; T. Altan, S. I. Oh, and H. L. Gegel, "Metal Forming," chap. 11, American Society for Metals, Metals Park, Ohio, 1983.

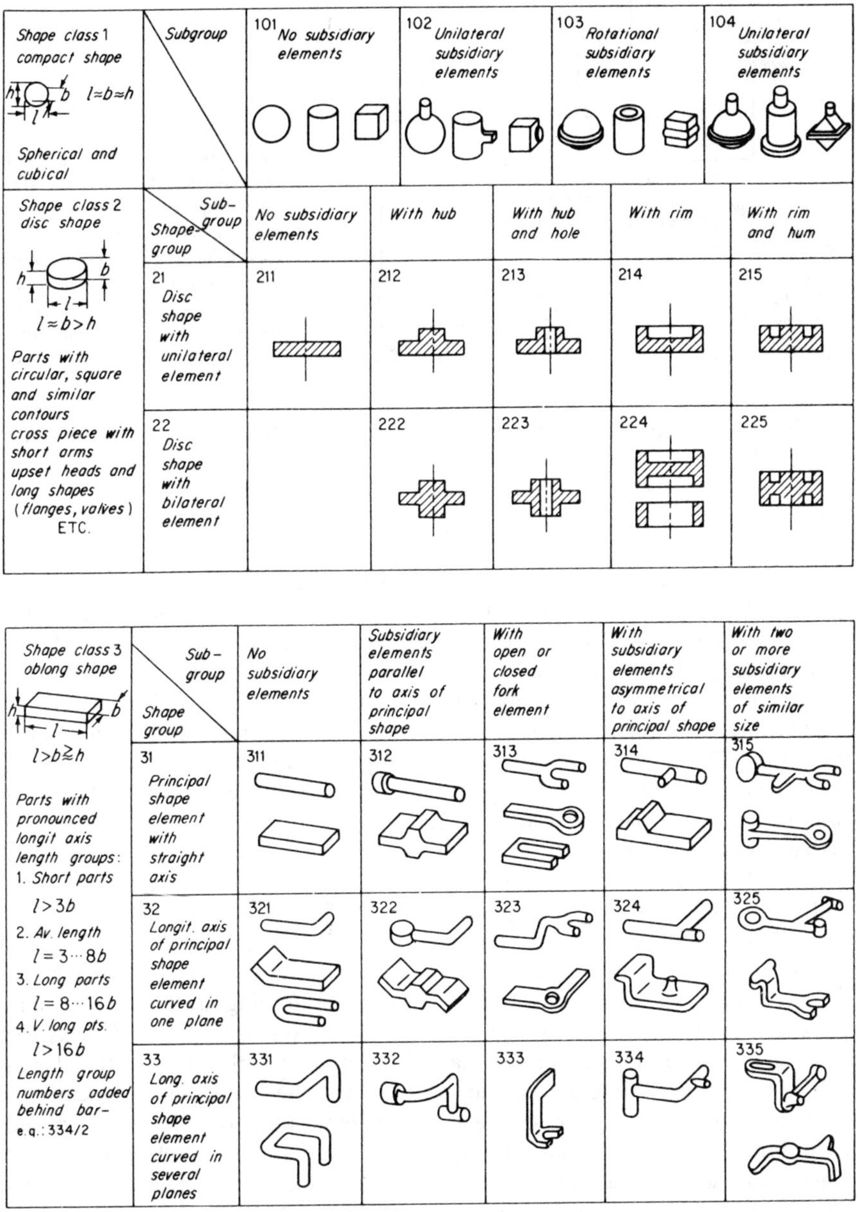

Shape class 1 compact shape $l \approx b \approx h$ Spherical and cubical	Subgroup	101 No subsidiary elements	102 Unilateral subsidiary elements	103 Rotational subsidiary elements	104 Unilateral subsidiary elements

Shape class 2 disc shape $l \approx b > h$ Parts with circular, square and similar contours cross piece with short arms upset heads and long shapes (flanges, valves) ETC.	Sub-group / Shape-group	No subsidiary elements	With hub	With hub and hole	With rim	With rim and hum
	21 Disc shape with unilateral element	211	212	213	214	215
	22 Disc shape with bilateral element		222	223	224	225

Shape class 3 oblong shape $l > b \gtrapprox h$ Parts with pronounced longit axis length groups: 1. Short parts $l > 3b$ 2. Av. length $l = 3 \cdots 8b$ 3. Long parts $l = 8 \cdots 16b$ 4. V. long pts. $l > 16b$ Length group numbers added behind bar– e.g.: 334/2	Sub-group / Shape group	No subsidiary elements	Subsidiary elements parallel to axis of principal shape	With open or closed fork element	With subsidiary elements asymmetrical to axis of principal shape	With two or more subsidiary elements of similar size
	31 Principal shape element with straight axis	311	312	313	314	315
	32 Longit. axis of principal shape element curved in one plane	321	322	323	324	325
	33 Long. axis of principal shape element curved in several planes	331	332	333	334	335

Figure 16-9 Shape classification for forging. *(After Spies. From "Forging Equipment, Materials, and Practices," p. 113, MCIC-HB-03, Battelle Memorial Laboratories, Columbus, Ohio, 1973.)*

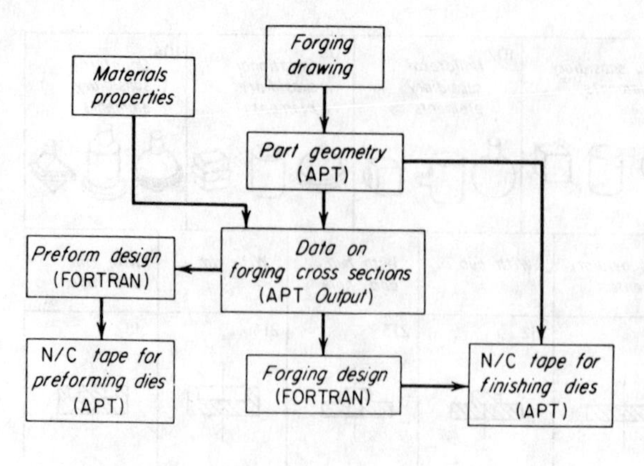

Figure 16-10 Flow diagram for computer-aided design *(CAD)* and computer-aided manufacturing (CAM) systems applied to closed-die forging. (*From "Forging Equipment Materials, and Practices," p. 140, MCIC-HB-03, Battelle Memorial Laboratories, Columbus, Ohio, 1973.*)

2. All concave radii on the preform should be larger than the radii on the final forged part.
3. The cross section of the preform should be higher and narrower than the final cross section, so as to accentuate upsetting flow and minimize extrusion flow.

Ideally, flow in the finishing step should be lateral toward the die cavity without additional shear at the die-workpiece interface. This type of flow minimizes friction, forging load, and die wear. An example of the use of basic principles in preform design has been given by Akgerman, Becker, and Altan.[1]

A milestone in metalworking is the use of *computer-aided design* (CAD) in establishing the proper design for preforming and finishing dies in closed-die forging. CAD[2] has been applied to rib-web type airframe forgings (Fig. 16-4) and to airfoil shapes, but it can be applied to any class shape for which there is a suitable volume of parts to justify the development work. The power of the CAD system can be appreciated from Fig. 16-10. Starting with a drawing of the final part, the CAD system defines this geometry in terms of points, planes, cylinders, and other regular geometric shapes using the APT computer language. APT is a specialized computer language for describing geometric changes produced in metal cutting that is at the heart of numerical controlled (N/C) machining. Next, the coordinates of the various cross sections of the forging are determined, and these are used to perform design calculations to establish such factors as the location of the neutral surface, the shape difficulty factor, the cross-sectional area and volume, the flash geometry, and the stresses, the loads, and the center of loading. An important aspect of this system is that it takes the part geometry and flash dimensions and generates the N/C tape for machining the electrodes in the sinking of the finishing dies by electric discharge machining. Thus, this system also involves *computer-aided manufacturing* (CAM). CAM is also used to machine the preforming dies.

[1] N. Akgerman, J. R. Becker, and T. Altan, *Metallurgia and Metal Forming*, vol. 40, pp. 135–138, May 1973.

[2] N. Akgerman and T. Altan, *Met. Eng. Q.*, vol. 13, pp. 26–28, February 1973; S. I. Oh, J. J. Park, S. Kobayashi, and T. Altan, *Trans. ASME J. Eng. Ind.*, vol. 105, pp. 251–258, 1983.

16-6 CALCULATION OF FORGING LOADS IN CLOSED-DIE FORGING

The prediction of forging load and pressure in a closed-die forging operation is quite a difficult calculation. There are three general approaches to the problem. The approach used in many forge shops is to estimate the forging load required for a new part from information available from previous forgings of the same material and similar shape. Slightly more sophistication is found in what might be called the empirical approach. Schey[1] has expressed the forging load as

$$P = \bar{\sigma} A_t C_1 \tag{16-23}$$

where A_t = cross-sectional area of the forging at the parting line, including the flash

C_1 = a constraint factor which depends on the complexity of the forging. C_1 has a value of 1.2 to 2.5 for upsetting a cylinder between flat dies. C_1 varies from 3 to 8 for closed-die forging of simple shapes with flash and from 8 to 12 for more complex shapes

The third approach is to use the *slab analysis*, suitably modified for the special situations found in closed-die forging. Although this level of analysis does not consider nonuniform deformation, when applied with consideration of the physical situation, it can give good agreement with experimental results.[2] The slab analysis for forging a plate in plane strain that was given in Sec. 16-3 has been extended to include forging between inclined die surfaces,[3] and the slab analysis for upsetting a disk (Sec. 15-7) has been extended to account for lateral flow between inclined dies and extrusion (longitudinal flow) into a rib or a shaft. The basic approach is to divide the actual forging into simple geometric shapes which can be treated by the slab analysis. The total forging load is the sum of the loads found for the components parts. Detailed examples of this approach are available in the technical literature.[3,4]

16-7 FORGING DEFECTS

If the deformation during forging is limited to the surface layers, as when light, rapid hammer blows are used, the dendritic ingot structure will not be broken down at the interior of the forging. Incomplete forging penetration can readily be detected by macroetching a cross section of the forging. The examination of a deep etch disk for segregation, dendritic structure, and cracks is a standard quality-control procedure with large forgings. To minimize incomplete penetration, forgings of large cross section are usually made on a forging press.

[1] J. A. Schey, "Principles of Forging Design," American Iron and Steel Institute, New York, 1964.
[2] N. Akgerman and T. Altan, *Trans. ASME, Ser. B: J. Eng. Ind.*, vol. 94, pp. 1025–1034, 1972.
[3] T. Altan and R. J. Fiorentino, *Trans. ASME, Ser. B: J. Eng. Ind.*, vol. 93, pp. 477–484, 1971.
[4] T. Altan, S. I. Oh, and H. L. Gegel, op. cit., pp. 159–169.

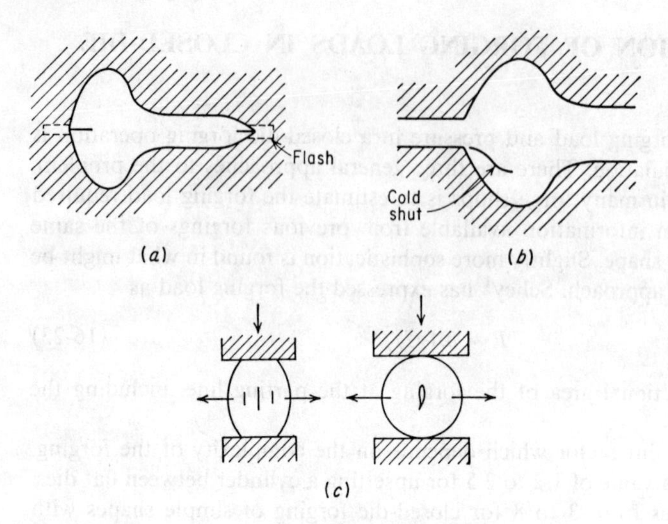

Figure 16-11 Typical forging defects. (*a*) Cracking at the flash; (*b*) cold shut or fold; (*c*) internal cracking due to secondary tensile stresses.

Surface cracking can occur as a result of excessive working of the surface at too low a temperature or as a result of hot shortness. A high sulfur concentration in the furnace atmosphere can produce hot shortness in steel and nickel. Cracking at the flash of closed-die forgings is another surface defect, since the crack generally penetrates into the body of the forging when the flash is trimmed off (Fig. 16-11*a*). This type of cracking is more prevalent the thinner the flash in relation to the original thickness of the metal. Flash cracking can be avoided by increasing the flash thickness or by relocating the flash to a less critical region of the forging. It also may be avoided by hot trimming or stress relieving the forging prior to cold trimming of the flash.

Another common surface defect in closed-die forgings is the cold shut, or fold (Fig. 16-11*b*). A cold shut is a discontinuity produced when two surfaces of metal fold against each other without welding completely. This can happen when metal flows past part of the die cavity that has already been filled or that is only partly filled because the metal failed to fill in due to a sharp corner, excessive chilling, or high friction. A common cause of cold shuts is too small a die radius.

Loose scale or lubricant residue that accumulates in deep recesses of the die forms scale pockets and causes underfill. Incomplete descaling of the workpiece results in forged-in scale on the finished part.

Secondary tensile stresses can develop during forging, and cracking can thus be produced.[1] Internal cracks can develop during the upsetting of a cylinder or a round (Fig. 16-11*c*), as a result of the circumferential tensile stresses. Proper

[1] S. L. Semiatin, Workability in Forging, "Workability Testing Techniques," chap. 8, American Society for Metals, Metals Park, Ohio, 1984.

design of the dies, however, can minimize this type of cracking. In order to minimize bulging during upsetting and the development of circumferential tensile stresses, it is usual practice to use concave dies. Internal cracking is less prevalent in closed-die forging because lateral compressive stresses are developed by the reaction of the work with the die wall.

The deformation produced by forging results in a certain degree of directionality to the microstructure in which second phases and inclusions are oriented parallel to the direction of greatest deformation. When viewed at low magnification, this appears as *flow lines*, or *fiber structure*. The existence of a fiber structure is characteristic of all forgings and is not to be considered as a forging defect. However, as was discussed in Sec. 8-15, the fiber structure results in lower tensile ductility and fatigue properties in the direction normal to it (transverse direction). To achieve an optimum balance between the ductility in the longitudinal and transverse directions of a forging, it is often necessary to limit the amount of deformation to 50 to 70 percent reduction in cross section.

16-8 POWDER METALLURGY FORGING

A new and rapidly growing area is the production of closed-die forgings from powder metallurgy preforms (P/M forging). The use of sintered P/M preforms rather than bar stock as the workpiece offers advantages of improved material utilization through reduction or elimination of machining, forming to final size in one forging stroke, uniformity of structure and reduced directionality of properties relative to conventionally forged parts.

Working with a sintered powdered metal preform introduces new aspects to the mechanics and metallurgy of plastic deformation. Because it contains a dispersion of interconnected voids, the deformation of a P/M preform is much different from a conventional fully dense workpiece. With a P/M preform the workpiece decreases in volume during plastic deformation as the porosity is closed up and eliminated by the act of plastic deformation. The presence of voids causes a significant decrease in local ductility which increases the likelihood of fracture during forging. However, the forming limit concept can be applied[1] to the design of P/M preforms to prevent fracture in forging. The presence of voids increases the surface area over which unfavorable oxidation or contamination reactions can occur.

The basic plasticity mechanics of a porous powder metallurgy preform can be described[2] by the following relations. The relationship between densification and plastic deformation is achieved through relating the plastic Poisson ratio ν to the

[1] C. L. Downey and H. A. Kuhn, *Trans. ASME, Ser. H., J. Eng. Mater. Technol.*, vol. 97, pp. 121–125, 1975.

[2] H. A. Kuhn and C. L. Downey, *Int. J. Powder Metall.*, vol. 7, p. 15–25, 1971.

fraction of theoretical density ρ/ρ_t.

$$\nu = 0.5\left(\frac{\rho}{\rho_t}\right)^2 \tag{16-24}$$

This relationship holds for hot- and cold-working, provided the preform has been sintered. For the frictionless compression of a cylinder the relative density change is given by

$$-\frac{d\rho}{\rho} = d\varepsilon_z + d\varepsilon_r + d\varepsilon_\theta \tag{16-25}$$

and from the definition of Poisson's ratio

$$d\varepsilon_r = d\varepsilon_\theta = -\nu\, d\varepsilon_z \tag{16-26}$$

so that

$$-\frac{d\rho}{\rho} = (1 - 2\nu)\, d\varepsilon_z \tag{16-27}$$

The substitution of Eq. (16-24) in Eq. (16-27) gives

$$-\frac{d\rho}{\rho} = (1 - \rho^2)\, d\varepsilon_z \tag{16-28}$$

which can be integrated to give

$$-\varepsilon_z = \ln\left[\frac{(\rho/\rho_i)^2(1 - \rho_i^2)}{(1 - \rho^2)}\right]^{1/2} \tag{16-29}$$

where ρ_i is the initial density of the cylinder. The more practical problem of compression of a disk between dies with friction has been analyzed[1] using the above approach and has been shown to give excellent agreement with experiment.

The classical theory of plasticity is based on the assumption of constancy of volume, which leads to the further condition that yielding is unaffected by the hydrostatic component of the stress state. A modification of the von Mises' yield criterion is needed for dealing with porous materials which densify with plastic deformation. Kuhn has shown that a workable yield criterion is

$$\sigma_0(\rho, \varepsilon) = \left[\frac{(\sigma_1 - \sigma_2)^2 + (\sigma_2 - \sigma_3)^2 + (\sigma_3 - \sigma_1)^2}{2} + (1 - 2\nu)(\sigma_1\sigma_2 + \sigma_2\sigma_3 + \sigma_3\sigma_1)\right]^{1/2} \tag{16-30}$$

The first term in Eq. (16-30) is the usual von Mises' criterion, and the second term accounts for the porosity through Poisson's ratio and Eq. (16-24). Several other yield functions for compressible P/M materials have ben proposed.[2]

[1] H. A. Kuhn and C. L. Downey, *Trans. ASME, Ser. H: J. Eng. Mater Technol.*, vol. 95, pp. 41–46, 1973.

[2] S. M. Doraivelu, H. L. Gegel, J. S. Gunasekera, J. C. Malas, and J. T. Morgan, *Int. J. Mech. Sci.*, vol. 26, pp. 527–535, 1984.

16-9 RESIDUAL STRESSES IN FORGINGS

The residual stresses produced in forgings as a result of inhomogeneous deformation are generally small because the deformation is usually carried out well into the hot-working region. However, appreciable residual stresses and warping can occur on the quenching of steel forgings in heat treatment.

Special precautions must be observed during the cooling of large steel forgings from the hot-working temperature. Large forgings are subject to the formation of small cracks, or *flakes*, at the center of the cross section. Flaking is associated with the high hydrogen content usually present in steel ingots of large size, coupled with the presence of residual stresses. In order to guard against the development of high thermal or transformation residual stresses, large forgings are very slowly cooled from the working temperature. This may be accomplished by burying the forging in ashes for periods up to several weeks or, in the controlled cooling treatment which is used for hot-rolled railroad rail and certain forgings, by transferring the hot forging to an automatically controlled cooling cycle which brings the forging to a safe temperature in a number of hours. The use of vacuum-degassed steel largely eliminates problems with flaking.

BIBLIOGRAPHY

Altan, T., F. W. Boulger, J. R. Becker, N. Akgerman, and H. J. Henning: "Forging Equipment, Materials, and Practices," Metals and Ceramics Information Center, MCIC-HB-03, October 1973. (Available through National Technical Information Service.)

Byrer, T. G. (ed.): "Forging Handbook," American Society for Metals, Metals Park, Ohio, 1985.

Geleji, A.: "Forge Equipment, Rolling Mills, and Accessories," (in English), Akademiai Kiado, Budapest, 1967.

Jensen, J. E. (ed.): "Forging Industry Handbook," Forging Industry Association, Cleveland, Ohio, 1970.

"Metals Handbook," 8th ed., vol. 5, Forging and Casting, American Society for Metals, Metals Park, Ohio, 1970.

"Open Die Forging Manual," 3rd ed., Forging Industry Association, Cleveland, Ohio, 1982.

Schey, J. A.: "Principles of Forging Design," American Iron and Steel Institute, New York, 1964.

Thomas, A. "DFRA Forging Handbook: Die Design," Drop Forging Research Association, Sheffield, England, 1980.

CHAPTER

SEVENTEEN

ROLLING OF METALS

17-1 CLASSIFICATION OF ROLLING PROCESSES

The process of plastically deforming metal by passing it between rolls is known as *rolling*. This is the most widely used metalworking process because it lends itself to high production and close control of the final product. In deforming metal between rolls, the work is subjected to high compressive stresses from the squeezing action of the rolls and to surface shear stresses as a result of the friction between the rolls and the metal. The frictional forces are also responsible for drawing the metal into the rolls.

The initial breakdown of ingots into blooms and billets is generally done by hot-rolling. This is followed by further hot-rolling into plate, sheet, rod, bar, pipe, rails, or structural shapes. The cold-rolling of metals has reached a position of major importance in industry. Cold-rolling produces sheet, strip, and foil with good surface finish and increased mechanical strength, at the same time maintaining close control over the dimensions of the product.

The terminology used to describe rolled products is fairly loose, and sharp limits with respect to dimensions cannot always be made for steelmaking terminology. A *bloom* is the product of the first breakdown of the ingot. Generally the width of a bloom equals its thickness, and the cross-sectional area is greater than 36 in^2. A further reduction by hot-rolling results in a *billet*. The minimum cross section of a billet is about $1\frac{1}{2}$ by $1\frac{1}{2}$ in. It should be noted that in nonferrous metallurgical terminology a billet is any ingot which has received hot-working by rolling, forging, or extrusion, or the term may refer to a casting which is suitable for hot-working, as an extrusion billet. A *slab* refers to a hot-rolled ingot with cross-sectional area greater than 16 in^2 and with a width that is at least twice the thickness. Blooms, billets, and slabs are known as *semifinished products* because they are subsequently formed into other mill products. The differentiation between *plate* and *sheet* is determined by the thickness of the product. In general, plate has a thickness greater than $\frac{1}{4}$ in, although there are exceptions to this limit, depending on the width. *Sheet* and *strip* refer to rolled products which generally have a thickness less than $\frac{1}{4}$ in. In general, strip refers to the rolled product with a width no greater than 24 in, while sheet refers to the product of greater width.

Generally, rolling starts with a cast ingot or an electroplated slab. However, dense sheet may be produced by rolling directly from powder. In *powder rolling*[1] a metal powder is introduced between the rolls and compacted into a "green strip," which is subsequently sintered and subjected to further hot-working and/or cold-working and annealing cycles. A major advantage of powder rolling is the elimination of the initial hot-ingot breakdown step, with a corresponding large economy in needed capital equipment. Other advantages may be the minimization of contamination in hot-rolling, and the production of sheet with very fine grain size or with a minimum of preferred orientation.

In conventional hot- or cold-rolling the main objective is to decrease the thickness of the metal. Ordinarily little increase in width occurs, so that the decrease in thickness results in an increase in length. *Roll forming* is a special type of cold-rolling in which strip is progressively bent into complex shapes by passing it through a series of driven rolls. The thickness of the metal is not appreciably changed during this process. Roll forming is particularly suited to producing long, molded sections such as irregular-shaped channels and trim. Another specialized use of rolling is *thread rolling*, in which a blank is fed between two grooved die plates to form the threads.

17-2 ROLLING MILLS

A rolling mill consists basically of rolls, bearings, a housing for containing these parts, and a drive for applying power to the rolls and controlling their speed. The forces involved in rolling can easily reach many millions of pounds. Therefore, very rigid construction is needed, and very large motors are required to provide the necessary power. When these requirements are multiplied several times for the successive stands of a large continuous mill, it is easy to see why a modern rolling-mill installation demands many millions of dollars of capital investment and many man-hours of skilled engineering design and construction.

Rolling mills can be conveniently classified with respect to the number and arrangement of the rolls (Fig. 17-1). The simplest and most common type of rolling mill is the *two-high* mill (Fig. 17-1*a*). Rolls of equal size are rotated only in one direction. The stock is returned to the entrance, or rear, of the rolls for further reduction by hand carrying or by means of a platform which can be raised to pass the work above the rolls. An obvious improvement in productivity results

[1] G. M. Sturgeon, "Metal Strip from Powder," Mills and Boon, London, 1972.

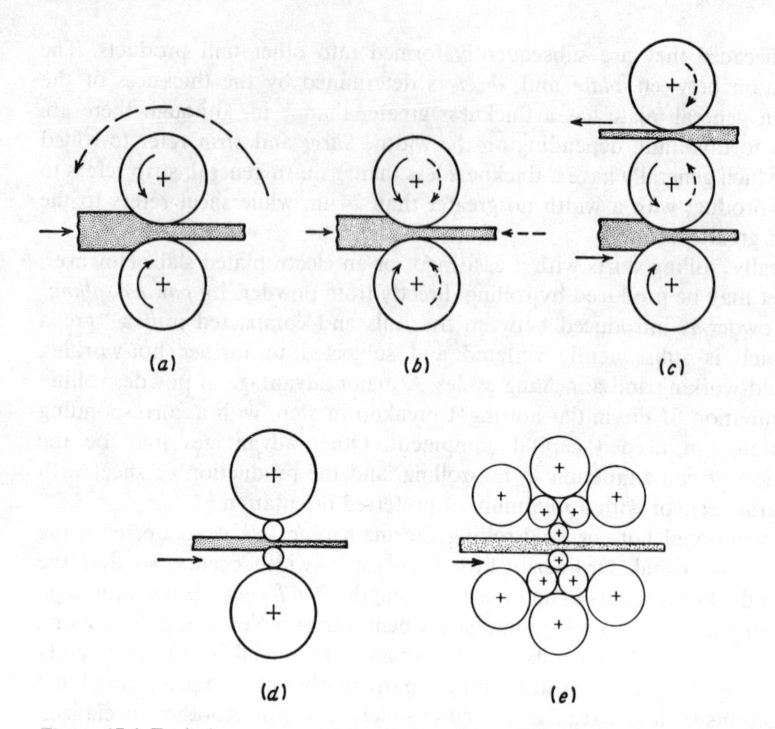

Figure 17-1 Typical arrangements of rolls for rolling mills. (*a*) Two-high, pullover; (*b*) two-high, reversing; (*c*) three-high; (*d*) four-high; (*e*) cluster.

from the use of a *two-high reversing mill*, in which the work can be passed back and forth through the rolls by reversing their direction of rotation (Fig. 17-1*b*). Another solution is the *three-high mill* (Fig. 17-1*c*), consisting of an upper and lower driven roll and a middle roll which rotates by friction.

A large decrease in the power required for rolling can be achieved by the use of small-diameter rolls. However, because small-diameter rolls have less strength and rigidity than large rolls, they must be supported by larger-diameter backup rolls. The simplest mill of this type is the *four-high mill* (Fig. 17-1*d*). Very thin sheet can be rolled to very close tolerances on a mill with small-diameter work rolls. The *cluster mill* (Fig. 17-1*e*), in which each of the work rolls is supported by two backing rolls, is a typical mill of this kind. The *Sendzimir mill* is a modification of the cluster mill which is very well adapted to rolling thin sheet or foil from high-strength alloys.

For high production it is common to install a series of rolling mills one after another in tandem (Fig. 17-2). Each set of rolls is called a *stand*. Since a different reduction is taken at each stand, the strip will be moving at different velocities at each stage in the mill. The speed of each set of rolls is synchronized so that each successive stand takes the strip at a speed equal to the delivery speed of the

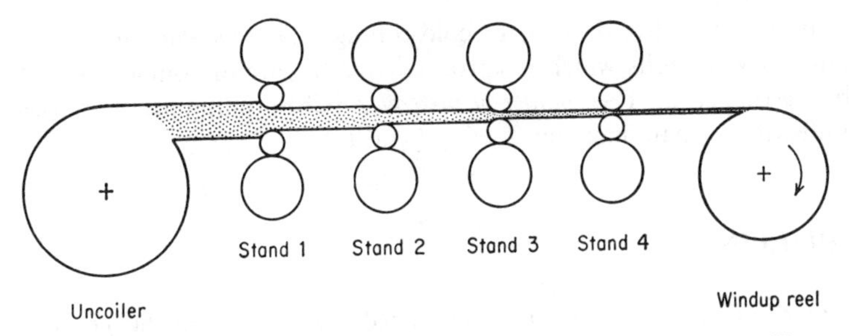

Figure 17-2 Schematic drawing of strip rolling on a four-stand continuous mill.

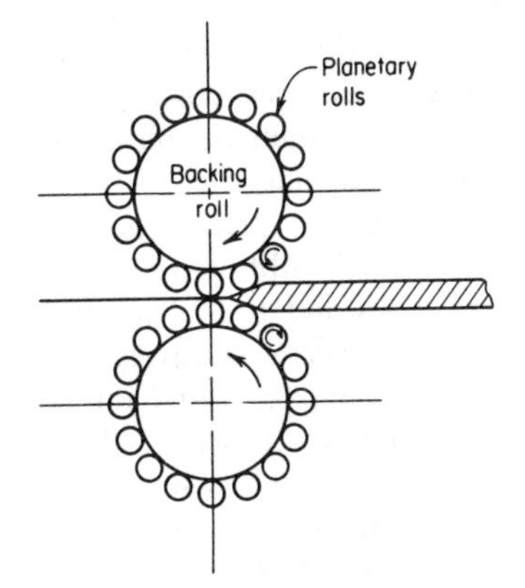

Figure 17-3 Arrangement of rolls in a planetary mill.

preceding stand. The uncoiler and windup reel not only accomplish the functions of feeding the stock to the rolls and coiling up the final product but also can be used to supply a *back tension* and a *front tension* to the strip. These added horizontal forces have several advantages that will be discussed later.

A different rolling mill design is the *planetary mill* (Fig. 17-3). This mill consists of a pair of heavy backing rolls surrounded by a large number of small planetary rolls.[1] The chief feature of the planetary mill is that it hot reduces a slab directly to strip in one pass through the mill. Each planetary roll gives an almost constant reduction to the slab as it sweeps out a circular path between the backing roll and the slab. As each pair of planetary rolls ceases to have contact with the workpiece another pair of rolls makes contact and repeats that reduction. The overall reduction is the summation of a series of small reductions by each pair of rolls in turn following each other in rapid succession. The action in the planetary mill is more like forging than rolling. It is necessary to use feed rolls to introduce the slab into the mill, and a pair of planishing rolls may be needed on the exit side to improve the surface finish.

[1] J. L. Giles and C. Gutteridge, *J. Iron Steel Inst. London*, vol. 211, pp. 9–12, 1973.

There are other innovative designs for cold-rolling. The *pendulum mill*[1] uses two small-diameter work rolls which reciprocate over the arc of contact to cold reduce a slab to a thin sheet. The *contact-bend-stretch* (CBS) rolling process[2] uses a four-high mill with a small-diameter floating bend roll.

17-3 HOT-ROLLING

The first hot-working operation for most steel products is done on the *primary roughing mill* (sometimes called blooming, slabbing, or cogging mills). These mills usually are two-high reversing mills with 24- to 54-in-diameter rolls. They are designated by the size of the rolls, as for example, a 45-in slabbing mill. The objective of this operation is the breakdown of the cast ingot into blooms or slabs for subsequent finishing into bars, plate or sheet. The initial breakdown passes often involve only small reductions. Heavy scale is removed initially by rolling the ingot while lying on edge, while the thickness is reduced by rolling after the ingot has been turned 90° so as to be lying flat. There is appreciable spreading of the ingot width in hot-rolling of ingots. To maintain the desired width and preserve the edges, the ingot is turned 90° on intermediate passes and passed through edging grooves in the rolls. A reversing primary mill has a relatively low production rate since the workpiece may be passed back and forth and turned from 10 to 20 times. Where high production rates are of prime concern, edging passes may be eliminated by using a *universal mill*. This type of mill essentially is two rolling mills, one with two large-diameter rolls and the other with vertical rolls which control the width at the same time the thickness is reduced. The production of slabs from cast ingots by hot-rolling can be eliminated by using *continuous casting* to produce the slab directly from the molten steel. Another method is to produce slabs by *bottom-pressure casting*.

Plates are produced by hot-rolling, either from reheated slabs or directly from ingots. *Sheared plate* is produced by rolling between straight horizontal rolls and then trimming all edges. *Mill edge* is the normal edge produced in hot-rolling between horizontal finishing rolls. Mill-edge plates have two mill edges and two trimmed edges. *Universal-mill plates* have been rolled on a universal mill and trimmed on the ends only.

The general distinction between *strip* and *sheet* is that strip usually is less than 24 in wide. However, irrespective of width, the equipment for producing these products is known as a *continuous hot-strip mill*. In the modern wide, hot-strip mill, reheated slabs are first passed through a scalebreaker mill, then through a *roughing train* of 4 four-high mills, followed by a *finishing train* of 6 four-high finishing mills. If the sheet to be produced is wider than the width of the slab, then the first stand in the roughing train is a *broadside mill* in which the width of the slab is increased by cross rolling. The roughing mills usually are

[1] K. Saxl, *Engineering*, vol. 195, pp. 494–495, 1963.
[2] L. F. Coffin, *J. Met.*, vol. 15, pp. 14–22, March 1967.

equipped with vertical edging rolls to control the width of the strip. High-pressure water jets are sprayed on the strip to remove scale. Following the last finishing stand there is either a flying shear to cut the strip to length or a coiler to produce continuous long lengths. In hot-rolling steel the slabs are heated initially at 2000 to 2400°F. The temperature in the last finishing stand varies from 1300 to 1600°F, but should be above the upper critical temperature to produce uniform equiaxed ferrite grains.

Because the nonferrous-metals industry deals with a diverse product mix, the equipment used for hot-rolling these materials is usually less specialized than equipment for hot-rolling of steel. The smaller ingot sizes and lower flow stresses found with most nonferrous alloys permit the use of smaller rolling mills. Two- and three-high mills are used for most hot-rolling of nonferrous alloys, although continuous four-high hot mills are used for aluminum alloys.

17-4 COLD-ROLLING

Cold-rolling is used to produce sheet and strip with superior surface finish and dimensional tolerances compared with hot-rolled strip. In addition, the strain hardening resulting from the cold reduction may be used to give increased strength. A greater percentage of rolled nonferrous metals is finished by cold-rolling compared with rolled-steel products. The starting material for cold-rolled steel sheet is pickled hot-rolled breakdown coil from the continuous hot-strip mill. Cold-rolled nonferrous sheet may be produced from hot-rolled strip, or in the case of certain copper alloys it is cold-rolled directly from the cast state.

High-speed four-high tandem mills with three to five stands are used for the cold-rolling of steel sheet, aluminum, and copper alloys. Generally, this type of mill is designed to provide both front and back tension. A continuous mill has high capacity and results in low labor costs. For example, the delivery speed of a five-stand continuous mill can reach 6,000 ft/min. However, this type of equipment requires a large capital investment and suffers further from lack of versatility. Four-high single-stand reversing mills with front and back tension are a more versatile installation. This type of mill is used often for the production of specialty items that vary widely in dimensions. However, it cannot compete with the continuous-tandem mill where large tonnages are involved.

The total reduction achieved by cold-rolling generally will vary from about 50 to 90 percent. In establishing the reduction in each pass or in each stand, it is desirable to distribute the work as uniformly as possible over the various passes without falling very much below the maximum reduction for each pass. Generally the lowest percentage reduction is taken in the last pass to permit better control of flatness, gage, and surface finish. One rational procedure[1] for developing cold-rolling schedules is to adjust the reduction in each pass so as to produce a constant rolling load.

[1] B. R. Oliver and J. E. Bowers, *J. Inst. Met.*, vol. 93, pp. 218–222, 1964–1965.

The elimination of the yield point from annealed steel sheet is an important practical problem since the existence of a yield-point elongation results in inhomogeneous deformation (stretcher strains) during deep drawing or forming. The usual practice is to give the annealed steel a final, small cold reduction, *temper rolling*, or *skin pass*, which eliminates the yield-point elongation. Temper rolling also results in an improved surface and in improved flatness. Other methods which are used to increase the flatness of rolled sheet are *roller leveling* and *stretcher leveling*. A roller-leveling machine consists of two sets of small-diameter rolls which are arranged so that the top and bottom rows are offset. When the sheet is passed into the leveler, it is flexed up and down and the sheet is straightened as it emerges from the rolls. The stretcher leveler consists of two jaws which grip the edges of the sheet and stretch it with a pure tensile force.

17-5 ROLLING OF BARS AND SHAPES

Bars of circular or hexagonal cross section and structural shapes like I beams, channels, and railroad rails are produced in great quantity by hot-rolling with grooved rolls (Fig. 17-4). Actually, the hot breakdown of an ingot into a bloom

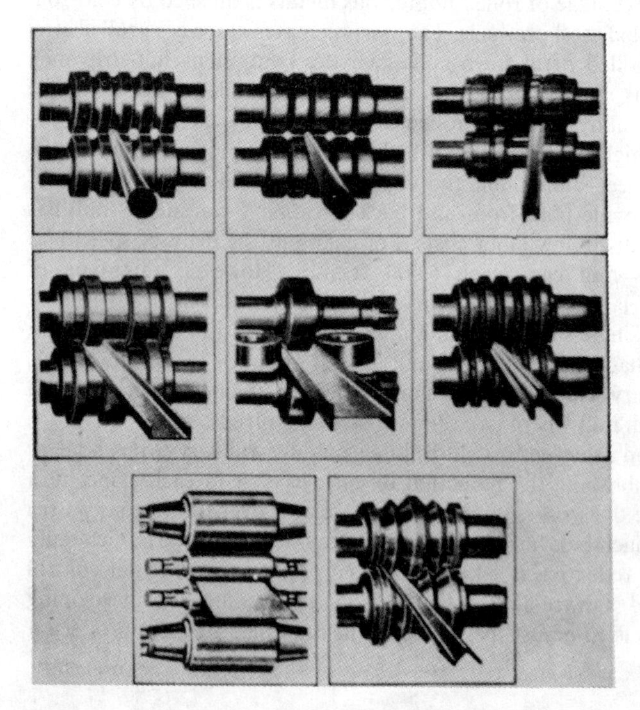

Figure 17-4 Rolling of bars and structural shapes. *(American Iron and Steel Institute.)*

falls in this category since grooved rolls are used to control the changes in shape during the blooming operation.

The rolling of bars and shapes differs from the rolling of sheet and strip in that the cross section of the metal is reduced in two directions. However, in any one pass the metal is usually compressed in one direction only. On the next pass it is rotated 90°. Since the metal spreads to a much greater extent in the hot-rolling of bars than in cold-rolling of sheet, an important problem in designing passes for bars and shapes is to provide allowance for the spreading.[1] A typical method of reducing a square billet to a bar is by alternate passes through oval and square-shaped grooves. The design of roll passes for structural shapes is much more complicated and requires extensive experience.[2] Because different metals spread different amounts, it is not generally possible to roll metals of widely different rolling characteristics on the same set of bar rolls.

A rolling mill designed to roll bars is known as a *bar mill*, or *merchant mill*. Most production bar mills are equipped with guides to feed the billet into the grooves and repeaters to reverse the direction of the bar and feed it back through the next roll pass. Mills of this type are generally either two- or three-high. A common installation consists of a roughing stand, a strand stand, and a finishing stand. It is common practice to arrange bar rolls *in train*; i.e., several mills are set close together, side by side, and the rolls in one stand are driven by connecting them to those of the adjacent stand.

17-6 FORCES AND GEOMETRICAL RELATIONSHIPS IN ROLLING

Figure 17-5 illustrates a number of important relationships between the geometry of the rolls and the forces involved in deforming a metal by rolling. A metal sheet with a thickness h_0 enters the rolls at the entrance plane XX with a velocity v_0. It passes through the roll gap and leaves the exit plane YY with a reduced thickness h_f. To a first approximation no increase in width results, so that the vertical compression of the metal is translated into an elongation in the rolling direction. Since equal volumes of metal must pass a given point per unit time, we can write

$$bh_0v_0 = bhv = bh_fv_f \tag{17-1}$$

where b = width of sheet

v = its velocity at any thickness h intermediate between h_0 and h_f

In order that a vertical element in the sheet remain undistorted, Eq. (17-1) requires that the exit velocity v_f must be greater than the entrance velocity v_0. Therefore, the velocity of the sheet must steadily increase from entrance to exit. At only one point along the surface of contact between the roll and the sheet is

[1] R. N. Wright, *Wire Technology*, vol. 7, pp. 88–92, 1980.

[2] W. Trinks, "Roll Pass Design," 2d ed., Penton Publishing Company, Cleveland, 1933; R. Stewartson, The Rolling of Rods, Bars, and Light Sections, *Metall. Rev.*, vol. 4, pp. 309–379, 1959.

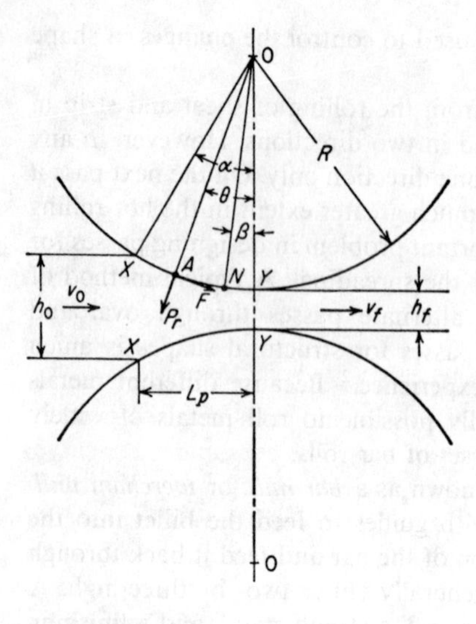

Figure 17-5 Forces acting during rolling.

the surface velocity of the roll v_r equal to the velocity of the sheet. This point is called the *neutral point*, or *no-slip point*. It is indicated in Fig. 17-5 by point N.

At any point along the surface of contact, such as point A in Fig. 17-5, two forces act on the metal. These are a radial force P_r and a tangential friction force F. Between the entrance plane and the neutral point the sheet is moving slower than the roll surface, and the frictional force acts in the direction shown in Fig. 17-5 so as to draw the metal into the rolls. On the exit side of the neutral point the sheet moves faster than the roll surface. The direction of the frictional force is then reversed so that it acts to oppose the delivery of the sheet from the rolls.

The vertical component of P_r is known as the *rolling load P*. The rolling load is the force with which the rolls press against the metal. Because this is also equal to the force exerted by the metal in trying to force the rolls apart, it is frequently called the *separating force*. The *specific roll pressure p* is the rolling load divided by the contact area. The contact area between the metal and the rolls is equal to the product of the width of the sheet b and the projected length of the arc of contact L_p.

$$L_p = \left[R(h_0 - h_f) - \frac{(h_0 - h_f)^2}{4} \right]^{1/2} \approx \left[R(h_0 - h_f) \right]^{1/2} \qquad (17\text{-}2)$$

Therefore, the specific roll pressure is given by

$$p = \frac{P}{bL_p} \qquad (17\text{-}3)$$

The distribution of roll pressure[1] along the arc of contact is indicated in Fig. 17-6. The pressure rises to a maximum at the neutral point and then falls off. The

[1] F. A. R. Al-Salehi, T. C. Firbank, and P. R. Lancaster, *Int. J. Mech. Sci.*, vol. 15, pp. 693–710, 1973.

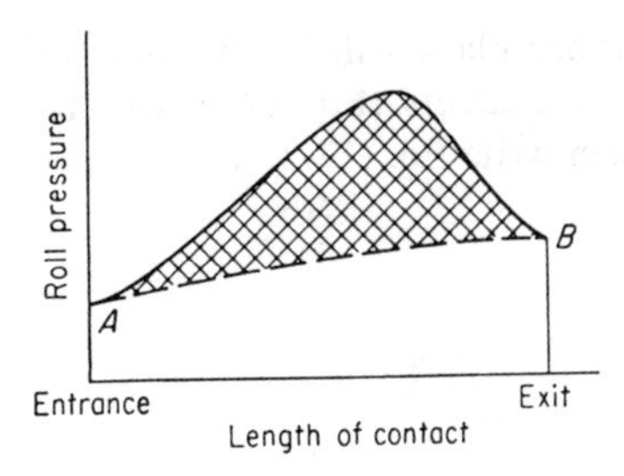

Figure 17-6 Distribution of roll pressure along the arc of contact.

fact that the pressure distribution does not come to a sharp peak at the neutral point, as required in theoretical treatments of rolling, indicates that the neutral point is not really a line on the roll surface but an area. The area under the curve is proportional to the rolling load, which for purposes of calculation acts at the center of gravity of the pressure distribution. Therefore, the shape of the pressure distribution is important because the location of the resultant rolling load with respect to the roll centers determines the torque and power required to produce the reduction. The area shown shaded in Fig. 17-6 represents the force required to overcome frictional forces between the roll and the sheet, while the area under the dashed line AB represents the force required to deform the metal in plane homogeneous compression. A similarity should be noted between the pressure distribution in rolling shown in Fig. 17-6 and the pressure distribution for compression between plates (Fig. 15-22).

The angle α between the entrance plane and the centerline of the rolls is called the *angle of contact*, or *angle of bite*. Referring to Fig. 17-5, the horizontal component of the normal force is $P_r \sin \alpha$, and the horizontal component of the friction force is $F \cos \alpha$. For the workpiece to enter into the throat of the roll the horizontal component of the friction force, which acts toward the roll gap, must be equal to or greater than the horizontal component of the normal force, which acts away from the roll gap. The limiting condition for unaided entry of a slab into the rolls is

$$F \cos \alpha = P_r \sin \alpha$$

$$\frac{F}{P_r} = \frac{\sin \alpha}{\cos \alpha} = \tan \alpha$$

but

$$F = \mu P_r$$

so that

$$\mu = \tan \alpha \tag{17-4}$$

The workpiece cannot be drawn into the rolls if the tangent of the contact angle exceeds the coefficient of friction. If $\mu = 0$, rolling cannot occur, but as μ increases progressively larger slabs will be drawn into the roll throat. In ingot breakdown by hot-rolling, where it is desired to achieve a large reduction in a short time, the rolls have grooves cut in them parallel to the roll axis to increase the effective value of μ.

For the same friction conditions, a large-diameter roll will permit a thicker slab to enter the rolls than will a small-diameter roll. This is because the angle subtended from the center of the roll to the entrance plane will be the same in both cases ($\tan\alpha$) but the lengths of the arcs of contacts (Eq. 17-2) will be appreciably different. Referring to Fig. 17-5, we can write Eq. (17-2) as

$$L_p \approx \sqrt{R\,\Delta h}$$

where Δh = the "draft" taken in rolling

$$\tan\alpha = \frac{L_p}{R - \Delta h/2} \approx \frac{\sqrt{R\,\Delta h}}{R - \Delta h/2} \approx \sqrt{\frac{\Delta h}{R}}$$

From Eq. (17-4), $\mu \geq \tan\alpha = \sqrt{\Delta h/R}$

or

$$(\Delta h)_{\max} = \mu^2 R \tag{17-5}$$

The high forces generated in rolling are transmitted to the workpiece through the rolls. Under these loading conditions there are two major types of elastic distortion. First, the rolls tend to bend along their length because the workpiece tends to separate them while they are restrained at their ends. This leads to the problems with thickness variation over the width that are discussed in Sec. 17-8. Second, the rolls flatten in the region where they contact the workpiece, so that the radius of curvature is increased from R to R'. The most widely used analysis for *roll flattening* is due to Hitchcock,[1] who represented the actual pressure distribution with elastically flattened rolls by an elliptical distribution. According to this analysis the radius of curvature increases from R to R'

$$R' = R\left[1 + \frac{CP'}{b(h_0 - h_f)}\right] \tag{17-6}$$

where $C = 16(1 - \nu^2)/\pi E$ is evaluated for the roll material ($C = 3.34 \times 10^{-4}$ in^2/ton for steel rolls) and P' is the rolling load based on the deformed roll radius. Since P' is a function of R', the exact solution of Eq. (17-6) requires a trial-and-error procedure.[2]

Example Determine the maximum possible reduction for cold-rolling a 12-in-thick slab when $\mu = 0.08$ and the roll diameter is 24 in. What is the maximum reduction on the same mill for hot rolling when $\mu = 0.5$?

$$\tan\theta_{\max} = \mu \qquad \alpha = \theta_{\max} = \tan^{-1}(0.08) = 4.6°$$

From Fig. 17-5 $\sin\alpha = L_p/R = \sqrt{R\,\Delta h}\,/R$, $\Delta h = 0.077$ in. Note that the same result would be obtained from Eq. (17-5).

$$(\Delta h)_{\max} = \mu^2 R = (0.08)^2(12) = 0.077 \text{ in}$$

For hot rolling

$$(\Delta h)_{\max} = (0.5)^2(12) = 3 \text{ in}$$

[1] J. H. Hitchcock, "Roll Neck Bearings," ASME, New York, 1935; see L. R. Underwood, "The Rolling of Metals," pp. 286–296, John Wiley & Sons, Inc., New York, 1950.

[2] E. C. Larke, "The Rolling of Strip, Sheet, and Plate," 2d ed., pp. 267–273, Chapman and Hall, Ltd., London, 1963.

17-7 SIMPLIFIED ANALYSIS OF ROLLING LOAD: ROLLING VARIABLES

The main parameters in rolling are:

The roll diameter
The deformation resistance of the metal as influenced by metallurgy, temperature, and strain rate
The friction between the rolls and the workpiece
The presence of front tension and/or back tension in the plane of the sheet

The rolling load is given by the roll pressure times the area of contact between the metal and the rolls, Eq. (17-3). Neglecting friction for the moment, the pressure is the yield stress of the material and the area of contact is the projected area of the arc of contact times the width of the metal in the roll gap. From Eq. (17-3)

$$P = pbL_p = \sigma_0' b\sqrt{R\Delta h} \tag{17-7}$$

The plane strain yield stress σ_0' is used when there is no change in the width (b) of the sheet. When spread occurs in rolling then the uniaxial yield stress should be used.

For the usual situation of plane-strain conditions the influence of friction on rolling can be introduced considering the previous analysis (Sec. 16-3) for the plane-strain compression of a slab. To a first approximation rolling is plane-strain compression in which a friction hill is generated. Thus, the mean deformation pressure is given by

$$\frac{\bar{p}}{\bar{\sigma}_0'} = \frac{1}{Q}(e^Q - 1) \tag{17-8}$$

where $Q = \mu L_p / \bar{h}$
$\bar{h}$ = the mean thickness between entry and exit from the rolls.

Referring to Eq. (17-3), the rolling load is given by

$$P = \bar{p}bL_p$$

and since $L_p \approx \sqrt{R\Delta h}$

$$P = \frac{2}{\sqrt{3}}\bar{\sigma}_0\left[\frac{1}{Q}(e^Q - 1)b\sqrt{R\Delta h}\right] \tag{17-9}$$

The factor $2/\sqrt{3}$ arises because flat rolling is a plane-strain situation, so that the flow stress should be the flow stress in plane strain.

Equation (17-9) shows that rolling load increases with roll diameter at a rate greater than $D^{1/2}$, depending on the contribution from the friction hill. The rolling load also increases as the sheet entering the rolls becomes thinner (due to the e^Q term). Eventually a point is reached where the deformation resistance of the sheet is greater than the roll pressure which can be applied and no further reduction in thickness can be achieved. This occur when the rolls in contact with the sheet are both severely elastically deformed. The roll diameter has an important influence on determining the minimum possible gage sheet that can be rolled with a particular mill. Both rolling load and the length of the arc of contact decrease with roll diameter. Therefore, with small-diameter rolls, properly stiffened against deflection by backup rolls, it is possible to produce a greater reduction before roll flattening becomes significant and no further reduction in sheet thickness is possible.

The mean flow stress for a rolling process can be determined directly from a plane-strain compression test. For cold-rolling it does not depend much on the strain rate or roll speed. However, as emphasized earlier, in hot-rolling changes in strain rate can produce significant changes in the flow stress of the metal.[1]

The friction between the roll and the metal surface is of great importance in rolling. We have already seen that a frictional force is needed to pull the metal into the rolls. However, Fig. 17-6 indicates that a large fraction of the rolling load comes from the frictional forces. The frictional contribution resides in the e^Q term in Eq. (17-8). High friction results in high rolling load, a steep friction hill, and great tendency for edge cracking. The friction varies from point to point along the contact arc of the roll.[2] However, because it is very difficult to measure this variation in μ, all theories of rolling are forced to assume a constant coefficient of friction. For cold-rolling[3] with lubricants μ varies from about 0.05 to 0.10, but for hot-rolling friction coefficients from 0.2 up to the sticking condition are common.

Example Calculate the rolling load if steel sheet is hot rolled 30 percent from a 1.5-in-thick slab using a 36-in-diameter roll. The slab is 30 in wide. Assume $\mu = 0.30$. The plane-strain flow stress is 20 ksi at entrance and 30 ksi at the exit from the roll gap due to the increasing velocity.

$$\frac{h_0 - h_1}{h_0} \times 100 = 30\% \qquad h_1 = 1.50 - 0.45 = 1.05 \text{ in} \qquad \Delta h = 0.45 \text{ in}$$

$$\bar{h} = \frac{h_0 + h_1}{2} = \frac{1.50 + 1.05}{2} = 1.275 \text{ in}$$

$$Q = \frac{\mu L_p}{\bar{h}} = \frac{(0.30)\sqrt{18 \times 0.45}}{1.275} = 0.67$$

$$\sigma_0' = \frac{20 + 30}{2} = 25 \text{ ksi}$$

$$P = 25\left[\frac{1}{0.67}(e^{0.67} - 1)30\sqrt{18 \times 0.45}\right] = 25(121.6) = 3039 \text{ kips}$$

What would be the rolling load if sticking friction occurs?

[1] P. M. Cook, *Proc. Conf. Properties of Materials at High Rates of Strain*, Institution of Mechanical Engineers, London, 1957, pp. 85–97.

[2] J. M. Capus and M. G. Cockcroft, *J. Inst. Met.*, vol. 90, pp. 289–297, 1961–1962.

[3] W. L. Roberts, *Blast Furn. Steel Plant*, vol. 56, pp. 382–394, 1968.

Continuing the analogy with compression in plane strain, from Eq. (16-15)

$$\bar{p} = \sigma_0'\left(\frac{a}{2h} + 1\right) = \sigma_0'\left(\frac{L_p}{4\bar{h}} + 1\right)$$

$$P = \sigma_0'\left(\frac{\sqrt{R\Delta h}}{4\bar{h}} + 1\right)b\sqrt{R\Delta h} = 25\left(\frac{2.846}{4 \times 0.45} + 1\right)30(2.846)$$

$$P = 5509 \text{ kips}$$

Example The previous example neglected the influence of roll flattening under the very high rolling loads. From Eq. (17-6) the deformed radius of a roll under load is $R' = R[1 + CP'/[b(h_0 - h_f)]]$. For a steel roll, $C = 3.34 \times 10^{-4}$ in^2/ton. Using $P' = 1{,}357$ tons from the previous example $R' = 18[1 + [3.34 \times 10^{-4}(1357)]/[30(0.45)]] = 8(1.034) = 18.612$ in. We now use R' to calculate a new value of P' and in turn another value of R'

$$Q = \frac{0.30\sqrt{18.612 \times 0.45}}{1.275} = 0.68$$

$$P'' = 25\left[\frac{1}{0.68}(e^{0.68} - 1)30\sqrt{18.612 \times 0.45}\right]$$

$$= 25(124.3) = 3108 \text{ kips} = 1387 \text{ tons}$$

$$R'' = 18\left[1 + \frac{3.34 \times 10^{-4}(1387)}{30(0.45)}\right] = 18(1.0343) = 18.617 \text{ in}$$

The difference between the two estimations of R' is not large, so we stop the calculation at this point.

The angle of bite can be used to establish μ through Eq. (17-4). However, the method is not very accurate and other techniques have been developed. We have already seen that the neutral point is the location on the arc of contact where the direction of the friction force changes. From the entry plane to the neutral point the friction force acts in the direction of roll rotation, while on the exit side of the neutral point it reverses direction. If back tension is applied gradually to the sheet, the neutral point shifts toward the exit plane. The total rolling load P and torque M_T (per unit of width b) are given by[1]

$$\frac{P}{b} = \int_0^{L_p} p\,dx \qquad \frac{M_T}{b} = \int_0^{L_p} (\mu p\,dx)R = \mu R\int_0^{L_p} p\,dx = \mu R\frac{P}{b}$$

thus, $$\mu = \frac{M_T}{PR} \qquad (17\text{-}10)$$

where μ is obtained by measuring the torque and the rolling load at constant roll speed and reduction with the proper back tension. The proper back tension to

[1] P. W. Whitton and H. Ford, *Proc. Inst. Mech. Eng. (London)*, vol. 169, p. 123, 1955.

bring the neutral point to the exit plane is achieved when the exit velocity of the sheet v_f is equal to the surface velocity of the rolls, $v_r = R\omega$. This can be expressed another way by stating that the *forward slip* S_f equals zero.

$$S_f = \frac{v_f - v_r}{v_r} \tag{17-11}$$

Measurement of forward slip for any value of back tension can be used to estimate μ through[1]

$$S_f = \frac{1}{4}\frac{r}{1-r}\left(1 - \frac{\alpha}{2\mu}\right)^2 \tag{17-12}$$

where $r = (h_0 - h_f)/h_0$ is the reduction and α is the angle of bite.

The minimum-thickness sheet that can be rolled on a given mill is directly related to the coefficient of friction. Since friction coefficients are much lower for cold-rolling than hot-rolling, thinner-gage sheet is produced by cold-rolling. The thickness of the sheet produced on a cold-rolling mill can also be decreased appreciably by increasing the rolling speed. This is attributable to the decrease in friction coefficient with increasing rolling speed.[2]

The presence of tension in the plane of the sheet can materially reduce the rolling load. Back tension may be produced by controlling the speed of the uncoiler relative to the roll speed and front tension may be created by controlling the coiler. The effect of sheet tension on reducing roll pressure p can be shown simply from a consideration of the von Mises' criterion in plane strain.

$$\sigma_1 - \sigma_3 = \frac{2}{\sqrt{3}}\bar{\sigma}_0$$

$$p - (-\sigma_h) = \frac{2}{\sqrt{3}}\bar{\sigma}_0$$

where σ_h is the horizontal sheet tension and compressive stresses are taken as positive.

$$p = \frac{2}{\sqrt{3}}\bar{\sigma}_0 - \sigma_h \tag{17-13}$$

Thus, the roll pressure is reduced in direct proportion to the tension in the plane of the sheet. This results in less wear of the rolls and improved flatness and uniformity of thickness across the width of the sheet. A study[3] of the effect of sheet tension has shown that back tension is about twice as effective in reducing the rolling load as front tension. The rolling load when tension is applied can be determined from Eq. (17-9) by replacing $\bar{\sigma}_0$ with $\bar{\sigma}_0 - \sigma_t$, where σ_t is the average of the front and back tension.

Nadai[4] developed a theory of rolling which permits the calculation of the effect of strip tension on the roll pressure distribution. As shown schematically in

[1] M. D. Stone, *Trans. ASME, Ser. D: J. Basic Eng.*, vol. 81, pp. 681–686, 1957.
[2] R. B. Sims and D. F. Arthur, *J. Iron Steel Inst. London*, vol. 172, pp. 285–295, 1952.
[3] W. C. F. Hessenberg and R. B. Sims, *J. Iron Steel Inst. London*, vol. 168, pp. 155–164, 1951.
[4] A. Nadai, *J. Appl. Mech.*, vol. 6, pp. A54–A62, 1939.

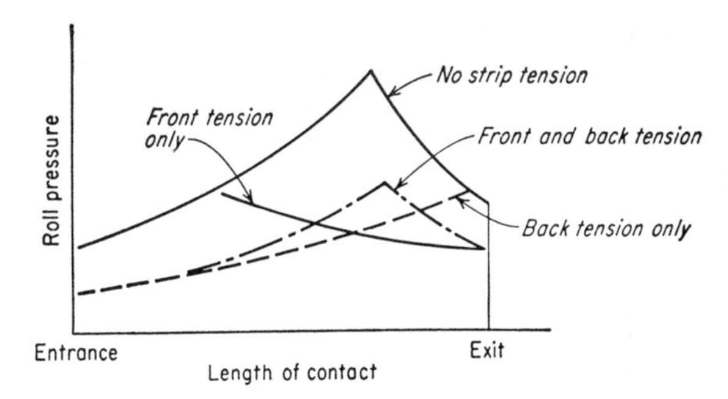

Figure 17-7 Effect of strip tension on the distribution of roll pressure.

Fig. 17-7, the addition of both front and back tension materially reduces the area under the curve, although there is little shift of the neutral point. If only back tension is applied, the neutral point moves toward the roll exit. If a high enough back tension is applied, the neutral point will eventually reach the roll exit. When this happens, the rolls are moving faster than the metal and they slide over the surface. On the other hand, if only front tension is used, the neutral point will move toward the roll entrance.

17-8 PROBLEMS AND DEFECTS IN ROLLED PRODUCTS

A variety of problems in rolling, leading to specific defects, can arise depending on the interaction of the plastically deforming workpiece with the elastically deforming rolls and rolling mill.[1] Under the influence of the high rolling forces the rolls flatten and bend, and the entire mill is elastically distorted. Because of *mill spring* the thickness of the sheet exiting from the rolling mill is greater than the roll gap set under no-load conditions. In order to roll to precise thickness it is necessary to know the elastic constant of the mill.[2] This is usually given in the form of a calibration curve (Fig. 17-8). The elastic constant for most screw-loaded rolling mills is 4,000 to 8,000 tons/in, while hydraulically loaded mills may exceed 10,000 tons/in.

Elastic flattening of the rolls with increasing roll pressure results in a condition where the rolls eventually deform more easily than the workpiece. Thus, for a given material and set of rolling conditions, there will be a minimum thickness below which the sheet can be reduced no further. It has already been seen that thinner-gage sheet can be produced with small-diameter rolls. More complete analysis of the problem[3] shows that the limiting thickness is nearly proportional to the coefficient of friction, the roll radius, the flow stress of the workpiece, and is inversely proportional to the elastic modulus of the rolls. For

[1] J. A. Schey, Workability in Rolling, "Workability Testing Techniques," Chap. 10, American Society for Metals, Metals Park, Ohio, 1984.

[2] E. A. Marshall and A. Shutt, *J. Iron Steel Inst. London*, vol. 204, pp. 837–841, 1966; K. Wiedemer, *Met. Technol.*, vol. 1, pp. 181–185, 1974.

[3] H. Ford and J. M. Alexander, *J. Inst. Met.*, vol. 88, pp. 193–199, 1959–1960.

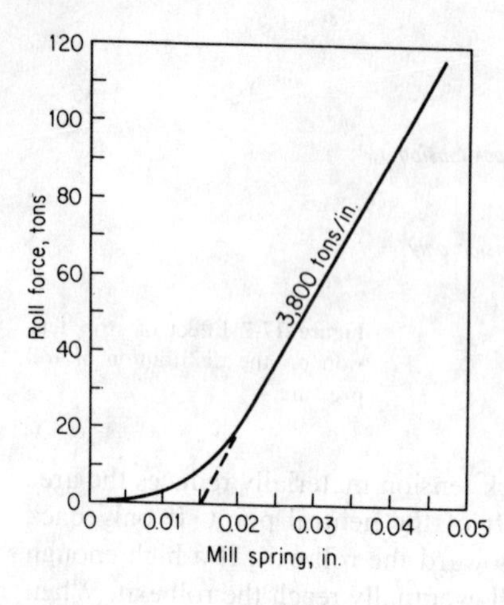

Figure 17-8 Typical calibration curve for the elastic constant of a rolling mill. *(From J. A. Schey, in "The Techniques of Materials Preparation and Handling," R. F. Bunshah, ed., vol. 1, pt. 3, chap. 34, p. 1452. John Wiley & Sons, Inc., New York, 1965.)*

steel rolls this is given by

$$h_{\min} = \frac{\mu R \bar{\sigma}_0'}{1{,}860} \tag{17-14}$$

where units are in pounds per square inch and inches. In general, problems with limiting gage can be expected when the sheet thickness is below $\frac{1}{400}$ to $\frac{1}{600}$ of the roll diameter.

The roll gap must be perfectly parallel, otherwise one edge of the sheet will be decreased more in thickness than the other, and since volume and width remain constant, this edge of the sheet elongates more than the other, and the sheet bows. There are two aspects to the problem of the shape of a sheet. The first pertains to uniform thickness over the width and along the length. This property of a sheet can be measured accurately, and it is subject to precise control with modern automatic gage control systems. The second important property of a sheet is *flatness*. It is difficult to measure this property accurately; particularly difficult are in-process measurements taken as the sheet moves through a continuous mill at high speed. The rolling process is very sensitive to flatness. A difference in elongation of one part in 10,000 between different locations in the sheet can give rise to waviness in a thin-gage sheet. Figure 17-9 shows how waviness (lack of flatness) develops. If the rolls deflect as in Fig. 17-9*a*, the edges of the sheet will be elongated to a greater extent in the longitudinal direction than the center, i.e., it has "long edges." If the edges were free to move relative to the center, we would have the situation shown in Fig. 17-9*b*. However, the sheet remains a continuous body and the strains readjust to maintain continuity. The result is that the center portion of the sheet is stretched in tension and the edges are compressed in the rolling direction Fig. 17-9*c*. The usual result is a wavy edge or edge buckle (Fig. 17-9*d*). Under other conditions of $\Delta = h/L$ the strain distribution produced by long edge could produce short "zipper breaks" or cracks in the center of the sheet (Fig. 17-9*e*). The obvious solution to roll bending is to contour the roll parallel to its axis so it is larger in the center than at the ends. Then when the rolls deflect,

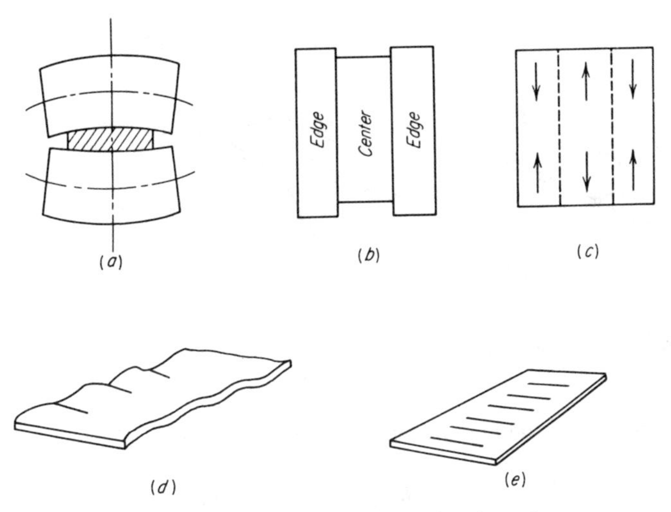

Figure 17-9 Consequences of roll bending to produce long edge.

they will present a parallel gap to the workpiece. Rolls having a ground *camber* or crown are usually used. Camber can also be produced thermally due to the thermal expansion. This corrective procedure suffers from the fact that the camber corrects for the roll deflection at only one value of roll force, and thus it may not be an effective measure for a range of rolling conditions. A better technique is to equip the rolling mill with hydraulic jacks which permit the elastic distortion of the rolls to correct for deflection under rolling conditions. Many modern sheet mills are equipped in this way. If the rolls have excess convex camber, the center of the sheet is elongated more than the edges. The strain distribution is the opposite of Fig. 17-9*c* and the sheet is said to have loose middle and tight edges. Such a sheet usually will contain center buckles.

It is generally considered[1] that sheet with bad shape is created in the hot-rolling step and the cold-rolling cannot correct bad shape completely. Moreover, shape problems are greatest when rolling thin strip (less than 0.010 in) because fractional errors in roll gap profile increase with decrease in thickness, producing larger internal stresses. Also, very thin sheet is less resistant to buckling. Mild shape problems may be corrected by *stretch leveling* the sheet in tension with a bridle roll between stands, or by bend flexing the sheet in a *roller-leveler*. Suggestions for a rolling mill with "flexible rolls" which would correct bad shape have been made[2] and a good beginning has been made[3] on a mechanics analysis of shape and flatness.

[1] T. Sheppard and J. M. Roberts, *Int. Metall. Rev.*, vol. 18, pp. 1–18, 1973.

[2] N. H. Polakowski, D. M. Reddy, and H. N. Schmeissing, *Trans. ASME, Ser. B: J. Eng. Ind.*, vol. 91, pp. 702–709, 1969.

[3] H. W. O'Connor and A. S. Weinstein, *Trans. ASME, Ser. B: J. Eng. Ind.*, vol. 94, pp. 1113–1123, 1972; D. J. McPherson, *Metall. Trans.*, vol. 5, pp. 2479–2499, 1974.

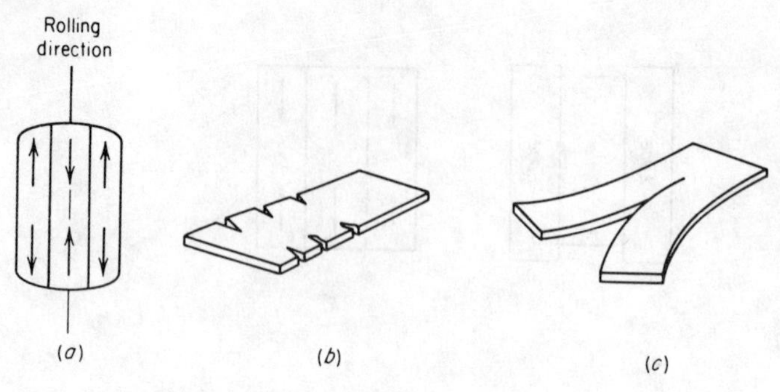

Figure 17-10 Defects resulting from lateral spread.

Problems with shape and flatness are brought about by inhomogeneities in deformation in the rolling direction of the sheet. Other forms of inhomogeneous deformation[1] can lead to problems with *cracking*. As the workpiece passes through the rolls all elements across the width experience some tendency to expand laterally (in the transverse direction of the sheet). The tendency for lateral spread is opposed by transverse friction forces. Because of the friction hill, these are higher toward the center of the sheet so that the elements in the central region spread much less than the outer elements near the edge. Because the thickness decrease in the center of the sheet all goes into a length increase, while part of the thickness decrease at the edges goes into lateral spread, the sheet may develop a slight rounding at its ends (Fig. 17-10*a*). Because there is continuity between the edges and center, the edges of the sheet are strained in tension, a condition which leads to *edge cracking* (Fig. 17-10*b*). Under severe conditions the strain distribution shown in Fig. 17-10*a* can result in a *center split* of the sheet (Fig. 17-10*c*).

Edge cracking also can be caused by inhomogeneous deformation in the thickness direction.[2] When the rolling conditions are such that only the surface of the workpiece is deformed (as in light reductions on a thick slab), the cross section of sheet is deformed into the shape shown in Fig. 17-11*a*. In subsequent passes through the rolls the overhanging material is not compressed directly but is forced to elongate by the neighboring material closer to the center. This sets up high secondary tensile stresses which lead to edge cracking. This type of cracking is experienced in initial ingot breakdown in hot-rolling where it was observed[3] that edge deformation like that shown in Fig. 17-11*a* occurred where $h/L_p > 2$. With heavy reductions, so that the deformation extends through the thickness of the sheet, the center tends to expand laterally more than the surfaces to produce barreled edges similar to those found in upsetting a cylinder (Fig. 17-11*b*). The

[1] I. Ya Tarnovskii, A. A. Pozdeyev, and V. B. Lyashkov, "Deformation of Metals During Rolling," Pergamon Press, New York, 1965; A. Jones and B. Walker, *Met. Technol.*, vol. 1, pp. 310–315, 1974.

[2] N. H. Polakowski, *J. Inst. Met.*, vol. 76, pp. 755–757, 1949–1950.

[3] P. A. Aleksandrov, in A. M. Samarin (ed.), "Contemporary Problems of Metallurgy," pp. 416–422, Consultants Bureau, New York, 1960.

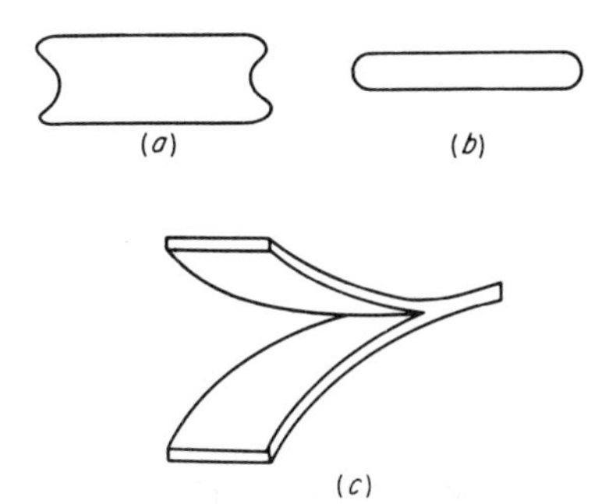

Figure 17-11 Edge distribution resulting from rolling with (*a*) light reduction, (*b*) heavy reduction, and (*c*) alligatoring.

secondary tensile stresses[1] created by the barreling are a ready cause of edge cracking. With this type of lateral deformation, greater spread occurs toward the center than at the surfaces so that the surfaces are placed in tension and the center is in compression. This stress distribution also extends in the rolling direction, and if there is any metallurgical weakness along the center line of the slab, fracture will occur there (Fig. 17-11*c*). This *alligatoring* type of fracture is accentuated if there is any curling of the sheet because one roll is higher or lower than the centerline of the roll gap.

Edge cracking is minimized in commercial rolling practice by employing vertical edge rolls which keep the edges straight and thus prevent a cumulative buildup of secondary tensile stresses due to barreling of the edge. Since most laboratory mills do not have edge rolls, a simple but time-consuming procedure to prevent edge cracking is to machine the edges square after each pass. A better procedure is to equip the mill with edge-restraining bars.[2] Materials with low ductility may be rolled without excessive cracking by "canning" on all sides with a material with a flow stress similar to that of the workpiece. The canning material minimizes thermal stresses and provides edge restraint and an increased degree of hydrostatic compression.

Defects other than cracks can result from defects introduced during the ingot stage of production or during rolling. Internal defects such as fissures are due to incomplete welding of pipe and blowholes. Longitudinal stringers of nonmetallic inclusions or pearlite banding in steels are related to melting and solidification practices. In severe cases these defects can lead to laminations which drastically reduce the strength in the thickness direction. Because rolled products usually have a high surface-to-volume ratio, the condition of the surface during all stages of production is of importance. In order to maintain high quality, the surface of billets must be conditioned by grinding, chipping, or burning with an oxygen lance to remove surface defects such as slivers, seams, and scabs. Scratches due to defective rolls or guides must be guarded against in cold-rolled sheet. Sometimes

[1] G. Cuminsky and F. Ellis, *J. Inst. Met.*, vol. 95, pp. 33–37, 1967; A. L. Hoffmanner, *Ibid.*, vol. 96, pp. 158–159, 1968.

[2] J. A. Schey, *J. Inst. Met.*, vol. 94, pp. 193–200, 1966.

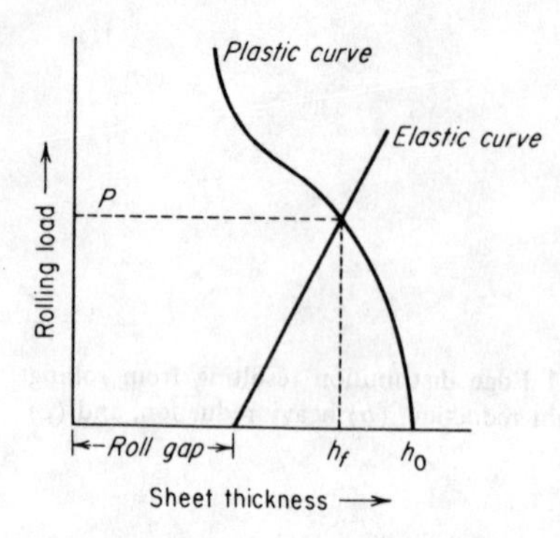

Figure 17-12 Characteristic elastic and plastic curves for a rolling mill.

problems arise with removing the rolling lubricant or because there is discoloration after heat treatment.

17-9 ROLLING-MILL CONTROL

The high throughput and production rate of modern continuous hot-strip and cold-rolling mills make it imperative that these mills be operated under automatic control. Of all the metalworking processes rolling is best suited for the adoption of automatic control because it is an essentially steady-state process in which the tooling geometry (roll gap) may be changed readily during the process. To date, the application of automatic control in rolling has been mostly concerned with control of sheet gage during the rolling of long continuous coils. This has required the development of on-line sensors to continuously measure sheet thickness. The two instruments most commonly used are the flying micrometer and x-ray, or isotope, gages which measure thickness by monitoring the amount of radiation transmitted through the sheet. More recently control procedures have been aimed at controlling strip shape[1] as well as thickness.

The problem of gage control can be understood by considering the characteristic curves for a rolling mill (Fig. 17-12). For a given set of rolling conditions, the rolling load varies with the final sheet thickness according to the *plastic curve*. This is essentially what would be obtained by the solution of Eq. (17-9). The *elastic curve* for mill spring is superimposed on the figure. This indicates that a sheet of initial thickness h_0 would have a final thickness h_f and the load on the mill would be P. The influence of changes in rolling variables can be visualized readily with this type of diagram. If the lubrication breaks down so that μ increases or the flow stress increases because the temperature decreases, the plastic curve will be raised (Fig. 17-13). The rolling load will be increased from P_1 to P_2 and the final thickness will be increased from h_{f1} to h_{f2}. Figure 17-13 shows that to maintain a constant thickness h_{f1} under these new conditions the roll gap

[1] A. J. Carlton, W. J. Edwards, P. W. Johnston, and N. Kuhn, *J. Aust. Inst. Met.*, vol. 18, pp. 22–36, 1973.

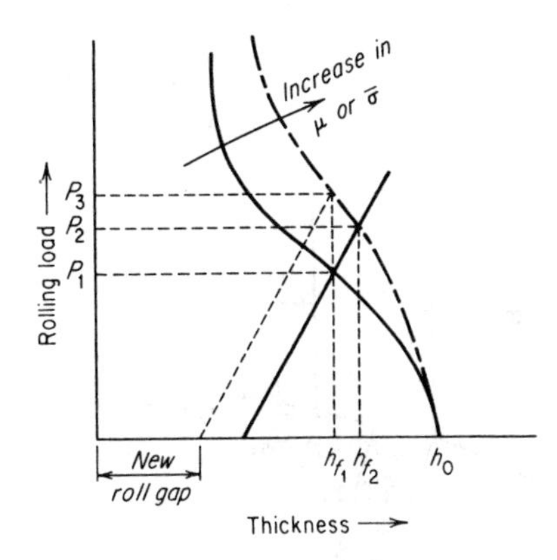

Figure 17-13 Use of characteristic curves to show changes in rolling conditions.

would have to be decreased. Moving the elastic curve to the left, in this way, further increases the rolling load to P_3. If, for example, there is an increase in the incoming sheet thickness, the plastic curve would move to the right relative to the elastic curve. If there is an increase in strip tension, the plastic curve would move to the left.

In a continuous hot mill the strip thickness is measured indirectly by measuring the rolling load and using the characteristic curve of the mill to establish thickness. The error signal is fedback to the rolling mill screws to reposition them so as to minimize the error. An x-ray gage is used after the last stand to provide an absolute measurement of sheet gage. In continuous cold-strip mills thickness is measured by x-ray gages. While the error in thickness following the first stand is usually fedback to adjust the gap setting on the first stand, gage control in subsequent stands usually is achieved by controlling the strip tension through controlling the relative roll speed in successive stands or the coiler speed. Gage control through control of strip tension has faster response time than control through change in roll setting.

17-10 THEORIES OF COLD ROLLING

Probably more work has been expended in developing a theoretical treatment of cold-rolling than for any other metalworking process. A theory of rolling is aimed at expressing the external forces, such as the rolling load and the rolling torque, in terms of the geometry of the deformation and the strength properties of the material being rolled.

The differential equation for the equilibrium of an element of material being deformed between rolls is common to all the theories of rolling. The derivation given below is based on the following assumptions.

1. The arc of contact is circular—no elastic deformation of the rolls.
2. The coefficient of friction is constant at all points on the arc of contact.

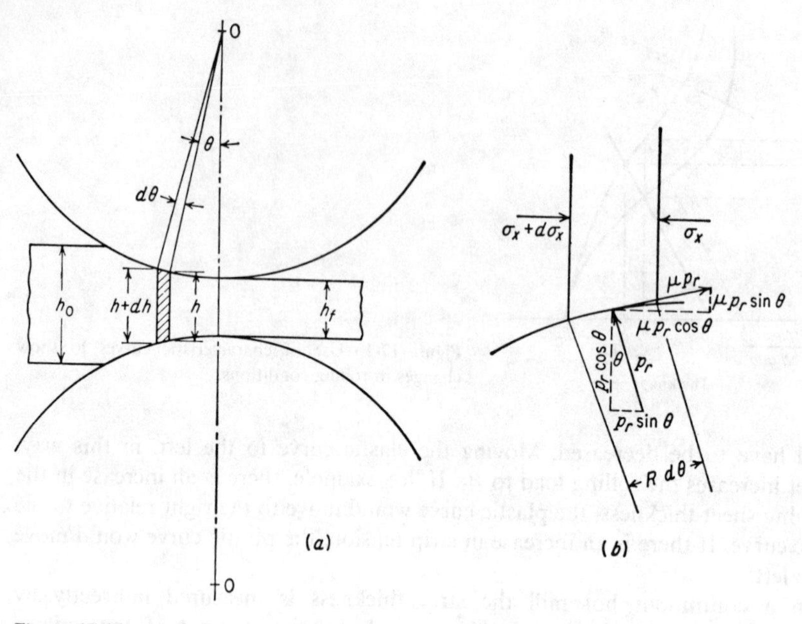

Figure 17-14 (*a*) Geometrical relationship for element undergoing plane-strain deformation by rolling; (*b*) stresses acting on an element.

3. There is no lateral spread, so that rolling can be considered a problem in plane strain.
4. Plane vertical sections remain plane; i.e., the deformation is homogeneous.
5. The peripheral velocity of the rolls is constant.
6. The elastic deformation of the sheet is negligible in comparison with the plastic deformation.
7. The distortion-energy criterion of yielding, for plane strain, holds.

$$\sigma_1 - \sigma_3 = \frac{2}{\sqrt{3}}\sigma_0 = \sigma_0'.$$

The stresses acting on an element of strip in the roll gap are shown in Fig. 17-14. At any point of contact between the strip and the roll surface, designated by the angle θ, the stresses acting are the radial pressure p_r and the tangential shearing stress $\tau = \mu p_r$. These stresses are resolved into their horizontal and vertical components in Fig. 17-14*b*. In addition, the stress σ_x is assumed to be uniformly distributed over the vertical faces of the element. The normal stress on one end of the element is $p_r R\, d\theta$, and the horizontal component of this force is $p_r R \sin\theta\, d\theta$. The tangential friction force is $\mu p_r R\, d\theta$, and its horizontal component is $\mu p_r R \cos\theta\, d\theta$. Taking the summation of the horizontal forces on the element results in

$$(\sigma_x + d\sigma_x)(h + dh) + 2\mu p_r R \cos d\theta - \sigma_x h - 2p_r R \sin\theta \, d\theta = 0$$

which simplifies to

$$\frac{d(\sigma_x h)}{d\theta} = 2p_r R(\sin\theta \pm \mu \cos\theta) \tag{17-15}$$

The positive sign applies between the exit plane and the neutral point, while the negative sign applies between the entrance plane and the neutral point. The minus and plus signs in Eq. (17-15) occur because the direction of the friction force changes at the neutral point. This equation was first derived by von Kármán[1] and is usually named for him.

The forces acting in the vertical direction are balanced by the specific roll pressure p. Taking the equilibrium of forces in the vertical direction results in a relationship between the normal pressure and the radial pressure.

$$p = p_r(1 \mp \mu \tan\theta) \tag{17-16}$$

The relationship between the normal pressure and the horizontal compressive stress σ_x is given by the distortion energy criterion of yielding for plane strain.

$$\sigma_1 - \sigma_3 = \frac{2}{\sqrt{3}}\sigma_0 = \sigma_0'$$

or

$$p - \sigma_x = \sigma_0' \tag{17-17}$$

where p is the greater of the two compressive principal stresses.[2]

The solution of problems in cold-rolling consists in the integration of Eq. (17-15) with the aid of Eqs. (17-16) and (17-17). Unfortunately, the mathematics is rather formidable, and various approximations must be made to obtain a tractable solution. Trinks[3] provided graphical solutions to von Kármán's equation, using the assumptions of constant yield stress and a parabolic arc of contact. The most complete solution of the rolling equations has been obtained by Orowan.[4] In this solution, allowance was made for the fact that the flow stress changes with θ owing to strain hardening. However, the complexity of the equations makes it necessary to obtain solutions by graphical integration.[5] Even though the equations for roll pressure have been expressed[6] in algebraic form, they are still too complicated for routine calculation of rolling problems.

[1] T. von Kármán, *Z. Angew. Math. Mech.*, vol. 5, pp. 139–141, 1925.

[2] In agreement with the literature on rolling, compressive stresses are taken as positive.

[3] W. Trinks, *Blast Furn. Steel Plant*, vol. 25, pp. 617–619, 1937; see also Underwood, op. cit., pp. 210–215.

[4] E. Orowan, *Proc. Inst. Mech. Eng. (London)*, vol. 150, pp. 140–167, 1943.

[5] A computer solution to Orowan's equations which also incorporates corrections for the deformation of the rolls has been given by J. E. Hockett, *Trans. Am. Soc-. Met.*, vol. 52, pp. 675–697, 1960.

[6] M. Cook and E. C. Larke, *J. Inst. Met.*, vol. 74, pp. 55–80, 1957.

Some simplification to this problem has been provided by Bland and Ford.[1] By restricting the analysis to cold-rolling under conditions of low friction and for angles of contact less than 6°, they were able to put $\sin\theta \approx \theta$ and $\cos\theta \approx 1$. Thus, Eq. (17-15) can be written

$$\frac{d(\sigma_x h)}{d\theta} = 2 p_r R'(\theta \pm \mu) \qquad (17\text{-}18)$$

It is also assumed that $p_r \approx p$, so that Eq. (17-17) can be written $\sigma_x = p_r - \sigma_0'$. By substituting into Eq. (17-18) and integrating,[2] relatively simple equations for the radial pressure result.

Roll entrance to neutral point:

$$p_r = \frac{\sigma_0' h}{h_0}\left(1 - \frac{\sigma_{xb}}{\sigma_{01}'}\right) e^{\mu(H_1 - H)} \qquad (17\text{-}19)$$

Neutral point to roll exit:

$$p_r = \frac{\sigma_0' h}{h_f}\left(1 - \frac{\sigma_{xf}}{\sigma_{02}'}\right) e^{\mu H} \qquad (17\text{-}20)$$

where $$H = 2\left(\frac{R'}{h_f}\right)^{1/2} \tan^{-1}\left[\left(\frac{R'}{h_f}\right)^{1/2} \theta\right]$$

and σ_{xb} = back tension
σ_{xf} = front tension

Subscript 1 refers to a quantity evaluated at the roll-entrance plane, and subscript 2 refers to a quantity evaluated at the roll-exit plane.

Equations (17-19) and (17-20) can be used to calculate the pressure distribution over the arc of contact of the rolls. The rolling load or total force P is the integral of the specific roll pressure over the arc of contact.

$$P = R'b \int_0^{\theta=\alpha} p \, d\theta \qquad (17\text{-}21)$$

where b = width of sheet
α = contact angle

This is best evaluated by graphical integration following a point-by-point calculation with Eqs. (17-19) and (17-20). Computational methods based on the Bland and Ford solution have been published.[3]

[1] D. R. Bland and H. Ford, *Proc. Inst. Mech. Eng. (London)*, vol. 159, pp. 144–163, 1948.
[2] See *ibid.*, or H. Ford, *Metall. Rev.*, vol. 2, pp. 5–7, 1957.
[3] P. W. Whitton, *J. Appl. Mech.*, vol. 23, pp. 307–311, 1956.

The modern digital computer removes the necessity for making simplifying assumptions to obtain solutions to von Kármán's equation. Alexander[1] removed all restrictions except the assumption of homogeneous deformation, and showed that while the previous approximate theories gave reasonable prediction of rolling load, they were not capable of predicting roll torque with any degree of precision.

17-11 THEORIES OF HOT-ROLLING

Theories of hot-rolling have not advanced to the state of knowledge that exists for cold-rolling because of the more difficult problem of accounting for inhomogeneous deformation and less well-defined friction conditions in hot-working. As in other hot-working processes, the flow stress for hot-rolling is a function of both temperature and strain rate (speed of rolls).

The strain rate for hot-rolling with sticking friction is given[2] by

$$\dot{\varepsilon} = \frac{v}{h} = \frac{2v_r \sin\theta}{h} = \frac{2v_r \sin\theta}{h_f + D(1 - \cos\theta)} \tag{17-22}$$

Evaluation of this equation shows that the maximum strain rate occurs near the entrance to the rolls. For equal percentage reduction, thin sheet will undergo much greater strain rates than thick slabs. The mean strain rate with sticking friction is given[3] by

$$\bar{\dot{\varepsilon}} = v_r\left(\frac{1}{R\Delta h}\right)\ln\frac{h_0}{h_f} \tag{17-23}$$

where $v_r = 2\pi Rn$ and n is in revolutions per second.

When slipping friction occurs the surface velocity of the metal cannot be assumed equal to the peripheral velocity of the rolls. Only at the neutral point does this situation exist. The strain rate for slipping friction is given[3] by

$$\dot{\varepsilon}_n = \frac{2v_r h_n \cos\beta \tan\theta}{\left[h_f + D(1 - \cos\theta)\right]^2} \tag{17-24}$$

where h_n is the metal thickness in the gap at the neutral point, and β and θ are given by Fig. 17-5.

Ford and Alexander[4] used slip-line field analysis of hot-rolling to develop equations for the rolling load and torque in hot-rolling of nonferrous alloys and steels.[5]

$$P = kbL_p\left(\frac{\pi}{2} + \frac{L_p}{h_0 + h_f}\right) \tag{17-25}$$

[1] J. M. Alexander, *Proc. R. Soc. London, Ser. A*, vol. 326, pp. 535–563, 1972.
[2] Larke, *op. cit.*, chap. 8.
[3] J. N. Harris, "Mechanical Working of Metals," pp. 144–148, Pergamon Press, New York, 1983.
[4] H. Ford and J. M. Alexander, *J. Inst. Met.*, vol. 92, pp. 397–404, 1963–1964.
[5] E. Gupta and H. Ford, *J. Iron Steel Inst. London*, vol. 205, pp. 186–190, 1967.

where k is the mean flow stress in pure shear $= \sigma_0/\sqrt{3}$. The torque is given by

$$M_T = kbL_p^2\left(1.60 + 0.91\frac{L_p}{h_0 + h_f}\right) \tag{17-26}$$

Denton and Crane[1] have pointed out that if Eq. (17-25) is rewritten slightly, it is similar to the equation for the force to forge a slab between perfectly rough platens.

Hot-rolling:

$$P = kbL_p\left[1.31 + 0.53\frac{L_p}{(h_0h_f)^{1/2}}\right]$$

Hot-forging:

$$P = kbw\left[1.5 + 0.5\frac{b}{h}\right]$$

where $(h_0h_f)^{1/2}$ is the geometric mean thickness of the strip.

A frequently used analysis for the rolling load in hot-rolling is due to Sims.[2] Using a shortened form of the Orowan equation and mathematical simplifications, similar to Bland and Ford, Sims developed the relationship

$$P = \sigma_0'b\left[R(h_0 - h_f)^{1/2}\right]Q_p \tag{17-27}$$

where Q_p is a complex function of the reduction in thickness and the ratio R/h_f. Values of Q_p may be obtained from Fig. 17-15 or from

$$Q_p = \sqrt{\frac{h_0}{4\Delta h}}\left[\pi \tan^{-1}\sqrt{\frac{\Delta h}{h_f}} - \sqrt{\frac{R}{h_f}}\ \ln\frac{h_n^2}{h_0h_f}\right] - \frac{\pi}{4} \tag{17-28}$$

17-12 TORQUE AND HORSEPOWER

Power is applied to a rolling mill by applying a torque to the rolls and by means of strip tension. The power is expended principally in four ways, (1) the energy needed to deform the metal, (2) the energy needed to overcome frictional forces in the bearings, (3) the energy lost in the pinions and power-transmission system, and (4) electrical losses in the various motors and generators. Losses in the windup reel and uncoiler must also be considered.

The total rolling load is distributed over the arc of contact in the typical friction-hill pressure distribution. However, the total rolling load can be assumed to be concentrated at a point along the arc of contact at a distance a from the line

[1] B. K. Denton and F. A. A. Crane, *J. Iron Steel Inst. London*, vol. 210, pp. 606–616, 1972.

[2] R. B. Sims, *Proc. Inst. Mech. Eng. (London)*, vol. 108, pp. 191–200, 1954; M. Tarokh and F. Seredynski, *J. Iron Steel Inst. London*, vol. 208, pp. 695–697, 1970.

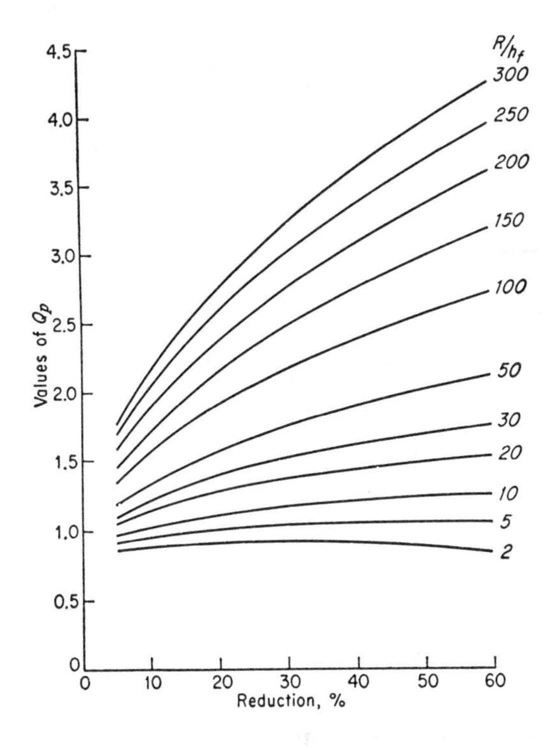

Figure 17-15 Values of Q_p for use with Eq. (17-26). *(From E. C. Larke, "The Rolling of Strip, Sheet, and Plate," Chapman & Hall, Ltd., London, 1957.)*

of centers of the rolls. In calculating the torque the problem is mainly how to calculate this moment arm. It is usual practice[1] to consider the ratio of the moment arm a to the projected length of the arc of contact.

$$\lambda = \frac{a}{L_p} = \frac{a}{\left[R(h_0 - h_f)\right]^{1/2}} \tag{17-29}$$

A typical value of λ is 0.5 for hot-rolling and 0.45 for cold-rolling.

The torque is equal to the total rolling load multiplied by the effective moment arm, and since there are two work rolls, the torque is given by

$$M_t = 2Pa \tag{17-30}$$

During one revolution of the top roll the resultant rolling load P moves along the circumference of a circle equal to $2\pi a$ (Fig. 17-16). Since there are two work rolls involved, the work done is equal to

$$\text{Work} = 2(2\pi a)P \text{ ft-lb} \tag{17-31}$$

Since horsepower is defined as the rate of doing work at 33,000 ft-lb/min, the

[1] Larke, *op. cit.*, chap. 11.

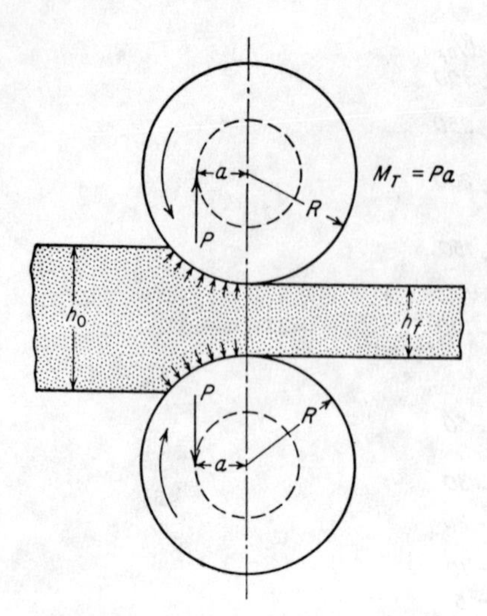

Figure 17-16 Schematic diagram illustrating roll torque.

horsepower needed to operate a pair of rolls revolving at N rpm is given by

$$\text{hp} = \frac{4\pi aPN}{33{,}000} \tag{17-32}$$

In SI units the power requirement expressed in kilowatts is given by

$$\text{kW} = \frac{4\pi aPN}{60{,}000} \tag{17-33}$$

where P is in newtons, a is in meters, and N is in rpm. The above equation expresses the horsepower required in deforming the metal as it flows through the roll gap. The horsepower needed to overcome friction in the bearings and the pinions must be determined separately.

The same basic equation for horsepower holds for hot-rolling, with the important condition that the equations which were given for determining the effective moment arm in cold-rolling are not applicable in hot-rolling. A procedure for determining the moment arm in hot-rolling, based on the work of Sims,[1] is given in detail by Larke.[2] Although the calculations are straightforward, they are too detailed for inclusion here.

Example A 12-in-wide aluminum alloy strip is hot-rolled in thickness from 0.80 to 0.60 in. The rolls are 40 in in diameter and operate at 100 rpm. The uniaxial flow stress for the aluminum alloy can be expressed as $\sigma = 20\varepsilon^{0.2}$

[1] Sims, *op. cit.*
[2] Larke, *op. cit.*, chap. 12.

(ksi). Determine the rolling load and the horsepower required for this hot reduction.

$$\varepsilon_1 = \ln\left(\frac{0.80}{0.60}\right) = 0.288 \qquad r = \frac{0.80 - 0.60}{0.80} = 0.25 \qquad R/h_f = \frac{20}{0.60} = 33.3$$

$$\bar{\sigma}_0 = \frac{k\int_0^{\varepsilon_1}\varepsilon^n\,d\varepsilon}{\varepsilon_1} = \left.\frac{k\varepsilon^{n+1}}{\varepsilon_1(n+1)}\right|_0^{\varepsilon_1} = \frac{k\varepsilon_1^{n+1}}{\varepsilon_1(n+1)} = \frac{k\varepsilon_1^n}{n+1}$$

$$= \frac{20(0.288)^{0.2}}{1.2} = 13 \text{ ksi}$$

From Eq. (17-27)
Rolling load:

$$P = \frac{2}{\sqrt{3}}(13)(12)[20(0.80 - 0.60)]^{1/2}(1.5) = 540 \text{ kips}$$

Horsepower:

$$\text{hp} = \frac{4\pi aPN}{33{,}000} \qquad a = 0.5\sqrt{R\Delta h} = 0.5\sqrt{20 \times 0.2} = 1 \text{ in}$$

$$a = \tfrac{1}{12} \text{ ft} = 0.083 \text{ ft}$$

$$\text{hp} = \frac{4\pi(0.083)(540{,}000)(100)}{33{,}000} = 1706 \text{ hp}$$

BIBLIOGRAPHY

"Elements of Rolling Practice," 2d ed., The United Steel Companies, Ltd., Sheffield, England, 1963.
Larke, E. C.: "The Rolling of Strip, Sheet, and Plate," 2d ed., Chapman and Hall, Ltd., London, 1963.
Roberts, W. L.: "Cold Rolling of Steel," Marcel Dekker, Inc., New York, 1978.
Roberts, W. L.: "Hot Rolling of Steel," Marcel Dekker, Inc., New York, 1983.
"Roll Pass Design," The United Steel Companies, Ltd., Sheffield, England, 1960.
Starling, C. W.: "The Theory and Practice of Flat Rolling," University of London Press, London, 1962.
"The Making, Shaping, and Treating of Steel," 8th ed., United States Steel Corporation, Pittsburgh, 1964.
Tselikov, A. I. and V. V. Smirnov: "Rolling Mills," Pergamon Press, New York, 1965.
Underwood, L. R.: "The Rolling of Metals," John Wiley & Sons, Inc., New York, 1950.
Wusatowski, Z.: "Fundamentals of Rolling," Pergamon Press, New York, 1969.

CHAPTER

EIGHTEEN

EXTRUSION

18-1 CLASSIFICATION OF EXTRUSION PROCESSES

Extrusion is the process by which a block of metal is reduced in cross section by forcing it to flow through a die orifice under high pressure. In general, extrusion is used to produce cylindrical bars or hollow tubes, but shapes of irregular cross section may be produced from the more readily extrudable metals, like aluminum. Because of the large forces required in extrusion, most metals are extruded hot under conditions where the deformation resistance of the metal is low. However, cold extrusion is possible for many metals and has become an important commercial process. The reaction of the extrusion billet with the container and die results in high compressive stresses which are effective in reducing the cracking of materials during primary breakdown from the ingot. This is an important reason for the increased utilization of extrusion in the working of metals difficult to form, like stainless steels, nickel-based alloys, and other high-temperature materials.

The two basic types of extrusion are *direct extrusion* and *indirect extrusion* (also called inverted, or back, extrusion). Figure 18-1*a* illustrates the process of direct extrusion. The metal billet is placed in a container and driven through the die by the ram. A dummy block, or pressure plate, is placed at the end of the ram in contact with the billet. Figure 18-1*b* illustrates the indirect-extrusion process. A hollow ram carries the die, while the other end of the container is closed with a plate. Frequently, for indirect extrusion, the ram containing the die is kept stationary, and the container with the billet is caused to move. Because there is no relative motion between the wall of the container and the billet in indirect extrusion, the friction forces are lower and the power required for extrusion is less than for direct extrusion. However, there are practical limitations to indirect

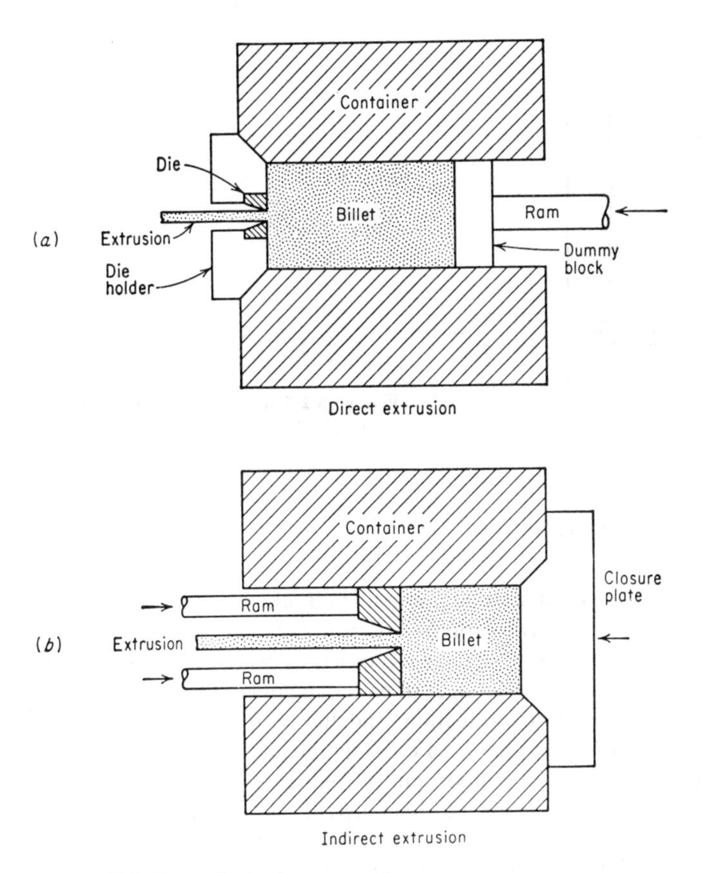

Figure 18-1 Types of extrusion.

extrusion because the requirement for using a hollow ram limits the loads which can be applied.

Tubes can be produced by extrusion by attaching a mandrel to the end of the ram. The clearance between the mandrel and the die wall determines the wall thickness of the tube. Tubes are produced either by starting with a hollow billet or by a two-step extrusion operation in which a solid billet is first pierced and then extruded.

Extrusion was originally applied to the making of lead pipe and later to the lead sheathing of cable. Figure 18-2 illustrates the extrusion of a lead sheath on electrical cables.

Impact extrusion is a process used to produce short lengths of hollow shapes, such as collapsible toothpaste tubes. It may be either indirect or direct extrusion, and it is usually performed on a high-speed mechanical press. Although the process generally is performed cold, considerable heating results from the high-speed deformation. Impact extrusion is restricted to the softer metals such as lead, tin, aluminum, and copper.

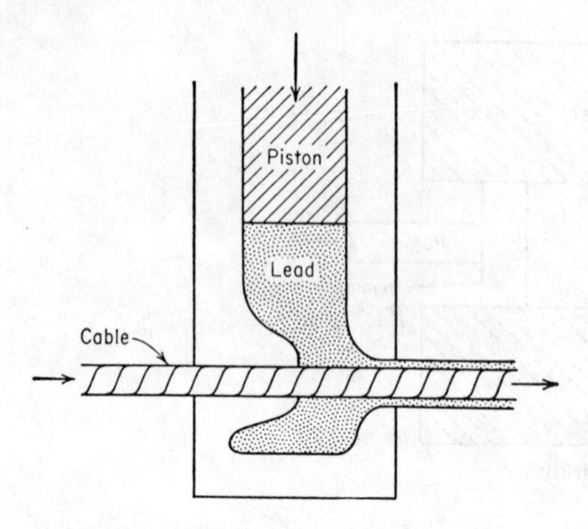

Figure 18-2 Extrusion of lead sheath on electrical cable.

18-2 EXTRUSION EQUIPMENT

Most extrusions are made with hydraulic presses. Hydraulic extrusion presses are classified into horizontal and vertical presses, depending upon the direction of travel of the ram. Vertical extrusion presses are generally built with capacities of 300 to 2,000 tons. They have the advantages of easier alignment between the press ram and the tools, higher rate of production, and the need for less floor space than horizontal presses. However, they need considerable headroom, and to make extrusions of appreciable length, a floor pit is frequently necessary. Vertical presses will produce uniform cooling of the billet in the container, and thus symmetrically uniform deformation will result. In a horizontal extrusion press the bottom of the billet which lies in contact with the container will cool more rapidly than the top surface, unless the extrusion container is internally heated, and therefore the deformation will be nonuniform. Warping of bars will result, and nonuniform wall thickness will occur in tubes. In commercial operations the chief use for vertical presses is in the production of thin-wall tubing, where uniform wall thickness and concentricity are required. Horizontal extrusion presses are used for most commercial extrusion of bars and shapes. Presses with a capacity of 1,500 to 5,000 tons are in regular operation, while a few presses of 14,000 tons capacity have been constructed.

The ram speed of the press can be an important consideration since high ram speeds are required in high-temperature extrusion where there is a problem of heat transfer from the billet to the tools. Ram speeds of 1,000 to 1,500 in/min may be used in extruding refractory metals, and this requires a hydraulic accumulator system with the press. At the other extreme, aluminum and copper alloys are prone to hot shortness so that the ram speed must be restricted to a few inches per minute. In this case, direct-drive pumping systems are adequate. Work has been done on presses equipped for preselected program control of ram speed in order to maintain a uniform finishing temperature.

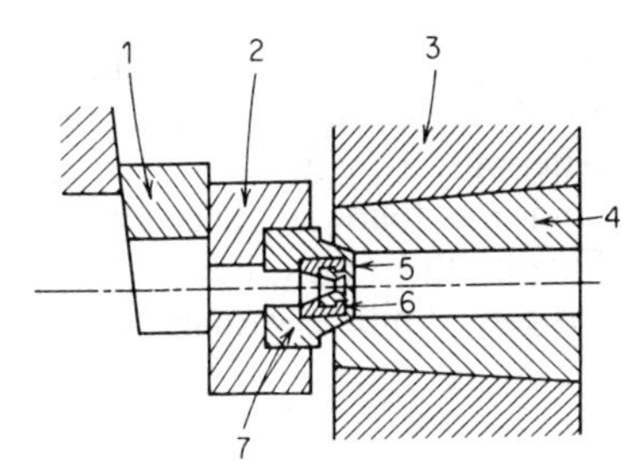

Figure 18-3 Typical arrangement of extrusion tooling. *(After Schey.)*

The dies and tooling used in extrusion must withstand considerable abuse from the high stresses, thermal shock, and oxidation. Figure 18-3 illustrates a typical extrusion tooling assembly. This assembly is designed for easy replacement of damaged parts and for reworking and reuse of components of the tooling. The die stack consists of the die (6), made from highly alloyed tool steel, which is supported by a die holder (5) and a bolster (7), all of which are held in a die head (2). This entire assembly is sealed against the container on a conical seating surface by the pressure applied by a wedge (1). Since the extrusion container must withstand high pressures, it is usually made in two parts. A liner (4) is shrunk into the more massive container (3) to produce a compressive prestress in the inside surface of the liner. The extrusion ram or stem is usually highly loaded in compression. It is protected from the hot billet by a follower pad immediately behind the billet. Since the liner and follower pad are subjected to many cycles of thermal shock, they will need to be replaced periodically.

There are two general types of extrusion dies. Flat-faced dies (FIg. 18-4*a*) are used when the metal entering the die forms a dead zone and shears internally to form its own die angle. A parallel land on the exit side of the die helps strengthen the die and allows for reworking of the flat face on the entrance side of the die without increasing the exit diameter. Dies with conical entrance angles (Fig. 18-4*b*) are used in extrusion with good lubrication. Decreasing the die angle increases the homogeneity of deformation and lowers the extrusion pressure, but beyond a point the friction in the die surfaces becomes too great. For most extrusion operations the optimum semidie angle (α) is between 45 and 60°.

In addition to the extrusion press, billet-heating facilities are needed, and for production operations, there should be automatic transfer equipment for placing

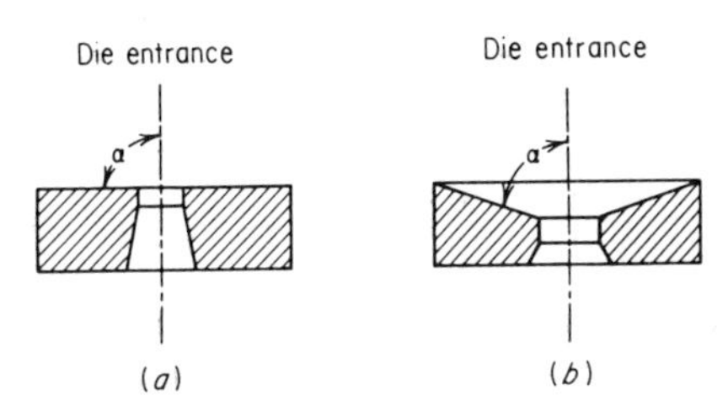

Figure 18-4 Typical extrusion dies. (*a*) Flat-faced (square) die; (*b*) conical die.

the heated billet in the container. Provision for heating the extrusion container may also be required. A hot saw is needed to cut off the extrusion so that the discard, or butt, can be removed from the die. Alternatively, extrusion may be done with a carbon block between the billet and the follower. This allows complete extrusion of the billet without a butt. Finally, there must be a runout table for catching the extrusion and a straightener to correct minor warpage in the extruded product.

18-3 HOT EXTRUSION

The principal variables which influence the force required to cause extrusion are (1) the type of extrusion (direct vs. indirect), (2) the extrusion ratio, (3) the working temperature, (4) the speed of deformation, and (5) the frictional conditions at the die and container wall.

In Fig. 18-5 the extrusion pressure is plotted against ram travel for direct and indirect intrusion. Extrusion pressure is the extrusion force divided by the cross-sectional area of the billet. The rapid rise in pressure during the initial ram travel is due to the initial compression of the billet to fill the extrusion container. For direct extrusion the metal begins to flow through the die at the maximum value of pressure, the *breakthrough pressure*. As the billet extrudes through the die the pressure required to maintain flow progressively decreases with decreasing length of the billet in the container. For indirect extrusion there is no relative motion between the billet and the container wall. Therefore, the extrusion pressure is approximately constant with increasing ram travel and represents the stress required to deform the metal through the die. While this appears to be an attractive process, in practice it is limited by the need to use a hollow ram which creates limitations on the size of the extrusion and the extrusion pressures which can be achieved. Therefore, most hot extrusion is done by the direct process. Finally, returning to Fig. 18-5, at the end of the stroke the pressure builds up rapidly and it is usual to stop the ram travel so as to leave a small discard in the container. This discard often contains defects which are unwanted in the product.

The *extrusion ratio* is the ratio of the initial cross-sectional area of the billet to the final cross-sectional area after extrusion, $R = A_0/A_f$. Extrusion ratios reach about 40 : 1 for hot extrusion of steel and may be as high as 400 : 1 for

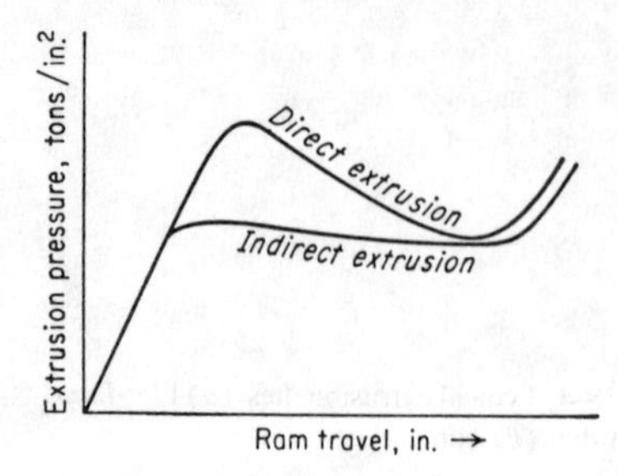

Figure 18-5 Typical curves of extrusion pressure vs. ram travel for direct and indirect extrusion.

aluminum. It is important to appreciate the distinction between the fractional reduction in area, $r = 1 - A_f/A_0$, Eq. (15-4), and the extrusion ratio R, $R = 1/(1 - r)$. For large deformations R is a more descriptive parameter. For example, the change in fractional reduction from 0.95 to 0.98 appears relatively minor, yet it corresponds to a change in area ratio from $R = 20:1$ to $R = 50:1$.

Because there is constancy of mass flow rate through the die, the velocity of the extruded product is the ram velocity $\times R$, so that quite high sliding velocities can be achieved along the die land. The extrusion pressure is directly related to the natural logarithm of the extrusion ratio, so that the extrusion force may be expressed as:

$$P = kA_0 \ln A_0/A_f \tag{18-1}$$

where k = the "extrusion constant," an overall factor which accounts for the flow stress, friction, and inhomogeneous deformation.

Most metals are extruded hot so as to take advantage of the decrease in flow stress or deformation resistance with increasing temperature. Since hot-working introduces the problems of oxidation of the billet and the extrusion tools and softening of the die and tools, as well as making it more difficult to provide adequate lubrication, it is advantageous to use the minimum temperature which will provide the metal with suitable plasticity. The upper hot-working temperature is the temperature at which hot shortness occurs, or, for pure metals, the melting point. Because of the extensive deformation produced in extrusion, considerable internal heating of the metal also results. Therefore, the top working temperature should be safely below the melting-point or hot-shortness range.

In the extrusion of steel[1] the billets are heated in the range 1100 to 1200°C, while the tooling is preheated to around 350°C. The extrusion pressures generally are in the range of 125,000 to 180,000 psi. The combination of high stresses and temperature place a severe demand on the glass lubrication system. The hot extrusion of lower-strength aluminum alloys is accomplished without any billet lubrication. The billet shears within itself near the container wall to create its own internal conical die surface. Metal deformation is very nonuniform, which results in a wide variation in heat-treatment response.

The interrelationship of temperature, deformation velocity, and deformation (reduction) to define conditions for successful crack-free hot-working was considered in Sec. 15-4 and illustrated with Fig. 15-15. This situation is very typical of hot extrusion.

Increasing the ram speed produces an increase in the extrusion pressure. A tenfold increase in the speed results in about a 50 percent increase in pressure. Greater cooling of the billet occurs at low extrusion speeds. When this becomes pronounced, the pressure required for direct extrusion will actually increase with increasing ram travel because of the increased flow stress as the billet cools. The higher the temperature of the billet, the greater the effect of low extrusion speed on the cooling of the billet. Therefore, high extrusion speeds are required with

[1] R. Cockroft, *Met. Mater.*, vol. 3, pp. 351–355, September 1969.

high-strength alloys that need high extrusion temperatures. The temperature rise due to deformation of the metal is greater at high extrusion speeds, and therefore problems with hot shortness may be accentuated.[1]

The selection of the proper extrusion speed and temperature is best determined by trial and error for each alloy and billet size. The interdependence of these factors is shown schematically in Fig. 15-15. For a given extrusion pressure, the extrusion ratio which can be obtained increases with increasing temperature. For any given temperature a larger extrusion ratio can be obtained with a higher pressure. The maximum billet temperature, on the assumption that there are no limitations from the strength of the tools and die, is determined by the temperature at which incipient melting or hot shortness occurs in the extrusion. The temperature rise of the extrusions will be determined by the speed of extrusion and the amount of deformation (extrusion ratio). Therefore, the curve which represents the upper limit to the safe extrusion region slopes upward toward the left. The worst situation is for extrusion at infinite speed, where none of the heat produced by deformation is dissipated. At lower extrusion speeds there is greater heat dissipation, and the allowable extrusion ratio for a given preheat temperature increases. The allowable extrusion range is the region under the curve of constant pressure and extrusion speed.

18-4 DEFORMATION, LUBRICATION, AND DEFECTS IN EXTRUSION

The pressure required to produce an extrusion is dependent on the way the metal flows in the container and extrusion die, and this is largely determined by the conditions of lubrication. Certain defects which occur in extrusion are directly related to the way the metal deforms during extrusion.

Figure 18-6 shows the characteristic types of deformation in extrusion. Nearly homogeneous deformation is shown in Fig. 18-6*a*. This would be typical of low container friction with a well-lubricated billet, hydrostatic extrusion in which the billet is surrounded with a pressurized fluid, or indirect extrusion (Fig. 18-6*d*). Deformation of the billet is relatively uniform until close to the die entrance. Figure 18-6*b* represents a case of increased container-wall friction, as shown by the severe distortion of the grid pattern in the corners of the die to produce a *dead zone* of stagnant metal which undergoes little deformation. Grid elements at the center of the billet undergo essentially pure elongation into the extruded rod, while elements near the sides of the billet undergo extensive shear deformation. The shear distortion requires an expenditure of energy which is not related to the change in external dimensions from billet to extruded product, and this is termed redundant work. For high friction at the container-billet interface, flow is concentrated toward the center and an internal shear plane develops (Fig.

[1] The heat balance during extrusion has been discussed by A. R. E. Singer and J. W. Coakham, *Metallurgia*, vol. 60, pp. 239–246, 1959; R. Akaret, *J. Inst. Met.*, vol. 95, pp. 204–211, 1967.

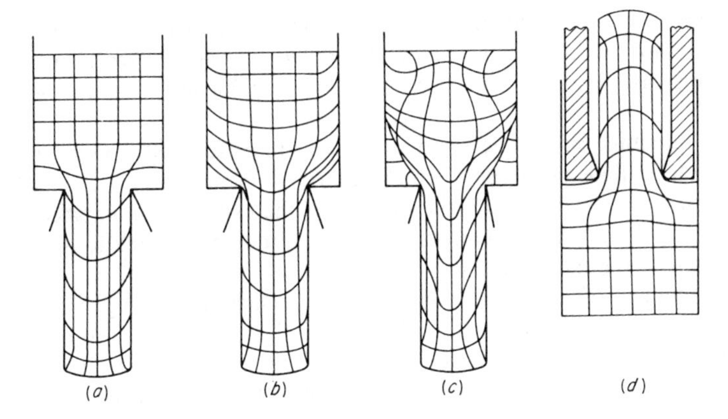

Figure 18-6 Patterns of metal deformation in extrusion. (*After Schey.*)

18-6*c*). This condition can also occur when the billet surface is chilled by a cold container. When sticking friction prevails between the billet and container, the metal will separate internally along a shear zone and a thin skin will be left in the container. Under these deformation conditions all of the surface of the extrusion is clean new material.

An effective hot-extrusion lubricant must have a low shear strength yet be stable enough to prevent breakdown at high temperature. For high-temperature extrusion of steel and nickel-based alloys the most common lubricant is molten glass, the Ugine-Sejournet process.[1] In this lubrication practice the billet is heated in an inert atmosphere and coated with glass powder before it enters the container of the press. The glass coating not only serves as a lubricant between the billet and the container wall, but it also serves as a thermal insulation to reduce heat loss to the tools. However, the main source of lubricant is a glass pad placed at the face of the die in front of the nose of the billet. During extrusion the hot billet progressively softens this glass pad and provides a lubricant film between the extrusion and the die, typically about 0.001 in thick. While it has been widely held that the viscosity of the glass is the chief criterion in selecting lubricants, more recent studies[2] have shown no correlation between the extrusion pressure or coating thickness and the glass viscosity. It is believed that successful lubrication depends on maintaining a reservoir of glass between the billet and the die. The coating thickness depends on the rate at which lubricant becomes available by melting or softening, rather than on its shear stress. A complex interaction exists between the optimum lubricant, the temperature, and the ram speed. If the speed is too low, the lubricant will produce thick coatings with low initial extrusion pressures, but this will rapidly exhaust the reservoir of lubricant and restrict the length that can be extruded. If the speed is too high, the coating

[1] J. Sejournet and J. Delcroix, *Lubr. Eng.*, vol. 11, p. 389, 1955.

[2] J. A. Rogers and G. W. Rowe, *J. Inst. Met.*, vol. 95, pp. 257–263, 1967.

may become dangerously thin. A modification of this technique is to apply thick films of glass only to the circumference of the billet to achieve hydrostatic lubrication.[1] In this case lubricant viscosity has a strong influence on extrusion pressure.

The lubricant film must be complete and continuous to be successful. Gaps in the film will serve to initiate shear zones which can develop into surface cracks. Also, lubricant film may be carried into the interior of the extrusion along shear bands to show up as longitudinal laminations in the product. Laminations of oxide can be created in the same way.

There are a number of other extrusion defects[2] which must be guarded against. Because of the inhomogeneous deformation in the direct extrusion of a billet, the center of the billet moves faster than the periphery. As a result, the dead metal zone extends down along the outer surface of the billet. After about two-thirds of the billet is extruded, the outer surface of the billet moves towards the center and extrudes through the die near the axis of the rod. Since the surface of the billet often contains an oxidized skin, this type of flow results in internal oxide stringers. This defect can be considered to be an internal pipe, and it is known as the *extrusion defect*. On a transverse section through the extrusion this will appear as an annular ring of oxide. The tendency toward the formation of the extrusion defect increases as the container wall friction becomes greater. If a heated billet is placed in a cooler extrusion container, the outer layers of the billet will be chilled and the flow resistance of this region will increase. Therefore, there will be a greater tendency for the center part of the billet to extrude before the surface skin, and the tendency for formation of the extrusion defect is increased.

One way of avoiding the extrusion defect is to carry out the extrusion operation only to the point where the surface oxide begins to enter the die and then discard the remainder of the billet. This procedure may have serious economic consequences since as much as 30 percent of the billet may remain at the point where the extrusion defect is encountered. An alternative procedure, which is frequently applied in the extrusion of brass, is to use a follower block which is slightly smaller than the inside diameter of the extrusion container. As the ram pushes the follower block forward, it scalps the billet and the oxidized surface layer remains in the container.

When extrusion is carried to the point at which the length of billet remaining in the container is about one-quarter its diameter, the rapid radial flow into the die results in the creation of an axial hole, or funnel, in the back end of the extrusion. This hole may extend for some distance into the back end of the extrusion, and therefore this metal must be discarded. The length of this defect can be reduced considerably by inclining the face of the ram at an angle to the ram axis.[3]

[1] K. M. Kulkarni, J. A. Schey, P. W. Wallace, and V. DePierre, *J. Inst. Met.*, vol. 100, pp. 33–39, 1972.

[2] R. N. Parkins, "Mechanical Treatment of Metals," pp. 232–239, American Elsevier Publishing Company, New York, 1968.

[3] An analysis of this defect based on the slip-field theory has been given by W. Johnson, *Appl. Sci. Res.*, vol. 8A, pp. 52–60, 228–236, 1959.

Surface cracking, ranging in severity from a badly roughened surface to repetitive transverse cracking called *fir-tree cracking* (Fig. 15-34*b*) can be produced by longitudinal tensile stresses generated as the extrusion passes through the die. In hot extrusion this form of cracking usually is intergranular and is associated with hot shortness. The most common cause is too high a ram speed for the extrusion temperature. At lower temperatures where hot shortness cannot occur transverse cracking is believed[1] to be caused by momentary sticking in the die land and the sudden building up of pressure, and then breakaway.

Center burst, or chevron cracking (Fig. 15-34*c*) can occur at low extrusion ratios. It has been shown[2] that this defect is related to the influence of frictional conditions on the zone of deformation at the extrusion die. In this instance high frictional restraint at the tool-billet interface produces a sound product while center bursts occur when the friction is low.

Because of the nonuniform deformation produced in extrusion, considerable variation in hot-worked structure and properties, or in properties after heat treatment, may be expected. A common problem is variation in structure and properties from front to back end of the extrusion in both the longitudinal and transverse directions. Regions of exaggerated grain growth often are found in hot extrusions. These coarse-grain regions appear on the surface or towards the center, depending on the deformation conditions.

18-5 ANALYSIS OF THE EXTRUSION PROCESS

Using the uniform deformation energy approach (Sec. 15-2), the plastic work of deformation per unit volume can be expressed for direct extrusion as

$$U_p = \bar{\sigma} \int d\varepsilon = \bar{\sigma} \int_{A_0}^{A_f} d \ln A = \bar{\sigma} \ln \frac{A_f}{A_0} = -\bar{\sigma} \ln R \tag{18-2}$$

The work involved is

$$W = U_p V = V \bar{\sigma} \ln R = pAL = \text{force} \times \text{distance} \tag{18-3}$$

where $\bar{\sigma}$ is defined as the effective flow stress *in compression* so that

$$p = \frac{V}{AL} \bar{\sigma} \ln R = \bar{\sigma} \ln R \tag{18-4}$$

This is an expression for the idealized extrusion pressure, since it considers neither friction nor redundant deformation. If we define the efficiency of the process η as the ratio of the ideal to actual energy per unit volume, we can express the actual extrusion pressure p_e by

$$p_e = \frac{p}{\eta} = \frac{\bar{\sigma}}{\eta} \ln R \tag{18-5}$$

It will be noted that this is just another way of expressing Eq. (18-1).

[1] R. J. Wilcox and P. W. Whitton, *J. Inst. Met.*, vol. 88, p. 145, 1959–1960.

[2] F J. Gurney and V. DePierre, *Trans. ASME, Ser. B: J. Eng. Ind.*, vol. 96, pp. 912–916, 1974.

Careful measurements of the forces in extrusion made by DePierre[1] show that the total extrusion force P_e is the sum of the die force P_d, the friction force between the container liner and the upset billet P_{fb}, and the frictional force between the container liner and the follower P_{ff}. Generally, $P_{ff} \approx 0$.

Assuming the billet frictional stress is equal to $\tau_i \approx k$, the ram pressure required by container friction is

$$p_f \frac{\pi D^2}{4} = \pi D \tau_i L$$

and

$$p_e = p_d + p_f = p_d + \frac{4\tau_i L}{D} \tag{18-6}$$

where τ_i = uniform interface shear stress between billet and container liner
L = length of billet in the container liner
D = inside diameter of the container liner

By directly measuring p_e and p_d, it is possible to obtain good values for the frictional shear stress.

A simple approach to determining die pressure p_d is to use a slab analysis to account for friction on extruding through a conical die.[2] Sachs[3] has performed this analysis for Coulomb sliding friction. The solution is directly analogous to that for wiredrawing through a conical die given in Sec. 19-3.

$$p_d = \sigma_{xb} = \sigma_0 \left(\frac{1+B}{B} \right) (1 - R^B) \tag{18-7}$$

where $B = \mu \cot \alpha$
α = semidie angle
R = extrusion ratio $= A_0/A$

While this analysis considers die friction, it does not allow for redundant deformation. However, extrusion probably is the most often studied problem using slip-line field theory.[4-8] Here the problem has been studied for plane-strain deformation without considering friction. All of these studies yield solutions of the form

$$p_d = \sigma_0 (a + b \ln R) \tag{18-8}$$

where typically $a = 0.8$ and $b = 1.5$ for axisymmetric extrusion.

[1] V. DePierre, *Trans. ASME, J. Lubr. Technol.*, vol. 92, pp. 398–405, 1970.

[2] For a 180° flat die it is assumed that the material shears internally to form an effective cone angle of $2\alpha = 90°$.

[3] O. Hoffman and G. Sachs, "Introduction to the Theory of Plasticity for Engineers," pp. 176–186, McGraw-Hill Book Company, New York, 1953.

[4] R. Hill, *J. Iron Steel Inst. London*, vol. 158, pp. 177–185, 1948.

[5] W. Johnson, *J. Mech. Phys. Solids*, vol. 4, pp. 191–198, 1956.

[6] A. P. Green, *ibid.*, vol. 3, pp. 189–196, 1955.

[7] T. F. Jordan and E. G. Thomsen, *ibid.*, vol. 4, pp. 184–190, 1956.

[8] W. Johnson and H. Kudo, "The Mechanics of Metal Extrusion," Manchester University Press 1962.

Extensive work using upper-bound analysis has also contributed to an understanding of the problem. Kudo[1] found the following expression for extrusion through rough square dies ($2\alpha = 180°$):

$$p_d = \sigma_0(1.06 + 1.55 \ln R) \tag{18-9}$$

Avitzur[2], produced a more generalized expression based on a spherical velocity field. This equation applies for lubricated extrusion through a die of semiangle α.

$$p_d = \frac{2\sigma_0}{\sqrt{3}}\left(\frac{\alpha}{\sin^2\alpha} - \cot\alpha\right) + \sigma_0[2f(\alpha) + m\cot\alpha]\ln(r_0/r_f) + 2m\left[\frac{L}{r_0} - \left(1 - \frac{r_f}{r_0}\right)\cot\alpha\right] \tag{18-10}$$

where m = interfacial friction factor
$f(\alpha)$ = complex function of semidie angle; $f(\alpha) = 1$ for small die angles
L = length of land on exit from die
r_0 = radius of billet
r_f = radius of extruded rod

A more recent upper-bound analysis[3] has used a different velocity field which produces good agreement with experiments on hydrostatic extrusion.

Based on the above theoretical background and experimental studies of hot extrusion, DePierre[4] used the following equation to describe die pressure:

$$p_d = \sigma_0(a + b\ln R) + mk\cot\alpha\ln R \tag{18-11}$$

where $m = \tau_i/k$ and a and b are evaluated as follows:

α	a	b
30	0.419	1.006
45	0.659	1.016
60	0.945	1.034

The flow distribution in extrusion has been measured[5] with billets gridded along the diametrial plane. Using the technique of visioplasticity,[6] it is possible to map out in detail the distribution of strain and strain rate and to calculate the variation of temperature and flow stress within the extrusion. Figure 18-7 shows the variation of *local* strain rate. Note that there are local maxima near the exit from the die on the surface, and along the centerline of the extrusion. The ambiguity of defining an average strain rate in a converging die was discussed in Sec. 15-5. The average strain rate for extrusion usually is defined by the time for

[1] H. Kudo, *Int. J. Mech. Sci.*, vol. 2, p. 102, 1960.
[2] B. Avitzur, *Trans. ASME, Sec. B: J. Eng. Ind.*, vol. 85, pp. 89–96, 1963; vol. 86, pp. 305–316, 1964.
[3] N. Ahmed, *Trans. ASME, Ser. B: J. Basic Eng.*, vol. 94, pp. 213–222, 1972.
[4] V. DePierre, op. cit.
[5] T. H. C. Childs, *Met. Technol.*, vol. 1, pp. 305–309, 1974.
[6] A. H. Shabaik and E. G. Thomsen, *Ann. CIRP*, vol. 17, pp. 149–156, 1969.

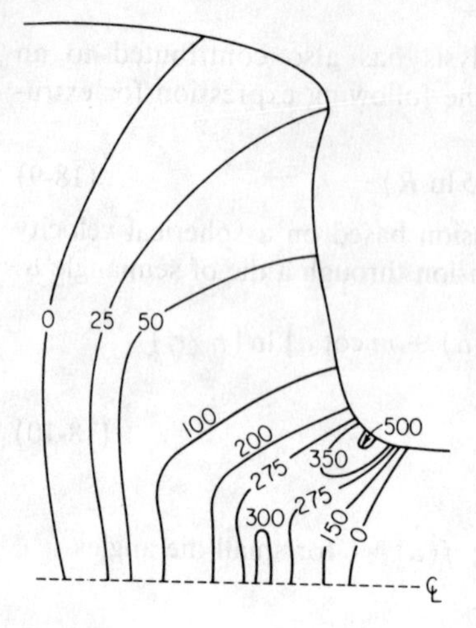

Figure 18-7 Strain-rate distribution in a partially extruded steel billet. $R = 16.5$, ram speed = 210 mm/s, average extrusion temperature = 1440° K. *(From T. H. C. Childs, Met. Technol., vol. 1, p. 306, 1974.)*

material to transverse through a truncated conical volume of deformation zone defined by the billet diameter D_b and the extrusion diameter D_e. For a 45° semicone angle,

$$V = \frac{\pi h}{3}\left(\frac{D_b^2}{4} + \frac{D_e^2}{4} + \frac{D_b D_e}{4}\right)$$

where

$$h = (D_b - D_e)/2$$

so

$$V = \frac{\pi}{24}\left(D_b^3 - D_e^3\right)$$

For a ram velocity v, the volume extruded per unit time is

$$v\frac{\pi D_b^2}{4}$$

and the time to fill the volume of the deformation zone is

$$\frac{V}{v\left(\pi D_b^2/4\right)} = \frac{D_b^3 - D_e^3}{6vD_b^2}$$

and if $D_b \gg D_e$

$$\frac{D_b^3 - D_e^3}{D_b^2} \approx D_b$$

and

$$t = D_b/6v$$

The time average mean strain rate is given by

$$\dot{\varepsilon}_t = \frac{\bar{\varepsilon}}{t} = \frac{6v \ln R}{D_b} \tag{18-12}$$

and for the general semidie angle α

$$\dot{\varepsilon}_t = \frac{6vD_b^2 \ln R \tan \alpha}{D_b^3 - D_e^3} \tag{18-13}$$

Example An aluminum alloy is hot extruded at 400°C at 2 in/s from 6-in diameter to 2-in diameter. The flow stress at this temperature is given by $\bar{\sigma} = 200(\dot{\varepsilon})^{0.15}$ (MPa). If the billet is 15 in long and the extrusion is done through square dies without lubrication, determine the force required for the operation.

Extrusion ratio $$R = \frac{6^2}{2^2} = 9$$

From Eq. (18-12)

$$\dot{\bar{\varepsilon}} = \frac{6v \ln R}{D_b} = \frac{6(2)\ln 9}{6} = 4.39 \text{ s}^{-1}$$

$$\bar{\sigma} = 200(4.39)^{0.15} = 200(1.25) = 250 \text{ MPa}$$

A dead-metal zone will form in the corners of the container against the die (see Fig. 18-6*c*). We can assume that this is equivalent to a die angle $\alpha = 60°$. Therefore, the extrusion pressure due to flow through the die, p_d, is from Eq. (18-7)

$$p_d = \sigma_0 \left(\frac{1 + B}{B} \right)(1 - R^B)$$

$$B = \mu \cot \alpha = 0.0577 \quad \text{where } \mu \text{ is assumed equal to } 0.1$$

$$p_d = 250(18.33)(1 - 1.174) \qquad \frac{1 + B}{B} = 18.33$$

$$p_d = 797 \text{ MPa}$$

The maximum pressure due to container wall friction will occur at breakthrough when $L = 15$ in. Aluminum will tend to stick to the container and shear internally.

$$\tau_i \cong k = \frac{\sigma_0}{\sqrt{3}} = \frac{250}{\sqrt{3}} = 144 \text{ MPa}$$

and from Eq. (18-6) the extrusion pressure is

$$p_e = p_d + \frac{4\tau_i L}{D} = 797 + 4(144)\left(\frac{15}{6} \right)$$

$$p_e = 797 + 1{,}440 = 2237 \text{ MPa} = 324{,}000 \text{ psi}$$

The extrusion load is

$$P = p_e A = 324{,}000 \frac{\pi}{4}(6)^2 = 9.16 \times 10^6 \text{ lb} = 4{,}580 \text{ metric tons}$$

18-6 COLD EXTRUSION AND COLD-FORMING

Cold extrusion is concerned with the cold forming[1] from rod and bar stock of small machine parts, such as spark plug bodies, shafts, pins, and hollow cylinders or cans. More properly, the subject should be expanded to include other cold-forming processes (such as upsetting, expanding, and coining) that are not strictly extrusion. Axisymmetric parts are particularly suited to cold-forming processes. Precision cold-forming can result in high production of parts with good dimensional control and good surface finish. Because of extensive strain hardening, it often is possible to use cheaper materials with lower alloy content. Extensive use is made of cold-formed low-alloy steels in the automotive industry.

The first major application of cold-forming of steel was the development during World War II of a process for making steel cartridge cases. The key was the development of a suitable lubricant system. For steel, a zinc phosphate conversion coating and soap is usually preferred. Other important factors are the use of a steel with high resistance to ductile fracture and the design of the tooling to minimize tensile-stress concentrations. The extensive literature on cold-forming has been reviewed by Watkins.[2]

18-7 HYDROSTATIC EXTRUSION

The concept of hydrostatic extrusion[3] was introduced in Figure 15-29. Because the billet is subjected to uniform hydrostatic pressure, it does not upset to fill the bore of the container as it would in conventional extrusion. This means that the billet may have a large length-to-diameter ratio (even coils of wire can be extruded) or it may have an irregular cross section. Since there is no container-billet friction, the curve of the extrusion pressure vs. ram travel is nearly flat, like that for indirect extrusion in Fig. 18-5. Because of the pressurized fluid, lubrication is very effective, and the extruded product has good surface finish and dimensional accuracy. Since friction is nearly absent, it is possible to use dies with a very low semicone angle ($\alpha \approx 20°$) which greatly minimizes the redundant deformation.

Because hydrostatic extrusion employs a pressurized fluid, there is an inherent limitation to hot-working with this process. A practical limit on fluid pressure of around 250,000 psi currently exists because of the strength of the container and the requirement that the fluid not solidify at high pressure. This limits the obtainable extrusion ratio for mild steel to less than 20 : 1, while for very soft metals like aluminum, it is possible to achieve extrusion ratios in excess of 200 : 1.

[1] C. H. Wick, "Chipless Machining," The Industrial Press, New York, 1961; H. D. Feldman, "The Cold Forging of Steel," Hutchinson & Co. (Publishers), Ltd., London, 1961; "Source Book on Cold Forming," American Society for Metals, Metals Park, Ohio, 1976.

[2] M. T. Watkins, *Int. Metall. Rev.*, vol. 18, September, December, 1973.

[3] H. L. D. Pugh (ed.), International Conference on Hydrostatic Extrusion, Institution of Mechanical Engineers, London, 1974.

The extrusion of wire is one of the areas actively being pursued. Because of the large amount of stored energy in a pressurized fluid, the control of the extrusion on the exit from the die may be a problem. However, this is solved by *augmented* hydrostatic extrusion in which an axial force is applied either to the billet or to the extrusion. The fluid pressure is kept at less than the value required to cause extrusion, and the balance is provided by the augmenting force. In this way, much better control is obtained over the movement of the extrusion.

A number of methods have been developed to increase the rate of production of hydrostatic extrusion or to place it on a continuous basis. Thick-film hydrostatic extrusion[1] minimizes the amount of pressurized fluid, and because the billets can be precoated with the hydrostatic medium, the production rate approaches that for conventional extrusion. Fuchs[2] developed a continuous extrusion process which uses viscous drag of a flowing polymer to feed the billet rod into an extrusion chamber and through the extrusion die. Problems with pressure control have led to a modified design[3] of a continuous gear-driven extruder. Alexander[4] has discussed continuous hydrostatic extrusion from a general point of view.

18-8 EXTRUSION OF TUBING

With modern equipment, tubing may be produced by extrusion to tolerances as close as those obtained by cold-drawing.[5] To produce tubing by extrusion, a mandrel must be fastened to the end of the extrusion ram. The mandrel extends to the entrance of the extrusion die, and the clearance between the mandrel and the die wall determines the wall thickness of the extruded tube. Generally, a hollow billet must be used so that the mandrel can extend to the die. In order to produce concentric tubes, the ram and mandrel must move in axial alignment with the container and the die. Also, the axial hole in the billet must be concentric, and the billet should offer equal resistance to deformation over its cross section.

One method of extruding a tube is to use a hollow billet for the starting material. The hole may be produced either by casting, by machining, or by hot piercing in a separate press. Since the bore of the hole will become oxidized during heating, the use of a hollow billet may result in a tube with an oxidized inside surface.

A more satisfactory method of extruding a tube is to use a solid billet which is pierced and extruded in one step in the extrusion press. With a modern

[1] R. J. Fiorentino, G. E. Meyer, and T. G. Byres, *Metallurgia and Metal Forming*, July, 1974, pp. 193–197; August 1974, pp. 210–213.

[2] F. J. Fuchs, *West. Electr. Eng.*, July–October 1971.

[3] J. A. Gumprecht and I. G. Histand, *Tech. Paper* MF74-223, Society of Manufacturing Engineers, 1974.

[4] J. M. Alexander, *Mater. Sci. Eng.*, vol. 10, pp. 70–74, 1972.

[5] F. W. Morris, *J. Inst. Met.*, vol. 90, pp. 101–106, 1961–1962.

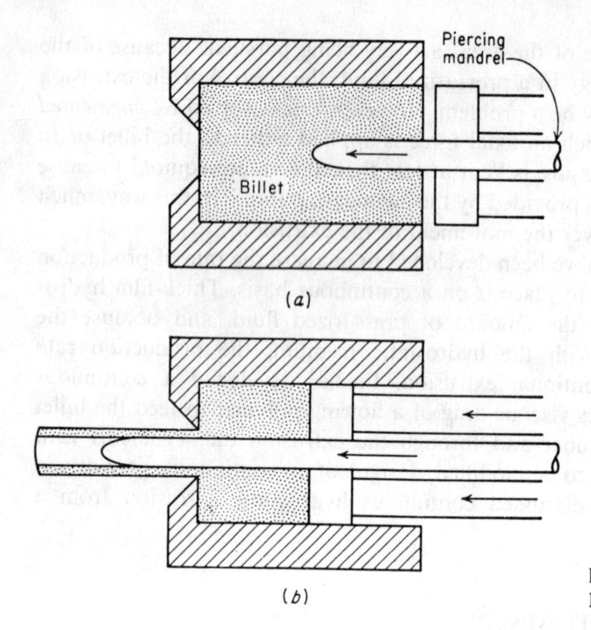

Figure 18-8 Tube extrusion. (*a*) Piercing; (*b*) extrusion.

extrusion press the piercing mandrel is actuated by a separate hydraulic system from the one which operates the ram. The piercing mandrel moves coaxially with the ram, but it is independent of its motion (Fig. 18-8). In the operation of a double-action extrusion press the first step is to upset the billet with the ram while the piercing mandrel is withdrawn. Next the billet is pierced with the pointed mandrel, ejecting a metal plug through the die. Then the ram advances and extrudes the billet over the mandrel to produce a tube.

A third method of extruding tubing, which is used with aluminum and magnesium alloys, is to use a solid billet and a porthole die with a standard extrusion ram without a mandrel. A sketch of a porthole die is shown in Fig. 18-9.

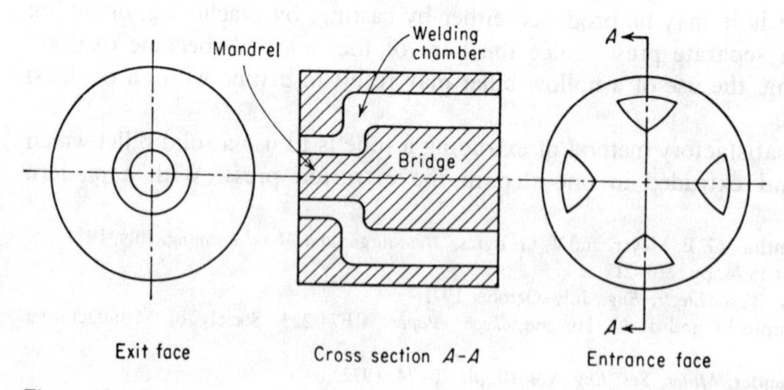

Figure 18-9 Porthole extrusion die.

The metal is forced to flow into separate streams and around the central bridge, which supports a short mandrel. The separate streams of metal which flow through the ports are brought together in a welding chamber surrounding the mandrel, and the metal exits from the die as a tube. Because the separate metal streams are joined within the die, where there is no atmospheric contamination, a perfectly sound weld is obtained. In addition to tubing, porthole extrusion is used to produce hollow unsymmetrical shapes in aluminum alloys.

18-9 PRODUCTION OF SEAMLESS PIPE AND TUBING

Pipe and tubing may be classified as seamless or welded, depending on the method of manufacture. Welded tubing is formed from strip and welded by hot-forming, fusion, or electric welding. As discussed in Sec. 18-8, extrusion is an excellent method of producing seamless pipe and tubing, especially for metals

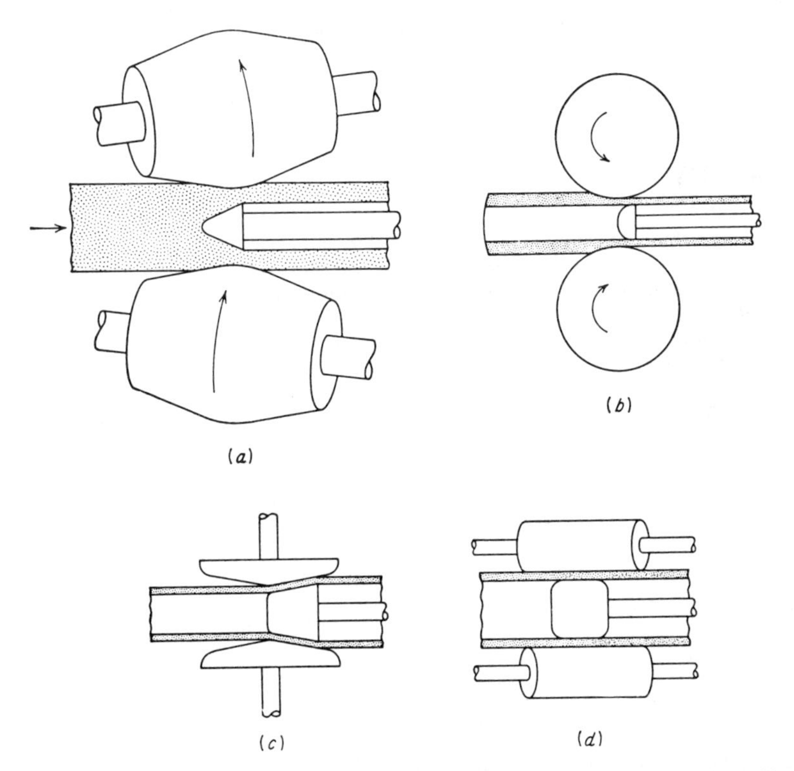

Figure 18-10 (*a*) Mannesmann mill; (*b*) plug rolling mill; (*c*) three-roll piercing mill; (*d*) reeling mill.

which are difficult to work. However, there are other well-established processes[1] for producing seamless pipe and tubing which generally are more economical than extrusion. Cold-working techniques for producing tubes are discussed in Chap. 19.

The Mannesmann mill (Fig. 18-10*a*) is used extensively for the rotary piercing of steel and copper billets. The process employs two barrel-shaped driven rolls which are set at an angle to each other. An axial thrust is developed as well as rotation to the billet. Because of the low arc of contact with the billet, Δ is high and tensile stresses develop along the axis of the billet. This assists in opening up the center of the billet as it flows around the piercing point to create the tube cavity. Piercing is the most severe hot-working operation customarily applied to metals.

The Mannesmann mill does not provide sufficiently large wall reduction and elongation to produce finished hot-worked tubes. Various types of plug rolling mills which drive the tube over a long mandrel containing a plug (Fig. 18-10*b*) have evolved. The Assel elongator,[2] which uses three conical driven rolls, has been widely adopted. This led to the development of three-roll piercing machines[3] (Fig. 18-10*c*) which produce more concentric tubes with smoother inside and outside surfaces than the older Mannesmann design. A *reeling mill* (Fig. 18-10*d*) which burnishes the outside and inside surfaces and removes the slight oval shape is usually one of the last steps in the production of pipe or tubing.

BIBLIOGRAPHY

Bishop, J. F. W.: The Theory of Extrusion, *Metall. Rev.*, vol. 2, pp. 361–390, 1957.

Chadwick, R.: The Hot Extrusion of Non-ferrous Metals, *Metall. Rev.*, vol. 4, pp. 189–255, 1959.

Chadwick, R.: Developments in Design and Application of Extrusion Presses for Metal Processing, *Int. Met. Rev.*, vol. 25, no. 3, pp. 94–136, 1980.

Laue, K. and H. Stenger: "Extrusion-Processes, Machinery, Tooling," American Society for Metals, Metals Park, Ohio, 1981.

Müller, E.: "Hydraulic Extrusion Presses" (in English), Springer-Verlag, Berlin, 1961.

Pearson, C. E. and R. N. Parkins: "The Extrusion of Metals," 2d ed., John Wiley & Sons, Inc., New York, 1960.

[1] T. Z. Blazynski, *Int. Met. Rev.*, vol. 22, pp. 313–322, 1977.

[2] C. E. Snee, *Iron Steel Eng.*, vol. 33, p. 124, 1956.

[3] W. L. Roberts, *Ibid.*, vol. 52, pp. 56–63, 1975.

CHAPTER

NINETEEN

DRAWING OF RODS, WIRES, AND TUBES

19-1 INTRODUCTION

Drawing operations involve pulling metal through a die by means of a tensile force applied to the exit side of the die. Most of the plastic flow is caused by compression force which arises from the reaction of the metal with the die. Usually the metal has a circular symmetry, but this is not an absolute requirement. The reduction in diameter of a solid bar or rod by successive drawing is known as *bar*, *rod*, or *wiredrawing*, depending on the diameter of the final product. When a hollow tube is drawn through a die without any mandrel to support the inside of the tube, this is known as *tube sinking*. When a mandrel or plug is used to support the inside diameter of the tube as it is drawn through a die, the process is called *tube drawing*. Bar, wire, and tube drawing are usually carried out at room temperature. However, because large deformations are usually involved, there is considerable temperature rise during the drawing operation.

19-2 ROD AND WIREDRAWING

The principles involved in the drawing of bars, rod, and wire are basically the same, although the equipment that is used is different for the different-sized products. Rods and tubes, which cannot be coiled, are produced on drawbenches (Fig. 19-1*a*). The rod is pointed with a swager, inserted through the die, and clamped to the jaws of the drawhead. The drawhead is moved either by a chain drive or by a hydraulic mechanism. Drawbenches with 300,000 lb pull and 100 ft of runout are available. Draw speeds vary from about 30 to 300 ft/min.

The cross section through a conical drawing die is shown in Fig. 19-1*b*. The entrance of the die (the bell) is shaped so that the wire entering the die will draw

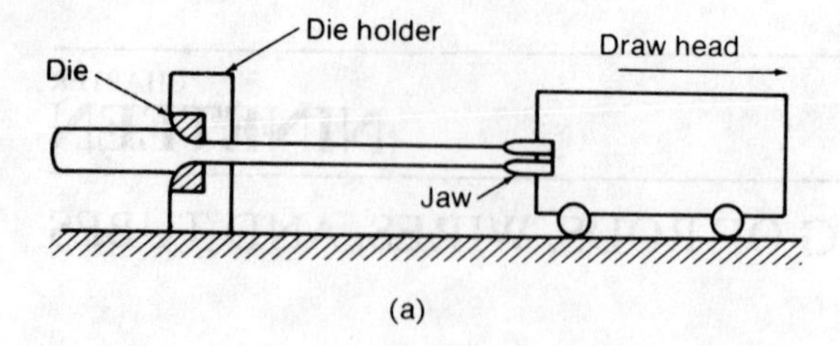

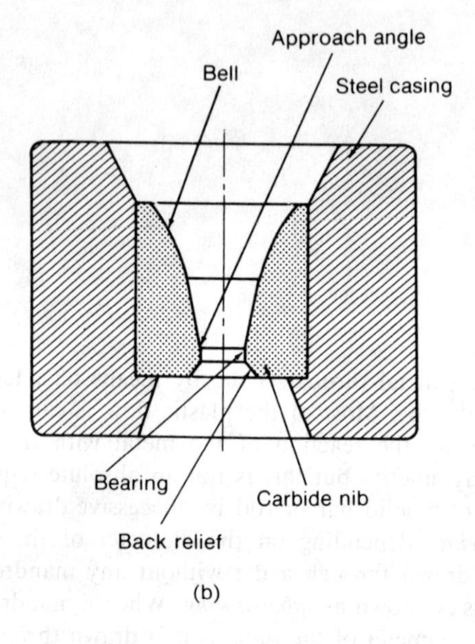

Figure 19-1 (*a*) Schematic drawing of a drawbench; (*b*) cross section of a drawing die.

lubricant with it. The shape of the bell causes the hydrostatic pressure to increase and promotes the flow of lubricant into the die. The *approach angle* is the section of the die where the actual reduction in diameter occurs. The half die angle α is an important process parameter. The *bearing region* does not cause reduction but it does produce a frictional drag on the wire. The chief function of the bearing region is to permit the conical approach surface to be refinished (to remove surface damage due to die wear) without changing the dimensions of the die exit. The *back relief* allows the metal to expand slightly as the wire leaves the die. It also minimizes the possibility of abrasion taking place if the drawing stops or the die is out of alignment. Most drawing dies are cemented carbide or industrial diamond (for fine wires). The die nib is encased for protection in a thick steel casing.

The distinction between wire and rod is somewhat arbitrary. In general the term wire refers to small diameter products under 5 mm which may be drawn rapidly on multiple-die machines.

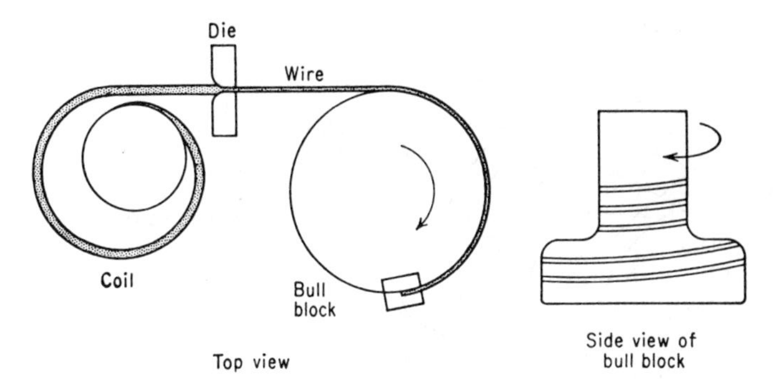

Figure 19-2 Wiredrawing equipment (schematic).

Wiredrawing usually starts with a coil of hot-rolled rod. The rod is first cleaned by pickling to remove any scale which would lead to surface defects or excessive die wear due to abrasion. The next step is to prepare the rod so that the lubrication is effective. With high-strength materials a soft surface coating such as copper or tin may be used as the lubricant. More typically, conversion coatings such as sulfates or oxalates may be applied to the rod. These are used in conjunction with a lubricant, typically soap, in dry drawing. In wet drawing the dies and the rod are completely immersed in an oil lubricant containing an EP additive.

When the rod diameter is sufficiently small to permit coiling, block drawing is usually employed because it allows the generation of long lengths in a much smaller floor space than required for a draw bench (Fig. 19-2). Because the area reduction per drawing pass is rarely greater than 30–35 percent many reductions are required to achieve the overall reduction. Multiple block machines with one die and one draw block for each reduction are common. Since the wire diameter will decrease after each pass the velocity and length of the wire will increase proportionately. Thus, the peripheral speed of each draw block must increase in turn if there is to be no slippage between the wire and the block. One way to achieve this is to equip each drawing block with its own electric motor with variable speed control. However, a more economical design is to use a single electrical motor to drive a series of stepped cones (Fig. 19-3). The diameter of each cone is designed to produce a peripheral speed equivalent to a certain size reduction. When precise agreement in wire surface velocity and block peripheral velocity is not achieved the wire slides on the blocks as they revolve, causing friction and evolution of heat. The drawing speed in multiple-die machines may reach 600 m/min for ferrous drawing, but with nonferrous drawing speeds up to 2000 m/min are common.

Heat generation is a major concern in drawing operations. Although most rod and wiredrawing is done cold, plastic deformation and friction can generate wire temperatures of several hundred degrees celsius. This heat is only partially

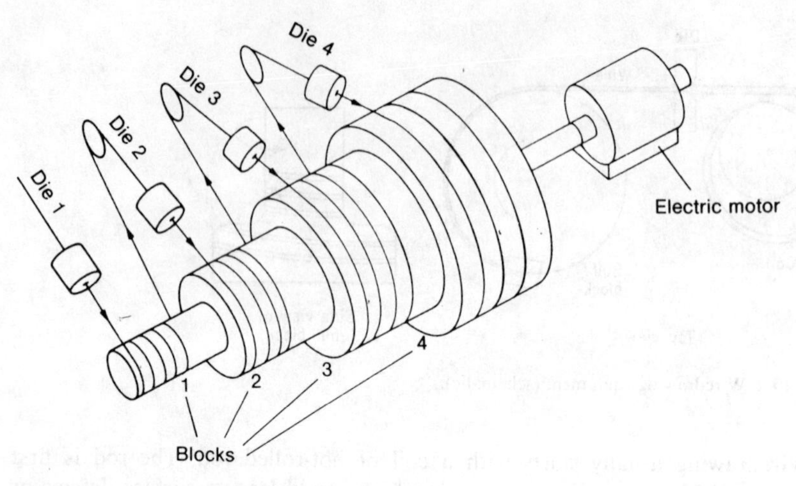

Figure 19-3 Stepped-cone multiple-pass wiredrawing. *(J. N. Harris, "Mechanical Working of Metals," p. 208, Pergamon Press, New York, 1983.)*

removed by interpass cooling, and since the dies extract little heat, they get quite hot.

Nonferrous wire and low-carbon steel wire are produced in a number of tempers ranging from dead soft to full hard. Depending on the metal and the reductions involved, intermediate anneals may be required. Steel wire with a carbon content greater than 0.25 percent is given a special *patenting* heat treatment. This consists in heating above the upper critical temperature and then cooling at a controlled rate or transforming in a lead bath at a temperature around 600°F to cause the formation of fine pearlite. Patenting produces the best combination of strength and ductility for the successful drawing of high-carbon music and spring wire.

Defects in rod and wire can result from defects in the starting rod (seams, slivers, and pipe) or from the deformation process itself.[1] The most common type of drawing defect is center burst, or chevron cracking (Fig. 15-34*c*). This also is called *cupping*. An upper-bound analysis[2] is capable of identifying the combinations of semidie angle and reduction for which less deformation energy is required if a hole forms at the centerline. This analysis predicts that center burst fracture will occur for low die angles at low reductions, and as α increases the critical reduction for freedom from center burst increases. For a given reduction and die angle, the critical reduction to prevent fracture increases with the friction.

[1] R. N. Wright, Workability in Extrusion and Wire Drawing, "Workability Testing Techniques," chap. 9, American Society for Metals, Metals Park, Ohio, 1984.

[2] B. Avitzur, "Metal Forming: Process and Analysis," pp. 172–176, 240–241, McGraw Hill Book Company, New York, 1968.

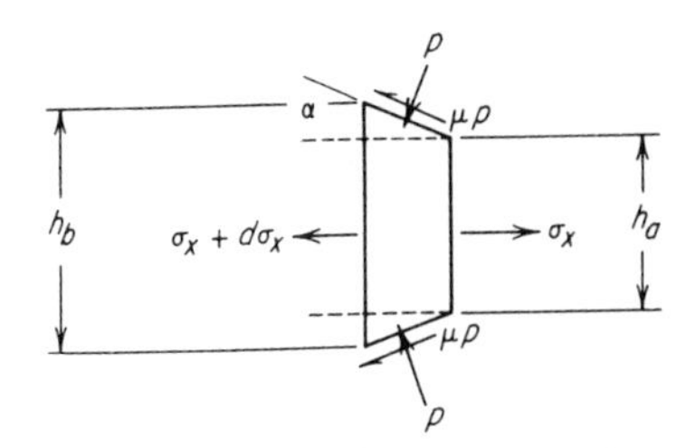

Figure 19-4 Stresses acting on an element of wire in plane-strain strip drawing.

19-3 ANALYSIS OF WIREDRAWING

Although wiredrawing appears to be one of the simplest metalworking processes, a complete analysis that enables calculation of the draw force to better than ±20 percent of the observed value is a rather difficult problem. We shall use wiredrawing as an example of how to add to the model to account for realistic conditions.

We have already seen that the uniform-deformation-energy method predicts a draw stress given by Eq. (15-13).

$$\sigma_{xa} = \bar{\sigma}_0 \ln \frac{A_b}{A_a} = \sigma_0 \ln \frac{1}{1-r} \tag{19-1}$$

This equation not only neglects friction, but it neglects the influence of transverse stresses and of redundant (shearing) deformation.

As a first step we shall consider the problem of strip drawing (see Sec. 15-2 and Fig. 15-3) where a Coulomb friction coefficient μ exists between the strip and the die. The friction stress μp opposes the motion of the strip through the die (Fig. 19-4). Referring to Eq. (15-6), the equilibrium of forces in the x direction is

$$\sigma_x\, dh + h\, d\sigma_x + 2p \tan\alpha\, dx + 2\mu p\, dx = 0 \tag{19-2}$$

and since $dh = 2\, dx \tan\alpha$,

$$\sigma_x\, dh + h\, d\sigma_x + p(1 + \mu \cot\alpha)\, dh = 0 \tag{19-3}$$

Since the yield condition for plane strain is $\sigma_x + p = \sigma_0'$ and $B = \mu \cot\alpha$, the differential equation for strip drawing is

$$\frac{d\sigma_x}{\sigma_x B - \sigma_0'(1+B)} = \frac{dh}{h} \tag{19-4}$$

If B and σ_0' are both constant, Eq. (19-4) can be integrated directly to give the draw stress σ_{xa}.

$$\sigma_{xa} = \sigma_0' \frac{1+B}{B}\left[1 - \left(\frac{h_a}{h_b}\right)^B\right] = \sigma_0' \frac{1+B}{B}\left[1 - (1-r)^B\right] \tag{19-5}$$

However, wiredrawing is conducted with conical dies. An analysis[1] following that for strip drawing but integrating around the circumference of the die results

[1] O. Hoffman and G. Sachs, "Introduction to the Theory of Plasticity for Engineers," McGraw-Hill Book Company, pp. 176–180, 1953.

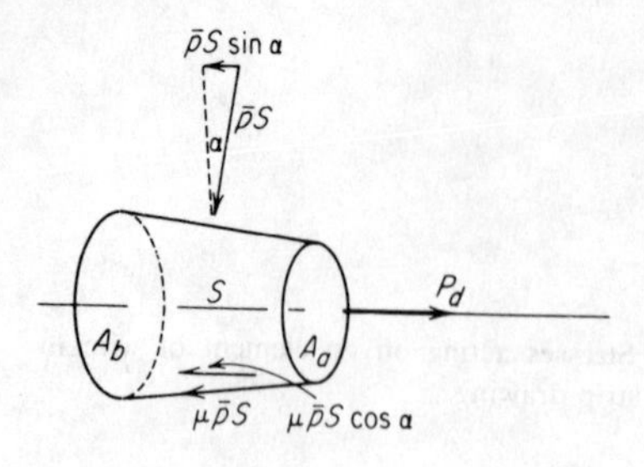

Figure 19-5 Forces acting on a conical element.

in

$$\sigma_{xa} = \sigma_0 \frac{1+B}{B}\left[1 - \left(\frac{D_a}{D_b}\right)^{2B}\right] \tag{19-6}$$

A similar, but slightly different analysis for wiredrawing with friction is given by Johnson and Rowe.[1] The surface area of contact between the wire and the die (Fig. 19-5) is given by

$$S = \frac{A_b - A_a}{\sin\alpha} \tag{19-7}$$

and the mean normal pressure on this area is $\bar{p}$. The forces acting in the axial direction are given in Fig. 19-5. The draw force P_d is balanced by the horizontal component of the frictional force and the horizontal component of the normal pressure.

$$P_d = \mu\bar{p}S\cos\alpha + \bar{p}S\sin\alpha \tag{19-8}$$

$$P_d = \bar{p}S(\mu\cos\alpha + \sin\alpha) = \bar{p}\frac{A_b - A_a}{\sin\alpha}(\mu\cos\alpha + \sin\alpha)$$

$$P_d = \bar{p}(A_b - A_a)(\mu\cot\alpha + 1) = \bar{p}(A_b - A_a)(1 + B) \tag{19-9}$$

In the absence of friction, $B = 0$ and

$$P_d = \bar{p}(A_b - A_a) = \bar{\sigma}_0 A_a \ln\frac{A_b}{A_a}$$

which is really Eq. (19-1). Therefore, the draw stress *with friction* is given by[2]

$$\sigma_{xa} = \frac{P_d}{A_a} = \bar{\sigma}_0 \ln\frac{A_b}{A_a}(1 + B) \tag{19-10}$$

Example Determine the drawing stress to produce a 20-percent reduction in a 10-mm stainless steel wire. The flow stress is given by $\sigma_0 = 1300\varepsilon^{0.30}$ (MPa).

[1] R. W. Johnson and G. W. Rowe, *J. Inst. Met.*, vol. 96, p. 105, 1968.
[2] J. G. Wistreich, *Proc. Inst. Mech. Eng. (London)*, vol. 169, pp. 654–665, 1955.

The die angle is 12° and $\mu = 0.09$

$$B = \mu \cot \alpha = 0.09/\tan 6° = 0.8571$$

$$\varepsilon_1 = \ln \frac{1}{1-r} = \ln \frac{1}{1-0.2} = 0.223$$

$$\bar{\sigma} = \frac{K\varepsilon_1^n}{n+1} = \frac{1300(0.223)^{0.30}}{1.30} = 637 \text{ MPa}$$

$$A_b = 10 \text{ mm} \qquad A_a = A_b - rA_b = A_b(1-r) = 10(0.8) = 8 \text{ mm}$$

From Eq. (19-6)

$$\sigma_{xa} = \bar{\sigma}\left(\frac{1+B}{B}\right)\left[\left(1 - \left(\frac{A_a}{A_b}\right)^B\right]$$

$$= 637\left(\frac{1.8571}{0.8571}\right)[1 - 0.8^{0.8571}] = 240 \text{ MPa}$$

From Eq. (19-10)

$$\sigma_{xa} = \bar{\sigma} \ln \frac{A_b}{A_a}(1+B)$$

$$= 637 \ln \frac{1.0}{0.8}(1.8571) = 264 \text{ MPa}$$

Note, that there is about a 10-percent difference between these expressions. However, they both predict a draw stress appreciably lower than the uniaxial flow stress.

If the wire is moving through the die at 3 m/s, determine the horsepower required to produce the deformation.

$$\text{Power} = \text{force} \times \frac{\text{distance moved}}{\text{time}}$$

Drawing force $P_d = \sigma_{xa} A_a = 240 \text{ N/mm}^2 \times \frac{\pi}{4}(8)^2 \text{ mm}^2 = 12.06 \text{ kN}$

$$\text{Power} = 12.06 \text{ kN} \times \frac{3 \text{ m}}{\text{s}} = 36.18 \text{ kW}$$

But, $\quad 1 \text{ hp} = 0.746 \text{ kW}$

$$\text{Horsepower} = \frac{36.18}{0.746} = 48.5 \text{ hp}$$

Equation (19-10) includes a term for uniform deformation energy and a term for frictional energy. The redundant deformation can be taken into consideration by including a factor ϕ which allows for the influence of redundant deformation in raising the flow stress of the material.

$$\sigma_{xa} = \phi\bar{\sigma}_0 \ln \frac{A_b}{A_a}(1+B) \tag{19-11}$$

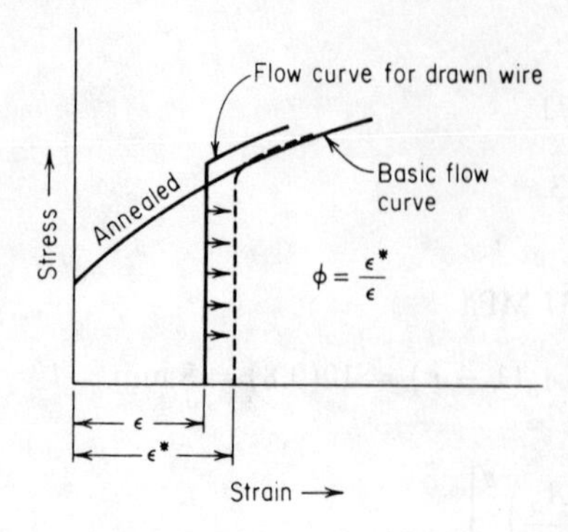

Figure 19-6 Procedure for determining redundant deformation of drawn wire.

The redundant work factor is defined by

$$\phi = f(\alpha, r) = \frac{\varepsilon^*}{\varepsilon} \tag{19-12}$$

where ϕ = the redundant work factor

ε^* = the "enhanced strain" corresponding to the yield stress of the metal which has been homogeneously deformed to a strain ε

The redundant work factor may be determined in a straightforward manner for drawn wires[1] as shown in Fig. 19-6. The flow curve of a drawn wire is superimposed on the flow curve for the annealed metal. The origin of the curve for the drawn metal is displaced along the strain axis by an amount equal to the drawing reduction, $\varepsilon = \ln(A_b/A_a) = \ln[1/(1-r)]$. The fact that the yield stress for the drawn wire is above the basic flow curve for the material is due to the redundant work that it received. To determine ϕ, the flow curve for the drawn metal is moved to the right to ε^* where the curves coincide. The area swept out in this procedure is the redundant work per unit of volume. An alternative approach[2] is to use upper-bound analysis to account for the redundant deformation by assuming simple yet reasonable deformation regions. Unfortunately, this leads to complex equations.

$$\sigma_{xa} = \sigma_0 \frac{\left\{2f(\alpha)\ln\frac{R_b}{R_a} + \frac{2}{\sqrt{3}}\left(\frac{\alpha}{\sin^2\alpha} - \cot\alpha\right) + 2\mu\left(\cot\alpha\left[1 - \ln\frac{R_b}{R_a}\right]\ln\frac{R_b}{R_a} + \frac{L}{R_a}\right)\right\}}{1 + 2\mu(L/R_a)} \tag{19-13}$$

where $f(\alpha)$ = a complex function of semidie angle

L = the length of the die land

R_b = the radius of the billet

R_a = the wire radius

[1] The use of microhardness measurements to determine ϕ is discussed by W. A. Backofen, "Deformation Processing," pp. 138–140, Addison-Wesley Publishing Company, Inc., Reading, Mass., 1972.

[2] B. Avitzur, *Trans. ASME J. Eng. Ind.*, vol. 86, pp. 305–316, 1964.

A less complicated, but still useful expression, that is based on an upper-bound analysis,[1] contains the redundant work term, $\frac{2}{3}\tan\alpha$.

$$\sigma_{xa} = \sigma_0\left[\left(1 + \frac{m}{\sin 2\alpha}\right)\ln\frac{R_b}{R_a} + \frac{2}{3}\tan\alpha\right] \tag{19-14}$$

where m = the friction factor, i.e., τ_i = km.

The concept of deformation-zone geometry, introduced in Sec. 15-8, provides a convenient method for treating the redundant work in wiredrawing.[2] For the drawing of round wire

$$\Delta = \frac{\alpha}{r}\left[1 + (1 - r)^{1/2}\right]^2 \tag{19-15}$$

where α = the approach semiangle, in radians
r = the drawing reduction

Commercial wiredrawing often employs α in the range 6 to 10° and drawing reductions of about 20 percent. Thus, the Δ values typically range from 2–3. Higher values of Δ correspond to lower reductions and higher die angles, while lower values correspond to higher reductions and lower die angles. Analysis of experimental data shows that the redundant work factor is related to Δ by[3]

$$\phi = C_1 + C_2\Delta \approx 0.8 + \frac{\Delta}{4.4} \tag{19-16}$$

The expression for draw stress given in Eq. (19-11) contains terms for homogeneous deformation, friction, and redundant work. The last two terms are functions of semidie angle α. The effect of die angle on these components of the total energy required to cause deformation is shown schematically in Fig. 19-7. The ideal work of plastic deformation U_p is independent of die angle. For a fixed coefficient of friction, the work required to overcome friction U_f decreases with increasing α as shown by Eqs. (19-7) and (19-8). On the other hand, the redundant work U_r increases with increasing α as shown by Eq. (19-14). The summation of these components of total energy U_T leads to a curve which has a minimum at some optimum die angle α^*. Increasing the reduction and the friction raise the optimum die angle.

The optimum value of α corresponds to a minimum in total energy of deformation or of draw stress. This also can be expressed in terms of an optimum value of Δ.

$$\Delta_{\text{opt}} \approx 4.9\left[\frac{\mu}{\ln(1/1 - r)}\right]^{1/2} \tag{19-17}$$

[1] W. F. Hosford and R. M. Caddell, "Metal Forming," pp. 162–163, Prentice-Hall, Inc., Englewood Cliffs, N.J., 1983.

[2] R. N. Wright, *Wire Technology*, vol. 4, pp. 57–61, 1976.

[3] J. G. Wistreich, *Proc. Inst. Mech. Eng.*, vol. 169, p. 654, 1955; R. M. Caddell and A. G. Atkins, *Trans. ASME J. Eng. Ind.*, vol. 90, pp. 411–419, 1968.

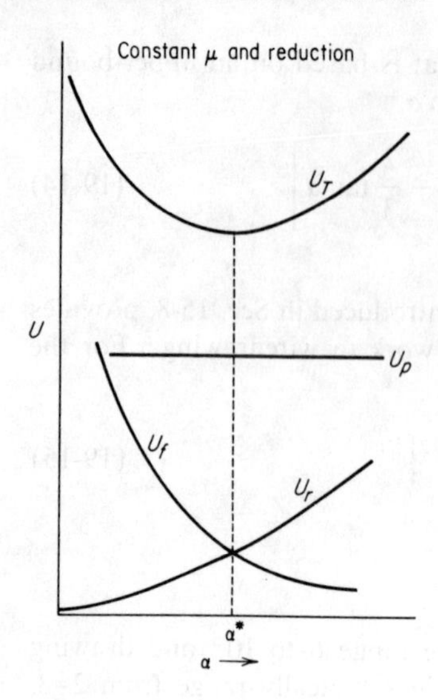

Figure 19-7 Components of total energy of deformation.

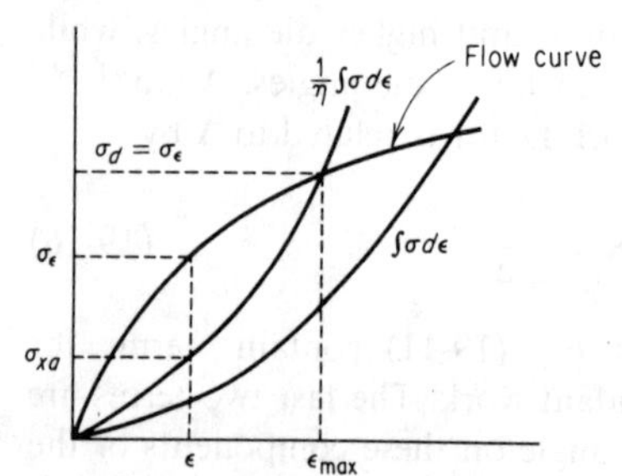

Figure 19-8 Development of limit on drawability. *(After Backofen.)*

We now shall examine the limit of drawability for steady-state wiredrawing.[1] Equation (19-11) can be expressed most simply by

$$\sigma_{xa} = \frac{1}{\eta}\int \sigma \, d\varepsilon \qquad (19\text{-}18)$$

where $\eta = U_p/U_T$ is the efficiency of the deformation process. Figure 19-8 plots the flow curve of the material, $\int \sigma \, d\varepsilon$ for an ideal process without friction or redundant work, and $1/\eta \int \sigma \, d\varepsilon$ for the actual drawing process. For a given strain $\varepsilon = \ln(A_b/A_a)$ produced by the die the values of draw stress σ_{xa} and flow stress σ_ε are as indicated. The weakest link is the wire which has exited from the die. On being deformed through the die the material is strain-hardened. For a severely strain-hardened material the formation of a neck is quickly followed by fracture. Therefore, the drawing limit is reached when $\sigma_d = \sigma_\varepsilon$ as shown in Fig. 19-8. If the

[1] W. A. Backofen, op. cit., pp. 227–230; R. M. Caddell and A. G. Atkins, *Trans. ASME, Ser. B: J. Eng. Ind.*, vol. 91, pp. 664–672, 1969.

material follows a power-law hardening relationship $\sigma_\varepsilon = K\varepsilon^n$, then Eq. (19-18) becomes

$$\sigma_d = \frac{K\varepsilon^{n+1}}{\eta(n+1)} = \frac{\sigma_\varepsilon \varepsilon}{\eta(n+1)} \tag{19-19}$$

Substituting the criterion for the maximum drawing strain in a single pass, that is, $\sigma_d = \sigma_\varepsilon$,

$$\varepsilon_{\max} = \eta(n+1) \tag{19-20}$$

Since $\varepsilon = \ln(A_b/A_a)$,

$$\left(\frac{A_b}{A_a}\right)_{\max} = e^{\eta(n+1)} \tag{19-21}$$

and by definition the reduction is $r = 1 - (A_a/A_b)$

$$r_{\max} = 1 - e^{-\eta(n+1)} \tag{19-22}$$

For wire subjected to repeated reductions through a series of dies, n will decrease toward zero and the allowable reduction will decrease. A comparison of the maximum strain in drawing with that in stretching (tension) shows the important role played by the strain-hardening exponent.

$$\left(\frac{\varepsilon(\text{drawing})}{\varepsilon(\text{stretch})}\right)_{\max} = \frac{\eta(n+1)}{n} \tag{19-23}$$

Typically the limiting strain in drawing is at least twice that in pure tension.

Example For the material drawn in the previous example, what is the largest possible reduction?

To a first approximation the limit on drawing reduction occurs when $\sigma_{xa} = \bar{\sigma}$

$$\sigma_{xa} = \sigma_0\left(\frac{1+B}{B}\right)\left[1-(1-r)^B\right]$$

$$637 = 637\left(\frac{1.8571}{0.8571}\right)\left[1-(1-r)^{0.8571}\right]$$

$$1 = 2.167\left[1-(1-r)^{0.8571}\right] \quad r = 0.51 \quad \text{or} \quad \varepsilon = \ln\frac{1}{1-r} = 0.71$$

$$\sigma_0 = 1{,}300\varepsilon^{0.30} = 1{,}300(0.7)^{0.30} = 1{,}173 \text{ MPa}$$

A better estimate is to let $\sigma_{xa} = \sigma_0$ at $\varepsilon = 0.71$, i.e., $\sigma_{xa} = 1173$ MPa

$$1173 = 637(2.167)1 - (1-r)^{0.8571}$$

$$r = 0.89$$

Note, that when there is no friction or redundant work ($\eta = 1$), even if there is no strain hardening ($n = 0$), Eq. (19-22) predicts that $r_{\max} = 1 - 1/e = 0.63$.

19-4 TUBE-DRAWING PROCESSES

Hollow cylinders, or tubes, which are made by hot-forming processes such as extrusion or piercing and rolling (see Chap. 18), often are cold finished by drawing. Cold-drawing is used to obtain closer dimensional tolerances, to produce better surface finishes, to increase the mechanical properties of the tube by strain hardening, to produce tubes with thinner walls or smaller diameters than can be obtained with hot-forming methods, and to produce tubes of irregular shapes.

The three basic types of tube-drawing processes are *sinking*, *plug drawing*, and *mandrel drawing*. Since the inside of the tube is not supported in tube sinking (Fig. 19-9*a*), the wall thickens slightly and the internal surface becomes uneven. Because the shearing at the entry and exit of the die is large, the redundant strain is higher for sinking and the limiting deformation is lower than for other tube-producing processes. Both the inner and outer diameters of the tube are controlled in drawing on a plug (Fig. 19-9*b*). The plug may be either cylindrical or conical. The plug controls the size and shape of the inside diameter and produces tubing of greater dimensional accuracy than in tube sinking. Because of the increased friction from the plug, the reduction in area seldom exceeds 30 percent. The situation where a carefully matched plug floats in the die throat is shown in Fig. 19-9*c*. Properly designed *floating plugs* can give a reduction in area of 45 percent, and for the same reduction the drawing loads are lower than for drawing with a fixed plug. An important feature of this design is that it is possible to draw and coil long lengths of tubing. However, tool design and lubrication can be very critical. Problems with friction in tube drawing are minimized in drawing with a long *mandrel* (Fig. 19-9*d*). The mandrel consists of a long hard rod or wire that extends over the entire length of the tube and is drawn through the die with the tube. In tube drawing with a moving mandrel the draw force is transmitted to the metal partly by the pull on the exit section and partly by the friction forces acting along the tube-mandrel interface. Since the mandrel is moving with a

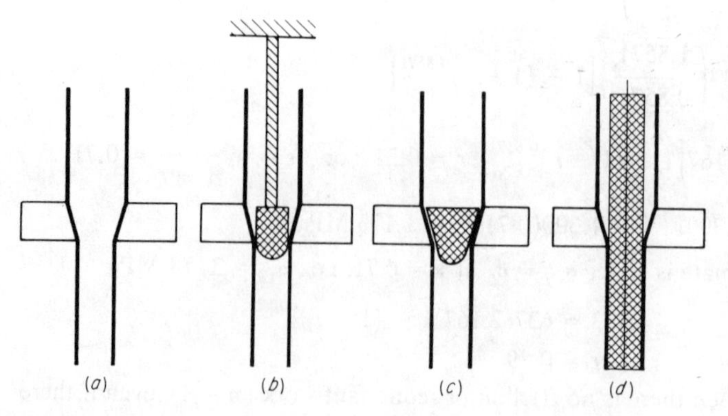

Figure 19-9 Methods of tube drawing. (*a*) Sinking; (*b*) fixed plug; (*c*) floating plug; (*d*) moving mandrel.

velocity equal to that of the tube on exiting from the die, and since this is higher than the velocity of the metal confined in the die channel, there is a *forward frictional drag* at the interface between the mandrel and the tube which tends to cancel the backward frictional drag between the stationary die and the tube. However, after drawing, the mandrel must be removed from the tube by rolling (reeling), which increases the tube diameter slightly and disturbs the dimensional tolerances.

19-5 ANALYSIS OF TUBE DRAWING

In the tube drawing with a plug or mandrel the greatest part of the deformation occurs as a reduction in wall thickness. Usually the inside diameter is reduced by a small amount equal to that needed to allow for insertion of the plug or mandrel before drawing. Thus, there is no hoop strain and the analysis[1] can be based on plane-strain conditions.

For tube drawing with a plug, the draw stress, by analogy with Eq. (19-5), can be expressed by

$$\sigma_{xa} = \sigma_0' \frac{1 + B'}{B'} \left[1 - \left(\frac{h_a}{h_b} \right)^{B'} \right] \tag{19-24}$$

where
$$B' = \frac{\mu_1 + \mu_2}{\tan \alpha - \tan \beta}$$

and μ_1 = coefficient of friction between tube and die wall
μ_2 = coefficient of friction between tube and plug
α = semicone angle of die
β = semicone angle of the plug; for a cylindrical plug, $\beta = 0$

In tube drawing with a floating plug the question of proper design of the plug is more important than the magnitude of the drawing stress. This has been discussed by Blazynski.[2]

In tube drawing with a moving mandrel, the friction forces at the stationary die-tube interface are directed toward the die entrance as for wiredrawing and tube drawing with a plug. However, as discussed earlier, the friction forces at the mandrel-tube interface are directed toward the exit of the die. Therefore, for a moving mandrel B' must be expressed as

$$B' = \frac{\mu_1 - \mu_2}{\tan \alpha - \tan \beta}$$

The draw stress is given by Eq. (19-24) with the proper value of B'. If $\mu_1 = \mu_2$, as

[1] Hoffman and Sachs, op. cit., pp. 190–195.

[2] T. Z. Blazynski, *Metall. Rev.*, no. 142, April, 1970; D. J. Smith and A. N. Bramley, *Proc. 13th Int. Machine Tool Design and Research Conf.*, Birmingham, England, 1972.

may often be the case, then $B' = 0$. The differential equation of equilibrium for this simple case is

$$h\,d\sigma_x + (\sigma_x + p)\,dh = 0$$

which integrates directly to the equation for ideal homogeneous deformation Eq. (19-1). However, it is entirely possible for the coefficient of friction on the mandrel μ_2 to exceed that at the die μ_1, so that B is negative. This situation results in a draw stress less than required by frictionless ideal deformation.

Tube sinking is the process by which the inside diameter of the tube is decreased. Although we have neglected this aspect of the deformation until now, some reduction in inside diameter is part of the tube-drawing process.

The stresses involved in tube sinking have been analyzed by Sachs and Baldwin[1] on the assumption that the wall thickness of the tube remains constant. The equation or the draw stress at the die exit is analogous with the equation describing the draw stress in wiredrawing. The cross-sectional area of the tube is related to the midradius r and the wall thickness h by $A \approx 2\pi rh$.

$$\sigma_{xa} = \sigma_0'' \frac{1+B}{B}\left[1 - \left(\frac{A_f}{A_b}\right)^B\right] \qquad (19\text{-}25)$$

The yield stress σ_0'' is taken equal to $1.1\sigma_0$ to account for the complex stresses in tube sinking. A more complete analysis of tube sinking has been given by Swift.[2]

19-6 RESIDUAL STRESSES IN ROD, WIRE, AND TUBES

Two distinct types of residual-stress patterns are found in cold-drawn rod and wire, depending on the amount of reduction. For reductions per pass of less than about 1 percent the longitudinal residual stresses are compressive at the surface and tensile at the axis, the radial stresses are tensile at the axis and drop off to zero at the free surface, while circumferential stresses follow the same trend as the longitudinal residual stresses. For larger reductions of commercial significance the residual-stress distribution is completely reversed from the first type of stress pattern. In this case the longitudinal stresses are tensile at the surface and compressive at the axis of the rod, the radial stresses are compressive at the axis, and the circumferential stresses follow the same pattern as the longitudinal stresses. The first type of residual-stress pattern is characteristic of forming operations where the deformation is localized in the surface layers.

The effect of die angle and the amount of reduction per pass on the longitudinal residual stress in cold-drawn brass wire was investigated by Linicus

[1] G. Sachs and W. M. Baldwin, Jr., *Trans. ASME*, vol. 68, pp. 655–662, 1946; also see Hoffman and Sachs, op. cit., pp. 252–255.

[2] H. W. Swift, *Philos. Mag.*, vol. 40, ser. 7, pp. 883–902, 1949.

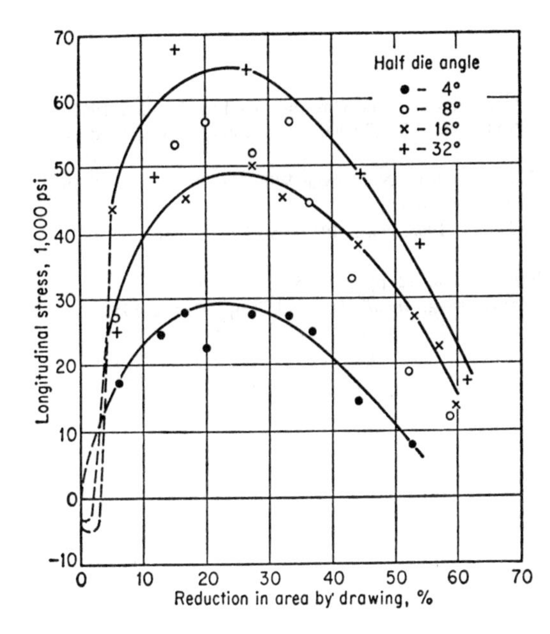

Figure 19-10 Longitudinal residual stress in cold-drawn brass wire. *(After Linicus and Sachs.)*

and Sachs.[3] Figure 19-10 shows that for a given reduction the longitudinal residual stress increases with the half-die angle. Maximum values of residual stress are obtained for reductions in the region of 15 to 35 percent.

For tubes produced by tube sinking, under conditions where the deformation is relatively uniform throughout the tube wall, the longitudinal residual stresses are tensile on the outer surface and compressive on the inner surface of the tube. The residual stresses in the circumferential direction follow the same pattern, while the stresses in the radial direction are negligible. Approximate measurements[1] of the circumferential stresses on the outer surface of sunk tubes indicate that the stresses increase with increasing reduction in diameter at the same rate at which the yield stress is increased by the cold-work.

Studies[2] of residual stresses in tubes produced by drawing over a plug and mandrel showed the same distribution of residual stresses as for tube sinking. An important result was that a substantial reduction in the level of residual stress could be produced by tandem drawing, whereby a small reduction (2 percent) was produced by a second die immediately following the main reduction. A systematic approach to developing tube-drawing schedules has been proposed[3] based on the criterion of developing zero circumferential residual stress.

[3] W. Lincius and G. Sachs, *Mitt. Dtsch. Materialprüfungsanst.*, vol. 16, pp.38–67, 1932; R. N. Wright, *Wire Tech.*, vol. 6, pp. 131–135, 1978.

[1] D. K. Crampton, *Trans. Metall. Soc. AIME*, vol. 89, pp. 233–255, 1930.

[2] S. K. Misra and N. H. Polakowski, *Trans. ASME, Ser. D: J. Basic Eng.* vol. 91, pp. 810–815, 1969.

[3] B. J. Meadows, and A. G. Lawrence, *J. Inst. Met.*, vol. 98, pp. 102–105, 1970.

BIBLIOGRAPHY

Bernhoeft, C. P.: "The Fundamentals of Wire Drawing," The Wire Industry Ltd., London, 1962.
Blazynski, T. Z.: "Metal Forming: Tool Profiles and Flow," Chaps. 4–6, John Wiley-Halsted Press, New York, 1976.
Collins, L. W., J. G. Dunleavy, and O. J. Tassi, (eds.): "Nonferrous Wire Handbook," vol. 1, 1977, vol. 2, 1981, Wire Association, Inc., Branford, Conn.
Dove, A. B. (ed.): "Steel Wire Handbook," vol. 1, 1968, vol. 2, 1969, vol. 3, 1972, vol. 4, 1980, Wire Association, Inc., Branford, Conn.
Rowe, G. W.: Wire Manufacture, *Int. Met. Rev.*, vol. 22, pp. 341–354, 1977.
Wistreich, J. G.: The Fundamentals of Wire Drawing, *Metall. Rev.*, vol. 3, pp. 97–142, 1958.

CHAPTER

TWENTY

SHEET-METAL FORMING

20-1 INTRODUCTION

The ability to produce a variety of shapes from flat sheets of metal at high rates of production has been one of the real technological advances of the twentieth century. This transition from hand-forming operations to mass-production methods has been an important factor in the great improvement in the standard of living which occurred during the period.

In essence, a shape is produced from a flat blank by stretching and shrinking the dimensions of all its volume elements in the three mutually perpendicular principal directions. The resulting shape is then the result of the integration of all the local stretching and shrinking of the volume elements. Attempts have been made to classify the almost limitless number of shapes which are possible in metal forming into definite categories depending on the contour of the finished part. Sachs[1] has classified sheet-metal parts into five categories.

1. Singly curved parts
2. Contoured flanged parts—including parts with stretch flanges and shrink flanges
3. Curved sections
4. Deep-recessed parts—including cups and boxes with either vertical or sloping walls
5. Shallow-recessed parts—including dish-shaped, beaded, embossed, and corrugated parts

[1] G. Sachs, "Principles and Methods of Sheet-Metal Fabricating," 2d ed., pp. 9–14, Van Nostrand Reinhold, New York, 1966.

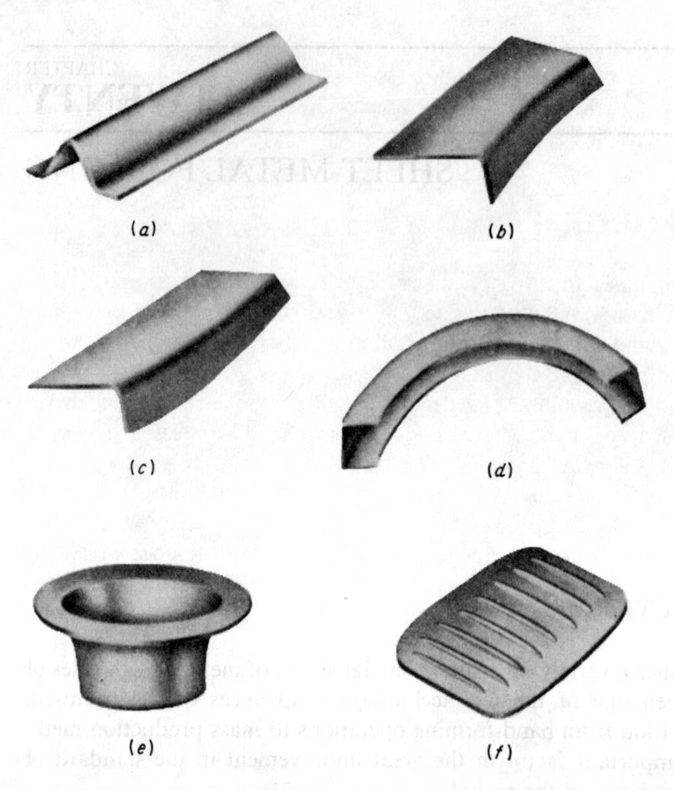

Figure 20-1 Typical formed shapes. (*a*) Singly curved; (*b*) stretch flange; (*c*) shrink flange; (*d*) curved sections; (*e*) deep-drawn cup; (*f*) beaded section.

Typical examples of these parts are shown in Fig. 20-1. Another classification system, developed in the automotive industry, groups sheet-steel parts into categories depending on the severity of the forming operation.[2] Severity of the operation is based on the maximum amount of bending or stretching in the part.

Still another way of classifying sheet-metal forming is by means of specific operations such as bending, shearing, deep drawing, stretching, ironing, etc. Most of these operations have been illustrated briefly in Fig. 15-1, and they will be discussed in considerably greater detail in this chapter.

We should note that unlike the *bulk-forming* deformation processes described in the earlier chapters, sheet forming is carried out generally in the plane of the sheet by tensile forces. The application of compressive forces in the plane of the sheet is avoided because it leads to buckling, folding, and wrinkling of the sheet. While in bulk-forming processes the intention is often to change the thickness or

[2] "Metals Handbook," 8th ed., vol. 1, pp. 319–335, American Society for Metals, Metals Park, Ohio, 1961.

lateral dimensions of the workpiece, in sheet-forming processes decreases in thickness should be avoided because they could lead to necking and failure. Another basic difference between bulk forming and sheet forming is that sheet metals, by their very nature, have a high ratio of surface area to thickness.

20-2 FORMING METHODS

The old method of hand forming of sheet metal is today used primarily as a finishing operation to remove wrinkles left by forming machines. In the metal-working industries hand forming is primarily limited to experimental work where only a few identical pieces are required.

Most high-production-volume sheet-metal forming is done on a press, driven by either mechanical or hydraulic action.[1] In mechanical presses energy is generally stored in a flywheel and is transferred to the movable slide on the downstroke of the press. Mechanical presses are usually quick-acting and have a short stroke, while hydraulic presses are slower-acting but can apply a longer stroke. Presses are usually classified according to the number of slides which can be operated independently of each other. In the *single-action press* there is only one slide, generally operating in the vertical direction. In the *double-action press* there are two slides. The second action ordinarily is used to operate the *hold-down*, which prevents wrinkling in deep drawing. A *triple-action press* is equipped with two actions above the die and one action below the die.

The basic tools used with a metalworking press are the *punch* and the *die*.[2] The punch is the convex tool which mates with the concave die. Generally the punch is the moving element. Because accurate alignment between the punch and die is usually required, it is common practice to mount them permanently in a *subpress*, or *die set*, which can quickly be inserted in the press. An important consideration in tooling for sheet-metal forming is the frequent requirement for a clamping pressure, or hold-down, to prevent wrinkling of the sheet as it is being formed. Hold-down can best be provided by a *hold-down ring*, which is actuated by the second action of a double-action press. However, by using mechanical springs or an auxiliary air cylinder, hold-down can be provided in a single-action press.

Frequently punches and dies are designed so that successive stages in the forming of the part are carried out in the same die on each stroke of the press. This is known as *progressive forming*. A simple example is a progressive blanking and piercing die to make a plain, flat washer (Fig. 20-2). As the strip is fed from left to right, the hole for the washer is first punched and then the washer is blanked from the strip. At the same time as the washer is being blanked from the

[1] C. Wick, J. T. Benedict, and R. F. Veilleux, "Tool and Manufacturing Engineers Handbook," 4th ed., vol. 2, chap. 5, Society of Manufacturing Engineers, Dearborn, Mich., 1984.

[2] Wilson, F. W. "Die Design Handbook," 2d ed., McGraw-Hill Book Company, New York, 1965; C. Wick, J. T. Benedict, and R. F. Veilleux, "Tool and Manufacturing Engineers Handbook," vol. 2, chap. 6, Society of Manufacturing Engineers, Dearborn, Mich., 1984.

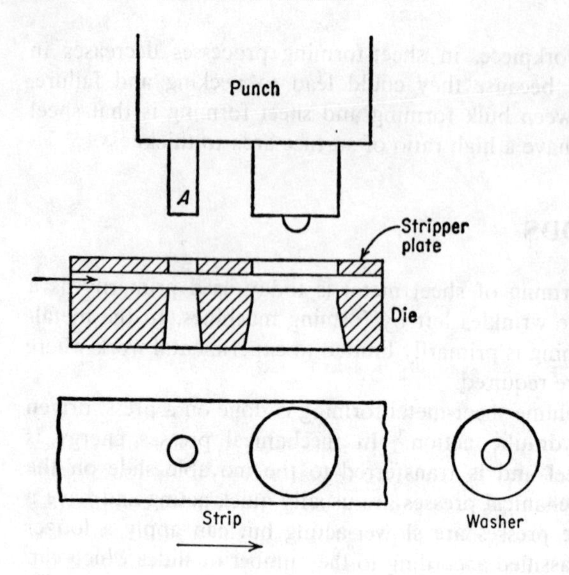

Figure 20-2 Progressive piercing and blanking die.

strip, the punch *A* is piercing the hole for the next washer. The stripper plate is used to prevent the metal from separating from the die on the up stroke of the punch.

Compound dies are designed to perform several operations on the same piece in one stroke of the press. Because of their complexity the dies are costlier and the operations somewhat slower than individual operations. Another strategy is to use *transfer dies*, where the part is moved from station to station within the press for each operation. The die materials depend on the severity of the operation and the required production run.[1] In aircraft work, where production runs are often small, tooling is frequently made from a zinc-base alloy called Kirksite or from wood or epoxy resins. For long die life, however, tool steel is required.

The *press brake* is a single-action press with a very long and narrow bed. The chief purpose of a press brake is to form long, straight bends in pieces such as channels and corrugated sheets. *Roll forming* (Chap. 17) is another common method of producing bent shapes in long lengths. The roll-forming process is also used to produce thin-wall cylinders from flat sheet.

Rubber hydroforming is a modification of the conventional punch and die in which a pad of rubber or polyurethane serves as the die. Rubber forming, or the *Guerin process*, is illustrated in Fig. 20-3. A form block (punch) is fastened to the bed of a single-action hydraulic press, and a thick blanket of rubber is placed in a retainer box on the upper platen of the press. When a blank is placed over the form block and the rubber forced down on the sheet, the rubber transmits a nearly uniform hydrostatic pressure against the sheet. A unit pressure of around 1,500 psi is sufficient for most parts, and higher local pressures can be provided

[1] "Metals Handbook," 8th ed., vol. 1, pp. 685–717, American Society for Metals, Metals Park, Ohio, 1961.

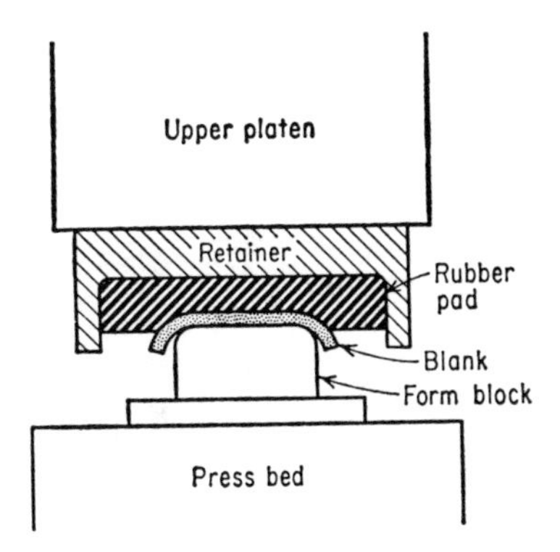

Figure 20-3 Rubber forming.

by auxiliary tooling.[1] The Verson-Wheelon process uses a soft rubber bag subjected to internal fluid pressure. Because the forming pressure is four to five times greater than in the Guerin process it can be used to form more complicated and deeper shapes. Rubber forming is used extensively in the aircraft industry. Shallow flanged parts with stretch flanges are readily produced by this method, but shrink flanges are limited because the rubber provides little resistance to wrinkling. Another limitation is that the blank tends to move on the form block unless holes for positioning pins are provided in the part.

A variety of methods are used to bend or to contour-form straight sections.[2] Cylindrical- and conical-shaped parts are produced with *bending rolls* (Fig. 20-4*a*). A three-roll bender is not very well suited to preventing buckling in thin-gage sheet. Often a fourth roll is placed at the exit to provide an extra adjustment in curvature. In three-point loading the maximum bending moment is at the midpoint of the span. This localization of strain can result, under certain circumstances, in the forming limit being reached at the midpoint before the rest of the part is bent to the proper contour. More uniform deformation along the length of the part is obtained with *wiper-type equipment*. In its simplest form this consists of a sheet which is clamped at one end against a form block; the contour is progressively formed by successive hammer blows, starting near the clamp and moving a short distance toward the free end with each blow. A wiper-type bender is sketched in Fig. 20-4*b*. In this case the form block or die has a nonuniform contour so that the wiper rolls must be pressed against the block with a uniform pressure supplied by a hydraulic cylinder. Still a third method of producing contours is by *wrap forming*. In wrap forming the sheet is compressed against a form block, and at the same time a longitudinal tensile stress is applied to prevent buckling and wrinkling (Fig. 20-4*c*). A simple example of wrap forming is the coiling of a spring around a mandrel. The stretch forming of curved sections is a special case of wrap forming.

A method of making tank heads, television cones, and other deep parts of circular symmetry is *spinning* (Fig. 20-5*a*). The metal blank is clamped against a

[1] Sachs, op. cit., pp. 424–455.
[2] *Ibid*., pp. 476–493.

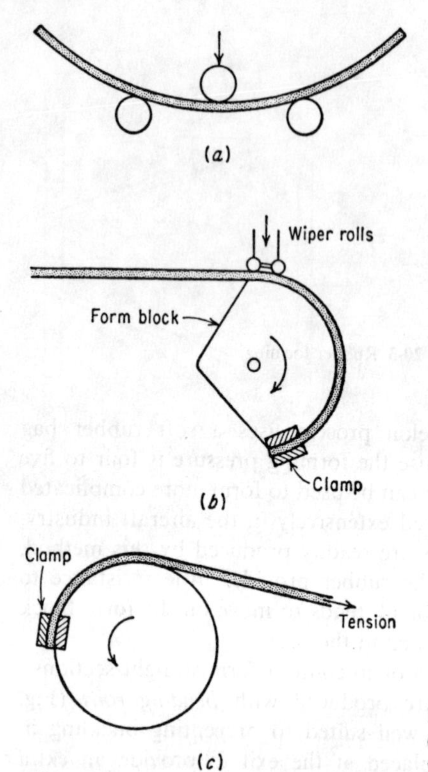

Figure 20-4 Methods of bending and contouring (*a*) Three-roll bender; (*b*) wiper-type benders; (*c*) wrap forming.

form block which is rotated at high speed. The blank is progressively formed against the block, either with a manual tool or by means of small-diameter work rolls. In the spinning process the blank thickness does not change but its diameter is decreased. The *shear-spinning* process, Fig. 20-5*b*, is a variant of conventional spinning. In this process the part diameter is the same as the blank diameter but the thickness of the spun part is reduced according to $t = t_0 \sin \alpha$. This process is also known as *power spinning*, *flowturning*, and *hydrospinning*. It is used for large axisymmetrical conical or curvilinear shapes such as rocket-motor casings and missile nose cones. Still a third variation of spinning is *tube spinning* in which a tube is reduced in wall thickness by spinning on a mandrel. The spinning tool can operate on either the outside or inside diameter of the tube.

Explosive forming[1] is well suited to producing large parts with a relatively low production lot size. The sheet-metal blank is placed over a die cavity and an explosive charge is detonated in water at an appropriate standoff distance from the blank. The shock wave propagating from the explosion serves as a "frictionless punch" to deform the blank.

[1] E. J. Bruno (ed.), "High Velocity Forming of Metals," Society of Manufacturing Engineers, Dearborn, Mich., 1968; A. A. Ezra, "Principles and Practice of Explosive Metalworking," Industrial Newspapers Ltd., London, 1973.

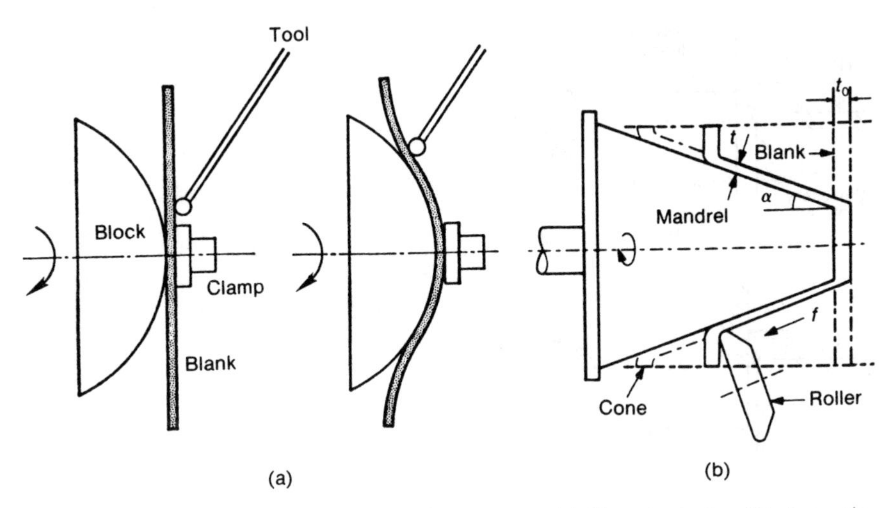

Figure 20-5 Schematic representation of spinning processes. (*a*) Manual spinning; (*b*) shear spinning.

20-3 SHEARING AND BLANKING

Shearing is the separation of metal by two blades moving as shown in Fig. 20-6*a*. In shearing, a narrow strip of metal is severely plastically deformed to the point where it fractures at the surfaces in contact with the blades. The fracture then propagates inward to provide complete separation. The depth to which the punch must penetrate to produce complete shearing is directly related to the ductility of the metal. The penetration is only a small fraction of the sheet thickness for brittle materials, while for very ductile materials it may be slightly greater than the thickness.

The clearance between the blades is an important variable in shearing operations. With the proper clearance the cracks that initiate at the edges of the blades will propagate through the metal and meet near the center of the thickness to provide a clean fracture surface (Fig. 20-6*a*). Note that even with proper clearance there is still distortion at a sheared edge. Insufficient clearance will produce a ragged fracture (Fig. 20-6*b*) and also will require more energy to shear the metal than when there is proper clearance. With excessive clearance there is greater distortion of the edge, and more energy is required because more metal must plastically deform before it fractures. Furthermore, with too large a clearance, burrs or sharp projections are likely to form on the sheared edge (Fig. 20-6*c*). A dull cutting edge also increases the tendency for the formation of burrs. The height of the burr increases with increasing clearance and increasing ductility of the metal. Because the quality of the sheared edge influences the formability of the part the control of clearance is important. Clearances generally range between 2 and 10 percent of the thickness of the sheet; the thicker the sheet the larger the clearance.

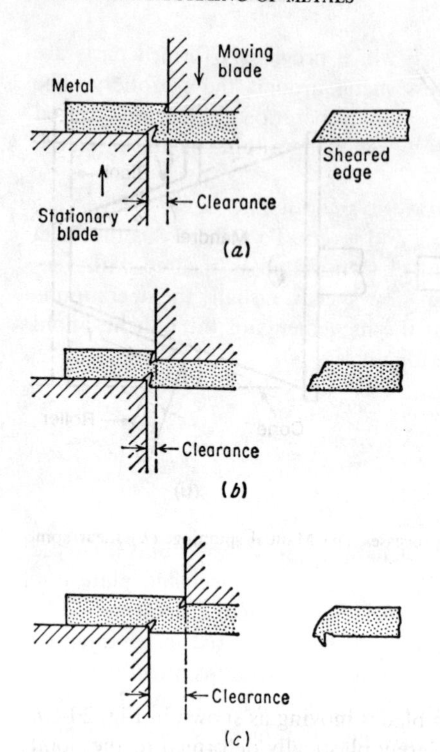

Figure 20-6 Shearing of metal. (*a*) Proper clearance; (*b*) insufficient clearance; (*c*) excessive clearance.

Neglecting friction, the force required to shear a metal sheet is the product of the length cut, the sheet thickness, and the shearing strength of the metal. Empirically, the maximum punch force to produce shearing is given by

$$P_{\max} \approx 0.7\sigma_u hL \tag{20-1}$$

where σ_u = the ultimate tensile strength
h = sheet thickness
L = total length of the sheared edge

The shearing force can be reduced appreciably by making the edges of the cutting tool at an inclined angle so that only a short part of the total length of cut is made at one time. The bevel of the cutting edge is called *shear*.

A whole group of press operations are based on the process of shearing. The shearing of closed contours, when the metal inside the contour is the desired part, is called *blanking*. If the material inside the contour is discarded, then the operation is known as *punching*, or *piercing*. Punching indentations into the edge of the sheet is called *notching*. *Parting* is the simultaneous cutting along at least two lines which balance each other from the standpoint of side thrust on the parting tool. *Slitting* is a shearing cut which does not remove any metal from the sheet. *Trimming* is a secondary operation in which previously formed parts are finished to size, usually by shearing excess metal around the periphery. The removal of forging flash in a press is a trimming operation. When the sheared edges of a part are trimmed or squared up by removing a thin shaving of metal, the operation is called *shaving*.

Fine blanking is a process in which very smooth and square edges are produced in small parts such as gears, cams, and levers. To achieve this the sheet metal is tightly locked in place to prevent distortion and is sheared with very small clearances on the order of 1 percent at slow speeds. Usually the operation is carried out on a triple-action press so that the movements of the punches, hold down ring, and die can be controlled individually.

20-4 BENDING

Bending is the process by which a straight length is transformed into a curved length. It is a very common forming process for changing sheet and plate into channel, drums, tanks, etc. In addition, bending is part of the deformation in many other forming operations. The definition of the terms used in bending are illustrated in Fig. 20-7. The bend radius R is defined as the radius of curvature on the concave, or inside, surface of the bend. For elastic bending below the elastic limit the strain passes through zero halfway through the thickness of the sheet at the neutral axis. In plastic bending beyond the elastic limit the neutral axis moves closer to the inside surface of the bend as the bending proceeds. Since the plastic strain is proportional to the distance from the neutral axis, fibers on the outer surface are strained more than fibers on the inner surface are contracted. A fiber at the mid-thickness is stretched, and since this is the average fiber, it follows that there must be a decrease in thickness (radial direction) at the bend to preserve the constancy of volume. The smaller the radius of curvature, the greater the decrease in thickness on bending.

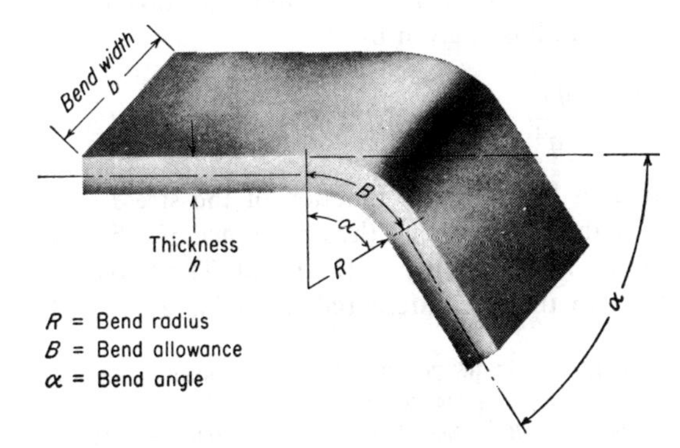

Figure 20-7 Definition of terms used in bending.

According to the theory of bending[1] the strain increases with decreasing radius of curvature. If the change in thickness is neglected, the neutral axis will remain at the center fiber and the circumferential stretch on the top surface e_a will be equal to the shrink on the bottom surface, e_b. The conventional strain at the outer and inner fibers is given by

$$e_a = -e_b = \frac{1}{(2R/h)+1} \tag{20-2}$$

Experiments show that the circumferential strain on the tension surface is considerably greater than that given by Eq. (20-2) for large values of h/R, while the strain on the compression surface is not very different from the strain predicted by the simplified equation.

For a given bending operation the bend radius cannot be made smaller than a certain value, or the metal will crack on the outer tensile surface. The *minimum bend radius* is usually expressed in multiples of the sheet thickness. Thus, a $3T$ bend radius indicates that the metal can be bent without cracking through a radius equal to three times the sheet thickness. Therefore, the minimum bend radius is a forming limit. It varies considerably between different metals and always increases with cold-working. Although some very ductile metals have a minimum bend radius of zero, indicating that they can be flattened upon themselves, it is general practice to use a bend radius of not less than $\frac{1}{32}$ in in order to prevent damage to punches and dies. For high-strength sheet alloys the minimum bend radius may be $5T$ or higher. The minimum bend radius is not a precise material parameter because it depends, among other things, on the geometry of the bending conditions.

The minimum bend radius for a given thickness of sheet can be predicted[2] fairly accurately from the reduction of area measured in a tension test q. If q is less than 0.2, then the shift in the neutral axis can be neglected and R_{min} is given simply by

$$\frac{R_{min}}{h} = \frac{1}{2q} - 1 \quad \text{for } q < 0.2 \tag{20-3}$$

When q is greater than 0.2, the shift in the neutral axis must be taken into consideration and the minimum bend radius is given by

$$\frac{R_{min}}{h} = \frac{(1-q)^2}{2q-q^2} \quad \text{for } q > 0.2 \tag{20-4}$$

The ductility of the outer fiber in bending is a function of the stress state acting on the surface. It is a well-established fact that the occurrence of a biaxial state of tension produces a decrease in the ductility of the metal. The biaxiality ratio σ_2/σ_1 of the transverse stress to the circumferential stress increases with

[1] J. D. Lubahn and G. Sachs, *Trans. ASME*, vol. 72, pp. 201–208, 1950; B. W. Shaffer and E. E. Unger, *Trans. ASME*, Ser. E: *J. Appl. Mech.*, vol. 27, pp. 34–40, 1960.

[2] J. Datsko and C. T. Yang, *Trans. ASME, Ser. B: J. Eng. Ind.*, vol. 82, pp. 309–314, 1960.

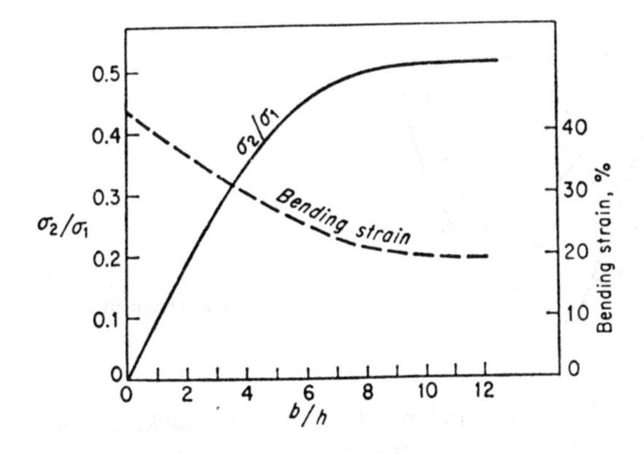

Figure 20-8 Effect of ratio of width to thickness on biaxiality and bend ductility in bending. *(From G. S. Sangdahl, E. L. Aul, and G. Sachs, Proc. Soc. Exp. Stress Anal., vol. 6, no. 1, p. 1, 1948.)*

increasing ratio of width to thickness, b/h. Figure 20-8 indicates that for low values of b/h the biaxiality is low because the stress state is practically pure tension, but as the width of the sheet increases relative to its thickness, the ratio σ_2/σ_1 increases until at approximately $b/h = 8$ the biaxiality reaches a saturation value of approximately $\frac{1}{2}$. Correspondingly, the strain to produce fracture in bending is a reverse function of the width-thickness ratio. In bending sheets with a high width-thickness ratio, the cracks will occur near the center of the sheet when the ductility is exhausted. However if the edges of the sheet are rough, edge cracking will occur. Frequently the minimum bend radius can be increased by polishing or grinding the edges of the sheet. In bending narrow sheets, the failure will occur at the edges because the biaxiality is quite low at the center of the width.

In addition to cracking on the tensile surface of a bend another common forming difficulty is springback. *Springback* is the dimensional change of the formed part after the pressure of the forming tool has been released. It results from the changes in strain produced by elastic recovery. When the load is released, the total strain is reduced owing to the elastic recovery. The elastic recovery, and therefore the springback, will be greater the higher the yield stress, the lower the elastic modulus, and the greater the plastic strain. For a given material and strain the springback increases with the ratio between the lateral dimensions of the sheet and its thickness.

Springback is encountered in all forming operations, but it is most easily recognized and studied in bending (Fig. 20-9). The radius of curvature before release of load R_0 is smaller than the radius after release of the load R_f. But, the bend allowance (Fig. 20-7) is the same before and after bending, so

$$B = \left(R_0 + \frac{h}{2}\right)\alpha_0 = \left(R_f + \frac{h}{2}\right)\alpha_f$$

Thus, the springback ratio, $K_s = \alpha_f/\alpha_0$, is given by

$$K_s = \frac{\alpha_f}{\alpha_0} = \frac{R_0 + h/2}{R_f + b/2} = \frac{2R_0/h + 1}{2R_f/h + 1} \tag{20-5}$$

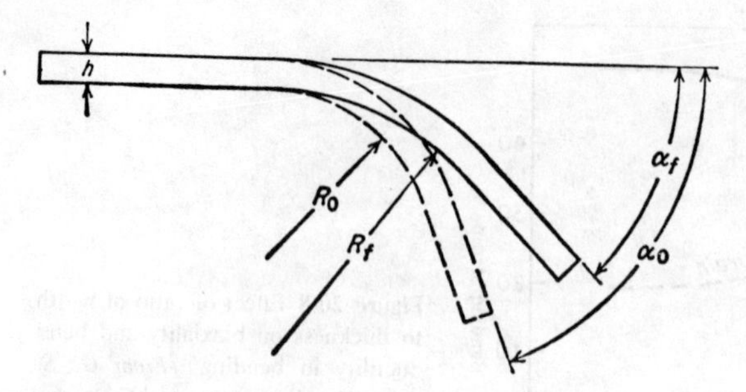

Figure 20-9 Springback in bending.

The springback ratio defined in this way is independent of sheet thickness and depends only on the ratio of bend radius to sheet thickness. Values are available[1] for aluminum alloys and austenitic stainless steels in a number of cold-rolled tempers. Other data,[2] which include a number of high-temperature alloys, indicate that to a first approximation the springback in bending can be expressed by

$$\frac{R_0}{R_f} = 4\left(\frac{R_0\sigma}{Eh}\right)^3 - 3\frac{R_0\sigma}{Eh} + 1 \qquad (20\text{-}6)$$

The commonest method of compensating for springback is to bend the part to a smaller radius of curvature than is desired (overbending) so that when springback occurs the part has the proper radius. The trial-and-error procedure of finding the proper die contour to correct for springback can be shortened somewhat by the use of the above equation, but the calculation is by no means a precise procedure. Furthermore, the correction to the die is valid only over a rather narrow range of yield stress. Other methods of compensating for springback are to bottom the punch in the die so as to produce a coining action and the use of high-temperature forming so as to reduce the yield stress.

The force P_b required to bend a length L about a radius R may be estimated from

$$P_b = \frac{\sigma_0 L h^2}{2(R + h/2)} \tan\frac{\alpha}{2} \qquad (20\text{-}7)$$

20-5 STRETCH FORMING

Stretch forming is the process of forming by the application of primarily tensile forces in such a way as to stretch the material over a tool or form block.[3] The process is an outgrowth of the stretcher leveling of rolled sheet. Stretch forming is

[1] Sachs, op. cit., p. 100.

[2] F. J. Gardiner, *Trans. ASME*, vol. 79, pp. 1–9, 1957; C. A. Queener and R. J. DeAngelis, *Trans. Am. Soc. Met.*, vol. 61, p. 757; 1968.

[3] R. D. Edwards, *J. Inst. Met.*, vol. 84, pp. 199–209, 1956; Sachs, op. cit., pp. 456–475; Wick, Benedict, and Veilleux, op. cit., pp. 7-13 to 7-22.

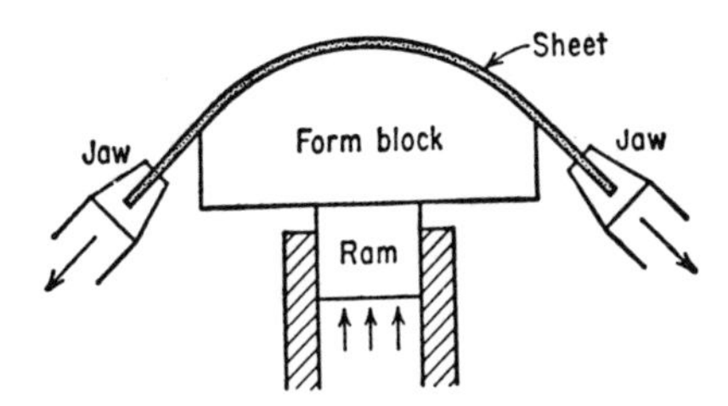

Figure 20-10 Stretch-forming operation.

used most extensively in the aircraft industry to produce parts of large radius of curvature, frequently with double curvature. An important consideration is that springback is largely eliminated in stretch forming because the stress gradient is relatively uniform. On the other hand, because tensile stresses predominate, large deformations can be obtained by this process only in materials with appreciable ductility.

Stretch-forming equipment consists basically of a hydraulically driven ram (usually vertical) which carries the punch or form block and two jaws for gripping the ends of the sheet (Fig. 20-10). No female die is used in stretch forming. The grips may be pivoted so that the tension force is always in line with the edge of the unsupported sheet, or they can be fixed, in which case a large radius is needed to prevent tearing the sheet at the jaws. In using a stretch-forming machine the sheet-metal blank is first bent or draped around the form block with relatively light tensile pull, the grips are applied, and the stretching load is increased until the blank is strained plastically to final shape. This differs from wrap forming (Sec. 20-2) since in the latter process the blank is first gripped and then while still straight is loaded to the elastic limit before wrapping around the form block.

Stretching commonly is found as a part of many sheet-forming operations. For example, in forming a cup with a hemispherical bottom, the sheet is stretched over the punch face. Most complex automotive stampings involve a stretching component.

For a strip loaded in tension we have seen (Sec. 8-3) that the limit of uniform deformation occurs at a strain equal to the strain-hardening exponent n. In biaxial tension, the necking which occurs in uniaxial tension is inhibited if $\sigma_2/\sigma_1 > \frac{1}{2}$, and instead the material develops *diffuse necking* which is not highly localized or readily visible to the eye. Eventually in the stretching of a thin sheet[1] plastic instability will occur in the form of a narrow *localized neck*. There will be a direction of zero-length increase inclined at an angle ϕ to the deforming axis (Fig. 20-11).

The normal strain ε_2' must be zero, for if it were not, material adjoining the edges of the band would have to deform and the band would spread out along X_1' and the band would grow into a diffuse neck. It can be shown (Prob. 20-3) that $\phi \approx \pm 55°$ for an isotropic material in pure tension. Moreover, the criterion for

[1] S. F. Keeler and W. A. Backofen, *Trans. Am. Soc. Met.*, vol. 56, pp. 25–48, 1963.

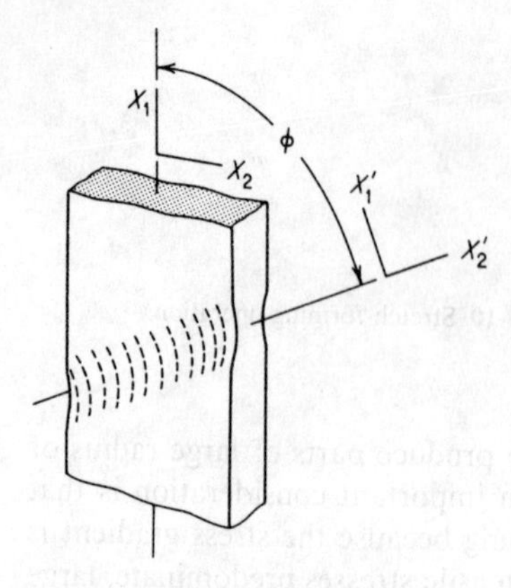

Figure 20-11 Localized necking in a strip loaded in tension. The normal strain along X_2' must be zero. *(From W. A. Backofen, "Deformation Processing," p. 205, Addison-Wesley, Publishing Company, Inc., Reading, Mass., 1972.)*

local necking was shown in Sec. 8-3 to be

$$\frac{d\sigma}{d\varepsilon} = \frac{\sigma}{2} \tag{20-8}$$

which arises from the fact that the area decreases with straining in local necking less rapidly than with diffuse necking. For power-law strain hardening, we have seen that for diffuse necking $\varepsilon_u = n$, but for local necking $\varepsilon_u = 2n$.

Diffuse necking usually will not constitute a limit to forming because the thinning is spread out over a fairly wide area of the sheet. Local necking is readily detected on exposed surfaces so that it represents a forming limit. Moreover, the formation of a local neck soon is followed by fracture of the sheet.

Local necking is very sensitive to the state of plastic strain,[1] usually denoted by the strain ratio $\rho = \varepsilon_2/\varepsilon_1$, where ε_1 is the algebraically larger of the principal strains in the plane of the sheet. If the sheet were deformed in pure shear, it would neither thicken nor thin and there would be no necking. At the other extreme, if the sheet is deformed in plane strain so that the width does not change, the local necking angle is $\phi = 90°$, and the strain for onset of both diffuse and local necking are equal. The dependence of necking strains on strain state is shown[2] in Fig. 20-12. ε_1^* is the strain for necking in the direction of the largest principal strain, and ε_u is the strain for diffuse necking in pure tension. If $\rho < 0$, both diffuse necking and local necking are possible, but both criteria coincide for plane-strain deformation. Theory predicts that only diffuse necking can occur if $\rho > 0$, since when both strains are tensile, there can be no zero-length change line. However, sheets tested under these conditions do thin locally and tear, generally at right angles to the ε_1 direction. For low-carbon steel[3] the critical strain for the onset of local necking is larger than ε_u and rises with increasing strain ratio in the manner shown by the dotted line in Fig. 20-12. Other materials (such as brass and austenitic stainless steel) do not show a rising necking limit

[1] R. Hill, *J. Mech. Phys. Solids*, vol. 1, p. 19, 1952.
[2] W. A. Backofen, *Metall. Trans.*, vol. 4, p. 2682, 1973.
[3] M. Azrin and W. A. Backofen, *Metall. Trans.*, vol. 1, pp. 2857–2865, 1970.

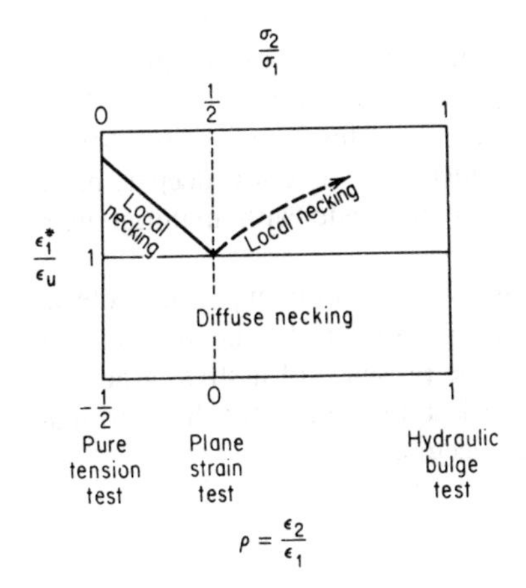

Figure 20-12 Variation of necking limits with strain ratio.

with strain ratio. The type of observations shown in Fig. 20-12 are the basis for the forming limit diagram discussed in Sec. 20-7. However, the theoretical basis for this behavior is still being developed.[1,2]

A common feature of a sheet-forming process is the existence of strain gradients. A strain gradient arises whenever the deformation is not uniform. The important role of strain hardening in modifying strain gradients can be shown by a simple example.[3] Consider a section of a sheet loaded in tension. The force is constant but the cross-sectional area increases with radial distance from some point, so that σr = constant. Therefore,

$$\frac{d\sigma}{dr} = -\frac{\sigma}{r}$$

and

$$\frac{d\sigma}{d\varepsilon}\frac{d\varepsilon}{dr} = \frac{d\sigma}{dr} = -\frac{\sigma}{r}$$

so the strain gradient is expressed by

$$\frac{d\varepsilon}{dr} = -\frac{\sigma}{(d\sigma/d\varepsilon)r} \tag{20-9}$$

If the strain hardening can be expressed by the power law $\sigma = K\varepsilon^n$, $d\sigma/d\varepsilon = nK\varepsilon^{n-1} = n(\sigma/\varepsilon)$ so

$$\frac{d\varepsilon}{dr} = -\left(\frac{\sigma}{r}\right)\frac{1}{n(\sigma/\varepsilon)} = -\frac{(\varepsilon/r)}{n} \tag{20-10}$$

[1] Z. Marciniak and K. Kuczynski, *Int. J. Mech. Sci.*, vol. 9, pp. 609–620, 1967.

[2] D. P. Koistinen and N. M. Wang (eds.), "Mechanical of Sheet Metal Forming," Plenum Press, New York, 1978; K. S. Chan, D. A. Koss, and A. K. Ghosh, *Met. Trans.*, vol. 15A, pp. 323–329, 1984.

[3] W. A. Backofen, "Deformation Processing," p. 211, Addison-Wesley Publishing Company, Inc., Reading, Mass., 1972.

Equation (20-10) shows that the strain gradient is reduced by greater strain hardening (larger n). Since the most highly strained region will have hardened the most, the load is passed on to the neighboring elements. This forces them to strain more and in so doing the strain gradient is reduced. As a result, deeper, more complex parts or overall greater reductions can be made with material with greater strain-hardening capacity.

The strain-rate sensitivity of the metal, m, also plays an important role in influencing the distribution of strain in a manner similar to the n value. A positive m reduces the localization of strain in the presence of a stress gradient. The higher the m value the more diffuse the neck, and the greater the deformation beyond the maximum load before fracture occurs.

20-6 DEEP DRAWING

Deep drawing is the metalworking process used for shaping flat sheets into cup-shaped articles such as bathtubs, shell cases, and automobile panels. This is done by placing a blank of appropriate size over a shaped die and pressing the metal into the die with a punch (Fig. 20-13). Generally a clamping or hold-down pressure is required to press the blank against the die to prevent wrinkling. This is best done by means of a *blank holder* or *hold-down ring* in a double-action press. Although the factors which control the deep-drawing process are quite evident, they interact in such a complex way that precise mathematical description of the process is not possible in simple terms. The greatest amount of experimental and analytical work has been done on the deep drawing of a flat-bottom cylindrical cup (Swift test) from a flat circular blank. The discussion of deep drawing which follows will be limited to this relatively simple situation.

In the deep drawing of a cup the metal is subjected to three different types of deformations. Figure 20-14 represents the deformation and stresses developed in

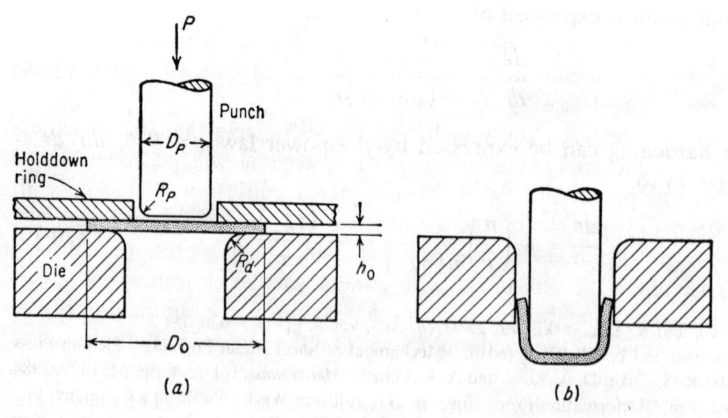

Figure 20-13 Deep drawing of a cylindrical cup. (a) Before drawing; (b) after drawing.

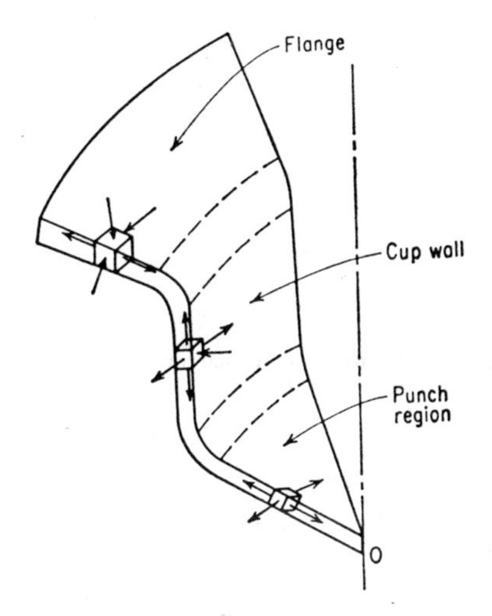

Figure 20-14 Stresses and deformation in a section from a drawn cup.

a pie-shaped segment of the circular blank during deep drawing. The metal at the center of the blank under the head of the punch is wrapped around the profile of the punch, and in so doing it is thinned down. The metal in this region is subjected to biaxial tensile stress due to the action of the punch. Metal in the outer portion of the blank is drawn radially inward toward the throat of the die. As it is drawn in, the outer circumference must continuously decrease from that of the original blank πD_0 to that of the finished cup πD_p. This means that it is subjected to a compressive strain in the circumferential, or hoop, direction and a tensile strain in the radial direction. As a result of these two principal strains there is a continual increase in the thickness as the metal moves inward. However, as the metal passes over the die radius, it is first bent and then straightened while at the same time being subjected to a tensile stress. This plastic bending under tension results in considerable thinning, which modifies the thickening due to the circumferential shrinking. Between the inner stretched zone and the outer shrunk zone there is a narrow ring of metal which has not been bent over either the punch or the die. The metal in this region is subjected only to simple tensile loading throughout the drawing operation.

If the clearance between the punch and the die is less than the thickness produced by free thickening, the metal in these regions will be squeezed, or *ironed*, between the punch and the die to produce a uniform wall thickness. In commercial deep drawing clearances about 10 to 20 percent greater than the metal thickness are common. Ironing operations in which appreciable uniform reductions are made in the wall thickness use much smaller clearances.

The force on the punch required to produce a cup is the summation of the ideal force of deformation, the frictional forces, and the force required to produce ironing (if present). Figure 20-15 illustrates the way in which these components of the total punch force vary with the length of stroke of the punch. The ideal force of deformation increases continuously with length of travel because the strain is

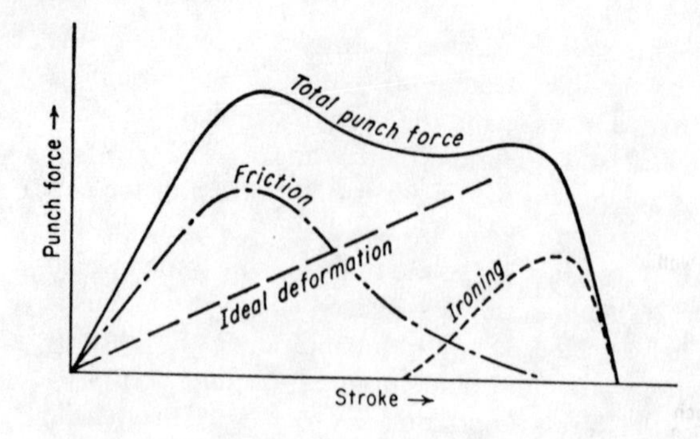

Figure 20-15 Punch force vs. punch stroke for deep drawing.

increasing and the flow stress is increasing owing to strain hardening. A major contribution to the friction force comes from the hold-down pressure. This force component peaks early and decreases with increasing travel because the area of the blank under the hold-down ring is continually decreasing. Any force required to produce ironing occurs late in the process after the cup wall has reached maximum thickness. An additional factor is the force required to bend and unbend the metal around the radius of the die. One measurement[1] of the work required in cupping showed that 70 percent of the work went into the radial drawing of the metal, 13 percent into overcoming friction, and 17 percent into the bending and unbending around the die radius.

From an analysis of the forces in equilibrium during the formation of a deep-drawn cup Sachs[2] has developed the following approximate equation for the total punch force as a function of the diameter of the blank, D_0, at any stage in the process,

$$P = \left[\pi D_p h(1.1\sigma_0)\ln\frac{D_0}{D_p} + \mu\left(2H\frac{D_p}{D_0}\right)\right]e^{(\mu\pi/2)} + B \qquad (20\text{-}11)$$

where P = total punch load
σ_0 = average flow stress
D_p = punch diameter
D_0 = blank diameter
H = hold-down force
B = force required to bend and restraighten blank
h = wall-thickness
μ = coefficient of friction

In Eq. (20-11) the first term expresses the ideal force required to produce the cup, and the second term is the friction force under the blank holder. The exponential

[1] H. W. Swift, *J. Inst. Met.*, vol. 82, p. 119, 1952.

[2] G. Sachs, "Spanlose Formung," pp. 11–38, Springer-Verlag OHG, Berlin, 1930; G. Sachs and K. R. Van Horn, "Practical Metallurgy," pp. 430–431, American Society for Metals, Metals Park, Ohio, 1940.

term considers the friction at the die radius, and the quantity B accounts for the force required to bend and unbend the sheet around this radius. A more complete and accurate analysis of the stresses and strains in the deep drawing of a cup has been presented by Chung and Swift[1] and Budiansky and Wang.[2]

The load in deep drawing is applied by the punch to the bottom of the cup and it is then transmitted to the sidewall of the cup. Usually failure occurs in the narrow band of material in the cup wall just above the radius of the punch which has undergone no radial drawing or bending but is subjected essentially to tensile straining. In this annular ring between the die wall and the punch the metal is subjected essentially to plane-strain stretching and thinning. Failure occurs by necking (followed by tearing) at a stress approximately equal to the tensile strength, increased by the plane-strain factor.

$$P_{\max} = \frac{2}{\sqrt{3}} s_u \pi D_p h \tag{20-12}$$

The applied stress in deep drawing is transmitted through a "weakest-link system" to a point of incipient failure which has undergone little strengthening by strain hardening. Thus, the strain-hardening ability of the material plays very little role in deep drawing, as opposed to wire drawing where n is very important. The *drawability* of a metal is measured by the ratio of the initial blank diameter to the diameter of the cup drawn from the blank (usually approximated by the punch diameter). For a given material there is a *limiting draw ratio* (LDR), representing the largest blank that can be drawn through a die D_p without tearing. The theoretical upper limit on LDR is[3]

$$\text{LDR} \approx \left(\frac{D_0}{D_p}\right)_{\max} \approx e^{\eta} \tag{20-13}$$

where η is an efficiency term to account for frictional losses. If $\eta = 1$, then LDR ≈ 2.7, while if $\eta = 0.7$, LDR ≈ 2. This agrees with experience that even with ductile metals it is difficult to draw a cup with a height much greater than its diameter.

Some of the practical considerations which affect drawability are:

Die radius—should be about 10 times sheet thickness.
Punch radius—a sharp radius leads to local thinning and tearing.
Clearance between punch and die—20 to 40 per cent greater than the sheet thickness.
Hold-down pressure—about 2 per cent of average of s_0 and s_u.
Lubricate die side to reduce friction in drawing.

[1] S. Y. Chung and H. W. Swift, *Proc. Inst. Mech. Eng.* (*London*), vol. 165, pp. 199–228, 1951; this theory has been reviewed in detail by J. M. Alexander, *Metall. Rev.*, vol. 5, pp. 349–411, 1960.

[2] B. Budiansky and N. M. Wang, *J. Mech. Phys. Solids*, vol. 14, p. 357, 1966.

[3] W. A. Backofen, op. cit., p. 233.

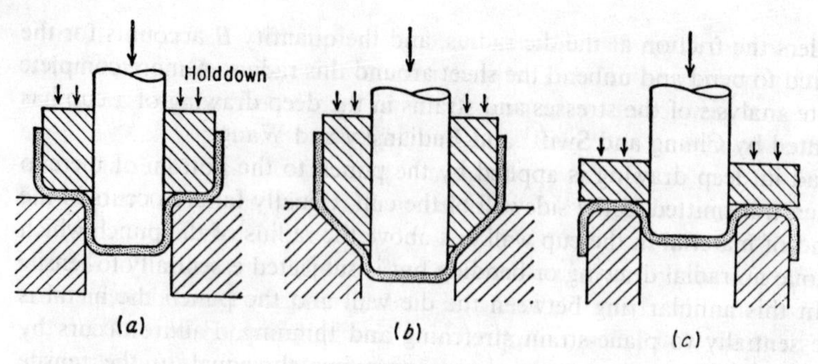

Figure 20-16 Redrawing methods (*a*) Direct redrawing; (*b*) direct redrawing with tapered die; (*c*) reverse redrawing.

Since the average maximum reduction in deep drawing is about 50 percent, to make tall slender cups (such as cartridge cases and closed-end tubes), it is necessary to use successive drawing operations. Reducing a cup or drawn part to a smaller diameter and increased height is known as *redrawing*.

The two basic methods of redrawing are *direct*, or *regular*, redrawing and *reverse*, or *indirect*, redrawing (Fig. 20-16). In direct redrawing the original outside surface of the cup remains the outside surface of the redrawn cup. Figure 20-16*a* illustrates direct redrawing by means of a hold-down ring. Note that the metal must bend twice and that it is bent and unbent at the punch and die radii. The high strain hardening that is encountered in the process shown in Fig. 20-16*a* is reduced somewhat by the design shown in Fig. 20-16*b*. Although the metal still goes through the same number of bends, the angle through which it bends is less than 90° and the punch load is reduced. The disadvantage of the design shown in Fig. 20-16*b* is that the first-stage drawn cup must be made with a tapered corner. This type of drawn cup cannot be produced in all metals without buckling. The thickness of the hold-down ring used in direct redrawing is determined by the percentage reduction of the redraw. For small reductions a hold-down ring cannot be used.

In reverse redrawing (Fig. 20-16*c*) the cup is turned inside out so that the outside surface of the drawn cup becomes the inside surface of the redrawn shell. Better control of wrinkling can be obtained in reverse redrawing because of the snug control of the metal around the die radius and the fact that there are no geometrical limitations to the use of a hold-down ring.

The reduction obtained by redrawing is always less than that obtainable on the initial draw because of the higher friction inherent in the redrawing process. Generally the reduction is decreased for each successive redrawing operation to allow for strain hardening. Greater reductions are, of course, possible if anealing is carried out between redraws. Most metals will permit a total reduction of 50 to 80 percent before annealing.

Redrawing operations may also be classified into drawing with appreciable decrease in wall thickness, called *ironing*, and drawing with little change in wall thickness, called *sinking*. The ironing process is basically the same as tube drawing with a moving mandrel. The predominant stress in ironing is the radial compressive stress developed by the pressure of the punch and the die. Redrawing without reduction in wall thickness is basically the same as tube sinking or tube drawing without a mandrel. The predominant stresses are an axial tensile stress from the action of the punch and a circumferential compression from the drawing in of the metal.

To improve drawability, the potential failure site near the bottom of the cup wall must be strengthened relative to the metal deforming by radial drawing near the top of the cup wall.[1] Some improvement can be obtained by shifting the failure site up the cup wall by mechanical factors such as gripping the cup more firmly around the punch by surrounding the cup with a pressurized rubber pad where the cup exits from the die. Roughening the punch or withholding lubrication to the punch may also help in this regard. It would also be possible to weaken the metal in the flange relative to the failure site by selectively heating the metal in the flange area. However, by far the greatest improvement in drawability comes about by the control of *crystallographic texture* in the sheet that is to be drawn. The correct texture gives the proper orientation of slip systems so that the strength in the thickness direction is greater than that in the plane of the sheet.

In Sec. 3-7 we saw that the resistance to through thickness thinning was measured by R, the plastic strain ratio[2] of width to thickness in a sheet. R measures the *normal anisotropy*. A large value of R denotes high resistance to thinning in the thickness direction (direction normal to the plane of the sheet).

$$R = \frac{\ln(w_0/w)}{\ln(h_0/h)} \tag{20-14}$$

where w_0 and w are the initial and final width and h_0 and h are the initial and final thickness. Since thickness measurements are difficult to make with precision on thin sheets, the equation can be rewritten using the constancy-of-volume relationship.

$$R = \frac{\varepsilon_w}{\varepsilon_h} = \frac{\varepsilon_w}{-(\varepsilon_w + \varepsilon_l)} = \frac{\ln w_0/w}{-\ln \dfrac{w_0 L_0}{wL}} = \frac{\ln w_0/w}{\ln \dfrac{wL}{w_0 L_0}} \tag{20-15}$$

Since most rolled sheets show a variation of elastic and plastic properties with orientation in the plane of the sheet, it is usual to allow for this planar anisotropy by $\bar{R}$ averaged over measurements taken at different angles to the rolling direction of the sheet.

$$\bar{R} = \frac{R_0 + 2R_{45} + R_{90}}{4} \tag{20-16}$$

[1] R. T. Holcomb and W. A. Backofen, *Sheet Met. Ind.*, vol. 43, pp. 479–484, 1966.
[2] M. Grumbach and G. Pomey, *ibid.*, vol. 43, pp. 515–529, 1966; ASTM Standard E517-74.

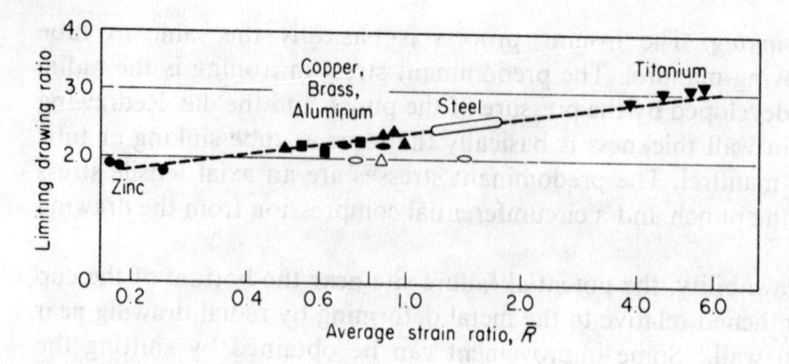

Figure 20-17 Correlation between limiting draw ratio and $\bar{R}$ for a wide range of sheet metals. *(From M. Atkinson, Sheet Met. Ind., vol. 44, p. 167, 1967.)*

Example A tension test on a special deep-drawing steel showed a 30-percent elongation in length and a 16-percent decrease in width. What limiting draw ratio would be expected for the steel?

$$\frac{L - L_0}{L_0} = 0.30 \qquad \frac{w - w_0}{w_0} = -0.16$$

$$\frac{L}{L_0} = 1.30 \qquad \frac{w}{w_0} = 1 - 0.16 = 0.84$$

$$R = \frac{\ln(w_0/w)}{\ln((w/w_0)(L/L_0))} = \frac{\ln(1/0.84)}{\ln(0.84 \times 1.30)} = \frac{\ln 1.190}{\ln 1.092} = 1.98$$

From Fig. 20-17, LDR ≈ 2.7.

Whiteley[1] first demonstrated the importance of $\bar{R}$ in drawability. An example of this type of correlation is given in Fig. 20-17. The ideal crystallographic orientation[2] to maximize $\bar{R}$ in bcc metals would be a sheet texture with ⟨111⟩ in the direction normal to the sheet and {111} oriented randomly in the plane of the sheet. Most steels fall in the range $R = 1$ to $R = 2$. The theoretical limit on $\bar{R}$ for bcc metals[3] appears to be 3. The best way to understand the role of $\bar{R}$ in deep drawing is with the use of a yield locus diagram. We need to relate the stress states at the failure site on the cup wall and in the flange to the yield locus. As Fig. 20-14 shows (neglecting the low contact pressure), in the cup wall both stresses are tensile and the deformation is essentially plane strain. This requires that $\sigma_1 = 2\sigma_2$, and on the yield locus for an isotropic sheet ($R = 1$) (Fig. 20-18),

[1] R. L. Whiteley, *Trans. Am. Soc. Met.*, vol. 52, pp. 154–169, 1960.
[2] D. V. Wilson, *Metall. Rev.*, no. 139, pp. 175–188, 1969.
[3] H. R. Piehler, Sc.D. thesis, Massachusetts Institute of Technology, Cambridge, 1967.

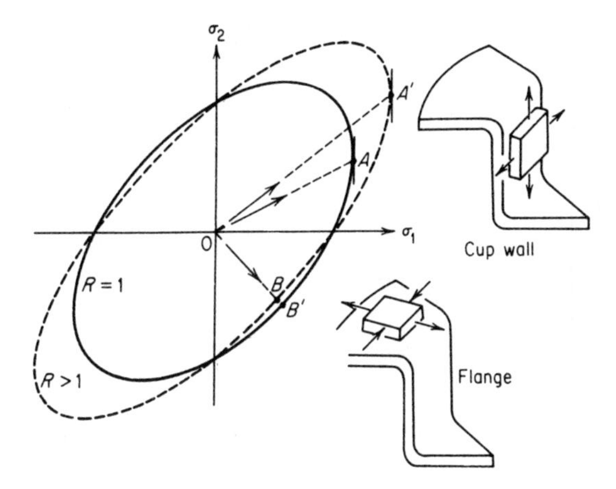

Figure 20-18 Comparison of the load paths on the yield locus for isotropic ($R = 1$) and crystallographically strengthened ($R > 1$) sheet.

it corresponds to the loading vector **OA**. The stress state in the flange is compression in the hoop direction and tension in the radial direction. If thickening can be neglected, the stress state approaches pure shear, vector **OB** on the yield locus. The yield locus for a sheet with $R > 1$ is shown dotted in Fig. 20-18. There hardly is any difference in the length of the load path (OB') for the pure shear stress state in the flange, but the load path (OA') for the plane-strain tension in the cup wall is increased appreciably. This shows that the cup wall is strengthened relative to the flange by texture control, and drawability is increased.

20-7 FORMING LIMIT CRITERIA

Because of the complexity of sheet-forming operations, simple mechanical property measurements made from the tension test are of limited value. Over the years a number of laboratory tests have been developed to evaluate the formability of sheet materials[1]. The Swift flat-bottom cup test[2] is a standardized test for deep drawing. The drawability is expressed in terms of the *limiting draw ratio*. In the Olsen and Erichsen tests[3] the sheet is clamped between two ring dies while a punch, usually a ball, is forced against the sheet until it fractures. The depth of the bulge before the sheet fractures is measured. These tests subject the sheet primarily to stretching, while the Swift test provides nearly pure deep drawing. However, most practical sheet-forming operations provide a combination of both

[1] B. Taylor, Sheet Formability, "Metals Handbook," 9th ed., vol. 8, pp. 547–570, American Society for Metals, Metals Park, Ohio, 1985.

[2] O. H. Kemmis, *Sheet Met. Ind.*, vol. 34, pp. 203, 251, 1957.

[3] "Metals Handbook," 8th ed., vol. 1, pp. 322–323, American Society for Metals, Metals Park, Ohio, 1961.

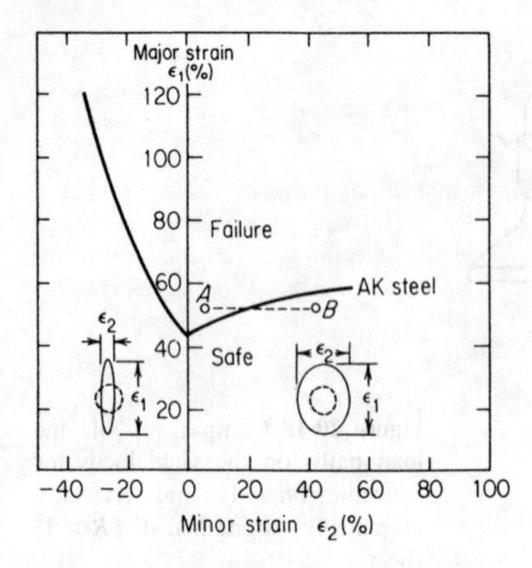

Figure 20-19 Keeler-Goodwin forming limit diagram.

biaxial stretching and deep drawing. The Fukui test,[1] which produces a conical cup using a hemispherical punch, provide a combination of both stretching and drawing.

A useful technique for controlling failure in sheet-metal forming is the *forming limit diagram*.[2] The surface of the sheet is covered by a grid of circles, produced by electrochemical marking.[3] When the sheet is deformed, the circles distort into ellipses. The major and minor axes of an ellipse represent the two principal strain directions in the stamping. The strain in these two directions is measured by the percentage change in the lengths of the major and minor axes. These strains, at any point on the surface, are then compared with the Keeler-Goodwin diagram for the material (Fig. 20-19). Strain states above the curve represent failure, those below do not cause failure. For example, point A is a failure, but if the strain distribution is altered (perhaps by changing a die radius), it could move to point B, which would not cause failure. The failure curve for the tension-tension region was determined by Keeler[4] and is nearly fixed for a variety of low-carbon steels. Other metals such as aluminum[5] have a different curve. The tension-compression region was first determined by Goodwin.[6]

Another approach to predicting sheet formability is the stretch-draw shape analysis.[7] The forming limit of the material is established with the Olsen test and the Swift cupping test. Then the part is broken down into simple shapes and the percentage of draw and stretch are calculated from the geometry. This places the

[1] J. C. Gerdeen and A. Daudi, *Trans. ASME J. Eng. Ind.*, vol. 105, pp. 276–281, 1983.

[2] S. P. Keeler, *Met. Prog.*, October 1966, pp. 148–153.

[3] S. Dinda, K. F. James, and S. P. Keeler, "How to Use Circle Grid Analysis for Die Tryout," American Society for Metals, Metals Park, Ohio, 1980.

[4] S. P. Keeler, *Sheet Met. Ind.*, vol. 42, pp. 683–691, 1965.

[5] S. S. Hecker, *Trans. ASME, Ser. H: J. Eng. Mater. Technol.*, vol. 97, pp. 66–73, 1975; J. D. Embury and J. L. Duncan, *Ann. Rev. Mater. Sci.*, vol. 11, pp. 505–521, 1981.

[6] G. M. Goodwin, *Metall. Ital.*, vol. 60, pp. 767–774, 1968.

[7] A. S. Kasper, *Met. Prog.*, May 1971, pp. 57–60; "Tool and Manufacturing Engineers Handbook," vol. 2, op. cit., pp. 1-24 to 1-33.

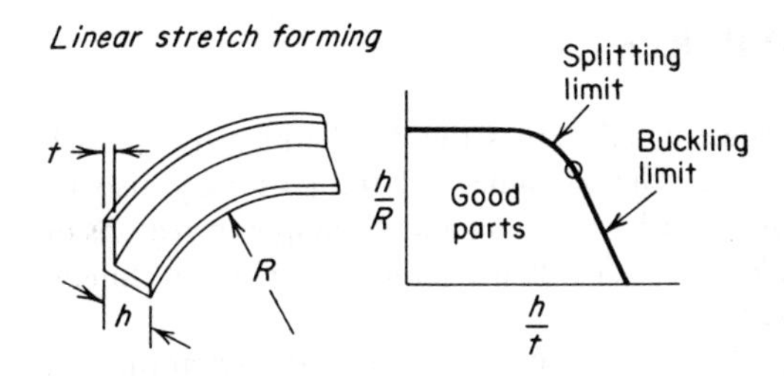

Figure 20-20 Formability chart for linear stretch forming. Each material would have to be calibrated, but the form of the chart would be the same. *(After Wood et al.)*

part on the forming limit diagram and the degree of severity of the part can be established.

Useful formability charts have been developed[1] for such processes as brake forming, joggling, dimpling, linear stretch forming, rubber forming, beading, and spinning. Generally, these are processes used in aerospace manufacture. Critical dimensional parameters were established for each process. An example for linear stretch forming is shown in Fig. 20-20.

Just as distributed computing has made the analysis of the deformation in bulk deformation processes a practical reality, so is computer-aided design and analysis making inroads in the field of sheet-metal forming.[2] The computer-aided engineering system is capable of predicting whether a particular sheet-metal part can be formed successfully based on the working drawings for the part. The computer analysis program consists of constitutive relations for the material, a program for analysis of the forming limit diagram (FLD), and a finite analysis program for determining the critical strains. The data for the FLD can either be experimentally determined curves for the sheet material or the FLD can be computed.[3]

Example A grid of 0.1-in circles is electroetched on a blank of sheet steel. After forming into a complex shape the circle in the region of critical strain is distorted into an ellipse with major diameter 0.18 in and minor diameter 0.08 in. How close is the part to failing in this critical region?

Major strain $$e_1 = \frac{0.18 - 0.10}{0.10} \times 100 = 80\%$$

Minor strain $$e_2 = \frac{0.08 - 0.10}{0.10} \times 100 = -20\%$$

When those coordinates are placed on the forming limit diagram for AK steel we see that the part is in imminent danger of failure.

[1] W. W. Wood et al., "Theoretical Formability," AFML-TR-64-411, January 1965.

[2] D. Lee, General Electric Corp., Corporate Research and Development, Report No. 82CRD131, May 1982.

[3] D. Lee and F. Zaverl, *Int. J. Mech. Sci.*, vol. 24, p. 154, 1982.

20-8 DEFECTS IN FORMED PARTS

The ultimate defect in a formed sheet-metal part is the development of a crack which destroys its structural integrity. The usefulness of the part may also be destroyed by local necking or thinning or by buckling and wrinkling in regions of compressive stress. Another troublesome defect is the failure to maintain dimensional tolerances because of springback.

In the deep drawing of a cup the most common failure is the separation of the bottom from the rest of the cup at the location of greatest thinning near the punch radius. This defect may be minimized either by reducing the thinning by using a larger punch radius or by decreasing the punch load required for the drawing operation. If radial cracks occur in the flange or the edge of the cup, this indicates that the metal does not have sufficient ductility to withstand the large amount of circumferential shrinking that is required in this region of the blank. This type of failure is more likely to occur in redrawing without annealing than on the initial draw.

Wrinkling of the flange or the edges of the cup results from buckling of the sheet as a result of the high circumferential compressive stresses. In analyzing this type of failure each element in the sheet can be considered as a column loaded in compression. If the blank diameter is too large, the punch load will rise to high values, which may exceed the critical buckling load of the column. Since column stability decreases with an increasing slenderness ratio, the critical buckling load will be achieved at lower loads for thin sheet. To prevent this defect, it is necessary to use sufficient hold-down pressure to suppress the buckling.

Since sheet-metal formed parts usually present a large surface area, they are particularly susceptible to surface blemishes which detract from the appearance of the part. Pronounced surface roughness in regions of the part which have undergone appreciable deformation is usually called *orange peeling*. The orange-peel effect occurs in sheet metal of relatively large grain size. It results from the fact that the individual grains tend to deform independently of each other, and therefore grains stand out in relief on the surface. This condition is best corrected by using finer-grain-size sheet metal so that the grains deform more nearly as a whole and the individual grains are difficult to distinguish with the eye.

Another serious surface defect that is commonly found in low-carbon sheet steel is the presence of *stretcher strains*, or "worms." This defect shows up as a flamelike pattern of depressions in the surface (Fig. 20-21). The depressions first appear along planes of maximum shear stress, and then, as deformation continues, they spread and join together to produce a uniform rough surface. The existence of stretcher strains is directly associated with the presence of a yield point in the stress-strain curve and the nonuniform deformation which results from the yield-point elongation (Fig. 20-22). The metal in the stretcher strains has been strained an amount equal to B in Fig. 20-22, while the remaining metal has received essentially zero strain. The elongation of the part is given by some intermediate strain A. As the deformation continues and the number of stretcher strains increase, the strain will increase until when the entire part is covered it has

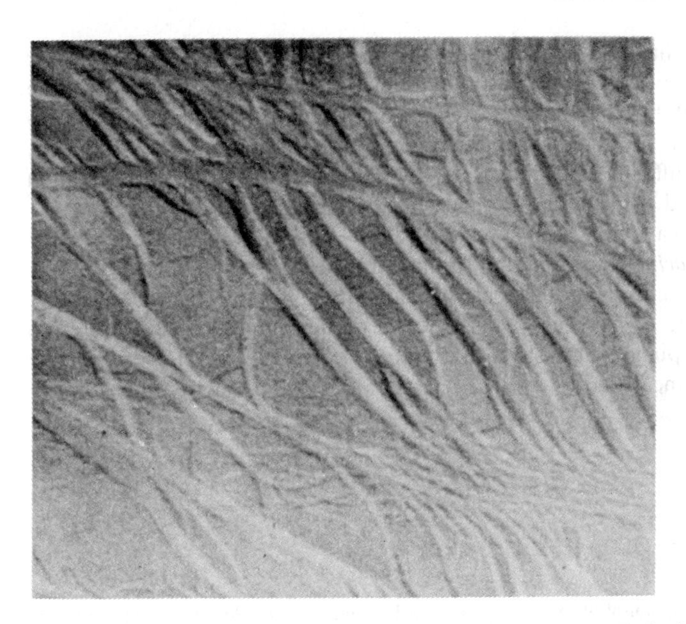

Figure 20-21 Stretcher strains in low-carbon steel sheet. *(Courtesy E. R. Morgan and Met. Prog., June, 1958, p. 89.)*

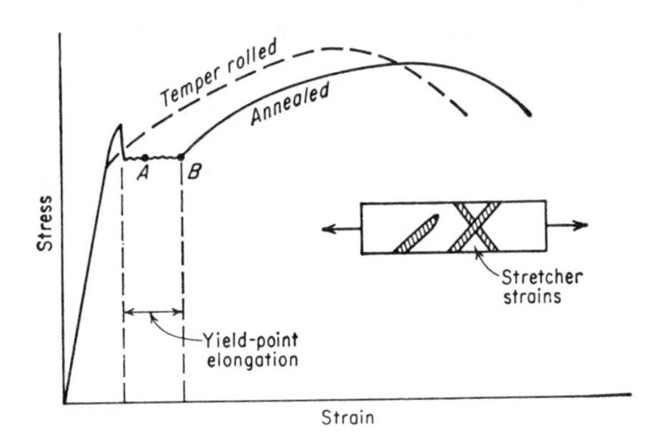

Figure 20-22 Relation of stretcher strains to stress-strain curve.

a strain equal to *B*. Beyond this strain the deformation is uniform and homogeneous. The main difficulty with stretcher strains, therefore, occurs in regions of the part where the strain is less than the yield-point elongation. The usual solution to this problem is to give the sheet steel a small cold reduction, usually $\frac{1}{2}$ to 2 percent reduction in thickness. Such a temper-rolling or skin-rolling treatment cold-works the metal sufficiently to eliminate the yield point. However, if the steel strain-ages during storage, the yield point will return and difficulties with stretcher strains will reappear.

The directionality in mechanical properties produced by rolling and other primary working processes can have an important effect on the fabricability of the metal. Mechanical fibering has little effect on formability, whereas crystallographic fibering or preferred orientation may have a large effect. Ordinarily, bending is more difficult when the bend line is parallel to the rolling direction than when the bend is made perpendicular to the rolling direction.

One of the ways that directionality shows up in deep drawing is the phenomenon of *earing*. Earing is the formation of a wavy edge on the top of a drawn cup which necessitates extensive trimming to produce a uniform top. Usually two, four, or six ears will be formed, depending on the preferred orientation in the plane of the sheet. Earing can be directly correlated[1] with the *planar anisotropy*, measured by ΔR

$$\Delta R = \frac{R_0 + R_{90} - 2R_{45}}{2} \tag{20-17}$$

BIBLIOGRAPHY

Alexander, J. M.: An Appraisal of the Theory of Deep Drawing, *Metall. Rev.*, vol. 5, pp. 349–411, 1960.

Eary, D. F. and E. A. Reed: "Techniques of Pressworking Sheet Metal," 2d ed. Prentice-Hall, Inc., Englewood Cliffs, N.J., 1974.

Keeler, S. P.: "Understanding Sheet Metal Formability, *Machinery*, February through July, 1968.

Koistinen, D. P. and N. M. Wang (eds.): "Mechanics of Sheet Metal Forming," Plenum Press, New York, 1978.

"Metals Handbook," 8th ed., vol. 4, Forming, American Society for Metals, Metals Park, Ohio, 1967.

Sachs, G.: "Principles and Methods of Sheet-Metal Fabricating," 2d ed., revised by H. E. Voegel, Reinhold Publishing Corporation, New York, 1966.

Strasser, F.: "Functional Design of Metal Stampings," Society of Manufacturing Engineers, Dearborn, MI, 1971.

Wick, C., J. T. Benedict, and R. F. Veilleux (eds.): "Tool and Manufacturing Engineers Handbook," 4th ed., vol. 2, chaps. 1–11, Society of Manufacturing Engineers, Dearborn, Mich., 1984.

[1] D. V. Wilson and R. D. Butler, *J. Inst. Met.*, vol. 90, p. 473, 1961–1962.

CHAPTER

TWENTY ONE

MACHINING OF METALS

21-1 TYPES OF MACHINING OPERATIONS

In the broadest sense there are two distinct classes of solid-state manufacturing processes. *Deformation processes* produce the required shape with the necessary mechanical properties by plastic deformation in which the material is moved and its volume is conserved. *Machining processes* produce the required shape by removal of selected areas of the workpiece through a machining process. Most machining is accomplished by straining a local region of the workpiece to fracture by the relative motion of the tool and the workpiece. Although mechanical energy is the usual input to most machining processes, some of the newer metal-removal processes employ chemical, electrical, or thermal energy.

The inexperienced reader should not be misled about the importance of machining processes to the industrial world because only one chapter is devoted to this subject. This choice partly reflects the author's own areas of interest and partly the fact that the deformation processes historically have been of more interest to the metallurgist, while machining processes have been of prime concern to mechanical and industrial engineers. There is no real rationale for this situation since, as will be seen shortly, machining processes depend critically on the tool material and the interaction of the machining forces with the workpiece material. In the space available only the briefest introduction to the technology of machining operations is possible. Emphasis is on fundamentals of the mechanics and the material response to more generalized machining processes.

Machining usually is employed to produce shapes with high dimensional tolerance, good surface finish, and often with complex geometry. Machining is a *secondary* processing operation since it usually is conducted on a workpiece that was produced by a primary process such as hot rolling, forging, or casting.

Something more than 80 percent of all manufactured parts must be machined before they are completed. The variety of machining processes and machine tools that can be utilized is very great. Since the development of machine tools paralleled the industrialization of our society, it is an old field with much specialized terminology and jargon. The student is referred to texts on machine tool practice[1] for details of machine tool construction and performance. Other reference sources[2] should be consulted for details on specific machining operations and selection of speeds, feeds, tool geometry and material, and cutting fluid.

Some generalization can be brought to this complex subject by considering how the workpiece is held in the machine and what the relative movements are between the workpiece and the tool. The ease and accuracy with which a surface can be machined depends on how well the surface can be matched to the movements of the machine and the cutting edge of the tool. Machine tools generate surfaces in two ways: (1) by using a *form tool* which matches the shape to be produced in the surface, e.g., a tool with a notch in its cutting edge produces a raised rib on the machined surface, and (2) feeding the cutting tool in and out along the length of the workpiece. Form tools lack flexibility and have high relative cost, but they are important in high-volume production. Machining by path cutting with a single-edge or multiple-edge tool requires the movement of the tool in two directions in order to generate shapes other than a simple cylinder. The *primary motion* is the main motion provided by the machine tool to cause relative motion between the tool and the workpiece so that the face of the tool approaches the workpiece. The *feed motion* is a motion provided to the tool or workpiece which when added to the primary motion leads to a repeated or continuous chip removal and the creation of a surface with the desired geometry. The feed motion absorbs a much smaller portion of the total power than the primary motion.

With only a few exceptions, machine tools may be grouped in to two broad categories: those that generate surfaces of rotation and those that generate flat or formed surfaces by linear movement. Figure 21-1*a* shows the relative motions needed to produce a machined surface on a turning machine such as a lathe. The principal parameters are the *cutting speed* v, the *depth of cut* d, and the *feed* f. To turn a cylindrical surface of length L_w requires L_w/f revolutions of the workpiece. If the number of revolutions of the workpiece per minute is n_w, then the time for one *pass* is $t = L_w/fn_w$. If the tool operates on the inside of a hollow cylinder (Fig. 21-1*b*), the process is called *boring*. Very heavy workpieces are machined on

[1] C. R. Hine, "Machine Tools and Processes for Engineers," 2d ed., McGraw-Hill Book Company, New York, 1959; S. Kalpakjian, "Manufacturing Processes for Engineering Materials," Addison-Wesley Publishing Company, Reading, Mass., 1984; R. A. Lindberg, "Processes and Materials of Manufacture," Allyn and Bacon, Inc., Boston, 1983.

[2] "Metals Handbook," 8th ed., vol. 3, Machining, American Society for Metals, Metals Park, Ohio, 1967; "Machining Data Handbook," 3d ed., Metcut Research Associates Inc., Cincinnati, Ohio, 1980; "Tool and Manufacturing Engineers Handbook," 4th ed., vol. 1, Society of Manufacturing Engineers, Dearborn, Mich., 1983.

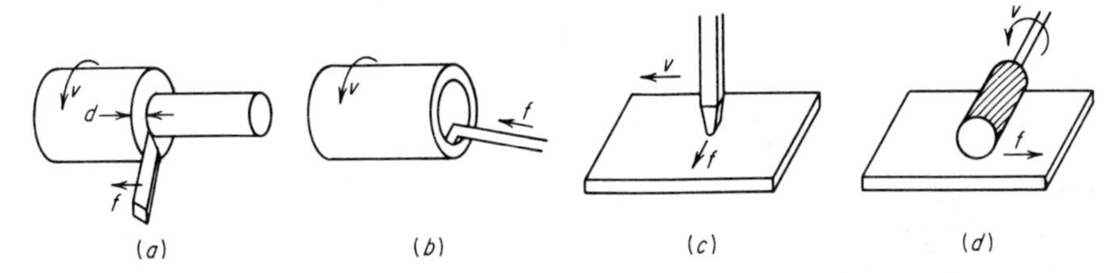

Figure 21-1 Schematic of some common machining operations. (*a*) Turning on a lathe; (*b*) boring; (*c*) shaping; (*d*) slab milling.

vertical boring mills where they are rotated about the vertical axis. *Shaping* is a process in which a single-point tool reciprocates over the workpiece in a linear manner (Fig. 21-1*c*). One of the most versatile processes for producing flat surfaces is *milling* (Fig. 21-1*d*). The milling cutter is a multiple-edge tool.

21-2 MECHANICS OF MACHINING

A simple model of two-dimensional machining (orthogonal cutting) is shown in Fig. 21-2. The tool is a single-point tool that is characterized by the *rake angle* α, the *clearance* (relief) angle θ, and the wedge angle ω. $\alpha + \omega + \theta = 90°$. The *rake face* of the tool is the surface over which the chip flows as it leaves the machined surface. The forces imposed on the tool create intense shearing action on the metal ahead of the tool. The metal in the chip is severely deformed on going from an undeformed chip of thickness t to the deformed chip with thickness t_c. The *chip thickness ratio* (also the cutting ratio) $r = t/t_c$ is an easily measured parameter which is typically around $r \approx 0.5$. The creation of the chip subjects the metal to large plastic shear strains ($\gamma = 2$ to 4) at strain rates above 10^3 s^{-1}. There is a localized region of intense shear in the vicinity of *OA*, but for simplicity this usually is represented by a well-defined shear plane *OA* occurring at a *shear angle* ϕ.

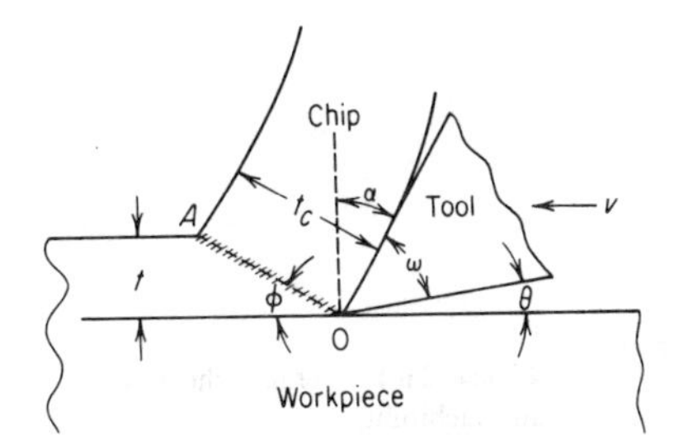

Figure 21-2 Idealized model for chip formation. Rake angle α as shown above is positive rake.

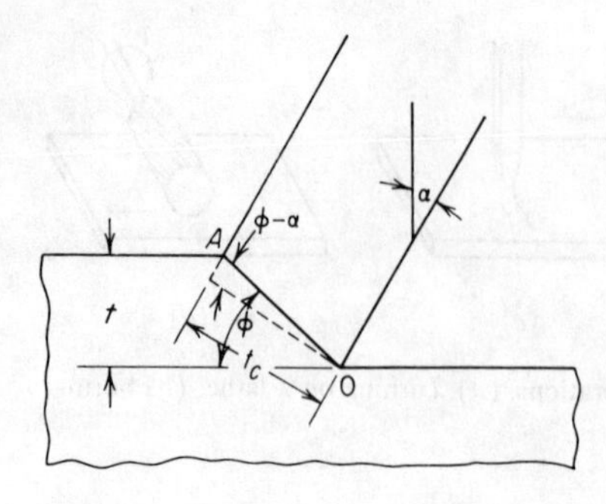

Figure 21-3 Relationships expressed in Eq. (21-1).

The relationship between rake angle, shear angle, and chip thickness ratio can be derived from Fig. 21-3.

$$r = \frac{t}{t_c} = \frac{OA \sin \phi}{OA \cos (\phi - \alpha)} \tag{21-1}$$

and on solving for ϕ

$$\tan \phi = \frac{r \cos \alpha}{1 - r \sin \alpha} \tag{21-2}$$

Thus, the chip thickness ratio r determined after the cut has been made may be used to establish the shear angle ϕ existing during the machining operation. Since volume is constant in plastic deformation, and chip width b is essentially constant, we could also obtain r from measurements of the ratio of chip length L_c to the length of the workpiece from which it came, L_w. If L_c is unknown, it can be determined by measuring the weight of chips W_c, and by knowing the density of the metal ρ, $L_c bd = W_c/\rho$.

If the deformation in cutting is assumed to be a simple block-like shearing process, we can estimate the shearing strain in machining using Fig. 21-4. An element originally defined by $ABCD$ is sheared to $EBCF$. The shear strain γ is the shear offset divided by the perpendicular distance between shear planes.

$$\gamma = \frac{AE}{OB} = \frac{AO}{OB} + \frac{OE}{OB} = \tan(\phi - \alpha) + \cot \phi \tag{21-3}$$

For chips without curl there are three velocities which must be specified in cutting. The *cutting speed* v is the velocity of the tool relative to the workpiece.

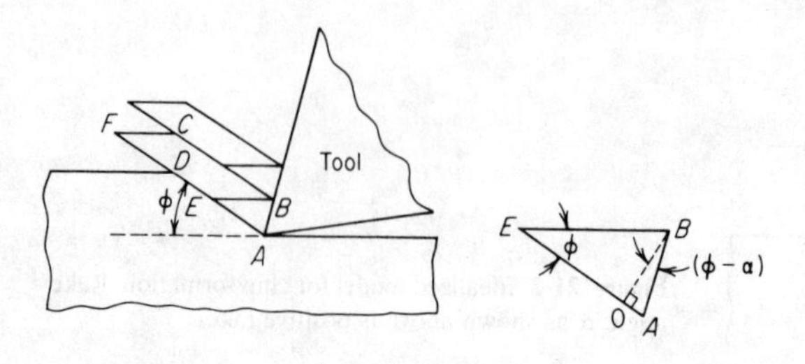

Figure 21-4 Average shear strain in machining.

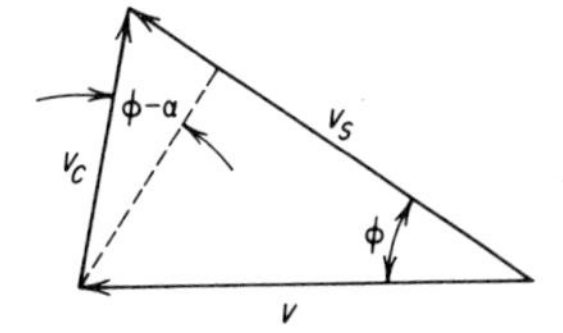

Figure 21-5 Velocity relationships in orthogonal machining.

The velocity of the chip relative to the tool face is the *chip velocity* v_c. The velocity of the chip relative to the work is the *shear velocity* v_s. From continuity of mass, $vt = v_c t_c$, so

$$r = \frac{t}{t_c} = \frac{v_c}{v} \tag{21-4}$$

The three velocity components give the kinematic relationship shown in Fig. 21-5, from which we see that the vector sum of the cutting velocity and the chip velocity is equal to the shear velocity vector. It can be shown from the geometry that

$$v_s = \frac{v \cos \alpha}{\cos(\phi - \alpha)} \tag{21-5}$$

Knowing v_s, we can establish the shear-strain rate in cutting.

$$\dot{\gamma} = \frac{d\gamma}{dt} = \frac{v_s}{(y_s)_{\max}} \tag{21-6}$$

where $(y_s)_{\max}$ is the estimate of the maximum value of the thickness of the shear zone, approximately 10^{-3} in. Using realistic values of $\phi = 20$, $\alpha = 5°$, $v = 600$ fpm and $(y_s)_{\max} \approx 10^{-3}$ in, we calculate $\dot{\gamma} = 1.2 \times 10^5\ \text{s}^{-1}$, which is several orders of magnitude greater than the strain rate usually associated with high-speed metal-working operations.

We are now ready to consider the forces and stresses acting in metal cutting.[1] In Fig. 21-6, P_R is the resultant force between the tool face and the chip and P'_R is the equal resultant force between the workpiece and the chip along the shear plane. These resultant forces can be resolved conveniently into several sets of components. An obvious choice is the component tangential, F_t, and normal, F_n, to the rake face of the tool. However, the force in metal cutting frequently is measured by a strain-gage toolpost dynamometer[2] which independently measures the horizontal (cutting) F_h and vertical (thrust) F_v forces in cutting. It can be shown that

$$\begin{aligned} F_t &= F_h \sin \alpha + F_v \cos \alpha \\ F_n &= F_h \cos \alpha - F_v \sin \alpha \end{aligned} \tag{21-7}$$

If the components of cutting force are known, then the coefficient of friction in

[1] M. E. Merchant, *J. Appl. Phys.*, vol. 16, pp. 207–318, 1945.

[2] E. G. Loewen, E. R. Marshall, and M. C. Shaw, *Proc. Soc. Exp. Stress Anal.*, vol. 8, p. 1, 1951; G. Boothroyd, *Engineer*, vol. 213, p. 351, 1962.

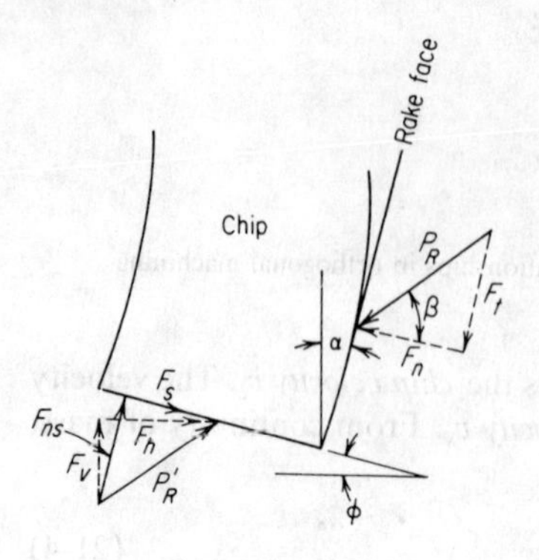

Figure 21-6 Force components in orthogonal cutting.

the tool face is given by

$$\mu = \tan \beta = \frac{F_t}{F_n} = \frac{F_v + F_h \tan \alpha}{F_h - F_v \tan \alpha} \tag{21-8}$$

Finally, the resultant force may be resolved parallel, F_s, and normal, F_{ns}, to the shear plane. In terms of the usual measured force components

$$\begin{aligned} F_s &= F_h \cos \phi - F_v \sin \phi \\ F_{ns} &= F_h \sin \phi + F_v \cos \phi \end{aligned} \tag{21-9}$$

The average shear stress is F_s divided by the area of the shear plane $A_s = bt/\sin \phi$.

$$\tau = \frac{F_s}{A_s} = \frac{F_s \sin \phi}{bt} \tag{21-10}$$

and the normal stress is

$$\sigma = \frac{F_{ns}}{A_s} = \frac{F_{ns} \sin \phi}{bt} \tag{21-11}$$

The shear stress in cutting is the main parameter affecting the energy requirement. Reasonable values of this shear stress have been predicted[1] by a dislocation model which considers high strain rates and very large strains.

Calculation of the shear stress in cutting from force measurements requires knowledge of the shear angle ϕ. This angle can be measured experimentally by suddenly stopping the cutting process and using metallographic techniques to determine the shear zone. However, it would be convenient to be able to predict ϕ from other more easily determined parameters. Merchant[2] assumed that the shear plane would be at the angle which minimizes the work done in cutting. This results in the expression

$$\phi = \frac{\pi}{4} + \frac{\alpha}{2} - \frac{\beta}{2} \tag{21-12}$$

[1] B. F. von Turkovitch, *Trans. ASME, Ser. B: J. Eng. Ind.*, vol. 92, pp. 151–157, 1970.
[2] M. E. Merchant, op. cit.

Experience in metal cutting research shows that, when machining different materials, a precise value of shear plane angle cannot be predicted. Rather, it is more realistic to predict a ϕ range which depends on the composition and heat treatment of the material. Based on an upper-bound model of the shear zone[1] a criterion for predicting ϕ has been developed[2] which accounts for strain hardening in the chip. First the lowest upper bound to cause shear in a severely strain-hardened material (of shear strength k) is calculated. This deformation requires a certain power input. This same power input is then applied to unstrained material (of shear strength k_0) and the angle of collapse instability is calculated. This is the shear plane angle for the softer material and it also predicts the shear plane angle for the strain-hardening material. The predicted shear plane angle ϕ_0 is given by

$$\cos(\phi_0 - \alpha)\sin\phi_0 = \frac{k_0}{k_1}\left[\cos\left(45 - \frac{\alpha}{2}\right)\sin\left(45 + \frac{\alpha}{2}\right)\right] \tag{21-13}$$

where $\alpha =$ rake angle

$k_0 = \sigma_0/\sqrt{3}$ and σ_0 is the yield strength of the material

$k_1 = s_u/\sqrt{3}$ and s_u is the tensile strength of the material

Equation (21-13) has shown excellent agreement with experimental values for different materials.

Example Determine the shear plane angle in orthogonal machining with a 6° positive rake angle for hot-rolled AISI 1040 steel and annealed commercially pure copper.

Hot rolled 1040 steel	$\sigma_0 = 60{,}000$ psi	$s_u = 91{,}000$ psi
Annealed copper	$\sigma_0 = 10{,}000$ psi	$s_u = 30{,}000$ psi

$$\cos(\phi_0 - 6)\sin\phi_0 = \frac{k_0}{k_1}\left[\cos\left(45 - \frac{6}{2}\right)\sin\left(45 + \frac{6}{2}\right)\right]$$

$$\cos(\phi_0 - 6)\sin\phi_0 = \frac{k_0}{k_1}(0.552)$$

$$\frac{1}{2}[\sin 6° + \sin(2\phi_0 - 6)] = \frac{k_0}{k_1}(0.552)$$

$$2\phi_0 = \left[\sin^{-1}\left(1.104\frac{k_0}{k_1} - 0.1045\right)\right] + 6$$

Note that since k_0/k_1 is a fraction we can use tensile values directly in the above equations.

[1] G. W. Rowe and F. Wolstencroft, *J. Inst. Met.*, vol. 98, pp. 33–41, 1970.

[2] P. K. Wright, *Trans. ASME J. Eng. Ind.*, vol. 104, pp. 285–292, 1982.

For hot-rolled 1040 steel:

$$2\phi_0 = \left[\sin^{-1}\left(1.104 \times \frac{60}{91}\right) - 0.1045\right] + 6 = \sin^{-1}(0.6234) + 6° = 44.6°$$

$\phi_0 = 22.3°$ experimental range is 23 to 29°

For commercial copper (annealed):

$$2\phi_0 = \left[\sin^{-1}\left(1.104 \times \frac{10}{30}\right) - 0.1045\right] + 6 = \sin^{-1}(0.2635) + 6° = 21.3°$$

$\phi_0 = 10.6°$ experimental range is 11 to 13.5°

This model does not allow for strain rate effects, which would raise k_0 more than k_1 and lead to somewhat higher values of ϕ_0.

The power (energy per unit time) required for cutting is $F_h v$. The volume of metal removed per unit time (metal-removal rate) is $Z_w = btv$. Therefore, the energy per unit volume U is given by

$$U = \frac{F_h v}{Z_w} = \frac{F_h v}{btv} = \frac{F_h}{bt} \tag{21-14}$$

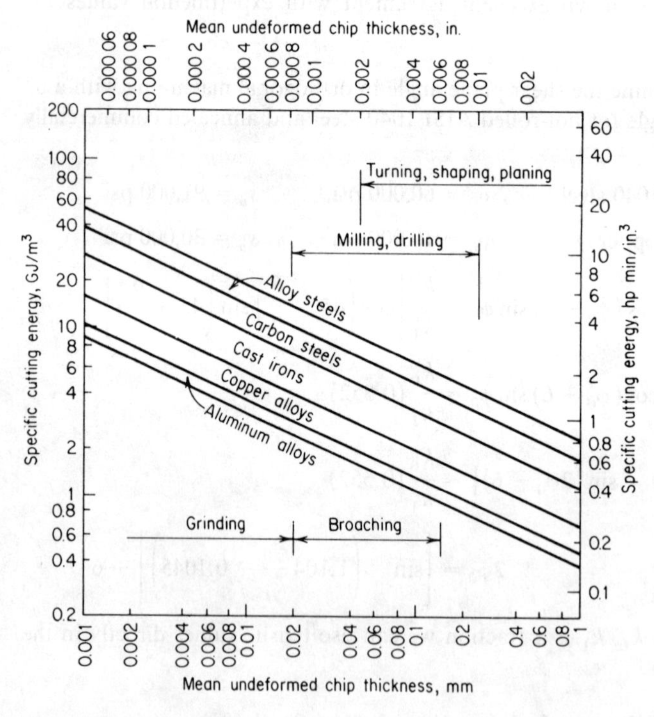

Figure 21-7 Force values of specific cutting energy for various materials and machining operations. *(From G. Boothroyd, "Fundamentals of Metal Machining and Machine Tools," p. 57, McGraw-Hill Book Company, New York, 1975.)*

where b is the width of the chip, and t is the undeformed chip thickness. This energy involved in removing a unit volume of metal often is referred to as the *specific cutting energy*. The specific cutting energy depends on the material being machined and also on the cutting speed, feed, rake angle, and other machining parameters. However, at high cutting speeds (greater than about 600 ft/min) it is independent of speed, and thus provides an approximate but useful indication of the forces required for machining. Figure 21-7 summarizes typical values of specific cutting energy for various materials and machining processes. The total energy for cutting can be divided into a number of components:

1. The energy required to produce the gross deformation in the shear zone,
2. The frictional energy resulting from the chip sliding over the tool face,
3. Energy required to curl the chip,
4. Momentum energy associated with the momentum change as the metal crosses the shear plane,
5. The energy required to produce the new surface area.

In a typical machining operation, the shear energy will be about 75 percent of the total, the frictional energy about 25 percent, and the energy associated with the other three components is negligible.

Example In an orthogonal cutting process $v = 500$ ft/min, $\alpha = 6°$, and the width of cut is $b = 0.40$ in. The underformed chip thickness is 0.008 in. If 13.36 g of steel chips with a total length of 20 in are obtained, what is the slip plane angle? If a toolpost dynamometer gives cutting and thrust forces of $F_h = 250$ lb and $F_v = 100$ lb, determine the percentage of the total energy that goes into overcoming friction at the tool-chip interface and the percentage that is required for cutting along the shear plane.

Thickness of chip

$$t_c = \frac{m}{\rho b L} = \frac{13.36 \text{ g}/(454 \text{ g/lb})}{(0.283 \text{ lb/in}^3) \times 0.40 \text{ in} \times 20 \text{ in}}$$

$$t_c = 0.013 \text{ in}$$

Chip thickness ratio

$$r = t/t_c = \frac{0.008}{0.013} = 0.615$$

From Eq. (21-2)

$$\tan\phi = \frac{r\cos\alpha}{1 - r\sin\alpha} = \frac{0.615\cos 6°}{1 - 0.615\sin 6°} = 0.6536$$

$$\phi = 33°$$

From Eq. (21-8)

$$\mu = \tan\beta = \frac{F_t}{F_n} = \frac{F_v + F_h \tan\alpha}{F_h - F_v \tan\alpha} = \frac{100 + 250\tan 6°}{250 - 100\tan 6°} = 0.527$$

$$\beta = 27.8°$$

From Figs. 21-5 and 21-6, the power required to overcome friction at the tool-chip interface is the product of F_t and v_c. The frictional specific energy is $U_f = F_t v_c/btv = F_t r/bt = (F_h \sin\alpha + F_v \cos\alpha)r/bt$. Likewise, the energy required for cutting along the shear plane is $U_s = F_s v_s/btv$ and the total specific energy is $U = U_f + U_s$.

$$\frac{\text{Friction energy}}{\text{Total energy}} = \frac{U_f}{U} = \frac{F_t v_c}{F_h v} = \frac{F_t}{F_h} r$$

From Fig. 21-6, $F_t = P_R \sin\beta$ and $P_R = P'_R = \sqrt{F_v^2 + F_h^2}$

$$P_R = \sqrt{(100)^2 + (250)^2} = 269.2 \text{ lb}$$

$$F_t = 269.2 \sin 27.8° = 125.6 \text{ lb}$$

$$\%\text{ friction energy} = \frac{125.6(0.615)}{250} \times 100 = 30.9\%$$

$$\frac{\text{Shearing energy}}{\text{Total energy}} = \frac{F_s v_s}{F_h v}$$

$$F_s = F_h \cos\phi - F_v \sin\phi = 250\cos 33° - 100 \sin 33° = 155.2 \text{ lb}$$

$$v_s = \frac{v\cos\alpha}{\cos(\phi - \alpha)} = \frac{500 \cos 6°}{\cos(33 - 6)} = 558 \text{ ft/min}$$

$$\%\text{ shearing energy} = \frac{155.2 \times 558}{250 \times 500} \times 100 = 69.1\%$$

The total energy per unit volume is

$$U = \frac{F_h v}{btv} = \frac{250 \times 500}{33{,}000} \times \frac{1}{0.40(0.008)500 \times 12} = 0.20 \frac{\text{hp min}}{\text{in}^3}$$

where 1 hp = 33,000 ft-lb/min. This analysis of energy distribution neglects two other energy requirements in cutting. A small *surface energy* is required to produce the new surfaces. There also is an energy associated with the *momentum change* as the metal crosses the shear plane. Although the momentum energy is negligible at ordinary cutting velocities, it can be significant in high-speed machining at cutting speeds above 25,000 ft/min.

The model of the machining process that has been presented so far is unrealistically simplified. Basically, the model considers two-dimensional (orthogonal) cutting in which a continuous chip is produced by intense shearing over

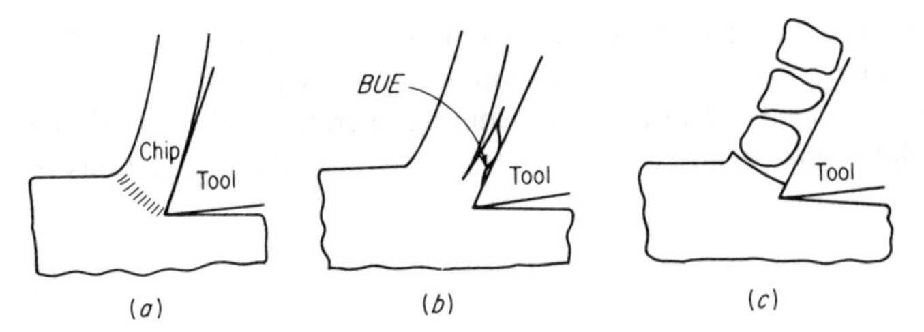

Figure 21-8 Types of machining chips. (*a*) Continuous chip; (*b*) chip with a builtup edge, BUE; (*c*) discontinuous chip.

a localized shear plane. We now shall consider some of the deviations from this simplified model.

Three general classifications of chips are formed in the machining process.[1] The *continuous chip* (type 2 chip) is characteristic of cutting ductile materials under steady-state conditions. Most basic research in metal cutting utilizes this type of chip (Fig. 21-8*a*). However, long continuous chips present handling and removal problems in practical operations. The usual practice is to include a *chipbreaker* on the rake face of the tool to deflect the chip and cause it to break into short lengths. Under conditions where the friction between the chip and the rake face of the tool is high the chip may weld to the tool face. The accumulation of chip material (Fig. 21-8*b*) is known as a *builtup edge* (BUE). The BUE (type 3 chip) acts as a substitute cutting edge and the major part of the contact between the chip and the tool is through the BUE. Periodically the BUE breaks off, part remaining on the back side of the chip and part welding to the machined surface. Thus, BUE is a source of rough surface finish. BUE is most prevalent with high feeds and low rake angle. For a set of machining parameters there usually is a cutting speed above which BUE does not occur. *Discontinuous chips* (type 1 chip) are formed in brittle materials which cannot withstand the high shear strains imposed in the machining process without fracture (Fig. 21-8*c*). Typical materials which machine with discontinuous chips are cast iron and cast brass, but this type of chip also may be produced in ductile materials machined at very low speeds and high feeds.

Forces other than those required to produce the sheared chip are involved in the machining process. No cutting tool is perfectly sharp, so that the tool edge "plows" its way through the work material.[2] For a large depth of cut the ploughing force is a small fraction of the total force, but for small values of depth of cut it becomes an appreciable fraction of the total cutting force. This leads to a "size effect", i.e., the specific cutting energy for processes that remove very thin chips, like grinding, is higher than for processes that produce thicker chips. Also difficult to model are the frictional forces between the chip and the rake face of

[1] For a more detailed classification of chips see G. Boothroyd, "Fundamentals of Metal Machining and Machine Tools," p. 186, McGraw-Hill Book Company, New York, 1975.

[2] P. Albrecht, *Trans. ASME, Ser B, J. Eng. Ind.*, vol. 82, pp. 348–362, 1960.

the tool. Along the area of contact there is a region of sticking friction as well as an area of sliding. The best analysis of the situation is given by Zorev.[1]

Because of the complexity of practical machining operations, the machining force F_h often is related empirically to the machining parameters by equations of the form

$$F_h = kd^a f^b \tag{21-15}$$

where d is the depth of cut, f is the feed, and k is a function of rake angle, decreasing about 1 percent per degree increase in rake angle.

21-3 THREE-DIMENSIONAL MACHINING

To this point we have considered only two-dimensional orthogonal machining in which the cutting edge is perpendicular to the cutting velocity vector. While a few practical machining operations (such as surface broaching, lathe cutoff operations, and plain milling) fall in this category, most practical machining operations are three dimensional. A comparison between two-dimensional and three-dimensional machining is given in Fig. 21-9. For orthogonal machining we see that rotating the tool around the x axis will change the width of the cut, and rotation around the y axis changes the rake angle. However, major changes in the cutting process occur if the tool is rotated about the z axis by an *angle of inclination* i. Now the chip no longer flows up the tool face along a path perpendicular to the cutting edge (path OA), but instead it follows the path OB which is inclined to the normal by the *chip flow angle* η_c. The chip flow angle may be obtained experimentally from the relation

$$\cos \eta_c = \frac{b_c}{b/\cos i} \tag{21-16}$$

where b_c is the width of the chip and b is the width of the workpiece that is being cut.

Three-dimensional machining can be treated as orthogonal machining if the process is analyzed on the plane defined by the cutting velocity v and the chip velocity v_c. The angle zOA is defined as the *normal rake angle* α_n. This is a geometric property of the tool that is independent of i, and for $i = 0$ it equals the rake angle for orthogonal cutting. The *effective rake angle* α_e is the angle between v_c and the normal to v in the plane through v and v_c. From the geometry of the situation, this is given by

$$\sin \alpha_e = \sin \eta_c \sin i + \cos \eta_c \cos i \sin \alpha_n \tag{21-17}$$

[1] N. N. Zorev, "Metal Cutting Mechanics," Pergamon Press, New York, 1966.

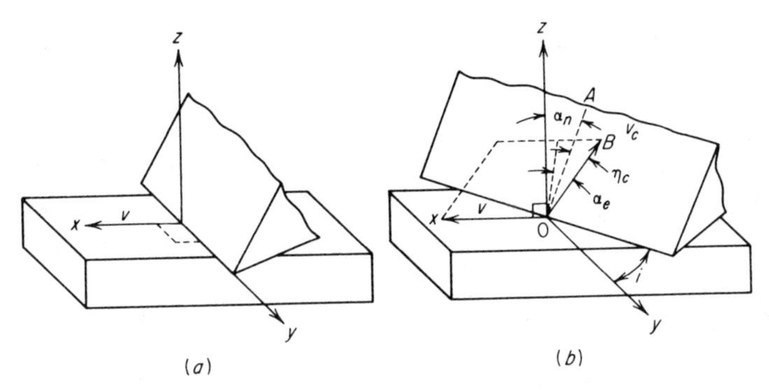

Figure 21-9 (*a*) Orthogonal cutting; (*b*) three-dimensional cutting.

Staebler has shown[1] that to a first approximation $\eta_c = i$, so that Eq. (21-17) becomes

$$\sin \alpha_e \approx \sin^2 i + \cos^2 i(\sin \alpha_n) \tag{21-18}$$

Since i and α_n may be determined fairly readily for a tool, α_e can be calculated and used in the equations for orthogonal machining to make predictions about three-dimensional cutting. From Eq. (21-18) we see that increasing the inclination of the cutting edge is equivalent to increasing the rake angle. From previous discussion of orthogonal cutting we have seen that this reduces the shear strain in the chip, and reduces the cutting force, energy per unit volume, and the tendency for a BUE.

The most common three-dimensional cutting tool is the ordinary lathe tool (Fig. 21-10). Like most practical three-dimensional tools, it has two cutting edges that cut simultaneously. The primary cutting edge is the side-cutting edge, while the secondary cutting edge is the end-cutting edge. The seven-digit code for describing the cutting tool geometry given in Fig. 21-10*b* is that of the American National Standards Institute (ANSI). Completely different methods for describing lathe tools are in use in Great Britain and Germany and international standards have been proposed to clarify this confusing situation.[2]

Milling and drilling are important machining operations that employ multiple-edge cutting tools.[3] The description of the tool geometry and the mechanics of the process are too detailed for discussion here.

[1] G. V. Staebler, *Proc. Inst. Mech. Eng. (London)*, vol. 165, p. 14, 1951.

[2] G. Boothroyd, op. cit., chap. 7; A. Bhattacharyya and I. Ham, "Design of Cutting Tools," chap. 2, Society of Manufacturing Engineers, Dearborn, Mich. 1969.

[3] A. Bhattacharyya and I. Ham, op. cit., pp. 124–149; E. J. A. Armarego and R. H. Brown, "The Machining of Metals," pp. 174–200, Prentice-Hall, Inc., Englewood Cliffs, N.J., 1969; "Tool and Manufacturing Engineers Handbook," 4th ed., vol. 1, Society of Manufacturing Engineers, Dearborn, Mich., 1983.

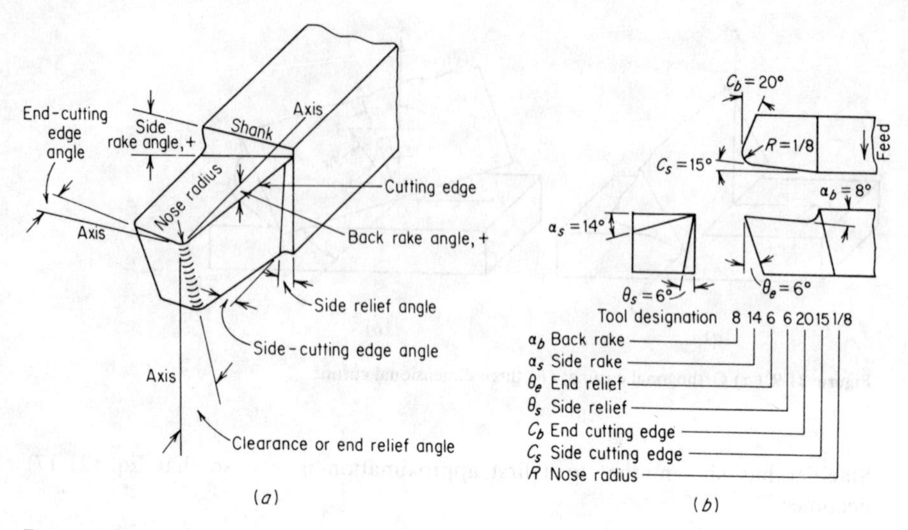

Figure 21-10 (*a*) Typical cutting tool terminology based on right-hand convention; (*b*) ANSI designation of a cutting tool. *(After S. Kalpakjian, "Mechanical Processing," p. 279, Van Nostrand Reinhold Company, New York, 1967.)*

21-4 TEMPERATURE IN METAL CUTTING

Although the vast majority of machining operations are conducted with the workpiece at ambient temperature, because of the large plastic strain and very high strain rate there is a significant temperature rise. This has an important bearing on the choice of tool materials, their useful life, and on the type of lubricant system required. The shearing energy required to form the chip is nearly all converted into heat. Heat also is generated by the rubbing under high pressure and velocity of the chip against the rake face of the tool. As a result of this rubbing action a secondary deformation zone develops in the chip adjacent to the chip-tool interface, and this also contributes to heat generation.

The heat transfer is strongly dependent on the cutting velocity. At very low cutting speeds there will be adequate time for conduction to occur, chiefly into the workpiece, and near isothermal conditions will prevail. At the other extreme, at very high cutting speeds there is little time for heat conduction and a near adiabatic condition will exist with high local temperatures in the chip. The usual machining conditions are somewhere between these two extremes.

Much data on temperature generation in metal cutting may be correlated with a dimensionless number $R_t = \kappa/vd$, where κ is the thermal diffusivity $= k/\rho c$, v is the cutting speed, and d is the depth of cut. R_t is called the *thermal number*. If all of the heat generated goes into the chip, the adiabatic temperature is given by

$$T_{ad} = \frac{U}{\rho c} \tag{21-19}$$

where U = the specific cutting energy
ρ = the density of the workpiece material
c = specific heat of workpiece

For lower velocities the temperature will be less than that given by Eq. (21-19). The approximate chip-tool interface temperature is given by the correlation[1]

$$\frac{T}{T_{ad}} = C\left(\frac{1}{R_t}\right)^p \tag{21-20}$$

where $C \approx 0.4$ and the exponent p ranges from $\frac{1}{3}$ to $\frac{1}{2}$. More complete analyses of the temperature in the chip have been given by Lowen and Shaw,[2] Rapier,[3] and Weiner.[4] The finite element method has been used[5] for calculating the temperature distributions in the chip and the tool.

21-5 CUTTING FLUIDS

We have seen that machining processes involve high local temperatures and high friction at the chip-tool interface. Thus most practical machining operations use a cutting fluid designed to ameliorate these effects. The primary functions of a cutting fluid are:

1. To decrease friction and wear,
2. To reduce temperature generation in the cutting area,
3. To wash away the chips from the cutting area,
4. To protect the newly machined surface against corrosion.

While the last two factors are important practical considerations, the first two factors are the prime functions of a cutting fluid. Important practical objectives which grow out of these functions are increased tool life, improved surface finish, reduced cutting forces and power consumption, and reduced thermal distortion of the workpiece. Cutting fluids usually are liquids, but they may be gases. Solid lubricants also play a role in improving machinability.

There are two basic types of liquid cutting fluids: petroleum-based nonsoluble fluids (straight cutting oils) and water-miscible fluids (soluble oils). Many additives are used in conjunction with each of these types of fluids to achieve a specific purpose. A cutting fluid of the first type may contain one or more parts of mineral oil, fatty oils, sulfur, or chlorine. Water-soluble oils may contain some

[1] N. H. Cook, "Manufacturing Analysis," p. 47, Addison-Wesley Publishing Company, Inc., Reading, Mass., 1966.

[2] E. G. Loewen and M. C. Shaw, *Trans. ASME*, vol. 76, pp. 217–231, 1954.

[3] A. C. Rapier, *J. Appl. Phys.*, vol. 5, p. 400, 1954.

[4] J. H. Weiner, *Trans. ASME*, vol. 77, p. 1331, 1955.

[5] M. G. Stevenson, P. K. Wright, and J. G. Chow, *Trans. ASME J. Eng. Ind.*, vol. 105, pp. 149–154, 1983; P. R. Dawson and S. Malkin, *Trans. ASME J. Eng. Ind.*, vol. 106, pp. 179–186, 1984.

combination of fatty oils, fatty acids, wetting agents, emulsifiers, sulfur, chlorine, rust inhibitors, and germicides in water. Since water has nearly twice the specific heat of any other cutting fluid, water-based fluids are better coolants than oil-based cutting fluids, and they tend to be used for high-speed machining. They also are more economical than oil-based fluids. Petroleum-based cutting fluids are better lubricants and are often preferred at lower cutting speeds. Cutting fluid selection is a field in which experience plays a major role.

The new surfaces created in cutting are clean and hot, and therefore they are active sites for chemical reaction. The sulfur and chlorine added to the cutting fluid react with the fresh metal surfaces to form sulfides or chlorides. Generally these are compounds with low shear strength and therefore they reduce friction. Iron chloride usually has a lower shear strength than iron sulfide, but the reaction time is longer to form the chloride compound. Thus, while chlorinated fluids work well at low speed and light loads, they are not as effective as sulfur compounds under severe conditions. Both sulfur and chlorine usually are added together to provide for effective lubrication over a range of machining conditions.

The cutting fluid reaches the chip-tool interface by a number of processes. For thin chips, it has been shown[1] that cutting fluid can diffuse through the highly distorted structure of the metal in the chip. In general the back of the chip is not in complete contact with the rake face of the tool so that low-surface-tension liquid can penetrate these cavities by capillary action. Since the chips are hot, the fluid often is a vapor, which can more easily penetrate in narrow cavities. Moreover, chlorine compounds have been shown[2] to prevent the welding shut of fine cracks in highly sheared metal, a mechanism which could introduce many paths to the chip-tool interface.

A form of solid lubrication is provided by certain *free-machining additions* to the work material. Lead added to brass or steel is present in the form of fine dispersed particles. When the leaded metal is cut, the lead spreads over the cut surface to provide a low-shear-stress interface.

21-6 TOOL MATERIALS AND TOOL LIFE

Cutting tools are subjected to high forces under conditions of high temperature and wear. Three main forms of wear occur in metal cutting. In *adhesive wear* the tool and the chip weld together at local asperities, and wear occurs by the fracture of the welded junctions. *Abrasive wear* occurs as a result of hard particles on the underside of the chip abrading the tool face by mechanical action as the chip passes over the rake face. The hard particles can arise from hard constituents of either the workpiece or the tool, or highly strain-hardened fragments of a BUE. Wear may also result from *solid-state diffusion* from the tool material to the workpiece at the high temperature and intimate contact existing at the interface

[1] C. Cassin and G. Boothroyd, *J. Mech. Eng. Sci.*, vol. 7, p. 67, 1965.
[2] E. Usui, A. Gujral, and M. C. Shaw, *Int. J. Mach. Tool Des., Res.*, vol. 1, p. 187, 1961.

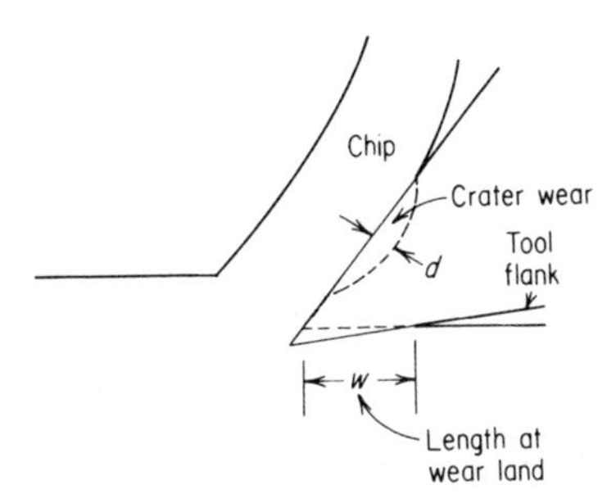

Figure 21-11 Crater wear and flank wear.

between the chip and the rake face. Thus, a good cutting tool material[1] must possess high hot hardness and wear resistance, with sufficient toughness to resist fracture or chipping. It should have good thermal shock resistance and low reactivity with the workpiece material. It should be easy to form, grind, and sharpen to the desired tool profile, and it should be economical.

Two main types of wear are observed in a cutting tool: (1) *flank wear* is the development of a *wear land* on the tool due to abrasive rubbing between the tool flank and the newly generated surface, and (2) *crater wear* is the formation of a circular crater in the rake face of the tool (see Fig. 21-11). Other modes of wear are chipping or fracture of the tool, rounding of the cutting edge, or plastic deformation of the tool.

The primary cause of crater wear is diffusion wear due to the high temperature developed at the interface between the chip and the rake face of the tool. Typically the crater forms at about the midpoint of the tool-chip contact area at a distance from the tool tip corresponding to the location of maximum temperature. The crater is formed by wear processes which are accelerated by the local softening of the tool material. Crater wear is followed by plotting crater depth vs. cutting time. A sharp increase in rate of crater wear often is observed above a certain temperature. This corresponds to a temperature at which the removal of atoms from the tool to the chip is accelerated due to a rapid rate of diffusion.[2] The predominant wear process depends on cutting speed. At higher speeds crater wear is predominant, while at lower speeds it is flank wear.

Flank wear usually is monitored by plotting the length of the wear land w against cutting time (Fig. 21-12). This typical wear curve shows an initial region where the sharp cutting edge quickly is broken down to establish a wear land, a linear region of constant wear rate, and finally, a region of accelerated wear. A similar curve could be obtained by measuring the depth of the crater.

The oldest cutting tool materials are the *high-carbon tool steels* containing from 0.7 to 1.5 percent carbon. These are chiefly water-hardening and limited in application to low cutting speeds because they fall rapidly in hardness above 300°F. Some improvement in hot hardness is achieved by alloying elements such as chromium, vanadium, tungsten, and cobalt.

[1] E. M. Trent, *Metall. Rev.*, no. 127, pp. 129–144, 1968; "Metals Handbook," 9th ed., vol. 3, pp. 470–477, 1980.

[2] B. M. Kramer and N. P. Suh, *Trans. ASME J. Eng. Ind.*, vol. 102, pp. 303–309, 1980.

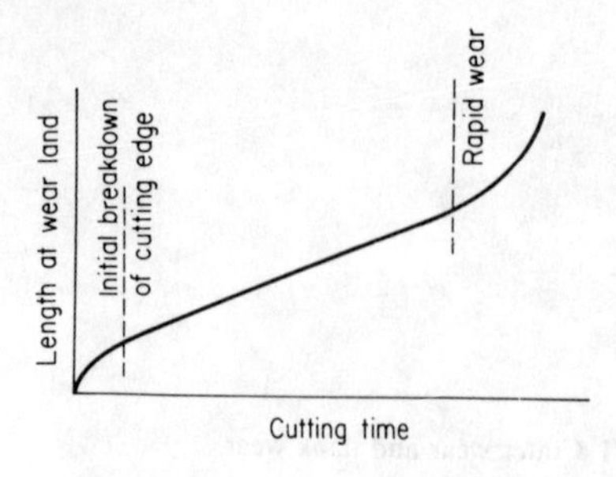

Figure 21-12 Typical wear curve for cutting tool.

High-speed tool steels retain their hot hardness to about 1000°F and can be used at cutting speeds about twice as fast as carbon tool steels. The classic example of high-speed steel is 18-4-1 (W-Cr-V), but since World War II the M-series steels, in which most of the expensive tungsten is replaced by molybdenum, have been the most common grades. The popular M-2 high-speed steel contains 6% W–4% Cr–2% V–5% Mo.

Cast nonferrous tools typically contain 40% Co–35% Cr–20% W. They are called nonferrous because they contain no iron. The structure of these materials is chiefly complex intermetallic compounds which retain useful hardness to about 1500°F. Because the materials are brittle, they must be cast to shape and used in situations where impact and vibration are not severe.

A major achievement in tool materials was the introduction of *cemented carbide* tools, which consist of heat-resistant refractory carbides embedded in a ductile metal matrix. They are produced by powder metallurgy using liquid-phase sintering. The most common system consists of tungsten carbide with a cobalt binder. To minimize crater wear when machining steel, titanium and tantalum carbide usually are combined with the tungsten carbide. Cemented carbides are useful at cutting temperatures up to 2000°F, and they can be used at speeds up to five times that used with high-speed steel. Cemented carbide tools are used in the form of inserts or tips that are brazed to a steel shank, or in the form of throwaway inserts with multiple cutting edges that are mechanically fastened to the tool post. Cemented carbides are brittle, and to prevent fracture the tool should be rigidly supported and run without vibration or chatter. Coated carbides consisting of a very thin layer of TiC or TiN over the WC-Co tool reduce the effects of adhesion and diffusion, and reduce crater wear.

Ceramic or *oxide tools* of sintered Al_2O_3 are harder than cemented carbides, possess better wear resistance, and have less tendency for the tool to weld to the chip. They also possess high compressive strength, but they are more brittle than WC. Ceramic tools can operate at cutting speeds two to three times that of cemented carbides in uninterrupted cuts where shock and vibration are minimized. Tool failure can be reduced by the use of rigid tool mounts and rigid machine tools. *Diamond* offers the highest hot hardness of any material. Synthetic diamonds were produced in the late 1950s, and cubic boron nitride (CBN) in the early 1970s, by high-pressure high-temperature processing. Until recently, diamond and CBN were used only for grinding, but in the midseventies "compax" tools in which these materials are sintered to a WC substrate were developed for machining.

Certain tool materials are chemically incompatible with certain workpiece materials because strong bonds tend to form between the chip and the tool. For example, poor results are obtained if Al_2O_3 is used to machine aluminum or titanium alloys. Also, poor tool life results when a diamond tool is used to machine steel because the iron on the face of the tool tends to dissolve carbon from the diamond. CBN is used as a substitute for diamond in machining steel.

Tool life is a very important parameter in metal cutting, but it is difficult to define without ambiguity. In the most general sense, tool life is determined by the point at which the tool no longer makes economically satisfactory parts. However, more specific criteria of tool life usually are used. Complete destruction of the tool when it ceases to cut is one such criterion. More typically, the tool life usually is defined in terms of an average or maximum allowable wear land. A typical value would be 0.3 mm. Less often, the tool life might be based on the degradation of the surface finish below some specified limit, or the increase in the cutting force above some value, or when the vibrational amplitude reaches a limiting value.

Cutting speed is the most important operating variable influencing tool temperature, and hence, tool life. The classic study by Taylor[1] established the relationship between cutting speed v and the time t to reach a wear land of certain dimensions as

$$vt^n = \text{constant} \tag{21-21}$$

Typical values of the exponent n are: high-speed steel, 0.1; cemented carbide, 0.2; ceramic tool, 0.4. Significant changes in tool geometry or machining parameters will change the value of the constant. The tool life used in this equation is the machining time between regrinding the tool, not the total life of the tool before it is discarded. While cutting speed is the major variable affecting tool life feed f and depth of cut d also are important parameters. Thus, the Taylor equation often is extended to the form

$$vt^n d^x f^y = \text{constant} \tag{21-22}$$

It is important to realize that the Taylor equation is completely empirical and, as with other empirical relationships, it is dangerous to extrapolate outside of the limits over which the data extend.

To summarize briefly, tool life will depend on the tool material, the machining parameters, and on such factors as the cutting fluid and the properties of the workpiece material. The ability of a material to be machined has been termed *machinability*.[2] A material has good machinability if the tool wear is low, the tool life is long, the cutting forces are low, the chips are well behaved and break into

[1] F. W. Taylor, *Trans. ASME*, vol. 28, pp. 31–350, 1907.

[2] B. Mills and A. H. Redford, "Machinability of Engineering Materials," Appl. Sci. Publ., Englewood, N.J., 1983.

small ringlets instead of long snarls, and the surface finish is acceptable. This illustrates the problem in defining machinability in a scientific sense; there are too many criteria for evaluating this elusive material property. Thus, there are no standard tests for evaluating machinability.

One of the most common methods of evaluating machinability is in terms of the cutting velocity for a specified tool life, e.g., 60 or 90 min. The machinability of any material is compared with a standard material. Thus, the *relative machinability* may be defined as (v_{60} material A)/(v_{60} standard material). However, defining machinability in this way essentially is equivalent to describing tool life. If the *tool* material changes, the relative machinability ratio would be expected to change. Moreover, this method does not consider the surface finish produced. Some success at predicting relative machinability from basic strength and ductility parameters has been obtained.[1]

Example A high-speed steel tool and cemented carbide tool have the following tool life relationships.

High-speed steel $\quad vt^{0.12} = 230$

Cemented carbide $\quad vt^{0.30} = 1{,}500$

What would be the increase in tool life for each tool material as a result of a 50 percent decrease in cutting speed?

For high-speed steel tool: $v_1t_1^{0.12} = v_2t_2^{0.12}$ but $v_2 = 0.5v_1$

$$\left(\frac{t_1}{t_2}\right)^{0.12} = 0.5 \qquad t_2 = t_1/(0.5)^{1/0.12} = \frac{t}{0.0031} = 322t_1$$

For cemented-carbide tool: $v_1t_1^{0.30} = v_2t_2^{0.30}$

$$\left(\frac{t_1}{t_2}\right)^{0.30} = 0.5 \qquad t_2 = t_1/(0.5)^{1/0.3} = \frac{t_1}{0.0994} = 10t_1$$

Failure of a tool in production can be costly and should be avoided. This leads to conservative estimates of tool life from curves like Fig. 21-12 and the general replacement of tools before they have used up their economic life. As a result, considerable research is being devoted to methods of on-line measurement of tool wear.[2] A variety of sensors has been employed but the most promising appear to be the use of frequency signature analysis as the worn tool flank vibrates from rubbing on the workpiece and the monitoring of acoustic emission from the stress waves generated during cutting. Another developing concept is the adaptive control of the machining process[3] in which the constraints imposed by the various tool failure mechanisms are mapped and optimized to give the safe working region for the control variables of cutting speed and feed rate.

[1] A. Henkin and J. Datsko, *Trans. ASME, Ser. B, J. Eng. Ind.*, vol. 85, pp. 321–328, 1963.
[2] S. Jetly, *Manufact. Eng.*, pp. 55–60, July 1984.
[3] D. W. Yen and P. K. Wright, *Trans. ASME J. Eng. Ind.*, vol. 105, pp. 31–38, 1983.

21-7 GRINDING PROCESSES

Grinding processes employ an abrasive wheel containing many grains of hard material bonded in a matrix. The action of a grinding wheel may be considered a multiple-edge cutting tool except the cutting edges are irregularly shaped and randomly spaced around the face of the wheel. Each grain removes a short chip of gradually increasing thickness, but because of the irregular shape of the grain there is considerable ploughing action between each grain and the workpiece. This results in progressive attritious wear of the grain in which the sharp edges become dull and the force on the grain increases because of increasing friction. Eventually the force on the grain becomes high enough either to tear the grain from the bonding matrix and expose a sharp fresh grain, or to fracture the worn grain to expose new cutting edges. In either case, a grinding wheel has a self-sharpening feature. However, under other conditions a wheel can become "loaded" with dull grains and/or fine adhering chips. Then the wheel must be *dressed* by passing a diamond-tip tool across the wheel face to generate new sharp cutting surfaces.

There are several aspects of the grinding process which make it distinctly different from other metal-removal processes. These differences chiefly are related to the different nature of the cutting tool in grinding. While most cutting processes are carried out with tools having positive rake angles, the shape of the abrasive grains results in a tool with large negative rake angles. Thus, there is a tendency for the grain to slide over the work rather than cut. The depth of cut in grinding usually is very small (10^{-4} in), and this results in very small chips that adhere readily to the wheel or the workpiece. The net effect is that the specific cutting energy for grinding is about 10 times greater than for turning or milling (see Fig. 21-7). In other cutting processes only about 5 percent of the energy input ends up in the finished surface, the bulk going into the chip. In grinding, greater than 70 percent of the energy goes into the finished surface. This results in considerable temperature rise and generation of residual stresses.

One of the chief applications of grinding is in *finishing operations*, where the ability to produce a highly finished surface to precise dimensions is important. The ability to grind a hard surface which cannot be machined with a conventional cutting tool also is important. Finishing processes which use abrasives more to produce a very smooth surface than to remove metal are *honing*, *lapping*, and *superfinishing*. However, there is one area of grinding processes used primarily to remove metal at high rate (abrasive machining). Examples are billet conditioning, rough grinding of castings, and abrasive cutoff operations.

A grinding wheel is a cutting tool whose properties and performance can be varied over wide limits by the selection and control of the abrasive grains[1] and the bonding matrix. Most commercial grinding wheels employ aluminum oxide Al_2O_3 or silicon carbide SiC as abrasive grain. Frequently these materials are alloyed with oxides of titanium, chromium, vanadium, zirconium, etc., to impart special properties. There is a growing use of diamond wheels for fine finishing. The

[1] R. Komanduri and M. C. Shaw, *Trans. ASME, Ser. H*, vol. 96, pp. 145–156, 1974.

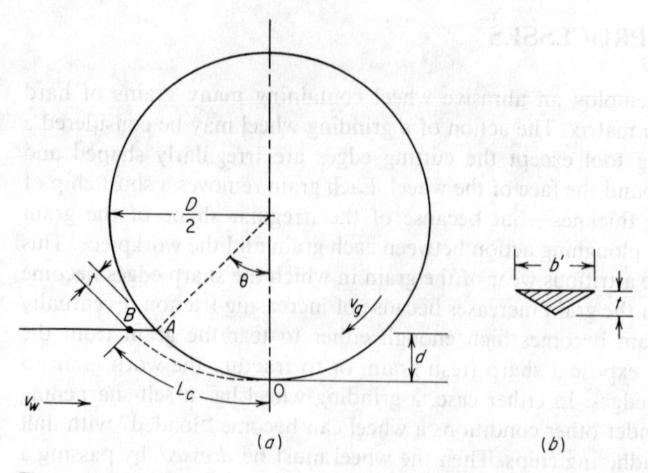

Figure 21-13 (*a*) Geometry in surface grinding; (*b*) approximate cross section of grinding chip.

performance of the wheel chiefly is controlled by the strength of the bond, which is a function of the type and amount of bonding material used. The most common bonding materials are a glassy *vitrified bond*, a *silicate bond* made from hardened water glass, a *resinoid bond* prepared from organic resins, and a *rubber bond*. The *grade* of a wheel expresses the degree to which the abrasive grains are bonded to the matrix. A soft wheel releases a worn grain easier than a hard wheel. The relative proportion of abrasive grain and binder, and the spacing between grains, is called the *structure* of the wheel. Structure is one of the variables that determines the number of cutting points per unit area on the face of the wheel and the clearance between grains for removing chips. Another variable relative to effective grain spacing is grain size. A wheel with an open structure provides for easy chip removal, but it will tend to act as a soft wheel compared with a wheel with a closed structure. The proper wheel for a grinding operation is one in which the bond permits a sharp grain to cut but allows a dull grain, which is acted on by a greater force, to be fractured or pulled from the wheel.

Figure 21-13 illustrates the geometry involved in surface grinding. The velocity of the grinding wheel is v_g and the velocity of the workpiece is v_w. Typically, grinding wheels move at very high cutting speeds (3,000 to 16,000 ft/min) while v_w might typically be 60 ft/min. An important variable in the grinding process is the *grain depth of cut t* (analogous to the undeformed chip thickness in ordinary metal cutting), which is distinct and much smaller than the wheel depth of cut (infeed) *d*. The metal removed by a grain is area *OAB* times the chip width *b*. The average length of chip L_c is

$$L_c \approx \frac{D}{2} \sin\theta \tag{21-23}$$

But
$$\cos\theta = \frac{(D/2) - d}{D/2} = 1 - \frac{2d}{D} \tag{21-24}$$

and
$$\sin^2\theta = 1 - \cos^2\theta = \frac{4d}{D} - \frac{4d^2}{D^2} \tag{21-25}$$

Neglecting the second-order term $4d^2/D^2$, and substituting Eq. (21-25) into Eq. (21-23), results in

$$L_c \approx \sqrt{Dd} \tag{21-26}$$

The total volume of chips removed per unit time V_{chip} is

$$V_{\text{chip}} = v_w bd \tag{21-27}$$

These chips tend to approximate a triangular cross section (Fig. 21-13*b*) where $r = b'/t$ is between 10 and 20. The average volume of a chip is $(b'tL_c)/4 = (rt^2L_c)/4$. The number of chips produced per unit time is

$$n = v_g bC \tag{21-28}$$

where C is the number of active grains on the wheel per square inch (typically 2,000).

The volume of material removed per unit time is $nV_{\text{chip}} = v_w bd$. Thus,

$$v_g bC\frac{(rt^2L_c)}{4} = v_w bd \tag{21-29}$$

and since $L_c \approx \sqrt{Dd}$,

$$t = 2\sqrt{\frac{v_w}{Crv_g}\sqrt{\frac{d}{D}}} \tag{21-30}$$

The grain depth of cut t given above is its maximum value. The mean value of t is $\bar{t} = t/2$. This analysis was for surface grinding, where the workpiece is a flat surface. Two other common grinding processes are the grinding of the external surface of a cylinder (cylindrical grinding) and grinding the inside cylindrical surface of a hollow cylinder (internal grinding). The above analysis can be used for these processes simply by substituting the equivalent diameter D_e for the wheel diameter D in Eq. (21-30).

$$D_e = \frac{D_g D_w}{D_w \pm D_g} \tag{21-31}$$

where D_g is the grinding wheel diameter, D_w is the workpiece diameter, and the plus sign is used for cylindrical grinding and the minus sign is used for internal grinding. For surface grinding, where $D_w = \infty$, we find that $D_e = D_g$.

The grain depth of cut plays an important role in grinding. It determines the area of contact between the chip and the grain, and thus the force on the abrasive grain. This, in turn, determines the bonding strength necessary to hold the grains in place. As t decreases the rake angle becomes more negative and the specific grinding energy increases. An increase in specific energy increases the grain-tip temperature, which causes the wear rate and the workpiece temperature to increase. Finally, the radial force on the grain increases with contact area ($t \times b$) and this results in greater elastic deflection of the wheel and the workpiece.

Example Compare the grain depth of cut (undeformed chip thickness) for fine grinding, where the object is to remove a small amount of material with a smooth surface finish, and stock removal grinding, where the objective is to remove a large amount of material.

Fine grinding	Stock removal grinding
$D_g = 200$ mm	$D_g = 400$ mm
$d = 0.05$ mm	$d = 5$ mm
$v_g = 30$ m/s	$v_g = 80$ m/s
$v_w = 0.3$ m/s	$v_w = 0.3$ m/s
$r = 20$	$r = 1$
$C = 2$ per mm^2	$C = 0.2$ per mm^2
$t = 2\left[\dfrac{v_w}{C_r v_g}\sqrt{\dfrac{d}{D_g}}\right]^{1/2} = 2\sqrt{\dfrac{0.3}{2 \times 20 \times 30}\sqrt{\dfrac{0.05}{200}}}$	$t = 2\sqrt{\dfrac{0.3}{0.2 \times 1 \times 80}\sqrt{\dfrac{5}{400}}}$
$t = 4\ \mu$m	$t = 0.09$ mm $= 90\ \mu$m

The specific cutting energy in grinding is

$$U = \frac{F_h v_g}{v_w b d} \tag{21-32}$$

where F_h = the tangential force on the wheel. As already discussed, U is appreciably higher for grinding than for turning. U is strongly dependent on t, and to a first approximation

$$U \propto \frac{1}{t} \tag{21-33}$$

If the grain cross section is assumed triangular, the force on a single abrasive grain F_g will be

$$F_g = U\left(\tfrac{1}{2}b't\right) = \tfrac{1}{2}Urt^2 \tag{21-34}$$

Substituting Eq. (21-33) into Eq. (21-34), we find

$$F_g \propto rt \propto \sqrt{\frac{rv_w}{Cv_g}\sqrt{\frac{d}{D}}} \tag{21-35}$$

Equation (21-35) illustrates the important point that the hardness of a grinding wheel depends on both the strength of the bond *and* the grinding conditions, since the greater the value of F_g the more easily the grains fracture or tear out. For a wheel of a given grade, the greater F_g the softer the wheel appears to be. Thus, a hard wheel can be made to grind softer by increasing the workpiece velocity or depth of cut, or by decreasing the wheel speed or the number of cutting points (move to a more open structure).

It was pointed out earlier that grinding requires a much larger energy input per unit volume than cutting, and that a far larger portion of this energy goes to raising the temperature of the workpiece. While a rigorous analysis is quite difficult[1] it has been established that the surface temperature T_w depends strongly on the energy per *unit surface area* that is ground.

$$T_w \propto \frac{F_g v_g}{v_w b} \propto Ud \qquad (21\text{-}36)$$

and since $U \propto 1/t$, because of the size effect,

$$T_w \propto \frac{d}{t} d^{3/4} \sqrt{\frac{v_g C}{v_w} \sqrt{d}} \qquad (21\text{-}37)$$

Temperatures of 3000°F or higher can be created in the ground surface. Since the time involved in producing a chip is of the order of microseconds, the peak temperature is very transitory. However, the temperature rise in grinding can lead to melting, or metallurgical changes such as the production of untempered martensite, or grinding cracks, or a visible oxidation of the surface known as grinding burn. Grinding carried out under improper conditions can result in high tensile residual stresses in the ground surface. Residual stresses can be minimized by using the proper grinding fluid and softer wheels at lower wheel speeds.

While the picture of grinding given above adequately describes many aspects of the process, it does not explain such important observations as the large thermal input into the finished surface, or why the specific cutting energy for fine grinding is at least 10 times greater than for metal cutting. Shaw[2] has presented a new approach in which the action of an abrasive grain is modeled with the plastic deformation under a Brinell hardness indentor (see Fig. 11-2). In this model the grinding chip is removed by a special type of extrusion rather than a concentrated shear process as an ordinary metal cutting.

Grinding ratio (G ratio) is the volume of material removed from the work per unit volume of wheel wear. It is a measure of ease of grinding or grindability. The higher the G ratio the easier a material is to grind. The G ratio depends on the grinding process and grinding conditions (wheel, fluid, speed, and feed) as well as the material. Values of G ratio range from about 2 to over 200.

Example A horizontal spindle surface grinder is cutting with $t = 5\ \mu$m and $U = 40$ GPa. Estimate the tangential force on the wheel if the wheel speed is

[1] S. Malkin, *Trans. ASME Ser. B., J. Eng. Ind.*, vol. 96, pp. 1184–1191, 1974; *Annal of CIRP*, vol. 27, pp. 233–236, 1978.

[2] M. C. Shaw, *Mech. Chem. Eng. Trans.*, vol. MC8, no. 1, pp. 73–78, May 1972.

30 m/s, the cross-feed per stroke is 1.2 mm, the work speed is 0.3 m/s, and the wheel depth of cut is 0.05 mm.

The metal removal rate is

$$M = v_w bd = 0.3(1.2 \times 10^{-3})(0.05 \times 10^{-3}) = 0.018 \times 10^{-6} \text{ m}^3/\text{s}$$

$$\text{Required power} = 40 \times 10^9 \text{ N/m}^2 \times 0.018 \times 10^{-6} \text{ m}^3/\text{s}$$

$$= 720 \text{ N-m/s} = 720 \text{ W}$$

$$\text{Power} = F_h v_g \qquad F_h = 720 \text{ N-m/s} \times 1/30 \text{ m/s} = 24 \text{ N}$$

21-8 NONTRADITIONAL MACHINING PROCESSES

A number of new material-removal processes[1] have been developed since World War II which mostly use forms of energy other than mechanical energy. The impetus for developing most of these processes was the search for better ways of machining complex shapes in hard materials. The various techniques are listed below in terms of the major energy source. However, space permits only a brief description of a few of these processes.

Source of energy	Name of process
Thermal energy processes	Electrical discharge machining, EDM
	Laser-beam machining, LBM
	Plasma-arc machining, PAM
Electrical processes	Electrochemical machining, ECM
	Electrochemical grinding, ECG
Chemical process	Chemical machining, CHM
Mechanical process	Ultrasonic machining, USM

Electrical discharge machining[2] (EDM) is a method for producing holes, slots, or other cavities in an electrically conductive material by the controlled removal of material through melting or vaporization caused by a high-frequency spark discharge. The workpiece (the anode) and the tool (the cathode) are immersed in a dielectric fluid with a spark gap of 0.0005 to 0.020 in. Important advantages of EDM are that it can produce deep holes in hard material without drifting, or can machine cavities of irregular contour. The metal-removal rate is independent of the hardness of the workpiece, but it does depend on thermal properties such as heat capacity and conductivity, melting point, and latent heat of melting and vaporization. Electrode wear is a problem which requires selection of the proper

[1] R. K. Springborn (ed.), "Non-Traditional Machining Processes," Society of Manufacturing Engineers, Dearborn, Mich., 1967.

[2] Electrical Discharge Machining, "Metals Handbook," op. cit., pp. 227–233.

electrode material for the workpiece material. Because high temperatures are attained in the spark, the metal is melted and then rapidly quenched by a mass effect when it resolidifies. The recast layer may be deleterious to fatigue properties.

Electrochemical machining[1] (ECM) is the controlled removal of metal by anodic dissolution in an electrolytic cell in which the workpiece is the anode and the tool is the cathode. It is similar to a reverse electroplating process. The electrolyte is pumped through the cutting gap while direct current is passed through the cell at low voltage to dissolve the metal from the workpiece. The rate of removal of material is proportional to the amount of current passing between the tool and the workpiece and is independent of the hardness of the workpiece. ECM is a cold process which results in no thermal damage to the workpiece. It results in a smooth burr-free surface. However, it is not suited for producing sharp corners or cavities with flat bottoms. *Electrochemical grinding* (ECG) is a combination of ECM and abrasive grinding in which most of the metal is removed by electrolytic action. It is used with hard carbides or difficult-to-grind alloys where wheel wear or surface damage must be minimized.

Chemical machining[2] (CHM) involves metal removal by controlled chemical attack with chemical reagents. The essential steps involve cleaning the surface, masking those areas which are not to be dissolved, attacking with chemicals, and cleaning. Chemical milling refers to chemical machining of large areas, such as aircraft structural parts. Chemical blanking is used for cutting or stamping parts from very thin sheet.

In *ultrasonic machining*[3] (USM) the tool is excited at around 20,000 cycles/sec with a magnetostrictive transducer while a slurry of fine abrasive particles is introduced between the tool and the workpiece. Each cycle of vibration removes minute pieces of the workpiece by fracture or erosion. USM mostly is used for machining brittle hard materials such as semiconductors, ceramics, or glass.

21-9 ECONOMICS OF MACHINING

The annual cost of operating machine tools in the United States (labor and overhead) exceeds $40 billion. Thus, there is a demonstrated need for understanding the cost elements in machining. In addition, the problem of determining the cutting speed to give minimum cost is a classic problem in cost optimization using Kelvin's law. If we consider for the moment only the costs involved in actual cutting, the machining time (and thus the cost of machining labor for one piece) will decrease with increasing speed or feed. However, tool wear will increase with speed and feed, and thus, there will be increased costs for tools and tool changing.

[1] Electrochemical Machining, "Metals Handbook," op. cit., pp. 233–240.

[2] Chemical Milling, "Metals Handbook," op. cit., pp. 240–249; J. W. Dini, *Int. Metall. Rev.*, vol. 20, pp. 29–55, March 1975.

[3] Ultrasonic Machining, "Metals Handbook," op. cit., pp. 249–251.

There will be an optimum speed which balances these opposing factors and results in minimum cost per piece.

The *total cost* of a machined piece is the sum of four costs:

$$C_u = C_m + C_n + C_c + C_t \tag{21-38}$$

where C_u = the total unit (per piece) cost
C_m = the machining cost
C_n = the cost associated with nonmachining time; including setup cost of mounting fixtures in the machine and preparing it for operation, time for loading and unloading parts in the machine, and idle machine time
C_c = the cost of tool changing
C_t = the tool cost per piece

Machining cost can be expressed by

$$C_m = t_m(L_m + O_m) \tag{21-39}$$

where t_m = the machining time per piece (including the time the feed is engaged whether or not the tool is cutting)
L_m = the labor cost of a production operator per unit time
O_m = the overhead charge for the machine, including depreciation, indirect labor, maintenance, etc.

For a simple turning operation, the machining time for one piece is

$$t_m = \frac{L}{fN} = \frac{LD}{12fv} \tag{21-40}$$

where L = the length of cut
f = the feed, in/rev
N = the rpm of the workpiece
D = the diameter of the workpiece, in
v = the cutting speed, ft/min

The *cost of nonmachining time* C_n is usually expressed as a fixed cost in dollars per piece. The *cost of tool changing* C_c can be expressed by

$$C_c = t_g\left(\frac{t_{ac}}{t}\right)(L_g + O_g) \tag{21-41}$$

where t_g = the time required to grind and change a cutting edge
t_{ac} = the actual cutting time per piece
t = the tool life for a cutting edge
L_g = the labor rate for a toolroom operator
O_g = the overhead rate for the toolroom operation

The Taylor equation for tool life can be written

$$vt^n = \text{constant} = K$$

or
$$t = \left(\frac{K}{v}\right)^{1/n} \tag{21-42}$$

$$t_{ac} = \frac{\pi L_a D}{12 f v}$$

where L_a is the actual length of cut and

$$C_c = t_g(L_g + O_g)\frac{\pi L_a D v^{(1/n)-1}}{12 f K^{1/n}} \tag{21-43}$$

The *tool cost* per piece can be expressed by

$$C_t = C_e \frac{t_{ac}}{t} \tag{21-44}$$

where C_e is the cost of a cutting edge, and t_{ac}/t is the number of tool changes required per piece.

The *total unit cost* is given by

$$C_u = C_n + C_m + C_c + C_t$$

or
$$C_u = C_n + (L_m + O_m)Av^{-1} + (L_g + O_g)t_g\frac{Av^{(1/n)-1}}{K^{1/n}} + C_e\frac{Av^{(1/n)-1}}{K^{1/n}} \tag{21-45}$$

where $A = (\pi LD)/12f$.

The optimum value of cutting speed to give minimum unit cost can be obtained by setting $\partial C_u/\partial v = 0$. The component costs are shown in Fig. 21-14. The minimum in the total unit cost curve occurs chiefly because tool wear increases rapidly with cutting speed. Improvements in cutting tool materials would push the optimum speed to higher values. Worked examples comparing

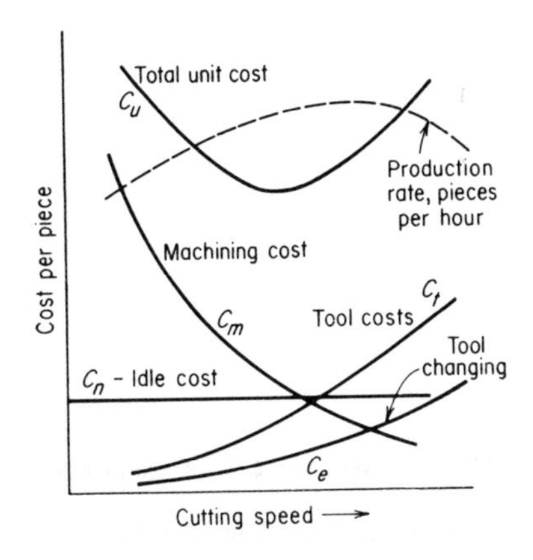

Figure 21-14 Variation of machining costs with cutting speed.

different tool materials have been presented by Boothroyd.[1] Detailed equations for costs for turning, milling, drilling, reaming, and tapping have been published.[2]

An alternative strategy to operating at the cutting speed to minimize unit cost is to operate at the cutting speed that results in maximum production rate This occurs at a higher value of cutting speed (Fig. 21-14). Still another alternative[3] is to operate at the value of cutting speed which maximizes profit.

BIBLIOGRAPHY

Armarego, E. J. A. and R. H. Brown: "The Machining of Metals," Prentice-Hall, Inc. Englewood Cliffs, N.J., 1969.

Bhattacharyya, A. and I. Ham: "Design of Cutting Tools," Society, of Manufacturing Engineers, Dearborn, Mich., 1969.

Boothroyd, G.: "Fundamentals of Metal Machining and Machine Tools," McGraw-Hill Book Company, New York, 1975.

Cook, N. H.: "Manufacturing Analysis," Addison-Wesley Publishing Company, Inc., Reading, Mass., 1966.

Kronenberg, M.: "Machining Science and Applications," Pergamon Press, New York, 1966.

"Metals Handbook," 8th ed., vol. 3, Machining, American Society for Metals, Metals Park, Ohio, 1967.

Shaw, M. C.: "Metal Cutting Principles," Oxford University Press, New York, 1984.

"Tool and Manufacturing Engineers Handbook," 4th ed., vol. 1, Society of Manufacturing Engineers, Dearborn, Mich., 1983.

Trent, E. M.: "Metal Cutting," 2d ed., Butterworth, Woburn, Mass., 1984.

[1] G. Boothroyd, op. cit., pp. 161–164.

[2] M. Field, N. Zlatin, R. Williams, and M. Kronenberg, *Trans. ASME, Ser. B, Eng. Ind.*, vol. 90, pp. 455–466, 585–596, 1968.

[3] A. M. Wu and D. S. Ermer, *Trans. ASME, Ser. B, J. Eng. Ind.*, vol. 88, pp. 435–442, 1966.

APPENDIX

A

THE INTERNATIONAL SYSTEM OF UNITS (SI)

Basic SI units appropriate to mechanical metallurgy

Quantity	Name	Symbol
Length	Meter	m
Mass	Kilogram	kg
Time	Second	s
Thermodynamic temperature	Kelvin	K
Amount of substance	Mole	mol

SI units derived from basic units

Quantity	Name	Symbol	Quantity expressed in other units
Area	Square meter	m^2	
Volume	Cubic meter	m^3	
Velocity	Meter per second	m/s	
Density	Kilogram per cubic meter	kg/m^3	
Concentration	Mole per cubic meter	mol/m^3	
Force	Newton	N	
Energy (work)	Joule	J	N-m
Power	Watt	W	J/s
Stress (pressure)	Pascal	Pa	N/m^2
Strain	Meter per meter	m/m	
Fracture toughness		$MPa\sqrt{m}$	
Viscosity (dynamic)	Pascal-second	Pa-s	
Surface tension	Newton per meter	N/m	
Specific heat	Joule per kilogram kelvin	J/(kg K)	

Decimal multiples and submultiples of SI units

Factor	Prefix	Symbol	Factor	Prefix	Symbol
10^9	giga	G	10^{-9}	nano	n
10^6	mega	M	10^{-6}	micro	μ
10^3	kilo	k	10^{-3}	milli	m
10^2	hecto	h	10^{-2}	centi	c
10^1	deka	da	10^{-1}	deci	d

Conversion factors for SI units

Quantity	To convert from	To	Multiply by
Angle	degree	rad	1.745×10^{-2}
Area	in^2	m^2	6.452×10^{-4}
	ft^2	m^2	9.290×10^{-2}
	in^2	mm^2	6.452×10^{2}
Density	lb/in^3	kg/m^3	2.768×10^{4}
	g/cm^3	kg/m^3	1.000×10^{3}
Energy	ft-lb	J	1.356
	Btu	J	1.054×10^{3}
	cal	J	4.184
	erg	J	1×10^{-7}
Force	lb_f	N	4.448
	kg_f	N	9.807
	dyne	N	1×10^{-5}
Fracture toughness	$ksi\sqrt{in}$	$MPA\sqrt{m}$	1.099
Length	angstrom	m	1×10^{-10}
	in	m	2.540×10^{-2}
	ft	m	3.048×10^{-1}
	micron (μm)	m	10^{-6}
	mil	m	2.540×10^{-5}
Mass	lb	kg	4.536×10^{-1}
	tonne (metric)	kg	1.000×10^{3}
	ton (short)		9.072×10^{2}
Power	hp	W	7.457×10^{2}
	Btu/h	W	2.931×10^{-1}
	ft-lb/min	W	2.260×10^{-2}
Stress (pressure)	lb_f/in^2	Pa	6.895×10^{3}
	ksi	MPa	6.895
	dyn/cm^2	Pa	1.000×10^{-1}
	kg_f/mm^2	Pa	9.806×10^{6}
Torque	ft-lb	N-m	1.356
Velocity	ft/min	m/s	5.080×10^{-3}
	rpm	rad/s	1.047×10^{-1}
Viscosity	poise	Pa-s	10^{-1}
Volume	in^3	m^3	1.639×10^{-5}
	ft^3	m^3	2.832×10^{-2}
	in^3	mm^3	1.639×10^{4}
Other conversions	erg/cm^2	J/m^2	1×10^{-3}
	$lb_f\text{-}ft/in^2$	J/m^2	2.101×10^{3}
	degree Celsius	kelvin (K)	$T_k = T_c + 273$
	degree Fahrenheit	degree Celsius	$T_c = (T_F - 32)/1.8$

Constants and conversion factors

1 angstrom unit, Å = 10^{-8} cm = 10^{-10} m
1 micron, μm = 10^{-3} mm = 10^4 Å = 10^{-6} m
1 centimeter = 0.39370 in
1 inch = 2.5400 cm
1 foot = 30.480 cm
1 radian = 57.29°
1 pound = 453.59 g
1 kilogram = 2.204 lb
1 dyne = 2.24×10^{-6} lb
1 erg = 1 dyne-cm = 10^{-7} joule
1 foot-pound = 1.3549 joules
1 calorie = 4.182 joules
1 kilocalorie = 10^3 cal
1 electron volt, eV = 1.602×10^{-12} erg = 23,050 cal/g-atom
Avogadro's number $n = 6.02 \times 10^{23}$ molecules per mole
Universal gas constant R = 1.987 cal/(deg)(mole) = 8.314 J/K mole
Boltzmann's constant $k = 1.380 \times 10^{-16}$ erg/(deg)(mole)
Planck's constant $h = 6.6234 \times 10^{-27}$ erg-s
1 atmosphere = 14.697 psi
1 pound per square inch = 7.04×10^{-4} kg/mm^2 = 6.93×10^4 dyn/cm^2
1 megapascal (MN/m^2) = 145 psi
1 square inch = 6.4516 cm^2
1 square foot = 929.03 cm^2
1 cubic inch = 16.387 cm^3
$\ln x = \log_e x = 2.3026 \log_{10} x$
The base of the natural system of logarithms = 2.718
kT at room temperature (300 K) = 4.8×10^{-10} erg = 1/40 eV
Lattice parameter: $a_0 \approx 3.5 \times 10^{-8}$ cm
Burgers vector: $\mathbf{b} \approx 2.5 \times 10^{-8}$ cm

APPENDIX

B

PROBLEMS

CHAPTER 1

1-1 An annealed-steel tensile specimen ($E = 30 \times 10^6$ psi) has a 0.505-in minimum diameter and a 2 in-gage length. Maximum load is reached at 15,000 lb, and fracture occurs at 10,000 lb.

(*a*) What is the tensile strength?

(*b*) Why does fracture occur at a lower load than maximum load?

(*c*) What is the deformation when a tensile stress of 15,000 psi is applied?

1-2 A wire 300 m long elongates by 3 cm when a tensile force of 200 N is applied. What is the modulus of elasticity (in MPa) if the diameter of the wire is 5 mm? Compare your answer with the result in units of psi.

1-3 Show that the deformation of a bar subjected to an axial load P is $\delta = PL/AE$.

1-4 The deflection of a cantilever beam of length L subjected to a concentrated load P at one end is given by $\delta = PL^3/3EI$, where I is the moment of inertia. Compare the relative deflections of steel, titanium, and tungsten beams. See Table 2-1 for necessary data.

1-5 A hollow copper cylinder (OD = 10 in and ID = 5 in) contains a solid steel core. Determine the compressive stresses in the steel and the copper when a force of 200,000 lb is applied with a press.

1-6 Discuss the relative factor of safety to be applied to the following situations: (1) boxcar coupling; (2) pressure vessel for nuclear reactor; (3) missile nose cone; (4) flagpole (5) automobile leaf spring. Consider each case in terms of the following factors:

(*a*) material reliability;

(*b*) type of loading;

(*c*) reliability of stress analysis;

(*d*) fabrication factors;

(*e*) influence of time;

(*f*) consequences of failure.

1-7 A cylinder of cast iron 0.50 in in diameter by 2 in long is tested in compression. Failure occurs at an axial load of 50,000 lb on a plane inclined 40° to the axis of the cylinder. Calculate the shearing stress on the plane of failure. The same cylinder of cast iron is pulled in tension. Give your estimate of the failure load and the plane of failure.

1-8 A bar of uniform cross section is subjected to an axial stress of 10 MN/m^2. θ is the angle between the transverse axis and a plane oblique to the longitudinal axis. Plot the normal stress and shear stress on this oblique plane over the range $\theta = 0°$ to $\theta = 90°$.

CHAPTER 2

2-1 Find the principal stresses and the orientation of the axes of principal stress with the x, y axes for the following situations:

(*a*) $\sigma_x = +50{,}000$ psi, $\sigma_y = +5{,}000$ psi, $\tau_{xy} = -8{,}000$ psi

(*b*) $\sigma_x = -60{,}000$ psi, $\sigma_y = +5{,}000$ psi, $\tau_{xy} = +25{,}000$ psi

2-2 Construct a Mohr's circle of stress for each of the plane-stress conditions given in Prob. 2-1.

2-3 The three-dimensional state of stress is given by

$$\sigma_x = +8{,}000 \text{ psi} \qquad \tau_{xy} = +2{,}000 \text{ psi}$$

$$\sigma_y = -4{,}000 \text{ psi} \qquad \tau_{yz} = +3{,}000 \text{ psi}$$

$$\sigma_z = +6{,}000 \text{ psi} \qquad \tau_{xz} = -5{,}000 \text{ psi}$$

Determine

(*a*) The total stress (magnitude and direction with x, y, z axes) on plane described by direction cosines $l = +1/\sqrt{2}$, $m = +\frac{1}{2}$, n = negative;

(*b*) magnitude of normal and shear stresses on this plane;

(*c*) principal stresses and direction cosines of the principal planes;

(*d*) maximum shear stress.

2-4 The state of stress at a point for a system of reference axes x, y, z is given by

$$\begin{pmatrix} 200 & 100 & 0 \\ 100 & 0 & 0 \\ 0 & 0 & 500 \end{pmatrix}$$

If a new set of axes x', y', z' is formed by rotating x, y, z 60° about the z axis, find the matrix of the stress tensor for the new axes x', y', z' through the same point. Use tensor concepts.

2-5 Determine the shear stress $\tau_{y'x'}$ for the x' axis inclined at $\theta = 30°$ to the x axis. The stress state is given by

$$\begin{bmatrix} \sigma_x & \tau_{xy} \\ \tau_{yx} & \sigma_y \end{bmatrix} = \begin{bmatrix} 270 & 320 \\ 320 & -210 \end{bmatrix} \text{MPa}$$

2-6 The strain produced by a temperature change is given by $e_{ij} = \alpha_{ij}(T - T_0)$, where α is the coefficient of thermal expansion.

(*a*) What is the rank of the thermal expansion tensor α?

(*b*) Because of the symmetry of cubic crystals α_{ij} is isotropic. Write the components of the α tensor for this case.

(*c*) In lower symmetry crystals, such as Sn and U, α_{ij} is not isotropic. What practical problems will this introduce?

2-7 Write a matrix giving the components of the hydrostatic (mean) strain tensor. Evaluate the first, second, and third invariants of the hydrostatic strain tensor.

2-8 If $E = 210$ GPa, $G = 79$ GPa, and $\nu = 0.33$, compute ε_z and ε_{xy} for the state of stress given in Prob. 2-5.

2-9 For the strain-gage arrangement shown in Fig. 2-17, $\alpha = 30°$, and $\beta = 40°$. If $e_a = 2.6 \times 10^{-3}$, $e_b = 0.4 \times 10^{-3}$, and $e_c = 1.4 \times 10^{-3}$, determine the principal strains by means of Mohr's circle.

2-10 On a plate of material ($E = 25 \times 10^6$ psi, $\nu = 0.25$) strain gages are arranged as shown. When the plate is loaded, the gages read $e_1 = 1{,}860 \times 10^{-6}$, $e_2 = 185 \times 10^{-6}$, and $e_3 = 1{,}330 \times 10^{-6}$.

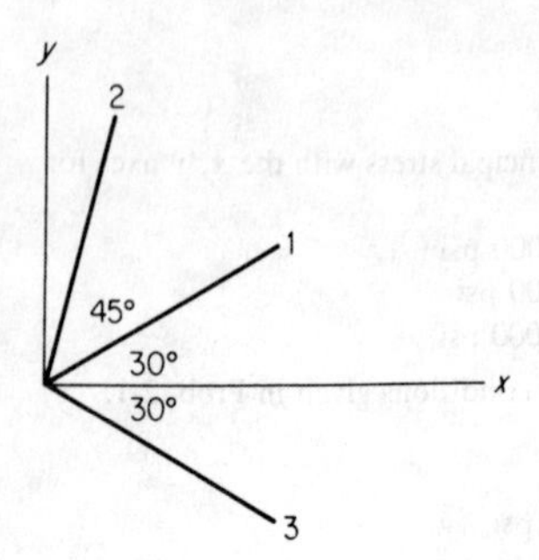

(*a*) What is the largest normal stress?
(*b*) What is the smallest normal stress?
(*c*) What is the largest shear stress?

2-11 In a nylon part (Young's modulus = 2,000 MPa and Poisson's ratio = 0.35) the normal strains at a free surface are measured along the three directions shown in Fig. 2-17, where $\alpha = \beta = 45°$. If

$$\varepsilon_a = 0.1 \times 10^{-3}, \qquad \varepsilon_b = -0.2 \times 10^{-3}, \qquad \varepsilon_c = 0.4 \times 10^{-3},$$

determine the following:
(*a*) The three principal strains and their directions;
(*b*) The magnitude of the greatest shear strain;
(*c*) The three principal stresses and their directions;
(*d*) The magnitude of the greatest shear stress;
(*e*) Draw Mohr's circle for the three-dimensional state of stress.

2-12 In an isotropic solid, a hydrostatic pressure produces a dilatation in which the strain in any direction is given by $\varepsilon = p(1 - 2\nu)/E$. However, in an aiosotropic solid the strain due to hydrostatic pressure is not equal in all directions.
(*a*) For a cubic crystal determine a relationship for strain in any direction in terms of E_x, ν_y, G_{xy}.
(*b*) Explain whether the dilatation can be calculated from $\Delta = \varepsilon_x + \varepsilon_y + \varepsilon_z$.
(*c*) Under the action of a hydrostatic pressure, what will be the shape of a face of a cube whose edges are originally parallel to the cubic crystal axes?

2-13 For a thin-wall cylindrical pressure vessel of internal diameter D and wall thickness h show that the stress in the circumferential direction is related to the internal pressure p by the equation $\sigma_t = pD/2h$. If the ends of the vessel are welded shut, what is the average longitudinal stress in the walls? Determine the ratio of circumferential to longitudinal stress. Why would these equations not hold for a thick-wall pressure vessel?

2-14 Strain gages on the outer surface of a pressure vessel read $e_l = 0.002$ in the longitudinal direction and $e_t = 0.005$ in the circumferential direction. Compute the stresses in these two principal directions: $E = 2.07 \times 10^5$ MPa and $\nu = 0.3$. What is the error if the Poisson effect is not taken into consideration?

2-15 It is found experimentally that a certain material does not change in volume when subjected to an elastic state of stress. What is Poisson's ratio for this material?

2-16 Determine the volume of a 10-cm copper sphere that is subjected to a fluid pressure of 12 MPa.

2-17 Assuming that atoms are hard elastic spheres, show that Poisson's ratio for a close-packed array of spheres is $\frac{1}{3}$.

CHAPTER 3

3-1 Determine the engineering strain e, the true strain ε, and the reduction in area q for each of the following situations:

(*a*) Extension from L to $1.1L$

(*b*) Compression from h to $0.9h$

(*c*) Extension from L to $2L$

(*d*) Compression from h to $\frac{1}{2}h$

(*e*) Compression to zero thickness

3-2 A 2-in-diameter forging billet is decreased in height (upset) from 5 to 2 in

(*a*) Determine the average axial strain and the true strain in the direction of compression.

(*b*) What is the final diameter of the forging (neglect bulging)?

(*c*) What are the transverse plastic strains?

3-3 A 1-in-thick plate is decreased in thickness according to the following schedule: 0.50, 0.25, 0.125 in. Compute the total strain on the basis of initial and final dimensions and the summation of the incremental strains, using

(*a*) conventional strain and

(*b*) true strain.

How does this show an advantage for the use of true strain in metal-forming work?

3-4 Show that constancy of volume results in $e_1 + e_2 + e_3 = 0$ and $\varepsilon_1 + \varepsilon_2 + \varepsilon_3 = 0$. Why is the relationship for conventional strain valid only for small strains but the relationship for true strains is valid for all strains?

3-5 A steel shaft ($\sigma_0 = 100{,}000$ psi) is subjected to static loads consisting of a bending moment of 200,000 in-lb and a torsional moment of 500,000 in-lb. Using a factor of safety of 2, determine the diameter of the shaft based on

(*a*) maximum-shear-stress theory and

(*b*) distortion-energy theory.

3-6 A hexagonal crystal has its basal slip plane normal to the α axis and the slip direction parallel to the β axis. If the stresses σ_{11} and σ_{22} are applied as shown in the sketch, develop and plot the two-dimensional yield locus for the 1, 2 system of axes. Let $k_{\alpha\beta}$ be the shear stress for yielding on the $\alpha\beta$ slip system.

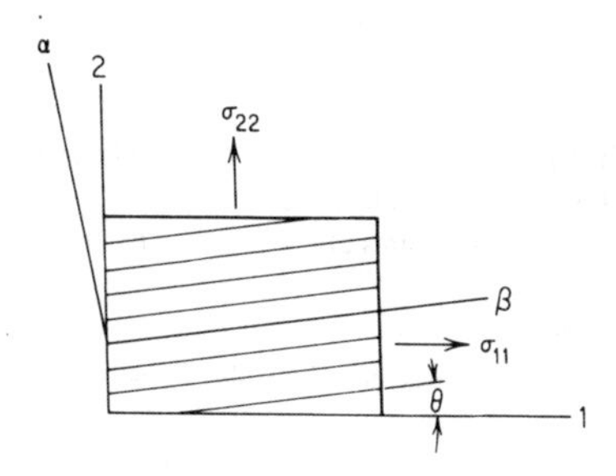

3-7 If the yield strength of a steel is 950 MPa, determine whether yielding will have occurred on the basis of both the von Mises and Tresca criteria. The stress state is given by

$$\begin{bmatrix} 0 & 0 & 300 \\ 0 & -400 & 0 \\ 300 & 0 & -800 \end{bmatrix} \text{MPa}$$

3-8 The yield stress of a tension specimen machined from a $\frac{1}{2}$-m wide by 0.6-cm thick copper sheet is 145 MPa. The sheet is rolled further with an applied tensile force in the plane of the sheet of 0.22 MN.

What roll pressure is needed to just cause yielding? Ignore friction between the metal and the rolls. For rolling, deformation occurs by plane-strain in which there is no increase in the width of the sheet.

3-9 If the yield strength of the steel used in the vessel in Prob. 2-13 is 30,000 psi, D is 12 in and h is $\frac{1}{4}$ in, what will be the value of internal pressure at which the tank will yield?

3-10 A thin-walled tube 10 in long has a wall thickness of 0.1 in and an initial outside radius of 2.0 in. The plastic stress-strain curve for the material is given by $\bar{\sigma} = 100{,}000\bar{\varepsilon}^{0.5}$. The tube is subjected to axial load P and torque T so that the stress ratio $\tau_{z\theta}/\sigma_z = 1$ at all times. The deformation is continuous until the effective stress equals 25,000 psi. Determine:

(*a*) The effective strain at the end of the deformation process
(*b*) The final dimensions of the tube
(*c*) The final axial load and torque

3-11 On the assumption that there is no change in width during the rolling of a sheet, derive the expression for the significant strain in terms of the change in thickness of the sheet. Also express in terms of percentage reduction.

3-12 A steel shaft transmits 400 hp at 200 rpm. The maximum bending moment is 8,000 ft-lb. The factor of safety is 1.5. Find the minimum diameter of solid shaft on the basis of

(*a*) maximum-shear-stress theory
(*b*) distortion-energy theory. ($\sigma_0 = 40{,}000$ psi, $E = 30 \times 10^6$, $\nu = 0.30$.)

3-13 A steel with a yield strength in tension of 42,000 psi is tested under a state of stress where $\sigma_2 = \sigma_1/2$, $\sigma_3 = 0$. What is the stress at which yielding will occur if it is assumed that

(*a*) the maximum-normal-stress theory holds,
(*b*) the maximum-shear-stress theory holds,
(*c*) the distortion-energy theory holds?

3-14 A material is tested under the state of stress $\sigma_1 = 3\sigma_2 = -2\sigma_3$. Yielding is observed at $\sigma_2 = 20{,}000$ psi.

(*a*) What is the yield stress in simple tension?
(*b*) If the material is used under conditions such that $\sigma_1 = -\sigma_3$, $\sigma_2 = 0$, at what value of σ_3 will yielding occur? (Assume distortion-energy criterion of yielding.)

3-15 Show that slip-line fields meet at 45° to a free surface on a frictionless surface.

CHAPTER 4

4-1 (*a*) Interstitials in the fcc lattice occupy the tetrahedral positions $(\frac{1}{4}, \frac{1}{4}, \frac{1}{4})$ and the octahedral positions $(\frac{1}{2}, \frac{1}{2}, \frac{1}{2})$. Indicate these interstitial positions on a sketch of the structure cell

(*b*) In the bcc lattice interstitials occupy sites at $(\frac{1}{2}, \frac{1}{4}, 0)$ and smaller sites at $(\frac{1}{2}, \frac{1}{2}, 0)$ and $(\frac{1}{2}, 0, 0)$. Show these sites on a sketch.

Compare the number of available interstitial sites in the fcc and bcc lattices.

4-2 For what geometrical conditions can a single crystal be deformed but not show slip lines on its surface?

4-3 Prove that the c/a ratio for an ideal hcp structure is $\sqrt{8/3}$.

4-4 How many atoms per square millimeter are there on a (100) face of a copper crystal? ($a_0 = 3.60$ Å.)

4-5 How many atoms per square millimeter are there on a (111) face of a copper crystal?

4-6 Verify the values for the distance between planes for fcc and bcc lattices given in Table 4-2.

4-7 (*a*) Show by means of a model or sketch that the intersection of {111} in an fcc structure cell produces an octahedron

(*b*) How many favorable slip systems are there for tensile straining along a [001] axis?
(*c*) How many for deformation along [111]?

4-8 A piece of copper has a measured density of 8.91. The lattice parameter of fcc copper is 3.6153 Å and its atomic mass is 63.51 g per g-mole. Calculate the percentage of vacancies in the pure copper.

4-9 If the planes shown in Fig. 4-15 are (111) planes what is the direction and length of the Burgers vector if the crystal is a bcc structure with a lattice parameter 3.303 Å.

4-10 Determine the critical resolved shear stress for an iron crystal which deforms by simultaneous slip on $(110)[\bar{1}11]$, $(110)[\bar{1}1\bar{1}]$, $(\bar{1}10)[111]$, and $(\bar{1}10)[11\bar{1}]$ when the tensile stress along [010] is 95.2 MPa.

4-11 The following data for zinc single crystals were obtained by Jillson[1]:

$90 - \phi$	λ	P, g
6.5	18	20,730
19.5	29	7,870
30	30.5	5.280
40	40	4,600
61	62.5	5,600
85	85.5	28,500

$A = 122 \text{ mm}^2$

(*a*) Establish that Schmid's law is obeyed.
(*b*) Demonstrate the orientation dependence of the yield stress by plotting σ_0 versus $\cos\phi \cos\lambda$.

4-12 If a material has a dislocation density of 10^{12} cm^{-2}, what is the average distance between dislocations?

4-13 If the critical resolved shear stress for yielding in aluminum is 240×10^3 Pa, calculate the tensile stress required to cause yielding when the tensile axis is [001].

4-14 A bicrystal with a simple cubic crystal structure is oriented as shown below. Which crystal will slip first if the slip system is $\{100\}\langle 100\rangle$?

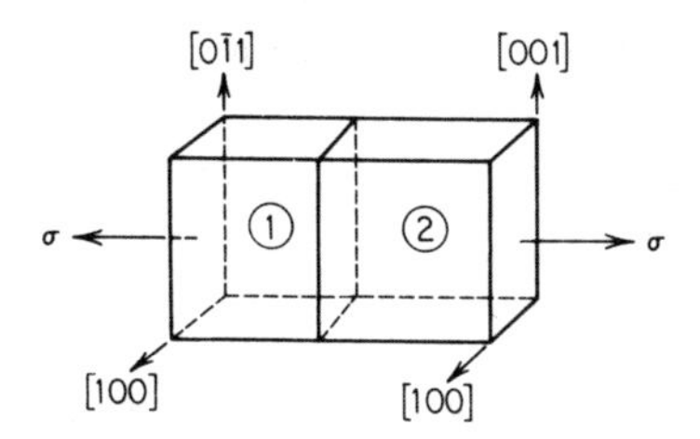

4-15 Along which crystallographic direction in an fcc crystal must the tensile axis be oriented to give multiple slip on
(*a*) four slip systems,
(*b*) six slip systems,
(*c*) eight slip systems?

4-16 Compare and contrast the deformation mechanisms of slip and twinning.

[1] D. C. Jilson, *Trans. AIME*, vol. 188, p. 1129, 1950.

CHAPTER 5

5-1 (*a*) Prove that the dislocation reaction $(a/2)[10\bar{1}] \rightarrow (a/6)[21\bar{1}] + (a/6)[11\bar{2}]$ is vectorially correct.

(*b*) Show that this reaction results in a decrease in strain energy.

5-2 What are the partial dislocations formed from $(a/2)[011]$ on the $(11\bar{1})$ in a face-centered cubic crystal?

5-3 What is the slip plane for the Lomer-Cottrell barrier formed by combining the $(a/2)[\bar{1}10]$ on (111) with the $(a/2)[101]$ on $(1\bar{1}\bar{1})$?

5-4 Construct a two-dimensional drawing of the (110) plane of a bcc lattice. By means of vector addition show that $(a/2)[\bar{1}\bar{1}1] + (a/2)[111] \rightarrow a[001]$ and that $(a/2)[\bar{1}\bar{1}1] + (a/2)[\bar{1}\bar{1}\bar{1}] \rightarrow a[\bar{1}\bar{1}0]$. Show that the last reaction is not energetically favorable.

5-5 Give the specific indices of the two {111} planes that contain the screw dislocation with the Burgers vector $a/2[\bar{1}01]$.

5-6 There is experimental evidence that [100] dislocations can form in alpha iron by the reaction $a/2[11\bar{1}] + a/2[1\bar{1}1] \rightarrow a[100]$. Show this reaction on a sketch of the iron unit cell and determine that the reaction is energetically favorable.

5-7 Using Fig. 5-13, show that the y components in Cottrell's equation $(a/2)[\bar{1}\bar{1}1] + (a/2)[111] \rightarrow a[001]$ cancel each other with a release of elastic energy. Show that this equals the energy gained by the x components of the reaction.

5-8 Express Eq. (5-4) in terms of the polar coordinates r and θ. By letting $\theta = 0$ show that the shear stress on the slip plane in the direction of the Burgers vector is equal to $\tau = \tau_0 b/r$.

5-9 Determine the equation for the energy of a mixed dislocation consisting of edge and screw components. Let θ be the angle between the Burgers vector and the dislocation line.

5-10 (*a*) Derive an expression for the volume change at a point in a crystal near an edge dislocation. What is the total volume change in the region surrounding the dislocation, not including the core?

(*b*) What is the volume change at a point near a screw dislocation? Explain the difference between this result and that for the edge dislocation.

5-11 The yield stress of most single crystals is $10^{-4}G$ or less. By using the concept of the Frank-Read source, develop an argument in which the yield stress is determined by the spacing of sources initially present in the crystal. Show that this leads to a reasonable estimate for the dislocation density. Give some alternative explanations for the yield stress.

5-12 (*a*) Estimate the strain energy of an edge dislocation where $r = 1$ cm, $r_0 = 10^{-7}$ cm, $G = 5 \times 10^{11}$ dyn/cm^2, $b = 2.5 \times 10^{-8}$ cm, and $\nu = \frac{1}{3}$. Express the result in electron volts per atomic plane. How much energy (electron volts) is required to produce 1 cm of dislocation line?

(*b*) Show that more than one-half the strain energy resides outside the core of the dislocation in the region $r = 10^{-4}$ to $r = 1$ cm.

5-13 (*a*) Show that the shear stress produced on a slip plane by an edge dislocation on a parallel plane at coordinates r, θ is given by

$$\tau_{xy} = \frac{Gb}{8\pi(1-\nu)} \frac{1}{h} \sin 4\theta$$

where h is the perpendicular distance between two parallel planes.

(*b*) What is the shear stress required for two edge dislocations to pass on parallel slip planes that are 1,000 Å apart in an annealed copper crystal? (Use the data in Prob. 5-12.)

5-14 If the binding energy U_i of carbon atoms to edge dislocations is about 0.2 eV, show that the dislocation sites become readily saturated at room temperature.

CHAPTER 6

6-1 Calculate the number of grains in the top surface of the head of a pin (0.1 in diameter) if the pin is made from steel with ASTM grain size 6.

6-2 What is the grain-boundary area in 1 in^3 of steel with ASTM grain size 6? Assume that the grains have a cubic shape.

6-3 If a grain-size count reveals 22 grains per square inch on a photomicrograph taken at a magnification of 250 ×, what is the ASTM grain-size number?

6-4 If the grain boundaries of an iron sample (ASTM grain size 6) are completely covered with a monatomic layer of oxygen, how much oxygen, in ppm, is present? (Radius of the oxygen atom is 0.6 Å.)

6-5 For a steel with ASTM grain size 3, what reduction in grain size would double the yield stress? Express as both average grain diameter and ASTM number.

6-6 The energy of a low-angle boundary is given by $E = E_0\theta(A - \ln\theta)$, where E is the energy per unit boundary area, E_0 is the elastic strain energy of a dislocation, θ is the angle of misfit, and A is a constant.

(*a*) Plot E/θ versus $\ln\theta$.

(*b*) Plot E versus θ.

6-7 The dislocation density ρ is the total length of dislocation line in the crystal volume. The units are number per square centimeter. Show that the relationship between ρ and the radius of curvature of a bent crystal, R, is given by $\rho = 1/Rb$.

Hint: Use Fig. 6-5.

6-8 Calculate the modulus of elasticity for a two-phase alloy composed of 50 vol percent cobalt, and 50 vol percent tungsten carbide on the basis of

(*a*) equality-of-strain and

(*b*) equality-of-stress hypotheses.

$E_{Co} = 30 \times 10^6$ psi; $E_{WC} = 100 \times 10^6$ psi. Compare the calculated values with the measured modulus of 54×10^6 psi.

6-9 For a unit cube containing a dispersed second phase of uniform spherical particles of radius r, the number of particles per unit volume N_v is related to the volume fraction of particles f by $N_v = f/\frac{4}{3}\pi r^3$. The average distance between particles is given by $\lambda = 1/(N_v)^{1/3} - 2r$.

(*a*) Show that $\lambda = 2r[(1/1.91f)^{1/3} - 1]$.

(*b*) For $f = 0.001, 0.01, 0.1$ and $r = 10^{-6}, 10^{-5}, 10^{-4}$, and 10^{-3} cm calculate λ. Plot f versus λ on log-log paper.

6-10 If we take the limit on matrix strength as $G/1{,}000$ (yield strength of the pure matrix) and $G/30$ (fracture stress of the matrix), show that the limits on interparticle spacing for effective dispersion hardening are between about 0.01 and 0.3 μm. To minimize low ductility, the volume fraction of the second phase should be less than about 15 percent. What restrictions does this place on particle size?

6-11 A SiC coated boron fiber (Borsic) is used to reinforce an aluminum matrix. Using the property data given below, determine the following properties for a composite containing 30 volume percent of fibers:

(*a*) E parallel to fiber axis;

(*b*) E perpendicular to fiber axis;

(*c*) strength of composite parallel to fiber.

	E	Tensile strength
Borsic fiber	55×10^6 psi	400,000 psi
Aluminum	10×10^6 psi	5,000 psi

6-12 If the shear strength of a grain boundary in iron is assumed to be $G/10$, determine the stress required to penetrate the grain boundary for
(*a*) a grain size of ASTM 2
(*b*) a grain size of ASTM 12.

6-13 Tests on alloy single crystals, such as 70/30 brass, show that the lattice rotation "overshoots" beyond the [001] – $[1\bar{1}\bar{1}]$ symmetry boundary before duplex slip occurs. On the basis of your knowledge of the effect of alloying on dislocation behavior, offer an explanation.

6-14 Considering that ductility is opposed to strength, propose on the basis of strengthening fundamentals the steps that can be taken to improve the ductility of an ordered Ti-Al alloy with a noncubic crystal structure.

CHAPTER 7

7-1 Estimate the theoretical fracture stress of iron if the surface energy is 1.2 J/m^2. How does this compare with the highest observed strength of heat-treated steel?

7-2 Using Griffith's equation for plane stress, determine the critical crack length for
(*a*) iron.
(*b*) zinc and
(*c*) NaCl if the following constants apply:

	σ_c, dyn/cm^2	γ, erg/cm^2	E, dyn/cm^2
Iron	9×10^9	1,200	20.5×10^{11}
Zinc	2×10^7	800	3.5×10^{11}
NaCl	2×10^7	150	5.0×10^{11}

7-3 What is the critical crack length in iron according to the Griffith-Orowan equation if $\gamma_p \approx 10^6$ erg/cm^2?

7-4 Show that the Cottrell mechanism for the creation of a microcrack by slip on intersecting planes is not energetically favorable in an fcc lattice.

7-5 Using Eq. (7-27), discuss how it qualitatively predicts the effect of the following factors on the ductile-to-brittle transition:
(*a*) material composition,
(*b*) impurities,
(*c*) melting practice,
(*d*) temperature,
(*e*) grain size,
(*f*) state of stress,
(*g*) rate of loading.

CHAPTER 8

8-1 The following data were obtained during the tension test of a low-carbon steel with a specimen having a 0.505-in diameter and a 2-in gage length.

Load, lb	Elongation, in	Load, lb	Elongation, in
500	0.00016	6,300	0.020
1,000	0.00032	7,000	0.060
1,500	0.00052	7,500	0.080
2,000	0.00072	8,500	0.12
2,500	0.00089	9,600	0.18
3,000	0.00105	10,000	0.26
3,500	0.00122	10,100	0.30
4,000	0.00138	10,200	0.50
4,500	0.00154	10,050	0.58
5,000	0.00175	9,650	0.62
5,500	0.00191	9,100	0.70
6,000	0.00204	8,100	0.76

Yield point, 6,200 lb; breaking load, 6,800 lb; final gage length, 2.87 in; final diameter 0.266 in.

(*a*) Plot the engineering stress-strain curve.

(*b*) Determine the (1) proportional limit; (2) modulus of elasticity; (3) lower yield point; (4) tensile strength; (5) fracture stress; (6) percentage elongation; (7) reduction of area.

(*c*) Plot the true-stress–true-strain curve up to maximum load.

(*d*) Determine $d\sigma/d\varepsilon$ at maximum load and the value of the strain-hardening exponent n.

8-2 Derive the relationship for the toughness of a metal whose true stress-strain curve obeys the power law $\sigma = K\varepsilon^n$.

8-3 (*a*) Show that $q = 1 - e^{-\varepsilon_f}$, where q is the reduction of area and ε_f is the true strain at fracture.

(*b*) If the true-stress–true-strain curve is given by $\sigma = K\varepsilon^n$, derive a relationship between the yield stress and the reduction in cross-sectional area.

8-4 For any finite value of e in the region of uniform tensile elongation, show that $d\sigma/d\varepsilon > ds/de$ according to the relationship $d\sigma/d\varepsilon = (ds/de)(1 + e)^2 + s(1 + e)$.

8-5 A useful measure of ductility for evaluating formability is the *zero-gage-length elongation* e_0. This represents the maximum possible elongation on the minimum possible gage length. Show that $e_0 = q/(1 - q)$. If q is measured at some point on the tensile specimen other than the neck, it gives a value of uniform reduction. Why?

8-6 The following data were obtained during the true-stress–true-strain test of a nickel specimen:

Load, lb	Diameter, in	Load, lb	Diameter, in
0	0.252	3,570	0.201
3,440	0.250	3,500	0.200
3,580	0.245	3,350	0.190
3,670	0.240	3,150	0.180
3,710	0.235	2,950	0.170
3,720	0.230	2,800	0.149

(*a*) Plot the true-stress–true-strain curve.

(*b*) Determine the following: (1) true stress at maximum load; (2) true fracture stress; (3) true fracture strain; (4) true uniform strain; (5) true necking strain; (6) ultimate tensile strength; (7) strain-hardening coefficient.

8-7 Will the superposition of lateral hydrostatic pressure affect the strain at which necking occurs in a tensile test? Explain.

8-8 For a material whose flow curve is given by $\sigma = K\varepsilon^n$, derive an expression for the tensile-yield ratio in terms of n and show how s_u/s_0 can be used to determine n.

8-9 Show that the tensile strength for a material that follows the power law is

$$s_u = K\left(\frac{n}{e}\right)^n$$

where $e = 2.718$.

8-10 The variation of percentage elongation X with gage length has been found empirically to follow the relationship $X = c\sqrt{A}/L_0 + b$, where c and b are constants. The following data were obtained for a steel plate that is 2 in wide and $\frac{1}{2}$ in thick:

L_0, in	2	4	6	10	12	14
X	40.0	30.6	27.5	25.7	25.0	24.3

Find b and c, and determine the percentage elongation for a 2- by $\frac{7}{8}$-in steel plate with a 10-in gage length.

CHAPTER 9

9-1 Estimate the Mohs' hardness for the following materials:
(*a*) steel file,
(*b*) chalk,
(*c*) pine plank,
(*d*) ball bearing,
(*e*) sapphire.

9-2 (*a*) Show by means of a sketch that in order to produce geometrically similar indentations in a Brinell test d/D must remain constant.
(*b*)

Diameter of indenter, mm	Diameter of indentation, mm	Load, kg
10	4.75	3,000
7	3.33	1,470
5	2.35	750
1.2	0.57	425

P/D^2 is constant for the above data. Determine the BHN, and show that it is approximately constant with load.

9-3 The following hardness data were obtained on copper by using a 10-mm ball indenter:

Load, kg	Diameter of indentation, mm		
	Annealed	$\frac{1}{4}$ hard	$\frac{1}{2}$ hard
500	4.4	3.2	2.9
1,000	5.4	3.9	3.7
1,5000	6.2	4.6	4.5
2,000		5.4	5.3
2,500		5.9	5.7

(*a*) Determine whether or not Meyer's law is obeyed.
(*b*) Determine the Meyer's law constants k, n', and C.
(*c*) Plot the BHN and Meyer hardness as a function of load for the $\frac{1}{4}$ hard copper.

9-4 If a $\frac{1}{4}$-in-diameter ball is used for the annealed copper in Prob. 9-3, calculate what the indentation would be at 500-, 1,000-, and 1,500-kg loads. Compare with the observed values.

Load, kg	Diameter of indentation, mm
500	4.0
1,000	4.6
1,500	5.5
2,000	5.9

Determine k and n', and compare with the values for a 10-mm ball.

CHAPTER 10

10-1 A 1-in-diameter hot-rolled steel bar was tested in torsion, with the following results:

Torque, in-lb	Number of $\frac{1}{4}$ turns of bar	Torque, in-lb	Number of $\frac{1}{4}$ turns of bar
6,700	1	10,600	12
7,400	2	11,000	15
8,200	3	11,400	18
8,700	4	11,800	24
9,100	5	12,400	32
9,700	7	12,600	38
10,200	9	12,800	39

If the length of the bar between chucks is 18 in, determine
(*a*) shear stress–shear-strain curve;
(*b*) modulus of rupture;
(*c*) shear-stress–shear-strain curve corrected for inelastic strain.
(*d*) If the twisting moment at the yield point was 4,500 in-lb and the angle of twist was 2.6°, determine the torsional yield stress, the modulus of elasticity, and the modulus of resilience.
(*e*) What is Poisson's ratio if $E = 29.0 \times 10^6$ psi?

10-2 Compare the torsional strength, stiffness, and weight of two steel shafts, one solid, the other hollow. The hollow shaft has twice the diameter of the solid shaft, but the solid cross-sectional area is the same.

10-3 Assume that a certain brittle material will fail at a tensile stress of 30,000 psi.
(*a*) For a bar with a $\frac{1}{4}$-in diameter, what torsional moment will cause failure? Sketch the failure.
(*b*) If this is an ideal brittle material ($\nu = \frac{1}{4}$), show that greater strain will be achieved in torsion than tension before failure occurs.

10-4 Explain why necking is observed in the tension test of a ductile material but not in the torsion test of the same material.

CHAPTER 11

11-1 A cylindrical pressure vessel 8 m long, 4 m in diameter exploded at an internal pressure of 10 MPa. The vessel is made from 20-mm-thick steel with the following properties: $E = 210$ GPa; $\sigma_0 = 1.2$ GPa; $\mathcal{G} = 120$ kJ/m^2. Show that a conventional strength of materials analysis is not consistent with the facts, but that the failure is compatible with the existence of a longitudinal crack about 10 mm long.

11-2 A ship steel has a value of $\mathcal{G}_c = 200$ lb/in.

(*a*) What is the fracture stress in a thin plate that is 12 in wide and that contains a central crack 0.5 in long?

(*b*) If the crack is 2 in long, what is the fracture stress?

(*c*) Increasing the plate thickness to 5 in reduces $\mathcal{G}_c$ to 100 lb/in. What is the fracture stress for a 0.5-in-long crack?

11-3 A steel plate is 12 in wide and $\frac{1}{4}$ in thick. There is a 1-in-long crack along each edge. If $K_{Ic} = 77.5$ ksi$\sqrt{\text{in}}$:

(*a*) Calculate the force required to propagate the crack the remaining 10 in across the width of the plate.

(*b*) Calculate the force required to break the plate in tension if there were no crack. Assume that the fracture strength is 100,000 psi.

(*c*) Calculate the force required to break the plate in tension if the steel had the theoretical cohesive strength.

11-4 For crack sizes of 4 and 7 mm, values of specimen compliance ($1/M$) of 1.485×10^{-5} mm/N and 1.635×10^{-5} mm/N were obtained for a single-edge notched plain carbon steel specimen at a critical load value of 10 kN. What is the fracture stress for a 10-mm-long crack in a wide plate?

11-5 A spherical pressure vessel is to be constructed by welding curved steel plates. The welds are inspected by radiography using a technique that is certain to detect any crack greater than 0.1 in long. Three grades of maraging steel with the properties given below are being considered. If the wall thickness of the vessel is 1 in and the diameter is 10 ft, which steel will give the greatest pressure capability?

Grade of steel	Yield strength, psi	K_{Ic}, psi$\sqrt{\text{in}}$
200	215,000	57,000
250	245,000	48,000
300	290,000	33,500

11-6 A compact tension specimen (Fig. 11-8) is tested according to ASTM standard E399. A type I load-displacement curve (Fig. 11-9) is obtained with $P_Q = 120$ kN. The specimen dimensions are $B = 5$ cm, $W = 10$ cm, and $a = 5$ cm. If the material yield stress is 600 MPa, are the conditions correct for a valid K_{Ic} measurement?

11-7 An internal crack has been detected with ultrasonics in the shell of a large cylindrical steel pressure vessel. The stresses in the region of the crack are: longitudinal tensile stress of 60,000 psi; circumferential tensile stress of 140,000 psi; radial compressive stress of 5,000 psi. The pressure vessel is so large that there is no appreciable variation of stress over the region containing the crack. The crack is oriented so that its normal makes an angle of 40° with the longitudinal direction and 60° with the circumferential direction.

(*a*) What is the normal stress on the plane of the crack? What is the shear stress in the plane of the crack?

(*b*) If K_{Ic} for the steel is 50 ksi$\sqrt{\text{in}}$, what is the smallest size crack that will cause fracture of the pressure vessel? State the assumptions made in arriving at this answer.

11-8 A cylindrical steel pressure vessel with a yield strength of 360 MPa is subjected to a hoop stress of 140 MPa. A tensile residual stress of 80 MPa also can be assumed to be present. The lowest service temperature of the vessel will be $-25°$C, under which conditions the K_{Ic} is 1,400 MPa$\sqrt{\text{mm}}$. The vessel will be designed according to a "leak-before-break" philosophy in which a detectable leak will occur before brittle fracture could occur. Determine the thickness of the pressure vessel based on fracture mechanics considerations.

CHAPTER 12

12-1 Peterson has shown that for an average value of $K_t = 2$ the notch sensitivity index is given by $q = 1/(1 + a/r)$, where a is a "particle size" dependent on the material and r is the notch radius. Kuhn and Hardrath[1] have related a with the tensile strength of the steel.

s_u, ksi	200	160	100	50
a	0.0004	0.001	0.0042	0.012

Using these data, plot a curve of q versus r for steels of different strengths.

12-2 Show that increasing the size of a fatigue specimen tested in bending decreases the stress gradient. Why would this result in a size effect?

12-3 Why are the notched fatigue properties of ordinary gray cast iron about equal to the unnotched properties?

12-4 Using the ideas about stress gradient, show that the increase in fatigue limit due to nitriding the surface of a steel shaft is given approximately by $1/(D/2\delta - 1)$, where δ is the depth of the nitrided layer and D is the diameter of the shaft. Would this equation be applicable for axial loading? Explain why nitriding is an effective means of counteracting stress raisers.

12-5 A steel shaft is subjected to a completely reversed bending moment of 6,000 in-lb. A transition section at one end contains a fillet with a 0.10-in radius. The steel has a tensile strength of 100,000 psi, and the fatigue limit is 45,000 psi. K_t for the fillet is about 2, and the dynamic factor of safety is 2.5. What is the minimum allowable diameter of the shaft?

12-6 A 5-cm-diameter shaft is subjected to a static axial load of 100 kN. If the yield stress of the material is 350 MN/m^2 and the fatigue limit ($R = -1$) is 260 MN/m^2, what is the largest completely reversed bending moment that can be applied? The factor of safety is 2, and K_f due to the presence of notches is 1.8.

12-7 A threaded bolt experiences a completely reversed axial load of 60,000 lb. The static factor of safety is 2.0, and the dynamic factor of safety is 3.0. $K_t = 2.0$. The root radius of the threads is 0.020 in.

(*a*) Compare the root area required for static and dynamic service for a high-strength bolt ($\sigma_0 = 180{,}000$ psi, $\sigma_u = 200{,}000$ psi).

(*b*) Repeat for a low-strength bolt ($\sigma_0 = 30{,}000$ psi, $\sigma_u = 50{,}000$ psi). Use data in Prob. 12-1.

(*c*) Determine the ratio of the areas for low-strength and high-strength bolts for the static and dynamic conditions.

12-8 A steam turbine used for peaking service experiences a thermal cycle from room temperature (70°F) to a service temperature of 1000°F several times each day. The *total* strain amplitude experienced by a turbine blade each cycle is $\varepsilon_a = \alpha(T_s - T_r)$. The properties of the blade material over the temperature interval are: 60,000-psi yield strength, 80,000-psi tensile strength, 30 percent reduction in area, elastic modulus 24×10^6 psi, and coefficient of thermal expansion 10^{-5} per °F. How many cycles can the blade withstand?

[1] NACA Tech. Note 2805, 1952.

12-9 A pressure cylinder with a 4-in. inside diameter and a $\frac{1}{4}$-in wall thickness is subjected to internal pressure that varies from $-p/4$ to p. The steel has a fatigue limit of 60,000 psi in completely reversed bending. Under a repeated stress cycle of zero to a maximum stress in tension the same steel can withstand 90,000 psi without failure in 10^7 cycles. What is the maximum internal pressure that can be withstood by the cylinder without fatigue failure before 10^7 cycles?

12-10 For a steel material, $s_0 = 30{,}000$ psi, $s_u = 50{,}000$ psi, $\varepsilon_f = 0.30$, and $E = 30 \times 10^6$ psi. What is the limit on the total cyclic strain range if the steel is to withstand 4.9×10^5 cycles?

12-11 A high-strength steel has a yield strength of 100 ksi and a fracture toughness (K_{Ic}) equal to 150 ksi$\sqrt{\text{in}}$. Based on the level of nondestructive inspection, the smallest size flaw that can be detected routinely, a_i, is 0.3 in. Assume that the most dangerous crack geometry in the structure is a single-edge notch so that $K_{Ic} = 1.12\sigma\sqrt{\pi a}$. The structure is subjected to cyclic fatigue loading in which $\sigma_{max} = 45$ ksi and $\sigma_{min} = 25$ ksi. The fatigue-crack growth rate for the steel is given by $da/dN = 0.66 \times 10^{-8}(\Delta K)^{2.25}$. With this information, estimate the fatigue life of the structure by

(*a*) integrating numerically using an increment of crack growth of 0.1 in for each calculation

(*b*) using Eq. (12-25). What changes could be made to increase the fatigue life? Which factor would be the most influential?

12-12 Make a Weibull plot for the fatigue limit from the data given in Fig. 12-5. Indicate the stress levels at which 90, 50, and 1 percent of specimens would fail.

12-13 A 5-cm-diameter shaft is subjected to a static bending moment of 900 N-m. What is the value of the maximum twisting moment varying from 0 to M_T to produce failure in 10^7 cycles? The yield strength is 620 MPa and the fatigue limit at 10^7 cycles in completely reversed stress is 400 MPa.

CHAPTER 13

13-1 Construct a three-dimensional plot showing the relationship between stress, strain, and time, for $T = T_m/2$.

13-2 Assuming that the mechanical equation of state is completely valid, show how creep curves could be constructed from stress-strain curves.

13-3 The following rupture times[1] were obtained for stress rupture tests on a steel alloy:

Stress, psi	Temperature, °F	Rupture time, hr
80,000	1080	0.43
80,000	1030	6.1
80,000	1000	22.4
80,000	975	90.8
10,000	1400	1.95
10,000	1350	6.9
10,000	1300	26.3
10,000	1250	84.7

Establish the validity of the Larson-Miller and Manson-Haferd parameters.

13-4 By means of sketches, show how isochronous stress-strain curves are derived from creep data.

[1] S. S. Manson, G. Succop, and W. F. Brown, Jr., *Trans. Am. Soc. Met.*, vol. 51, p. 924, 1959.

13-5 Steady-state creep rate can be expressed by

$$\dot{\varepsilon}_s = B\sigma^n e^{-Q/kT}$$

Apply this relationship to the creep of a steel support rod in a boiler operating at 1000° F. The rod is stressed in tension to 8,000 psi and its creep elongation must not exceed 10 percent. Using the data given below, evaluate the constants in the creep equation and estimate the lifetime of the rod.

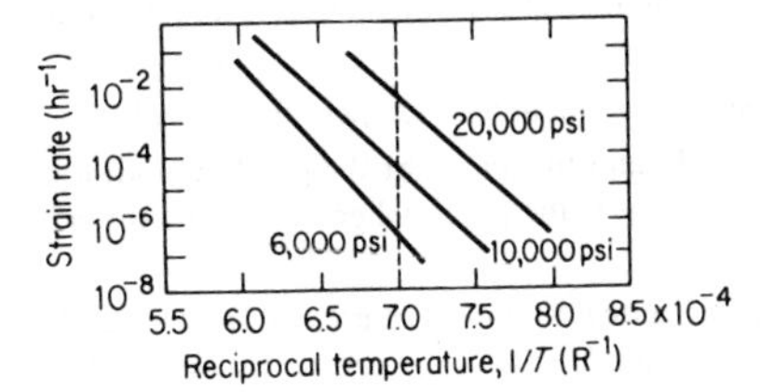

13-6 From Fig. 13-20,
(*a*) determine the stress required to produce rupture in 2,000 h at 1600° F;
(*b*) determine the rupture time at 1400° F for the same stress.

13-7 Using the data in Table 13-3, compute the value of Q.

13-8 Explain how the Larson-Miller parameter can be used as a figure of merit for elevated temperature materials. If we have two materials described by $P_A = T(C_A + \log t)$ and $P_B = T(C_B + \log t)$, which material would be stronger
(*a*) if $C_A < C_B$ and the curves of σ vs. P coincide;
(*b*) if $C_A = C_B$ and the curve of σ vs. P is higher for A than for B.

13-9 The following data[1] were obtained in creep tests at 1500° F on an austenitic high-temperature alloy.

Stress, psi	Minimum creep rate, %/h
10,000	0.00008
15,000	0.0026
20,000	0.025
30,000	2.0
40,000	30
50,000	320

Use these data to evaluate Eq. (13-31).

13-10 A tie rod 10 ft long is made from the austenitic alloy whose creep properties are given in Prob. 13-9. At 1500° F it is subjected to an axial load of 8,000 lb. The yield stress at 1500° F is 60,000 psi, and the stress for rupture in 1,000 h is 20,000 psi. Using a factor of safety of 3.0, establish the necessary cross-sectional area of the tie rod based on the following considerations:
(*a*) yield stress at 1500° F;
(*b*) 1,000 h rupture life at 1500° F;
(*c*) an allowable creep rate of 1 percent per 10,000 h.

13-11 A thin-wall pressure vessel of an austenitic alloy has a 1-in wall thickness and an 18-in inside diameter. The vessel operates at 1500° F. Find the allowable internal pressure if the maximum

[1] N. J. Grant and A. G. Bucklin, *Trans. Am. Soc. Met.*, vol. 42, p. 720, 1950.

allowable increase in diameter is 0.2 in over a 2-year period. Assume steady-state conditions, and use the values of B and n' derived from the data in Prob. 13-9.

CHAPTER 14

14-1 Sketch the transition-temperature curves for a plain carbon steel tested in tension, torsion, and notched impact.

14-2 The Izod test does not lend itself to testing specimens as a function of temperature. Why?

14-3 The weight of the pendulum of an impact tester is 45 lb, and the length of the pendulum arm is 32 in. If the arm is horizontal before striking the specimen, what is the potential energy of the tester? What is the striking velocity?

14-4 A weight W falls vertically downward (without friction) along a thin rod of area A and length L until it is brought to rest by striking a flange at the end of the rod. Using energy considerations, derive an expression for the maximum stress in the rod produced by the weight falling through a distance h. Assume perfectly elastic impact. Show that the sudden release of load from a height $h = 0$ produces twice the stress that would result from a gradual application of load.

14-5 Outline the steps that can be taken to eliminate hydrogen embrittlement.

14-6 From an engineering design viewpoint what steps can be taken to minimize stress-corrosion cracking?

14-7 The critical impact velocity for a steel is 160 ft/s, and the longitudinal sound velocity is 19,500 ft/s. If $E = 30 \times 10^6$ psi and the density is 0.32 lb/in^3, what is the fracture stress under these impulsive-loading conditions?

CHAPTER 15

15-1 Discuss the relative ease of hot-work and cold-work for the following metals: aluminum, zirconium, lead, SAE 4340 steel, molybdenum. Consider the following factors: melting point, flow stress, oxidation behavior, brittleness.

15-2 When a compressive force of 400 tons is applied to the top surface of a well-lubricated cube it just causes plastic flow. The cube is 3.16 in on each edge. What force would be required to produce flow if the faces of the cube other than the top face were constrained by die forces of 100 and 200 tons?

15-3 A circular plate is compressed by inclined dies, as shown below. The angle of the dies α is small and sliding friction occurs at the surface with coefficient of friction μ. Derive the differential equation for normal pressure. What is the significance of the case when $\alpha = \mu$?

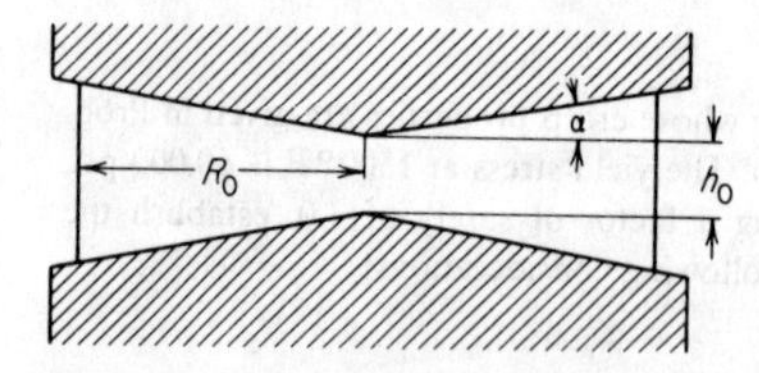

15-4 Calculate the ideal work of deformation per unit volume in drawing a wire through a die. Express in terms of the strain in the longitudinal direction.

15-5 The reduction per pass is given by $q_n = (h_{n-1} - h_n)/h_{n-1}$, where n is the number of the pass. Derive an expression between the total reduction $Q_n = (h_0 - h_n)/h_0$ and the summation of the reductions per pass q_n.

15-6 Determine $p/2k$ for a 50 percent plane-strain extrusion using the slip-line field given below.

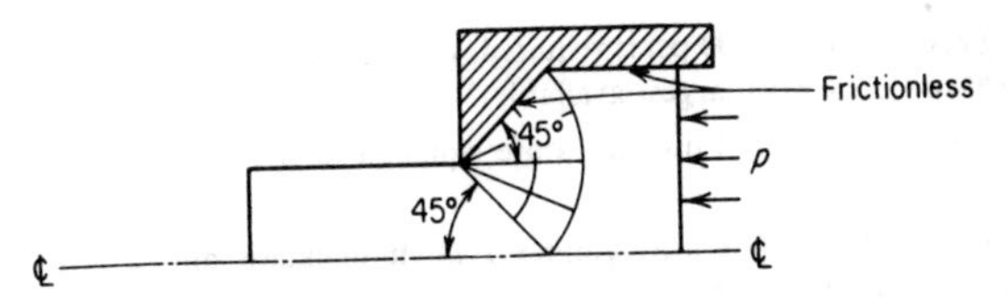

15-7 Determine $p/2k$ for a 50 percent extrusion (frictionless) using the upper-bound field given below.

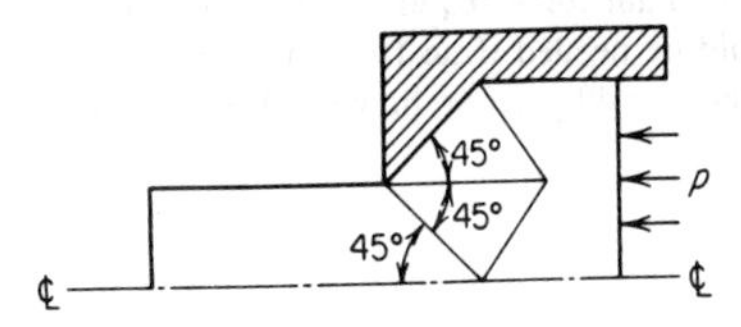

15-8 A frictionless uniaxial compression test of a copper cylinder gave the following results:

True stress, ksi	4	10	16	24	32	38	40	41.4	42
Reduction in height, %	0	2.5	5	10	20	30	40	50	60

(a) Plot the true-stress–true-strain curve in compression;

(b) Construct an engineering tensile stress-strain curve up to the ultimate tensile stress from these data.

15-9 Show how a value of flow stress independent of friction can be obtained with the compression test using an extrapolation method.

15-10 Derive the expressions for the flow stress and the strain in the plane-strain compression test.

15-11 Compare the advantages and disadvantages of the tension test, the compression test, the plane-strain compression test, and the torsion test for determining the flow stress for metalworking processes.

15-12 In upsetting cylindrical specimens two types of cracks are observed:

(a) vertical cracks which are parallel to the direction of compression and run into the specimen normal to its free surface, and

(b) inclined cracks at approximately 45° to the direction of compression. Both types of fracture occur in the plane of maximum shear stress, which is a diametral plane at midheight. By applying Mohr's circle in three-dimensions show the two different stress states that must exist to produce these different fractures.

CHAPTER 16

16-1 Plot a curve of p_{av}/σ_0 versus the parameter $2\mu a/h$ for a rectangular slab and a cylinder for values of the parameter from 0 to 5.

16-2 A 3,000-lb power forging hammer has a total nominal energy of 35,000 ft-lb. If the blow efficiency is 40 percent and the forging load builds up from $P/3$ at the beginning of the stroke to P at the end of the stroke, what is the total forging load for

(a) a stroke of 0.2 in,

(b) a stroke of 0.6 in.

16-3 What would be the effect of the section thickness h on the forging load if the friction were reduced to zero?

16-4 Assuming that an average forging pressure of 50,000 psi is required, what is the largest closed-die forging that could be made on a 35,000-ton press (the largest available)?

16-5 Show that the strains in the transverse and radial directions are equal for a thin circular disk compressed in the z direction.

16-6 Compare the average pressure needed to compress a cylinder 1 in in thickness and 2 ft in diameter with one $\frac{1}{8}$ in in thickness and 2 ft in diameter.

16-7 An SAE 1040 steel at the forging temperature has a yield stress of 6,000 psi. A right-circular cylinder 3 in high and 1 in in diameter is to be upset to half height between flat dies.

(*a*) If the coefficient of friction is 0.4, what is the maximum force required for the upsetting?

(*b*) How much extra force is required over what would be needed if no friction were present?

(*c*) If it takes 3 s to produce the forging and the efficiency is 40 percent, how much power must be available in order to do the job?

CHAPTER 17

17-1 Using the simplified theory of rolling, plot curves for the variation of rolling load with roll diameter, coefficient of friction, and mean strip thickness.

17-2 Sheet steel is reduced from 0.160 to 0.140 in with 20-in-diameter rolls having a coefficient of friction equal to 0.04. The mean flow stress in tension is 30,000 psi. Neglect strain hardening and roll flattening.

(*a*) Calculate the roll pressure at the entrance to the rolls, the neutral point, and the roll exit.

(*b*) If $\mu = 0.40$, what is the roll pressure at the neutral point?

(*c*) If 5,000-psi front tension is applied in Prob. 17-2*a* what is the roll pressure at the neutral point?

17-3 Calculate the minimum gage possible in hot rolling low-carbon steel strip at 700°F on a 23-in mill with steel rolls. ($\sigma_0' = 35{,}000$ psi, $\mu = 0.4$, $E = 20 \times 10^6$ psi.) What is the minimum gage for cold rolling on the same mill? ($\sigma_0' = 60{,}000$ psi, $\mu = 0.10$, $E = 30 \times 10^6$ psi.)

17-4 Explain the reason for designing rolling mills with complex clusters of rolls such as in Fig. 17-1*e*.

17-5 In hot rolling with an 18-in rolling mill a groove was placed on the roll so as to put an imprint on the sheet every revolution. If the distance between marks on the sheet was 62.1 in, what was the forward slip? What is the coefficient of friction in rolling from 0.30 to 0.20 in in thickness?

17-6 List three basic requirements for a good rolling lubricant.

17-7 The residual stresses produced in rolling are proportional to $(h_f/L)^2$, where L is the length of the arc of contact.[1] From the geometry of rolling show that $(h_f/L)^2 = (h_f/R)[(100 - q)/q]$, where $q = [(h_0 - h_f)/h_0]100$.

17-8 On a certain mill the rolling load for a 30 percent reduction is 19.5 tons/in of width. What is the rolling load when a back tension of 10 tons/in² and a front tension of 8 tons/in² are applied? Assume that $\sigma_0' = 35$ tons/in².

17-9 (*a*) Determine the rolling schedule for cold-rolling annealed copper and mild steel strips from $16 \times \frac{1}{8}$ in to 16×0.100 in with 6-in diameter rolls on a four-high mill. The rolling load of the mill may not exceed 100 tons. Use a simplified rolling theory. The following mean flow stresses were obtained with a plane-strain compression test: copper, 20 ksi; mild steel, 35 ksi.

(*b*) Determine the horsepower required to drive the mill in rolling the copper strip if it exits at 600 ft/min,

[1] R. McC. Baker, R. E. Ricksecker, and W. M. Baldwin, Jr., *Trans. AIME*, vol. 175, pp. 337–354, 1948.

(c) What would be the maximum temperature reached in rolling copper strip if there are no heat losses?

17-10 A metal strip, 0.160 in thick, is cold-rolled in one pass under the conditions given below. Determine the mill constant and the roll gap setting to produce a strip 0.125 in thick.

Rolling load, lb	Roll gap setting, in	Thickness of strip, in
50,000	0.143	0.150
81,000	0.132	0.145
101,000	0.122	0.138
135,000	0.098	0.119
57,000	0.070	0.095

17-11 Using the simplified rolling equation, Eq. (17-7), sketch the pressure distribution in the roll gap
(a) as a function of μ for constant Δh;
(b) as a function of reduction for constant μ.

CHAPTER 18

18-1 Derive Eq. (18-6) using a slab analysis.

18-2 Using Eq. (18-6), suggest a simple method for measuring the frictional shear stress if only breakthrough pressure, and not die pressure p_d, can be measured.

18-3 In the hot extrusion of aluminum at 600° F, $\sigma_0 = 12{,}000$ psi.
(a) For a 12-in-diameter billet, 36 in long, what is the breakthrough pressure required to extrude a 3-in-diameter bar if $\mu = 0.10$?
(b) What is the required extrusion pressure at the end of the stroke?
(c) What capacity press would be needed for this extrusion?

18-4 For extrusion through a flat die the strain rate is given approximately by $\dot{\varepsilon} = (6v/D)\ln R$, where v is the ram velocity, D is the billet diameter, and R is the extrusion ratio. For the conditions given in Prob. 18-3 compare the strain rates and the time the metal is in the die for a 2 in/s and 10 in/s ram speed.

18-5 Starting with Eq. (18-8), derive an expression for the adiabatic temperature rise in extrusion as a function of the extrusion ratio. See Fig. 15-15 for the trend.

18-6 The following equation expresses the pressure for the extrusion of aluminum bars:

$$p = \sigma_0(0.47 + 1.2\ln R)e^{4\mu L/D}$$

Billets 8 in in diameter and 16 in long are extruded into $\frac{3}{4}$-in-diameter bars. In order to increase the length of the product by 10 ft, would it be more economical in terms of pressure to increase the billet length or the diameter? (Assume that $\mu = 0.10$.)

CHAPTER 19

19-1 Plot a curve of the ratio of draw stress to flow stress vs. reduction for $B = 0$, 1, and 2.

19-2 Plot a curve of the ratio of draw stress to flow stress versus semiangle die for the conditions $\mu = 0.05$, $r = 0.4$, and $D_a = 0.2$ in.

19-3 For a value of $B = 1$, plot a curve similar to Prob. 19-1 which shows the effect of back pull on draw stress. Use ratios of back stress to flow stress of 0.0, 0.4, and 0.8.

19-4 Show that the strain rate of wire being pulled through a die can be expressed by $\dot{\varepsilon} = (2 \tan \alpha)/r(dx/dt)$. What is the maximum strain rate for a 0.040-in-diameter wire pulled at 2,000 ft/min through a 10° semiangle die?

19-5 Compare the force required to give a 30 percent reduction in a 0.5-in-diameter wire in a 10° semiangle die with the force required to produce the same reduction in a tube blank of 0.500-in outside diameter, 0.400-in inside diameter using a cylindrical mandrel. Assume that $\mu = 0.08$ for both cases.

19-6 Determine the maximum reduction by drawing for
(*a*) a non-strain-hardening material;
(*b*) annealed low-carbon steel ($n = 0.26$);
(*c*) annealed copper ($n = 0.5$).

19-7 A common phenomenon is "die ringing," circumferential wear at the die entrance. Show that the maximum die pressure occurs at this location.

CHAPTER 20

20-1 Derive Eq. (20-2) for the fiber strain in a bent sheet. What is the basic assumption in this derivation?

20-2 Show how minimum bend radius may be predicted by the tensile reduction of area, Eq. (20-3).

20-3 Use Mohr's circle of strain to establish that a local neck forms at 55° to the tensile axis in the uniaxial straining of an isotropic sheet.

20-4 The flow curves for two metals are given by

$$\text{Metal } A\text{: } \sigma = 89{,}000\,\varepsilon^{0.5}$$

$$\text{Metal } B\text{: } \sigma = 27{,}000\,\varepsilon^{0.1}$$

Both metals initially are loaded to a stress of 20,000 psi and a strain of 0.05. If the stress now is increased to 23,000 psi, what is the strain in each metal? What is the importance of this example to sheet-metal forming?

20-5 Estimate the punch load required for a deep-drawn cup with a 2-in diameter and 0.040-in wall if it is made from low-carbon steel, $\sigma_0 = 30{,}000$ psi, $\sigma_u = 50{,}000$ psi. The blank diameter is 4 in, and the overall coefficient of friction is 0.08. Estimate the hold-down pressure and force required for bending and unbending, from other considerations.

CHAPTER 21

21-1 List 10 specific or specialized machining operations or machine tools. Give examples of the type of part or geometry produced by each.

21-2 In orthogonal cutting of a low-carbon steel the specific cutting energy is 4 GJ/m³. The undeformed chip thickness is 0.2 mm and the chip width is 5 mm at a cutting speed of 1.1 m/s. The rake angle of the tool is 5°. Assume coefficient friction between tool face and chip is 0.7. Determine
(*a*) the cutting force,
(*b*) the average shear stress in the shear plane,
(*c*) the normal stress on the shear plane,
(*d*) the average shear strain in cutting,
(*e*) the average shear-strain rate.

21-3 The unit power required for turning with a feed of 0.010 in/rev is given in units of hp/(in^3)/(min). For machining gray cast iron the unit power is 1.0, and for machining heat-treated alloy steel at R_c 40 it is 2.0 hp/(in^3)(min). Compare the time required for rough turning 2-in-diameter × 10 in long cylinders of cast iron and steel to a diameter of 1.8 in. The lathe is rated at 10 hp and has a 70 percent efficiency.

21-4 Describe two methods for measuring the temperature generated in metal cutting.

21-5 Specify the conditions of grinding that would produce a fine finish on the ground surface.

21-6 In billet conditioning by grinding the specific cutting energy is 6,900 MPa. If the metal removal rate is 1,080 cm^3/min, estimate the power required for the operation. If the wheel speed is v_g = 80 m s^{-1}, what is the tangential force on the wheel?

21-7 A 1200-mm by 150-mm surface is to be milled on a gray cast-iron casting using a face milling cutter with 12 teeth and a 200-mm diameter. The depth of cut will be 3 mm and the feed per tooth of the cutter is 0.30 mm. The cutting speed is 70 m/min and the unit power is 1.36 kW/cm^3 per s^{-1}. Let the machine efficiency be 0.80 and provide a 25-percent allowance for tool wear. Determine the power at the milling machine drive motor to perform this operation.

ANSWERS TO SELECTED PROBLEMS

CHAPTER 1

1-1 (*a*) 75,000 psi;
(*b*) Because necking reduces the diameter at a point along the gage length;
(*c*) 0.001 in

1-2 $E = 102$ GPa

1-4 1 : 2 : 3.4

1-5 $\sigma_{c_u} = 2{,}116$ psi; $\sigma_{s_t} = 3{,}835$ psi

1-7 Shearing stress = 125 ksi

CHAPTER 2

2-1 (*a*) $\sigma_{max} = 51.4$ ksi; $\sigma_{min} = 3.6$ ksi; $\theta = -9.8°$
(*b*) $\sigma_{max} = 13.5$ ksi; $\sigma_{min} = -68.5$ ksi; $\theta = -18.7°$

2-3 (*a*) 10.66 ksi;
(*b*) $\sigma = 10.651$ ksi, $\tau = 0.462$ ksi
(*c*) $\sigma_1 = 12.1$ ksi, $\sigma_2 = 3.6$ ksi, $\sigma_3 = -5.7$ ksi
(*d*) $\tau_{max} = 8.9$ ksi

2-5 -47.84 MPa

2-8 $\varepsilon_z = -0.000952$; $\varepsilon_{xy} = 0.00202$

2-9 $e_1 = 0.0058$; $e_2 = 0.0002$

2-10 (*a*) $\sigma_{max} = 58{,}000$ psi;
(*b*) $\sigma_{min} = 0$;
(*c*) $\tau_{max} = 29{,}000$ psi

2-14 $\sigma_L = 7.96 \times 10^2$ MPa; $\sigma_T = 12.7 \times 10^2$ MPa

2-15 522.94 cm^3

CHAPTER 3

3-1

	e	ε	q
(*a*)	0.10	0.095	0.0909
(*b*)	−0.10	−0.105	−0.111
(*c*)	1.0	0.693	0.50
(*d*)	−0.5	−0.693	−1.0
(*e*)	−1.0	$-\infty$	undefined

3-2 (*a*) $e_c = -0.6$, $\varepsilon = -0.916$;
(*b*) $D = 3.16$ in;
(*c*) $\varepsilon_1 = \varepsilon_2 = 0.458$

3-5 (*a*) 4.78 in;
(*b*) 4.59 in

3-7 $\sigma_1 = 100$, $\sigma_2 = -400$, $\sigma_3 = -900$. Yielding occurs based on Tresca criterion but not with von Mises.

3-9 $p = 1443$ psi

3-11 $\bar{\varepsilon} = 2/\sqrt{3}\ \ln(1 - r)$

3-14 (*a*) 60,000 psi; (*b*) 45,100 psi

CHAPTER 4

4-4 1.54×10^{13} atoms/mm^2

4-7 (*b*) Eight slip systems; (*c*) Six slip systems

4-8 0.22 percent

4-10 38.9 MPa

4-13 0.587 MPa

CHAPTER 5

5-2 $a/2[011] \rightarrow a/6[112] + a/6[\bar{1}21]$

5-3 $a/2[\bar{1}10] + a/2[101] \rightarrow a/2[011]$

5-4 $\pm(111)$ and $\pm(\bar{1}1\bar{1})$

5-12 (*a*) 37.4×10^7 eV

5-13 (*b*) 7.46×10^7 dyne/cm^2

CHAPTER 6

6-1 2500 grains

6-3 ASTM 8.1

6-4 20 ppm

6-8 (*a*) 65×10^6 psi;
(*b*) 46×10^6 psi

6-11 (*a*) 23.5×10^6 psi;
(*b*) 13.2×10^6 psi;
(*c*) 123,000 psi

CHAPTER 7

7-1 31 GPa or 4.5×10^6 psi
7-3 2.5×10^{-2} cm

CHAPTER 8

8-1 (*b*) (1) 30,000 psi; (2) 29.4×10^6 psi; (3) 31,000 psi; (4) 51,000 psi
(*d*) (1) 52,200 psi; (2) $n = 0.223$
8-6 (*b*) (1) 89,600 psi; (2) 161,000 psi (uncorrected) (3) 1.05; (4) 0.183; (5) 0.871;
(6) 74,700 psi; (7) 0.17
8-10 $b = 21$; $c = 37.5$

CHAPTER 9

9-3 (*b*) Annealed $k = 3.2$, $n' = 3.5$; $\frac{1}{4}$ hard and $\frac{1}{2}$ hard
$k = 25.1$, $n' = 2.6$

CHAPTER 10

10-1 (*b*) 49,000 psi;
(*d*) yield stress 22,900 psi, $G = 11.2 \times 10^6$ psi
(*e*) Poisson's ratio 0.295
10-2 For equal shearing yield strength the hollow shaft has 3.5 times the torque carrying capacity. Hollow shaft will be 7 times stiffer than solid shaft of same weight.

CHAPTER 11

11-2 (*a*) 87.4 ksi; (*b*) 43.2 ksi; (*c*) 65.5 ksi
11-3 (*a*) 107,000 lb; (*b*) 250,000 lb; (*c*) 6×10^6 lb
11-4 $\mathscr{G}_c = 25$ N/mm; $\sigma = 406$ MPa
11-8 10.5 mm

CHAPTER 12

12-5 $D = 1.88$ in

12-6 6.27×10^6 N-m

12-9 11,160 psi

12-10 2.34×10^{-3}

12-13 5,300 N-m

CHAPTER 13

13-6 (a) 30,000 psi;
(b) 6.4×10^5 h

13-7 89,125 cal/mole

13-9 $B = 1.92 \times 10^{-42}$; $n' = 9.4$

13-10 (a) 0.4 in^2;
(b) 1.2 in^2;
(c) 2.33 in^2

13-11 1263 psi

CHAPTER 14

14-3 13 ft/sec

14-4 $\sigma = W/A[1 + \sqrt{1 + 2h(AE/WL)}]$

14-7 373,000 psi

CHAPTER 15

15-2 540 tons

15-3 $\dfrac{p}{\sigma_0} = \dfrac{\tan\alpha}{\mu} - \left(\dfrac{\tan\alpha}{\mu} - 1\right)\left[\dfrac{1 + D/h_0(r/R_0)\tan\alpha}{1 + (D/h_0)\tan\alpha}\right]^{-\mu/\tan\alpha}$

15-4 $U_p = \bar{\sigma}\varepsilon$

15-6 $p/2k = \frac{1}{2} + \pi/4 = 1.29$

15-7 $p/2k = 1.5$

CHAPTER 16

16-2 (a) 630 tons; (b) 210 tons

16-3 Forging pressure would be independent of h

16-4 If $C_1 = 3$, $A = 3.3$ ft^2; if $C_1 = 12$, $A < 1$ ft^2

16-6 Thin cylinder $\bar{p}/\sigma_0 = 1 \times 10^6$; thick cylinder $\bar{p}/\sigma_0 = 2.6$

CHAPTER 17

17-2 (*a*) At entrance and exit $p_r = 34{,}640$ psi (neglecting strain hardening); at neutral point $p_r = 37{,}420$ psi
(*b*) $p_r = 67{,}700$ psi
(*c*) $p_r = 32{,}000$ psi

17-3 Hot rolling $h_{\min} = 0.016$ in, cold-rolling $h_{\min} = 0.058$ in

17-5 $\mu = 0.5$

17-8 14.5 ton/in

17-9 (*b*) 180 hp; (*c*) $\Delta T = 9°\text{C}$

CHAPTER 18

18-3 (*a*) 120,000 psi; (*b*) 38,300 psi; (*c*) 6,500 tons

18-4 $v = 2$ in/s, $t = 1$ s; $v = 10$ in/s, $t = \frac{1}{5}$ s

18-6 It would reduce pressure to increase D to get greater length but it may not be economical to use a different size container.

CHAPTER 19

19-4 3520 s^{-1}

19-5 For wiredrawing $\sigma_{xa}/\sigma_0 = 0.48$; for tube drawing $\sigma_{xa}/\sigma_0 = 0.67$.

19-6 (*a*) 0.63; (*b*) 0.72; (*c*) 0.78

CHAPTER 20

20-4 Metal A: $\varepsilon = 0.07$, $\Delta\varepsilon = 0.02$
Metal B: $\varepsilon = 0.24$, $\Delta\varepsilon = 0.19$
Shows that low n material develops steep strain gradients

20-5 8950 lb. Force to tear cup wall is about 14,000 lb.

CHAPTER 21

21-2 (*a*) 4 kN; (*b*) 846 MN/m^2; (*c*) 844 MN/m^2;
(*d*) 3.9; (*e*) $4.4 \times 10^5\ \text{s}^{-1}$

21-3 For cast iron, $t = 0.9$ min; for steel, $t = 1.8$ min

21-6 Power 124 kW; tangential force 1,550 N

21-7 8.6 hp

NAME INDEX

SUBJECT INDEX